With the recent advent of large, ground-based telescopes and space telescopes, it is now possible to study in detail stars outside our galaxy – in neighbouring galaxies in the so-called Local Group. The VIII Canary Islands Winter School of Astrophysics gathered leading experts from around the world to review this exciting new area of research – extragalactic stellar astrophysics. This volume presents eight specially written articles based on the meeting, reviewing how the study of stars in nearby galaxies can be used to understand stellar and galactic structure and evolution in general.

This book covers all aspects of extragalactic stellar astrophysics: stellar physics, stellar winds, stellar evolution, the use of photometric and spectroscopic techniques for studying extragalactic stars, stellar populations, chemical evolution, star formation histories and the calibration of the extragalactic distance scale.

This timely volume provides graduate students and researchers with an invaluable introduction to and reference on the new subject of extragalactic stellar astrophysics.

CAMBRIDGE CONTEMPORARY ASTROPHYSICS

Stellar Astrophysics for the Local Group

CAMBRIDGE CONTEMPORARY ASTROPHYSICS

Series editors
José Franco, Steven M. Kahn, Andrew R. King and Barry F. Madore

Titles available in this series

Gravitational Dynamics, *edited by O. Lahav, E. Terlevich and R. J. Terlevich* (ISBN 0 521 56327 5)

High-sensitivity Radio Astronomy, *edited by N. Jackson and R. J. Davis* (ISBN 0 521 57350 5)

Relativistic Astrophysics, *edited by B. J. T. Jones and D. Marković* (ISBN 0 521 62113 5)

Advances in Stellar Evolution, *edited by R. T. Rood and A. Renzini* (ISBN 0 521 59184 8)

Relativistic Gravitation and Gravitational Radiation, *edited by J.-A. Marck and J.-P. Lasota* (ISBN 0 521 59065 5)

Instrumentation for Large Telescopes, *edited by J. M. Rodríguez Espinosa, A. Herrero and F. Sánchez* (ISBN 0 521 58291 1)

Stellar Astrophysics for the Local Group, *edited by A. Aparicio, A. Herrero and F. Sánchez* (ISBN 0 521 63255 2)

Stellar Astrophysics for the Local Group

VIII Canary Islands Winter School of Astrophysics

Edited by
A. APARICIO
Instituto de Astrofísica de Canarias

A. HERRERO
Instituto de Astrofísica de Canarias

F. SÁNCHEZ
Instituto de Astrofísica de Canarias

PUBLISHED BY THE PRESS SYNDICATE OF THE UNIVERSITY OF CAMBRIDGE
The Pitt Building, Trumpington Street, Cambridge CB2 1RP, United Kingdom

CAMBRIDGE UNIVERSITY PRESS
The Edinburgh Building, Cambridge CB2 2RU, United Kingdom
40 West 20th Street, New York, NY 10011–4211, USA
10 Stamford Road, Oakleigh, Melbourne 3166, Australia

© Cambridge University Press 1998

This book is in copyright. Subject to statutory exception
and to the provisions of relevant collective licensing agreements,
no reproduction of any part may take place without
the written permission of Cambridge University Press.

First published 1998

Printed in the United Kingdom at the University Press, Cambridge

A catalogue record for this book is available from the British Library

ISBN 0 521 63255 2 hardback

Contents

Participants   x
Preface   xiii
Acknowledgements   xv

Fundamentals of Stellar Evolution Theory: Understanding the HRD
C. Chiosi

Introduction	1
Basic stellar evolution	2
Physical causes of violent ignition, explosion, and collapse	11
Two basic ingredients: nuclear reactions and opacities	16
Stellar winds: observational and theoretical hints	18
Classical evolution of massive stars with mass loss	24
Convection: the major uncertainty	32
Passing from theory to observations	47
Globular clusters	51
Old open clusters	61
Young rich clusters of the LMC	62
The HRD of supergiants stars: open problems	63
Modelling AGB & Carbon stars: recent results	72
Cepheid stars: mass discrepancy and mixing	76
References	79

Observations of the Most Luminous Stars in Local Group Galaxies
P. Massey

Introduction	95
Introducing the unevolved luminous stars	98
Finding main-sequence luminous stars in the Local Group: methodology for a hard problem	108
Finding the evolved descendants of massive stars: LBVs, WRs, and RSGs	116
Secrets of star formation as revealed by luminous stars	124
Secrets of stellar evolution revealed by luminous stars	130
Summary: what to take away from all this	144
References	145

Quantitative Spectroscopy of the Brightest Blue Supergiant Stars in Galaxies
R.P. Kudritzki

Introduction	149
Atmospheres of luminous hot stars	158
Methods of spectral diagnostics	177
The X-ray emission of O-stars	195
IR - diagnostics of blue supergiants with extreme mass-loss	201
The most massive stars in the Local Group	213
Stellar abundances in Local Group galaxies and beyond	214
The Wind Momentum - Luminosity Relationship and extragalactic distances	232

Extragalactic stellar astronomy - a vision	254
Appendix A. Oscillator strengths distribution function of hydrogen	256
References	256

Calibration of the Extragalactic Distance Scale
B.F. Madore & W.L. Freedman

Introduction to the lectures	263
Cepheids	264
Brief summary of the observed properties of cepheid variables	266
Simple physical considerations	267
Observational considerations	271
Advances driven by new technology	275
CCDs and multiwavelength coverage	277
Obtaining accurate cepheid distances	278
Local Group galaxies	280
Beyond the Local Group	283
The Hubble constant	288
The future	288
Contrasting aspects of the PL and PLC	289
A reddening-free formulation of the PL relation	291
Comments on reddening determinations	295
Comparisons with other distance indicators	299
The key project	300
Other ground-based work	303
Helium core flash and the tip of the red giant branch as a primary distance indicator	305
The ideal distance indicator	305
Some history concerning the red giant branch	306
Concerns and technical issues	309
An overview of the theoretical underpinnings: core helium ignition	312
Recent applications of the TRGB method	313
The scorecard	315
Discussion	318
Implications of the Hipparcos observations of galactic Cepheids	319
Comparison with V–band period–luminosity relations	319
Multiwavelength period–luminosity relations	320
Discussion	324
Implications of a cepheid distance to the Fornax cluster	327
NGC 1365 and the Fornax cluster	328
HST observations	329
Cepheids in NGC 1365	330
The Hubble constant	331
The Hubble constant at Fornax	333
The nearby flow field	334
Beyond Fornax: the Tully-Fisher relation	336
Beyond Fornax: other relative distance determinations	338
Beyond Fornax: type Ia supernovae	339
Cosmological implications	340
Conclusions	341
References	343

Dwarf Galaxies
G.S. Da Costa

Introduction	351
Prelude: results from standard stellar evolution	352
"Old" populations in the Magellanic Clouds	358
Local Group dE and dSph galaxies	363
Local Group dIrr galaxies	388
Dwarf galaxies beyond the Local Group	396
Summary	401
References	402

Resolved Stellar Populations of the Luminous Galaxies in the Local Group
M. Mateo

Introduction	407
Photometric techniques	408
Star clusters in the Local Group	415
The old and intermediate-age populations in luminous LG galaxies	423
"Young" field star populations in luminous LG galaxies	429
Variable star populations in LG galaxies	433
Beyond the Local Group	438
Epilogue	444
Appendix: stellar photometry examples using DoPHOT	445
References	452

Chemical Evolution of the ISM in Nearby Galaxies
E.D. Skillman

Introduction and purpose	457
Abundances from HII regions	459
Simple chemical evolution	468
Abundance patterns in dwarf galaxies	472
Abundance patterns in spiral galaxies	489
Self-consistent star formation histories	504
Summary	518
References	518

Populations of Massive Stars and the Interstellar Medium
C. Leitherer

Introduction	527
Regions of high-mass star formation	528
Massive stars in resolved populations	543
Evolutionary synthesis of unresolved high-mass populations	556
Release of mass and energy by massive stars	569
Massive stars and the dynamics of the ISM	585
References	598

Participants

Andronova, A.	Pulkovo Observatory (Russia)
Aparicio Juan, A.	Instituto de Astrofísica de Canarias (Spain)
Becker, S.R.	Universitäts-Sternwarte München (Germany)
Bosch, G.	Institute of Astronomy, Univ. Cambridge (UK)
Bremmes, T.	Astronomical Institute of Basel (Switzerland)
Cairós Barreto, L.M.	Instituto de Astrofísica de Canarias (Spain)
Castellani, M.	Osservatorio di Monteporzio (Italy)
Chiosi, C.	Universita di Padova (Italy)
Cole, A.A.	University of Wisconsin-Madison (USA)
Cordero Gracia, M.	Universidad Complutense de Madrid (Spain)
Da Costa, G.	Mt. Stromlo & Siding Spring Obs. (Australia)
Dorado de Cáceres, M.	LAEFF-INTA (Spain)
Gallart Gallart, C.	Carnegie Observatories of Washington (USA)
García Navas, J.	VILSPA (Spain)
Girardi, L.	Instituto de Fisica- UFRGS (Brasil)
Gouliermis, D.	National Observatory of Athens (Greece)
Gummersbach, C.A.	Landessternwarte Heidelberg (Germany)
Hansen Ruiz, C.	Royal Greenwich Observatory (UK)
Herrero Davó, A.	Instituto de Astrofísica de Canarias (Spain)
Hidalgo Gámez, A.M.	Astronomiska Observatoriet Uppsala (Sweden)
Hurley-Keller, D.A.	University of Michigan (USA)
Johnson, K.	J.I.L.A., University of Colorado (USA)
Kohle, S.	Radioastron. Institut Univ. Bonn (Germany)
Kudritzki, R.P.	Universitäts-Sternwarte München (Germany)
Larsen, S.	Niels Bohr Institut (Denmark)
Leitherer, C.	Space Telescope Science Institute (USA)
Lourenso Prieto, S.	Instituto de Astrofísica de Canarias (Spain)
Madore, B.	Infrared Processing and Analysis Center, CalTech (USA)
Marigo, P.	Universita di Padova (Italy)
Martínez Delgado, D.	Instituto de Astrofísica de Canarias (Spain)
Massey, P.	Kitt Peak National Observatory (USA)
Mateo, M.	University of Michigan (USA)
Moitinho de Almeida, A.	Instituto de Astrofísica de Andalucía (Spain)
Möller, C.	Universitäts-Sternwarte Göttingen (Germany)
Monteverde Hernández, M.I.	Instituto de Astrofísica de Canarias (Spain)
Musella, I.	Osservatorio Astronomico di Capodimonte (Italy)
Pedraz Marcos, S.	Universidad Complutense de Madrid (Spain)
Portinari, L.	Universita di Padova (Italy)
Prins, S.	University of Amsterdam (The Netherlands)
Rekola, R.	Turola Observatory (Finland)
Rosenberg González, A.	Osservatorio Astronomico di Padova (Italy)
Roth, M.M.T.	Astrophysikalisches Institut Postdam (Germany)
Royer, F.	DASGAL- Observatoire de Paris-Meudon (France)
Royer, P.	Institut d'Astrophysique de Liège (Belgium)
Santillán, A.J.	Universidad Nacional Autónoma de México (México)
Santolamazza, P.	Osservatorio Astronomico di Capodimonte (Italy)
Skillman, E.	University of Minesota (USA)
Smartt, S.J.	The Queen's University of Belfast (UK)

Tantalio, R.	Universita di Padova (Italy)
Taresch, G.	Universitäts-Sternwarte München (Germany)
Thomas, D.	Universitäts-Sternwarte München (Germany)
van Loon, J. T.	ESO, Garching bei München (Germany)
Vassiliadis, E.	Instituto de Astrofísica de Canarias (Spain)
Vega Beltrán, J.C.	Osservatorio Astronomico di Padova (Italy)
Vilchez Medina, J.M.	Instituto de Astrofísica de Canarias (Spain)
Villamariz Cid, M.R.	Instituto de Astrofísica de Canarias (Spain)
Zurita Muñoz, A.	Instituto de Astrofísica de Canarias (Spain)

Preface

The goal of the Canary Islands Winter School of Astrophysics, organized by the Instituto de Astrofísica de Canarias (IAC), is to bring together each year advanced graduate students, recent postdocs and interested scientists with a group of leading experts in a particular area of astrophysics. The one held in 1996 in La Laguna (Tenerife, Spain) was devoted to the stellar content of the Local Group and the application of its study to more distant galaxies.

The idea of using the Local Group as a typical case and as a first step towards understanding the more distant Universe has its origins in the possibility of arriving at a detailed knowledge of the properties of its constituent galaxies and their stars. We are still making progress in acquiring a detailed knowledge of the Local Group, but we realize that the unknowns far outweight the knowns, and this is precisely the reason why study of the Local Group is still, and will continue to be, useful. As the results from the Hubble Space Telescope are coming in, we are witnessing a rapid advance in terms of quantity of information. What only a few years ago was no more than vague, often erroneous, conjecture concerning the properties of the nearest galaxies is now becoming irrefutable evidence, which in its turn raises new questions on aspects that were previously beyond our grasp. This change currently under way has also been aided by large ground-based telescopes, such as the WHT on La Palma, and especially the Keck I and II telescopes on Hawaii, and will be reinforced by the new technological achivements represented by the new generation of 8-10 m telescopes (from the VLT to the 10 m GTC, and the LBT, Gemini, Subaru, HET, etc., in between), together with rapid advances in detector size and sensitivity.

This is therefore a fitting moment to review what we know and do not know about the Local Group, to recognize our present limitations and identify areas where we might begin to glimpse an answer.

Why stellar astrophysics? Stars are born from the gaseous medium of galaxies; they evolve in a manner which depends mainly on their mass and eventually they die, returning part of their constituent material to the gaseous medium from which they came; but this material now has a different composition and dynamics. This irreversible process is the main driver of the evolution of most galaxies. If we knew how many stars of each age and chemical composition a galaxy has (i.e., its star formation history), we could, by making use of what we know concerning the processes that affect stars, understand what the galaxy is really like and how it evolves. What we manage to unravel concerning the conditions and the way in which stars are formed, the details of their evolution and the processed material which they return to the interstellar medium, will drive our knowledge of galaxies.

One of the most important applications of extragalactic stellar observations is the measurement of distances in the Universe. The Cepheids provide one of the standard measuring rods that enable us to construct a cosmic distance scale. This standard distance candle has to be continously reviewed and updated, and great efforts are dedicated to the refinement of this method. But progress continues, and new techniques appear that may complement the Cepheid method. The *wind momentum–luminosity relation*, whose fundamentals are explained in chapter 3, is one of them.

For a long time, many classical applications of observations of stars, such as studies of ages and populations, stellar evolution, abundances, detailed interaction with the surrounding medium, etc., were limited by the faintness of extragalactic stars. At the same time, it was known that galaxies different from the Milky Way offered different conditions for stellar formation and evolution so that including them in studies already carried out on our Galaxy would permit a significant advance in our understanding of these fields. At present, the technological developments referred to above are overcoming many of these difficulties, and a new era of stellar astrophysics may soon open up for us.

This whole conjunction of positive aspects encouraged us to suggest the topic of this book and to work enthusiastically towards bringing about the meeting. Trying to find the best list of topics and the best people to teach and review them, we brought together eight specialists in various aspects of the problem, which range from stellar evolution to stellar population synthesis as applied to distant galaxies; from the physics of stellar atmospheres to the properties of galaxies and the interstellar medium and the extragalactic distance scale.

We have no doubt that all the effort involved has greatly benefited all the participants, and we would like to extend this experience to all who are interested through the publication of these proceedings.

Artemio Herrero, Antonio Aparicio
La Laguna, Tenerife
Noviembre, 1997

Acknowledgements

It is a pleasure to acknowledge the participation of the eight lecturers in the School. Their skill in presenting the material in a clear and concise way, and the spirited discussions and presentations of the 48 students all contributed towards making the School a very enjoyable event. We are also indebted to Lourdes González, Nieves Villoslada and Campbell Warden for their work during the preparation and organization of the School.

Fundamentals of Stellar Evolution Theory: Understanding the HRD

By CESARE CHIOSI

Department of Astronomy, University of Padua, Vicolo Osservatorio 5, 35122 Padua, Italy

We summarize the results of stellar evolution theory for single stars that have been obtained over the last two decades, and compare these results with the observations. We discuss in particular the effect of mass loss by stellar winds during the various evolutionary stages of stars that are affected by this phenomenon. In addition, we focus on the problem of mixing in stellar interiors calling attention on weak aspects of current formulations and presenting plausible alternatives. Finally, we survey some applications of stellar models to several areas of modern astrophysics.

1. Introduction

The major goal of stellar evolution theory is the interpretation and reproduction of the Hertzsprung-Russell Diagram (HRD) of stars in different astrophysical environments: solar vicinity, star clusters of the Milky Way and nearby galaxies, fields in external galaxies. The HRD of star clusters, in virtue of the small spread in age and composition of the component stars, is the classical template to which stellar models are compared. If the sample of stars is properly chosen from the point of view of completeness, even the shortest lived evolutionary phases can be tested. The HRD of field stars, those in external galaxies in particular, contains much information on the past star formation history. In these lectures, no attempt has been made to cover all the topics that could be addressed by a report on the progress made in understanding the HRD. Rather we have selected a few topics on which in our opinion most effort has concentrated over the past few years. Among others, the subject of the extension of convective regions in real stars was vividly debated with contrasting appraisals of the problem. Accordingly, various scenarios for the evolution of stars were presented and their far-reaching consequences investigated. Similarly, much effort was dedicated to understanding the evolution of globular cluster stars, and to quantifying the effect of important parameters (see below) in the aim of clarifying whether an age spread is possible.

The lectures are organized as follows; 2. a summary of basic stellar evolution theory, whenever possible updated to include the most recent results; 3. a summary of the physical causes determining violent ignition of a nuclear fuel, core collapse, and explosions. 4. a summary of recent results on relevant nuclear reaction rates and opacities. 5. a critical discussion of stellar winds and their implications on stellar models. 6. a review of the evolution of massive stars under the effect of mass loss. 7. a discussion of several problems related to convective instability and mixing in stellar interiors (semiconvection, overshoot, and diffusion) together with a summary of the evolutionary results under different mixing schemes. 8. a summary of the problems concerning the transformation of the theoretical HRD into the observational color-magnitude diagram. 9. a description of the results obtained for the luminosity functions, age, age-metallicity relation, age spread, and *second parameter* of globular clusters; 10. a discussion of the old open clusters as a means for calibrating the extension of convective cores in the range of low-mass stars; 11. similarly for the rich young clusters of the Large Magellanic Cloud (LMC) but in the range of intermediate-mass stars. 12. a modern description of the properties of supergiant stars in the Milky Way and LMC together with current understanding of their evolution in light of the problems raised by SN1987A. 13. a summary of the most recent developments in modeling the evolution of AGB stars, their luminosity function, and the formation of Carbon stars; 14. a summary of the recent progress made in modeling the pulsation of the Cepheid stars and the specific topics of the shape of the instability strip and mass discrepancy.

For more information the reader is referred to the many review articles that have appeared over the years, including Iben & Renzini (1983, 1984), Iben (1985), Castellani (1986), Chiosi & Maeder (1986), Hesser (1988), Renzini & Fusi-Pecci (1988), Iben (1991), VandenBerg (1991), Demarque *et al.* (1991), Fusi-Pecci & Cacciari (1991), Chiosi *et al.* (1992), Maeder & Conti (1994), Stetson *et al.* (1996), VandenBerg *et al.* (1996), and others quoted in the text.

2. Basic stellar evolution

Independently of the chemical composition, stars can be loosely classified into three categories according to their initial mass, evolutionary history, and final fate: low-mass stars, intermediate-mass stars, and massive stars. Various physical causes concur to define the three groups and related mass limits:

1. The existence of a natural sequence of nuclear burnings from hydrogen to silicon.

2. The amount of energy liberated per gram by gravitational contraction which is increasing with stellar mass.

3. The tendency of the gas in the central regions to become electron degenerate at increasing density.

4. The existence of threshold values of temperature and density in the center for each nuclear step.

5. The relation between these threshold values and the minimum stellar or, more precisely, core mass at which each nuclear burning can start, and the fact that the minimum core mass for a given nuclear burning is not the same for electron degenerate and non-degenerate gas.

6. Finally, the explosive nature of a nuclear burning in a degenerate gas.

Because the evolutionary path of a star in the HRD is a natural consequence of the interplay between those physical processes, they will be the main guide-lines of our summary of the stellar evolution theory.

2.1. *Low-, Intermediate-, and High-Mass Stars: definition*

By low mass stars we define those which shortly after leaving the main sequence toward the red giant branch (RGB), develop an electron degenerate core composed of helium. When the mass (M_{He}) of the He core has grown to a critical value ($0.45 \div 0.50 M_\odot$), the precise value depends on the composition, star mass, and input physics), a He-burning runaway is initiated in the core (He-flash), which continues until electron degeneracy is removed. The maximum initial mass of the star (otherwise called M_{HeF}) for this to occur is about $1.8 \div 2.2 M_\odot$, depending on the initial chemical composition. Within the same mass range we distinguish the stars lighter than $M_{con} \simeq 1.2 \div 1.3 M_\odot$ that burn hydrogen in a radiative core from the more massive ones doing it in a convective core. Furthermore, it is worth recalling that stars lighter than about $0.5\ M_\odot$ cannot proceed to central He-ignition because they fail to reach the threshold value for the He-core burning. Stars more massive than M_{HeF} are classified either as intermediate-mass or massive stars. In turn we distinguish the intermediate-mass stars from the massive ones by looking at the stage of carbon ignition in the core. By intermediate mass we mean those stars which, following core He-exhaustion, develop a highly degenerate carbon-oxygen (C-O) core, and as asymptotic giant branch (AGB) stars experience helium shell flashes or thermal pulses. The AGB phase is terminated either by envelope ejection and formation of a white dwarf ($M_{HeF} \leq M_i \leq M_w$) or carbon ignition and deflagration in a highly degenerate core once it has grown to the Chandrasekhar limit of 1.4 $M_\odot$. The limit mass M_w is regulated by the efficiency of mass loss by stellar wind during the RGB and AGB phases (see Iben & Renzini 1983). This point will be discussed in more detail below. The minimum mass of the C-O core, below which carbon ignition in non degenerate condition fails and the above scheme holds, is 1.06 $M_\odot$ corresponding to an initial mass from 7 to 9 $M_\odot$, depending on the chemical composition. This particular value of the initial mass is known as M_{up}. Finally, massive stars are those that ignite carbon nonviolently and through a series of nuclear burnings proceed either to the construction of an iron core and subsequent photodissociation instability with core collapse and supernova explosion ($M_i \geq M_{mas}$), or following a more complicated scheme undergo core collapse and supernova explosion ($M_{up} \leq M_i \leq M_{mas}$). M_{mas} is about 12 $M_\odot$.

Figure 1 shows the evolutionary path in the HRD of model stars of 0.8 $M_\odot$, 5 $M_\odot$, 20 $M_\odot$, and 100 $M_\odot$ which can be considered to be representative of the three categories. These evolutionary tracks have the chemical composition [Z=0.008, Y=0.250] The thick portions of each track approximately indicate the regions of slow evolution, where the majority of stars are observed. The various evolutionary phases discussed in the text are shown as appropriate for the star mass. The reader should refer to this figure to locate on the HRD the particular evolutionary phase under discussion.

2.2. Core and Shell H-Burning Phases

The core H-burning main sequence phase of stars lighter than M_{con} is characterized by the gradual formation of a small He core at the center and the buildup of a smooth chemical profile from a He-rich core to the outer unprocessed layers. The luminosity steadily increases while the star climbs along the zero-age main sequence itself departing significantly from it toward cooler T_{eff} only at the very end of the phase. The duration of the core H-burning phase strongly decreases with increasing mass of the star going from about 15×10^9 yr for a typical 0.7 $M_\odot$ star to about 1×10^9 yr for a typical 1.7 $M_\odot$ star.

After the main sequence phase, the H-exhausted core temporarily cools as electron degeneracy sets in, and the energy liberated by gravitational contraction flows out by electron conduction, delaying the increase in central temperature required to ignite helium in the core. As a low-mass star reaches the base of the RGB, the central temperature reaches a minimum approximately equal to the temperature of the H-burning shell. Thereafter, the mass of the helium core grows under the action of the H-burning shell, the core contracts, and temperatures in the core and H-burning shell increase. The luminosity of the star is proportional to the increase in the shell temperature. The rate at which matter is added to the core by the H-burning shell, and consequently the rate of release of gravitational energy and heating of the core, are proportional to the luminosity. The star climbs the RGB (Hayashi line), while convection in the outer layers gets deeper and deeper, eventually reaching those layers that were nuclearly processed in previous stages and generating a discontinuity in the chemical profile (first dredge-up). The steady outward migration of the H-burning shell forces the external convection to recede. The ascent of the RGB is temporarily slowed when the H-burning shell reaches the discontinuity in the chemical profile. Owing to the electron degeneracy, all low-mass stars, independently of initial mass, build up an helium core of approximately the same mass. When this core has grown to about $0.45 \div 0.50 M_\odot$, violent He-burning starts off-center because neutrino emission has cooled the innermost regions (Thomas 1967, Mengel & Sweigart 1981). As the nuclear burning progresses inwards, degeneracy is removed, so that a quiescent nuclear burning in the core begins. The RGB phase is terminated. Because this stage occurs at essentially identical core masses, the maximum luminosity of the RGB is almost the same, independent of the initial mass and chemical composition of the star. The duration of the RGB phase depends on stellar mass going from about 2.0×10^9 yr for a typical 0.7 $M_\odot$ star to about 2.7×10^8 yr for a typical 1.7 $M_\odot$ star.

The evolution of stars lighter than M_{HeF} but heavier than M_{con} is basically similar to the above scheme, although toward the upper mass end it reflects in many respects the evolution of intermediate-mass stars.

In intermediate- and high-mass stars, the main sequence core H-burning phase is characterized by the formation of a convective core, a steady increase in luminosity and radius, and a decrease of the T_{eff}. The size of the convective core, which is customarily fixed by the classical Schwarzschild (1958) criterion ($\nabla_R = \nabla_A$, with the usual meaning of the symbols), increases with stellar mass, whereas the duration of the core H-burning phase decreases with increasing mass owing to the overwhelming effect of the increasing luminosity. The main sequence core H-burning lifetime goes from several 10^8 yr to a few 10^6 yr as the mass of the star increases from about 2 $M_\odot$ to 100 $M_\odot$. Massive stars may be affected by semiconvective instability (thereinafter the H-semiconvection) and mass loss by stellar wind. Semiconvection has long been the characterizing feature of the structure of massive stars evolved at constant mass, whereas to date the most salient signature of the evolution of massive stars is the occurrence of mass loss by stellar wind (see Chiosi & Maeder 1986).

After exhausting central hydrogen while on the main sequence, intermediate- and high-mass stars up to say 15 $M_\odot$ (the evolution is more complicated for the most massive ones) evolve rapidly to the red giant (supergiant) region, burning hydrogen in a thin shell above a rapidly contracting and heating core, composed essentially of helium. As they approach the Hayashi line, a convective envelope develops whose base extends inward until it reaches layers in which hydrogen has been converted into helium and carbon into nitrogen via the CNO cycle. As a consequence, the surface abundance of those elements varies in a detectable way (first dredge-up). H-burning in the shell not only provides the bulk of the stellar luminosity but also adds matter to the H-exhausted core which continues to grow. When temperature and density in the core reach suitable values, helium is ignited.

The question as to why stars become red giants has been debated for many years without a satisfactory answer. Renzini (1984) identifies the physical cause for the rapid expansion of the envelope to red giant dimensions in a thermal instability in the envelope, which is primarily determined by the derivatives of the opacity in the middle temperature region (see also Iben & Renzini 1984). Applegate (1988) finds that a radiative envelope in which a Kramers' opacity law holds cannot transport a luminosity larger than a critical value. He argues that the transition to red giant structure is triggered by the star's luminosity exceeding the critical value. Weiss (1989a) reanalysing the criterion introduced by Renzini (1984) concludes that the opacity is not the main cause. Renzini et al. (1992) consider the envelope expansion caused by its thermal instability, whereas according to Iben (1993) there is no simple physical explanation to why stars become red giants. In contrast, Renzini & Ritossa (1994) with numerical experiments document the physical nature of the thermal instability causing the envelope expansion (but for low mass stars). The red giant problem still exists.

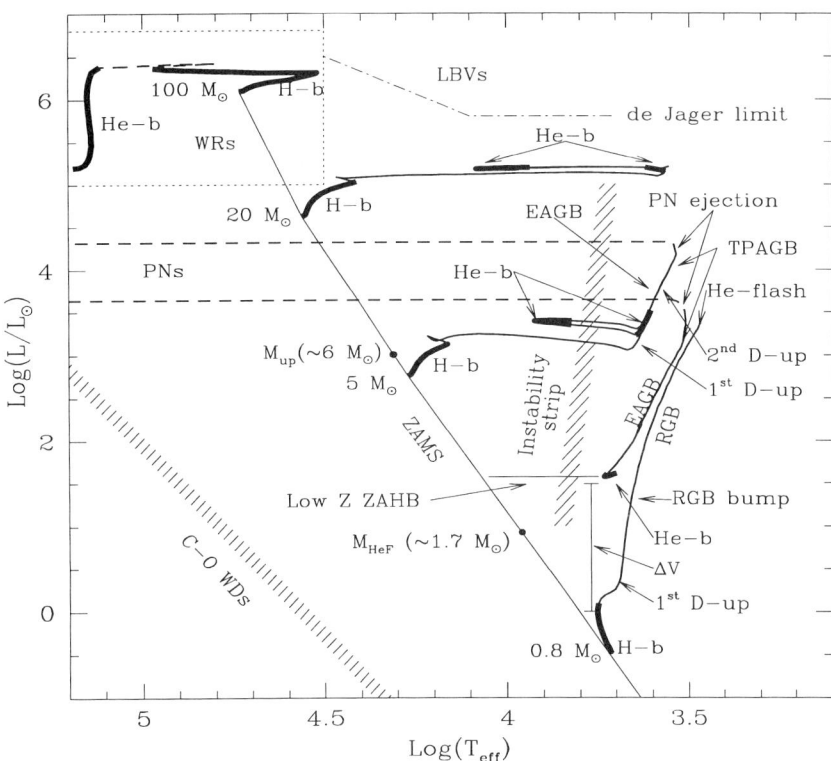

FIGURE 1. The evolutionary paths in the HRD of model stars of composition [Z=0.008, Y=0.25] and of initial mass 0.8 $M_\odot$, 5 $M_\odot$, 20 $M_\odot$, and 100 $M_\odot$. The models are calculated with the overshoot scheme for central convection. M_{HeF} and M_{up} are the masses separating low-mass stars from intermediate-mass stars, and the latter from the massive ones, respectively. For low- and intermediate-mass stars the tracks go from the zero-age main sequence (ZAMS) to the end of the asymptotic giant branch (AGB) phase, whereas for the massive stars they reach the stage of C-ignition in the core. Massive stars include the effect of mass loss by stellar wind. H-b and He-b stand for core H- and He-burning, respectively. He-flash indicates the stage of violent ignition of central He-burning in low-mass stars at the tip of the red giant branch (RGB). The main episodes of external mixing (1st and 2nd dredge-up) are indicated by 1st D-up and 2nd D-up, respectively. The AGB phase is separated into early stages (E-AGB) and thermally pulsing regime (TP-AGB) of the He-burning shell. For low- and intermediate-mass stars we show the stage of planetary nebula (PN) ejection, the region where PN stars are observed, and the white dwarf (WD) cooling sequence. The horizontal line labeled ZAHB indicates the locus of the zero-age horizontal branch − core He-burning models − of low-mass stars with composition typical of globular clusters. The shaded vertical band shows the instability strip of Cepheid and RR Lyrae stars. In the region of massive stars, we show the de Jager limit, the location of the blue luminous variables (LBVs) and Wolf-Rayet stars (WRs). Finally, the thick portions of the tracks indicate the stages of slow evolution, where the majority of stars are observed.

2.3. Core He-Burning Phase

The development of the He-flash at the top of the RGB has been the subject of many quasi static as well as dynamical studies aimed at understanding whether the violent burning may acquire explosive characteristics or induce some sort of mixing (see the detailed discussion by Iben & Renzini 1984). Arguments exist for excluding both the total disruption of the star (type II-like supernova) and a substantial mixing between the inner core and the outer envelope (see Renzini & Fusi-Pecci 1988). In fact, type II-like supernovae are not seen in elliptical galaxies but post core He-flash stars do exist (HB and AGB), whereas mixing and consequent dredge-up of carbon would produce a kind of red HB star that is not observed. Following He-flash at the termination of the RGB, stars lighter than M_{HeF} quiescently burn helium in a convective core. Their position on the HB depends on several factors, among which the metallicity and the mass of the H-rich envelope dominate, the latter reduced by mass loss from the red giant precursor. For metal-rich stars, the core He-burning is confined to a narrow region or clump near the RGB about 3 magnitudes below the RGB tip, whereas for metal-poor stars the evolution covers a much broader range of T_{eff}s to the blue of the RGB at approximately the same luminosity as for the metal-rich ones. Under favorable circumstances (sufficiently low metal content or high enough mass loss for high metallicity stars) the HB can intersect the instability strip, giving rise to the RR Lyrae pulsators. The luminosity of the HB stars is determined primarily by the composition (helium abundance) of their main sequence progenitors. The nearly constant luminosity and duration of the He-burning phase (approximately 10^8 yr) reflect the convergency of precursor stars of different initial mass toward a common value for helium core mass (M_{He}). However, as the contribution to the luminosity by the H-burning shell may depend on the mass of the envelope, blue HB models can be less luminous.

The occurrence of mass loss from RGB stars cannot yet be derived from a satisfactory theory, but is basically justified by the observations (e.g. Renzini 1977; Chiosi 1986; Chiosi et al. 1992, and below). Since the amount of mass lost by a star depends both on the mean mass-loss rate, customarily expressed by empirical relationships as a function of the stellar parameters (e.g. Reimers 1975), and the duration of the phase in which mass loss is supposed to occur, the observational rates and lifetimes along the RGB are such that mass loss plays an important role only in stars with mass smaller than about 1 $M_\odot$, hence typical of old globular clusters (see Iben 1974, Renzini 1977, and Iben & Renzini 1983 for details). All low-mass stars more massive than about 1 $M_\odot$ remain in the clump for the entire core He-burning phase.

In intermediate- and high-mass stars, core He-burning ignites in nondegenerate conditions as soon as the central temperature and density are approximately equal to 10^8 K and 10^4 g cm^{-3}, respectively. This requires a minimum core mass of 0.33 $M_\odot$. Since M_{He} increases with the initial mass of the star because of the larger convective core on the main sequence, the mean luminosity of core He-burning phase increases with stellar mass. During the core He-burning phase, hydrogen continues to burn in a shell at about the same rate as it did during the main sequence phase. The rate at which helium is burnt in the convective core determines the rate at which the star evolves. Typically, the lifetime in the core He-burning stage is about 20 to 30% of the main sequence lifetime, being longer in models of smaller mass.

The slow evolution during core He-burning of intermediate-mass stars takes place in two distinct regions of the HRD, a first near the Hayashi line and a second at higher T_{eff}s and luminosities. The early stages of core He-burning take place in the first region. Subsequently, when the energy released by the burning core (which is increasing) equals the energy released by the H-burning shell (which is decreasing), a rapid contraction of the envelope readjusts the outer layers from convective to radiative and the star moves to the second region, where the remaining part of the core He-burning phase occurs. This causes the blue loops. The precise modeling and lifetime of the second phase depend on the stellar mass, chemical composition, nuclear reaction rates [$^{12}C(\alpha,\gamma)^{16}O$ in particular], extension of the convective core, opacity, mass loss along the RGB, inward penetration of the outer convection during the RGB stages (first dredge-up), and other physical details. For any choice of composition, as the stellar mass decreases, the location of the blue loop region moves toward the Hayashi line, eventually merging with the red giant region. Thus, for an assigned chemical composition, core He-burning breaks into two bands, one roughly corresponding to the locus of the Hayashi line or red giant stars, and another that breaks off the red giant band at low luminosities and moves toward the blue with increasing luminosity (the so-called blue band). The mean slope of this band is determined by a complicated interplay among the above physical factors which cannot be established a priori. The blue band of the core He-burning models may intersect the instability strip of the Cepheid stars.

Finally, the location of the core He-burning phase of stars more massive than say 15 $M_\odot$ is highly uncertain because it is dominated entirely by the effect of mass loss and convective overshoot (see Chiosi & Maeder 1986). These stars will be discussed in more detail below.

The core He-burning phase of intermediate-mass stars (toward the lower mass end) and low-mass stars on the HB is known to be affected by two types of convective instability: in early stages by a semiconvective mixing similar to that encountered by massive stars, and in late stages by the so-called breathing convection. They will be examined in more detail later on. Suffice it to recall that semiconvection prolongs the He-burning lifetime by approximately a factor of two in low-mass stars, i.e. for stars on the HB, whereas it has a negligible effect in intermediate-mass stars. Breathing convection determines a moderate increase in the lifetime (about 20%), whereas it gives origin to much larger C-O cores in all stars of this mass range.

The result of core He-burning, which turns helium into carbon, oxygen, and traces of heavier species, is the formation of a C-O core whose dimensions are determined by M_{He} and, once again, on the physical model adopted to describe central convection and its efficiency. Low-mass stars form C-O cores of approximately equal mass, whereas all other stars build C-O cores whose sizes increase with stellar mass.

After core He-exhaustion, the structure of the stars is composed of a C-O core, a He-burning shell, and an H-rich envelope at the base of which an H-burning shell is active. However, in massive stars, mass loss by stellar wind may be so strong at this stage that the entire envelope is lost even during the completion of the core He-burning phase (see below).

2.4. Later Evolutionary Phases

From the point of view of understanding the HRD, the evolutionary phases beyond the core He-burning of stars more massive than M_{up} are scarcely relevant because of their very short lifetime, hence low probability of detection, were it not for the final supernova explosion. Therefore, their evolution will not be described here (see Chiosi 1986; Woosley 1986, 1988; and Woosley & Weaver 1986).

Following the exhaustion of central helium, low- and intermediate-mass stars evolve through the AGB phase. The AGB phase is separated into two parts: the early AGB or E-AGB, which lasts until the H-burning shell is re-ignited (see below), and the thermally pulsing AGB or TP-AGB (see below), which lasts until the H-rich envelope is lost via a normal giant wind (low-mass progenitors) or via a "superwind" (intermediate-mass progenitors).

As the abundance of helium in central regions goes to zero, the He-exhausted core contracts and heats up while the H-rich envelope expands and cools. Cooling in the layers external to the C-O core is so effective that the H-burning shell extinguishes. In the HRD the stars evolve running almost parallel to the RGB, and once again the base of the convective envelope penetrates inward. According to Iben & Renzini (1983) there is a limiting mass (4.6 M$_\odot$ for solar composition) above which external convection eventually reaches layers processed by the CNO cycle. This means that fresh helium and fresh nitrogen are brought to the surface (second dredge-up). Eventually, the expansion of the envelope is halted by its own cooling and the envelope re-contracts, the luminosity decreases, and matter at the base of the convective envelope heats up. Ultimately, the H-burning shell is re-ignited, forcing the envelope convection to move outward in mass ahead of the H-burning shell. This terminates the E-AGB. In the meantime, the matter in the C-O core reaches such high densities that the electrons there become degenerate. Electron conduction causes the core to become nearly isothermal, while neutrino cooling carries away the gravitational energy liberated by the material added to the core by the outward migration of the He- and H-burning shells. Therefore, the temperature in the core tends to remain close to the temperature in the He-burning shell (about 10^8 K), well below the threshold value for carbon ignition.

Following the re-ignition of the shell H-burning, nuclear burning in the He-shell becomes thermally unstable (for a more detailed discussion see Iben & Renzini 1983, 1984; and Iben 1991). In brief, the nuclear burning does not occur at a steady rate, but the two shells, one H and the other He, alternate as the major source of energy. For 90% of the time the He-burning shell is inactive and the H-burning shell is the major source of energy. However, as the mass of the He-rich zone below the H-burning shell increases, the density and temperature at the base of this zone increase until the rate of energy production by the $3\alpha - {}^{12}$C reaction becomes higher than the rate at which energy can be carried out by radiative diffusion. As originally discovered by Schwarzschild & Harm (1965), a thermonuclear runaway occurs. A thin convective layer is generated on the top of the He-burning shell. At first the energy goes into raising the temperature of and expanding the matter in and near the burning zone and the material is pushed away in both directions. Matter at the base of the H-burning region is pushed out and cools to such low temperatures that the H-burning shell is temporarily extinguished. Eventually, matter at the He-burning region begins to cool as it overexpands and the rate of burning there drops dramatically. The convective layer disappears and a steady state is reached in which He-burning occurs quiescently at a slowly decreasing rate as the He-burning shell actually runs out of fuel. This quiescent phase lasts for about 10% of the time elapsing between successive outbursts. The material propelled outward falls back and the H-burning shell eventually re-ignites.

During this phase, material processed into the intershell region can be brought into the outer convective envelope and exposed to the surface. The so-called third dredge-up can then take place. In AGB stars of large C-O core mass (hence with large initial mass) the dredge-up can occur easily. But in AGB stars of small C-O core mass (hence with small initial mass) this is possible only if extra mixing is forced into the intershell region. The goal is achieved either by means of semiconvection induced by the more opaque C-rich material deposited in the intershell region by the tiny convective shell ahead of the flashing He-burning shell or by crude overshoot of convective elements from the convective shell itself. The mechanism of semiconvection has been proposed by Iben & Renzini (1982) following a suggestion by Sackmann (1980). Convective overshoot has subsequently been used by Hollowell (1988) and Hollowell & Iben (1988, 1989). In both discussions C-rich material is deposited in more external layers where it can be easily engulfed by the external convection during the subsequent cycle. This is the basic mechanism to convert an M giant into a carbon star (C star).

In the classical theory of the TP-AGB phase (a deep revision of this scheme will be presented in section 13), the luminosity of the star increases linearly with the mass of the H-exhausted core (Paczynski 1970a,b). Among others, popular relations (slightly different according to the mass range of the progenitor stars) are

$$L = 2.38 \times 10^5 \mu^3 Z_{CNO}^{0.04} \times (M_c^2 - 0.0305 M_c - 0.18021) \qquad (2.1)$$

where L is the luminosity in solar units, M_c is the mass of the H-exhausted core (in solar units), μ is the molecular weight, Z_{CNO} is the abundance (by mass) of CNO elements in the envelope. This relation applies to stars with core mass in the range $0.5 M_\odot \leq M_c \leq 0.66 M_\odot$ (Boothroyd & Sackmann 1988a,b).. i.e. in the low mass stars range. For stars with core mass $M_c \geq 0.95 M_\odot$ it is replaced by

$$L = 1.226 \times 10^5 \times \mu^2 \times (M_c - 0.46) M^{0.19} \qquad (2.2)$$

originally from Iben & Truran (1978). A linear interpolation is adopted for stars with $0.66 M_\odot \leq M_c \leq 0.95 M_\odot$. This secures that the TP-AGB stars brighten in M_{bol} at a constant rate (see Renzini 1977 and Iben & Renzini 1983 for details).

Given that C-ignition in highly degenerate conditions requires a C-O core mass of 1.4 $M_\odot$ (the Chandrasekhar limit), and considering the effect of mass loss, the minimum initial mass of the star, M_w, for C-ignition to occur is estimated in the range 4 to 6 $M_\odot$, depending on the adopted mass-loss rates, evolutionary lifetimes, and chemical composition (Iben & Renzini 1983). Stars lighter than the above limit will fail C-ignition and, by losing the H-rich envelope will become C-O white dwarfs (WDs) with a modest increase of the C-O core mass during the TP-AGB phase (about 0.1 $M_\odot$). In a very low-mass star (0.8÷1.0) $M_\odot$, the ejection of the envelope may be completed even before the H-burning shell is re-ignited and the thermal pulsing regime begins. Direct observational evidence for the existence of the TP-AGB phase and third dredge-up is given by properties of long-period variables (LPVs) with enhanced strength of the ZrO band. In fact, Zr is formed by s-processing in the convective He-burning shell during a shell flash and is dredged up to the surface (Wood et al. 1983).

However, even if intermediate-mass stars with initial mass in the range $M_w \leq M \leq M_{up}$ could experience deflagrating C-ignition, this does not occur for several reasons (Iben 1985, 1991). In short, as we infer from the density of matter in planetary nebula (PN) shells, the estimated outflow rates from OH/IR sources, the several nearby C stars, the paucity of C stars in rich clusters of the LMC (like NGC 1866), and finally the luminosity function of carbon stars in the same galaxy (Reid et al. 1990), there must be some fast mechanism, which on a very short time scale (10^3 yr) terminates the TP-AGB phase soon after it begins with a modest increase in the mass of the C-O core with respect to the initial value. The sudden termination of the AGB phase of all intermediate-mass stars has been long attributed to a sort of "superwind" (see Renzini & Voli 1981, and Iben & Renzini 1984), the physical interpretation of which is not yet understood. The manifestation of the superwind could be the OH/IR phenomenon. Estimates of the mass-loss rates from AGB stars and speculations about the physical nature of the superwind have been made by many authors among whom we recall Baud & Habing (1983), Bedijn (1988), van der Veen (1989), and Bowen & Willson (1991). Computations of AGB models including the effect of mass loss are still rare. Recent calculations are by Wood & Vassiliadis (1991) and Vassiliadis & Wood (1993) who identify the superwind in the combined effect of large amplitude radial pulsations and radiation pressure on grains.

The arguments presented above also suggest that the maximum mass of WDs is 1.1 $M_\odot$, considerably lower than the Chandrasekhar mass of 1.4 $M_\odot$, and slightly larger than the value of the C-O core mass at the start of the TP-AGB for a star with initial mass equal to M_{up} (see Iben 1991, and Weidemann 1990).

2.5. Planetary Nebulae and White Dwarfs

The main parameters governing the evolution from the AGB to the WD stage through the planetary nebula (PN) phase are the precise stage in a thermal cycle at which the final ejection of the H-rich envelope occurs and the amount of H-rich material which is left on the surface of the remnant at the termination of the ejection phase. Summarizing the results of many authors (Schoenberner 1979, 1981, 1983, 1987; Iben 1984, 1989; Iben & MacDonald 1985, 1986; Iben et al. 1983; Renzini 1979, 1982; Wood & Faulkner 1986), three evolutionary schemes are possible: 1. the ejection of the envelope occurs during the quiescent H-burning interpulse phase and the mass dM_{He} of the helium layer between the C-O core and the bottom of the H-rich envelope is "small", i.e. in the range 0.2 to 0.8 dM_H, where dM_H is the mass processed by the H-burning shell between He-shell flashes on the AGB (for a 0.6 $M_\odot$ core, dM_H is about 0.01 $M_\odot$); 2. the ejection occurs in the same stage but dM_{He} is greater than 0.8 dM_H; or 3. the ejection of the H-rich envelope occurs during a He-shell flash or shortly thereafter.

(1) In the first case, following the loss of the envelope, the star evolves blueward at about constant luminosity sustained by the H-burning shell at the base of the residual envelope. The surface temperature gets higher at decreasing mass of the H-rich surface layer. At $T_{eff} \geq 30,000$ K, the flux emitted by the central star ionizes the surrounding nebula and the complex appears as a PN. The time scale is of the order of 10^4 yr for a 0.6 $M_\odot$ star. In this phase, stellar winds from the central stars are known to occur at rates of about $10^{-9} \div 10^{-7}$ $M_\odot$/yr (Perinotto 1983, Cerruti-Sola & Perinotto 1985). When the mass of the H-rich surface layer falls below 10^{-2} dM_H, the H-burning shell extinguishes. The surface layers contract and the luminosity – which is a complicated consequence of gravitational energy release, cooling of non degenerate ions, and neutrino losses (e.g. D'Antona & Mazzitelli 1990) – drops dramatically. Gravitational diffusion becomes so strong that heavy elements sink and hydrogen, if any is left, becomes the dominant element at the surface. Ultimately, the star settles onto the cooling sequence of WDs for the given mass and composition. This model approximates well the characteristics of observed DA-WDs.

(2) In the second case, a final He-shell flash is possible. Following the extinction of the H-burning shell as in the previous case, helium ignites in a shell and the star is pushed back to the tip of the AGB. There the same mechanism that removed the H-rich envelope when the star left the AGB for the first time is likely to operate for a second time, forcing the departure from the AGB. However, in this case the luminosity of the star is sustained by the He-burning shell and departure from the AGB requires that mass loss continues until the residual mass of the H-rich material is less than 10^{-5} $M_\odot$. Evolving to high T_{eff} once again the PN is re-excited, and stellar winds from the central stars cause the loss of all remaining H-rich matter at the surface. The duration of this phase is about three times longer than in the previous case. Eventually He-burning ceases and gravitational sinking of heavy elements makes helium the dominant element at the surface. Finally, the star settles onto the WD sequence. This model nicely corresponds to non-DA-WDs.

(3) In the third case, the H-rich envelope is ejected during a He-shell flash when the intershell region contains the smallest amount of mass. Departed from the AGB, the luminosity, sustained by the He-burning shell, fades to a minimum as the star evolves to higher T_{eff}s. At a certain point, hydrogen is re-ignited and the luminosity increases again at almost constant T_{eff}. The subsequent evolutionary track lies close to the corresponding H-burning track (Iben 1984, Wood & Faulkner 1986). In coincidence with H-re-ignition, a marked slowdown of the evolutionary rate occurs. AGB stars becoming WDs through

this scheme, after about 10^7 yr from the extinction of the He-burning shell for a typical 0.6 $M_\odot$ star, may undergo a final H-burning runaway leading to an outburst which exhibits the characteristics of a slow nova (Iben & MacDonald 1985, 1986). There are some reasons to believe that planetary nebulae are the descendents of AGB stars via the ejection of the envelope during helium shell flashes (Iben 1991). Whether the slow nova episode is a common feature has not yet been tested.

The structure and evolution of WDs has been reviewed by D'Antona & Mazzitelli (1990), whereas the problems with the masses, mass distribution, and evolutionary status of these stars have been described by Weidemann (1990), to whom the reader should refer for further details. The distribution of single WDs with respect to the mass is based upon the position in the HRD and mass-radius relationship. The distribution is strongly peaked at M=0.55÷0.60 $M_\odot$, it extends downward at 0.4 $M_\odot$, and falls off exponentially beyond M=0.6 $M_\odot$ (Iben 1991, Weidemann 1990).

3. Physical causes of violent ignition, explosion, and collapse

He-Flash. How does the degenerate core react to the input of energy by the onset of He-burning ? In the case of a perfect gas, as a result of a small increase in the energy production, temperature and density increase, and to maintain hydrostatic equilibrium the core is forced to expand decreasing the density, hence the temperature. This ultimately regulates the rate of burning. On the contrary, in a degenerate gas pressure is not sensitive to temperature changes and core expansion does not follow the temperature rise. The rate of burning increases dramatically and the result is a thermal runaway (He-flash). A great deal of theoretical work has been devoted to clarify the nature of instability leading to a thermal runaway, to determine where in the core He-flash begins, and to follow the growth of a flash in hydrodynamical approximation to check whether partial or total disruption of the star may occur and/or whether partial or total mixing between incipient core convection and envelope convection may take place. All these topics have been discussed by Iben & Renzini (1984) in great detail. Suffice to recall here that with current neutrino energy losses, the maximum temperature in the core, and in turn He- ignition, occurs away from the centre at a mass fraction which increases with the core mass. The mass of the core at the stage of He-ignition depends on the initial chemical composition, stellar mass and neutrino energy losses. In a simplified picture of the He-flash, two factors tend to moderate the strength of the He-flash: first as temperature increases, the degree of degeneracy recedes and the expansion of the core begins; second, following the rapid increase in the core luminosity, convection sets in diluting the liberated energy over larger and larger masses. However, model computations in quasi static approximation show that following a major He-flash which removes degeneracy in that part of the core lying above the ignition layer, several minor flash episodes occur closer and closer to the centre (Mengel & Gross 1976; Despain 1981; 1982; Mengel & Sweigart 1981). By the time the last flash reaches the centre and degeneracy is completely removed, a significant fraction of helium has been converted into carbon (about 5%). There is the interesting possibility that some carbon is dredged up into the convective envelope, which could be somewhat related to elemental abundance peculiarities shown by red giants (cf. Iben & Renzini 1984). Dynamic flash calculations by different authors (Cole & Deupree 1980, 1981; Deupree & Cole 1981; Deupree 1986; Edwards 1969), which predict either total or partial disruption of the star, or in less extreme cases, a deflagrating wave sweeping across the star, leading to substantial conversion of helium into carbon, and distortions of the H-burning shell location by buoyant bubbles of nuclearly processed material rising from the core with possible mixing of this material into the H-burning shell, are not entirely convincing, and at some extent, also in contradiction with the observational evidence. In the light of these considerations, the general consensus is that the only effect of the He-flash is to remove degeneracy in the core and ultimately initiate the quiescent core He-burning phase.

Carbon detonation or deflagration. The evolution of central cores ensuing C-ignition in highly degenerate conditions has long been the subject of a great deal of theoretical work (cf. the reviews by Mazurek & Wheeler 1980; Sugimoto & Nomoto 1980). Since the early work of Rose (1969) and Arnett (1969), it was soon evident that if C-ignition occurs when the core is still not too concentrated, the energy liberated would be high enough to disrupt the star. Paczynski (1970a,b) also demonstrated that all stars which develop highly degenerate cores have virtually identical internal structures differing only in the envelope, hence in the total mass. This occurs because of the unique density structure of degenerate matter and a feedback between neutrino cooling and the accretion of mass onto the core from overlying H- and He-burning shells which control the temperature distribution (Barkat 1971). Evolutionary calculations of single constant mass stars show that C-ignition occurs at a central density of about 3×10^9 gr/cm^3 and a central temperature of about 3×10^8 K in a core of 1.39 M$_\odot$ just short of the Chandrasekhar limit (Paczynski 1970a,b; 1971). C-ignition occurs when the nuclear energy generation rate exceeds the neutrino loss rate. The initial phase of C-burning takes place in a central convective core and there is no immediate thermal runaway. However, very soon carbon flash grows into an explosive burning. In early studies of this phenomenon (Arnett 1968, 1969; Bruenn 1971) it was supposed that a detonation would be generated propagating through the core (cf. Mazurek & Wheeler 1980; Woosley 1986). A detonation is a shock-induced burning which propagates into the unburnt medium and its description is intimately connected with the physics of shock waves (cf. Mazurek & Wheeler 1980; Sugimoto & Nomoto 1980; Woosley, 1986, for all details). The result of those calculations was that all the fuel turned into nuclear statistical equilibrium abundances (NSE) and the liberated nuclear energy exceeded the binding energy of the star, thus disrupting and dispersing the whole core. Such a model of carbon detonation raised two problems: i) Shortage of pulsar progenitors (Gunn & Ostriker 1970; Ostriker *et al.* 1974). These in fact should be generated only by stars more massive than M_{up}. ii) Overproduction of iron group elements in the galaxy. In fact, if stars in the mass range $M_w < M < M_{up}$ experience carbon detonation and each star ejects about 1.4 M$_\odot$ of iron group elements, then these should have been produced by about six times more than expected (Ostriker et al 1974; Arnett 1974c). Although the problems have come to seem less severe for several reasons (as it will be discussed later on), yet many efforts have been made to avoid total disruption of the star in the C-detonation model. One way out was to postpone C-ignition at such high densities that β-processes lead to re-implosion of the core, thus trapping the newly formed iron group elements in a neutron star (cf. Sugimoto & Nomoto 1980; Mazurek & Wheeler 1980; Iben 1982a). However, as discussed by Sugimoto & Nomoto (1980), this seems unlikely to occur. Another possibility was seen in the convective URCA processes, which may take energy away with neutrino pair production (Paczynski 1972). In Paczynski's picture the following cycle should occur: convection sweeps material to higher density regions triggering the reactions $e^- + (Z, A) \Rightarrow (Z - 1, A) + \nu_e$. The convection then carries material to low density regions where the Fermi energy is less and beta decay occurs spontaneously $(Z - 1, A) \Rightarrow (Z, A) + e^- + \nu_e$. The net result is a restoration of the original composition but the loss of a neutrino pair. The neutrinos emitted by the URCA process basically carry away electron Fermi energy. The fluctuations in the Fermi energy are supplied by convection which in turn is driven by

C-burning. Thus the flux of URCA neutrinos in a sense limits the burning. However, in order for this mechanism to work everything has to be tuned very finely because the convective and beta processes time scales should be much shorter than the life time of C-burning. Furthermore, another complication was pointed out by Mazurek (1972) and Bruenn (1973), i.e. the URCA process tends to add energy to the thermal reservoir of electrons. In fact, in the high density part of the cycle, electrons are selectively captured from below the Fermi sea. The liberated levels are filled by other electrons that made the transition downward from the top of the Fermi sea, thus releasing the excess energy as heat. In the low density half of the cycle, beta decay produces suprathermal electrons which thermalize and heat up the gas. Thus, while URCA neutrinos remove a large quantity of energy, which ultimately comes from C-burning, the process seems to require a net heating of matter which would promote runaway. Finally, the consistency of the detonation assumption was questioned. If on one hand carbon detonation is a reasonable outcome, in that once initiated it appears to represent a self-consistent solution to the planar shock, on the other hand the major concern is with the initiation of the detonation itself. First of all it has to be checked whether or not explosive burning may grow into a detonation. In this context it is important to consider that the propagation of a shock through an inhomogeneous, spherical medium inhibits the strengthening of the shock itself (Ono 1960; Sugimoto & Nomoto 1980). Secondly, the energy released in C-burning is rather small compared to the Fermi energy and thus the overpressure that results from burning is rather mild. The spherical symmetry imposes zero velocity at the center and therefore a considerable volume must burn before self-consistent velocities are attained. A careful analysis of those effects indicated that detonation likely does not occur (Mazurek *et al.* 1974, 1977; Nomoto *et al.* 1976; Chechetkin *et al.* 1977). On the contrary, heat is carried outward by convection and conduction, while fresh fuel is carried inward by Rayleigh-Taylor instability, yielding what is called a "deflagration front". The primary characteristics of a deflagration front is that it propagates subsonically (cf. Mazurek & Wheeler 1980; Sugimoto & Nomoto 1980; Woosley 1986). The outcome depends largely on the rate at which burnt and unburnt materials come together, the limit of rapid transport tending toward detonation again. Buchler & Mazurek (1975) found that a deflagrating star ejects its H and He-rich envelope, but retains a partially burnt core that eventually proceeds to iron. Nomoto's *et al.* (1976) deflagration model, on the other hand, completely disrupted the star though less iron was produced. The results are still very uncertain, going from disruption, to the production of a quasi static iron core which lives for 10^3 yr before collapsing, to the creation of a stable white dwarf if the core mass is brought below the Chandrasekhar limit by mass loss driven by shock ejection's of material during the oscillations that are found to develop in several cases (Buchler & Mazurek 1975).

Electron capture instability. This phenomenon is typical of stars in the mass range $8 \div 12 M_\odot$. Following C-burning in non degenerate conditions, degeneracy gets very strong in the oxygen-neon magnesium core. Within this mass range two subgroups can be recognized, namely $M_{up} < M < M_{ec}$ and $M_{ec} < M < M_{mas}$, where M_{ec} and M_{mas} denote the minimum initial mass for stars being able to undergo core collapse by electron capture and to proceed to the formation of an iron core by the onion skin model, respectively.

(a) *The range* $M_{up} < M < M_{ec}$. Model stars in this range of mass have been studied by Miyaji *et al.* (1980) and Nomoto (1981, 1983a,b, 1984a,b,c,d), and the resulting supernova explosion has been investigated by Hillebrandt *et al.* (1984). Following non degenerate C-burning, an O+Ne+Mg core is formed. During the contraction phase of the core, Ne is not ignited as the mass interior to the He-burning shell is smaller than the critical mass for Ne ignition (1.37 $M_\odot$). Then the core becomes strongly degenerate while a temperature inversion is built up by efficient neutrino cooling. The core mass is gradually increased toward the Chandrasekhar limit by the double shell burning phase. A carbon layer is built up at the top of the core but compressional heating is not enough to ignite carbon in a shell. Once the mass interior to the H-burning shell is above 1.375 $M_\odot$, electron capture on ^{24}Mg and ^{20}Ne begins. As the number of free electrons decreases, the core starts contracting on the rapid time scale of electron captures. When the central density is about 2.5×10^{10} gr/cm^3, oxygen burning is ignited and a deflagration front incinerates the material into NSE composition. The propagation velocity of the burning front due to convective energy transport is much smaller than the infall velocity of the material and therefore almost a stationary front is formed. Eventually a full collapse sets in, further accelerated by electron captures on NSE elements, once the mass of the NSE core has become greater than its own Chandrasekhar limit. At the time of core bounce the NSE core contains about 1.1 $M_\odot$ (Hillebrandt et al. 1984). A supernova (type II) explosion releasing a total energy of about 2×10^{51} ergs and a neutron star remnant of about 1.2 $M_\odot$ is likely to occur.

(b) *The mass range* $M_{ec} < M < M_{mas}$. The evolution in this mass range is complicate and sensitive to the stellar mass. These stars undergo non degenerate C-burning and form O+Ne+Mg cores whose mass is in the range 1.37 to 1.5 $M_\odot$. The core mass is therefore high enough to ignite Ne, yet the core is semidegenerate and the degree of degeneracy depends on the ratio of the core mass to the Chandrasekhar mass. In such a core, combined effects of degeneracy and neutrino cooling produce a temperature inversion. This leads to an off center Ne-ignition. Because of electron degeneracy, shell Ne-burning is unstable to a flash (Nomoto 1984b). Subsequent behaviour of shell Ne-burning is sensitive to the stellar mass and crucial for the ultimate fate of the star (Woosley et al. 1980). For stars of about 12 $M_\odot$, to which a helium core mass of about 3 $M_\odot$ corresponds, a neon flash is first started off center ($M_r = 0.3 M_\odot$) and subsequent Ne and O-burning form a layer composed of ^{28}Si, ^{30}Si and ^{34}S. The Ne-burning layer (with $T = 2 \times 10^9$ K) propagates inwards all the way down to the center to form a Si-S core. Ne-burning is ignited layer by layer so that the released energy in one flash is too small to induce major dynamical effects (Nomoto 1985). For slightly lower masses (11.2 $M_\odot$ with $M_{He} = 2.8 M_\odot$), the inward Ne-burning shell reaches such high density layers ($\rho > 10^8$ gr/cm^3) that Ne-shell flashes are so explosive as to cause dynamical effects (Woosley et al. 1980). Inward propagation of Ne-burning is induced by compressional heating due to gravitational contraction of the O+Ne+Mg core rather than heat conduction. However, this may not be true for even smaller stellar masses (or core masses), as whether Ne-burning shell reaches the center depends on the competition between neutrino cooling and compressional heating. In other words, the central temperature could start decreasing before reaching Ne-ignition. Therefore, Ne and O-burning could be quenched by neutrino cooling (stars in the range 10 to 11 $M_\odot$). Further evolution after central Ne and O-burning also depends on the mass of the Si-S core relative to the Chandrasekhar mass. It is important to note that the relative number of free electrons is reduced as low as 0.48 during O-burning even before appreciable electron captures start, because O-burning produces copious amounts of neutron-rich elements (^{30}Si, ^{34}S). During the contraction of the Si-S core, this number decreases even still due to electron captures (Thielemann & Arnett 1985). Therefore, electron degeneracy may not become strong and Si-ignition is expected to occur with the formation of an iron core.

Iron break-down. This phenomenon is the starting cause of the final collapse of a massive star, i.e. in the range $M_{mas} < M < M_{vms}$, with M_{mas} and M_{vms} of about $13 \div 15 M_\odot$ and $100\ M_\odot$ respectively. This class of stars is perhaps the most studied, models having been computed to the very latest stages for a wide range of masses, chemical compositions and considerable variety of input physics (Rakavy *et al.* 1967; Paczynski 1970a,b; Arnett 1977; Sparks & Endall 1980; Lamb *et al.* 1976; Weaver *et al.* 1978; Weaver & Woosley 1980; Woosley & Weaver 1982a,b, 1983, 1985; Weaver *et al.* 1982, 1985, 1986; Woosley *et al.* 1984). The general consensus is that these stars carry out the whole natural sequence of nuclear reactions. The dominant feature of these models is the important role played by neutrino cooling in the interiors as originally pointed out by Rakavy *et al.* (1967). The strong neutrino cooling promotes degeneracy in the core and postpones the ignition of each successive fuel. The tendency for a given central burning stage to be convective is diminished. Since fuel is not swept in from a large volume, the cores are restricted in mass, the corresponding to each successive fuel being somewhat smaller than in a previous one. The growth of the mass of successive cores is also inhibited by the overlying burning shells. Because the cores are partially degenerated and cooled by neutrino losses, they reach the point of ignition by evolving to higher densities and hence close to the Chandrasekhar mass. This tends to develop cores of about $1.4\ M_\odot$ independently of the total mass of the star. Although there are differences in details, all models computed give similar results. Successive fuels are burned terminating in the development of an iron core. This core is surrounded by burning shells spaced by layers of inert fuel. The resulting core structure depends mainly on the mass of the He core built up during the core He-burning phase. There is a general agreement on the ultimate fate of stars which develop iron cores. The iron cannot undergo exothermic nuclear reactions. On the contrary, subject to the intense pressure of the overlying layers, the iron core is heated and undergoes endothermic breakdown into He, neutrons and protons. Furthermore, the high temperature and density promote electron captures. Both effects concur to remove pressure support at the center and the core begins to collapse on a dynamical time scale.

Electron pair instability. It occurs in very massive stars confined in the mass range $M_{vms} < M < M_{sms}$, i.e. from about 100 to $5 \times 10^4 M_\odot$. They are found to be dynamically stable against general relativity effects, pulsationally unstable during core H and He-burning phases and finally to suffer pair instability during core oxygen burning. Souffrin (1960) demonstrated that there is a region near $T = 2 \times 10^9$ K and $\rho < 10^6$ gr/cm^3 where copious electron-positron pair production lowers the adiabatic index Γ_1 below 4/3. Fowler & Hoyle (1964) first suggested that a new type of supernova can occur for massive stars going pair unstable. When enough of the core has $\Gamma_1 < 4/3$, it begins to collapse on a time scale of minutes. As it heats, oxygen burning sets in releasing energy, heating the gas still further and burning more oxygen until the equation of state stiffens, the collapse stops and reverses itself. Many numerical calculations show that the final outcome is ultimately regulated by the mass of the oxygen (carbon- oxygen) core M_{co}. If the mass of the oxygen core is greater than some critical value, M_{cr}, the heavy element core formed later collapses to a black hole. For $M_{co}^* < M_{co} < M_{cr}$ complete thermonuclear disruption occurs. In stars whose oxygen core is close to the lower end, M_{co}^*, oxygen-burning does not set in until the center is too tightly bound to be destroyed by nuclear energy release. The inner part of the star, then burns on up to iron and perhaps joins the fate of less massive stars. The critical core masses are estimated to be $M_{co}^* = 30 M_\odot$ and $M_{cr} = 110 M_\odot$. The initial mass of the star corresponding to the above values are about $100 M_\odot$ and $300\ M_\odot$ respectively. The above scheme is substantiated by model computations of Woosley & Weaver (1982a,b), Ober et al. (1983), Bond et al. (1982, 1984), Bond (1984), Woosley et al. (1984).

Explosive H-burning and general relativity instability. Supermassive stars are those that collapse directly to black holes via general relativity effects which raise the critical adiabatic index Γ_1 to $4/3+O(GM/Rc^2)$ or suffer total disruption due to explosive H-burning. The mass ranges in which the two alternatives are expected to occur are $M > 10^5 M_\odot$ for total collapse and $M < 10^5 M_\odot$ for explosive H-burning, depending upon the mass, initial metal content, and degree of rotation. Supermassive stars have been studied by Appenzeller & Fricke (1973), Appenzeller & Tscharnuter (1973), Fricke (1973, 1974), Fuller et al. (1982a,b). Their evolution can be summarized briefly as follows: Stars having such great mass are general relativity unstable (Iben 1963; Fowler 1966). After spending only a few thousand years contracting as gravity provides the radiation leaving their surface, the stars reach a critical radius (smaller than 10^{14} cm for an object of $5 \times 10^5 M_\odot$), where the first order relativistic correction to Newtonian gravity renders the star, which has Γ_1 nearly equal to $4/3$, unstable to continue collapse. Were it not for the presence of unburnt material, this collapse would continue indefinitely until the star becomes a black hole. However, since hydrogen is present, it is possible for nuclear reactions during the implosion to release sufficient energy to power an explosion. Fuller et al. (1982a,b) similarly to Appenzeller & Fricke (1972), Appenzeller & Tscharnuter (1973) and Fricke (1973, 1974) but not to Ober (1981), found that non rotating stars in this mass range will collapse to black holes with no reversal of implosion unless the initial metallicity of the star is a substantial fraction of the solar value ($Z > 0.005$). The results are quite sensitive to metallicity since the rate of H-burning in the CNO cycle is known to depend on the abundance of CNO elements. Lower burning rates do not release enough energy during collapse to reverse the implosion (cf. Woosley et al. 1984 for more details). Quite different results may be expected if rotation is considered (Fricke 1974). In particular, higher central temperatures may be reached by objects that have very low initial metallicity, and explosion is facilitated by the centrifugal potential.

Central Conditions. In concluding this section we present the plane of central conditions (temperature versus density) in which the hydrostatic history of stars of any initial mass is illustrated (Fig. 2). The same figure also visualizes the various regions of instability and the loci of successive nuclear burnings.

4. Two basic ingredients: nuclear reactions and opacities

Nuclear reactions. Although the rates of the major reactions involved in H and He-burnings are sufficiently known, the works of Fowler et al. (1967, 1975) and Harris et al. (1983) show that many of them have changed over the years and that some uncertainty is still possible. This is particularly true for the $^{12}C(\alpha,\gamma)^{16}O$ reaction whose cross section has been varied several times. The measurements by Kettner et al. (1982) and the analysis of Langanke & Koonin (1982) increased the cross section by a large factor (from 3 to 5) with respect to the Fowler et al. (1975) estimate. Subsequent revision of the rate (Fowler 1984, Caughlan et al. 1985) set the increase at about $2 \div 3$ times the old value. The most recent revision of this problem (Caughlan & Fowler 1988) has lowered the rate to nearly the same value as in Fowler et al. (1975). The effects of varying the rate of this reaction on evolutionary models have been known since the early study by Iben (1972). Specifically, the larger the cross section, the greater is the extent to which carbon is converted into oxygen, and the further the loop extends to the blue before rapid core contraction and envelope expansion set in and the evolution proceeds back to the red.

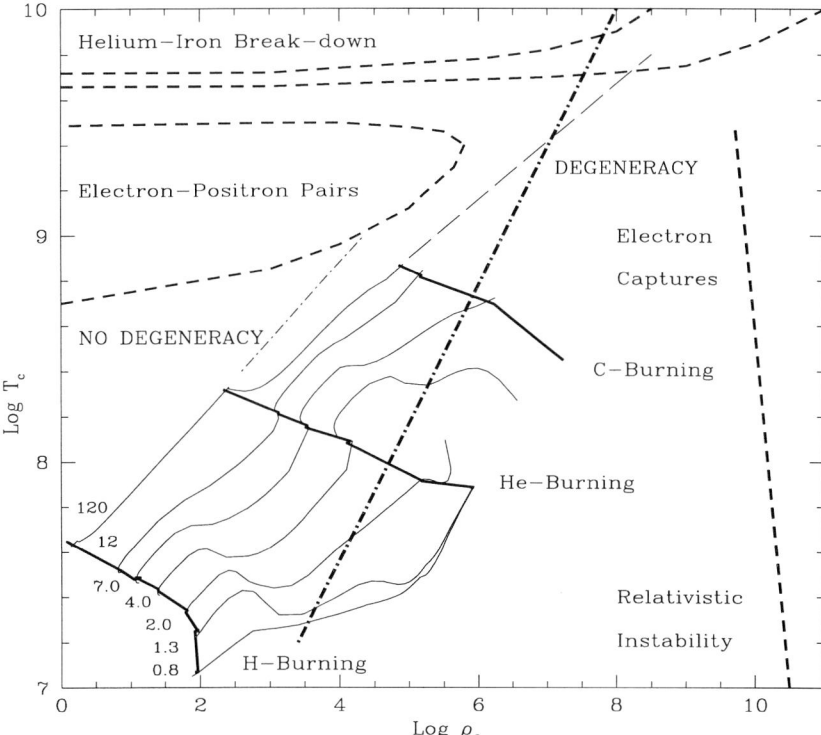

FIGURE 2. The path in the ρ_c versus T_c plane followed by the cores of stars with different initial mass during their evolution. The thick dashed lines delimit the regions of dynamical instability due to nuclear dissociation, electron-positron production, electron capture and relativistic instability as indicated. The heavy dotted long-dashed line shows the boundary between electron degenerate and non degenerate conditions. The thick solid lines show the loci of H-, He- and C-ignition (burning). The thin solid lines are the evolutionary paths of the 0.8, 1.3, 2.0 4.0, 7.0, 12 and 120 $M_\odot$ stars. These stellar models are with mass loss and mild core overshoot (see the text for more details). The 120 $M_\odot$ star evolved in presence of mass loss avoids the electron-pair instability region and goes towards the helium-iron break-down. In contrast the same star evolved at constant mass would go pair unstable. The path is schematically shown by the thin dotted-long-dashed line

The core He-burning lifetime is increased by a few percent. In massive stars, however, the above effects are blurred by mass loss and convective overshoot (see below). Finally, the abundance of carbon can get so low that the C-burning phase is actually missing, with profound consequences on later evolution of these stars (e.g. Woosley 1986, 1988; Woosley & Weaver 1986).

Opacities. For more than twenty years, almost all evolutionary calculations were made with the Los Alamos Opacity Libraries (LAOL) based on the work of many groups (Cox & Stewart 1965, 1970a,b; Cox & Tabor 1976; Magee et al. 1975; Huebner et al. 1977; Weiss et al. 1990). Occasionally, other opacities calculated by T. R. Carson (1976, unpublished) were used (Carson & Stothers 1976, 1988; Carson et al. 1981; Stothers 1976; Stothers & Chin 1977, 1978). The role played by each particular source of opacity in building up the total radiative opacity has been discussed by Iben & Renzini (1984), to whom we refer. The region in the temperature range 5×10^5 to 5×10^6 K, where the bound-bound and bound-free transitions of elements from carbon to iron dominate the opacity, is the one suffering from the highest uncertainty. The high number of electrons in each elemental species, the large number of ionization and excitation stages, the nonhydrogenic structure of the electronic configuration, and the distortions of this induced by nearby electrons and ions that are difficult to model, all conspire to make the opacity in this temperature range very difficult to calculate, and therefore subject to continuous upgrading. This is the main reason why the characteristics of stellar models that are very sensitive to the so-called middle temperature opacity are still highly uncertain and a matter of debate. LAOL opacities were based on the hydrogenic atomic model, whereas Carson's opacities stood on the hot "Thomas-Fermi" approximation. The two opacities were quite similar except for the region of the CNO ionization where in the Carson opacity a pronounced bump was present. Although Carson's opacities were retracted by Carson et al. (1984), various reasons suggested that an increase of this type was indeed necessary. For example the possibility that opacity enhancements could be responsible for the pulsation of β-Cephei stars was examined by Stellingwerf (1978). Simon (1982) noticed that by increasing the opacity of the metals by factors of $2 \div 3$ for $10^5 K \leq T \leq 10^6 K$ in models of classical Cepheids, the observed period ratios could be reproduced for masses and luminosities in agreement with those of evolutionary models. Bertelli et al. (1984) introduced an opacity bump in the CNO ionization region of the LAOL opacity and studied the effects on the location in the HRD of models with core overshoot and mass loss. The opacity peak was set at $\log T = 5.80$ and the opacity was increased by a factor of $2 \div 3$. These models were particularly successful in explaining the overall properties (main sequence extension, lifetimes, etc) of massive stars. The suggestion by Simon (1982) was confuted by Magee et al. (1984) who claimed that such opacity increase was incompatible with atomic physics. However, the opposite conclusion was reached by Iglesias et al. (1987,1990) and Iglesias & Rogers (1991a,b) who presented new opacity calculations using improved atomic physics showing that a significant increase in the opacity (bump-like structure) is present at about $\log T = 5.38$. Using the opacity calculations of Iglesias et al. (1987) as a guide, Andreasen & Petersen (1988), Andreasen (1988), and Petersen (1989, 1990) artificially enhanced the LAOL opacity by factor α in the range $1.5 \times 10^5 \leq T \leq 8 \times 10^5$ K. Adopting $\alpha = 2.5$ they reproduced the period ratios for double mode Cepheids of population I, whereas using $\alpha = 1.5$ they resolved the mass anomalies for the lower metallicity RR Lyrae stars. The enhancement and mass range proposed for the population I Cepheids is in close agreement with the suggestion of Bertelli et al. (1984) for supergiant stars. Those suggestions for an opacity enhancement are confirmed by the recent opacity calculations of by at least two independent groups, i.e. the OPAL project at the Livermore Laboratory (Iglesias & Rogers 1991a,b; Rogers & Iglesias 1992, 1993; Iglesias et al. 1992, 1995; Iglesias & Rogers 1996 and references) and the OP project (Seaton 1987, 1991; Berrington et al. 1987). The new opacities, the OPAL in particular, have become of general use and almost the totality of stellar models in literature are calculated with the most recent release of opacity tabulations.

5. Stellar winds: observational and theoretical hints

It is now widely recognized that many stars have an emergent mechanical as well as radiative flux indicating that they are losing mass at some rate. Luminous main sequence OB stars, OB giants and supergiants, M giants and supergiants, Wolf-Rayet (WR) stars, AGB stars, and central stars of planetary nebulae, lose mass at much higher rates that do affect their evolution. Since the theory of stellar winds is magistrally presented by

Kudritzski (this volume), I concentrate on a few aspects more tightly related to stellar models, i.e. the use of the theoretical or empirical relationships for Ṁ as a function of basic stellar parameters. The most recent compilation of mass loss rates of stars of any spectral type and luminosity class is by de Jager *et al.* (1988), who derived empirical relationships for Ṁ as a function of luminosity and effective temperature.

5.1. *Mass loss from early type stars and Wolf Rayet stars*

In recent years, quantitative mass-loss theory applicable to massive early type stars has been developed starting from the seminal studies of Lucy & Solomon (1970), Castor *et al.* (1975), Abbott (1982), Pauldrach *et al.* (1986), and Owocki *et al.* (1988), which allows one to calculate the mass-loss rate for a given star.

Early type stars. Essentially both theoretical and empirical studies of stellar winds from early type stars lead to two types of dependencies for Ṁ on stellar parameters. The mass-loss rate is a function of the sole luminosity

$$\log \dot{M} = \alpha + \beta \log L \qquad (5.3)$$

where α depends on the units in usage and $\beta = 1 \div 2$. This kind of dependence is primarily suggested by the data and is one of the main results of the radiation pressure theory in its various versions. However as the data show a large scatter at any given luminosity which cannot be accounted for by uncertainties of the measuring procedure, other more complex relations are proposed in which Ṁ is a function of luminosity, mass, radius, metallicity, and other stellar quantities

$$\log \dot{M} = \alpha + \beta \log L + \gamma \log M + \delta \log R + \zeta \log(Z/Z_\odot) \qquad (5.4)$$

with obvious meaning of the symbols.

A compilation of mass loss rate prescriptions (somewhat out of date) can be found in Chiosi & Maeder (1986) and Chiosi (1986b). Despite their similarity, the various relations are not strictly equivalent, and when used in evolutionary computations, they lead to substantially different results(cf. Chiosi & Maeder 1986).

LBV stars. Considerable uncertainties remain in the mass loss rates for stars near the so-called Humphreys-Davidson limit (LBV candidates) and WR stars. The limit runs from $\log(L/L_\odot) = 6.8$ at $T_{\text{eff}} = 40,000$ K to $\log(L/L_\odot) = 5.8$ at $T_{\text{eff}} = 15,000$ K and it stays constant at lower T_{eff} (Humphreys & Davidson 1979; Humphreys 1989, 1992). This limit has been interpreted as the *generalized* Eddington limit with the flux-mean opacity, lines included, as compared to the classical one with only electron scattering opacity (Lamers & Fitzpatrick 1988). The limit was shown to agree with the observed one in the Milky Way and LMC. Subsequent investigations indicate that the Eddington limit increases again at low T_{eff} since the opacity decreases considerably there (Lamers & Noordhoek 1993). The saddle-shape of the Eddington limit in the HRD is more and more pronounced (lower luminosities) at increasing metallicity. Keeping in mind that stars begin their evolution at the blue side of the HRD and tend to move redward, it is nowadays understood that those hitting the Eddington limit suffer from strong mass loss pushing them back to the blue. After several attempts to cross the Eddington limit, each time with strong mass loss, stability and blueward evolution are restored. Stars in this stage are currently associated to the LBV phenomenon. Stars remaining fainter than the saddle of the Eddington limit (approximately $\log(L/L_\odot) = 5.8$) can reach the red supergiant region. In principle red stars brighter than the above limit could exist with a stable atmosphere. Their absence is ascribed to the fact that they cannot reach this area because they met already earlier the Eddington limit during their redward evolution. Given these premises, the general question remains about the physical origin of the outbursts in LBV and in hypergiant stars. The topic is highly controversial and several models have been proposed. In brief, the most striking properties is the strong density inversion in the outer layers, where a thin gaseous layer floats upon a radiatively supported zone. This is due to the peak in the opacity which forces supra-Eddington luminosities in some layers. Three different conclusions are drawn: 1. The Rayleigh-Taylor instability occurs washing out the density inversion; 2. Super-Eddington luminosities drive an outward acceleration and induce mass loss without a density inversion; 3. Strong convection and turbulence develop and the inversion is maintained. How all this translates into enhancing the mass loss rate is not understood. The empirical solution is to adopt the rate of about $10^{-3} M_\odot/\mathrm{yr}$ for all phases reasonably associable with LBV stars.

WR stars. For $\dot{M}$ in WR stars, the average observed rates (Abbott et al. 1986; Conti 1988) have often been used. However these rates yield stellar models whose masses and luminosities are too high with respect to the observational ones (Schmutz et al. 1989). It has been suggested that for WNE and WC stars $\dot{M}$ ought to depend on the mass. For instance Langer (1989) gives the relation

$$\dot{M}_{WR} = (0.6 - 1.0) \times 10^{-7} \left(\frac{M}{M_\odot}\right)^{2.5} \; M_\odot \; \mathrm{yr}^{-1} \tag{5.5}$$

where the first coefficient applies to WNE and the second to WC stars. Similar dependence has been inferred from binary WR, and models of stellar winds (Turolla et al. 1988; Bandiera & Turolla 1990; Schaerer & Maeder 1992). There is at present no indications of a mass dependence of $\dot{M}$ for WNL stars. The physical cause of the high mass loss rates in WR stars is not understood. On the notion that pure He stars with mass in excess of about 16 $M_\odot$ are vibrationally unstable (Noels & Masarel 1982, Noels & Magain 1984), it has been suggested that WR stars, WN and WC in particular owing to their inner structure (cf. Chiosi & Maeder 1986, Maeder & Conti 1994), go vibrationally unstable and this causes an increase of their mass loss rate. The strong winds from WR stars have also been attributed to multi-scattering and purely radiative processes (Pauldrach et al. 1988, Cassinelli 1991), radiation and turbulence (Blomme et al. 1991), or radiation and Alfven waves (Dos Santos et al. 1993). The main problem remains to explain why the wind momentum of WR stars may be up to 30 times the photon momentum (Barlow et al. 1981, Cassinelli 1991, Lucy & Abbott 1993).

Finally, it is worth recalling that according to current empirical mass-loss rates, massive stars lose much more mass than is expected from the theoretical mass-loss rates, in particular during the core H-burning phase. The reason for the discrepancy is not known.

5.2. *Mass loss rates from late type stars*

The problems related to detection and modeling of mass loss in evolved cool stars have been reviewed by Lafon & Berruyer (1991). As for the mechanism powering the wind from late type stars, the high mass-loss rates observed in Mira stars, which are in excess of $10^{-8} M_\odot/\mathrm{yr}$ and often as large as $10^{-4} M_\odot/\mathrm{yr}$, together with the high luminosity of these stars suggested that the mass-loss mechanism is the radiation pressure on the gas. However, standard opacities of the atmospheric layers were too small to effectively accelerate the gas against gravity. The inclusion of H_2O to the opacity (Alexander et al. 1989) could alleviate the problem (Elitzur et al. 1989). Subsequently, the key mechanism to transfer momentum from radiation field to gas was identified in the opacity due to graphite (in carbon stars) and dust grains formed by coagulation of oxides and carbides of heavier elements (Si, Mg). Since the opacity of dust is very high, the radiation force on dust can overcome gravity and the momentum of the dust is imparted to the gas which is dragged along. This mechanism was first suggested by Kwok (1975) and has been applied to AGB stars by many authors (Gail & Sedlmayr 1987 and references therein). However, the main problem with this model is that dust forms at large distances from the star, where the density is too low to build a significant wind. Therefore, it was suggested that another mechanism should exist extending the stellar atmospheres and increasing the density at radii where dust can form.

Bowen (1988) has shown that stellar pulsation can enhance the atmospheric density scale height and can drive a wind together with radiation forces on dust. The connection between pulsation and mass loss has been thoroughly discussed by Willson (1988) to whom we refer. Often, useful dependencies of the mass-loss rates on basic stellar parameters are given (see for instance Volk & Kwok 1988). This mechanism cannot be applied to those giants showing substantial mass loss and no evidence of circumstellar dust.

To this aim, another mechanism was advanced, in which sound waves are responsible for the mass loss. The sound waves are generated either by convection in the mantle of the star or by pulsation at high eigenmodes. Models of this type were applied to AGB stars by Pijpers (1990), Pijpers & Hearn (1989), and Pijpers & Habing (1989).

There has been much speculation about the nature and cause of the fast mass loss otherwise called superwind (e.g. Iben & Renzini 1983, Iben 1987). Recent work on the subject is by Bowen & Willson (1991) who calculated large grids of dynamical atmosphere models for Mira-like stars. They found that as a natural consequence of evolutionary changes in stellar parameters, the mass-loss rate increases as an approximately exponential function of time and the final evolution is characterized by a powerful wind that resembles the kind of superwind first advocated by Renzini & Voli (1981).

Recently, Bloecker (1995) has suggested a relationship to evaluate the mass rate, based on the dynamical calculations of the atmosphere of Mira-like stars by Bowen (1988). Always in this context, Vassiliadis & Wood (1993) have suggested another semi-empirical formalism relating the mass loss rate $\dot{M}$ to the pulsational period P. It has been derived from observational determinations of $\dot{M}$ for Mira variables and pulsating OH/IR stars both in the galaxy and the LMC. The notable feature of this prescription is the onset of the superwind which develops naturally on the AGB, instead of the artificial sudden transition that is needed if a Reimers-like law for $\dot{M}$ is used. Two distinct phases of mass loss are considered. For periods shorter than 500 days $\dot{M}$ increases exponentially with P (Mira phase), while beyond this limit the mass-loss rate is practically constant at values typical of the superwind (few times 10^{-5} $M_\odot$ yr^{-1}). The relations are:

$$\log \dot{M} = -11.4 + 0.0123 P, \qquad P \leq 500 \qquad (5.6)$$

$$\dot{M} = 6.07023 \; 10^{-3} \frac{L}{c v_{exp}}, \qquad P \geq 500 \qquad (5.7)$$

Here, $\dot{M}$ is given in units of $M_\odot$ yr^{-1}, the stellar luminosity L is expressed in $L_\odot$, the pulsation period P in days, c is the light speed (in km s^{-1}) and v_{exp} (in km s^{-1}) denotes the terminal velocity of stellar wind. In eq. (5.7) $\dot{M}$ is the maximum mass loss rate which is obtained by equating the final mass momentum flux $\dot{M} v_{exp}$ to the momentum flux of the entire stellar luminosity, according to the radiation-driven-wind theory (Castor et al. 1975). The wind expansion velocity v_{exp} (in km s^{-1}) is calculated in terms of the pulsation period P (in days):

$$v_{\text{exp}} = -13.5 + 0.056 \text{P} \qquad (5.8)$$

with the additional constraint that v_{exp} lies in the range $3.0 - 15.0$ Km s^{-1}, the upper limit being the typical terminal velocity detected in high mass-loss rate OH/IR stars. The pulsation period P is derived from the period-mass-radius relation [eq. (4) in Vassiliadis & Wood (1993)], with the assumption that variable AGB stars are pulsating in the fundamental mode:

$$\log \text{P} = -2.07 + 1.94 \; \log \text{R} - 0.9 \; \log \text{M} \qquad (5.9)$$

where the period P is given in days; the stellar radius R and mass M are expressed in solar units. This relations has been successfully applied by Marigo *et al.* (1996a, 1997) in their semi-analytical models of AGB stars (see also section 13).

Finally, concerning the rate of mass loss from red supergiant stars, Salasnich *et al.* (1997) have noticed that the empirical law of de Jager *et al.* (1988) underestimates the rate of mass loss in the RSG phase as compared to the data of Reid *et al.* (1990), based on the IRAS infrared fluxes of 15 RSG variable stars in the LMC.

$$\log \dot{\text{M}} = 1.32 \times \log \text{P} - 8.17 \qquad (5.10)$$

where P is period (in days). Salasnich *et al.* (1997) combined the above relation with the empirical one by Feast (1991) between the absolute bolometric magnitude and period

$$\text{M}_{\text{bol}} = -2.38 \times \log \text{P} - 1.46 \qquad (5.11)$$

obtained from IRAS data, and suggested the following relation for $\dot{\text{M}}$

$$\log \dot{\text{M}} = -8.17 + 0.554 \times [2.5 \times \log(\frac{\text{L}}{\text{L}_\odot}) - 6.18] \qquad (5.12)$$

The adoption of this mass loss rate in models of massive stars yield results able to improve our understanding of the HRD of supergiant stars and provide an alternative scenario for the formation of the low luminosity WR stars to be presented in section 12 below.

5.3. *Evolutionary arguments for mass loss from low and intermediate mass stars*

Single stars less massive than about $10 \text{M}_\odot$ do not lose much mass while on the main sequence, but develop strong winds during the RGB and AGB phases. Stellar model calculations show that the evolutionary characteristics of red giant and AGB models, having degenerate cores, are practically insensitive to the envelope mass. This applies in particular to the evolutionary rates along the RGB and AGB. Therefore the decrease in stellar mass due to mass loss by stellar wind can be obtained integrating the equation

$$d\text{M} = \dot{\text{M}} \times dt_{\text{ER}} \qquad (5.13)$$

provided that laws for $\dot{\text{M}}$ and evolutionary rate "dt_{ER}" as a function of basic stellar parameters are assigned. While the evolutionary rate is well known from the body of stellar model calculations, the mass loss rate suffers from the uncertainties discussed above.

5.3.1. *Mass loss along the RGB*

Perhaps the most convincing argument determining the amount of mass to be lost by low mass stars along the RGB is that constant mass models cannot account for the morphology of horizontal branches (HB) in globular clusters. In fact, the mass of HB stars must be less than 0.63 $M_\odot$ for Z=0.001 and Y=0.230 in order to populate the zero age horizontal branch (ZAHB) blueward of the red edge of the instability strip. The argument of constant mass and blueward evolution during the HB phase is very weak and can be easily confuted (cf. Renzini 1977). Furthermore, not all stars in HB have the same mass but a little dispersion is required. Typical values of $\Delta M = M_{RGB} - M_{HB}$ and related dispersion σ are 0.2 $M_\odot$ and 0.025 $M_\odot$ for a typical main sequence mass of about 0.8 $M_\odot$. Various suggestions have been made to identify the origin of the HB mass dispersion. Fusi-Pecci & Renzini (1975) argued that small differences from one star to another in the maximum luminosity attainable during RGB, before core He-flash occurs, may result into differences in ΔM. Furthermore, differences in the core rotation among stars in a given cluster could explain the differences in the flash luminosity. Detailed studies of the effect of mass loss along the RGB phase have been carried out by Fusi-Pecci & Renzini in a series of papers using the Reimers (1975) formulation

$$\dot{M} = \eta \ 4 \times 10^{-13} \frac{L}{gR} \quad M_\odot yr^{-1} \tag{5.14}$$

(where all quantities are in solar units) to assign the mass loss rate. The evolutionary rate dt_{ER} was derived from Rood (1973) who gave useful expressions relating the time spent by a RGB star above a given luminosity, as functions of the stellar luminosity, mass and chemical composition.

The study of Fusi- Pecci & Renzini (cf. Renzini 1977) set very sharp constraints on the parameters η. The most important result is that η must be confined in a very narrow range of values in order to match the requirement imposed by the HB morphology. The parameter η was found in the range 0.40 ± 0.05. Evolutionary models calculated in occurrence of mass loss according to the Reimers (1975) relationship confirmed the theoretical predictions (Schoenberner 1979; Scalo 1976; Renzini & Voli 1981).

No similar constraints can be posed on η from the evolution of intermediate mass stars, which are expected to lose only a tiny fraction of mass during their red giant phase (first ascent along the Hayashi line following core H-exhaustion). This problem has been discussed by Renzini (1977, 1981a,b).

Finally, whether the mass loss rate for RGB stars is sensitive to metallicity is neither theoretically nor observationally well understood. Arguments are given, however, that at least for globular clusters the Z dependence must be very mild, otherwise the correlation between HB morphology and Z would be opposite to that observed (Renzini 1981a,b). Theoretical considerations on the nature of UV fluxes in elliptical galaxies by Greggio & Renzini (1990) suggest the following dependence

$$\dot{M} = \eta \ 4 \times 10^{-13} (\frac{L}{gR}) \times (1 + \frac{Z}{Z_{cr}}) \quad M_\odot yr^{-1} \tag{5.15}$$

with $Z_{cr} \simeq Z_\odot$. Recent studies on old open clusters (age of a few Gyr) and high metallicity have suggested that the mass loss rate during the RGB must strongly increase with the metallicity (Tripicco et al. 1993; Liebert et al. 1994). This claim has been however confuted with a number of counter-arguments by Carraro et al. (1996). The question is still open.

5.3.2. *Mass loss along the AGB*

Also in this case, the morphology of globular clusters strongly suggests that mass loss must occur along the AGB. There in fact, no stars brighter than the RGB tip luminosity are seen. In a typical old globular cluster with red giant (and/or turn off) mass of $0.8 M_\odot$

(the actual value depending also on the chemical composition), an HB mass of about $0.6 M_\odot$, one would expect a maximum AGB luminosity of $\log(L/L_\odot) = 4.2$, while the observed luminosity is about $\log(L/L_\odot) = 3.3$, very close to the RGB tip luminosity. It goes without saying that the maximum luminosity is the one obtained from the luminosity-core mass relation in which $M_{co} = M_{HB}$ is assumed (cf. Renzini 1977 for all details). On the other hand, on the basis of the lifetime, about $0.02 N_{HB}$ stars (N_{HB} being the number of HB stars) with luminosity up to 2.5 mag above the He-flash luminosity should be present in globular clusters if mass loss does not occur. On the contrary, there is no trace of such very bright AGB stars in well populated globular clusters. This means that mass loss must prevent stars from reaching luminosities significantly above the He-flash luminosity during AGB evolution. Approximately $0.1 M_\odot$ must be lost by AGB stars to match the observational constraints. If mass loss is regulated by the Reimers (1975) relation this would imply $\eta \gg 0.3$.

For younger clusters or equivalently stars of higher initial mass, the situation is more uncertain as a sharp observational constraint is missing. In clusters of the LMC there is a lack of bright AGB stars in the luminosity interval $-6 \geq M_{bol} \geq -7$, contrary to what observed in the field. Furthermore the observed C-stars are fainter than predicted by the theory (see below). For long time these facts have been considered basic constraints on modeling the evolution of AGB stars and setting limits on the amount of mass that has to be lost during the AGB phase. In the standard scenario the structure and evolution of AGB stars from intermediate mass progenitors is similar to that in the lower range of mass. At this stage we neglect the complicacy of envelope burning, which deeply alters the scheme we are going to discuss. Considerations similar to those holding for globular clusters should apply also to this case setting the maximum luminosity attainable by an AGB stars at the value corresponding to $M_{co} = 1.4 M_\odot$, i.e. $\log(L/L_\odot) = 4.7$ or $M_{bol} = -7$. Using the Reimers mass loss rate with the same η found to hold for classical globular clusters, not enough mass can be lost to prevent these stars from becoming very luminous, and the lack of cluster AGB stars in luminosity interval $-6 \geq M_{bol} \geq -7$ cannot be explained. The solution was found rather than in the increase of the mass loss rates all along the AGB in the concept of superwind (see Iben & Renzini 1983, 1984), blowing off the residual envelope on a very short time scale at the very end of the AGB phase. The amount of mass lost by an individual star is maximum at some critical value of the initial mass M_w. For $M < M_w$, the wind evaporates the H-rich envelope of the star before the CO core reaches the Chandrasekhar limit, while for $M > M_w$ the CO core reaches the Chandrasekhar limit and the star explodes. In both cases the AGB phase is terminated.

Nowadays, the understanding of the luminosity function of AGB stars (in clusters and field), and the concept of superwind have been deeply revised. We will come back to this point later on.

6. Classical evolution of massive stars with mass loss

While low and intermediate mass stars are occasionally affected by mass loss (tip of the RGB and AGB phases), massive stars are the only case in which the entire evolution is strongly affected by mass loss. Nowadays, mass loss by stellar wind is always included in all model calculations. They may differ in other physical details but all agree on the need of mass loss. Therefore, it is wise to present here a short summary of somewhat old models (with semiconvection) in which a first rationalization of the effect of mass loss was attempted. See for more details the reviews by Chiosi (1982), de Loore (1982),

Chiosi & Maeder (1986), Chiosi et al. (1992a,b), and Maeder & Conti (1994). In the following, we will be concerned only with the main general results.

6.1. *The Core and Shell H-Burning Phases*

Evolutionary models in the core and shell H-burning phases have been calculated using a large variety of relations for $\dot{M}$. In most models, similar effects of mass loss have been recognized. These are physically connected as follows:

1) The progressive reduction of the stellar mass makes the central temperature increase less rapidly than for constant-mass evolution, thus the mass of the convective core decreases more rapidly as evolution proceeds. However, the core mass fraction is larger in a star evolving with mass loss.

2) As a result of the smaller core, the luminosity of stars evolving with mass loss is lower than for constant-mass evolution. However, the star is always overluminous for its current mass. As a consequence of the lower luminosity, the extension of semiconvection and/or intermediate full convection at the top of the H-burning shell is much smaller than in constant-mass models (Chiosi et al. 1978). The reduction is proportional to the mean mass loss rate. This fact makes less important the whole problem of semiconvective instability, one of the major uncertainties in the structure of massive stars.

3) As a result of the lower luminosity, the main sequence lifetime is somewhat increased by increasing mass loss, though both the smaller convective core and the lack of sufficient semiconvective feeding (leading to less fuel to burn), would concur to shorten the lifetime.

4) For moderate mass loss, there is a slight main sequence widening as a result of the larger core mass fraction. In the case of heavy mass loss (i.e. loss sufficient to expose nuclearly processed material at the surface during main sequence evolution), there is a main sequence narrowing for the higher masses due to the lower surface hydrogen content.

5) The mass of the He core at the end of the core H-burning phase is smaller, and the chemical structure of the models shallower, than in constant-mass stars. Finally, a tiny intermediate convective zone may develop on top of the H-burning shell, when the temperature criterion against convection and moderate mass loss rates are used in the previous stages.

6.2. *The Core He-Burning Phase and the formation of WR stars*

Core He-burning. The core He-burning phase of model stars evolved with mass loss is characterized by an apparently erratic location in the HRD, which is the result of the extreme sensitivity of the models to many structural properties. Several criteria for blueward and/or redward movements in the HRD have been suggested (Chiosi 1981a, b; Falk & Mitalas 1981; Maeder 1981a,c). Broadly speaking, we may distinguish four internally competing effects: i) the classical mirror expansion that responds to core contraction (cf. Renzini 1984a for a recent discussion); ii) the increasing size of the He core by outward shifting of the H-burning shell and the increasing size of the fractional mass of the He core by mass loss, which favours redward evolution up to some critical value and blueward evolution afterward (Giannone et al. 1968); iii) the homogenization of the envelope by large intermediate convective zones, which (if present) tends to limit the increase in the stellar radius; iv) a large luminosity-to-mass ratio, which favours envelope expansion. In particular, point (ii) is regulated by mass loss during the post main sequence stages and more precisely at low effective temperatures. It is clear that several plausible evolutionary scenarios are generated by the competition between the various factors discussed above. Many sets of models of massive stars predict a ratio t_{He}/t_H of the lifetimes in the helium and hydrogen burning phases going from about 0.08 to 0.20 or even 0.30 depending on the adopted mass loss rates, mass range, and

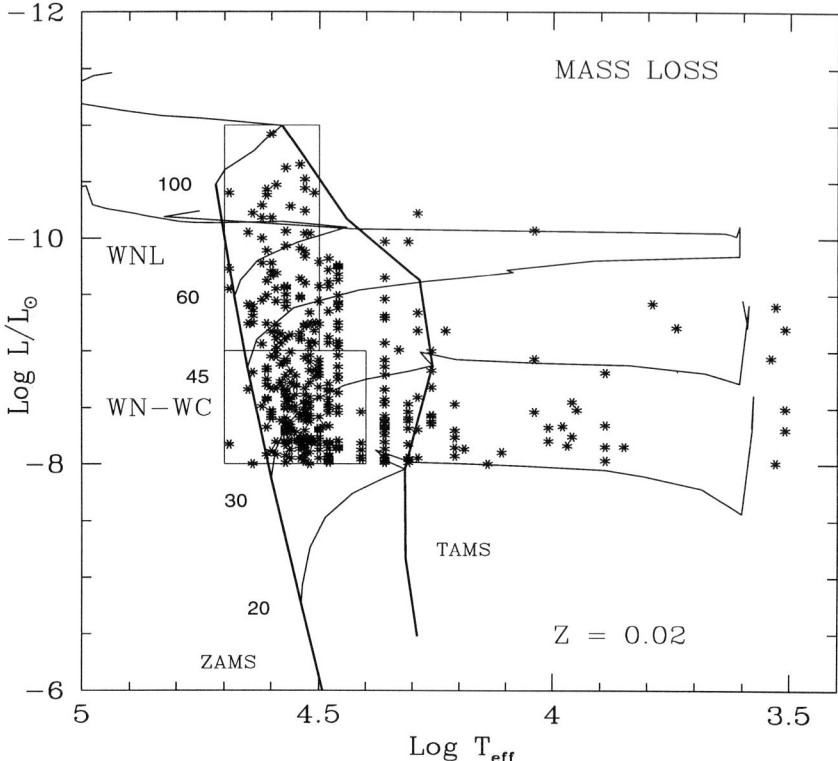

FIGURE 3. A semi-artistic HRD for massive stars in presence of mass loss by stellar wind (and convective overshoot). The models are calculated according to the Bressan et al. (1981) prescription. The asterisks are the stars from the Humphreys & McElroy (1984) catalog limited to those more luminous than $M_{bol} = -8$. The thick lines encompass the band of core H-burning models from the zero age main sequence (ZAMS) to the termination of the main sequence (TAMS). The boxes schematically show the location of WNL, WN and WC stars according to the old data (cf. Conti 1976)

other model assumptions, i.e. opacities, nuclear reaction rates (the $^{12}C(\alpha,\gamma)^{16}O$ in particular), size of the convective core, location and size of the H-burning shell and even the treatment of the external layers. See Chiosi & Maeder (1986) for more details and references. A typical, somewhat out of date HRD for massive stars evolved with mass loss under plausible mass loss rates during core H- and He-burning phases is shown in Fig. 3.

WR stars. WR stars are commonly interpreted as central He-burning objects that have lost the main part of their hydrogen-rich envelope, and in consequence show products of different burning stages. See Chiosi & Maeder (1986), Lamers et al. (1991), Maeder & Conti (1994) for recent reviews on the subject. The surface composition is used to assign the evolutionary stage of these objects and to classify these stars. WR stars, whose spectra are dominated by strong emission lines, are generally grouped into three sequences according to their spectra: WN, WC and WO. The spectra of the WN stars exhibit transition lines of He and N ions with little evidence of C; those of the WC stars predominantly show lines of He and C with little evidence of N; finally the spectra of the rarer WO stars are dominated by lines of O. The WN stars are commonly separated into two further groups according to whether or not evidence of significant H is detected. Broadly speaking, separation corresponds to that between early (WNE–little or no H) and late (WNL–some significant H) WN stars. Furthermore the CNO ratios are typical of nuclear equilibrium in WN stars.

The above interpretation is sustained by the continuity of the abundances in the sequence O, Of, WNL, WNE, WC, WC and WO (cf. Lamers et al. 1991) that nicely corresponds to a progression in peeling off the outer material from evolving massive stars. WR stars have low average masses, i.e. in the range 5 to 10 $M_\odot$ (Abbott & Conti 1987) and obey the mass-luminosity relation for He stars (Smith & Maeder 1989). They occur in young clusters and associations with ages smaller than 5-6 10^6 yr (Humphreys & McElroy 1984, Schild & Maeder 1984). WR stars are located in the HRD in the same region of luminous OB type stars (Hamann et al. 1993). Finally, the number ratios WR/O and WN/WC are consistent with the theoretical expectation in galaxies of different metallicity.

As far as the evolutionary sequence generating WR stars from their natural progenitors, namely the O and B type stars, is concerned various scenarios have been proposed over the years. They differ in the physical mechanism that in addition to mass loss by stellar wind concurs to produce luminous, chemically anomalous objects of high effective temperature and mass loss rate. The simplest scheme was suggested long ago by Conti (1976) starting from an original idea put forward by Smith (1973). In this picture, via the mechanism of mass loss, normal O type stars evolve to Of and WNL stars, which in turn may evolve into WNE and finally into WC stars. The so-called Conti scenario suffered from a number of difficulties (cf. Chiosi et al. 1978, Chiosi 1981a,b; 1982a,b; Bressan et al. 1981, and the reviews by Chiosi & Maeder 1986, and Maeder & Conti 1994) that led to a revision of this scheme. As first proposed by Chiosi et al. (1978) and Chiosi (1981a,b; 1982a,b), three mass ranges can be singled out for the progenitor stars with different avenues for the formation of WR stars. Stars lighter than M_1 evolve through the sequence O stars – blue supergiant stars (BSG) – red supergiant stars (RSG), whether or not a loop is present is irrelevant to the present purposes, and do not form WR stars of any type. Stars in the range $M_1 < M < M_2$ follow the sequence O – BSG – RSG – BSG – WNE – WC – (WO ?). This is known as the red supergiant channel in which mass loss along the Hayashi line plays an important role. Finally, stars heavier than M_2 follow the Conti scenario, i.e. the sequence O- Of - WNL -(WNE ?) - (WC ?) - (WO ?). Typical values for the limiting masses are $M_1 \simeq 25 M_\odot$ and $M_2 = 60 M_\odot$ (these mass limits depend however on the mass loss rates and other physical details). This scheme has been refined over the years by many authors (cf. Chiosi & Maeder 1986, and Maeder & Conti 1994 for more details and referencing), and it is referred to as the "Maeder Scenario".

A) Initial Masses Larger than 60 $M_\odot$

The winds responsible for peeling during the main sequence and early shell H-burning phases are high enough to remove all stellar outer layers, leaving a bare He core. This prevents any evolution toward the red supergiant stage, and therefore the stars always keep on the left half of the HRD [see effect (ii) above]. Zones that were initially in the core are exposed at the stellar surface, where they lead to large spectroscopically observable changes of chemical abundances. During the blue supergiant phase, these stars are located in the region of the LBV, where very strong winds are observed. As a result of the very high mass loss rates [of the order of $10^{-3} M_\odot$/yr or more (Lamers 1985)], the remaining part of the envelope is ejected and the star evolves directly as a bare core, likely identifiable as a WR star.

B) Initial Masses Between 60 $M_\odot$ and 25 $M_\odot$

In this range, mass loss on and near the main sequence is not high enough to remove all the outer envelope, and thus the star rapidly becomes a red supergiant because the intermediate convective zone is small or absent [effect (iii)]. The star spends part of the He-burning phase as a red supergiant, where it can expect a lifetime longer than that calculated from constant-mass models. In this mass range, the high stellar winds in the red supergiant stage progressively remove the outer envelope, and the star then evolves to the blue due to effect (ii) above (Chiosi *et al.* 1978; Maeder 1981b,c). It is worth emphasizing, however, that whether or not this will actually occur is entirely determined by the competition between the time scale of mass loss (rate) and the nuclear burning time scale. With insufficient mass loss, a star may not be able to leave the red supergiant region. If a blue loop occurs, a star may then spend part of the He phase as a blue supergiant. It is interesting to note that such a blue supergiant could be differentiated from supergiants on their first crossing by their pulsation properties (Lovy *et al.* 1984) and surface abundances. If the mass loss is high enough, the star may lose all the envelope and become a good candidate for the WR stage (Chiosi *et al.* 1978; Maeder 1981b,c). This will likely occur in the mass range 40 to 60 $M_\odot$. In this range of mass we face the complex situation that mass loss on or near the main sequence favours the formation of red supergiants, whereas heavy mass loss in the red may shorten the lifetime of red supergiants, since the star turns either into a blue supergiant and/or a WR star.

C) Initial Masses Below 25 $M_\odot$

Below about 25 $M_\odot$ the mass loss in the blue or the red is never large enough to remove the outer layers and to produce a definite blueward motion. After the main sequence the star may either become a blue supergiant and later on a red supergiant, or it may first become a red supergiant and then undergo a blue loop during which the Cepheid instability strip may be crossed on an appreciable time scale. Later, the star becomes a red supergiant again. The reasons for which a star will follow one or the other of the two schemes have been already amply discussed. However, below the limit of about 20 $M_\odot$, the red-blue-red scheme is always followed by stars of any initial mass. It is worth noting however, that in all cases the blue extension of the loop is significantly reduced by mass loss (Bertelli *et al.* 1985; Lauterborn *et al.* 1971; Maeder 1981c). In case of heavy mass loss the loops may be entirely suppressed.

Remarks. It is worth commenting on the growth of the He core during the He phase. Two alternatives exist: either the core remains substantially smaller than in the classical case, or it grows to a mass comparable to that of constant-mass models. The former case is typical of models calculated with the Schwarzschild (1958) criterion in the intermediate convective zone (if present), and low mass loss rates during the core H-burning phase. In such a case, the H-burning shell is topped by a fully convective zone that prevents the shell from migrating outwards and thus keeps the He core small (Chiosi *et al.* 1978). High mass loss rates and/or the adoption of the Ledoux (1947) criterion make the intermediate convective layer very small (or even nonexistent). The H-burning shell cannot therefore undergo any significant replenishment, and thus it migrates outward, which increases the mass of the He core (Maeder 1981b).

6.3. *Effects of Mass Loss on the Advanced Stages*

After the He-burning phase, the evolutionary time scales are so short that further mass loss (if it proceeds at current rates) is insignificant. Core contraction should lead to core collapse at essentially a constant total mass. Thus the effects of mass loss on late stages are those already present at the end of the He-burning phase.

In contrast with the huge effects of mass loss on evolutionary tracks, mass loss scarcely

For M > 60M$_\odot$ A l w a y s   B l u e

O star - Of - BSG - LBV - WN - WC - (WO) - SN

For 25M$_\odot$ < M < 60M$_\odot$ B l u e - R e d - B l u e

O star - BSG - YSG - RSG - WN - (WC) - SN: High M's

O star - BSG - YSG - RSG - WN - ——— - SN: Low M's

For M < 25M$_\odot$ B l u e - R e d

O star - (BSG) - RSG - YSG - Cepheid - RSG - SN

TABLE 1. SCHEMATIC EVOLUTION OF MASSIVE STARS – Legend: BSG - Blue supergiant; RSG - Red supergiant; YSG - Yellow supergiant; SN - Supernova; WC - Wolf Rayet of type C; WN - Wolf Rayet of type N; WO - Wolf Rayet of type O; LBV - Luminous blue variables

influences the course of central conditions up to the advanced stages (Brunish & Truran 1982b; Chiosi *et al.* 1978; Maeder 1981a,c; Maeder & Lequeux 1982). In spite of this, the effects of mass loss cannot be ignored.

We have already reported that as a result of mass loss, the mass of He core left over at the end of the H-burning phase is smaller than in constant mass models. Following central H-burning, the gravitational contraction of the core builds up a high density contrast throughout the star. Owing to this, the effect of mass loss on the high density He core become negligible (this is even more true in subsequent, higher density stages). Therefore the mass of the He-C-O core at the end of the He-burning phase is almost independent of mass loss for a large range of mass loss rates. Starting from the end of the H-burning constant and variable-mass evolution are seen to follow nearly the same path in the T_c vs ρ_c diagram. Moreover, a comparison of the internal distribution of chemical elements and the run of temperature and density shows identical results in models with or without mass loss. Even for moderate reduction of the He- C-O core in the WR stars of type WC (cf. Chiosi & Maeder 1986), the departures from the standard path in the T_c vs ρ_c diagram are very small (Maeder & Lequeux 1982). However, there is an important effect to consider, i.e. that the final fate of a massive star much depends on the size of its original He core, which is basically determined by what happend during the core H-burning phase. It can be easily understood that as long as the initial mass of the star is far from to M_{mas} and M_{vms} (this latter the mass limit for electron pair instability to occur), its final evolution will be the same as in the constant mass case, provided that the scaling to a smaller mass of the He core is taken into account. In contrast stars with mass close to the limits may have a different evolution. The problem is not relevant for stars with mass $\simeq M_{mas}$ because in the range $10 - 15M_\odot$ very little mass is lost during the core H-burning phase, so that the whole evolution remains unchanged. In contrast, the evolution may be different for stars near M_{vms}. As already shown in Fig. 2, this is the case of the 120 $M_\odot$ star, which evolved in presence of mass loss avoids the electron-pair instability region and goes towards the He-Fe break-down. In contrast the same star evolved at constant mass would go pair-unstable.

Finally, another important effect of mass loss (via the size of the He-core and stellar winds themselves) is on the net yields of chemical elements (cf. Arnett 1978b, Chiosi & Caimmi 1979, Maeder 1983c, and Chiosi 1986b for all details).

6.4. *Drawbacks of the classical scheme*

Although the above evolutionary scenario substantially agrees with the observational data, a closer scrutiny reveals that there are many points of severe uncertainty, which have spurred revisions of the above scheme.

Starting with the early studies of Stothers & Chin (1977) and Cloutman & Whitaker (1980), star counts in different areas of the HRD of supergiant stars by Bressan *et al.* (1981), Meylan & Maeder (1982), Bertelli *et al.* (1984), Vanbeveren (1987), and Tuchman & Wheeler (1989, 1990) indicate that in spite of the successful understanding the connection between O, Of, B through M, WR stars, and LBV stars, number frequencies do not fully agree with the theoretical expectation (see Chiosi & Maeder 1986, Maeder & Conti 1994 and references). *It seems as if there are too many post main sequence stars as compared to the main sequence ones.* Let us cast the problem as follows:

(1) The evolution of massive stars proceeds at nearly constant luminosity with a modest increase passing from the core H- to core He-burning phase. There can be a decrease in the luminosity during the latest stages of central He-burning if the models are subjected to heavy mass loss.

(2) The width of main sequence band, i.e the range of $T_{\rm eff}$ spanned by core H-burning models, depends on the kind of models in usage. It gets larger, i.e. its low temperature edge gets hotter or cooler at varying three basic ingredients of stellar models: mass loss rates, opacity, and size of the inner mixed core. The comparison of the observed and predicted main sequence width can be used to assess the success or failure of a certain class of models.

(3) The core He-burning phase splits in two stages one in the blue and one in the red side of the HRD, whose relative duration is a sensitive function of the star structure.

(4) The ratio of main sequence to post main sequence lifetime is of the order of 10 for obvious reasons (nuclear energy release of the two main evolutionary phases). Changing stellar models scarcely affects the above ratio.

(5) On the observational side there are three main requirements: star catalogs including all evolutionary stages, good completeness of the data, and good transformations from colors and magnitudes to $T_{\rm eff}$ and luminosities.

(a) There are several catalogs of blue, red supergiants, and WR stars. Unfortunately not always the same source of data, including all spectral types from O to M, LBVs and WR stars, can be used so that to perform the above analysis data from different sources are patched together. This is an important point to consider, because if M supergiants LBV and WR stars are the descendent of early type supergiants, their effect on the total and relative number of stars must be taken into account. Limiting the analysis to the subset of early type stars may lead to misleading conclusions.

(b) Assessing the completeness of the samples under consideration is a cumbersome affaire. This is particularly severe for the early type (main sequence) stars first because of their intrinsically difficult detectability, second because any shortage of such stars would immediately reflect on the contribution of the dominant population. Indeed 10% uncertainty in the number of stars on the main sequence corresponds to a population almost equalling the total population in the post main sequence phases.

(c) The translation of the observational data (mainly visual magnitudes and colors or spectral types) into the theoretical plane requires good calibrations of (B-V) - SP - $T_{\rm eff}$ - BC (cf. Massey, this volume).

(6) In order to compare the theoretical prediction from the lifetimes of the various phases with the observational number frequencies, we need to guess *a priori* the correspondence between the evolutionary phase and distinct groups of stars. While this is easily feasible in single clusters and associations, where main sequence stars are clearly separated from the evolved ones and the latter in turn are easily assignable to blue and red stages of core He-burning, this is not the case with the HRD of supergiant stars where a continuous distribution of stars is observed and stars in different evolutionary stage often fall into the same area.

(7) The test consists in checking whether the equality below is verified

$$\frac{t_H}{t_{He}} = \frac{N_{MS} + \Delta N_{MS}}{\sum_J (M_J + \Delta N_J)} \quad (6.16)$$

where N_{MS} and ΔN_{MS} are the numbers of detected and undetected main sequence stars, respectively. Likewise, the summation terms $(M_J + \Delta N_J)$ stand for $(N_{BSG} + \Delta N_{BSG})$, $(N_{RSG} + \Delta N_{RSG})$, $(N_{WR} + \Delta N_{WR})$, and $(N_{LBV} + \Delta N_{LBV})$ indicating blue supergiants, red supergiants, WR stars, and LBV stars, respectively. In the discussion below we attribute the greatest incompleteness to the group of main sequence stars and assume that all the other type are almost complete (their ΔN is set zero). It is worth to remind the reader that the above star numbers depend on the luminosity range, because as already seen not all types of star are present at each luminosity interval.

(8) Given these premises, when star counts are performed as a function of spectral type in suitable luminosity strips where stars from O to M and WR type are present, it seems as if in order to satisfy the above requirement of the main sequence to post main sequence lifetime ratio, stars from O to about A type must be assigned to the core H-burning phase. In other words, if only stars falling into the band drawn in the HRD by standard models (with semiconvection and mass loss) are identified as core H-burners, the embarrassing result is met that a large fraction (that sometimes amounts to about 40% depending to the adopted observational sample, degree of completeness, evolutionary models) of the stars should be in post main sequence phase. In brief, there is some conviction that the core H-burning phase should stretch to lower T_{eff} than commonly accepted (Stothers & Chin 1977, Bressan et al. 1981, Bertelli et al. 1984, Chiosi & Maeder 1986, Chiosi et al. 1992a,b).

(9) Other methods based on ratios of star counts in different T_{eff} bins but limited to subsets of stars, e.g. among blue supergiants alone or blue and red supergiants or all stars comprised between age and luminosity intervals (cf. Massey et al. 1995) do not provide the right answer. In fact what they actually tell is either the evolutionary rate across limited portions of the HRD (the rate is basically driven by nuclear burning and it is scarcely dependent on external phenomena determining the spectral appearance of a star) or simply reflect the fact that, within the uncertainty, the number of stars per luminosity interval is roughly proportional to the total lifetime. This indeed is the result of the Massey et al. (1995) analysis, which owing to the intrinsic nature of the proposed test cannot prove whether or not the main sequence band has to be wider than predicted. Indeed looking at Figs. 18 and 19 of Massey (this volume) aimed to show that previous samples on which the claim of a wider main sequence is based were highly incomplete (as surely was the case), we may notice that the problem is still there. In Fig. 18 we see a large number of stars in the region comprised between the red edge of the main sequence band and the blue edge of the blue loop. In principle this region should be almost void of stars because of the underlying evolutionary rate (we will come back again to this point). Second, in Fig. 19 we may notice that in the luminosity interval bracketed by the

15 and 25 $M_\odot$ tracks of Schaerer et al. (1994) — incidentally these models are in a much better situation as compared to the classical ones as they have overshoot and modern opacities, which are known to widen the main sequence band — there are about 15 - 17 stars in the main sequence band and about 10 stars beyond this. Taking these numbers as face values, the ratio is 1.5 to 1.7 (or even lower if WR stars are to be included), i.e. well below the expected value of about 10. Either incompleteness is still severe or there is something still missing with the stellar models (all stellar models in general, not only those of Schaerer et al. in particular as they are as good as many others in literature).

Even if the above conclusions can be strongly criticized for many reasons (cf. the discussion in Massey et al. 1995), yet they spurred much theoretical work to find plausible physical causes yielding a wider main sequence band. Stothers & Chin (1977) advocated a strong enhancement in the CNO opacity. Bertelli et al. (1984) and Nasi & Forieri (1990) investigated the effects of mass loss by stellar wind, atmospheric effects on the stellar radius caused by mass loss, convective overshoot from the inner core, and finally a suitable increase in the standard opacity in the region of the CNO ionization. Bottom line of their reasoning was to look for plausible physical phenomena stretching the main sequence band up to the point it could merge the region of core He-burning thus eliminating the blue gap in the HRD and explaining the large percentage of stars observed over there. Although models filling the blue gap could be found, none of these was considered as fully adequate because of several drawbacks: for instance the required opacity enhancement and overshoot efficiency were indeed too extreme. As a matter of facts, although the new radiative opacities by the Livermore group (cf. Iglesias et al. 1992) turned out to be larger than the classical ones (Huebner et al. 1977) their effect on massive stars is small (see Bressan et al. 1993; Fagotto et al. 1994a,b,c; Schaerer et al. 1993a,b; Schaller et al. 1992). The same is true for the atmospheric effects caused by stellar winds. The question of large core overshoot is still a matter of debate and will be examined below.

7. Convection: the major uncertainty

In the classical approach, the Schwarzschild criterion provides the simplest evaluation of the size of the convectively unstable regions and the MLT simplifies the complicated pattern of motions therein by saying that full, and instantaneous mixing of material takes place. In this scheme, well known inconsistencies are known to develop at the border of the convective regions. Various attempts have been made to cure the above difficulties, among which we recall H- and He-semiconvection, breathing convection, overshoot, and diffusion. Let us cast the problem as follows:

(1) What determines the extension of the convectively unstable regions (either core or envelope or both) together with the extension of the surrounding regions formally stable but that in a way or another are affected by mixing? In other words how far convective elements can penetrate into formally stable regions?

(2) What is the thermodynamic structure of the unstable and potentially unstable regions?

(3) What is the time scale of mixing? Instantaneous or over a finite (long) period of time? What is the mechanism securing either full or partial homogenization of the unstable regions?

Over the past decade different answers to above questions have been suggested and in turn different types of stellar model have been calculated.

(a) In massive stars, the inconsistency at the border of the formal convective core set by the Schawrzschild criterion (layer at which the acceleration imparted by the buoyancy force to convective elements) is eliminated by supposing extra slow mixing over a finite region so that the neutrality condition is maintained (H-semiconvection, see below).

(b) The same at core of intermediate and low mass stars in the central He-burning phase (He-semiconvection, see below).

(c) In both schemes above, the chemical profile of the semiconvective regions is obtained either by solving the neutrality condition or by means of a more complicated approach based on diffusive algorithms.

(d) The acceleration condition is replaced by a velocity condition, i.e. the extension of the convective regions is set at the layer where the velocity rather than acceleration of convective elements vanishes. The region above the formal Schwarzschild core is called region of overshoot. The problem now is to know whether this region has an adiabatic or radiative structure and whether mixing through this is instantaneous or over a finite time scale. The Bressan *et al.* (1981) models consider it as adiabatic and instantaneously mixed. The extension of the overshoot region is a matter of vivid debate. Nowadays it has settled to a sizable fraction of the local pressure scale height. For the particular case of overshoot from the H-burning convective cores in stars of low mass, the additional (reasonable) assumption is made that it must vanish at decreasing convective core, i.e. in stars approaching M_{con}. We will refer to stellar models of this type of mixing as those with straight overshoot (Bressan *et al.* 1981, Bertelli *et al.* 1985).

(e) Within the same scheme, there are models trying to take into account that mixing actually requires a suitable time scale to occur. To this aim straight mixing is abandoned, and the more appropriate diffusive approach is adopted (cf. Deng *et al.* 1996a,b). The efficiency of diffusion (or equivalently the time scale of it) seeks to incorporate physical processes known to occur in laboratory hydrodynamics, such as intermittence and stirring, and varies as function of the local properties of the overshoot region. In this context, the thermodynamics structure of these layers plays a secondary role even if a radiative stratification ought to be preferred. Other more physically grounded but by far more complicated formulations of the problem (cf. Xiong 1986; Grossman *et al.* 1993) have not yet been included in stellar model calculations.

7.1. *Hydrogen Semiconvection*

During the core H-burning phase of massive stars on the main sequence, radiation pressure and electron scattering opacity give rise to a large convective core surrounded by an H-rich region, which is potentially unstable to convection if the original gradient in chemical abundance is maintained, but stable if suitable mixing is allowed to take place. Theoretical models picture this region undergoing sufficient mixing until the condition of neutrality is restored, but carrying negligible energy flux. The gradient in chemical abundance depends on which condition is used to achieve neutrality, either Schwarzschild (1958) or Ledoux (1947). The former condition tends to give smoother chemical profiles and in some cases leads to the onset of a fully intermediate convective layer. It is worth recalling that the Ledoux criterion is a stronger condition favoring stability with respect to the Schwarzschild criterion. Similar instability occurs also during the early shell H-burning stages. The effects of H-semiconvection on the evolution of massive stars have been summarized by Chiosi & Maeder (1986) and most recently by Chiosi *et al.* (1992a,b).

7.2. *Helium Semiconvection*

As He-burning proceeds in the convective core of stars of any mass, the C-rich mixture inside the core becomes more opaque than the C-poor material outside; therefore the radiative temperature gradient increases within the core. The resulting superadiabaticity at the edge of the core leads to a progressive increase (local convective overshoot) in the size of the convective core during the early stages of He-burning (Schwarzschild 1970, Paczynski 1971, Castellani *et al.* 1971a,b). Once the convective core exceeds a certain size, the continued overshooting is no longer able to restore the neutrality condition at the border due to a characteristic turn-up of the radiative gradient. The core splits into an inner convective core and an outer convective shell. As further helium is captured by the convective shell, this latter tends to become stable, leaving behind a region of varying composition in which $\nabla_R = \nabla_A$. This type of mixing is called He-semiconvection. The extension of the semiconvective region varies with the star mass, being important in low- and intermediate-mass stars up to say 5 $M_\odot$, and negligible in more massive stars. Various algorithms have been devised to treat semiconvection (Castellani *et al.* 1971b; Demarque & Mengel 1972; Sweigart & Demarque 1972; Gingold 1976; Robertson & Faulkner 1972, Sweigart & Gross 1976, 1978; Castellani *et al.* 1985; Lattanzio 1986, 1987b, 1991; Fagotto 1990). In all computed models, when $Y_c \leq 0.1$, the convective core may undergo recurrent episodes of rapid increase followed by an equally rapid decrease until it engulfs the whole semiconvective region. Castellani *et al.* (1985) have designated this phase as "breathing pulses of convection". Semiconvection increases the core He-burning lifetime (by approximately a factor of two), whereas breathing convection increases the mass of the C-O core leftover at the end of He-burning phase. This fact will greatly shorten the early AGB phase. The only exception to this scheme are models calculated by Gingold (1976) in which for some chemical compositions the breathing convection phase is apparently missing. The reason for its absence has never been understood. Models with semiconvection alone and models with semiconvection plus breathing convection have different predictable effects on the expected ratio of the number of AGB stars to the number of HB stars in well studied globular clusters. Renzini & Fusi-Pecci (1988), comparing the above ratio with Gingold's (1976) models, consider semiconvection as a true theoretical prediction and argue that breathing convection is most likely an artifact of the idealized algorithm used in describing mixing (see also Chiosi 1986). Given that breathing convection is a consequence of the time-independent treatment of semiconvection, and that both are based on local descriptions of mixing, the question arises whether nonlocal, e.g. full convective overshoot (see below), and/or time-dependent mixing may overcome the above difficulties.

7.3. *Convective overshoot*

The argument for the occurrence of convective overshoot is that the traditional criteria for convective stability look at the locus where the buoyancy acceleration vanishes. Since it is very plausible that the velocity of the convective elements is not zero at that layer, these will penetrate (overshoot) into regions that are formally stable. If the physical ground of convective overshoot is simple, its formulation and efficiency are much more uncertain. This uncertainty is reflected in the variety of solutions and evolutionary models that have been proposed over the years. Major contributions to this subject are from Shaviv & Salpeter (1973), Maeder (1975), Cloutman & Whitaker (1980), Bressan *et al.* (1981), Stothers & Chin (1981, 1990), Matraka *et al.* (1982), Doom (1982a,b;1985), Bertelli *et al.* (1985, 1986a,b), Bressan *et al.* (1986), Xiong (1983, 1986, 1989, 1990), Langer (1986), Baker & Kuhfuss (1987), Renzini (1987), Maeder & Meynet (1987, 1988, 1989, 1991) Aparicio *et al.* (1990), Alongi *et al.* (1992, 1993), and Maeder (1990). In those studies the overshoot distance at the edge of the convective core has been proposed between zero and about $2 \times H_P$ (pressure scale height). As many evolutionary results depend on the extension of the convective regions, this uncertainty is very critical. In addition to the convective core, overshoot may occur at the bottom of the convective envelope during the various phases in which this develops, such as on the RGB. The effect of envelope overshoot on stellar models of low- and intermediate-mass stars has been studied by Alongi *et al.* (1992), whereas that for high-mass stars by Chiosi *et al.* (1992a,b).

A substantial improvement to the theory of non local convection occurred with the studies of Xiong (1983, 1986, 1989, 1990), Canuto (1992, 1993), Canuto & Mazzitelli (1991), Canuto *et al.* (1991, 1994a,b; 1996a,b), Grossman *et al.* (1993), Grossman (1996), and Grossman & Taam (1996). In brief, Xiong (1983, 1986, 1989, 1990) added three differential equations governing turbulent variables and used the diffusion approximation to treat non locality. Furthermore he described locally the dissipation of turbulent energy with the aid of the local pressure scale height and assumed turbulence to be isotropic. Canuto (1992, 1993), Canuto *et al.* (1991, 1994a,b; 1996a,b) went one step further increasing the number of differential equations governing turbulent variables, considered the more general case of anisotropic flow, and relaxed the diffusion approximation. Grossman *et al.* (1993) and Grossman (1996) developed a Boltzmann transport theory for the evolution of turbulent fluid elements and derived the equations for the hydrodynamic evolution of the high order correlations of velocity and temperature and presented simulations of nonlocal convection. Applications of the Canuto and collaborators theory to stellar models are by D'Antona & Mazzitelli (1994) and Mazzitelli *et al.* (1995).

The improvement in the quality of turbulence modeling has cast light on a controversial subject. i.e. the relation between the temperature stratification in the overshoot region and the extension of this latter. According to Renzini (1987), the overshoot zone is small if the temperature gradient is radiative there, large if adiabatic. In contrast, Xiong (1983, 1986, 1989, 1990) shows that overshoot is a very complicated process, in which different physical quantities have different distances of penetration, and finds that the overshoot region at the border of the convective core can be very large and radiative at the same time. Similar conclusion has recently been reached by Grossman (1996) who finds that the temperature stratification in the overshoot region is nearly radiative, and that the velocities of turbulent elements penetrating into it have exponential decay over many e-folding distances. In this context see also Zahn (1992) who favors the adibatic solution. The response of the stellar structure to passing from adiabatic to radiative temperature stratification has been studied by Deng (1992). Finally, all the above models of non local convection seem to converge to the conclusion that the size of the convective region is significant.

Despite the above achievements, fully hydrodynamical descriptions of non local convection have not yet been incorporated into stellar models calculations at least at the same level of popularity as the classical MLT.

7.4. *Modeling convective overshoot*

Bressan's et al (1981) method is particularly suited for model calculations (see also Maeder 1975). The method rests on the study of Shaviv & Salpeter (1973) who found that independently of the theory used to follow convective motions, the stellar temperature gradient can be taken as adiabatic up to the point where convective motions vanish and radiative elsewhere. In fact, the time scale necessary to establish the adiabatic value is rather short (5×10^2 yr), smaller than other evolutionary time scales. In the framework of the MLT of convection, the acceleration imparted to a convective element formed at the radial distance r_1 is

$$v_r \frac{dv_r}{dr} = -g \frac{\Delta \rho}{\rho} \quad (7.17)$$

where g is the local gravity acceleration and $\Delta \rho$

$$\Delta\rho = \int_{r_1}^{r} \left(\frac{d\rho^*}{dr'} - \frac{d\rho}{dr'}\right) dr' \tag{7.18}$$

is the density excess. With the aid of the equation of state we obtain

$$\Delta\rho = \int_{r_1}^{r} \left(\frac{\rho\chi_T}{T\chi_\rho}\right)\left[\frac{dT^*}{dr'} - \frac{dT}{dr'}\right] dr' + \int_{r_1}^{r} \left(\frac{\rho\chi_\mu}{\mu\chi_\rho}\right)\frac{d\mu}{dr'} dr' \tag{7.19}$$

Quantities marked with an asterisk refer to convective elements, while χ_T, χ_ρ and χ_μ are the temperature, density and molecular weight exponents in the pressure equation of state. The convective flux carried by elements originating at the level r_1 and vanishing at the distance r, is

$$F_c = Kc_P \rho v_r \int_{r_1}^{r} \left[\frac{dT^*}{dr'} - \frac{dT}{dr'}\right] dr' \tag{7.20}$$

where $K = 2$ takes the contribution from both rising and descending elements into account. With some manipulations we obtain

$$\left(\frac{1}{3}\right)\frac{dv^3}{dr} = \left(\frac{1}{K}\right)\frac{(g\chi_T F_c)}{(T\chi_\rho c_P \rho)} - \frac{(g\chi_\mu)}{(\mu\chi_\rho)}\Delta\mu v_r \tag{7.21}$$

where $\Delta\mu = \mu(r) - \mu(r_1)$. The convective flux F_c can be derived from the condition

$$F = F_r + F_c \tag{7.22}$$

where the total flux F and the radiative flux F_r are known. Starting from the layer r_1, the above equation is numerically integrated up to the distance $r = r_1 + \Lambda$, where Λ is the distance travelled by the elements before losing their identity and dissolving into the surrounding medium. The distance Λ travelled by convective elements is given by $\Lambda = \lambda H_P$. At any given distance r, all convective elements originated in the range of radial distances $r - \Lambda$ are present with different velocities. This allows us to derive the maximum velocity as a function of r. Finally, the border of the convective core is set at the layer where $v_{MAX}(r) = 0$. This modeling of convective overshoot is often referred to as the ballistic scheme. With the above formalism, small discontinuities or steep gradients in molecular weight would constitute almost insuperable barriers against the penetration of convective elements into stable radiative regions. This is due to the fact that even small discontinuities of molecular weight impart a negative acceleration to convective elements, which is greater by orders of magnitude than the acceleration originating from the temperature terms of eq. (7.19). However, the arguments suggested by Castellani et al. (1971a,b), Maeder (1975), and Renzini (1977) seem to indicate the possibility that barriers of molecular weight can be eroded on very short time scales, thus allowing further extension of the convective core (see also the section below). This problem is particularly important during the core He-burning phase when the convective core grows outward in mass. If the core shrinks, as it occurs during the core H-burning phase, the term of eq. (7.19) depending on the molecular weight can be neglected. In such a case the velocity of convective elements is given by

$$\left(\frac{1}{3}\right)v^3 = \left(\frac{1}{K}\right)\int_{r_1}^{r} \frac{(g\chi_T F_c)}{(T\chi_\rho c_P \rho)} dr' \tag{7.23}$$

From the above relation, it is evident that the layer at which the velocity v vanishes does not depend on the constant K. The contribution to eq. (7.17) by dissipative forces can be taken into account by properly adding a term dependent on a suitable power of velocity. It is worth recalling that among others in literature, the Bressan *et al.* (1981) model of convective overshoot is the one going closer to the results of Xiong (1990 and references) and Grossman *et al.* (1993) as pointed out by Zahn (1992).

7.5. *Can a gradient in molecular weight be eroded by convection ?*

Perhaps the key question to be addressed prior to any other consideration is whether the convective elements can penetrate a region with a gradient in molecular weight. This situation can occur either at the border of a growing convective core or at the base of a sinking envelope.

A growing convective core is present during the core H-burning phase of low mass stars ($1.1 M_\odot < M < 1.6 M_\odot$) and the central He-burning phase of stars of any mass. The second alternative is typical of stars ascending the Hayashi line. In both cases the convective regions expand across a gradient and/or barrier in molecular weight.

If the analysis is limited to radial motions along which the restoring force is most effective (Renzini 1977), a gradient in molecular weight can fully inhibit the propagation of convective motions. In fact a rigid cylindrical test element would penetrate across the gradient over the distance

$$d_p^2 = \frac{\lambda v_0^2}{g(1 - \mu^o/\mu^i)} \quad (7.24)$$

where λ is the mixing length (a fraction of the local pressure scale height), g is the local gravity, μ^o and μ^i are the mean molecular weight for matter outside and inside of the discontinuity respectively. This penetration distance is ~ 3 orders of magnitude smaller than λ, so that overshoot is virtually zero.

However as the element is not a rigid body, it is hard to conceive that it would immediately stop. Similarly, it is difficult to accept that an element moving with the large speed of the bulk motion, keeps its shape when it matches the barrier. We can picture the real situation imagining that the element will change shape and move along another direction. Since the total mass of the element must be conserved, the only possible motion is along the tangential direction, along which no buoyancy force, and negligible viscous force are experienced. If the tangential motion turns out to be turbulent, mixing along this direction would be easier than along the vertical direction (e.g. Zahn 1992).

Denoting with v_0 and l_0 the radial velocity and the dimension of the element, respectively, an estimate of the tangential velocity v_t is provided by the mass conservation

$$\pi(\frac{l_0}{2})^2 v_0 \approx 2\pi(\frac{l_0}{2}) d_p v_t \quad (7.25)$$

from which we get

$$v_t \approx \frac{1}{4} \frac{l_0}{d_p} v_0 \quad (7.26)$$

It is clear from this equation that larger v_t is to be expected for larger values of l_0 and v_0, and for smaller d_p. To derive the above estimate, we have assumed the density to be constant. It follows that in presence of a discontinuity in molecular weight, the tangential velocity is greatly amplified (e.g. Renzini 1977).

Because the matter above the transition level is basically at rest, the tangential motion will surely create shears at the interface. The question arises whether these shears are strong enough to trigger an instability. To check this possibility we look at the Richardson number for a stratified fluid (Zahn 1987). In presence of a gradient in molecular weight the Richardson number is

$$J_r = \frac{g}{H_P} \frac{d\ln\mu/d\ln P}{(dv_t/dz)^2} \qquad (7.27)$$

The Richardson condition says that the interface is stable if $J_r > \frac{1}{4}$ everywhere. Rewriting equation (7.27) in our formalism we obtain

$$J_r = gd_p \frac{\mu^i - \mu^o}{\bar{\mu}v_t^2}. \qquad (7.28)$$

Adopting typical estimates of the various quantities at the border of the convective core of a He-burning star, namely $g \sim 2 \times 10^6$, $v_0 \sim 3 \times 10^4$ cm s^{-1}, $\mu^i \sim 1.36$, $\mu^o \sim 1.33$, we get $J_r \sim 2 \, 10^{-3}$, which is smaller than the critical value by 2 orders of magnitude! Therefore the region is highly unstable and mixing is likely to occur.

With the aid of the above relations we tray to estimate the distance d_p over which the Richardson instability can occur. To this aim we look at the ratio d_p/l_0 which must satisfy the condition

$$(\frac{d_p}{l_0}) \leq [\frac{v_0^2 \times 0.25}{16g(\Delta\mu/\mu)H_P}]^{1/3} \qquad (7.29)$$

Inserting the same values for the various physical quantities as above we get

$$(\frac{d_p}{l_0}) \simeq 3 \times 10^{-4} \div 3 \times 10^{-3} \qquad (7.30)$$

depending on the value adopted for $\Delta\mu/\mu$. The distance d_p can be interpreted as the thickness of the region in which mixing caused by shear instability can erode the chemical profile.

7.6. *A simple theory of diffusive mixing in stellar interiors*

Deng *et al.* (1996a,b) have proposed a simple theory of mixing in stellar interiors, which seeks to amalgamate overshoot into a diffusive scheme. Key points of the model are: (i) In contrast to the MLT of convection, in which mixing is caused by the instantaneous disintegration of the largest scale elements at the end of their lifetime, it is thought of as the result of all scales thus in closer agreement with the evidence from laboratory fluid dynamics. (ii) The goal is achieved by introducing the concept of *characteristic scale most effective for mixing*. (iii) Two important phenomena known to occur in laboratory hydrodynamics, i.e. *intermittence* and *stirring*, are taken into account (cf. Deng *et al.* 1996a for details and referencing). (iv) The extension of the region potentially interested by mixing is derived from the velocity method of Bressan *et al.* (1981) for ($\Lambda = H_P$), which implies an extended overshoot region (cf. Xiong 1990 and Grossman 1996)). (v) The formalism makes use of the MLT to calculate a few important quantities, such the the maximum scale and associated velocity within a convective region. (vi) The adiabatic temperature stratification is assumed.

The diffusion coefficient $\mathcal{D}$ is expressed as

$$\mathcal{D} = \frac{1}{3}F_i F_s v_d L \qquad (7.31)$$

where, v_d is the velocity of a suitable mean effective scale driving mixing, and L is the dimension of the region interested by diffusion. The characteristic velocity v_d is not known a priori, but it is derived from properly analyzing the physical conditions under which mixing occurs. The factor F_i accounts for *intermittence*, while the factor F_s accounts for *stirring*.

We start with defining the following quantities: L_j, the linear dimension of the unstable region, which is labelled L_c for the inner convective core, L_{ov} for the overshoot region, L_e for the external convective envelope, L_s for the intermediate convective shell; l_0, the dimension of the largest eddy in the region under consideration; v_0, the velocity of l_0; l_d, the characteristic scale length of mixing; and v_d, the velocity of l_d.

Using the so-called β-model of intermittence (elements become less volume filling at decreasing scales, cf. Deng et al. 1996a for details), the intermittence factor is

$$F_i = (\frac{l_d}{l_0})^{3/2} \qquad (7.32)$$

The stirring term stems from the fact that eddies in a turbulent field work as a rigid stick stirring the material in a mixer and inducing smaller scale motions. However if the element size is comparable to that of the container the mixing efficiency is much lower.

$$F_s = (\frac{L - l_0}{l_0})^3 \qquad (7.33)$$

for $l_0 \leq L$.

With the aid of these simple ideas, the following prescription is formulated: *Inner Core with Nuclear Burning.* Homogenization in the core follows from the very fast motion of the largest convective elements securing that all the material has the same probability of being exposed to nuclear reactions. In such a case, the diffusion coefficient for the inner convective core is simply given by

$$\mathcal{D} = \frac{1}{3} v_0 L_c \qquad (7.34)$$

where L_c is the dimension of the unstable region and v_0 is the velocity of the largest element in it. *Regions of Overshoot.* At any layer inside the regions of overshoot, there is a natural maximum scale for turbulent elements which is set by the distance l_X between the current position X and the outer border of the overshoot region. Let v_X be its corresponding characteristic velocity. Considering that each element in the overshoot region can be at the same time either the off-spring of a bigger element within the same region or the off-spring of another element coming from the convective region underneath, the diffusion coefficient takes the expression

$$\mathcal{D} = \frac{1}{3} \left(\frac{l_d}{l_0}\right)^{\frac{5}{3}} \left(\frac{l_X}{L_{ov}}\right)^{\frac{5}{3}} (\frac{L_{ov}}{l_0} - 1)^3 v_0^S L_{ov} \qquad (7.35)$$

where v_0^S is the velocity at the transition layer between the fully unstable and the overshoot region ($\nabla_R = \nabla_A$). *The External Envelope without Nuclear Burning.* In this region the diffusion coefficient is simply given by

$$\mathcal{D} = \frac{1}{3}(L_e/l_0 - 1)^3 (\frac{l_d}{l_0})^{\frac{5}{3}} v_0 L_e \qquad (7.36)$$

In order to apply the above formalism to stellar models, one needs to determine the characteristic scale l_d best driving the mixing process. In a turbulent region, all scales from the maximum one equal to the dimension of the unstable zone itself down to that of the dissipative processes, i.e. the Kolmogorov micro-scale, are present. If l_d is equal to the Kolmogorov micro-scale l_k, essentially no mixing will occur because $l_K \sim 10^2$cm. In contrast, assuming l_d to be equal to l_0 (largest scale in the turbulence region) almost instantaneous mixing will take place. The effective scale l_d lies in between these two extreme values. Unfortunately, no theory can be invoked to fix the effective scale l_d a priori so that this must be considered as a sort of parameter.

Deng et al. (1996a) find that the following relations between l_d and l_0 leads to incomplete mixing over a time scale comparable with the evolutionary time scale

$$l_d = P_{dif} \times 10^{-5} l_0 \tag{7.37}$$

where l_0 is expressed in units of H_P and P_{dif} is a fine tuning parameter of the order of unity. The analysis of Deng et al. (1996a) indicates that $P_{dif} \ll 0.4$ leads to standard semiconvective models, $P_{dif} \gg 0.4$ leads to models with full homogenization of the overshoot region, whereas $P_{dif} \simeq 0.4$ leads to models with incomplete mixing over a time scale comparable to the evolutionary time scale. This formalism allows us to recover all possible cases in literature at varying P_{dif}.

7.7. Another diffusive prescription

Salasnich et al. (1997) have presented models of massive stars in which new prescriptions for diffusive mixing and mass loss by stellar wind (as already reported in section 5.2) are adopted. The diffusive algorithm is based on the comparison between the evolutionary timescale and the growth rate of overstability (see below). The diffusion coefficient $\mathcal{D}$ varies according to the region under consideration

(a) In the core, i.e. the region defined by the Schwarzschild criterion,

$$\mathcal{D} = (v \times L_c)/3 \tag{7.38}$$

where v is the turbulent velocity and L_c is the core radius. This choice ensures complete homogenization of the material interior to L_c.

(b) In the overshoot region, following the studies of Xiong (1990) and Grossman (1996) the velocity of the turbulent elements is assumed to decline exponentially with the distance from the border of the core

$$\mathcal{D} = H_P v_0 \times \exp[-r/(\alpha_1 H_P)]/3, \tag{7.39}$$

where r is the distance from the classic border, v_0 is the velocity at the border, and H_P is the pressure scale height. The parameter α_1 is introduced because the original formulation by Xiong (1989, 1990) and Grossman (1996) yields complete homogenization over too wide a region of the star, and the resulting stellar models are unable to match the observational data.

(c) Following central H-exhaustion an extended region with a gradient in molecular weight develops in which a convective zone may arise owing to the so-called *oscillatory convection or overstability* (Kato 1966). In such a case the diffusion coefficient is taken from Langer et al. (1985)

$$\mathcal{D} = L^2/t_{growth} \tag{7.40}$$

where t_{growth} is the growth timescale of the oscillatory perturbations. Complete mixing is ensured if $t_{growth} < t_{evol}$ (time step between two successive models). The time scale t_{growth} is expressed as a fraction α_2 of the time scale of heat dissipation over a distance equal to the wavelength of the perturbation. α_2 therefore controls the mixing efficiency in the intermediate convective region. The Schwarzschild or Ledoux neutrality condition is recovered for small and large values of α_2, respectively.

7.8. Stellar models with straight overshoot

The core H-burning phase of all stars possessing a convective core on the zero-age main sequence (M $\geq$ M$_{con}$) is affected by convective overshoot. Because of the larger cores, the models run at higher luminosities and live longer than the classical ones. They also extend the main sequence band over a wider range of T_{eff}s, this trend increasing with stellar mass (e.g. Alongi et al. 1992, 1993; Bertelli et al. 1985, 1986a,b; Maeder & Meynet 1987, 1988, 1989, 1991). Massive stars (M $\geq$ 40M$_\odot$) would spread all across the HRD, were it not for the contrasting effect of mass loss (see Bressan et al. 1981). The mass range M$_{con} \leq$ M $\leq$ 2M$_\odot$, where the onset of the convective core takes place gradually, deserves particular attention because the time scale required to establish equilibrium in the CN-CNO cycle is a significant fraction of the total H-burning lifetime. The convective core starts small, grows to a maximum, and then recedes as usual, independently of the model – either classical or overshoot – used to define the extension of the core. Within a given overshoot scheme, the growth of the core against a gradient in molecular weight is difficult to model. The morphology of the turnoff in the HRD of old clusters (age of a few 10^9 yr) suggests that overshoot cannot exceed a certain extent (Aparicio et al. 1990; Alongi et al. 1993; Bertelli et al. 1992; Maeder & Meynet 1989). The central H-burning phase of stars lighter than M$_{con}$ is clearly not affected by core overshoot.

In intermediate- and high-mass stars, the overluminosity caused by overshoot during the core H-burning phase still remains during the shell H- and core He-burning phases because of the larger size of the H-exhausted core, M$_{He}$. As a consequence of the higher luminosity, the lifetime of the He-burning phase (t$_{He}$) gets shorter in spite of the larger mass of the convective core. This, combined with the longer H-burning lifetime, t$_H$, makes the ratio t$_{He}$/t$_H$ fairly low (from 0.12 to 0.06 when the stellar mass varies from 2 M$_\odot$ to 9 M$_\odot$). The lifetime ratio is about a factor of 2 to 3 lower than in classical models of the same mass.

Since all low mass stars possess nearly identical helium core masses, the inclusion of convective overshoot leads to results similar to those obtained with the classical semi-convective scheme (Bressan et al. 1986).

Models of intermediate-mass stars evolved with core overshoot alone produce luminosity functions of main sequence stars that agree much better with the observational data for rich clusters (Chiosi et al. 1989a,b), however they hardly match the extension of the blue loops observed in the same clusters (Alongi et al. 1992) because they possess less extended blue loops on the HRD. To overcome this difficulty Alongi et al. (1992) considered the effect of envelope overshoot in addition to that of core overshoot. Envelope overshoot does not alter the scheme determined by core overshoot, but simply makes possible the occurrence of extended blue loops.

Since the evolution of massive stars is heavily dominated by mass loss, the effects of convective overshoot alone are more difficult to single out. These will be examined in greater detail below.

Due to the larger masses of the He and C-O cores left over at the end of core H- and He-burning phases, respectively, the critical masses M$_{up}$ and M$_{HeF}$ are about 30% smaller than in classical models (Barbaro & Pigatto 1984; Bertelli et al. 1985; Bertelli et al. 1986a,b). The impact of this result on the observational front is straightforward.

Models incorporating core overshoot all along their evolutionary history have not yet been evolved into the TP-AGB regime, however we may foresee a behavior qualitatively similar to that of classical models (Chiosi et al. 1987). We have already reported that

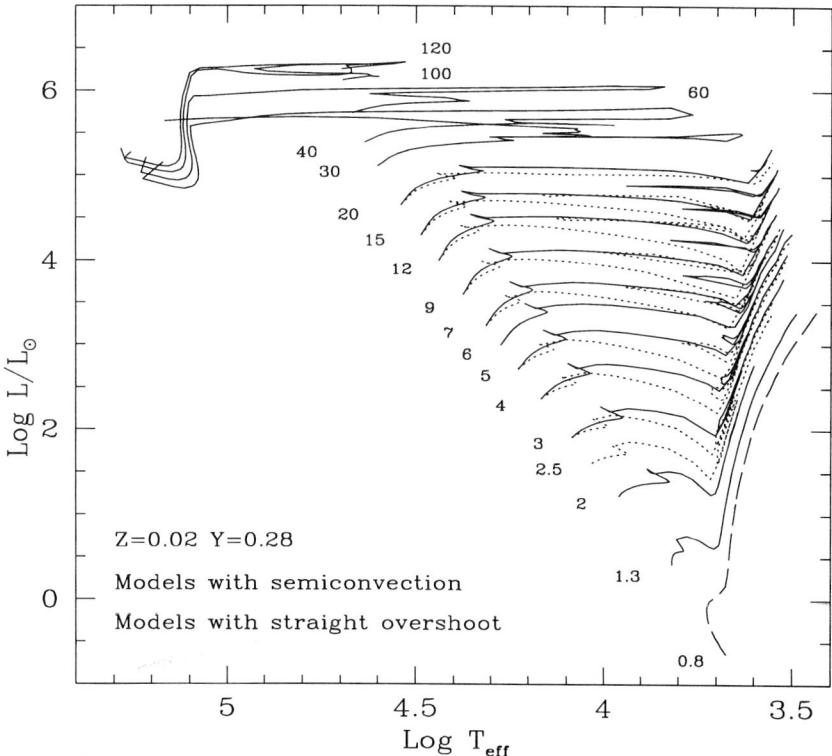

FIGURE 4. The HRD of models with full overshoot (solid lines) and semiconvection (dotted lines). Massive stars (M ≥ 10M$_\odot$) are calculated in presence of mass loss by stellar wind. In the models with overshoot the transition masses M$_{up}$ and M$_{HeF}$ are 5 and 1.6 M$_\odot$, respectively. In contrast, in semiconvective models M$_{up}$ ≃ 9M$_\odot$ and M$_{HeF}$ ≃ 2.2M$_\odot$. The chemical composition is [Z=0.02, Y=0.28]. All the models are from the Padua library

overshoot from the convective shell that follows a thermal pulse has occasionally been adopted to improve upon the explanation of C stars (Hollowell 1988; Hollowell & Iben 1988, 1989, 1990).

The path in HRD of models with convective overshoot as compared to those with semiconvection is shown in Fig. 4 limited to the case of solar composition. Massive stars are calculated taking mass loss by stellar wind into account. Finally, we also display a sequence of low mass (0.8 M$_\odot$) for the sake of comparison. All the models are from the Padua library.

7.9. Stellar models with Deng's et al. diffusion

In this section we present the stellar models of Deng et al. (1996a,b) giving major emphasis to the main properties of the HRD, lifetimes, and lifetime ratios. Two chemical compositions are considered, i.e. [Z=0.008, Y=0.25] suited to the LMC supergiant stars, and [Z=0.020, Y=0.28] typical of the same stars in the solar vicinity. The stellar models span the mass range 5 to 100 M$_\odot$ and go from the zero age main sequence up to the stage of core He-exhaustion.

The HRD. The evolutionary path in the HRD of diffusive models with chemical composition [Z=0.008, Y=0.250] is shown both in Figs. 5 and 6, which also display the

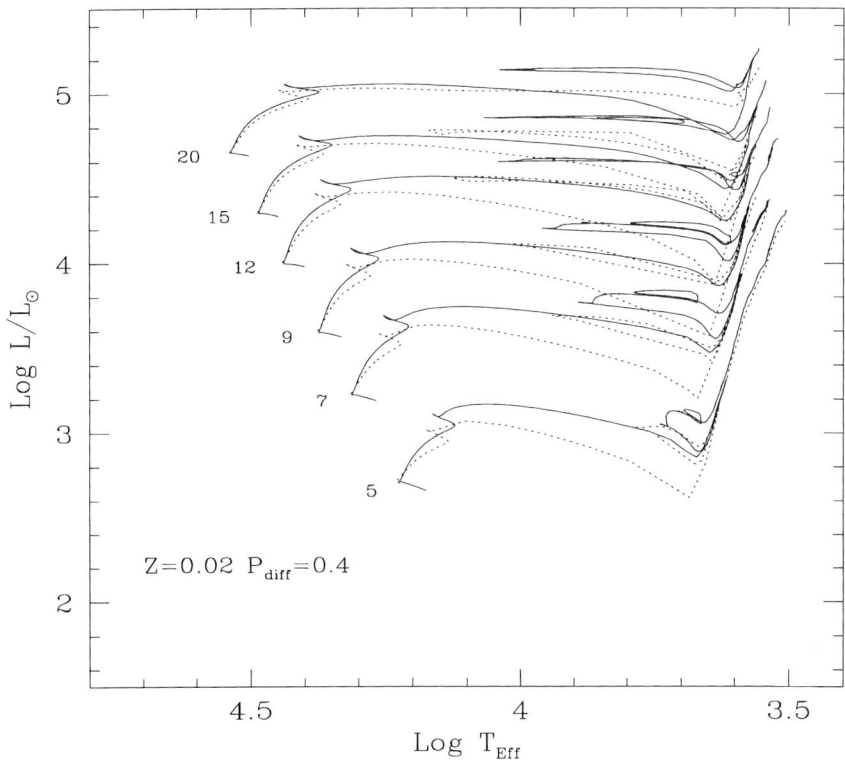

FIGURE 5. The HRD of models with diffusion (solid lines) according to Deng's et al. (1996a,b) formalism and mass loss by stellar wind (limited to the case of massive stars) as compared to classical semiconvective models (dotted lines) of the Padua library. The chemical composition is [Z=0.02, Y=0.28]. The diffusive models are calculated with $P_{dif} = 0.4$

correspondent semiconvective and straight overshoot models, respectively. For masses lower than about 30 $M_\odot$, all sequences perform wide loops, whose extension however gets narrower at increasing metallicity. Starting with the 30 $M_\odot$ star, the evolution is dominated by mass loss so that the well behaved loops are destroyed. We notice the typical blue - red - blue evolution leading to the formation of WR stars of different morphological type. It is worth of interest to compare the HRD of the present diffusive models with the HRD of models calculated with the same input physics but with different schemes of mixing, i.e. the semiconvective models with the Schwarzschild criterion (SE-models), presente Figs. 5, and the overshoot models with full homogenization of this region (FO-models) by Alongi et al. (1993), Bressan et al. (1993) and Fagotto et al. (1994a,b,c). shown in 6. It is soon evident that the path in the HRD of diffusive models is much akin to that of FO-models, the major difference being the wider extension of the loops. Compared to the SE-models, diffusive models have a wider main sequence band and much brighter luminosity during the core He-burning phase. However, they possess loops of comparable extension. Similar remarks can be made for the lifetimes and the lifetime ratios (see the data of Table 2 and 3 and the discussion below). This result is even more remarkable considering the different values of λ that are adopted for the FO- and diffusive models, i.e. $\lambda = 0.5$ and $\lambda = 1$, respectively (see Deng et al. 1996a,b for more details). The cause of it resides in the partial mixing induced by diffusion in the

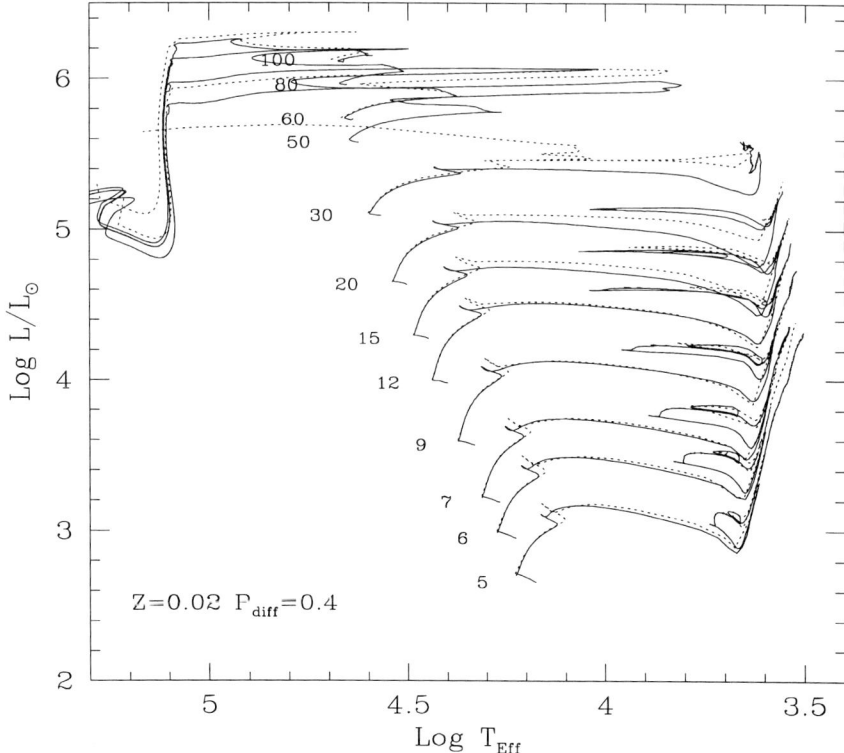

FIGURE 6. The HRD of models with diffusion (solid lines) according to Deng's et al. (1996a,b) formalism and mass loss by stellar wind (limited to the case of massive stars) as compared to straight overshoot models (dotted lines) from the Padua library. The chemical composition is [Z=0.02, Y=0.28]. The diffusive models are calculated with $P_{dif} = 0.4$

overshoot region. Indeed, despite the bigger penetration distance, the diffusive models develop smaller cores at the end of H-burning phase. As a consequence of this, diffusive models near the end of the main sequence phase possess slightly higher T_{eff}s and lower luminosities.

Lifetimes. Table 2 contains the lifetimes of the central H- and He-burning phases, τ_H and τ_{He}, respectively, together with the ratio τ_{He}/τ_H for all the diffusive models under consideration, whereas Table 3 compares the lifetimes of diffusive models with those of FO- and SE-models limited to stars up to 20 $M_\odot$. The case of more massive stars is treated separately in relation to the WR stars. For the sake of clarity we list also the lifetimes τ_H and τ_{He} together with the ratio τ_{He}/τ_H of SE- and FO-models taken from the Padua library. The comparison is made looking at the following ratios: $^H R_{ov} = (\tau_H)_{dif}/(\tau_H)_{ov}$, $^{He} R_{ov} = (\tau_{He})_{dif}/(\tau_{He})_{ov}$, $^H R_{se} = (\tau_H)_{dif}/(\tau_H)_{se}$, and $^{He} R_{se} = (\tau_{He})_{dif}/(\tau_{He})_{se}$. With respect to the FO-models the diffusive models have nearly identical τ_H and significantly longer τ_{He} (about 30%). The constancy of τ_H's and the different extension of the overshoot regions in the two types of model, implied by the different choice for λ (i.e. 1 here and 0.5 in the FO-models) means that for the adopted P_{dif} the mixing rate in diffusive models is about 50% of the rate in the FO-models. The longer τ_{He} of diffusive models can be understood as a result of their lower luminosity and almost equal dimension of the core and hence equal amount of fuel. This implies that the mixing

M	τ_H	τ_{He}	τ_{He}/τ_H	M	τ_H	τ_{He}	τ_{He}/τ_H
Z=0.008, Y=0.250				Z=0.020, Y=0.280			
5	105.83	14.664	0.138	5	101.27	19.210	0.189
6	71.64	8.833	0.123	6	67.99	9.586	0.141
7	51.56	5.810	0.113	7	48.72	6.103	0.125
8	39.65	4.441	0.112				
9	31.52	3.115	0.099	9	28.55	3.233	0.113
12	18.09	1.933	0.107	12	17.17	1.889	0.110
15	13.40	1.308	0.097	15	12.31	1.301	0.105
20	9.19	0.971	0.105	20	8.42	0.764	0.09
25	7.20	0.763	0.106				
30	6.35	0.635	0.100	30	6.20	0.78	0.13
40	5.16	0.57	0.11	40	4.97	0.74	0.15
60	4.13	0.50	0.12	60	4.06	0.64	0.16
80	3.72	0.43	0.12	80	3.57	0.59	0.16
100	3.46	0.39	0.11	100	3.35	0.61	0.18

TABLE 2. Lifetimes (in units of 10^6 years) of the core H- and He-burning phases for diffusive models. Masses are in solar units.

rate during the core He-burning phase is higher than in the FO-models. Furthermore, as a consequence of the above properties, the ratio τ_{He}/τ_H of the present models is about 20-50% higher than that of old FO-models with $\lambda = 0.5$. In contrast, with respect to the SE case, diffusive models have a much longer τ_H and a slightly shorter τ_{He}. The longer τ_H is caused by the bigger core, while the nearly identical τ_{He} is caused by the higher luminosity which is only partially compensated by the larger core and hence amount of fuel. It follows from the above results that the ratios τ_{He}/τ_H of diffusive models are comprised between those of FO- and SE-models. This is a theoretical prediction that can be tested by means of star counts in the HRD of rich clusters like those of the LMC and SMC with accurate photometric data. Although the question is not fully settled, recent analyses (Chiosi et al. 1995) favor models having lifetimes ratios τ_{He}/τ_H much similar to those of the FO- and diffusive models.

Blue to red supergiant ratios. In Table 4 we give the ratio of the main sequence to the post main sequence lifetimes (shortly indicated by MS/PMS) together with the ratio (BSG/RSG) of the core He-burning lifetime spent in the blue and red side of the HRD, limited to stars with initial mass up to 20 $M_\odot$. Keeping the discussion very short, the ratio MS/PMS increases from about 5 to 10 as the mass of the star goes from 5 to 20 $M_\odot$ with little dependence on the metallicity. The variation of the ratio MS/PMS with the mass simply reflects the different relative duration of the core He-burning, whereas its scarce sensitivity to the metallicity reflects the small effect of Z on τ_H. The ratio BSG/RSG is more metal dependent, for Z=0.008 it goes from about 2 to 6-8 over the same range of mass, whereas for Z=0.02 it remains close to unity but for the 20 $M_\odot$ star where it drops to about 0.6. Before drawing any conclusion from these numbers, we prefer to look at the detailed number frequencies derived from star counts in the HRD and the corresponding theoretical predictions. The analysis is postponed to section 12.

The Blue gap in the HRD. Table 4 gives the position in the HRD of the bluest stage of the loop (referred to as LOOP) and the reddest stage of the core H-burning band (the

	Overshoot			Classical Mixing			Comparison Ratios			
M	τ_H	τ_{He}	τ_{He}/τ_H	τ_H	τ_{He}	τ_{He}/τ_H	$^H R_{ov}$	$^{He} R_{ov}$	$^H R_{se}$	$^{He} R_{se}$
				Z=0.008	Y=0.250					
5	104.75	10.197	0.097	81.206	18.930	0.233	1.010	1.438	1.303	0.775
6	71.536	5.6120	0.078	57.051	10.171	0.178	1.001	1.574	1.256	0.868
7	52.336	3.5966	0.069	42.559	6.1211	0.144	0.985	1.615	1.211	0.949
9	32.387	1.9975	0.062	26.877	3.0546	0.114	0.973	1.560	1.173	1.020
12	19.830	1.1906	0.060	16.722	1.6371	0.098	0.912	1.624	1.082	1.181
15	14.015	0.8776	0.063	11.970	1.1234	0.094	0.956	1.490	1.200	1.164
20	9.430	0.6451	0.068	8.3121	0.7766	0.093	0.974	1.505	1.106	1.250
				Z=0.020	Y=0.280					
5	103.81	10.94	0.105	81.580	19.41	0.239	0.976	1.756	1.241	0.990
6	68.924	5.566	0.081	–	–	–	0.986	1.722	–	–
7	49.595	3.523	0.071	40.415	5.881	0.146	0.982	1.732	1.205	1.038
9	29.861	1.884	0.063	24.925	2.772	0.111	0.956	1.716	1.145	1.166
12	17.765	1.104	0.062	12.205	1.507	0.087	0.967	1.711	1.407	1.254
15	12.501	0.811	0.065	10.904	1.016	0.093	0.985	1.604	1.129	1.280
20	8.545	0.605	0.071	7.550	0.711	0.094	0.998	1.516	1.130	1.290

TABLE 3. Comparison of the lifetimes of diffusive, standard overshoot and semiconvective models. (see the text for more details). $^H R_{ov} = (\tau_H)_{dif}/(\tau_H)_{ov}$, $^{He} R_{ov} = (\tau_{He})_{dif}/(\tau_{He})_{ov}$, $^H R_{se} = (\tau_H)_{dif}/(\tau_H)_{se}$, and $^{He} R_{se} = (\tau_{He})_{dif}/(\tau_{He})_{se}$.

so-called termination main sequence, TAMS). The extension of the TAMS toward the red gives an idea of the widening of the main sequence band (MSB) induced by diffusion to be compared with the corresponding locus for SE- and FO-models (cf. the results by Alongi et al. 1993; Bressan et al. 1993; and Fagotto et al. 1994a,b,c). TAMS and LOOP determine the expected width of the blue gap in the HRD. It is soon evident that the evolutionary sequences with diffusion have a narrower blue gap than the SE- and FO-sequences of the same composition and input physics. See also the study by Chiosi et al. (1995) in which the observational HRD of NGC 330, a blue young cluster of the SMC, has been compared to diffusive, SE- and FO-models. Despite the advantages offered by the diffusive models, the problem of the many stars in the blue gap still remains.

General remarks. To summarize, the key points of the above discussion are:

(a) Diffusive models simultaneously share the key properties of FO- and SE-models, i.e. wide main sequences and extended loops. The lifetimes and lifetime ratios of diffusive models are closer to FO- rather than SE-models. Although these results are based on the use of the Schwarzschild criterion and $P_{dif} = 0.4$, similar results are obtained using the Ledoux criterion. The alternative is meaningful only for massive stars.

(b) The main sequence band (MSB) of diffusive models is only slightly narrower and much broader than those of FO-models and SE-models, respectively.

(c) The loops are in general much more extended than those of the FO-models, and comparable in size to those of the SE-models. Therefore, diffusion rules out a point of embarrassment of the FO-models, i.e. their narrow blue loops.

(d) Finally, we like to emphasize that by varying P_{dif} the results obtained with full

M/M$_\odot$	MS/PMS	BSG/RSG	LOOP		TAMS	
			logL/L$_\odot$	logT$_{eff}$	logL/L$_\odot$	logT$_{eff}$
		Z=0.008	Y=0.250			
5	6.221	2.092	3.2847	3.8910	3.0978	4.1589
6	7.246	1.939	3.6100	3.9740	3.3910	4.2050
7	7.874	2.076	3.8636	4.0325	3.6321	4.2413
8	8.197	2.243	4.0729	4.0729	3.8442	4.2694
9	9.174	1.671	4.2454	4.1040	4.0178	4.2938
12	8.547	4.060	4.5899	4.2002	4.4018	4.3478
15	10.090	1.890	4.8546	4.2194	4.6829	4.3798
20	9.351	1.539	5.1387	4.2439	4.9963	4.4111
		Z=0.020	Y=0.280			
5	4.505	1.152	3.0496	3.7394	3.0417	4.1225
6	6.623	1.683	3.4623	3.8301	3.3636	4.1693
7	7.407	1.263	3.7668	3.8916	3.6231	4.2073
9	6.757	1.528	4.2056	3.9597	4.0243	4.2629
12	9.174	1.037	4.6001	4.0413	4.4382	4.3140
15	10.000	0.710	4.8570	4.0677	4.6997	4.3500
20	10.638	0.288	5.1381	4.0389	5.0119	4.3735

TABLE 4. The post-MS to MS lifetime ratio (PMS/MS) together with the maximum extension of the blue loop (LOOP) and the location of the reddest stage of the main sequence band (TAMS) in the HRD, limited to stellar tracks up to 20 M$_\odot$

overshoot and simple semiconvection are recovered. As shown by Deng *et al.* (1996a,b) with P$_{dif} \simeq 1$ the full overshoot case is obtained, while P$_{dif} \simeq 0.1$ the pure semiconvective case is recovered.

7.10. *Stellar models with Salasnich's et al. diffusion*

Salasnich *et al.* (1997) models calculated with $\alpha_1 = 0.009$ and $\alpha_2 = 100$ (values indicated by a preliminary analysis of the problem) are shown in Fig. 7 limited to the stars of 18, 20 and 30 M$_\odot$ and composition [Z=0.02, Y=0.28].

These models are best suited to explain the missing blue gap in the HRD and the origin of faint WR stars (see below). In brief, they possess very extended blue loops which may reach the main sequence region. Since in this phase the models spend $\sim 50\%$ of the He-burning lifetime, **they are able to solve for the first time the mystery of the missing blue gap.**

8. Passing from theory to observations

There is no branch of modern astrophysics in which the results of stellar evolution theory cannot find immediate application. Stellar models are required to understand the genetic connection between stars of different morphological type, to infer the age of stellar systems (the stellar clock is perhaps the best clock to our disposal), to synthesize integrated spectra, magnitudes and colours of stellar aggregates of different complexity going from star clusters to galaxies, to derive the initial mass function, to explain

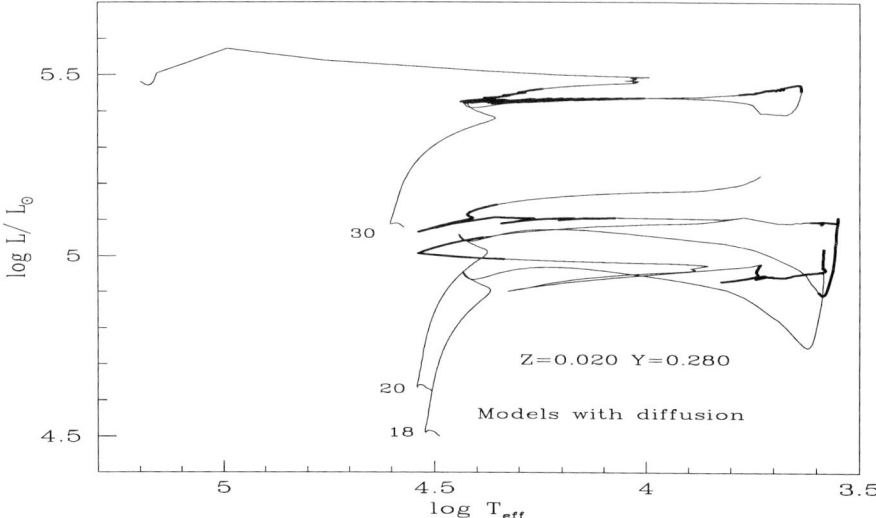

FIGURE 7. HRD of 18, 20, and 30$M_\odot$ stars with composition [Z=0.02, Y=0.28], the Salasnich et al. (1997) diffusive algorithm, and the new mass loss rates for RSG stars. The thick portion of each track indicates the regions of slow evolution (high detectability) during the core He-burning phase. Note how the blue loops may extend to high T_{eff} in the regions of the missing blue gap and WR stars. See the text for more details

the abundances of elemental species observed at surface of stars and in the interstellar medium, to understand the past history of chemical enrichment in galaxies of different morphological type, to predict properties of supernovae and planetary nebulae, to predict the ionizing flux of HII regions, to calibrate standard candles of the distance ladder, to address questions of cosmological nature such as the age of the universe and the formation processes and evolution of galaxies, to understand their evolution with the redshift. Owing to the tremendous impact of results of stellar models on any of the above areas, much effort has been invested on continuously refining this *basic tool*, in order to provide as accurate stellar models as possible so that not only satisfactory understanding of the pattern of stellar properties is possible but also the use of these in other branches of astrophysics is solidly grounded.

8.1. *Basic requirements: adequacy, accuracy, and calibration*

Stellar models must satisfy at least three prerequisites, (i) adequacy of the input physics; (ii) accuracy of the numerical calculations; finally, (iii) calibration of three main parameters intervening in stellar models, i.e. mixing length, mass loss efficiency, and extension of convective regions. In addition to this, there is another aspect to consider when stellar models are used to build extended libraries of isochrones, tabulations of luminosity functions, etc, that are at the base of many astrophysical applications. This is the homogeneity of the stellar grids, i.e. models calculated with the same input physics, numerical technique, and calibrating method. Patching together stellar tracks from different sources should be avoided by as much as possible. While opacities, nuclear reaction rates, equation of state are less of a problem because all modern libraries of stellar models are calculated using the state of art of this important physical ingredients, problems still remain with mixing and mass loss on which we have already reported in previous sections.

The calibration of the mixing length (ML) is an important step because the MLT contains the ratio of the ML to the local pressure scale height as a parameter ($\alpha = \Lambda/H_P$), whose variations affect the structure of the outer envelope (temperature and radius) but not the structure of the inner core and hence the luminosity of the star. The calibration consists in forcing the model temperatures and radii to match their observational counterparts. It must be kept in mind that there is no *a priori* reason to expect that a single value of α holds for every situation. The first calibrator of α is the Sun: solar models are imposed to reproduce the solar radius. This calibration yields $\alpha = 1.5 \div 1.7$ when the classical MLT is adopted. The same value is able to fit the slope of the main sequence of young clusters (VandenBerg & Bridges 1984), and the main sequence and lower RGB of galactic globular clusters (Bell & VandenBerg 1987). Subdwarf stars of known trigonometric parallaxes, metallicity, and color are second calibrator. Agreement is found for $\alpha = 1.6$, consistent with the previous determinations (cf. Buonanno et al. 1989). Finally, the location of the RGB is another possible calibrator of α. Fixed the ML, the temperature and slope of the RGB depends on the abundance of low ionization potential elements (Fe, Si, and Mg). Therefore this calibration is tightly related to the metallicity scale in use. This method (cf. Renzini & Fusi-Pecci 1988 for all details) yields $\alpha \simeq 1.6$, which is once more consistent with the previous determinations.

8.2. *Theory of SSP*

A simple stellar population (SSP) is defined as an assembly of coeval, initially chemically homogeneous, single stars, whose properties are described by three parameters, i.e. the age t, the chemical composition [Z,Y], and the initial mass function (IMF). The observational analogs of SSP are star clusters as to a first good approximation they satisfy the above requirements. The existence of binary stars can be taken into account. Galaxies are not SSPs, as they contain stars of any age, and chemical composition, and the IMF can have varied as a function of time and physical environment. However, even in this case much can be learned from SSPs as the complex populations of a galaxy can always be conceived as a convolution of many SSPs. Although some general ideas on how SSP should behave have long been known (cf. Tinsley 1980), the systematic presentation of the SSP theory is due to the seminal papers by Renzini (1981c) and Renzini & Buzzoni (1983, 1986).

The number of stars N_j in any post main sequence evolutionary stage j of a SSP is given by

$$N_j = B(t) \times L_T \times t_j \quad (8.41)$$

where t_j is the duration of the phase, L_T is the integrated luminosity of the SSP, and $B(t)$ is the *specific evolutionary flux* of the SSP. The above equation can be derived from very simple considerations. For a SSP of age t, the turnoff mass M_{TO}, i.e. the mass of the star at the core H-exhaustion stage, separates stars on the main sequence ($M < M_{TO}$) from stars in post main sequence stages ($M > M_{TO}$). The mass M_{TO} is a function of age and composition $M_{TO}(t, Z, Y)$. The rate (number of stars per year) at which stars leave the main sequence is

$$b(t) = \phi(M_{TO}) \times \dot{M}_{TO} \quad (8.42)$$

where $\phi(M)$ is the IMF and $\dot{M}_{TO}$ is derived from stellar models. The function $b(t)$ gives not only the rate at which stars leave the main sequence but also the rate at which stars enter or leave any subsequent evolutionary stage, i.e. it is the evolutionary flux of the

SSP. Owing to the narrow range of masses in post main sequence stages, this flux is about constant. The number of stars in a given evolutionary stage is therefore proportional to

$$N_j = b(t) \times t_j \tag{8.43}$$

The total luminosity of the SSP is given by

$$L_T = L_{MS} + L_{PMS} = \int_{M_{inf}}^{M_{TO}} L(M)\phi(M)dM + \sum_j n_j L_j \tag{8.44}$$

with obvious meaning of the symbols. The term $n_j L_j$ can be written as $b(t)t_j L_j$ where $t_j L_j$ is the total energy radiated by stars in the stage j. This energy is simply proportional to the amount of fuel F_j burned during the phase. The above equation is the so-called **Fuel Consumption Theorem**: *The contribution of stars in any post main sequence phase to the integrated luminosity of a SSP is directly proportional to the amount of fuel burnt during that stage.* The quantities F_j are the products of stellar model calculations; they are functions of the mass and hence age. Defining the specific evolutionary flux, B(t), as the ratio of the evolutionary flux b(t) to the total integrated luminosity of a SSP, $B(t) = b(t)/L_T$, after some algebraic manipulations one gets the basic relation (8.41) above. B(t) is nearly constant over a wide range of ages, going from 0.5×10^{-11} to 2×10^{-11} stars per year per solar luminosity as the age increases from 10^7 to 10^{10} years. Furthermore it is nearly independent of the IMF, and depends very weakly on the chemical composition. See Renzini & Buzzoni (1983, 1986) for all other details. Another interesting consequence of the *Fuel Consumption Theorem* is the relative contribution of a given phase to the total luminosity. Re-arranging relation (8.44) one gets

$$\frac{L_j}{L_T} \simeq 9.7 \times 10^{10} \times \frac{F_j(M_{TO})}{L_T} \tag{8.45}$$

The study of relation (8.45) at varying age and hence M_{TO} of SSPs has given origin to the long debated question of the so-called *Phase Transitions* in SSPs (cf. Renzini & Buzzoni 1986). Relations (8.41) and (8.45) or equivalently the *Fuel Consumption Theorem* are powerful tools to understand the properties of star clusters and more complex stellar aggregates (galaxies), to immediately figure out what kind of studies are feasible, and finally to correctly plan observations and comparison between theory and observations. It provides the most direct way of comparing theoretical lifetimes to star counts and constraining the minimum requirements to be satisfied by the data sample in order that the comparison be meaningful (cf. the discussion of this topic by Renzini & Fusi-Pecci 1988).

8.3. Isochrones

Perhaps the most complete and popular library of theoretical isochrones nowadays in usage is by Bertelli *et al.* (1994). Another, equally widely used library of isochrones is that of the Yale group (Green *et al.* 1987) which, however, being specifically designed for globular clusters, has a smaller age range. Finally, a modern library of isochrones suited for studies of globular clusters has been produced by VandenBerg *et al.* (1997). In the following we make fairly extensive use of the Padua isochrones, obviously because they are immediately at hand, but also because they have proved to perform reasonably well in wide range of circumstances. The Bertelli *et al.* (1994) isochrones stand on stellar models calculated with convective overshoot from the core, mass loss by stellar wind, and from the zero age main sequence up to the formation of white dwarfs or carbon ignition in the core as appropriate for the initial mass of the star. Particularly useful is the theoretical luminosity function along each isochrone which allows direct comparison with the observational luminosity function of star clusters.

The theoretical isochrones are translated into observational Color-Magnitude Diagrams (CMD) to provide the basic tool for the interpretation of CMD of star clusters and complex stellar assemblies. The details of the conversion in use are not given here. They are amply discussed by Bertelli et al. (1994) to whom the reader should refer.

In addition to this, Bertelli et al. (1994) present useful relationship as function of the age and chemical composition for characteristic loci of the CMD, i.e. the turnoff (TO), the bottom of the RGB (B_{RGB}), tip of the RGB (T_{RGB}), medium core He-burning (M_{Heb}), termination of the AGB phase (T_{AGB}) to be directly compared with observational data.

8.4. Single and composite CMDs

This library of isochrones and companion integrated magnitudes and colors have been used to study CMD of individual clusters of any age (Vallenari et al. 1994, 1997, and references therein), integrated colors of clusters in the LMC (Girardi et al. 1995), the past history of star formation in the fields of the LMC (Bertelli et al. 1992, Vallenari et al. 1994, 1996a,b), the stellar content of the Galactic Bulge (Bertelli et al. 1996), the structure of the Milky Way towards the galactic center (Ng et al. 1996), and finally the complex CMDs for the stellar content of galaxies, e.g the study by Gallart et al. (1996) of the old and intermediate age populations in the local Group dwarf galaxy NGC 6822.

No details of these studies are given here. We limit ourselves to show for the sake of illustration first two groups of isochrones with different composition in the M_V versus (V-I) CMD of Fig. 8 (left: [Z=0.0004, Y=0.23]; right; [Z=0.02, Y=0.28]), and second the four steps of the analysis by Gallart et al. (1996) in which using a highly sophisticated crowding technique and a recent version of the synthetic HRD simulator by Bertelli (1997, unpublished) the CMD of NGC 6822 is analyzed. This is shown in the four panels of Fig. 9: panel (a) is a synthetic CMD, panel (b) is the same after simulations of crowding (no differential reddening), panel (c) with crowding and differential reddening; finally panel (d) the observed CMD. With the aid of suitably defined ratios of star counts both in the theoretical and observational CMD, and assumptions for the time dependence of the past star formation rate, Gallart et al. (1996) successfully modeled the past star formation history of this galaxy.

9. Globular Clusters

The recent revolution in photometric techniques (CCD detectors) dramatically improved the quality of the color-magnitude diagrams (CMD) of globular clusters (GCs) thus allowing comparisons with theoretical models for low-mass stars of unprecedented sophistication. Since an exhaustive referencing to the impressive list of high quality CCD-CMDs is impossible and beyond the scope of these lectures, we limit ourselves to a few illustrative cases: 47 Tuc (Hesser et al. 1987), M3 (Buonanno et al. 1987), and M62 (McClure et al. 1987), and M92 (Stetson & Harris 1988). A very informative review on the data on GCs is by Hesser (1988).

There also exists an equally impressive list of theoretical studies for low-mass stars at varying basic parameters: mass, helium abundance Y, metallicity Z -- this latter separated into three components [CNO/H], [α]/H] (α-elements), and [Fe/H] and their relative proportions -- mixing length in the outer convective layer, opacities, nuclear reactions rates, mixing process, diffusion processes, equation of state, neutrino energy losses, mass loss by stellar wind, etc. Most of these models are calculated all the way

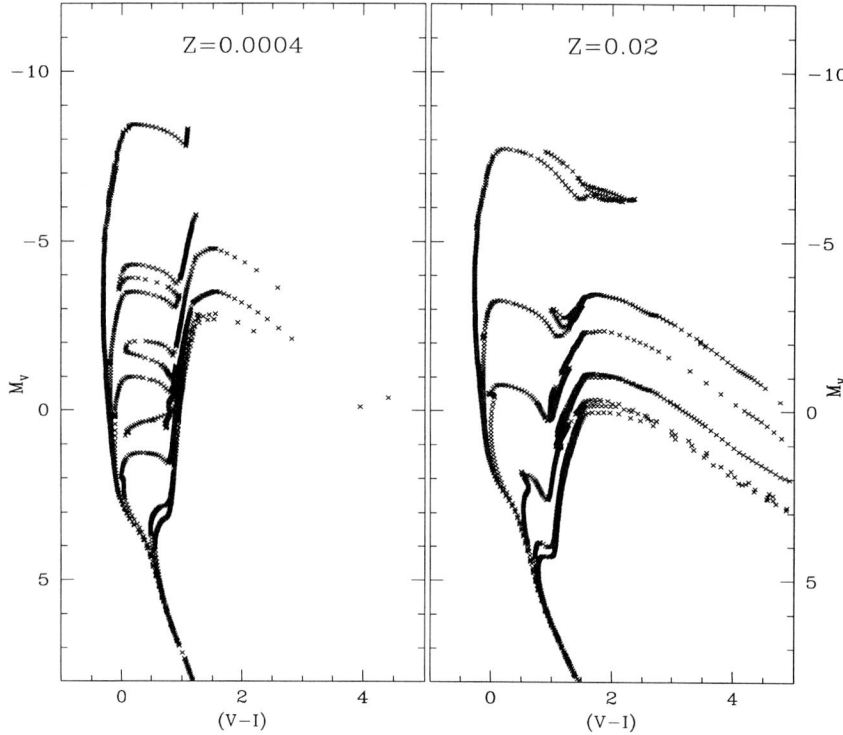

FIGURE 8. The M_V versus (V-I) CMD of isochrones (single stellar populations) with different age and composition as indicated. Left: [Z=0.0004, Y=0.23], right: [Z=0.02, Y=0.28]. From bottom to top, ages are 15, 2.0, 0.4, 0.1, and 0.01 Gyr. Note the different morphology of the AGB phase.

from the main sequence to the latest stages, thus making available homogeneous sets of evolutionary tracks and isochrones. The most recent reviews on the subject are by Renzini & Fusi-Pecci (1988), VandenBerg (1991), Stetson et al. (1996), and VandenBerg et al. (1996) to whom we refer for an exhaustive referencing.

Among the studies presenting extensive grids of stellar models we recall VandenBerg (1983, 1985), VandenBerg & Bell (1985), Sweigart (1987), Caputo et al. (1987), Sweigart et al. (1987, 1990), Chieffi & Straniero (1989), Straniero & Chieffi (1991), Bencivenni et al. (1989), Lee & Demarque (1990), Bergbusch & VandenBerg (1992), Bertelli et al. (1990), Alongi et al. (1993), Bressan et al. (1993), Fagotto et al. (1994a,b,c), VandenBerg (1992), Claret (1995), Claret & Gimenez (1995a,b), Alexander et al. (1996), Cassisi et al. (1997), Salaris et al. (1993).

As thoughtfully discussed by Renzini & Fusi-Pecci (1988) and recalled in the previous sections, to be safely used in the interpretation of the CMDs the evolutionary models must be tested for accuracy in the input physics and adequacy of the physical assumptions, and finally calibrated using known reference objects (see also VandenBerg 1991, Fusi-Pecci & Cacciari 1991, VandenBerg et al. 1996). Among the various parameters, the calibration of the mixing length in the outer convective layer is particularly important because it affects the luminosity and T_{eff} at the turnoff and position of the RGB. As has been recognized over the years the dominant error in the derivation of ages is the luminosity at the turn-off

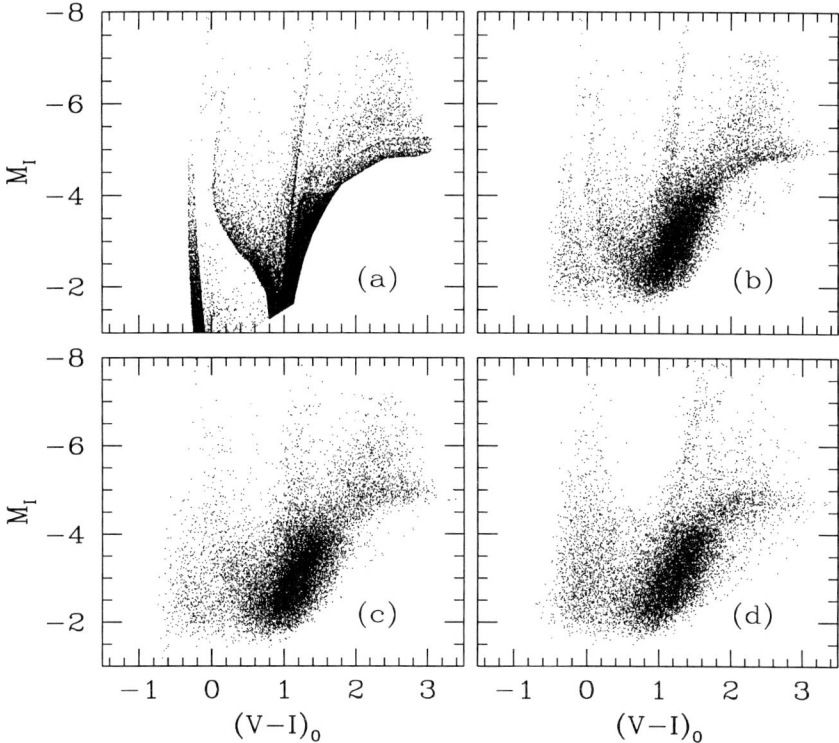

FIGURE 9. Simulations of crowding effects and differential reddening. In (a) the synthetic CMD is displayed; (b) shows the same synthetic CMD after correction for crowding (no reddening); while in (c) both differential reddening and crowding are taken into account; finally (d) shows the observed CMD of NGC 6822. Reproduced from Gallart et al. (1996)

and the uncertainty in the distance scale (see below). Proffitt (1993), Chaboyer (1995), Chaboyer & Kim (1995), and Mazzitelli et al. (1995) have recently examined the impact of changing the basic input physics (nuclear reaction rates, opacities, equation of state, mixing length, convection theory, etc,) on globular clusters ages. See also Chieffi et al. (1995) for a recent suggestion that the mixing length parameter is a weak function of the metallicity, and the exhaustive discussion of all the above topics by VandenBerg et al. (1996).

9.1. *Luminosity Functions, Mass Functions, and Star Counts*

Luminosity functions (LFs) of the main sequence to very faint magnitudes allow one to test the low-mass main sequence models and to derive information on the present-day mass function (PMF) from which a guess on the initial mass function (IMF) can be obtained. Stars with $M \leq 0.6 M_\odot$ are characterized by complex physics which makes it difficult to calculate adequate models. The main reasons are in the adopted model atmospheres, the low temperature molecular opacities, the formation and dissociation of the H_2 molecules, the failure of the ideal gas approximation in the equation of state and adiabatic gradient – specifically the presence of coulomb interactions (Copeland et al. 1970), the coexistence of partial degeneracy and ionization (Magni & Mazzitelli 1979), the dominant role by overadiabatic convection (D'Antona & Mazzitelli 1994). The models calculated by VandenBerg et al. (1983), D'Antona & Mazzitelli (1986, 1993), D'Antona (1987), Burrows et al. (1993), Baraffe et al. (1995), and Chabrier et al. (1996) although differing in many details of the input physics, predict a sudden flattening of the mass-luminosity relation and steepening of the main sequence below $0.5 M_\odot$. Both are due to the effect of the H_2 molecule rather than to the models becoming fully convective. The LF is also expected to steepen at the same luminosity (mass) as perhaps indicated by the observations (see Richer & Fahlman 1991). However, at the present, data are not accurate enough to discriminate among current evolutionary models (see the discussion in Renzini & Fusi-Pecci 1988).

From the earliest studies on the LFs in clusters with different [Fe/H] it was soon evident that the slopes of the LFs were dissimilar, in the sense that metal-rich clusters have LFs with a flatter slope. Interpreted with the aid of theoretical models, because a purely empirical mass-luminosity relation for Population II stars is still lacking, the LFs were transformed into PMFs by means of the usual power-law representation $\phi(m)dm = m^{-(1+x)}$, and a correlation between x and [Fe/H] was suggested (McClure et al. 1986, 1987). Specifically, the metal-rich clusters should possess a PMF flatter than that of the metal-poor ones. This result has potentially far-reaching consequences, in particular if interpreted in terms of a dependence of the IMF on the metallicity, and therefore it must be examined in detail. First, almost none of the existing LFs extend below 0.5 $M_\odot$; this limit depends however on the adopted stellar models. An exception is given by the LFs presented by Richer & Fahlman (1991), which reach about 0.2 $M_\odot$. Second, the LF obtained in a given portion of the cluster must be corrected to become representative of the whole cluster (see Pryor et al. 1986). If internal dynamical evolution and tidal stripping are unimportant, the PMF is simply the IMF because the H-burning lifetime of these stars is much longer than the Hubble time. However, this is not the case because many GCs show clear evidence of internal dynamical interaction and mass segregation (see Pryor et al. 1986; also Richer & Fahlman 1991 and references therein) with more massive stars sinking toward the center. Furthermore, tidal stripping may be very efficient (Stiavelli et al. 1991). Therefore, as a result of evaporation and stripping, GCs may have lost a large fraction of their low-mass stars with consequent lowering of the slope x. Finally, it has been shown that if observations are taken for very low mass stars at large distances from the cluster core, the observed PMF is similar to the IMF in the absence of extensive tidal stripping (Richer et al. 1991). Therefore the observed PMF in these regions constitutes a lower limit to the IMF. With these limitations in mind, Ortolani et al. (1989), Capaccioli et al. (1991), and Piotto (1991) reached the following conclusions: (a) the PMF varies from cluster to cluster; (b) for intermediate and low metallicity GCs there is no correlation between x and [Fe/H]; (c) the most metal-rich clusters have a flatter PMF; and finally, (d) the PMF slope seems to correlate with the position of the clusters with respect to the galactic gravitational potential. Specifically, the PMF becomes steeper and steeper at increasing galactocentric distance R_G and height z_G above the galactic plane. The simulations by Stiavelli et al. (1991) suggest that tidal disk shocking could be responsible for the observed correlations.

Luminosity functions for the upper main sequence stars (at the turnoff) have been occasionally proposed as age calibrators (Paczynski 1984, Ratcliff 1987, Demarque 1988). However, for the reasons discussed by Renzini (1986, 1988) this method cannot give ages with a precision better than $\pm 5 \times 10^9$ yr.

Star counts of RGB stars in GCs (Hesser et al. 1987, Fusi-Pecci et al. 1990) have confirmed the existence of a bump in the differential LF, or equivalently a knee in the cumulative LF, whose origin was first pointed out by Iben (1968). As amply discussed by Renzini & Fusi-Pecci (1988), the luminosity of the bump identifies the mass coordinate of the bottom of the homogeneous envelope and, therefore, the maximum penetration of the external convection during the RGB phase. Fusi-Pecci et al. (1990), discussing the relationship between the magnitude M_V of RR Lyrae stars and [Fe/H] (see below), noticed that the observed M_V of the bump is about 0.415 ± 0.07 mag fainter than predicted by current models. Three causes of disagreement were indicated – old opacities, more efficient envelope convection, and initial abundance of helium – to which different abundances of α-elements, like Ne, Mg, Si and S can be added (F. Ferraro 1991, private communication). Alongi et al. (1992) argued that opacity and helium abundance cannot rule out the discrepancy, whereas a more efficient mixing at the base of the convective envelope is plausible. Indeed, they found that envelope overshoot of about 0.7 of a pressure scale height H_P can reconcile theory with observations. They also considered the bump luminosity to be an additional constraint to the otherwise uncertain efficiency of this phenomenon.

Star counts in RGB, HB and AGB phases are considered as probes of stellar structure as they reflect the duration of the underlying evolutionary phases hence the adequacy of the physical assumptions in model calculations (Buzzoni et al. 1983, Buonanno et al. 1985). Four ratios can be constructed:

$$R = \frac{t_{HB}}{t_{RGB}} \simeq \frac{N(HB)}{N(RGB)} \tag{9.46}$$

$$R' = \frac{t_{HB}}{(t_{RGB} + t_{AGB})} \simeq \frac{N(HB)}{[N(RGB) + N(AGB)]} \tag{9.47}$$

$$R_1 = \frac{t_{AGB}}{t_{RGB}} \simeq \frac{N(AGB)}{N(RGB)} \tag{9.48}$$

$$R_2 = \frac{t_{AGB}}{t_{HB}} \simeq \frac{N(AGB)}{N(HB)} \tag{9.49}$$

where all symbols have their usual meaning. Specifically, t_{RGB}, t_{HB}, and t_{AGB} are lifetimes, whereas N(RGB), N(HB), and N(AGB) are star counts. In particular, N(RGB) is the number of RGB stars brighter than the RR Lyrae luminosity, and N(AGB) is the number of AGB stars up to 2.5 magnitude above the RR Lyrae luminosity. The ratios R, R', and R_1 can be used to trace back the helium abundance (Iben 1968) and convective mixing during the HB phase (Castellani et al. 1971a,b). The ratio R_2 is particularly sensitive to mixing in HB, and has been found to be able to discriminate among different types of mixing. This ratio is 0.15 ± 0.01 in well studied GCs (Buzzoni et al. 1983, Buonanno et al. 1985). Contrary to the claim by Renzini & Fusi-Pecci (1988), models with semiconvective mixing cannot match this ratio because several breathing pulses of the convective core prolong the core He-burning lifetime and shorten the AGB lifetime. On the contrary, models with convective overshoot match the above ratio if a plausible efficiency is assumed (Bressan et al. 1986, Chiosi 1986).

9.2. *Determining the ages of globular clusters: premises*

Determining the age of GCs is a complex game (see Demarque et al. 1991, and Fusi-Pecci & Cacciari 1991), which requires a knowledge of many parameters, such as the helium content Y, metallicity [M/H], CNO abundance [CNO/H], distance modulus, and reddening.

Abundance of Helium. Since GC stars are too cool to allow direct spectroscopic measurements of the abundance by mass of helium Y, less direct methods are used. A general assumption is that helium abundance in GCs reflects primordial nucleosynthesis as GCs are among the oldest objects in the Universe. In general, the helium abundance is estimated from various sources among which we recall: the big-bang nucleosynthesis (Boesgaard & Steigman 1985, Olive et al. 1991, Krauss & Romanelli 1990, Mathews et al. 1993), the extragalactic HII regions (Davidson & Kinman 1985), the empirical determinations of the pregalactic conditions (Pagel et al. 1992, Izotov et al. 1994, Olive & Steigman 1995), the R-method (Buzzoni et al. 1983; Buonanno et al. 1985; Ferraro et al. 1992, 1993), the morphology of observed HB (Lee et al. 1990, Dorman et al. 1991), and the red edges of the RR Lyrae instability strip (Bono et al. 1995). All these concur to a common value. The adopted helium abundance is $Y = 0.235 \pm 0.005$ and it is generally assumed to be constant throughout the halo clusters, even if helium abundance has been often considered a candidate for the second parameter (see below).

Metallicity. The metallicity is usually referred to the observed abundance [Fe/H] and the nowadays accepted metallicity scale is from Zinn & West (1984). The majority of GCs have [Fe/H] values from -1.0 to -2.3 dex with typical uncertainty of 0.15 dex. However, recalling that in metal-poor stars, the abundances of Ne, Mg, Si, and S are significantly enhanced with respect to [Fe/H] (Nissen et al. 1985), [Fe/H] $\neq$ [M/H], so that [Fe/H] alone is not fully representative of the real content of heavy elements. Useful compilations of [M/H] are from Pilachowski (1984), Webbink (1985), and Hesser & Shawl (1985). The observation of CMDs showing sequences of virtually undetectable width indicates a uniform abundance of heavy elements within the stars of a particular cluster (cf. Stetson 1993, Folgheraiter et al. 1993). Two exceptions exist: ωCen and M22 which show star to star differences of about 0.5 dex).

Abundances of α-elements. It is well known that the morphology of the turnoff greatly depends on the abundance of oxygen. The controversy of the oxygen enhancement in cluster stars as measured by [O/Fe] is still far from being solved. Given [O/Fe]=0 for the Sun by definition, the questions are whether [O/Fe] is different for the halo stars and whether it varies with [Fe/H]. Observing giant stars in GCs, Pilachowski et al. (1983) obtained [O/Fe]=0.25 regardless of [Fe/H]. A similar estimate is given by Gratton & Ortolani (1989) who find [O/Fe]=0.40 (see also Sneden et al. 1991, McWilliam et al. 1992). Therefore, giant stars in GCs seem enhanced in oxygen relative to the Sun with no correlation to [Fe/H]. However, the question arises whether this result holds for all the stars in a cluster, main sequence included, or whether it is limited to giants. An enhancement of the oxygen abundance in giant stars, resulting from inner processes can be excluded on the basis of stellar evolution theory. In dwarf field stars, the ratio [O/Fe] for [Fe/H] $\leq$ -1 is more controversial. Current estimates for [O/Fe] go from 0.4 to 0.7 dex with no variations with [Fe/H]. The opposite conclusion was reached by Abia & Rebolo (1989) who claimed that [O/Fe] varies from 1.2 for [Fe/H]=-2.3 to 0.6 for [Fe/H]=-1. This result was criticized by Spite & Spite (1991) who argued that the results were too sensitive to the adopted atmosphere parameters. In general, either [O/Fe]=0 or [O/Fe]=0.5 to 0.7 is adopted.

Reddening. Since there are many GCs with low color excess ($E_{B-V} \leq 0.1$) spanning a broad range of metallicities (up to [Fe/H]=-1), reddening is not a serious problem in finding the intrinsic color of the turnoff.

Distances. The distance scale of GCs is another topic of strong controversy (see Renzini 1991; Chaboyer 1995, Bolte & Hogan 1995). There are three standard candles from which globular cluster distances are derived: namely, the nearby sub-dwarfs, the RR Lyrae stars, and thanks to HST the white dwarfs. *Subdwarfs.* The nearby subdwarfs (metal poor stars with halo kinematics) close enough to have measurable parallaxes allows (i) to derive absolute M_V so that accurate testing of theoretical zero-age main sequences at varying [Fe/H] in the low metallicity regime is possible; (ii) under the assumption that these stars are similar to unevolved stars in globular clusters, to tie the relative cluster distances into the local distance scale. This topic has been discussed in great detail by VandenBerg et al. (1996) to whom the reader should refer. *RR Lyrae*. Modern

determinations of the distance modulus reduce to comparing the apparent magnitudes of the RR Lyrae stars or HB stars with the corresponding absolute visual magnitudes. There are several independent methods to obtain the absolute visual magnitudes of RR Lyrae stars, $M_V(RR)$ (see the discussion by Renzini & Fusi-Pecci 1988 and references therein), which ultimately lead to assess whether or not a correlation between $M_V(RR)$ and [Fe/H] exists and try to fix the slope and zero point of this relation (see also Fusi-Pecci & Cacciari 1991). The zero points is of critical importance for determining the age of the oldest clusters, whereas the slope bears very much on the resulting age-metallicity relation (see below) The method based on the pulsational properties of RR Lyrae stars proposed by Sandage (1982, 1986) gives $\Delta M_V(RR)/\Delta[Fe/H] = 0.35$. Specifically, in his investigation of the Oosterhoff effect Sandage (1982, 1986) determined the slope of the above relation from a argument known as the period shift. Assuming that the light curve shapes (rise time) and amplitude of RR Lyrae stars are unique functions of T_{eff}, the periods are found to increase with the metallicity according to $\Delta \log P = -0.12\Delta[Fe/H]$. The slope of the $M_V(RR) - [Fe/H]$ relation follows from the pulsation theory assuming that the mass of RR Lyrae is the same irrespective of [Fe/H]. The zero point is derived from an average of Baade-Wesselink and main sequence fitting magnitudes of RR Lyrae stars: $M_V = 0.89 \pm 0.05$ at [Fe/H]=-1.4 (see Renzini & Fusi-Pecci 1988). Similar slope and zero point, however with a larger range of uncertainty, were recently obtained by Sandage (1990a) by means of a more sophisticated analysis of the problem. [See also the review by Sandage & Cacciari 1990)]. Sandage's (1982, 1990a) conclusions seem to be supported by the observational studies of Buonanno et al. (1990) and Longmore et al. (1990). However, standard calculations of ZAHB models (Sweigart et al. 1987) are not able to predict any appreciable period shift, unless helium is anticorrelated with [Fe/H] which likely does not occur [see the discussions by Fusi-Pecci & Renzini (1988), Fusi-Pecci & Cacciari (1991), and Lee (1991a)]. To cast light on the problem, a different analysis was performed by Lee et al. (1987), Lee & Demarque (1990), Lee et al. (1990), and Lee (1990) who included off-ZAHB evolution and a possible dependence of the T_{eff}-pulsation amplitude relation on [Fe/H]. Calculations of synthetic HBs with solar [O/Fe] give $\Delta M_V(HB)/\Delta[Fe/H] = 0.17$ and $M_V(HB)=0.70$ at [Fe/H]=-1.4. This value for the slope seems to be supported by the observational work of Cacciari et al. (1989), Liu & Janes (1990), Fusi-Pecci et al. (1990), and Sarajedini & Lederman (1991). However, this new analysis implies the period shift relation $\Delta \log P = -0.04\Delta[Fe/H]$, whose slope is much lower than the Sandage (1982) value. Owing to the far-reaching implications of the $M_V(RR)$-[Fe/H] relation on the age problem (see below), this topic is a matter of debate. For more details on the subject the reader should refer to Fusi-Pecci & Renzini (1988), Sandage & Cacciari (1990), Lee (1991a) and VandenBerg et al. (1996) for an updated discussion of this topic. *White Dwarfs*. This method rests on fitting to the

white dwarf cooling sequences the observational position of WDs in the CMD (cf. Fusi-Pecci & Renzini 1979). Preliminary HST data for M4 by Richer et al. (1995) give a distance very similar to that derived by Richer & Fahlman (1984) on the assumption that $M_V(HB) = 0.84$.

Rotation & Diffusion Rotation may affect the structure of a low mass star in several ways: first there is evidence of rotation at significant rates (Peterson 1985), perhaps there are signatures in the observational surface abundances (see Kraft 1994 and references therein), it may induce deep mixing thus altering the surface abundance of H (Langer & Hoffman 1995). However as far as the fundamental question whether rotation can affect the age predicted by standard, non rotating models, according to Deliyannis et al. (1989) the answer is probably not (cf. VandenBerg et al. 1996 for more details). Under the action of gravity, in GC stars helium can sink inward relative to hydrogen. This process may affect the age in two ways. First the lower relative central abundance of hydrogen decreases the main sequence lifetime. Second a higher relative hydrogen abundance in the envelope results in a larger radius (lower T_{eff}) without changing the RGB position. The main sequence turnoff is redder, thus implying that a lower age is required to fit a given cluster (Deliyannis et al. 1990, Proffitt et al. 1990, Proffitt & VandenBerg 1991, J. Richer et al. 1992, Chaboyer et al. 1992, Chaboyer & Demarque 1994). The reduction in age is estimated to be about $10 \div 20\%$.

Methods. Given a good CMD, most likely obtained with a CCD detector, ages can be derived by means of the classical isochrone fitting (IF) method, the ΔV method, and the $\Delta(B-V)$ method.

9.3. Isochrone fitting

In the IF method, all the parameters discussed above are necessary. Therefore, the ages obtained from isochrone fitting are by far the most uncertain (see also the arguments given by Renzini & Fusi-Pecci 1988 and Fusi-Pecci & Cacciari 1991). Most studies have assumed solar [O/Fe] and find ages going from $10 \div 12 \times 10^9$ yr for a cluster like Pal 12 ([Fe/H]=-1.1; Sarajedini & King 1989, Straniero & Chieffi 1991) to $16 \div 18 \times 10^9$ yr for clusters like M92 and M68 ([Fe/H]=-2.1; King et al 1988, Stetson et al. 1989, Straniero & Chieffi 1991). If [O/Fe] varies with [Fe/H], this age range is less clear. If helium diffusion is included, an age reduction of 2×10^9 yr is possible, as estimated by Chaboyer et al. (1992) for the cluster NGC 288.

9.4. The ΔV method

The ΔV method rests on the fact that the turnoff magnitude becomes fainter as a cluster evolves, while the HB luminosity is virtually constant. ΔV is the magnitude difference between the turnoff and the HB at the turnoff color. This method is independent of reddening. Furthermore, the magnitude of RR Lyrae stars and turnoffs are likely scarcely dependent on [O/Fe] and helium diffusion. The disadvantage with this method is that not all GCs possess RR Lyrae stars, and some HBs are not horizontal. Furthermore, the turnoff is almost vertical, which makes uncertain the definition of the turnoff magnitude as well. It requires an assumption for the helium abundance. Finally, there is the effect of the controversial relations $M_V(RR)$-[Fe/H] and [CNO]-[Fe/H] (see above). On the observational side, ΔV does not correlate with [Fe/H] (Buonanno et al. 1989 and references therein), but there is some scatter in ΔV at given metallicity. According to Buonanno et al. (1989) $\Delta V \simeq 3.54$. An overview of the possible alternatives one may have from the different combinations of the slopes and zero points of the above relations is given by Fusi-Pecci & Cacciari (1991) to whom we refer in the summary below. With [CNO/Fe]=0 and $\Delta M_V(RR)/\Delta$[Fe/H] = 0.35, all clusters are coeval and no significant age-metallicity relation exists. With [CNO/Fe]=0 but $\Delta M_V(RR)/\Delta$[Fe/H] = 0.20, the cluster ages decrease with increasing metallicity by about 4×10^9 yr (Sarajedini & Demarque 1990, Sarajedini & King 1989, Lee et al. 1990, Sandage & Cacciari 1990). The same is true if $M_V(RR)$ is independent of [Fe/H] and equal to the classical value of 0.6 (Sarajedini & King 1989). Because of the period shift effect, this alternative is less probable. The zero point of the $M_V(RR)$-[Fe/H] is however crucial to setting the scale of the absolute ages, but unfortunately it is uncertain. With [CNO/Fe] ≥ 0 the situation is more complicated. If [CNO/Fe] ≥ 0 but independent of [Fe/H], $\Delta M_V(RR)/\Delta[Fe/H] = 0.35$, and $\Delta V = 3.54$, then all clusters are coeval but the absolute ages decrease with respect to the case with [CNO/Fe]=0 by a quantity depending on the degree of CNO-enhancement (e.g. with [CNO/Fe]=0.3 the ages are decreased by about 1×10^9yr). Keeping constant all other relations but letting [CNO/Fe] increase with decreasing [Fe/H], the condition $\Delta V = 3.54$ does not imply coevality of GCs. Due to the differential enhancement of [CNO], all metal-poor clusters are younger, while the age of the metal-rich ones are only marginally decreased. However, this combination of slope and abundances may lead to the following two indications against intuition: Y anticorrelates with [Fe/H] and the metal-poor clusters are younger than the metal-rich ones. If $\Delta M_V(RR)/\Delta[Fe/H] = 0.20$ and $\Delta V = 3.54$ the following cases are possible. With [CNO/Fe] ≥ 0 but constant at varying [Fe/H] all clusters have ages decreasing with increasing [Fe/H] – the metal-rich ones are the youngest. The absolute ages depend on the degree of [CNO]-enhancement as above. If [CNO] increases with decreasing [Fe/H], all the clusters may be coeval for a suitable difference in the [CNO]-enhancement between metal-rich and metal-poor clusters. Finally, the inclusion of helium diffusion in model calculations would lead to even lower ages without changing the above scheme.

9.5. $\Delta(B-V)$ method

The $\Delta(B-V)$ method is based on the color difference between the turnoff and the base of the RGB (Sarajedini & Demarque 1990, VandenBerg et al. 1990). This color difference decreases as the cluster age increases. Assuming that the mixing length used in stellar models is calibrated, the method is independent of distance, reddening, photometric zero point, helium abundance, and, to first order it seems to be insensitive to variations in [Fe/H]. The major uncertainties are with the transformations from T_{eff} to colors, the degree of helium diffusion, and [O/Fe], all of these affecting the turnoff color. $\Delta(B-V)$ is reduced by an increased age, an enhancement in oxygen abundance, and helium diffusion. According to VandenBerg et al. (1990) this method is particularly suited to determine relative ages as the determination of absolute ages is affected by the uncertainties in the convection theory and color transformations. Using the revised Yale isochrones (Green et al. 1987) and assuming [O/Fe]=0, Sarajedini & Demarque (1990) and Sarajedini (1991) find that the age of the oldest clusters is about 18×10^9 yr and indicate that GCs span an age range of at least 2.5×10^9 yr. Enhancement of [O/Fe] and/or helium diffusion would reduce the age by about 2×10^9 yr (see Proffitt & VandenBerg 1991). As far as the age spread among clusters with similar metallicity is concerned VandenBerg et al. (1990) give the following indication. The most metal-poor clusters ([M/H]=-2.1) are uniform in age within 0.5×10^9 yr; clusters with [M/H]=-1.6 are also coeval though some age spread cannot be excluded; finally the most metal-rich clusters, [M/H] ≥ -1.3, appear to encompass a significant range. This indicates that the age spread increases with the metallicity as expected if the collapse of the halo was of prolonged rather than of short duration ($\leq 1 \times 10^9$ yr).

9.6. Ages, age spread, and age-metallicity relation

Absolute ages, age spread, and age-metallicity relation of GCs are significant to cosmology and galaxy formation. The oldest GCs set a lower limit to the age of the Universe, whereas the age spread and age-metallicity relation, if real, not only could be a solution to the problem of the second parameter controlling the morphology of the CMDs of GCs, but also constrain the time scale and mechanism of halo formation. Long ago Searle & Zinn (1978) made the hypothesis that age is the second parameter driving the morphology of HBs (the metallicity is the first). Other second parameter candidates, such as Y, [CNO/Fe], or core rotation have been considered that could also account for the observed differences (see Renzini 1977; Lee 1991a,b) but to date only the age seems to provide an explanation compatible with both the standard theory of stellar evolution and the observed distribution of RR Lyrae stars. Two ideal clusters for testing the possibility that the age is the second parameter are NGC 288 and NGC 362. These clusters have similar [Fe/H] (-1.4 and -1.28 respectively; Zinn 1985) but totally different HBs. Another pair is provided by NGC 6397 (see Demarque et al. 1991) and Ruprecht 106 (Buonanno et al. 1990) with [Fe/H]=-1.9. Analyzing these pairs, Sarajedini & Demarque (1990) and Demarque et al. (1991) find the age difference of about 3×10^9 yr for the first pair and 4×10^9 yr for the second one. They also argue that the age is the second parameter of GCs. A similar conclusion was reached by Bolte (1989). VandenBerg & Stetson (1992) using the Δ(B-V) method argue that the pair NGC 362÷NGC 288 shows a difference in Δ(B-V) consistent either with an age difference of 2×10^9 yr or with a difference in [O/Fe] of 0.6, and consider premature the identification of the age as the second parameter. See also the more recent studies on relative ages and age gradients by Stetson et al. (1996) and Richer et al. (1996) and references quoted therein.

Because an important characteristic of the second parameter phenomenon is its systematic variation with the galactocentric distance (Searle & Zinn 1978), Lee (1991b) sought for a global interpretation of the available information correlating [Fe/H], [CNO/Fe], HB type, galactocentric distance, and relative ages of GCs. In the Lee (1991b and references therein) scenario, very likely the age is the second parameter that has the largest influence in determining the HB morphology, and the clusters in the inner halo ($R_G \leq 8$ Kpc) are in the mean several billion years older than the outer halo clusters. At the same time, arguments are given that run counter to the hypothesis that helium abundance, core rotation, or [CNO/Fe] abundance are the second parameter. If this interpretation is correct, it lends support to the idea of prolonged phase of Halo formation, possibly involving mergers and accretion of large fragments with independent dynamical and nucleosynthetic histories (Larson 1990). It is worth recalling that, as pointed out by Sandage (1990b), a significant age spread among GCs does not contradict the picture of Galaxy formation suggested long ago by Eggen et al. (1962).

Although absolute ages are less important from the point of view of interpreting the CMD of GCs, they are a key constraint to the minimum age of the Universe. The above discussion has clarified that the absolute age depends very strongly on the accuracy and adequacy of both observational parameters and stellar models (see the discussion by VandenBerg 1991, Stetson et al. 1996, VandenBerg et al. 1996). Therefore, the absolute ages are subject to change as soon as one of the basic parameters is improved. The ages estimated by Sarajedini & King (1989) for a selected sample of GCs show that their distribution peaks at about $16(\pm 2) \times 10^9$yr, with wings going down to 10×10^9 yr and up to 20×10^9 yr. The recent revision of the whole problem by Stetson et al. (1996) and VandenBerg (1996) have somewhat quenched the enthusiasm for the age being a good second parameter candidate and changed the limits for the absolute ages. According to Stetson et al. (1996) there is no substantial body of evidence indicating that the age is is the dominant second parameter, nor for a significant spread in age among clusters of given metallicity or for systematic variations of the age as a function of the Galactocentric distance. According to VandenBerg et al. (1996), a careful assessment of all uncertainties affecting the age problem suggests that the most metal-poor (presumably the oldest objects) of the globular clusters in the Galaxy have ages near 15 Gyr. Ages below 12 Gyr and above 20 Gyr appear to be unlikely.

10. Old open clusters

The old open clusters, whose ages range from say 1 to $7 \div 8 \times 10^9$ yr, trace most of the history of the Galactic Disk. Therefore, the correct ranking of old open clusters as a function of age, chemical composition, and kinematical properties, is of paramount importance to understanding the process of star formation in the Galactic Disk. Furthermore, having turnoff masses between 1 $M_\odot$ and 2 $M_\odot$, they are probes of stellar structure in that mass range, in which the transition from radiative to convective cores on the main sequence, from pp chain to CNO cycle for the core H-burning phase, and from very bright RGBs as in M67 to much less evident RGBs as in the Hyades, occur. We will limit ourselves to discuss problems related to the structure of these stars. As first pointed out by Barbaro & Pigatto (1984), the interpretation of the CMD of these clusters (e.g. NGC 2420, NGC 3680, IC 4651, King 2, King 11, M67, etc) in terms of the classical models encountered some difficulties that could be solved by invoking a certain amount of convective overshoot during the main sequence core H-burning phase and hence older ages with respect to those from classical models. The main signatures are the detailed shape of the main sequence turnoff, the shape of the RGB, the clump of red stars (most likely core He-burners), and the number of stars brighter than the main sequence at the beginning of the subgiant branch with respect to the main sequence stars (see Mazzei & Pigatto 1988; Maeder 1990; Antony-Twarog et al. 1988; 1989, 1990; Andersen & Nordstrom 1991; Aparicio et al. 1990). Another type of evidence comes from small samples of stars and eclipsing binaries for which good determinations of mass, radius, luminosity, and abundances are available (Andersen et al. 1990; Napiwotzki et al. 1991, Nordstrom et al. 1996), falling near the turnoff of some of these clusters. Specifically, binary stars with small convective cores (M = $1.2 M_\odot$) are very well fitted by standard models, while at slightly larger masses $1.5 M_\odot \leq M \leq 2.5 M_\odot$, the moderately evolved binaries require a certain amount of convective overshoot. This agrees with the scheme proposed by Aparicio et al. (1990), Maeder &Meynet (1991), and Alongi et al. (1991b), in which the efficiency of convective overshoot during the core H-burning phase of stars in this mass range is suggested to gradually increase with the star mass. A similar study was made by Napiwotzki et al. (1991) for the somewhat younger cluster NGC 2301 (estimated age of a few 10^8 yr) for which a careful determination of T_{eff}s and gravities for the brightest members of the cluster were available. Since four out of five stars fall beyond the limit for the core H-burning phase of classical models, a cooler turnoff of the main sequence seems to be required. This was attributed to substantial overshoot. The result was criticized by Brocato & Castellani (1993) who claimed that their recent models with the classical scheme (Castellani et al. 1990, 1991) possess the required extension in the CMD. However, the main sequence extension in the Brocato & Castellani (1993) HRD is not too different from the classical one shown by Napiwotzki et al. (1991). This implies once again that with classical models too many stars are in the short-lived phase of shell H-burning. Further support to this scheme comes from the careful analysis of the CMD of IC 4651 by Bertelli et al. (1992) who adopted both classical and overshoot models using the same input physics (Fagotto 1990; Alongi et al. 1992).

A recent determination of ages for a selected sample of old Galactic clusters, including the Sandage (1988) list, is by Carraro (1991) and Carraro & Chiosi (1994). This study collects the most recent CMD of each cluster, adopts the compilation of metallicities by Friel & Janes (1991), makes use of both classical and overshoot models calculated by the Padova group (Alongi et al. 1992, 1993: Bressan et al. 1993; Fagotto et al. 1994a,b,c), and finally relies on the synthetic CMD technique (see Chiosi et al. 1989b) instead of the simple isochrone fitting to estimate reddening, distance modulus, and age at the same time. The cluster ages span from 0.9×10^9 yr for NGC 2477 to 8×10^9 yr for NGC 6791.

11. Young rich clusters of the LMC

The young rich clusters of the Large Magellanic Cloud (LMC) are classical templates to which the results of stellar evolution theory for intermediate-mass stars are compared. Because of the large number of stars contained in these clusters, it is possible to make meaningful comparisons even for the shortest lived evolutionary phases. A powerful workbench is NGC 1866, a type III cluster in the classification of Searle et al. (1980), whose total mass is estimated in the range $3.6 \div 5 \times 10^5 M_\odot$. This cluster is well populated throughout the various evolutionary phases, exhibits an extended loop of giant stars, and is rich in Cepheids (Walker 1987, Welch et al. 1991). First attempts to interpret the CMD of NGC 1866 date from Arp (1967), Hofmeister (1969), and Robertson (1974). Becker & Mathews (1983) using the Robertson (1974) CMD noticed two important features. First, for the observed luminosity of the giants, there are too many stars above the predicted main sequence turnoff, whose number is a significant fraction of the number of giant stars. Second, the predicted ratio of post main sequence stars to the main sequence stars was about four times the observed one. Bertelli et al. (1985) using the same CMD concluded that only models with convective overshoot could overcome the difficulty. More recent CCD data (Chiosi et al. 1989a, Brocato et al. 1990) and new stellar models reopened the question whether models with overshoot ought to be preferred to the classical ones or to those with semiconvection. Chiosi et al. (1989b), using both the standard models by Becker (1981), which do not incorporate any special treatment of central convection, and models with overshoot by Bertelli et al. (1985, 1986a,b), gave the following results: the turnoff mass and age were $5 M_\odot$ and 70×10^6 yr with the former models, and $4 M_\odot$ and 200×10^6 yr with the latter, respectively. Because the evolutionary tracks alone could not show unambiguously which of the evolutionary schemes was correct, Chiosi et al (1989b) made use of the integrated luminosity function of the main sequence stars normalized to the number of giants (NILF) as a way to achieve the goal. This is possible because the NILF simply reflects the ratio of core He- to H-burning lifetimes, which depend on the stellar models in use. The conclusion was that models with substantial core overshoot reproduced the observed NILF, whereas classical models failed. Similar analysis, repeated using models with semiconvection (Lattanzio et al. 1991), reached identical conclusions. The analysis was extended to other clusters of the LMC, like NGC 1831 (Vallenari et al. 1991) and NGC 2164 (Vallenari et al. 1992) with similar results, i.e. models with core overshoot provided a good fit to the CMD and luminosity functions at the same time. Brocato et al. (1990) obtained CCD photometry of NGC 1866 and analyzed the CMD and LF following the method outlined by Chiosi et al. (1989b) but using the models without convective overshoot calculated by Castellani et al. (1990, 1991). They came to the conclusion that core overshoot is not required, if one makes use of modern opacities (see Castellani et al. 1990, 1991). Although the opacity may lower the ratio of core H- to He-burning lifetimes from 0.33 (Becker 1981) to 0.23, this value is still far from that indicated by the observations (0.10) or given by models with overshoot. Bressan (1990) clearly showed that Brocato's et al. (1990) conclusion was entirely due to the different luminosity function, a point of embarrassment because similar observing and reducing techniques were used. The reason for the disagreement is likely to be the different number of red giants in the two samples. Bencivenni et al. (1991), analyzing the much younger cluster NGC 2004 (the age and turnoff mass of which are a few 10^6 yr and about $20 M_\odot$, respectively), claim that arguments can be given against the existence of convective overshoot. However, we recall that in this mass range the inclusion of convective overshoot does not bring a significant difference with respect to the classical models (e.g. the core He- to H-burning lifetime ratio is modestly changed). Furthermore, the evolution is dominated by mass loss, which was not taken into account by Bencivenni *et al.* (1991). Many observational tests of convective overshoot in intermediate-mass stars have been critically scrutinized by Stothers (1991). Defining the ratio d/H_P of the effective convective overshoot distance beyond the classical Schwarzschild boundary to the local pressure scale height H_P as an index characterizing published models, Stothers (1991) comes to the conclusion that $d/H_P \leq 0.4$ is likely an upper limit to this phenomenon. This estimate is comparable to the value adopted by Maeder & Meynet (1991 and references therein), and Alongi *et al.* (1992, 1993) in their recent model calculations. Despite the net improvement of observational CMDs and LFs, the situation appears to be rather confused simply reflecting the difficulty of producing accurate and adequate stellar models with correct lifetimes for the core H- and He-burning phases.

12. The HRD of supergiant stars: open problems

Despite the many successful achievements, all existing stellar models have serious difficulties in simultaneously reproducing the whole pattern of properties of luminous stars. For the purposes of the present lectures, it is wise to review here several observational aspects of the HRD of supergiant stars that are still a matter of vivid debate together with their present day theoretical understanding.

12.1. *Star counts across the HRD*

Sources of the HRDs. The most recent HRD for supergiant stars in the solar vicinity is by Blaha & Humphreys (1989), whereas those for the LMC and SMC are by Fitzpatrick & Garmany (1990), and Massey *et al.* (1995). Although these samples are based on different selection criteria and suffer from a certain degree of incompleteness, these HRDs show common features that can be used to constrain theoretical models. The discussion below will be limited to the HRD of Milky Way and LMC and to stars brighter than $M_{bol} = -7.5$, which is a sort of lower limit above which incompleteness is less of a problem.

The luminosity boundary. The upper boundary to the luminosity of supergiant stars is clearly visible both in the HRD for the Milky Way and LMC. Since we have already reported on the current interpretation of this boundary, no other detail is added to the previous discussion.

The missing blue gap in the HRD. There is no evidence of the expected gap (scarcity of stars) in HRD between core H- and He-burning phase. The gap is the observational counterpart of the very rapid evolution across the HRD following core H-exhaustion and prior to stationary core He-burning. In contrast, we observe a continuous distribution of stars all across the region comprised between the red side of the so-called main sequence band (core H-burning phase) and the region of stationary core He-burning. The stars in the gap have several puzzling features suggesting an advanced stage of evolution, more advanced than inferred from their position in the HRD. Indeed, many of them show evidence of CNO processed material and He-rich composition at the surface. In fact, the group of OBN/OBC stars (Walborn 1988) and the He-rich objects near the main sequence (Kudritziski *et al.* 1983, 1989; Bohannan *et al.* 1986) are considered typical examples of stars, in addition to WR stars, whose interpretation is based on the exhibition at the surface of CNO and 3α processed matter. Similarly, many B type supergiants in the LMC (Fitzpatrick & Bohannan 1992) have their surface contaminated by the CNO processed material.

The ledge. At somewhat lower luminosities the density of stars in the HRD shows a distinct decrease red-ward of $3.9 \leq \log T_{\rm eff} \leq 4.2$ and the density drop-off forms a diagonal line otherwise called the "ledge" (Fitzpatrick & Garmany 1990), going to lower luminosities at decreasing $T_{\rm eff}$. According to Fitzpatrick & Garmany (1990) the most plausible explanation of the ledge is that stars of initial mass up to about 40-50 $M_\odot$ during the core He-burning phase either perform an extended blue loop in the HRD before core He-exhaustion or slowly evolve from blue to red and only at the very end of the He-burning phase quickly move to the red supergiant region (the classical case A or case B scenarios proposed long ago by Chiosi & Summa 1970 for constant mass models). Leaving aside the old constant mass models, extended loops are easy to get with the semiconvective scheme and the adoption of the Ledoux criterion ($\nabla_R = \nabla_A + \nabla_\mu$). The situation is less clear with the Schwarzschild criterion ($\nabla_R = \nabla_A$), because in many cases it gives rise to an intermediate fully convective region that interacts with the H-burning shell. Semiconvective schemes with criteria intermediate between Ledoux and Schwarzschild also exist that render the picture even more intrigued. Depending on the mass, chemical composition, and other physical details, extended loops, slow blue-red evolution, and even pure red evolution are possible. Most recent semiconvective models with a variety of physical assumptions are by Brocato & Castellani (1993), Bressan et al.(1993), Fagotto et al. (1994a,b,c), Langer (1986, 1991), Langer et al. (1990), Mowlavi & Forestini (1994), Stothers & Chin (1991, 1992) to whom the reader should refer for all details. Tuchman & Wheeler (1989, 1990) analyzing the distribution of stars in the HRD favored the slow blue-red evolution and excluded models with core overshoot. Indeed, models with mass loss and core overshoot alone (Maeder & Meynet 1987, 1988, 1989, 1991; Maeder 1990; Bressan et al 1993, Fagotto et al. 1994a,b,c; Schaerer et al. 1993a,b; Schaller et al. 1992) are less suited to explain the ledge because they tend to have rather short loops or no loops at all. However, the inclusion of overshoot from bottom of the convective envelope (Alongi et al. 1992) favors extended blue loops and yields stellar models that are able to reproduce the shape of the blue ledge (Alongi et al. 1993, Bressan et al. 1993). Moreover these models are able to account for the observations of CNO-processed and He-rich material at the surface of some He-rich objects near the main sequence (Kudritzski et al. 1983, Bohannan et al. 1986).

The Red Gap in the HRD. The population of red supergiants is distinctly separated from all remaining stars in the HRD by the so-called red Hertzsprung gap (RHG) showing that the stars must cross the region between the late G supergiants and the M supergiants on a very short time scale. The maximum luminosity attained by red supergiants is about $M_{\rm bol} = -9.5$. There are a few differences between the population of Galactic and LMC supergiants that can be ascribed to the different chemical composition.

Relative Frequencies of stars across the HRD. Do the relative frequencies of stars across the HRD agree with the theoretical prediction for the lifetimes and evolutionary rates? To clarify the question we present here some simple star counts using the catalog of supergiant stars in the Milky Way and LMC by Blaha & Humphreys (1989), the same adopted by Langer & Maeder (1995) for a similar purpose so that comparison is possible. In this context, although the use of the Massey et al. (1995) catalog would be more appropriate to perform star counts on the blue side of the HRD, its use is hampered by the lack of red supergiants. It must be said from the very beginning that we do not include the effect of binary stars (cf. Vanbeveren 1995), even if a binary-free sample

Milky Way. LUMINOSITY INTERVAL: $-7.5 \geq M_{bol} \geq -9$

Main Sequence up to O9.5: classical constant mass models

Total	O-O9.5	B	A	F-G	K-M	N_B/N_R	B_l	R_l	B_l/R_l
375	233	121	10	2	9	40	61	9	7
%	0.621	0.323	0.027	0.005	0.024				

Main Sequence up to B0.5: models with mass loss and overshoot or diffusion

Total	O-B0.5	B	A	F-G	K-M	N_B/N_R	B_l	R_l	B_l/R_l
375	295	59	10	2	9	40	61	9	7
%	0.787	0.157	0.027	0.005	0.024				

Main Sequence up to B3: ideal case

Total	O-B3	B	A	F-G	K-M	N_B/N_R	B_l	R_l	B_l/R_l
375	340	14	10	2	9	40	24	9	1.6
%	0.907	0.037	0.027	0.005	0.024				

Large Magellanic Cloud. LUMINOSITY INTERVAL: $-7.5 \geq M_{bol} \geq -9$

Main Sequence up to O9.5: classical constant mass models

Total	O-O9.5	B	A	F-G	K-M	N_B/N_R	B_l	R_l	B_l/R_l
639	117	375	54	21	72	8	328	72	4.5
%	0.183	0.587	0.085	0.033	0.113				

Main Sequence up to B0.5: models with mass loss and overshoot or diffusion

Total	O-B0.5	B	A	F-G	K-M	N_B/N_R	B_l	R_l	B_l/R_l
639	194	298	54	21	72	8	328	72	4.5
%	0.304	0.466	0.085	0.033	0.113				

Main Sequence up to B3: ideal case

Total	O-B3	B	A	F-G	K-M	N_B/N_R	B_l	R_l	B_l/R_l
639	445	47	54	21	72	8	122	72	4.5
%	0.696	0.974	0.085	0.033	0.113				

TABLE 5. Raw star counts in the HRD according to different extensions of the MS band. The last case of each group shows how far the MS band should extend in the HRD in order to get the desired ratio $N_{PMS}/N_{MS} \simeq 0.1$ in the ideal circumstance of a complete sample of main sequence stars ($\Delta N_{MS} = 0$). See the text for more details

of cluster supergiants (cf. Stothers & Chin 1994) would perhaps be more appropriate. Furthermore, to minimize the effects of incompleteness we select the luminosity interval $-7.5 \geq M_{bol} \geq -9$, which approximately correspond to main sequence stars in the mass interval 25 to 40 $M_\odot$. Finally, the correspondence between spectral type and T_{eff} is according to Humphreys & McElroy (1984). The counts are performed according to the following criteria:

(a) Stars are grouped per evolutionary phase and spectral type. In the first group (indicated by O-O9.5, O-B0.5, and O-B3 according to the case under consideration) we include all stars formally attributable to the main sequence phase, whereas in the remaining groups from B to K-M we sample stars according to their spectral type without further distinction.

(b) Three cases of main sequence extensions are considered: (1) the main sequence does not go beyond O9.5 as in the classical constant mass models; (2) the main sequence extends up to the spectral type B0.5 as in the present models with mass loss and diffusive mixing or in previous ones with mass loss and standard overshoot (there is a little dependence on metallicity that can be neglected here); (3) the main sequence stretches at least up to the spectral type B3.

The results of star counts are presented in Table 5. The first row of each case contains the counts, while the second row shows the relative percentages. Analyzing the data of Table 5 we notice what follows.

(a) There is a clear shortage of main sequence stars (first group), which is more pronounced for the LMC sample. Indeed, recalling that the ratio N_{MS}/N_{PMS} cannot be too different from 10, it means that the group of stars supposedly in the main sequence phase should amount to about 80-90% of the total. In contrast, if the MS band extends only up to the spectral type O9.5, the percentage is too low. The situations gets better when the MS band extends to B0.5, however also in this case it is not fully satisfactory as the WR stars should be added to our counts for the groups of PMS objects.

(b) The above shortage of main sequence stars can be caused by photometric incompleteness or a wrong limit to the red-ward extension of the MS band.

(c) If photometric incompleteness is the cause, it appears to be more severe than customarily accepted. The problem still remains even taking into account the analysis of completeness presented by Massey et al. (1995). Looking at their Fig. 6 in the luminosity interval of interest here and for $T_{eff} \geq 4.4$ we count 12 new stars as compared to 3 already known: the incompleteness factor is about 4. If this correction is applied to the case of LMC supergiants with main sequence expected to stretch up to the spectral type B0.5 (mass loss and overshoot or diffusion) we get $N_{MS}/N_{PMS} = 970/450 \simeq 2 << 10$. The problem is still there, further aggravated by the too many stars in the spectral types B, which roughly correspond to the Hertzsprung-Russell gap in the HRD (see above).

(d) If the red limit of the MS band is the cause, it requires some yet unknown physical process. Possible ways out of the above dilemma have already been discussed and will not be repeated here.

(e) Finally, since in our counts we have not included WR stars, the above results are a sort of lower limit. The difficulty with WR is that one needs some *a priori* information about the mass range of the progenitors. In any case, as WR stars are likely in the core He-burning phase (at least the WNE, WC and WO), their inclusion in star counts makes the shortage of massive stars in the main sequence group even more severe.

The N_B/N_R Ratios. The number ratio of blue to red supergiants (N_B/N_R) has long been taken as a probe of the inner structure of a massive star. The bottom line of the reasoning is that if N_B/N_R can be identified with the ratio of core He-burning lifetimes in the blue and red side of the HRD, the former would reflect the structure of the star. The most recent analysis of the N_B/N_R ratio and its dependence on the galaxy metallicity (the ratio seems to increases with the metallicity) is by Langer & Maeder (1995). They conclude that no current stellar model is able to cope with this trend. Indeed, using models with convection and semiconvection according to the Ledoux criterion satisfactory results were possible for for Z=0.002 but not for Z=0.02. They argue that a mixing efficiency in between the Schwarzschild and the Ledoux criterion might be appropriate to explain the trend of the above number ratio with the metallicity. However, the use of the N_B/N_R to test stellar interiors is hampered by the basic uncertainty whether or not it relates only to the ratio of core He-burning lifetimes in the two areas of the HRD. Recalling that N_B is evaluated including stars from O to A, it goes without saying that part of the H-burning phase should be included and therefore the above conclusions changed. Indeed the authors notice that there is an excess of B and A type stars, as if the main sequence were more extended than given by theoretical predictions, which is exactly the conclusion reached long ago by Stothers & Chin (1977), Bressan et al. (1981) and Bertelli et al. (1984). Mowlavi & Forestini (1994) analyzing the distribution of supergiant stars in the HRD concluded that semiconvective models ought to be preferred to the overshoot ones. To illustrate the above points we present here some evaluations of the N_B/N_R ratio. The total number of blue stars N_B is the sum of the numbers from O to G included, while the total number of red stars is the same but for the stars in the spectral classes K and M. This is the same definition used by Langer & Maeder (1995), however we differ for the luminosity interval. They consider all stars brighter than $M_{bol} = -7.5$, whereas we prefer to confine the counts in the luminosity interval $-9 \leq M_{bol} \leq -7.5$. The ratios N_B/N_R obtained according to the above definition are listed in Table 5. The ratio N_B/N_R considerably varies going from the Milky Way to LMC which may reflect both a different degree of incompleteness among the two samples (most likely the very early type stars) and different intrinsic frequency of red supergiants (see below). On the theoretical side adopted definition of N_B/N_R is

$$\frac{N_B}{N_R} = \left(\frac{\tau_H + \tau_{He}^B}{\tau_{He}^R}\right) \qquad (12.50)$$

with obvious meaning of the symbols. As the ratio N_B/N_R defined in this way is somewhat contaminated by main sequence stars, it cannot be straightforwardly compared to the ratio of lifetimes τ_{He}^B/τ_{He}^R.

To improve upon this point we try to single out from the sample those stars actually belonging to the core-He burning phase. To this aim we have to make use of stellar models to evaluate the maximum extension toward the blue of the core He-burning loops. Using Deng et al. (1996) models with mass loss and diffusion (which have the bluest extension) we notice that blue loops in the mass interval under consideration extend up the spectral type B1. Accordingly we identify as stars in the blue phase of the loop those in the spectral types from B1 to late G included, and as stars in the red stages of the loop those in the spectral types from K to M. Their total number is named B_l and R_l, respectively. If the main sequence band is supposed to extend to the spectral type B3, then B_l will be given by the stars from B4 to G9. These new counts and their ratios are also listed in Table 5.

Despite the crudeness of our approach, N_B/N_R in LMC is lower than in the Milky Way (see also Langer & Maeder 1995). Perhaps it mainly reflects the much higher number of K-M stars in the LMC as compared to that in the Galactic sample. Scaling the percentages to an equal total number of stars in the two populations, the LMC sample contains at least four times more K-M stars than the Galactic sample. This is also visible in the different values for the B_l/R_l ratios. The difference in the number of red supergiants provides hints to explain the observational trend of N_B/N_R with the environment metallicity. In conditions of relatively higher metallicity fewer red supergiants are formed. Considering the effects of metallicity on the core He-burning phase of stellar models calculated with mass loss, i.e. narrower blue loops and BSG/RSG ratios of about 0.6 to 1.2, the only possibility we can foresee is a much higher efficiency of mass loss perhaps combined with a deeper envelope mixing during the red stages of central He-burning so that red supergiants are quickly turned into blue objects (cf. Salasnich *et al.* 1997). This should not occur in conditions of lower metallicity where red supergiants can survive for longer times.

Finally, we would like to remark that the discrepancy of the star frequencies across the HRD and N_B/N_R ratios has long been masked by the apparently successful models for WR stars and the progenitor of SN1987A. However, it remains an essential and unsolved problem.

12.2. *The puzzling nature of Wolf-Rayet stars*

All existing scenarios for the formation of WR stars (cf. Maeder 1994) suffer from a basic difficulty as far as the exact location of the stellar models in the HRD and the comparison with the observational data for the WR stars are concerned (see Hamann *et al.* 1993). In brief, the models cannot explain the low luminosity and relatively low effective temperatures assigned to many WR stars. According to the current understanding of the formation of WR stars via the mechanism of mass loss already amply discussed in these lectures, stars initially more massive than about $40M_\odot$ (the precise value depends on the initial chemical composition and the efficiency of mass loss) do not evolve to the red side of the HRD, but soon after central H-exhaustion or even in the middle of this phase reverse their path in the HRD towards higher effective temperatures, first at constant luminosity and later at decreasing luminosity (Schaller *et al.* 1992; Charbonnel *et al.* 1993; Schaerer *et al.* 1993a,b; Meynet *et al.* 1994; Maeder & Meynet 1994; Alongi *et al.* 1993; Bressan *et al.* 1993; Fagotto *et al.* 1994a,b,c). During all these evolutionary stages, the stellar models have the pattern of surface chemical abundances typical of the WR objects. The correspondence between theoretical surface abundances and WR subclass is based on the Maeder & Meynet (1994) prescription: WNL: $0 \leq X_s \leq 0.3$, WNE: $X_s = 0$, WC: $C_s > N_s$ WO: $(C_s/12 + O_s/16)/Y_s/4 > 1$ where X, Y, C, N and O are the hydrogen, helium, carbon, nitrogen and oxygen abundances by mass, respectively, and s stands for the surface chemical abundances.

However, while the surface abundances and the decrease in the luminosity passing from WNL to WNE and WC stars indicated by the data (Hamann *et al.* 1993) are matched by the models, their effective temperatures are much hotter than those possessed by real WR stars. The situation is best shown in the HRD of Fig. 10, displaying the data by Hamann *et al.* (1993) for Galactic WN stars and the evolutionary path of diffusive models of Deng *et al.* (1996a,b) both with Z=0.020 and Z=0.008 limited to the range of massive stars. Similar results are obtained using FO- and SE-models. See for instance Bressan *et al.* (1993), Fagotto *et al.* (1994a,b,c). As a conclusion all current models of massive, no matter whether with diffusive or standard overshoot, cannot easily account for the low luminosities and relatively low effective temperatures assigned to many WR stars at the same time.

As a matter of facts, the models of massive stars that can evolve towards the luminosity range where the majority of the WR stars are found, do it during the latest stages of core He-burning when they have lost a large fraction or even the entire envelope and,

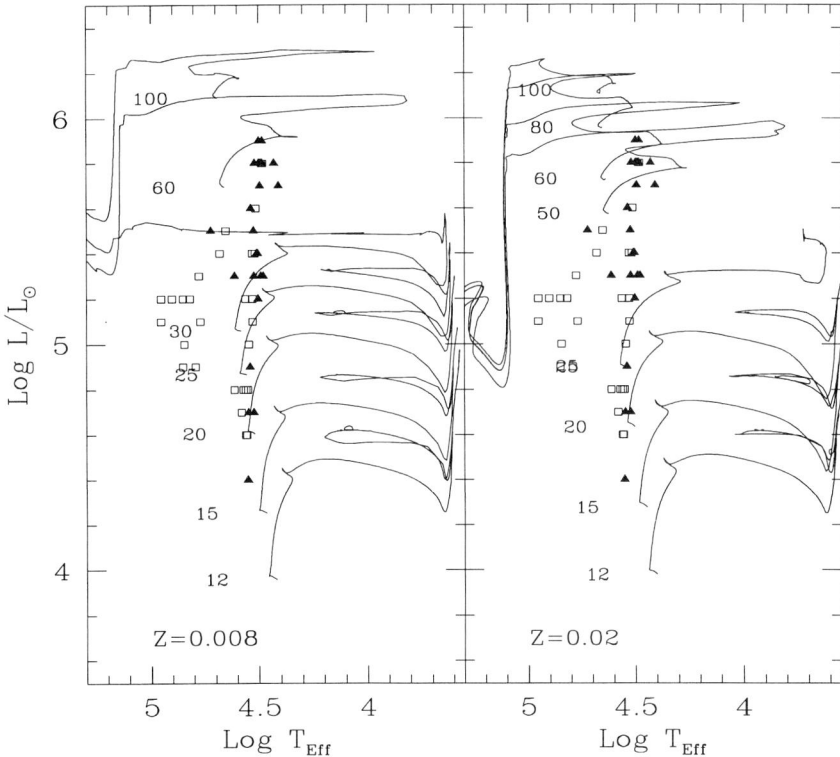

FIGURE 10. The HRD of massive stars with mass loss by stellar wind and Deng et al. (1996a,b) diffusion for two chemical compositions: [Z=0.008, Y=0.25] (left panel) and [Z=0.02, Y=0.28] (right panel) superposed to the data of Hamann et al. (1993) for galactic WR stars. Solid triangles stand WNs with detection of hydrogen at the surface, whereas open squares are the hydrogen free object. The initial mass (in solar units) of the stellar tracks is indicated along each sequence. These stellar models are calculated assuming the de Jager et al. (1988) rate of mass loss.

therefore, as bare cores possess a very high effective temperature. During all previous evolutionary stages they run at much higher luminosities.

It is often argued that the discrepancy in the effective temperature can be cured by applying the well known correction to the model effective temperatures. This takes into account the departure from hydrostatic equilibrium and the optical thickness of an expanding atmosphere. Indeed the photosphere of an expanding dense envelope can be different from that of a hydrostatic model, i.e. it can be located at larger radii and hence cooler effective temperatures (Bertelli et al. 1984, Maeder 1987, Hamann et al. 1993). However, among the galactic WN stars studied by Hamann et al. (1993), only WNE stars with strong emission lines show photospheric radii larger than the hydrostatic ones. So that the above correction does not apply to the majority of WR stars.

Finally, since the minimum initial mass of the models able to reach the so-called WR configuration is greater than about $40M_\odot$, there is the additional difficulty that the relative number of WR stars with respect to the progenitor OB type stars exceeds the expectation based on the possible duration of the WR phase and initial mass function (cf. Chiosi et al. 1992a,b for details).

An interesting possibility is offered by the diffusive models of Salasnich et al. (1997) in

$M^i/M_\odot$	Δt_{WNL}	M^i_{WNL}	Δt_{WNE}	M^i_{WNE}	Δt_{WC}	M^i_{WC}	Δt_{WO}	M_f or M^i_{WO}
			$Z=0.008$	$Y=0.250$				
30	0.0264	12.65	0.0786	12.05	0.3293	10.36	0.0000	6.87
40	0.0892	18.53	0.0315	16.39	0.3958	14.83	0.0354	7.55
60	0.0739	27.43	0.0105	25.06	0.3352	23.48	0.0930	9.38
80	0.1000	38.79	0.0049	35.10	0.1926	33.34	0.1416	13.50
100	0.3912	56.32	0.0032	43.62	0.0926	41.67	0.1532	20.30
			$Z=0.020$	$Y=0.280$				
30	0.0253	13.40	0.0596	12.42	0.6747	10.32	0.0000	4.28
40	0.0363	17.91	0.0220	16.34	0.7124	14.64	0.0000	4.49
60	0.0420	28.14	0.0066	25.99	0.6278	24.27	0.0000	5.16
80	0.4018	47.51	0.0032	36.51	0.5166	34.56	0.0530	5.98
100	0.6407	60.51	0.0028	43.57	0.5681	40.86	0.0352	5.69

TABLE 6. The lifetimes (in units of 10^6 yr) of massive stars during the various stages of the WR sequence WNL, WNE, WC and WO (see the text for more details). In addition to this we list the value of the current mass of the stars at the beginning of each sub-stage. The last column shows either the final mass (M_f) at the C-ignition stage if the WO stage is missing, or the mass at the start of the WO stage (M_{WO}) if present. All masses are in solar units.

which a different (much higher) efficiency of mass loss during the RSG stages as compared to the standard de Jager et al. (1988) relationship is adopted. In brief, Salasnich's et al. (1997) models soon after leaving the RSG region, exhibit surface H-abundance $X_H = 0.43$, which is close to the value of 0.40 assumed by Maeder (1990) to mark the transition from O-type to WNL stars; and after spending about half of the core He-burning lifetime in the region of the Hertzsprung-Russell gap are able to reach the zone where the faint WR are observed. The situation is shown in Fig. 11.

These models open a new avenue for the formation of faint single WR stars ($\log(L/L_\odot) \sim 4 - 4.5$). Contrary to what suggested in the past, these stars are not generated by the most massive ones ($M > 60 M_\odot$) evolving "*vertically*" in the HRD but can be generated by less massive progenitors (say 18-20 $M_\odot$) evolving "*horizontally*" in the HRD provided that they suffered from significant mass-loss during the RSG phase. Soon after the RGB stages, the models are structurally similar to WNL stars, but they become visible as such only near the main sequence, when the effective temperature is high enough to allow the spectral signatures of WR stars. Finally, with this new scenario both WR and RSG stars can co-exist in the same cluster.

12.3. *The Progenitor of the Supernova 1987A.*

The occurrence of SN1987A in LMC from a blue progenitor has spurred many questions about the evolution of massive stars. The basic requirements are a blue progenitor ($\log T_{\rm eff} = 4.0$) of initial mass of about 20 $M_\odot$ (Arnett et al. 1989), which underwent significant surface enrichment of He and C/N elements shortly before explosion (Fransson et al. 1989), and finally followed a blue-red-blue evolution shortly after core He-exhaustion. Many physical effects, not yet fully understood, may intervene to determine the final location of supernova progenitors. Current models with core overshoot, mass loss and chemical compositions in the range appropriate for the LMC can start C-burning

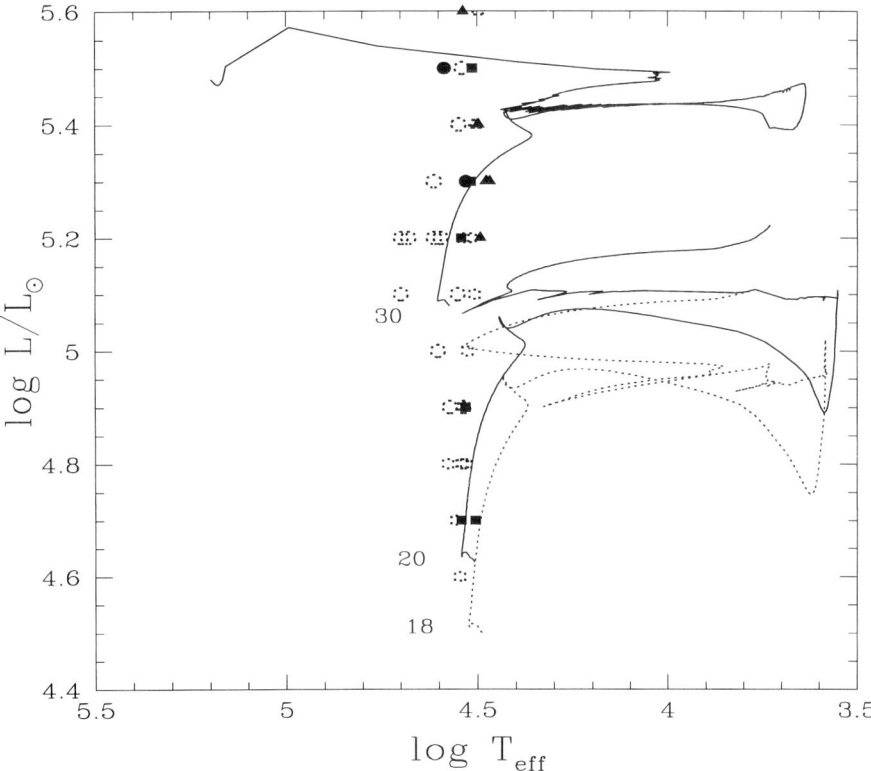

FIGURE 11. The HRD of massive stars with mass loss by stellar wind and Salasnich et al. (1996) diffusion. The chemical composition is [Z=0.008, Y=0.25]. The tracks are compared with the data of Hamann et al. (1993) for galactic WR stars. Solid triangles stand for WNL stars, whereas filled squares are WNE with weak HeII emission lines, filled circles are the WNE with strong HeII emission lines. In general filled symbols mean that hydrogen has bee detected. Open symbols have the same meaning as before, but indicate that no hydrogen has been detected. The initial mass (in solar units) of the stellar tracks is indicated along each sequence. These stellar models are calculated assuming the revised mass-loss rate during the RSG stages.

in the blue only with an envelope mass less than deduced for the SN1987A progenitor. Semiconvective models with the Ledoux criterion (Woosley 1988, Weiss 1989) and constant mass evolution lead to a blue progenitor. However, when a moderate mass loss is included, the final blue location occurs during the mid core He-burning. A treatment of semiconvection intermediate between the Ledoux and the Schwarzschild criterion (Langer et al. 1990, Langer 1991) accounts for the blue progenitor. However, first the goal is achieved with some fine tuning of the parameters, second it does not simultaneously account for the observed He and C/N surface enrichment. To cope with this difficulty, Langer et al. (1989a,b) have invoked rotational diffusion. Though reasonable, the main drawback of these models with semiconvection and/or rotationally induced diffusion is the fine tuning of the various parameters needed to hit the target. Enhanced mixing in the external layers of a red supergiant at the end of the He-burning phase has been envisaged by Saio et al. (1988). The bottom line of the reasoning resides in the possibility that the intermediate convective shell and the external convective envelope merge together, thus easily explaining the observed anomalies in the surface abundances and the blue location of the progenitor at the same time.

13. Modeling AGB & Carbon stars: recent results

In recent years the classical scenario for the structure and evolution of TP-AGB stars, the formation of C-stars and their associated luminosity functions has been deeply changed thanks to the flourishing of new theoretical models and improved observational data. Several breakthroughs have concurred to better clarify the subject.

(*a*) The different luminosity functions of cluster and field AGB stars of the LMC: while in clusters there are no AGB stars brighter than about $M_{bol} = -6$, much brighter objects are seen in the field such as the oxygen-rich LPV variables, e.g. Mira and OH/IR stars, falling in the range $M_{bol} = -6 \div -7$ (cf. Smith *et al.* 1995; Wood *et al.* 1983, 1992; Zijstra *et al.* 1996).

(*b*) The luminosity functions of C-type stars in LMC and SMC spanning the magnitude range $M_{bol} = -6.5$ down to $M_{bol} = -3.5$ and $M_{bol} = -3$, respectively (cf. Reid *et al.* 1990; Costa & Frogel 1996).

(*c*) The adoption of better algorithms for the mass loss rate, the so-called superwind phase in particular. In this context the empirical mass loss rate by Vassiliadis & Wood (1993) has proved to be particularly useful.

(*d*) The many studies (cf. Lattanzio 1991; Boothroyd & Sackmann 1988; Hollowell 1988, Straniero *et al.* 1995, and references) aimed at improving current treatments of mixing and opacity in the intershell region.

(*e*) The inclusion and refinements of envelope burning in most massive TP-AGB stars, which led to a revision of the classical core-mass-luminosity relation and more specifically brought into evidence the breakdown of it at high luminosities. Massive TP-AGB stars in fact leave the core-mass-luminosity relation and rapidly evolve at much higher luminosities because of the additional energy generation from the deepest layers of the envelope (Bloecker & Schoenberner 1991; Lattanzio 1992; Boothroyd & Sackmann 1992; Bloecker 1995; Marigo *et al.* 1997).

(*f*) The existence of flash-driven luminosity variations (well established by detailed numerical models) which concur to force a TP-AGB star to deviate from the classical core-mass-luminosity relation. The main deviations from the light curve occur at the peak-flash luminosity following a thermal pulse and during the subsequent long-lived luminosity dip. The former corresponds to a sudden and brief increase of the luminosity over its quiescent value (the outer layers are forced to expand). The latter reflects the declining of the He-burning luminosity that causes the stars to spend as much as 20–30% of the interpulse lifetime at a luminosity about a factor of 2 lower than predicted by the core-mass-luminosity relation.

(*g*) The much improved situation as far as the starting models of the TP-AGB phase are concerned. These are customarily taken from detailed stellar model calculations over ample ranges of intial masses with fine mass spacing.

(*h*) Finally, the very detailed semi-analytical approaches (Groenewegen & de Jong 1993, Marigo *et al.* 1996a, 1997) which first have clarified the role played by key parameters of the TP-AGB evolution and second have much improved upon the pioneering models of Renzini & Voli (1981).

In the following, we summarize the results of the recent studies by Marigo *et al.* (1996a,b; 1997) who have investigated the TP-AGB phase of low and intermediate-mass stars from the first pulse till the complete ejection of the envelope, paying particular attention to

(1) inter-shell nucleosynthesis and convective dredge-up;

(2) envelope burning in the most massive AGB stars ($M \geq 3 - 4 M_\odot$);

(3) mass loss by stellar wind;

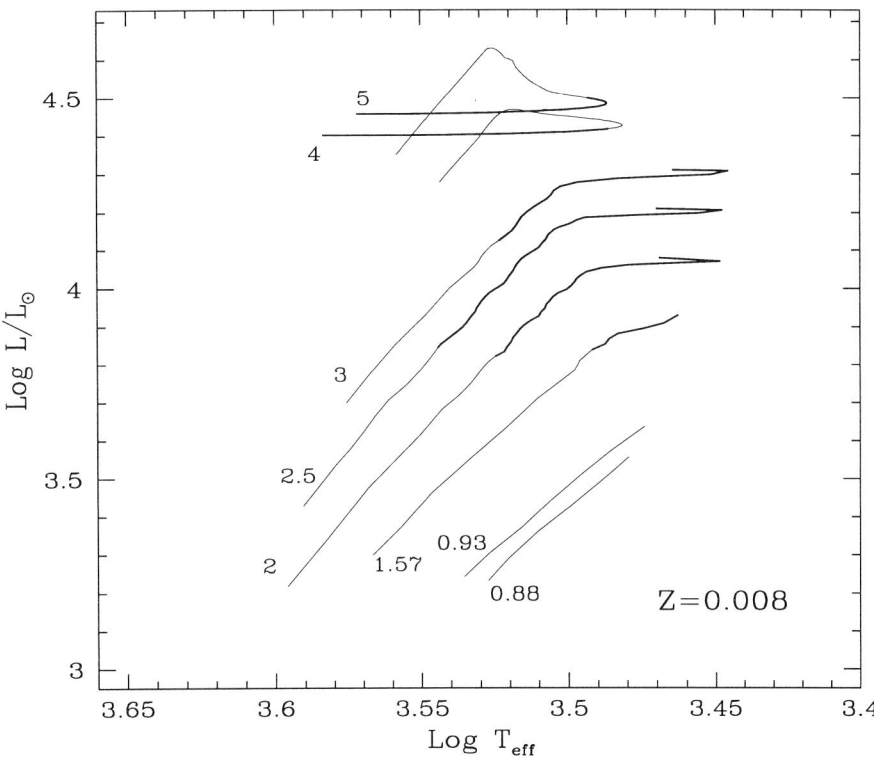

FIGURE 12. Evolutionary tracks of low and intermediate-mass stars during the TP-AGB stages. The initial chemical composition is [Z=0.008, Y=0.25]. Each track corresponds to a different value of the initial mass (in solar units) as indicated. Notice the different behaviour passing from the no-envelope burning to envelope-burning regime ($M \geq 3M_\odot$). For most of the phase massive TP-AGB stars are O-rich (C/O $\leq$ 1), thin portion of the tracks, and eventually become C-rich objects (C/O $\geq$ 1), thick portion of the tracks. The transition to the C-class occurs at $M_{bol} \simeq -6.5 \div -6.3$ in agreement with the bright end of the C-star luminosity function of the LMC.

(4) the formation of carbon stars.

Omitting all details of the analysis we focus here on a few major assumptions and results.

Mass loss. Mass loss by stellar winds during the TP-AGB phase is according to the semi-empirical relation by Vassiliadis & Wood (1993) that has already been anticipated in section (5.2). In addition to the self-accelerating behaviour closely mimicking superwind during the latest stages, in presence of envelope burning this relation strongly favors the early ejection of the outer envelope because of the increased luminosity.

Third dredge-up. The analytical treatment of the 3rd dredge-up involves two parameters: M_c^{min}, the minimum core mass for convective dredge-up, and λ the fractionary core mass increment during the previous interpulse period dredged up to the surface. The calibration ($M_c^{min} = 0.58M_\odot$ and $\lambda = 0.65$) is constrained on the luminosity function of C-stars in the LMC shown in Fig. 14.

Envelope burning. In agreement with previous studies (Boothroyd & Sackmann 1992 and references therein), in massive TP-AGB stars ($M \geq 3 - 4M_\odot$) deep and hot-bottom

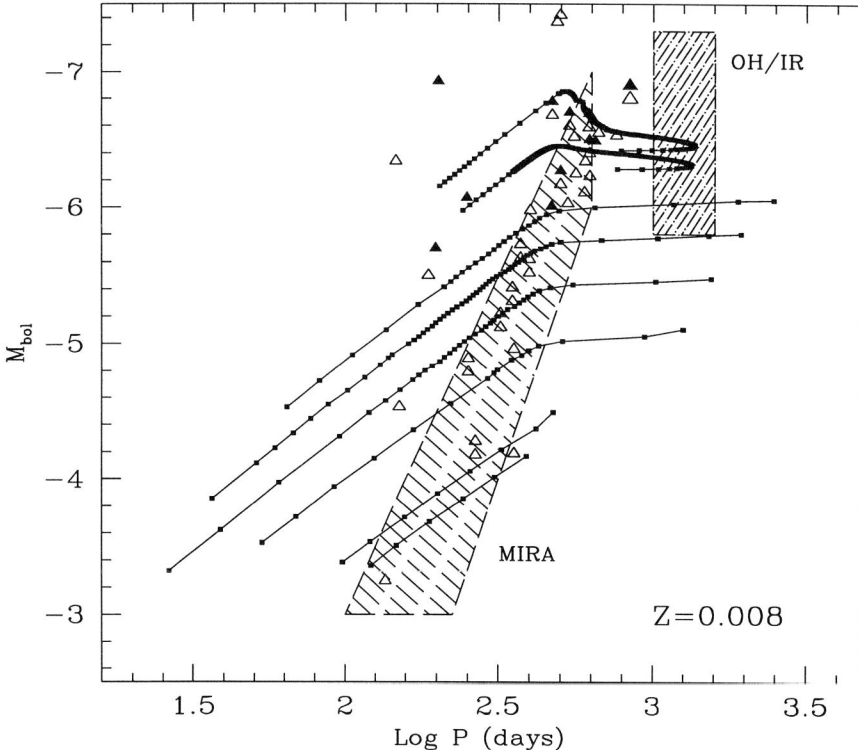

FIGURE 13. The same tracks as in Fig. 12, but in the Period-Luminosity Diagram. The models predictions are compared with observational data by Wood et al. (1992) for Miras and OH/IR stars (shaded areas and open triangles). In addition to this, the positions of Li-strong LPV stars by Smith & Lambert (1990) and Smith et al. (1995) are indicated (filled triangles). All observed LPVs are O-rich stars. This can be understood as the effect of envelope burning which keeps the ratio C/O $\leq$ 1 for most of AGB stars.

convective envelopes ($T_B \geq 60 - 100 \times 10^6 K$) are possible so that efficient envelope burning can occur.

Chemical abundances. As far as the chemical surface abundances are concerned, the rapid conversion of ^{12}C into ^{13}C and then into ^{14}N via the first reactions of the CNO cycle, can delay and even prevent the formation of C-stars. Moreover, the production of ^7Li possibly occurs by means of electron captures on ^7Be nuclei carried from the hot regions of the envelope into cooler layers (T $< 3 \times 10^6 K$) before the reaction $^7Be(p,\gamma)^8B$ proceeds (Cameron & Fowler 1971).

HRD and PLD. In Figs. 12 and 13, we show the path in the HRD and Period-Luminosity Diagram (PLD) of low and intermediate mass stars and compare them with observational data as indicated. The corresponding luminosity function of C-stars derived from these models is shown Fig. 14 together with the data for the LMC.

LMC C-stars progenitors. Examining the available data for AGB stars in the LMC clusters, Marigo et al. (1996b) address the question about the mass interval of low- and intermediate-mass stars which eventually evolve into C-stars during the TP-AGB phase. They combine the data compiled by Frogel et al. (1990) – near infrared photometry and spectral classification for luminous AGB stars in clusters – with the ages for individual

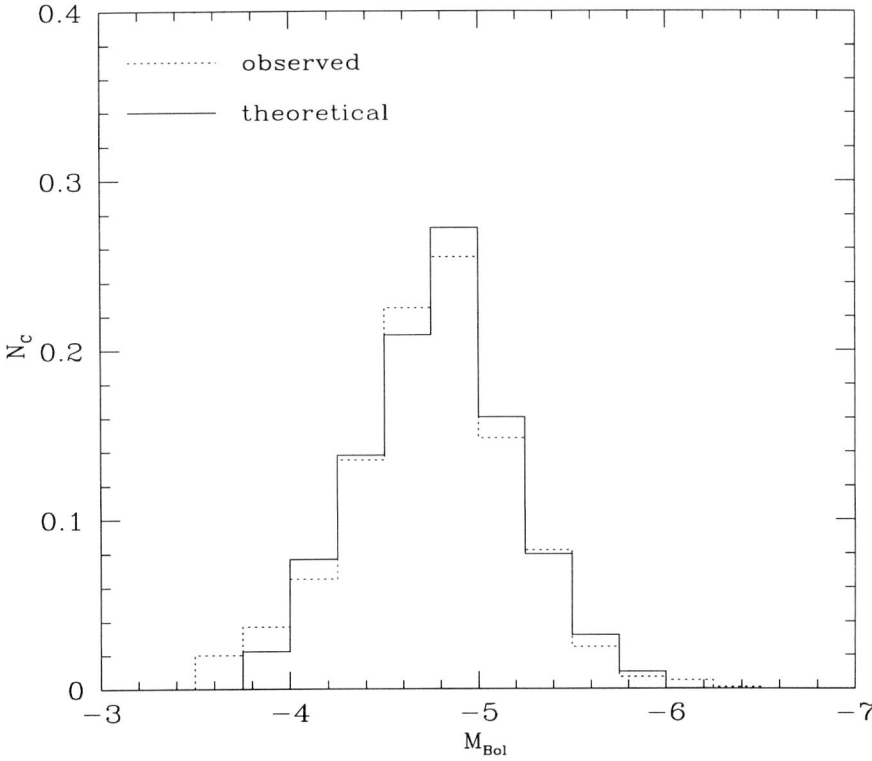

FIGURE 14. The luminosity function of C-star in the LMC (dotted line) taken from Groenewegen & de Jong (1993) as compared to the theoretical prediction (solid line)

clusters derived from independent methods (Girardi et al. 1995). The resulting distribution of C-stars in the $M_{bol} - \log(age)$ plane evidences that the upper and the lower limits of the mass range for the formation of C-stars cannot be derived from cluster data. The explanation of this resides in the presence of two different periods of quiescence in the cluster formation history of the LMC, shaping the age (and progenitor mass) distribution of C-stars. The most recent of these quiescence episodes could also explain the lack of very luminous AGB stars (with $-6 > M_{bol} > -7$) in the clusters, contrary to what observed in the field. Finally, they compare the distribution of C-stars in the $M_{bol} - \log(age)$ diagram with models of AGB evolution which were previously constrained to reproduce the observed luminosity function of C-stars in the field. These models provide a good description of the relative frequency of M- versus C-stars. Notice that the progenitors of C-stars have masses in the range $1.5 \div 2 M_\odot \leq M \leq 4 M_\odot$ which is consistent with previous analyses. Bryan et al. (1990) studying the luminosity function of AGB stars indicate that these objects are smaller than $4 M_\odot$. Claussen et al. (1987) deduced a mass range for the progenitors between $1.2 \div 1.6 M_\odot$. On the basis of kinematical properties Barnbaum et al. (1991) consider the existence of a population of C stars that is more massive than described by Claussen et al. (1987). The initial mass of C stars in this group should be above $2.5 \div 4 M_\odot$.

C-Stars in the Galactic Bulge. Finally, we like to comment on the old discovery of Azzopardi et al. (1985, 1988) of a group of C stars in the Galactic Bulge whose

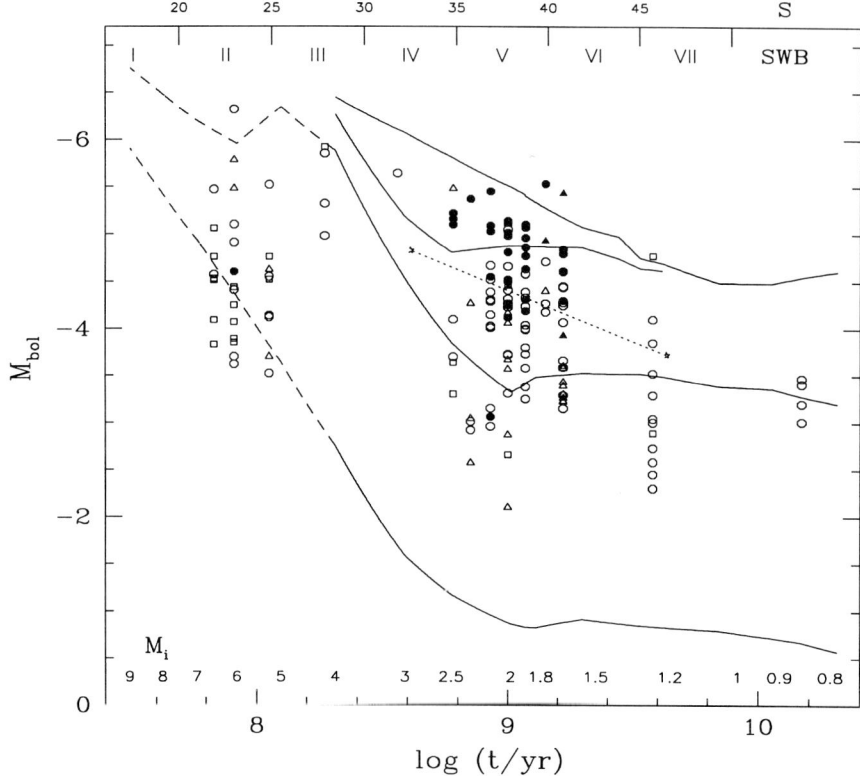

FIGURE 15. Absolute bolometric magnitudes of LMC AGB stars as a function of the age of the parent cluster. The data are from Frogel et al. (1990). The distance modulus to the LMC is $(m - M)_o = 18.5$. The ages are taken from Girardi et al. (1995). Open circles refer to M stars; filled circles correspond to C stars; open and filled triangles indicate stars that are presumably M and C objects; open squares refer to stars whose spectral type has not been assigned. The lines represent the theoretical models by Marigo et al. (1996a,b). From bottom to top, the set of four solid lines refer to the start of the E-AGB, onset of the TP-AGB, transition from O-rich (M-type) to C-rich stars, and end of the TP-AGB.

interpretation has long been a puzzle. Specifically, the bulge C stars are bluer and intrinsically much fainter compared to most other known C stars. In view of their location they should be metal-rich (Azzopardi et al. 1988). At first sight these C stars run counter to current models. According to Westerlund et al. (1991) the stars in the Galactic Bulge should be old, possess a mass of about 0.8 $M_\odot$, and have a metallicity in the range 0.1 solar to a few times solar (Rich 1988). Recently, a very interesting way out has been put forward by Ng (1997) who suggests that the C stars toward the Galactic Bulge actually belong to the Sagittarius dwarf galaxy. He finds that these stars have a metallicity comparable to the LMC, with an age between $0.1 \div 1$ Gyr.

14. Cepheid Stars: mass discrepancy and mixing

In recent years, there have been a considerable number of photometric studies of Cepheids in the field of the LMC and SMC (e.g. Caldwell & Coulson 1986 and references therein) and there is currently much observational effort being put into the search for, and study of, Cepheids in the rich star clusters of the Magellanic Clouds (e.g. Mateo et al 1990, 1991; Welch *et al.* 1991). Because they lie at the same distance, the Cepheids in the Magellanic Cloud clusters are basic to two important topics of astronomy: the understanding of pulsation theory itself and stellar evolution theories in general, and the testing of current calibrations of the cosmic distance scale through the calibration of the period-luminosity-color (PLC) relation (Sandage 1958, Sandage & Tammann 1968, Schmidt 1984, Feast & Walker 1987, Walker 1988, van den Bergh 1989; Madore & Freedman 1991). Furthermore, modern observations are done increasingly towards the red using BVRcIc photometry rather than the more traditional BV photometry (see Madore & Freedman 1991). Most of the theoretical modeling of the Cepheid pulsation rests on the pioneer work of Iben & Tuggle (1972a,b; 1975) and Becker *et al.* (1977). See the reviews by Becker (1985) and Chiosi (1989, 1990). However, these calculations covered a limited range of masses, were based on old models for intermediate-mass stars, and were made mostly at solar metallicity (Z=0.02). Recent models of Cepheid stars were calculated by Chiosi *et al.* (1993) in the mass range 3 $M_\odot$ to 12 $M_\odot$ and with chemical abundances appropriate for the solar vicinity, LMC, and SMC, i.e. Y=0.25 and Y=0.30, and Z=0.016, Z=0.008, and Z=0.004 (e.g. Russell & Bessell 1989). In addition to this, Chiosi *et al.* (1993) analyzed the response of pulsation to different schemes for the evolution of intermediate-mass stars, i.e. for classical models, models with mild core overshoot, and models with full core overshoot. For each model, three modes of pulsation were calculated: fundamental, first overtone and second overtone. They adopted the radiative opacities by Huebner *et al.* (1977) plus the molecular contribution by Alexander (1975), and Alexander *et al.* (1983) according to the prescription by Bessell *et al.* (1989) and the revision by P. R. Wood (1990, unpublished). Finally, the luminosities and T_{eff}s of the models were converted to magnitudes and colors in the BVRI passbands with the aid of either the Green *et al.* (1987) scale or tables amalgamating data from Bell & Gustafsson (1978), Gustafsson & Bell (1979), Buser & Kurucz (1978), and R. Buser (1989, unpublished). For more details see Chiosi *et al.* (1993).

The blue edges of the instability strips of Chiosi *et al.* (1993) agree with the corresponding ones of Iben & Tuggle (1972a,b; 1975), whereas the red edges have a different inclination whose slope varies with the metallicity. Red edges not running parallel to the blue ones have been suggested by Fernie (1990) for the Galactic Cepheids and are perhaps confirmed by the observational study of Mateo *et al.* (1991 and references) of Cepheids in LMC clusters. Chiosi *et al.* (1992a), comparing the Fernie (1990) empirical instability strip to theoretical predictions obtained from the Chiosi *et al.* (1993) Cepheid models and the synthetic CMD technique, showed that both the edges and the distribution of stars within the strip could be reproduced.

Chiosi *et al.* (1993) also presented the period-luminosity (PL) and PLC relationships in the BVRI passbands for the three harmonics and the various compositions. The PL relations agree well with the observational ones (see Feast & Walker 1987) and their zero points are nearly independent of the chemical composition as indicated by the analysis of observational data by Madore & Freedman (1991). The PLC relations have the period term in good agreement with the observational determinations (see Feast & Walker 1987, Madore & Freedman 1991), whereas the color term is larger than estimated by Caldwell & Coulson (1986) and closer to the early estimate by Sandage (1958). The reason for the difference is not understood.

14.1. *Mass Discrepancy of the Cepheid Stars*

It has long been debated whether the masses determined from stellar evolution theory agree with those derived from pulsation theory (see Iben 1974; Iben & Tuggle 1972a,b, 1975; Cox 1980, 1985). In general, pulsational masses (M_{pul}) are estimated to be 30 to

40% lower than evolutionary masses (M_{evol}) of the same luminosity. The mass discrepancy problem can be reduced to the following causes, each of which is affecting the masses in question in a different way (see the reviews by Becker 1985; Cox 1980, 1985; Pel 1985; and Iben & Tuggle 1972a,b, 1975): (a) significant mass loss at some point between the main sequence and the Cepheid stage could decrease M_{evol}; (b) uncertainties in the determination of the distance of the Cepheid stars would affect largely M_{pul} and to a lesser extent M_{evol}; (c) uncertainties in the conversion from colors to T_{eff}s would affect both M_{pul} and M_{evol}; (d) inadequacy of the pulsation theory which would obviously reflect on M_{pul}; finally, (e) inadequacies of current stellar models which bear on the determination of M_{evol}.

The effect of mass loss was studied by Willson (1988) and Willson & Bowen (1984), who suggested that stellar winds, somehow triggered by the pulsational instability itself, should occur while the star is within the instability strip. Evolutionary calculations by Brunish & Willson (1987), including mass loss during the Cepheid stage, confirm that Cepheids are trapped in the instability strip, mass loss can continue over a relatively long time scale, and the total mass is significantly reduced, whereas the luminosity is about constant. The rates of mass loss required by these model calculations are of the order of $7 \times 10^{-9} M_\odot$/yr for a 5 $M_\odot$ star and 2×10^{-7} $M_\odot$/yr for a 7 $M_\odot$ star. Unfortunately, efforts to observe Cepheid winds directly have given so far inconclusive results (see Willson 1988). Finally, the amount of mass loss along the Hayashi line is negligible according to the current mass-loss rates (de Jager et al. 1988).

As far as the calibration of the distance scale and the conversion from colors to T_{eff}s are concerned the reader is referred to Pel (1985) for further details.

We have already recalled that period ratios at evolutionary masses and luminosities can be explained by an enhancement in the opacity (Simon 1982, 1987), like that found in modern opacity calculations.

As already discussed above, convective overshoot alters the mass-luminosity relationship of core He-burning models. Thus, at any given initial mass, the tracks cross the instability strip at higher luminosity than classical models, or conversely, at any given luminosity the corresponding Cepheid mass is significantly lower (Matraka et al 1982, Bertelli et al. 1985). Once again, the star clusters of the LMC with Cepheids are the ideal workbench, because all the stars lie at the same distance and membership is less of a problem. This topic has been examined in great detail by Chiosi et al. (1992b) using the Cepheid stars and CMD of NGC 2157 (Mateo et al. 1990). On the one hand, the fit of the CMD with theoretical simulations based either on classical models or models incorporating core overshoot leads to accurate determination of the M_{evol} of the Cepheid stars, together with the age and chemical compositions. On the other hand, the use of a well calibrated relation between mass-period-luminosity-color (MPLC) for Cepheid stars with the chemical composition suited to the cluster in question, allows a good determination of M_{pul}, M_{evol}, and distance modulus to the LMC at the same time. This analysis indicates that the problem of mass discrepancy likely originates from the adoption of classical models, i.e. without overshoot, to derive M_{evol}, and from the lack of sufficient accuracy in the determination of the distance which bears on both M_{pul} and M_{evol}. The resulting distance modulus to the LMC is $(m - M)_o = 18.5 \pm 0.1$ in agreement with the determination by Panagia et al. (1991) based on the circumstellar ring observed by the Hubble Space Telescope around supernova 1987A in the LMC.

It is a pleasure to thank my collaborators and students for their contribution to my understanding of stellar evolution, invaluable help, and hard work over the years:

Maurizio Alongi, Gianpaolo Bertelli, Alessandro Bressan, Giovanni Carraro, Licai Deng, Franco Fagotto, Paola Marigo, Emma Nasi, Laura Portinari, Sergio Ortolani, Bernardo Salasnich, Rosaria Tantalo, and Antonella Vallenari. I want to thank Artemio Herrero and Antonio Aparicio for their efforts to make the Winter School successful. Special thanks to Francisco Sanchez for the very friendly hospitality at the IAC. This research was supported by the Ministry of University and Scientific and Technological Research (MURST), the National Council of Research (CNR-GNA), and the Italian Space Agency (ASI).

REFERENCES

ABBOTT, D. C. 1982 *ApJ* **259**, 282.
ABBOTT, D. C., BIEGING, H., CHURCHWELL, E., TORRES, A.V. 1986 *ApJ* **303**, 239.
ABBOTT, D.C., CONTI, P.S. 1987 *ARA&A* **25**, 113.
ABIA, C., & REBOLO, R. 1989 *ApJ* **347**, 186.
ALEXANDER, D. R. 1975 *ApJS* **29**, 363.
ALEXANDER, D. R., BROCATO, E., CASSISI, S., CASTELLANI, V., CIACIO, F., DEGL'INNOCENTI, S. 1996 *A&A* submitted.
ALEXANDER, D. R., JOHNSON, H. R., & RYMPA, R. C. 1983 *ApJ* **273**, 773.
ALEXANDER, D. R., AUGASON, G. C., JOHNSON, H. R. 1989 *ApJ* **345**, 1014.
ALONGI, M., BERTELLI, G., BRESSAN, A., CHIOSI, C. 1992 *A&A* **224**, 95.
ALONGI, M., BERTELLI, G., BRESSAN, A., CHIOSI, C., FAGOTTO, F., GREGGIO, L., NASI, E. 1993 *A&AS* **97**, 851.
ANDERSEN, J., NORDSTROM, B. 1991 *ApJ* **363**, L33.
ANDERSEN, J., NORDSTROM, B., CLAUSEN, J.V. 1990. *ApJ* **363**, L33.
ANDREASEN, G. K. 1988 *A&A* **201**, 72.
ANDREASEN, G. K., PETERSEN, J. O. 1988 *A&A* **192**, L4.
ANTONY-TWAROG, B. J., MUKHERJEE, K., CALDWELL, N., TWAROG, B. A. 1988 *AJ* **95**, 1453.
ANTONY-TWAROG, B. J., TWAROG, B. A., SHODAN, S. 1989 *AJ* **98**, 1634.
ANTONY-TWAROG, B. J., KALUZNY, J., SHARA, M. M., TWAROG, B. A. 1990 *AJ* **99**, 1054.
APARICIO, A., BERTELLI, G., CHIOSI, C., GARCIA-PELAYO, J. M. 1990 *A&A* **240**, 262.
APPENZELLER, I. 1986 In *Luminous Stars and Associations in Galaxies*, ed. P. S. Conti, C. de Loore, & E. Kontizas, (Dordrecht: Reidel), p. 139.
APPENZELLER, I., FRICKE, K. J. 1972 *A&A* **21**, 285.
APPENZELLER, I., TSCHARNUTER, W. 1973 *A&A* **25**, 125.
APPLEGATE, J. H. 1988 *ApJ* **329**, 803.
ARNETT, W. D. 1968 *Nature* **219**, 1344.
ARNETT, W. D. 1969 *Astrophys. Space Sci.* **5**, 180.
ARNETT, W. D. 1974 *ApJ* **191**, 727.
ARNETT, W. D. 1977 *ApJS* **35**, 145.
ARNETT, W. D. 1978 *ApJ* **219**, 1008.
ARNETT, W. D., BAHCALL, J. N., KIRSHNER, R. P., WOOSLEY, S. E. 1989 *ARA&A* **27**, 629.
ARP, H. 1967 *ApJ* **149**, 91.
AZZOPARDI, M., LEQUEUX, J., REBEIROT, E. 1985 *A&A* **145**, L4.
AZZOPARDI, M., LEQUEUX, J., REBEIROT, E. 1988 *A&A* **202**, L27.
BAKER, N. H., KUHFUSS, R. 1987 *A&A* **185**, 117.
BANDIERA, R., TUROLLA, R. 1990 *A&A* **231**, 85.

BARAFFE, I., CHABRIER, G., ALLARD, F., HAUSCHILDT, P. H. *ApJ* **446**, L35.
BARBARO, G., PIGATTO, L. 1984 *A&A* **136**, 355.
BARKAT, Z. 1971 *ApJ* **163**, 433.
BARNBAUM, C., KASTNER, J. H., ZUCKERMAN, B. 1991 *AJ* **102**, 289.
BARLOW, M.J., SMITH, L.J., WILLIS, A.J. 1981 *MNRAS* **196**, 101.
BAUD, B., HABING, H. J. 1983 *A&A* **127**, 73.
BECKER, S. A. 1981 *ApJS* **45**, 478.
BECKER, S. A. 1985 In *Cepheids, Theory and Observations*, ed. B.F. Madore, (Cambridge: Cambridge Univ. Press), p. 104.
BECKER, S. A., IBEN, I. JR., TUGGLE, R. S. 1977 *ApJ* **218**, 633.
BECKER, S. A., MATHEWS, G. J. 1983 *ApJ* **270**, 155.
BEDIJN, P. J. 1988 *A&A* **205**, 105.
BELL, R. A., GUSTAFSSON, B. 1978 *A&AS* **34**, 229.
BELL, R. A., VANDENBERG, D. A. 1987 *ApJS* **63**, 335.
BENCIVENNI, D., CASTELLANI, V., TORNAMBE', A., WEISS, A. 1989 *ApJS* **71**, 109.
BENCIVENNI, D., BROCATO, E., BUONANNO, R., CASTELLANI, V. 1991 *AJ* **102**, 137.
BERGBUSCH, P. A., VANDENBERG, D. A. 1992 *ApJS* **81**, 163.
BERRINGTON, K. A., BURKE, P. G., BUTLER, K., SEATON, M. J., STOREY, P. J., et al. 1987 *J. Phys.B* **20**, 6397.
BERTELLI, G., BETTO, F., BRESSAN, A., CHIOSI, C., NASI, E., VALLENARI, A. 1990 *A&A* **85**, 845.
BERTELLI, G., BRESSAN, A., CHIOSI, C. 1984 *A&A* **130**, 279.
BERTELLI, G., BRESSAN, A., CHIOSI C. 1985 *A&A* **150**, 33.
BERTELLI, G., BRESSAN, CHIOSI, C. 1992 *ApJ* **392**, 522.
BERTELLI, G., BRESSAN, A., CHIOSI, C., ANGERER, K. 1986a *A&AS* **66**, 191.
BERTELLI, G., BRESSAN, A., CHIOSI, C., ANGERER, K. 1986b In *The Age of Star Clusters*, ed. F. Caputo. *Mem. Soc. Astron. Ital.* **57**, 427.
BERTELLI, G., BRESSAN, A., CHIOSI, C., FAGOTTO, F., NASI, E. 1994 *A&AS* **106**, 275.
BESSELL, M. S., BRETT, J. M., SCHOLZ, M., WOOD, P. R. 1989 *A&AS* **77**, 1.
BLAHA, C., HUMPHREYS, R.M. 1989 *AJ* **98**, 1598.
BLANCO, B. M., BLANCO, V. M., MCCARTHY, M. F. 1978 *Nature* **271**, 638.
BLANCO, B. M., MCCARTHY, M. F., BLANCO, V. M. 1980 *ApJ* **242**, 948.
BLOECKER, T. 1995 *A&A* **299**, 755.
BLOECHER T., SCHOENBERNER, D. 1991 *A&A* **244**, L43.
BLOMME, R., VANBEVEREN, D., VAN RENSBERGEN, W. 1991 *A&A* **241**, 479.
BOER, B., DE JAGER, C., NIEUWENHUIJZEN, H. 1988 *A&A* **195**, 218.
BOESGAARD, A., STEIGMAN, G. 1985 *ARA&A* **23**, 319.
BOHANNAN, B., ABBOT, D.C., VOELS, A.A., HUMMER, D.G. 1986 *ApJ* **308**, 728.
BOLTE, M. 1989 *AJ* **97**, 1688.
BOLTE, M., HOGAN, C. J. 1995 *Nature* **376**, 399.
BOND, J. R. 1984 In *Stellar Nucleosynthesis*, eds. C. Chiosi & A. Renzini, (Dordrecht: Reidel), p. 297.
BOND, J. R., ARNETT, W. D., CARR, B. J. 1982 In *Supernovae: A Survey of Current Research*, eds. M. J. Rees & R. J. Stoneham, (Cambridge: Cambridge Univ. Press), p. 303.
BONO, G., CAPUTO, F., CASTELLANI, V., MARCONI, M., STAIANO, L., STELLINGWERF, R. F. 1995 *ApJ* **442**, 159.
BOND, J. R., ARNETT, W. D., CARR, B. J. 1984 *ApJ* **280**, 825.

BOOTHROYD, A. I., SACKMANN, I. J. 1988a *ApJ* **328**, 632.
BOOTHROYD, A. I., SACKMANN, I. J. 1988b *ApJ* **328**, 641.
BOOTHROYD, A. I., SACKMANN, I. J. 1988c *ApJ* **328**, 653.
BOOTHROYD, A. I., SACKMANN, I. J. 1988d *ApJ* **328**, 671.
BOWEN, G. H. 1988 *ApJ* **329**, 299.
BOWEN, G. H., WILLSON, L. A. 1991 *ApJ* **375**, L53.
BRESSAN, A. 1990 In *Chemical and Dynamical Evolution of Galaxies*, eds. F. Ferrini, P. Franco, & F. Matteucci (Pisa: Giardina).
BRESSAN, A., BERTELLI, G., CHIOSI, C. 1981 *A&A* **102**, 25.
BRESSAN, A., BERTELLI, G., CHIOSI, C. 1986 In *The Age of Star Clusters*, ed. F. Caputo. *Mem. Soc. Astron. It.* **57**, 411.
BRESSAN, A., BERTELLI, G., CHIOSI, C. 1991 *Bull. Am. Astron. Soc.* **23 (2)**, 969.
BRESSAN, A., FAGOTTO, F., BERTELLI, G., CHIOSI, C. 1993 *A&AS* **100**, 647.
BROCATO, E., BUONANNO, R. CASTELLANI, V., WALKER, A. R. 1990 *ApJS* **71**, 25.
BROCATO, E., CASTELLANI, V. 1993 *ApJ* **410**, 99.
BRUENN, S. W. 1971 *ApJ* **168**, 203.
BRUENN, S. W. 1973 *ApJ* **183**, L125.
BRUNISH, W. M., GALLAGHER, J. S., TRURAN, J. W. 1986 *AJ* **91**, 598.
BRUNISH, W. M., TRURAN, J. W. 1982a *ApJ* **256**, 247.
BRUNISH, W. M., TRURAN, J. W. 1982b *ApJS* **49**, 447.
BRUNISH, W. M., WILLSON, L. A. 1987 In *Stellar Pulsation*, ed. A. N. Cox, W. M. Sparks, & S. G. Starrfield (Springer-Verlag), p. 27.
BRYAN, G. L., VOLK, K., KWOK, S. 1990 *ApJ* **365**, 301.
BUCHLER, J. R., MAZUREK, T. J. 1975 *Mem. Soc. Roy. Sci. Liege* **8**, 435.
BUONANNO, R., BUSCEMA, G., FUSI-PECCI, F., RICHER, H. B., FAHLMAN, G. G. 1990 *AJ* **100**, 1811.
BUONANNO, R., CORSI, C. E., FERRARO, I., FUSI-PECCI, F. 1987 *A&AS* **67**, 327.
BUONANNO, R., CORSI, C. E., FUSI-PECCI, F. 1985 *A&A* **145**, 97.
BUONANNO, R., CORSI, C. E., FUSI-PECCI, F. 1989 *A&A* **216**, 80.
BURROWS, A., HUBBARD, W. B., SAUMON, D., LUNINE, J. I. 1993 *ApJ* **406**, 158.
BUSER, R., KURUCZ, R. L. 1978 *A&A* **70**, 555.
BUZZONI, A., FUSI-PECCI, F., BUONANNO, R., CORSI, C. E. 1983 *A&A* **128**, 94.
CACCIARI C., CLEMENTINI, G., BUSER, R. 1989 *A&A* **209**, 154.
CALDWELL, J. A. R., COULSON, I. M. 1986 *MNRAS* **218**, 223.
CAMERON, A. G. W., FOWLER, W. A. 1971 *ApJ* **164**, 111.
CANUTO, V. 1992 *ApJ* **392**, 218.
CANUTO, V. 1993 *ApJ* **416**, 331.
CANUTO, V., CHENG, Y., HARTKE, G. I., SCHILLING, O. 1991 *Phys. Fluids A* **3**, 1633.
CANUTO, V., MAZZITELLI, I. 1991 *ApJ* **370**, 295.
CANUTO, V. MINOTTI, F. O., RONCHI, C., YPMA, R. M., ZEMAN, O. 1994a . *J. Atmos. Sci.* **51** , 1605.
CANUTO, V., MINOTTI, F. O., SCHILLING, O. 1994b *ApJ* **425**, 303.
CANUTO, V., GOLDMAN, I., MAZZITELLI, I. 1996a *ApJ* in press.
CANUTO, V., CALOI, V. D'ANTONA, F., MAZZITELLI, I. 1996b *Apj*, submitted.
CAPACCIOLI, M., ORTOLANI, S., PIOTTO, G. P. 1991 *A&A* **244**, 298.
CAPUTO, F., DE STEFANIS, P., PAEZ, E., QUARTA, M. L. 1987 *A&AS* **68**, 19.
CARPAY, J., DE JAGER, C., NIEUWENHUIJZEN, H., MOFFATT, A. 1989 *A&A* **216**, 143.

CARRARO, G. 1991. Master's thesis. Univ. Padua, Italy.

CARRARO, G., CHIOSI, C. 1994 A&A **287**, 761.

CARRARO, G., GIRARDI, L., BRESSAN, A., CHIOSI, C. 1996 A&A **305**, 849.

CARSON, T. R., HUEBNER, W. F., MAGEE, N. H. JR., MERTZ, A. L. 1984 ApJ **283**, 466.

CARSON, T. R., STOTHERS, R. 1976 ApJ **204**, 461.

CARSON, T. R., STOTHERS, R. 1988 ApJ **328**, 196.

CARSON, T. R., STOTHERS, R., VERMURY, S. K. 1981 ApJ **244**, 230.

CASSINELLI, J. P. 1991 In *Wolf-Rayet Stars and Interrelations with Other Massive Stars in Galaxies*, eds. K. A. van der Hucht & B. Hidayat, (Dordrecht: Kluwer), p. 289.

CASSISI, S., CASTELLANI, M., CASTELLANI, V. 1997 Preprint.

CASTELLANI, V. 1986 *Fundamentals of Cosmic Physics* **9**, 317.

CASTELLANI, V., CHIEFFI, A.. PULONE, L., TORNAMBE', A. 1985 ApJ **296**, 204.

CASTELLANI, V., CHIEFFI, A., STRANIERO, O. 1990 ApJS **74**, 463.

CASTELLANI, V., CHIEFFI, A., PULONE, L. 1991 ApJS **76**, 911.

CASTELLANI, V., GIANNONE, P., RENZINI, A. 1971a *Astrophys. Space Sci.* **10**, 340.

CASTELLANI, V., GIANNONE, P., RENZINI, A. 1971b *Astrophys. Space Sci.* **10**, 355.

CASTOR, J. I., ABBOTT, D. C., KLEIN, R. I. 1975 ApJ **195**, 157.

CAUGHLAN, G. R., FOWLER, W. A. 1988 *At. Data Nucl. Data Tables* **40**, 283.

CAUGHLAN, G. R., FOWLER, W. A., HARRIS, M., ZIMMERMANN, B. 1985 *At. Data Nucl. Data Tables* **32**, 197.

CERRUTI-SOLA, M., PERINOTTO, M. 1985 ApJ **291**, 237.

CHABOYER, B. 1995 ApJ **444**, L9.

CHABOYER, B., DELIYANNIS, C.P., DEMARQUE, P., PINSONNEAULT, SARAJEDINI, A. 1992 ApJ **388**, 372.

CHABOYER, B., DEMARQUE, P. 1994 ApJ **433**, 510.

CHABOYER, B., KIM, Y-C. 1995 ApJ **454**, 767.

CHABRIER, G., BARAFFE, I., PLEZ, B. 1996 ApJ **459**, L91.

CHIEFFI, A. STRANIERO, O. 1989 ApJS **71**, 47.

CHIEFFI, A. STRANIERO, O., SALARIS, M. 1995 ApJ **445**, L39.

CHIOSI, C. 1981a In *Effects of Mass Loss on Stellar Evolution*, eds. C. Chiosi & R. Stalio, (Dordrecht: Reidel), p. 229.

CHIOSI, C. 1981b In *The Most Massive Stars*, eds. S. D'Odorico, D. Baade & K. Kjar, (Garching: ESO), p. 27.

CHIOSI, C. 1982 In *Wolf Rayet Stars: Observation, Physics and Evolution*, eds. C. de Loore & A. J. Willis, (Dordrecht: Reidel), p. 323.

CHIOSI, C. 1986a In *Spectral Evolution of Galaxies*, eds. C. Chiosi & A. Renzini, (Dordrecht: Reidel), p. 237.

CHIOSI, C. 1986b In *Nucleosynthesis and Stellar Evolution*, 16th Saas-Fee Course, ed. B. Hauck, A. Maeder, G. Meynet, (Geneva: Geneva Obs), p. 199.

CHIOSI, C. 1989 In *The Use of Pulsating Stars in Fundamental Problems of Astronomy*, ed. E. G. Schmidt, (Cambridge: Cambridge Univ. Press), p. 19..

CHIOSI, C. 1990 PASP **102**, 412.

CHIOSI, C. 1991 In *Confrontation between Stellar Pulsation and Evolution*, ed. C. Cacciari & G. Clementini, *ASP Conf. Ser.* **11**, 158.

CHIOSI, C., BERTELLI, G., BRESSAN, A. 1987 See Kwok & Pottasch 1987, p. 239.

CHIOSI, C., BERTELLI, G., BRESSAN, A. 1992a In *Instabilities in Evolved Super and Hypergiants*, ed. C. de Jager & H. Nieuwenhuijzen, (Amsterdam: North Holland), p. 145.

CHIOSI, C., BERTELLI, G., BRESSAN, A. 1992b ARA&A **30**, 235.

CHIOSI, C., BERTELLI, G., BRESSAN, A., NASI, E. 1986 *A&A* **165**, 84.
CHIOSI, C., BERTELLI, G., MEYLAN, G., ORTOLANI, S. 1989a *A&AS* **78**, 89.
CHIOSI, C., BERTELLI, G., MEYLAN, G., ORTOLANI, S. 1989b *A&A* **219**, 167.
CHIOSI, C., CAIMMI, R. 1979 *A&A* **80**, 234.
CHIOSI, C., MAEDER, A. 1986 *ARA&A* **24**, 329.
CHIOSI, C., NASI, E., SREENIVASAN, S. R. 1978 *A&A* **63**, 103.
CHIOSI, C., PIGATTO, L. 1986 *ApJ* **308**, 1.
CHIOSI, C., SUMMA, C. 1970 *Astrophys. Space Sci.* **8**, 478.
CHIOSI, C., VALLENARI, A., BRESSAN, A., DENG, L., ORTOLANI, S. 1995 *A&A* **293**, 710.
CHIOSI, C., WOOD, P. R., CAPITANIO, N. 1993 *ApJS* **86**, 541.
CHIOSI, C., WOOD, P. R., BERTELLI, G., BRESSAN, A. 1992a *ApJ* **387**, 320.
CHIOSI, C., WOOD, P. R., BERTELLI, G., BRESSAN, A., MATEO, M. 1992b *ApJ* **385**, 285.
CLARET, A. 1995 *A&AS* **109**, 441.
CLARET, A., GIMENEZ, A. 1995 *A&AS* **114**, 549.
CLAUSSEN, M. J., KLEINMMANN, S. G., JOYCE, R. R., JURA, M. 1987 *ApJS* **65**, 385.
CLOUTMAN, L. D., WHITAKER, R. W. 1980 *ApJ* **237**, 900.
COLE, P. W., DEUPREE, R. G. 1980 *ApJ* **239**, 284.
COLE, P. W., DEUPREE, R. G. 1981 *ApJ* **247**, 607.
CONTI, P. S. 1976 *Mem. Soc. Roy. Sci. Liege* **6 - 9**, 193.
CONTI, P. S. 1988 In *O-stars and WR stars, NASA Sp-497* ed. P.S. Conti & A.B. Underhill, (Washington: NASA), p. 81.
COPELAND, H., JENSE, J. O., JORGENSEN, H. E. 1970 *A&A* **5**, 12.
COX, A. N. 1980 *ARA&A* **18**, 15.
COX, A. N. 1985 In *Cepheids, Theory and Observations*, ed. B.F. Madore, (Cambridge: Cambridge Univ), p. 126.
COX, A. N., STEWART, J. N. 1965 *ApJS* **11**, 22.
COX, A. N., STEWART, J. N. 1970a *ApJS* **19**, 243.
COX, A. N., STEWART, J. N. 1970b *ApJS* **10**, 261.
COX, A. N., TABOR, J. E. 1976 *ApJS* **31**, 271.
D'ANTONA, F. 1987 *ApJ* **320**, 653.
D'ANTONA, F., MAZZITELLI, I. 1986 *ApJ* **296**, 502.
D'ANTONA, F., MAZZITELLI, I. 1990 *ARA&A* **28**, 139.
D'ANTONA, F., MAZZITELLI, I. 1994 *ApJS* **90**, 467.
DAVIDSON, K. 1987 *ApJ* **317**, 760.
DAVIDSON, K., KINMAN, T. D. 1985 *ApJS* **58**, 321.
DE JAGER, C. 1984 *A&A* **138**, 246.
DE JAGER C., NIEUWENHUIJZEN H., VAN DER HUCHT K. A. 1988 *A&AS* **72**, 295.
DE KOTER, A., DE JAGER, C., NIEUWENHUIJZEN, H. 1988 *A&A* **200**, 146.
DELIYANNIS, C. P., DEMARQUE, P., KAWALER, S. D. 1990 *ApJ* **73**, 21.
DELIYANNIS, C. P., DEMARQUE, P., PINSONNEAULT, M. H. 1989 *ApJ* **347**, L73.
DEMARQUE, P. 1988 In *Globular Cluster Systems in Galaxies*, ed. J. A. Grindlay & A. G. Philip, (Dordrecht: Kluwer), p. 121.
DEMARQUE, P., DELIYANNIS, C. P., SARAJEDINI, A. 1991 In *Observational Tests of Inflation*, ed. T. Shank, (Durhan: England).
DEMARQUE, P., MENGEL, J. C. 1972 *ApJ* **171**, 583.
DENG, L. 1992 PHD. Thesis, ISAS, Trieste, Italy.
DENG, L., BRESSAN, A., CHIOSI, C. 1996a *A&A* **313**, 145.

DENG, L., BRESSAN, A., CHIOSI, C. 1996b *A&A* **313**, 159.

DESPAIN, K. H. 1981 In *Physical Processes in Red Giants*. ed. I. Iben Jr. & A. Renzini, (Dordrecht: Reidel), p. 173.

DESPAIN, K. H. 1982 *ApJ* **253**, 11.

DEUPREE, R. G. 1986 *ApJ* **303**, 649.

DEUPREE, R. G., COLE, P. W. 1981 *ApJ* **294**, L35.

DOOM, C. 1982a *A&A* **116**, 303.

DOOM, C. 1982b *A&A* **116**, 308.

DOOM, C. 1985 *A&A* **142**, 143.

DORMAN, B., LEE, Y-W., VANDENBERG, D. A. 1991 *ApJ* **366**, 115.

DOS SANTOS, L. C., JATENCO-PEREIRA, V., OPHER, R. 1993 *ApJ* **410**, 732.

EDWARDS, A. C. 1969 *MNRAS* **146**, 445.

EGGEN, O. J., LYNDEN-BELL, D., SANDAGE, A. 1962 *ApJ* **136**, 748.

ELITZUR, M., BROWN, J. A., JOHNSON, H. R. 1989 *ApJ* **341**, L95.

FAGOTTO, F. 1990 Master Thesis, Univ. Padua, Italy.

FAGOTTO, F., BRESSAN, A., BERTELLI, G., CHIOSI, C. 1994a *A&AS* **104**, 365.

FAGOTTO, F., BRESSAN, A., BERTELLI, G., CHIOSI, C. 1994b *A&AS* **105**, 29.

FAGOTTO, F., BRESSAN, A., BERTELLI, G., CHIOSI, C. 1994c *A&AS* **105**, 39.

FALK, H. J., MITALAS, R. 1981 *MNRAS* **196**, 225.

FEAST, M. W., WALKER, A. R. 1987 *ARA&A* **25**, 345.

FEAST M. W. 1991 In *Instabilities in Evolved Super and Hyper-Giants*, eds. C. de Jager & H. Nieuwenhuijzen, (Amsterdam: North Holland).

FERNIE, J. D. 1990 *ApJ* **354**, 295.

FERRARO, F. R., CLEMENTINI, G., FUSI-PECCI, F., SORTINO, R., BUONANNO, R. 1992 *MNRAS* **256**, 391.

FERRARO, F. R., CLEMENTINI, G., FUSI-PECCI, F., VITELLO, E., BUONANNO, R. 1992 *MNRAS* **264**, 273.

FITZPATRICK, E. L., GARMANY, C. D. 1990 *ApJ* **363**, 119.

FOLGHERAITER, E. L., PENNY, A. J., GRIFFITHS, W. K. 1993 *MNRAS* **264**, 991.

FOWLER, W. A. 1966 In *High Energy Astrophysics*, ed. L. Gratton, (New York: Academic Press), p. 313.

FOWLER, W. A. 1984 *Rev. Mod. Phys.* **56**, 149.

FOWLER, W. A., CAUGHLAN, G. R., ZIMMERMANN, B. A. 1967 *ARA&A* **5**, 525.

FOWLER, W. A., CAUGHLAN, G. R., ZIMMERMANN, B. A. 1975 *ARA&A* **13**, 69.

FOWLER, W. A., HOYLE, F. 1964 *ApJS* **9**, 201.

FRANSSON, C., CASSATELLA, A., GILMOZZI, R., KIRSHNER, R. P., PANAGIA, et al. 1989 *ApJ* **336**, 429.

FRICKE, K. J. 1973 *ApJ* **183**, 941.

FRICKE, K. J. 1974 *ApJ* **189**, 535.

FRIEL, E. D., JANES, K. A. 1991 See Janes, 1991, p. 569.

FROGEL, J. A., MOULD, J., BLANCO, V. M. *ApJ* **352**, 96.

FULLER, G. M., FOWLER, W. A., NEWMAN, M. J. 1982 *ApJ* **252**, 715.

FUSI-PECCI, F., RENZINI, A. 1975 *A&* **39**, 413.

FUSI-PECCI, F., RENZINI, A. 1979 In *Astronomical Use of the Space Telescope*, eds. F. Macchetto, F. Pacini & M. Tarenghi, (Geneva: ESO), p. 181.

FUSI-PECCI, F., CACCIARI, C. 1991 In *New Windows to the Universe*, ed. F. Sanchez & M. Vasquez, (Cambridge: Cambridge Univ. Press), p. 364.

Fusi-Pecci, F., Ferraro, F. R., Crocker, D. A., Rood, R. T., Buonanno, R. 1990 *A&A* **238**, 95.

Gail, H. P., Sedlmayr, E. 1987 *A&A* **171**, 197.

Gallart, C., Aparicio, A., Bertelli, G., Chiosi, C. 1996 *AJ* **112**, 1950.

Garmany, C. D, Conti, P. S., Chiosi, C. 1982 *ApJ* **263**, 777.

Giannone, P., Kohl, K., Weigert, A. 1968 *Z. Ap* **68**, 107.

Gingold, R. A. 1976 *ApJ* **204**, 116.

Gratton, R. G., Ortolani, S. 1989 *A&A* **211**, 41.

Grossman, S. A. 1996 *MNRAS* **279**, 305.

Grossman, S. A., Taam, R. E. 1996 *MNRAS*, in press.

Grossman, S. A., Narayan, R., Arnett, D. 1993 *ApJ* **407**, 284.

Green, E. M., Demarque, P., King, C. R. 1987 In *The Revised Yale Isochrones and Luminosity Functions* (New Haven: Yale Univ. Obs).

Greggio, L., Renzini, A. 1990 *ApJ* **364**, 35.

Groenewegen, M. A. T. & de Jong, T. 1993 A&A **267**, 410.

Gunn, J.E., Ostriker, J.P. 1970. *ApJ* **160** 979.

Gustafsson, B., Bell, R. A. 1979 *A&A* **74**, 313.

Hamann, W. R., Koesterke, L., Wessolowski, U. 1993 *A&A* **274**, 397.

Hanson, R. B. 1979. In *The HR Diagram*, eds. A. G. D. Philip, & D. S. Hayes, (Dordrecht: Reidel), p. 154.

Harris, M. J., Fowler, W. A., Caughlan, G. R., Zimmermann, B. A. 1983 *ARA&A* **21**, 165.

Hesser, J. E. 1988 In *Progress and Opportunities in Southern Hemisphere Optical Astronomy*, eds. V. M. Blanco & M. M. Phillips, ASP Conf. Ser. **vol. 1**, p. 161.

Hesser, J. E., Harris, W. E., VandenBerg, D. A., Allwright, J. W. B., Shott, P., Stetson, P. B. 1987 *PASP* **99**, 739.

Hesser, J. E., Shawl, S. J. 1985 *PASP* **97**, 465.

Hillebrandt, W., Nomoto, K., Wolff, R.G. 1984 *A&A* **133**, 175.

Hofmeister, E. 1969 *A&A* **2**, 143.

Hollowell, D. E. 1987 See Kwok & Pottasch, 1987, p. 239.

Hollowell, D. E. 1988 PHD. Thesis. Univ. Illinois.

Hollowell, D. E., Iben, I. Jr. 1988 *ApJ* **333**, L2.

Hollowell, D. E., Iben, I. Jr. 1989 *ApJ* **340**, 966.

Hollowell, D. E., Iben, I. Jr. 1990 *ApJ* **349**, 208.

Huebner, W. F., Mertz, A. L., Magee, N. H. Jr., Argo, M. F. 1977 *Astrophys. Opacity Library Los Alamos*, no. **6760-M**.

Humphreys, R. M. 1989 In *Physics of Luminous Blue Variables*, eds. K. Davidson, A. J. F. Moffatt & H. J. G. L. M. Lamers, (Dordrecht: Kluwer), p. 3.

Humphreys, R. M. 1992 In *Instabilities in evolved super- and hyper-giant stars*, eds. C. de Jager & H. Nieuwenhuijzen, (Amsterdam: North-Holland), p. 13.

Humphreys, R. M., Davidson, K. 1979 *ApJ* **232**, 409.

Humphreys, R. M., Davidson, K. 1984 *Science* **223**, 343.

Humphreys, R. M., McElroy, D. B. 1984 *ApJ* **284**, 565.

Iben, I. Jr. 1963 *ApJ* **138**, 1090.

Iben, I. Jr. 1968 *Nature* **220**, 143.

Iben, I. Jr. 1972 *ApJ* **178**, 433.

Iben, I. Jr. 1974 *ARA&A* **12**, 215.

Iben, I. Jr. 1975a *ApJ* **196**, 525.

IBEN, I. JR. 1975b *ApJ* **196**, 549.
IBEN, I. JR. 1976 *ApJ* **208**, 165.
IBEN, I. JR. 1981 *ApJ* **246**, 278.
IBEN, I. JR. 1982 *ApJ* **253**, 248.
IBEN, I. JR. 1984 *ApJ* **277**, 333.
IBEN, I. JR. 1985 *Q.J.R.A.S.* **26**, 1.
IBEN, I. JR. 1987 See Kwok & Pottasch, 1987, p. 175.
IBEN, I. JR. 1988 In *Progress and Opportunities in Southern Hemisphere Optical Astronomy*, eds. V. M. Blanco & M. M. Phillips, *ASP Conf. Ser.* **vol. 1**, p. 220.
IBEN, I. JR. 1989 In *Evolution of Peculiar Red Giants*, eds. H. R. Johnson & B. Zuckerman, (Cambridge: Cambridge Univ. Press), p. 205.
IBEN, I. JR. 1991 *ApJS* **76**, 55.
IBEN, I. JR. 1993 *ApJ* **415**, 767.
IBEN, I. JR., KALER, J. B., TRURAN, J. W., RENZINI, A. 1983 *ApJ* **264**, 605.
IBEN, I. JR., MACDONALD, J. 1985 *ApJ* **296**, 615.
IBEN, I. JR., MACDONALD, J. 1986 *ApJ* **301**, 164.
IBEN, I. JR., RENZINI, A. 1982 *ApJ* **259**, L79.
IBEN, I. JR., RENZINI, A. 1983 *ARA&A* **21**, 271.
IBEN, I. JR., RENZINI, A. 1984 *Phys. Rep.* **105 (6)**, 329.
IBEN, I. JR., TRURAN, J. 1978 *ApJ* **220**, 980.
IBEN, I. JR., TUGGLE, R. S. 1972a *ApJ* **173**, 135.
IBEN, I. JR., TUGGLE, R. S. 1972b *ApJ* **178**, 441.
IBEN, I. JR., TUGGLE, R. S. 1975 *ApJ* **197**, 39.
IGLESIAS, C. A., ROGERS, F. J. 1991a *ApJ* **371**, 408.
IGLESIAS, C. A., ROGERS, F. J. 1991b *ApJ* **371**, L73.
IGLESIAS, C. A., ROGERS, F. J. 1996 *ApJ* **464**, 943.
IGLESIAS, C. A., ROGERS, F. J., WILSON, B. G. 1987 *ApJ* **322**, L45.
IGLESIAS, C. A., ROGERS, F. J., WILSON, B. G. 1990 *ApJ* **360**, 221.
IGLESIAS, C. A., ROGERS, F. J., WILSON, B. G. 1992 *ApJ* **397**, 717.
IGLESIAS, C. A., WILSON, B. G., ROGERS, F. J., GOLDSTEIN, W. H., BAR-SHALOM, A., OREG, J. 1995 *ApJ* **445**, 855.
IZOTOV, Y. I., THUAN, T. X., LIVOVETSKY, V. A. 1994 *ApJ* **435**, 647.
JANES, K., ED. 1991 *The Formation and Evolution of Star Clusters*, *ASP Conf. Ser.* **vol. 13**.
JURA, M., KLEINMANN S. G. 1990 *ApJS* **73**, 769.
KATO, S. 1966 *PASJ* **18**, 374.
KETTNER, K. U., BECKER, H. W., BUCHMAN, L., GORRES, J., KRAVINKEL, H., et al. 1982 *Z. Phys. A - Atoms and Nuclei* **308**, 73.
KING, C. R., DEMARQUE, P., GREEN, E. M. 1988 In *Calibration of Stellar Ages*, ed. A. G. D. Philip, (Schenectady: L. Davis), p. 211.
KRAFT, R. P. 1994 *PASP* **106**, 553.
KRAUSS, L. M., ROMANELLI, P. 1990 *ApJ* **358**, 47.
KUDRITZKI, R. P., GABLER, R., GROTH, H. G., PAULDRACH, A. W., PULS, J. 1989 In *Physics of Luminous Blue Variables*, eds. K. Davidson, A.F.J. Moffatt, & H. J. G. L. M. Lamers, (Dordrecht: Reidel), p. 67.
KUDRITZKI, R. P., SIMON, K. P., HAMANN, W. R. 1983 *A&A* **118**, 245.
KWOK, S. 1975 *ApJ* **198**, 583.
KWOK, S., POTTASCH, S. R., ED. 1987 *Late Stages of Stellar Evolution*, (Dordrecht: Reidel).

LAFON, J. P. J., BERRUYER, N. 1991 *A&A Rev.* **2**, 249.

LAMB, S. A., IBEN, I. JR., HOWARD, W. M. 1976 *ApJ* **207**, 209.

LAMERS, H. J. G. L. M. 1981 *ApJ* **245**, 593.

LAMERS, H. J. G. L. M. 1986 In *Luminous Stars and Associations in Galaxies*, eds. P.S. Conti, C. de Loore & E. Kontizas, (Dordrecht: Reidel), p. 157.

LAMERS, H. J. G. L. M., FITZPATRICK, E. 1988 *ApJ* **324**, 279.

LAMERS, H. J. G. L. M., MAEDER, A., SCHMUTZ, W., CASSINELLI, J.P. 1991 *ApJ* **368**, 538.

LAMERS, H. J. G. L. M., NOORDHOEK, R. 1993 In *Massive Stars and Their Lives in the Interstellar Medium*, eds. J. P. Cassinelli & E. Churchwell, *ASP Conf. Ser.* **vol. 35**, 517.

LANGANKE, K., KOONIN, S. E. 1982 *Nucl. Phys. A* **410**, 334.

LANGER, N. 1986 *A&A* **164**, 45.

LANGER, N. 1989a *A&A* **210**, 93.

LANGER, N. 1989b *A&A* **220**, 135.

LANGER, N. 1991 *A&A* **252**, 669.

LANGER, N., EL EID, M. F., BARAFFE, I. 1989 *A&A* **224**, L17.

LANGER, N., EID M. F. E., FRICKE, K. J. 1985 *A&A* **145**, 179.

LANGER, G. E. HOFFMAN R.D. 1995 *PASP* **107**, 1177.

LANGER, N., MAEDER, A. 1995 *A&A* **295**, 685L.

LARSON, R. 1990 *PASP* **102**, 709.

LATTANZIO, J. C. 1986 *ApJ* **311**, 708.

LATTANZIO, J. C. 1987a See Kwok & Pottasch, 1987, p. 235.

LATTANZIO, J. C. 1987b *ApJ* **313**, L15.

LATTANZIO, J. C. 1988a In *Evolution of Peculiar Red Giant Stars*, eds. H. R. Johnson & B. K. Zuckerman, (Cambridge: Cambridge Univ. Press), p. 131.

LATTANZIO, J. C. 1988b In *Origin and Distribution of the Elements*, ed. G. J. Mathews (Singapore: World Scientific), p. 398.

LATTANZIO, J. C. 1989 *ApJ* **344**, L25.

LATTANZIO, J. C. 1991 *ApJ* **76**, 215.

LATTANZIO, J. C., VALLENARI, A., BERTELLI, G., CHIOSI, C. 1991 *A&A* **250**, 340.

LAUTERBORN, D., REFSDAL, S., WEIGERT, A. 1971 *A&A* **10**, 97.

LEDOUX, P. 1947 *ApJ* **94**, 537.

LEE, Y. W. 1990 *ApJ* **363**, 159.

LEE, Y. W. 1991a See Janes, 1991, p. 205.

LEE, Y. W. 1991b *ApJ* **367**, 524.

LEE. Y. W., DEMARQUE, P. 1990 *ApJ* **73**, 709.

LEE, Y. W., DEMARQUE, P., ZINN, R. J. 1987 In *The Second Conference on Faint Blue Stars*, eds. A. G. D. Philip, D. S. Hayes & J. W. Liebert, (Schenectady: L. Davies), p. 137.

LEE, Y. W., DEMARQUE, P., ZINN, R. J. 1990 *ApJ* **350**, 155.

LEQUEUX, J. 1990 In *From Red Giants to Planetary Nebulae, Which Path for Stellar Evolution?*, eds. M. O. Mennessier & A. Omont, (Gif-sur-Yvette: Ed. Frontieres), p. 271.

LIEBERT, J., SAFFER, R. A., GREEN, E. M. 1994 *AJ* **107**, 1408.

LIU, T., JANES, K. A. 1990 *ApJ* **354**, 273.

LONGMORE, A. J., DIXON, R., SKILLEN, I., JAMESON, R. F., FERNLEY, J. A. 1990 *MNRAS* **247**, 695.

LOVY, D., MAEDER, A., NOELS, A., GABRIEL, M. 1984 *A&A* **133**, 307.

LUCY, L. B., ABBOTT, D. C. 1993 *ApJ* **405**, 738.

LUCY, L. B., SOLOMON, P. M. 1970 *ApJ* **159**, 879.

MADORE, B., FREEDMAN W. L. 1991 *PASP* **103**, 933.

MAEDER, A. 1975 *A&A* **40**, 303.

MAEDER, A. 1981a *A&A* **101**, 385.

MAEDER, A. 1981b *A&A* **102**, 401.

MAEDER, A. 1981c In *The Most Massive Stars*, eds. S. D'Odorico, D. Baade & K. Kjar, (Garching: ESO), p. 173.

MAEDER, A. 1983a *A&A* **120**, 113.

MAEDER, A. 1983b *A&A* **120**, 130.

MAEDER, A. 1983c In *Primordial Helium*, eds. P.A. Shaw, D. Kunt & K. Kjar, (Garching: ESO), p. 89.

MAEDER, A. 1990 *A&AS* **84**, 139.

MAEDER, A. CONTI, P. S. 1994 *ARA&A* **32**, 227.

MAEDER, A., LEQUEUX, J. 1982 *A&A* **114**, 409.

MAEDER, A., MERMILLIOD, J. C. 1981 *A&A* **93**, 136.

MAEDER, A., MEYNET, G. 1987 *A&A* **182**, 243.

MAEDER, A., MEYNET, G. 1988 *A&AS* **76**, 411.

MAEDER, A., MEYNET, G. 1989 *A&A* **210**, 155.

MAEDER, A., MEYNET, G. 1991 *A&AS* **89**, 451.

MAGEE, N. H. MERTZ, A. L., HUEBNER, W. F. 1975 *ApJ* **196**, 617.

MAGEE, N. H. MERTZ, A. L., HUEBNER, W. F. 1984 *ApJ* **283**, 264.

MAGNI, G., MAZZITELLI, I. 1979 *A&A* **72**, 134.

MARIGO, P., BRESSAN, A., CHIOSI, C. 1996a *A&A* **313**, 545.

MARIGO, P., BRESSAN, A., CHIOSI, C. 1997 *A&A* submitted.

MARIGO, P., GIRARDI, L., CHIOSI, C. 1996b *A&A* **316**, L1.

MASSEY, P., LANG, C. C., DEGIOIA-EASTWOOD, K., GARMANY. C. 1995 *ApJ* **438**, 188.

MATEO, M., OLSZEWSKI, E., MADORE, B. F. 1990 *ApJ* **107**, 203.

MATEO, M., OLSZEWSKI, E. W., MADORE, B. F. 1991 In *Confrontation between Stellar Pulsation and Evolution*, eds. C. Cacciari & G. Clementini, *ASP Conf. Ser.* **vol. 11**, p. 214.

MATHEWS, G. I., SCHRAMM, D. N., MEYER, B. S. 1993 *ApJ* **404**, 476.

MATRAKA, B., WASSERMANN, C., WEIGERT, A. 1982 *A&A* **107**, 283.

MAZUREK, T. J. 1972 PHD. Thesis, Yeshiva University.

MAZUREK, T. J., MEYER, D. L., WHEELER, J. C. 1977 *ApJ* **215**, 518.

MAZUREK, T. J., TRURAN, J. W., CAMERON, A. G. W. 1974 *Astr. Space Sci.* **27**, 261.

MAZUREK, T. J., WHEELER, J. C. 1980 *Fundamentals of Cosmic Physics* **5**, 193.

MAZZEI, P., PIGATTO, L. 1988 *A&A* **193**, 148.

MAZZEI, P., PIGATTO, L. 1989 *A&A* **213**, L1.

MAZZITELLI, I., D'ANTONA, F., CALOI, V. 1995 *A&A* **302**, 384.

MCCLURE, R. D., VANDENBERG, D. A., BELL, R. A., HESSER, J. E., STETSON, P. B. 1987 *AJ* **93**, 1144.

MCCLURE, R. D., VANDENBERG, D. A., SMITH, G. H., FAHLMAN, G. G., RICHER, et al. 1986 *ApJ* **307**, L49.

MCWILLIAM, A., GEISLER, D., RICH, R. M. 1992 *PASP* **104**, 1193.

MENGEL, J. G., GROSS, P. G. 1976 *Astrophys. Space Sci.* **41**, 407.

MENGEL, J. G., SWEIGART, A. V. 1981. In *Astrophysical Parameters for Globular Clusters*, eds. A. G. D. Philip, & D. S. Hayes, (Schenectady: L. Davis), p. 277.

MERMILLIOD, J. C., MAEDER, A. 1986 *A&A* **158**, 45.

MEYLAN, G., MAEDER, A. 1982 *A&A* **108**, 148.

MEYNET, G. 1991 In *Instabilities in Evolved Super and Hypergiants*, eds. C. de Jager & H. Nieuwenhuijzen, (Amsterdam: North Holland).

MIYAJI, S., NOMOTO, K., YOKOI, K., SUGIMOTO, D. 1980 *PASJ* **32**, 303.

MOULD, J., AARONSON, M. 1982 *ApJ* **263**, 629.

NAPIWOTZKI, R., SCHOENBERNER, D., WEIDMANN, V. 1991 *A&A* **243**, L5.

NASI, E., FORIERI, C. 1990 *Astrophys. Space Sci.* **166**, 229.

NG, Y. K. 1997 *A&A*, submitted.

NG, Y. K., BERTELLI, G., CHIOSI, C., BRESSAN, A. 1996 *A&A* **310**, 771.

NISSEN, P. E., EDVARDSSON, B., GUSTAFSSON, B. 1985 In *Production and Distribution of C, N and O Elements*, eds. I. J. Danziger, F. Matteucci & K. Kjar, (Garching: ESO), p. 131.

NOELS, A., MAGAIN, E. 1984 *A&A* **139**, 341.

NOELS, A., MAZAREL, C. 1982 *A&A* **105**, 293.

NOMOTO, K. 1981 In *Fundamental Problems in the Theory of Stellar Evolution*, eds. D. Sugimoto, D. Q. Lamb & D. N. Schramm, (Dordrecht: Reidel), p. 295.

NOMOTO, K. 1983 In *Supernova Remnants and Their X-Ray Emission*, eds. J. Danziger & P. Gorenstein, (Dordrecht: Reidel), p. 139.

NOMOTO, K. 1984a In *Stellar Nucleosynthesis*, eds. C. Chiosi & A. Renzini, (Dordrecht: Reidel), p. 239.

NOMOTO, K. 1984b *ApJ* **277**, 791.

NOMOTO, K. 1984c In *Problems of Collapse and Numerical Relativity*, eds. D. Bancel & M. Signore, (Dordrecht: Reidel), p. 89.

NOMOTO, K. 1984d In *Stellar Nucleosynthesis*, eds. C. Chiosi & A. Renzini, (Dordrecht: Reidel), p. 205.

NOMOTO, K. 1985 In *12th Texas Symposium on Relativistic Astrophysics* (New York: Acad. Sci).

NOMOTO, K., SUGIMOTO, D., NEO, S. 1976 *Astrophys. Space Sci.* **39**, L37.

NORDSTROM, B., ANDERSEN, J., ANDERSEN, M. I. 1996 *A&AS* **118**, 407.

OBER, W. W., EL EID, W., FRICKE, K. J. 1983 *A&A* **119**, 61.

ONO, Y. 1960 *Progr. Theor. Phys. Kyoto* **24**, 825.

OLIVE, K. A., STEIGMAN, G. 1995 *ApJS* **97**, 49.

OLIVE, K. A., STEIGMAN, G., WALKER, T. P. 1991 *ApJ* **380**, L1.

OSTRIKER, J. P., RICHSTONE, D. O., THUAN, T. X. 1974 *ApJ* **188**, L87.

ORTOLANI, S., PIOTTO, G.P., CAPACCIOLI, M. 1989 *The Messenger* **56**, 54.

OWOCKI, S. P., CASTOR, J. I., RYBICKI, G. B. 1988 *ApJ* **335**, 914.

PACZYNSKI, B. 1970a *Acta Astron.* **20**, 47.

PACZYNSKI, B. 1970b *Acta Astron.* **20**, 287.

PACZYNSKI, B. 1971 *Acta Astron.* **21**, 417.

PACZYNSKI, B. 1972 *Astrophys. Lett.* **11**, 53.

PACZYNSKI, B. 1984 *ApJ* **284**, 670.

PAGEL, B. E. J., SIMONSON, E. A., TERLEVICH, R. J., EDMUNDS, M. G. 1992 *MNRAS* **255**, 325.

PANAGIA, N., GILMOZZI, R., MACCHETTO, F., ADORF, H. M., KIRSHNER, R. P. 1991 *ApJ* **380**, L23.

PAULDRACH, A., PULS, J., KUDRITZKI, R. P. 1986 *A&A* **164**, 86.

PAULDRACH, A., PULS, J., KUDRITZKI, R. P. 1988 In *O-Stars and WR stars, NASA SP-497*, eds. P. S. Conti & A. B. Underhill, (Washington: NASA), p. 173.

PEL, J. W. 1985 In *Cepheids, Theory and Observations*, ed. B. F. Madore (Cambridge: Cambridge Univ. Press), p. 1.

PERINOTTO, M. 1983 In *Planetary Nebulae*, ed R. D. Fowler, (Dordrecht: Reidel), p. 323.
PETERSEN, J. O. 1989 *A&A* **226**, 151.
PETERSEN, J. O. 1990 *A&A* **238**, 160.
PETERSON, R. C. 1985 *ApJ* **294**, L35.
PILACHOWSKI, C. A. 1984 *ApJ* **281**, 614.
PILACHOWSKI, C. A., SNEDEN, C., WALLERSTEIN, G. 1983 *ApJS* **52**, 241.
PIOTTO, G. P. 1991 See Janes, 1991, p. 200.
PIJPERS, F. P. 1990 PHD. Thesis, Univ. Leiden.
PIJPERS, F. P., HABING H. J. 1989 *A&A* **215**, 334.
PIJPERS, F. P., HEARN, A. G. 1989 *A&A* **209**, 198.
PITERS, A., DE JAGER, C., NIEUWENHUIJZEN, H. 1988 *A&A* **196**, 115.
PROFFITT, C. R. 1993 In *Inside the Stars*, eds. W. W. Weiss & A. Baglin *ASP Conf. Ser.* **40**, 451.
PROFFITT, C. R., MICHAUD, G., RICHER, J. 1990 In *Cool Stars, Stellar Systems, and the Sun*, ed. G. Wallerstein, *ASP Conf. Ser.* **vol. 9**, p. 351.
PROFFITT, C. R., VANDENBERG D. A. 1991 *ApJS* **77**, 473.
PRYOR, C., SMITH, G. H., MCCLURE, R. D. 1986 *AJ* **92**, 1358.
RAKAVY, G., SHAVIV, G., ZINAMON, Z. 1967 *ApJ* **150**, 131.
RATCLIFF, S. J. 1987 *ApJ* **318**, 196.
REID, N., TINNEY, C., MOULD, J. 1990 *ApJ* **348**, 98.
REIMERS, D. 1975 *Mem. Soc. R. Sci. Liege* **6** (8), 369.
RENZINI, A. 1977 In *Advanced Stages of Stellar Evolution*, eds. P. Bouvier & A. Maeder, (Geneva: Geneva Obs), p.151.
RENZINI, A. 1979 In *Stars and Stellar Systems*, ed. B. E. Westerlund, (Dordrecht: Reidel), p. 155.
RENZINI, A. 1981a In *Physical Processes in Red Giants*, eds. I. Iben & A. Renzini, (Dordrecht: Reidel), p. 431.
RENZINI, A. 1981b In *Physical Processes in Red Giants*, eds. I. Iben & A. Renzini, (Dordrecht: Reidel), p. 165.
RENZINI, A. 1981c *Ann. Phys. Fr.* **6**, 87.
RENZINI, A. 1982 In *Wolf Rayet Stars*, eds. C. de Loore & A. J. Willis, (Dordrecht: Reidel), p. 413.
RENZINI, A. 1984a In *Observational Tests of the Stellar Evolution Theory*, eds. A. Maeder & A. Renzini, (Dordrecht: Reidel), p. 21.
RENZINI, A. 1984b In *Stellar Nucleosynthesis*, eds. C. Chiosi & A. Renzini, (Dordrecht: Reidel), p. 99.
RENZINI, A. 1986 In *Stellar Populations*, eds. C. Norman, A. Renzini & M. Tosi, (Cambridge: Cambridge Univ. Press), p. 73.
RENZINI, A. 1987 *A&A* **188**, 49.
RENZINI, A. 1988 In *Globular Cluster Systems in Galaxies*, eds. J. A. Grindlay & A. G. Philip, (Dordrecht: Kluwer), p. 443.
RENZINI, A. BUZZONI, A. 1983 *Mem. Soc. astron. It.* **54**, 335.
RENZINI, A. BUZZONI, A. 1986 *In Spectral Evolution of Galaxies*, eds. C. Chiosi, & A. Renzini, (Dordrecht: Reidel), p. 135.
RENZINI, A., GREGGIO, L., RITOSSA, C. 1992 *ApJ* **400**, 280.
RENZINI, A., FUSI-PECCI, F. 1988 *ARA&A* **26**, 199.
RENZINI, A., RITOSSA, C. 1994 **ApJ 433**, 293.
RENZINI, A., VOLI, M. 1981 *A&A* **94**, 175.
RICH, R. M. 1988 *AJ* **95**, 828.

RICHER, H. B., FAHLMAN, G. G. 1984 ApJ **277**, 227.

RICHER, H. B., FAHLMAN, G. G. 1991 See Janes, 1991, p. 120.

RICHER, H. B., FAHLMAN, G. G., BUONANNO, R., FUSI-PECCI, F., SEARLE, L., THOMPSON, I. B. 1991 ApJ **381**, 147.

RICHER, H. B., FAHLMAN, G. G., IBATA, R. A., STETSON, P. B., BELL, R. A., et al. 1995 ApJ **451**, L17.

RICHER, H. B., HARRIS, W., FALHMAN, G. Q., et al. 1996 **ApJ 463**, 602.

RICHER, J., MICHAUD, G., PROFFITT, C. R. 1992 ApJS **82**, 329.

ROBERTSON, J. W. 1974 ApJ **191**, 67.

ROBERTSON, J. W., FAULKNER, D. J. 1972 ApJ **171**, 309.

ROGERS, F. J., IGLESIAS, C. A. 1992 ApJ, **401**, 361.

ROGERS, F. J., IGLESIAS, C. A. 1993 ApJ, **412**, 375.

ROOD, R. T. 1973 ApJ **184**, 815.

ROSE, W. K. 1969 ApJ **155**, 491.

RUSSELL, S. C., BESSELL, M. S. 1989 ApJ **70**, 865.

SACKMANN, I. J. 1980 ApJ **241**, L37.

SALARIS, M., CHIEFFI, A., STRANIERO, O. 1993 ApJ **414**, 580.

SALASNICH, B., BRESSAN, A., CHIOSI, C. 1997 A&A, submitted.

SANDAGE, A. 1958 ApJ **127**, 513.

SANDAGE, A. 1982 ApJ **252**, 553.

SANDAGE, A. 1986 ARA&A **24**, 421.

SANDAGE, A. 1988 In *Calibration of Stellar Ages*, ed. A. G. Philip, (Schenectady: L. Davis), p. 43.

SANDAGE, A. 1990a ApJ **350**, 631.

SANDAGE, A. 1990b J. R. Astron. Soc. Can. **84 (2)**, 70.

SANDAGE, A., CACCIARI, C. 1990 ApJ **350**, 645.

SANDAGE, A., TAMMANN, G. A. 1969 ApJ **157**, 683.

SARAJEDINI, A. 1991 In *Precision Photometry, Astrophysics of the Galaxy*, ed. A. G. D. Philip, (Schenectady: L. Davies).

SARAJEDINI, A., DEMARQUE, P. 1990 ApJ **365**, 219.

SARAJEDINI, A., KING, C. R. 1989 AJ **98**, 1624.

SARAJEDINI, A., LEDERMAN, A. 1991 See Janes, 1991, p. 293.

SCALO, J. M. 1976 ApJ **206**, 215.

SCHALLER, G., SCHAERER, D., MEYNET, G., MAEDER, A. 1992 A&AS **96**, 296.

SCHAERER, D., CHARBONNEL, C., MEYNET G., MAEDER, A., SCHALLER, G. 1993a A&AS **102**, 339.

SCHAERER, D., MAEDER, A. 1992 A&A **263**, 129.

SCHAERER, D., MEYNET G., MAEDER, A., SCHALLER, G. 1993b A&AS **98**, 523.

SCHMIDT, E. G. 1984 ApJ **285**, 501.

SCHOENBERNER, D. 1979 A&A **79**, 108.

SCHOENBERNER, D. 1981 A&A **103**, 119.

SCHOENBERNER, D. 1983 ApJ **272**, 708.

SCHOENBERNER, D. 1987 See Kwok & Pottasch, 1987, p. 337.

SCHMUTZ, W., HAMANN, W. R., WESSELOWSKI, U. 1989 A&A **210**, 236.

SCHWARZSCHILD, M. 1958 *Structure and Evolution of the Stars*, (Princeton: Princeton Univ. Press).

SCHWARZSCHILD, M. 1970 Q. J. R. Astron. Soc. **11**, 12.

SCHWARZSCHILD, M., HARM, R. 1958 *ApJ* **142**, 855.
SEARLE, L., ZINN, R. J. 1978 *ApJ* **225**, 357.
SEARLE, L., WILKINSON, A., BAGNUOLO, W. G. 1980 *ApJ* **239**, 803.
SEATON, M. J. 1987 *J. Phys. B* **20**, 6363.
SEATON, M. J. 1991 *J. Phys. B* **23**, 3255.
SHAVIV, G., SALPETER, E. E. 1973 *ApJ* **184**, 191.
SIMON, N. R. 1982 *ApJ* **260**, L87.
SIMON, N. R. 1987 In *Pulsation and Mass Loss in Stars*, eds. R. Stalio & L. A. Willson, (Dordrecht: Reidel), p. 27.
SKINNER, C. J., WHITEMORE, B. 1988 *MNRAS* **231**, 169.
SMITH, L. F. 1973 In *Wolf Rayet Stars and High Temperature Stars*, eds. M. K. V. Bappu & J. Sahade, (Dordrecht: Holland), p. 15.
SMITH, V. V., LAMBERT, D.L. 1990 *Apj* **361**, L69.
SMITH, L. F., MAEDER, A. 1989 *A&A* **211**, 71.
SMITH, V. V., PLEZ, B., LAMBERT, D. L. 1995 *ApJ* **441**, 735.
SNEDEN C., KRAFT, R. P., PROSSER, C. F., LANGER, G. E. 1991 *AJ* **102**, 2001.
SPARKS, W. H., ENDAL, A. S. 1980 *ApJ* **237**, 130.
SPITE, M., SPITE, F. 1991 *A&A* **252**, 689.
STELLINGWERF, R. F. 1978 *AJ* **83**, 1184.
STETSON, P. B. 1993 In *The Globular Cluster-Galaxy Connection*, eds. G. H. Smith & J. P. Brodie, *ASP Conf. Ser.* **48**, 14.
STETSON, P. B., HARRIS, W. E. 1988 *AJ* **96**, 909.
STETSON, P. B., VANDENBERG, D. A., BOLTE, M. 1996 *PASP* **108**, 560.
STETSON, P. B., VANDENBERG, D. A., BOLTE, M., HESSER, J. E., SMITH, G. H. 1989 *AJ* **97**, 1360.
STIAVELLI, M., PIOTTO, G. P., CAPACCIOLI, M., ORTOLANI, S. 1991 See Janes, 1991, p. 449.
STOTHERS, R. 1976 *ApJ* **209**, 800.
STOTHERS, R. 1985 *ApJ* **298**, 521.
STOTHERS, R. 1991 *ApJ* **383**, 820.
STOTHERS, R., CHIN, C. W. 1977 *ApJ* **211**, 189.
STOTHERS, R., CHIN, C. W. 1978 *ApJ* **225**, 939.
STOTHERS, R., CHIN, C. W. 1981 *ApJ* **247**, 1063.
STOTHERS, R., CHIN, C. W. 1990 *ApJ* **348**, L21.
STOTHERS, R., CHIN, C. W. 1991 *ApJ* **374**, 288.
STOTHERS, R., CHIN, C. W. 1994 *ApJ* **426**, 143.
STRANIERO, O., CHIEFFI, A. 1991 *ApJS* **76**, 525.
STRANIERO, O., GALLINO, R., BUSSO, R., CHIEFFI, A., RAITERI C., LIMONGI C., SALARIS, M. 1995 *ApJ* **440**, L85.
SUGIMOTO, D., NOMOTO, K. 1980 *Space Sci. Rev.* **25**, 155.
SWEIGART, A. V. 1987 *ApJS* **65**, 95.
SWEIGART, A. V., DEMARQUE, P. 1972 *A&A* **20**, 445.
SWEIGART, A. V., GREGGIO, L., RENZINI, R. 1990 *ApJ* **364**, 527.
SWEIGART, A. V., GROSS, P. G. 1976 *ApJS* **32**, 367.
SWEIGART, A. V., GROSS, P. G. 1978. *ApJS* **36**, 405.
SWEIGART, A. V., RENZINI, A., TORNAMBE', A. 1987 *ApJ* **312**, 762.
THIELEMANN, F. K., ARNETT, W. D. 1985 *ApJ* **295**, 604.
TINSLEY, B. M. 1980 In *Fundamentals of Cosmic Physics* **5**, 287.

THOMAS, H. C. 1967 *Zs. Ap.* **67**, 420.

TRIPICCO, M. J., DORMAN, B., BELL, R. A. 1993 *AJ* **106**, 618.

TUCHMAN, J., WHEELER, J. C. 1989 *ApJ* **344**, 835.

TUCHMAN, J., WHEELER, J. C. 1990 *ApJ* **363**, 255.

TUROLLA, R., NOBILI, L., CALVANI, M. 1988 *ApJ* **324**, 899.

VALLENARI, A., CHIOSI, C., BERTELLI, G., MEYLAN, G. ORTOLANI, S. 1991 *A&AS* **87**, 517.

VALLENARI, A., CHIOSI, C., BERTELLI, G., MEYLAN, G., ORTOLANI, S. 1992 *AJ* **105**, 1100.

VALLENARI, A., APARICIO, A., FAGOTTO, F., CHIOSI, C., ORTOLANI, S., MEYLAN, G. 1994, *A&A* **284**, 447.

VALLENARI, A., CHIOSI, C., BERTELLI, G., ORTOLANI, S. 1996a *A&A* **309**, 358.

VALLENARI, A., CHIOSI, C., BERTELLI, G., ORTOLANI, S. 1996b *A&A* **309**, 367.

VALLENARI, BETTONI, D., CHIOSI, C. 1997 *A&A* submitted.

VANBEVEREN, D. 1987 *A&A* **182**, 207.

VANBEVEREN, D. 1995 *A&A* **294**, 107.

VANDENBERG, D. A. 1983 *ApJS* **51**, 29.

VANDENBERG, D. A. 1985 *ApJS* **58**, 781.

VANDENBERG, D. A. 1991 See Janes, 1991, p. 183.

VANDENBERG, D. A. 1992 *ApJ* **391**, 685.

VANDENBERG, D. A., BELL, R. A. 1985 *ApJS* **58**, 561.

VANDENBERG, D. A., BOLTE, M., STETSON, P. B. 1996 *ARA&A* **34**, 461.

VANDENBERG, D. A., BRIDGES, T. J. 1984 *ApJ* **278**, 679.

VANDENBERG, D. A., BOLTE, M., STETSON, P. B. 1990 *AJ* **100**, 445.

VANDENBERG, D. A., HARTWICK, F. D. A., DAWSON, P., ALEXANDER, D. R. 1983 *ApJ* **266**, 747.

VANDENBERG, D. A., STETSON, P.B. 1991 *AJ* **102**, 1043.

VANDENBERG, D. A., SWENSON, F. J., ROGERS, F. J., IGLESIAS, C. A., ALEXANDER D. R. 1997 In preparation.

VAN DEN BERGH, S. 1989 *A&ARev.* **vol. 1**, 111.

VAN DER VEEN, W. E. C. J. 1989 *A&A* **210**, 127.

VAN DER VEEN, W. E. C. J., RUGERS, M. 1989 *A&A* **226**, 183.

VASSILIADIS, E., WOOD, P. R. 1993 *ApJ* **413**, 641.

VOLK, K., KWOK, S. 1988 *ApJ* **331**, 435.

WALKER, A. R. 1987 *MNRAS* **225**, 627.

WALKER, A. R. 1988 In *Extragalactic Distance Scale*, eds. S. van den Bergh & C. J. Pritchett, *ASP Conf. Ser.* **vol. 4**, p. 89.

WEAVER, T. A. 1980 In *Energy and Technology Review*, (Laurence Livermore Laboratory), UCRL: 52000-80-2

WEAVER, T. A., WOOSLEY, S. E. 1980 *Ann. N. Y. Acad. Sci.* **336**, 335.

WEAVER, T. A., WOOSLEY, S. E., FULLER, G. M. 1982 *Bull. A. A. S.* **14**, 957.

WEAVER, T. A., WOOSLEY, S. E., FULLER, G. M. 1985 In *Numerical Astrophysics*, eds. J. Centrella, J. LeBlanc & R. Bowers, (Boston: Jones & Bartlett), p. 374.

WEBBINK, R. F. 1985 In *Dynamics of Star Clusters*, eds. J. Goodman & P. Hut, (Dordrecht: Reidel), p. 541.

WEIDEMANN, V. 1990 *ARA&A* **28**, 103.

WEISS, A. 1989a *A&A* **209**, 135.

WEISS, A. 1989b *ApJ* **339**, 365.

WEISS, A., KEADY, J. J., MAGEE, N. H. 1990 *At. Data and Nucl. Data Tables* **45**, 209.

WELCH, D. L., MATEO, M., COTE', P., FISHER, P., MADORE, B. 1991 *AJ* **101**, 490.

WESTERLUND, B. E., LEQUEUX, J., AZZOPARDI, M., REBEIROT, E. 1991 *A&A* **244**, 367.

WILLSON, L. A. 1988 In *Pulsation and Mass Loss in Stars*, eds. R. Stalio, & L. A. Willson, (Dordrecht: Reidel), p. 285.

WILLSON, L. A., BOWEN, G. H. 1984 *Nature* **312**, 429.

WOOD, P. R., BESSELL, M. S., FOX, M. W. 1983 *ApJ* **272**, 99.

WOOD, D. O. S., CHURCHWELL, E. 1989 *ApJ* **340**, 265.

WOOD, P. R., FAULKNER, D. J. 1986 *ApJ* **307**, 659.

WOOD, P. R., VASSILIADIS, E. 1991 In *Highlights of Astronomy*, **vol. 9**.

WOOD, P. R., WHITEOAK, J. B., HUGHES S. M. G., BESSEL, N. S., GARDNER, F. F., HYLAND, A. R. 1992 *ApJ* **397**, 552.

WOOSLEY, S. E. 1986 In *Nucleosynthesis and Stellar Evolution*, 16th Saas-Fee Course, eds. B. Hauck, A. Maeder & G. Meynet, (Geneva: Geneva Observatory), p. 1.

WOOSLEY, S. E. 1988 *ApJ* **330**, 218.

WOOSLEY, S. E., AXELROD, R. S., WEAVER, T. A. 1984 In *Stellar Nucleosynthesis*, eds. C. Chiosi & A. Renzini, (Dordrecht: Reidel), p. 262.

WOOSLEY, S. E., WEAVER, T. A. 1982a In *Supernovae: A Survey of Current Research*, eds. M. J. Rees & R. J. Stoneham, (Dordrecht: Reidel), p. 79.

WOOSLEY, S. E., WEAVER, T. A. 1982b In *Essays in Nuclear Astrophysics*, eds. C. A. Barnes, D. D. Clayton & D. N. Schramm, (Cambridge: Cambridge Univ. Press), p. 337.

WOOSLEY, S. E., WEAVER, T. A. 1985 In *Nucleosynthesis and Its Implications on Nuclear and Particle Physics*, eds. J. Audouze & T. van Than, (Dordrecht: Reidel).

WOOSLEY, S. E., WEAVER, T. A. 1986a *ARA&A* **24**, 205.

WOOSLEY, S. E., WEAVER, T. A. 1986b In *Radiation Hydrodynamics in Stars and Compact objects*, eds. D. Mihalas & K.H. Winkler, (Dordrecht: Reidel).

WOOSLEY, S. E., WEAVER, T. A., TAAM, R. E. 1980 In *Type I Supernovae*, ed. J. C. Wheeler, (Austin: Univ. of Texas), p. 96.

XIONG, D. R. 1983 *A&A* **150**, 133.

XIONG, D. R. 1986 *A&A* **167**, 239.

XIONG, D. R. 1989 *A&A* **213**, 176.

XIONG, D. R. 1990 *A&A* **232**, 31.

ZAHN, J. P. 1987 In *Instabilities in Luminous Early Type Stars*, eds. C. de Loore & H. J. G. L. M. Lamers, (Dordrecht: Reidel), p. 143.

ZAHN, J. P. 1992 *A&A* **265**, 115.

ZINN, R. J. 1985 *ApJ* **293**, 424.

ZINN, R. J., WEST, M. J. 1984 *ApJ* **55**, 45.

ZIJSTRA, A. A., LOUP, C., WATERS, L. B. F. M., WHITELOCK, P. A., VAN LOON J. TH., GUGLIELMO, F. 1996 *MNRAS* **279**, 32.

Observations of the Most Luminous Stars in Local Group Galaxies

By PHILIP MASSEY

Kitt Peak National Observatory, National Optical Astronomy Observatories†,
P. O. 26732, Tucson, AZ 85726-6732 USA

In these lectures I discuss both the difficulties and scientific rewards of observing the most luminous stars in nearby galaxies. Main-sequence, *bolometrically luminous* O-type stars are not to be confused with the *visually brightest* stars in these systems; the first stars encountered in V are late-type B or early-A stars of "modest" mass (15-25 $\mathcal{M}_\odot$), while the most luminous and massive stars (85-100 $\mathcal{M}_\odot$) are several visual magnitudes fainter than these due to their extreme temperatures and bolometric corrections. Because of their high temperatures, the observable colors of main-sequence luminous stars tell you nearly nothing about the actual effective temperature and bolometric corrections: for this, you need spectroscopy. The evolved descendents of the massive stars include the Luminous Blue Variables, Wolf-Rayet stars, and red supergiant (RSG) populations of these galaxies, and issues concerning selection-effects governing their detection are described. Finally, we discuss the results obtained over the past decade in determining the initial mass function (IMF), as well as what the Local Group galaxies are telling us about massive star evolution as a function of metallicity. The massive star IMF appears to be constant for OB associations in the SMC, LMC, and Milky Way (despite the factor of 10 difference in metallicity) with an IMF slope $\Gamma \sim -1.3$ (Salpeter): stars of similarly high mass are found in all three galaxies. There is a substantial population of *field* massive stars, born far from any OB associations, and the IMF slope of these stars is considerably steeper, $\Gamma \sim -4$, although stars of equally high mass are also found in the field, including isolated O3 and O4 stars. This suggests that in regions where the star-formation activity is high the IMF may be biased towards higher mass stars; this is just what is expected on theoretical grounds due to the relation between gas temperature and fragment mass (Larson 1985, 1986). Massive star evolutionary models reproduce the distribution of stars in the H-R diagram of the LMC and SMC very well, once incompleteness is taken into account. We find that the relative proportion of WC and WN stars in Local Group galaxies correlate very well with metallicity, in accord with the "Conti scenario", in which one expects that the luminosity and mass limit for becoming a WC star will depend upon how much mass-loss a star experiences. The exception is the starburst galaxy IC 10, which has both low metallicity and a high WC/WN ratio, but again this is consistent with the suggestion that high star formation "vigor" favors the production of the highest mass stars by heating up the gas. Studies of coeval populations can be used to diagnose the mass range that turn into Wolf-Rayet stars of various types, while studies of mixed-age populations of massive stars in the Local Group can be used to determine the mass-limit for becoming WR stars. The foreground contamination of Galactic red dwarfs seen against NGC 6822, M 31, and M 33 is emphasized, and a method for distinguishing true red supergiants is described. M 31 lacks the most luminous RSGs, despite a very strong WR population, a result also consistent with the predictions of the Conti scenario: rather than becoming RSGs, moderately high-mass stars are able to become Wolf-Rayet stars thanks to higher-mass loss rates due to their higher initial metallicity.

When Artemio Herrero Davó first contacted me about giving a series of lectures at the Winter School, he wrote that he would like for me to "treat the observational techniques of extragalactic spectroscopy, the problems of classification, the [content of the] stellar associations, and the Wolf-Rayet stars in the Local Group." It was an offer that proved

† NOAO is operated by the Association of Universities for Research in Astronomy, Inc. (AURA) under cooperative agreement with the National Science Foundation.

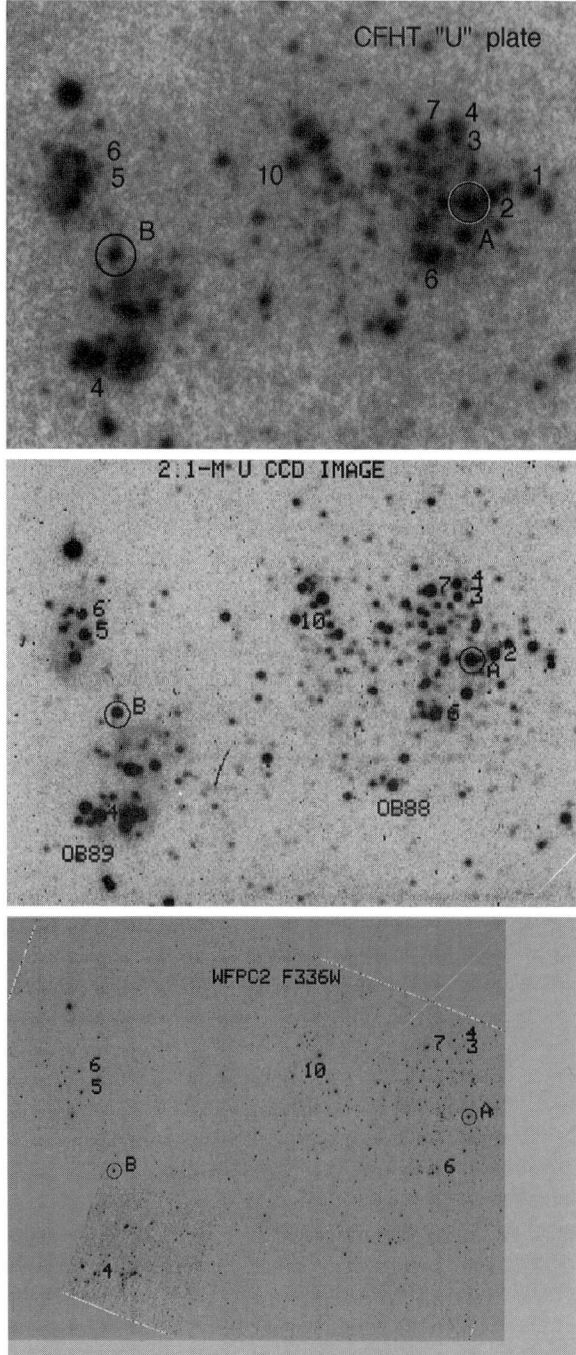

FIGURE 1. Three U-band images of the OB88/89 region in M 33. All three of these were "state-of-the-art" in their day, and show the improvement that has taken place during the past 15 years. The top image is from a CFHT plate taken by Wendy Freedman and Barry Madore circa 1981. The middle image was obtained at the 2.1-m with an RCA CCD during September 1986. Both ground-based image were taken in sub-arcsecond conditions. The bottom image was obtained with the WFPC2 camera on *HST* earlier this year.

irresistible, as this field is one that has been near and dear to my heart since I was the same age as the participants at this School. I got hooked on the idea of doing "galactic astronomy in extragalactic environments" when I was a graduate student in the late 1970's. It was during this time period that Roberta Humphreys' series "Studies of Luminous Stars in Nearby Galaxies" was first appearing (Humphreys 1978, 1979a, 1979b, 1980a, 1980b, 1980c; Humphreys & Davidson 1979). These papers contained spectra of actual extragalactic stars in galaxies beyond the Magellanic Clouds: seemingly individual stars in NGC 6822, IC 1613, M 31, and M 33. I learned everything that I could about the stellar populations in these galaxies by starting with the excellent review papers by Sidney van den Bergh (van den Bergh 1972, 1975; a more recent treatment can be found in van den Bergh 1991). Seminal works were appearing yearly by Paul Hodge (Hodge 1977, 1978, 1979, 1980). I knew I wanted in on this action! Then, in 1980, the last year I was in Boulder, my thesis advisor Peter Conti happened to mention that he and Bruce Bohannan were going to Kitt Peak to *look for Wolf-Rayet stars in M 33* using a low-dispersion objective "grism" at the prime focus of the Mayall 4-m. "You've GOT to take me along!" I argued. Peter was genuinely surprised, and not at first inclined to agree. But, I managed to be enough of a nuisance for the next few months until it was just easier for them to give in. I've never asked if they regretted it or not, but it was there that all the fun really began.

In Fig. 1 I show just how this field has changed since then, using three images which were each "state of the art" in their time. All three are U images of a small region in M 33. The first was taken with the Canada-France-Hawaii Telescope in sub-arcsecond conditions, using a photographic plate. (The image was taken by Wendy Freedman, who kindly allowed me to copy the plate for my own work.) The middle image was taken with an RCA CCD on the Kitt Peak 2.1-m (also in sub-arcsecond conditions), and was the basis for the photometry given in Massey et al. (1995a). The third is a recent image obtained with the "PC" chip of WFPC2 on the *Hubble Space Telescope*. One of the very comforting things that this comparison shows is that stars that we concluded were probably single from good, ground-based imaging generally continued to appear single even at *HST* resolution! As Rolf Kudritzki has emphasized in his lectures, generally we also will have *spectral* indications if a star is composite; that generally, we have not been fooling ourselves. Star OB88-6 (on the right) doesn't look quite stellar in either of the ground-based images, and on the WFPC2 image we find it is indeed multiple; similarly, star OB89-4 (at lower left) is a real mess in all three images. A star like OB88-10, though, which appeared stellar on the plate and on the ground-based CCD frame really looks single even on the WFPC2 image. Star OB89-7 is the only exception within this image: a close examination of the WFPC2 frame reveals that it is a very close double.

What was it that was so appealing to me to be doing studies that were traditionally "galactic" in nearby galaxies? After all, these extragalactic massive stars are far less bright and much more difficult to work on.

Well, for one thing, reddening greatly confuses things in our own Galaxy. Our knowledge of Galactic Wolf-Rayet stars becomes incomplete at distances of only a kpc or so. Therefore we can't really say much about how the numbers and properties of massive stars change in the Milky Way with, say, metallicity. (This situation is slowly improving thanks to IR array photometry and K-band classification of hot stars; for example, Margaret Hanson's work on the M17 region [Hanson 1995; Hanson et al. 1997].) In a galaxy like M 33 you can identify WR stars across the entire face, and hence see how their surface density and properties change with distance from the nucleus. Furthermore, if you believe that some change correlates with M 33's metallicity gradient, you can check this by seeing if the correlation holds when you look at other Local Group galaxies whose metallicities match that at a few different places within M 33. But, I think a more fundamental, philosophical reason is that astronomers never get to do controlled experiments. In order to understand the evolution of massive stars, what one *really* would like to do is start with two balls of gas with the same composition, one of 40 $\mathcal{M}_\odot$ and the other of 80 $\mathcal{M}_\odot$—then stand back and see what happens after a few million years. Will they both pass through a Luminous Blue Variable (LBV) phase? Will each of them turn into Wolf-Rayet (WR) stars? If they do, will they both become WC stars, or will the lower mass one explode as a supernova while still a WN star, owing to its smaller luminosity and hence lower mass-loss rate? If you could repeat this experiment, this time using two balls of 60 $\mathcal{M}_\odot$, but of differing metallicity, you would then have pretty much answered all the stellar evolution questions I've been trying to answer.

Instead, we have to use "our little grey matter" and take our laboratories as we find them. The galaxies of the Local Group that contain significant massive star populations span about a factor of 10 in metallicity. If we want to know how metallicity affects, say, the initial mass function (IMF) or massive star evolution, we have the ability to answer this from the resolved stellar content of these galaxies. This is really possible only because the Local Group does contain galaxies not only of differing metallicity but also multiple example with the *same* metallicity—else, we would have a hard time disentangling IMF differences from the affect of metallicity on massive star evolution, say. (For example, if WC stars come only from the very highest mass stars in regions of low metallicity, but can come from a stars of slightly lower masses in regions of higher metallicity, then the relative proportion of WC and WN stars will depend both on the metallicity and the initial population of stars.) As such, the galaxies of the Local Group indeed do make great laboratories for the study both of star formation and stellar evolution of massive, luminous stars.

The first three sections will be aimed at delineating some of the key issues involved in studying the most luminous stars in galaxies of the Local Group, while the next two sections will present the main results of these studies to date. The final section contains a summary of what I hope that each of the students at this Winter School will take away from these two weeks together.

1. Introducing the unevolved luminous stars

One of the great mysteries to me at the time of reading these early papers presenting spectroscopy of "luminous" stars in distant members of the Local Group was why these stars were *invariably* of late-B and A-type supergiants. My work as a graduate student was on O-type stars, stars which covered many superlatives: the hottest, the shortest-lived, the most massive, and, by extension, *the most luminous*, and yet these beautiful spectra being published by Humphreys and others—although touted as being of the most luminous stars—were never of the kind of stars that I worked on. Why not? As it turns out, the answer to this question is also tied to why it is difficult to tell very much about the most massive stars from photometry alone: the very high temperatures of high-mass, main-sequence stars result in most of their luminosity being in the far ultraviolet. Their high temperatures also render their optical colors (and observable UV colors) nearly useless for actually telling much about the stars themselves.

Many things that we've absorbed from the language and physics of intermediate-mass stars simply don't apply to highly luminous stars. For instance, we all "know" that the mass-luminosity relationship goes something like $L \sim M^{3.5}$, with a corresponding lifetime of $\tau \sim M/L \sim M^{-2.5}$. But this is not the case for high-mass stars! If it were, a star of 100 $\mathcal{M}_\odot$ would live only $10^{-5} \times 10^{10} = 100{,}000$ years, and we probably wouldn't see any

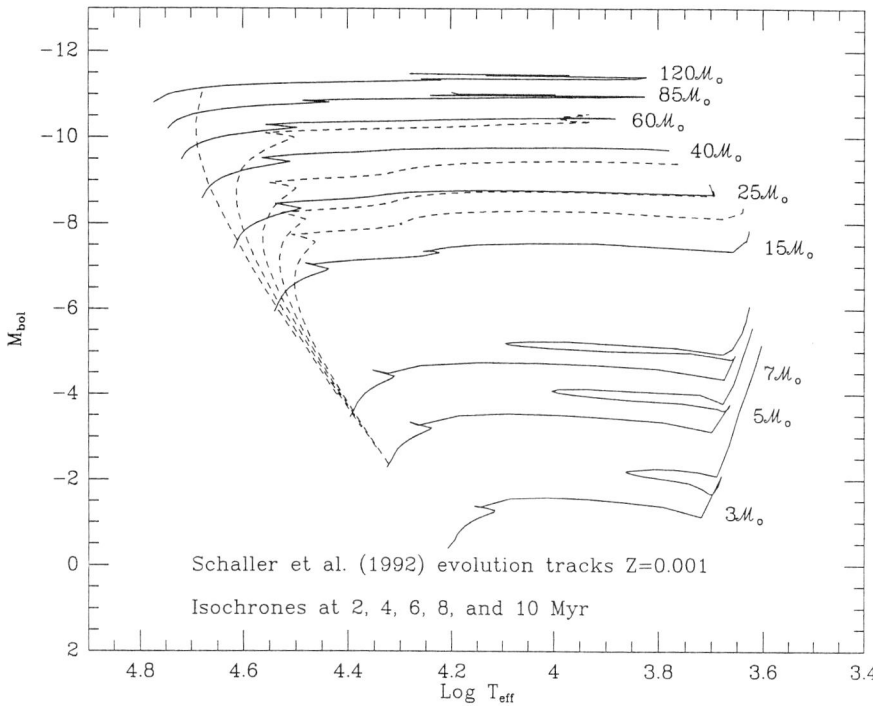

FIGURE 2. The evolutionary tracks of Schaller et al. (1992) are shown for an SMC-like metallicity. The H-burning main-sequence extends from the beginning of the tracks on the left to the first "kink" in each track. The tracks are labeled with the initial stellar masses ("initial" as mass-loss occurs during this evolution). The dashed lines are isocrones at 2 Myr intervals.

of them. Instead, the M-L relationship is less steep at the high-mass end, with $L \sim M^{2.0}$ or so, and all highly massive stars are expected to last at least a couple of Myr on the H-burning main-sequence. (Why is this? The reason has to do with the fact that all the atoms are ionized in very hot stars, so the main source of opacity is electron scattering. As a result, the mean free path has little temperature dependence. See problem 8.1 in Shu 1982.) As we will see later in this section, the colors of these stars don't tell us much about their temperatures. Even odder, the visual absolute magnitude of a star does not give much indication about its bolometric luminosity, in the sense that two stars with the same M_V in M 33 (and with identical colors) might easily be a factor of 10 different (2.5 mags) in bolometric luminosity and hence different in mass by factors of 3 or more. There is even a very significant nomenclature problem: "supergiant" OB stars are in fact *main-sequence* objects, if what we mean by "main-sequence" is that a star is still burning H in its core. For G-type stars this statement certainly isn't true—our sun is a H-burning G2 V, while most G-type luminosity class "III" (giant) stars are *shell* H-burning (and He-burning) objects that come from masses less than the sun. G-type stars of luminosity class "I" (supergiants) are true He-core burning objects that have come from a range of higher masses. Thus it is common for stellar evolution people to use the term "dwarf" (meaning luminosity class "V") and "main-sequence" interchangeably, while the term "supergiant" implies "He-burning". But, in contrast to this, most OB stars of *all* luminosity classes—including supergiants—are in fact core-H burning objects. (The

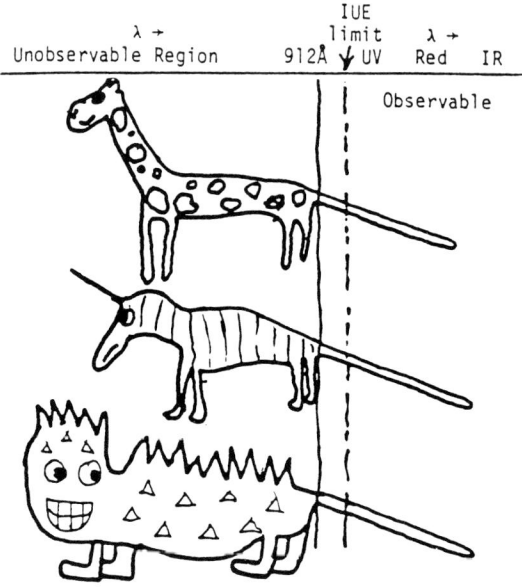

FIGURE 3. The only part of a hot star's spectral energy distribution that is accessible to us tells us only about the *tails* of the Raleigh-Jeans distribution, but not much about the beast itself. From Conti (1986), *Luminous Stars and Associations in Galaxies*, Dordrecht: Reidel, p. 199.

progenitor of SN1987A in the LMC was a He-burning B supergiant, but comparisons of the actual numbers of B supergiants with that predicted by stellar evolutionary theory suggests that the fraction of these objects is small; see Massey et al. 1995c.)

As we all learned as undergraduates, the most fundamental property of a star—the thing that everything else depends upon—is its *mass*, so let me introduce the unevolved luminous stars via this quantity. The mass and age (and, to a far lesser extent, composition) determines the star's effective temperature, luminosity, and hence radius and surface gravity—in other words, the star's location in the "physical" H-R diagram ($\log T_{eff}$ vs. M_{bol}). As we can see in the evolutionary tracks shown in Fig. 2, this mapping is unique—if we can determine a star's effective temperature and luminosity, we then also know its mass and its age.

We see that the regime of "high mass stars" (which we take to be $> 10 M_\odot$) is the realm of bolometric luminosities of $M_{bol} < -5$. (The sun has an $M_{bol} = 4.75$, so we are saying that these stars are nearly 10 magnitudes more luminous than the sun, or a factor of 10^4.) The highest mass stars reach bolometric magnitudes of -11 or -12, or about a factor of about *5 million times* more luminous than the sun.

The *zero-age* main sequence (ZAMS) can be inferred from where the evolutionary tracks begin in Fig. 2. We see that the temperatures we're talking about range from 60,000 °K ($\log T_{eff} = 4.78$) to 30,000 °K ($\log T_{eff} = 4.48$). Let me pose the question: what can we learn about the temperature of a star from its colors? Fig. 3 answers this in cartoon form: not much!

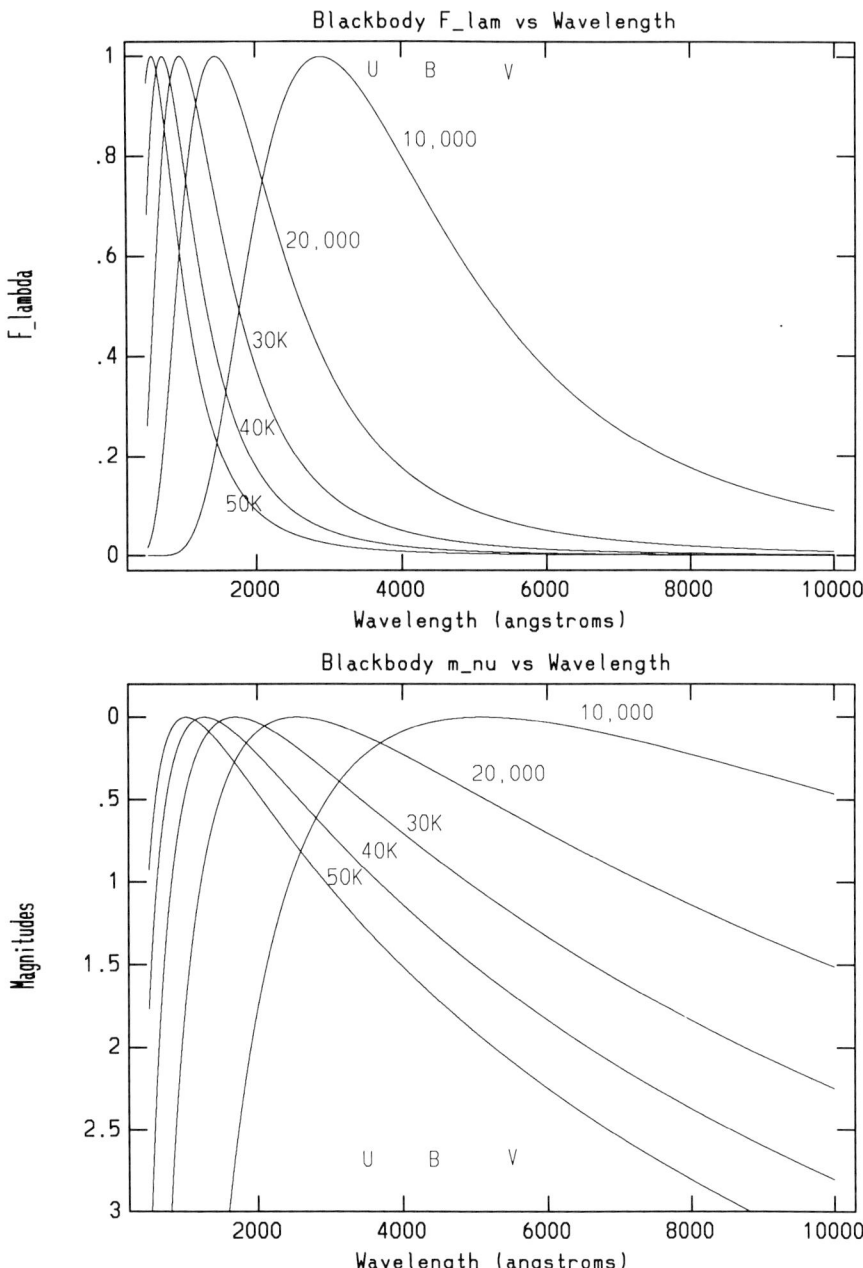

FIGURE 4. The flux as a function of wavelength for different temperature black-bodies. The units in the upper one are F_λ; in the bottom, I have plotted normalized spectrophotometric magnitudes.

Temperature	$m_U - m_B$	$m_B - m_V$	$U - B$	$B - V$
50,000 °K	−0.32	−0.37	−1.19	−0.33
40,000 °K	−0.30	−0.36	−1.17	−0.32
30,000 °K	−0.26	−0.33	−1.08	−0.30
20,000 °K	−0.19	−0.26	−0.81	−0.24
10,000 °K	+0.07	−0.03	0.00	0.00

TABLE 1. Blackbody spectrophotometric colors as a function of temperature, as well as representative Johnson UBV colors of "real" stars.

What do I mean by this? If we look at Fig. 4, we see what we would find by looking at blackbodies in this temperature regime. Most of the energy is in the UV—actually, the far-UV, below the Lyman limit (912 Å), a region that will always be shrouded by H absorption. We see in the upper plot the flux F_λ as a function of wavelength for various temperatures. In the bottom plot, we show these fluxes in "spectrophotometric" magnitude units, where

$$m_\nu = -2.5 \log_{10} F_\nu - 48.60$$

where $F_\nu (= F_\lambda \times c/\lambda^2)$ is in cgs units. (The constant was determined by Hayes & Latham 1975 by comparing the spectral energy distribution of Vega to a laboratory standard.) The Johnson UBV system has a somewhat different zero-point for U, B, and V (such that the colors of an A0 V star are zero), but how the colors change with temperature will be much the same. (I'm going through this in some detail just to show that this is something you can work out for yourself just from basic physics, it is not some subtlety having to do with stellar atmospheres.) If we make a table of these "spectrophotometric blackbody" colors we find (Table 1) that there is *simply no way* we could distinguish a 50,000 °K star from a 40,000 °K star, or a 40,000 °K star from one of only 30,000 °K, based upon its UBV colors. As we will see shortly, these effective temperatures span *the entire range of O3 to B0 in spectral types, and correspond to the entire range of masses we're discussing here, from the most massive (120 $M_\odot$) down to 10 $M_\odot$*. However, by the time we get to the temperatures characteristic of A-type stars (around 10,000 °K), color is actually a very accurate gauge of effective temperatures.

The fact that the colors are essentially *useless* for judging the effective temperature of an O-type star is important because it means that we really cannot guess the luminosity of a star (and hence its mass) without additional information. Remember that the bolometric magnitude of a star is related to the flux integrated over all wavelengths (or frequencies):

$$\int_0^\infty f_\lambda \, d\lambda = \int_\infty^0 f_\nu \, d\nu.$$

The zero-point of the bolometric magnitude system is consistent with underpinnings of the Johnson UBV system; i.e., an A0 V star has a bolometric magnitude (M_{bol}) equal to its absolute visual magnitude (M_V). In general, though,

$$M_{bol} = M_V + BC,$$

where the BC is the bolometric correction (with $BC = 0.0$ for an A0 V star).

As we can imagine from Fig. 4, the bolometric correction is indeed a very strong function of effective temperature for stars this hot, changing by 1.5 mags just from B0 to O3. We show in Fig. 5 how the BC changes with temperature.

One of the paradoxes that crop up when studying luminous stars is the one alluded

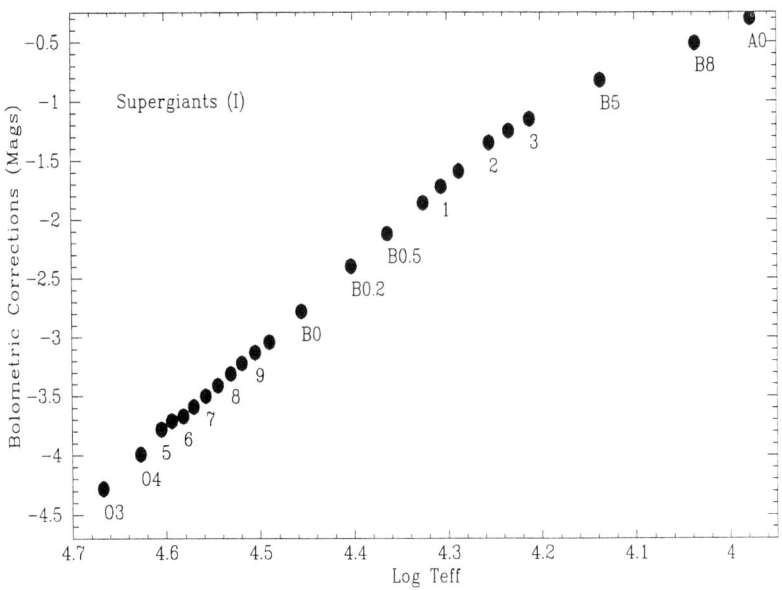

FIGURE 5. The bolometric correction as a function of temperature; see Massey et al. (1995c) and references therein.

to at the beginning of this section: that two stars can have indistinguishable colors and absolute visual magnitudes, but substantially different luminosities and masses. The M_V's of O3 V stars and B0 I stars are nearly identical (see Table 1-21 in Conti 1988), their $B - V$ colors differ by at most 0.1 and yet the O3 V star has a $BC = -4.3$ while the B0 I has a $BC = -2.8$, a difference of 1.5 mag. The mass of an O3 V star ($\log T_{eff} = 4.686$) with $M_V = -5.8$ (and hence $M_{bol} = -10.1$) is $85 \mathcal{M}_\odot$. The mass of a B0 I star ($\log T_{eff} = 4.456$) with $M_V = -5.8$ is that corresponding to an $M_{bol} = -8.6$, or about 25 $\mathcal{M}_\odot$. So, we see that these two nearly identical-looking stars (at least at B and V) have masses that differ by factors >3. Of course, we have exaggerated slightly here—if we have U photometry in addition to B and V then you really *can* tell the difference—barely—between these two objects if you do *very* good photometry. But, we could repeat this example, using instead an O7 V and O9 I star (whose values for M_V overlap), and whose Q values really are absolutely identical. (Q is a reddening-free index derived from UBV photometry, as I'll describe shortly.) In this case, we literally can't tell the placement of these stars in the HRD to better than 1.5 mag, and hence whether the mass is 25 or 60 $\mathcal{M}_\odot$.

There is, however, a way out. Spectroscopy answers the question of effective temperatures very neatly for these stars, and once the temperatures are known, the bolometric corrections are known via stellar atmosphere calculations; once the bolometric corrections are known (along with M_V), the mass can be inferred from stellar evolutionary tracks.

I show in Fig. 6 the spectra of O and B stars, where the data have been taken from

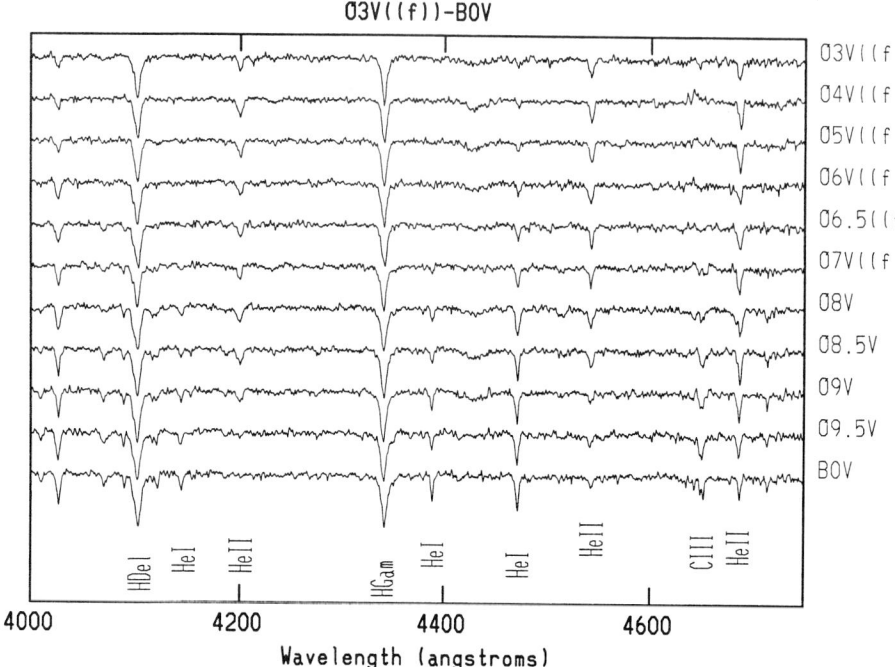
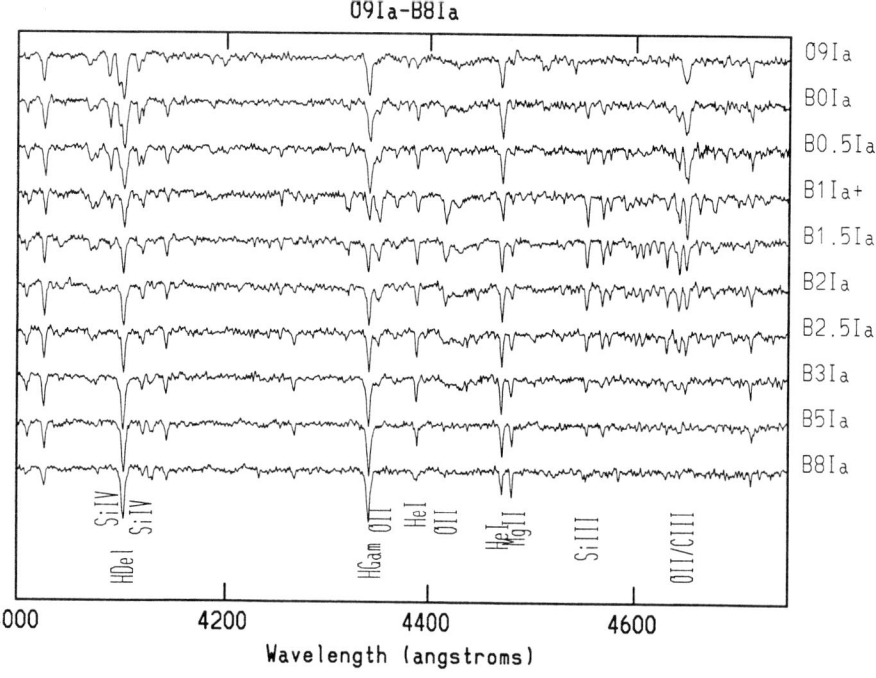

FIGURE 6. Spectral classification of OB stars is discussed extensively in Walborn & Fitzpatrick (1990), the source for the data in this figure.

	V			III			I		
Type	$\log T_{eff}$	BC	M_V	$\log T_{eff}$	BC	M_V	$\log T_{eff}$	BC	M_V
O3	4.686	−4.31	−5.6	4.667	−4.28	−6.0	4.648	−4.28	−6.5
O4	4.667	−4.27	−5.4	4.647	−4.06	−6.0	4.627	−3.99	−6.5
O5	4.646	−4.06	−5.2	4.626	−3.98	−6.0	4.605	−3.78	−6.5
O5.5	4.635	−3.98	−5.2	4.616	−3.85	−5.9	4.594	−3.71	−6.5
O6	4.625	−3.90	−5.2	4.604	−3.83	−5.7	4.582	−3.67	−6.5
O7	4.603	−3.77	−4.9	4.581	−3.66	−5.6	4.558	−3.50	−6.5
O8	4.580	−3.65	−4.6	4.556	−3.49	−5.1	4.531	−3.31	−6.5
O9	4.555	−3.48	−4.4	4.531	−3.31	−5.1	4.505	−3.13	−6.1
O9.5	4.543	−3.39	−4.0	4.517	−3.21	−5.1	4.490	−3.04	−6.0
B0	4.471	−3.00	−3.8	4.481	−2.96	−5.0	4.456	−2.78	−6.0

TABLE 2. The effective temperatures (T_{eff}), bolometric corrections (BC), and absolute visual magnitude (M_V) for O and early B stars of luminosity classes V, III, and I. The values are taken adapted from Chlebowski & Garmany (1991) and Conti (1988) with some smoothing.

the Walborn & Fitzpatrick (1990) atlas. For the O-type stars, the primary classification criterion are the He I to He II ratios, and in particular He I $\lambda 4471$ and He II $\lambda 4542$. For the early B stars (through about B2) the primary criterion is the Si IV $\lambda 4089$ to Si III $\lambda 4553$ ratio. Beyond B2 you can use Si III to Si II $\lambda 4128$ if the signal-to-noise is high enough, but the Mg II $\lambda 4481$ to He I $\lambda 4471$ ratio gains in importance, and is the main classification criterion from B5 to A0. Now, the nice thing about these temperature diagnostics is that they are independent of metallicity, at least through B2. In other words, we are using the relative strengths of different ions of the same species (He II to He I; Si IV to Si III), and so even if Si itself is much less abundant in one star than another, the ratio of Si IV to Si III winds up peaking at about the same effective temperature. Clearly this sort of thing isn't true for something like Mg II to He I, but the good news is that the UBV colors become a reasonable indicator of temperature past B2, and so we're home free.

Well, almost. Does an O9.5 III star in the LMC have the same T_{eff} and bolometric correction as one in the Milky Way or the SMC? Surprisingly, we don't know the answer to this. The data in Table 2 is based primarily on the old (zero metallicity) non-LTE models of Auer & Mihalas (1970) as calibrated with O-type spectral type by Conti (1973). This was the first non-LTE-based temperature scale for O stars, and is still the best available, despite newer atmosphere models that contain many more lines. Vacca, Garmany, & Shull (1996) have attempted to revisit this calibration using modern models, but the dispersion at a given spectral type is appallingly large, and a more critical study needs to be done. So, although there may be systematic errors in (any) adopted temperature scale, these can always be taken out by subsequent recalibration, and—as long as what one is doing is self-consistent— relative comparisons of HRDs (as we will do in Secs. 4 and 5) should remain unaffected.

Just exactly how much better do we do by using the spectral type information? We illustrate the answer to this in Fig. 7, where we show in the top half the errors that result from a misclassification of one spectral subtype and one spectral luminosity class. The error bars are slanted simply because an error in effective temperature also results in an error in the bolometric correction—and, indeed as we've been saying, in terms of determining the masses (which depend only a little on the effective temperature in the HRD), this is by far the most significant error. An error of one luminosity class is nearly parallel to the error produced in being off by one spectral subtype. For comparison, we

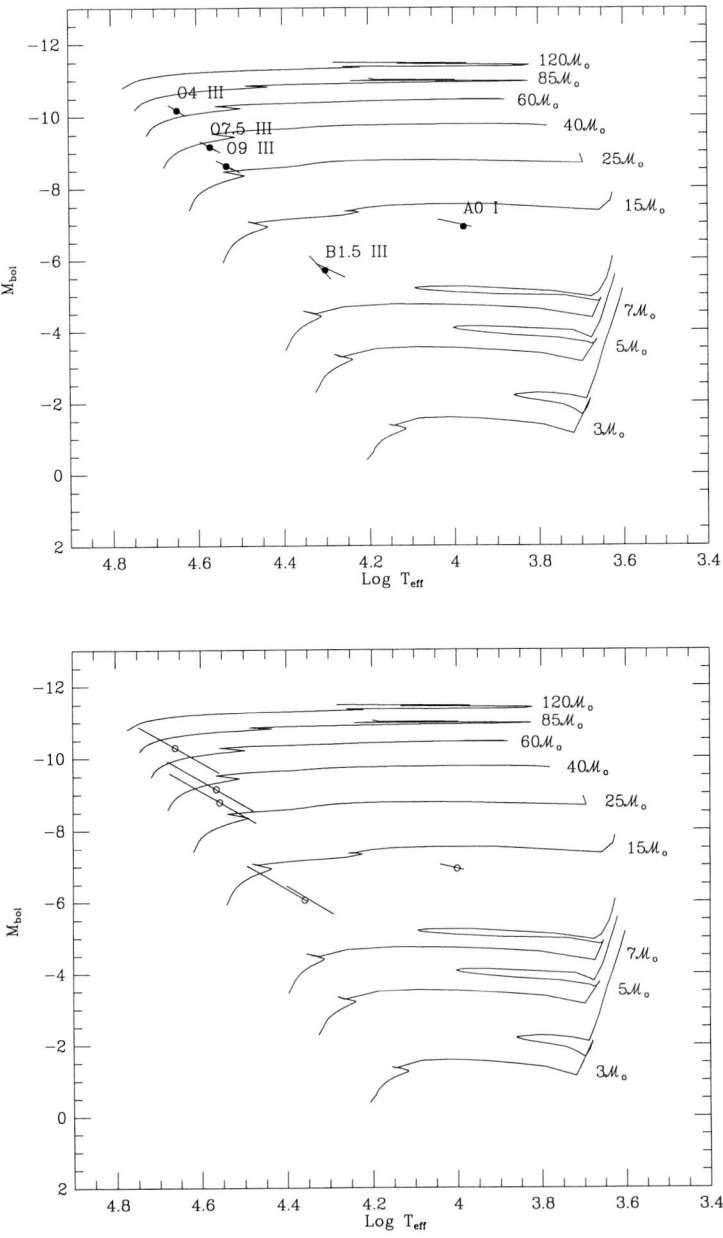

FIGURE 7. The errors in the HRD are shown. In the top figure, I've used spectral information to assign effective temperatures and BCs, and assumed an error of one spectral subtype and one luminosity class. In the bottom figure, I've used photometry to determine the star's placement via the reddening-free index Q, and the error bars correspond to a 0.02 mag uncertainty in the colors. This figure is based upon one appearing in Massey et al. (1995c).

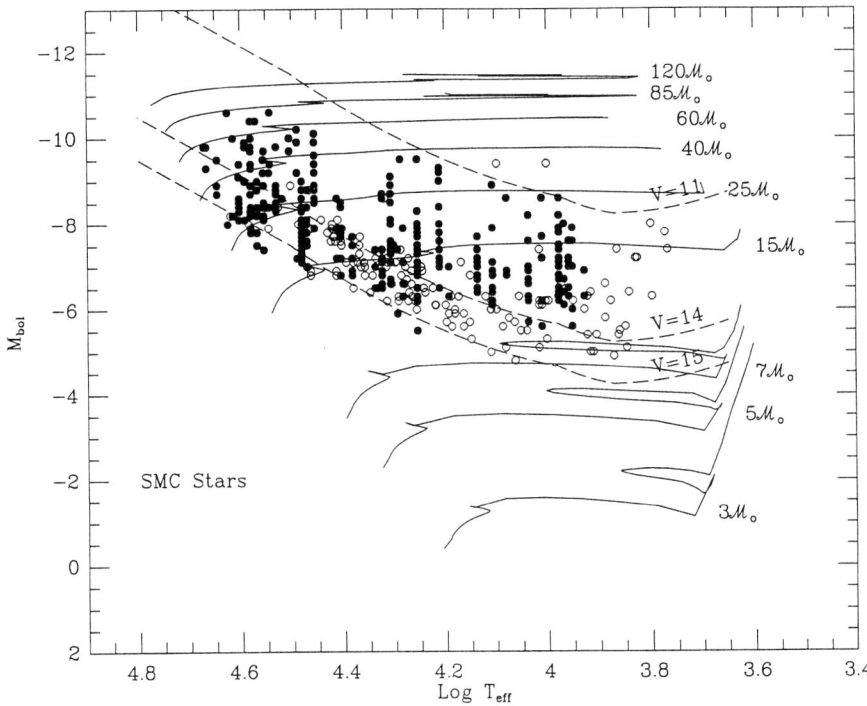

FIGURE 8. The distribution of stars in the SMC serves to illustrate that the *visually* brightest stars are late-B and early-A type supergiants of "modest" mass, not the far more luminous and massive hotter stars. Stars on the left side of this diagram are missing as a result of catalogs for stars in the SMC being incomplete beyond $V \approx 13$. The figure is taken from Massey et al. (1995c). An analogous figure for the LMC will be found in Fig. 18.

show in the lower part of the figure the resulting errors in the HRD due to finite errors in photometry, where the error bars correspond to a very modest uncertainty of 0.02 mag in $(B-V)_o$ and $(U-B)_o$. In the later case, we clearly don't know much of anything in terms of mass.

1.1. HR diagrams: the difference between "luminous" and "bright"

Before leaving this section, let me come back to the point that so puzzled me as a graduate student: why was it that all these nice spectra shown by early work on "hot stars" in nearby galaxies were invariably of A-type or late-B? Fig. 8 shows the distribution of stars in the HRD for the SMC from Massey et al. (1995c). We have added lines of constant V magnitude in this figure, using the bolometric corrections as a function of temperature.

So what stars are the brightest? I think this diagram is very interesting and revealing, explaining a great deal of how various workers have been misled in interpreting the data for nearby galaxies. The brightest stars are the 15-25 $\mathcal{M}_\odot$ stars with $\log T_{eff} \approx 4$, which are indeed the early A-type stars. Oh, sure, there are a few stars of higher mass near the $V \sim 11$ line, going on up to maybe 40 $\mathcal{M}_\odot$. But to find main-sequence stars of high mass would require digging down another couple of magnitudes visually because of the strong bolometric corrections.

Why don't we instead see 100 $\mathcal{M}_\odot$ A-type supergiants rather than 20 $\mathcal{M}_\odot$ A-type supergiants? The answer to this is a little more subtle: the lifetimes beyond the main-sequence are very short for all of these stars, and there are proportionally so many more of the 20 $\mathcal{M}_\odot$ stars. As we will see in Sec. 4, we expect that for every star of mass 70-120 $\mathcal{M}_\odot$ born in the field, there will be about 500 stars born between 15 and 25 $\mathcal{M}_\odot$.

1.2. Exercises for the student

(a) One day I was walking up to the Hill in Boulder to have lunch with my erstwhile thesis advisor Peter Conti, and a colleague who was visiting from abroad, Dr. "X". Dr. "X" was telling us that he was about to go take spectra of O stars in NGC 6822 using the fabulous ESO spectrograph. Well, this worried me quite a bit, as I had in the works a 3-year plan to get good *UBV* CCD photometry and then slowly begin accumulating spectra. "How did you pick out the candidates?" I asked him. "Oh, I used the POSS Sky Survey prints, and just picked out anything that was bright and blue." Assume that, at best, Dr. "X" was able to separate stars into "blue" and "not blue" corresponding to $B - V = 0.0$. What do you suppose he found?

(b) Sometimes astronomers think of "spectral types" and masses as one and the same—that you mentally think of O5 V stars as having some mass, and O5 I's some other mass. This is fine within some limits, but shouldn't be carried to an extreme. Using the numbers in Table 2, and the HRDs previously shown, determine what mass corresponds to an O7 III star. *Hint: the question really should be "What range of masses correspond to an O7 III star?"*

2. Finding main-sequence luminous stars in the Local Group: methodology for a hard problem

Studying the H-burning luminous stars in Local Group galaxies is *hard*, for the two reasons discussed above: (a) the bolometric corrections conspire in such a way that the most luminous stars are not the brightest stars, and (b) the colors don't tell us enough to get the bolometric corrections by themselves. Add to this one complication we haven't discussed yet: reddening.

In our favor, however, has been the revolution in detectors and instruments in the past decade, both in imaging and spectroscopy. CCDs now abound, and multi-object fiber feeds make it possible to obtain spectra of tens or hundreds of stars over $\approx 1°$, and of course *HST* has completely changed how we image and obtain spectra in "crowded" fields. Let us begin by considering the observational quantities we obtain with these new instruments, and how we connect the observational quantities to the physical quantities we want to know: effective temperature, luminosity, and mass.

We saw in Table 1 that the $B - V$ colors of all hot stars are about the same. So, in the "classic" color-magnitude diagram (CMD) of V vs. $B - V$, a young (<5 Myr), coeval population will be nearly a vertical line—if there is no reddening.

What will be the effects of reddening in such a diagram? For a given color excess $E(B - V)$ there will be extinction $A_V = 3.1 \times E(B - V)$. So the effect will be to move stars to the right (larger $B - V$) and lower (fainter V) in the CMD. If the reddening isn't constant, the effect will be to smear things out in the CMD.

The top section in Fig. 9 shows the V, $B - V$ CMD of the η Car cluster (Tr14/16) taken from Massey & Johnson (1993), and indeed this is just what we see. The filled circles are spectroscopically confirmed members of the cluster, and they do not form a vertical line—there is appreciable width to their distribution in this diagram. The HRD of this cluster will be shown in Sec. 5, in Fig. 22.

What would happen if we used $U - B$ as our color rather than $B - V$? Although the

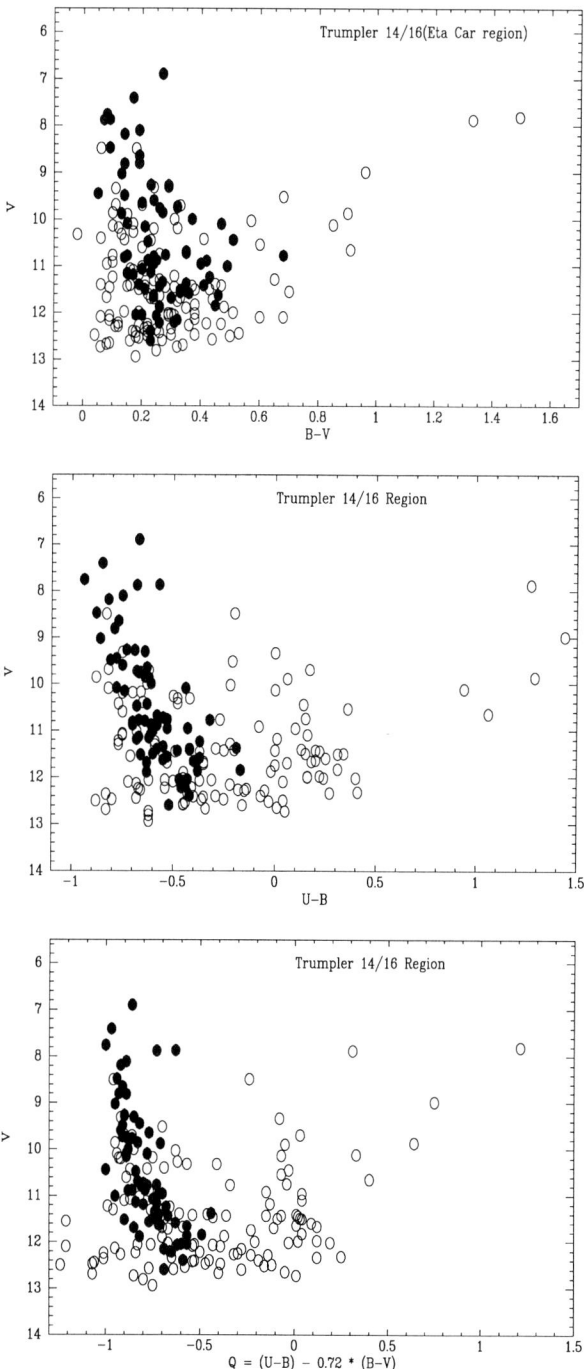

FIGURE 9. Three CMDs of the η Car cluster Tr14/16, based upon Massey & Johnson (1993).

interstellar extinction is considerably greater at U than at V [$A_U = 4.8 \times E(B-V)$], the baseline between U and B is shorter than between B and V, and so the color excess at $U-B$ is actually less than that at $B-V$: in fact, $E(U-B)/E(B-V) \approx 0.72$ for typical reddening in the Milky Way. Yet at the same time, $U-B$ is more sensitive to temperature than $B-V$ (Table 1), and so we do much better in constructing a CMD using $U-B$. Such a figure is shown in the middle section of Fig. 9.

But wait! We can do even better than this. If we know that the slope $E(U-B)/E(B-V) = 0.72$ (and we can substantiate this by taking spectra, and comparing the intrinsic colors to the observed colors), we can construct a *reddening-free index*, Q as follows:

$$Q = (U-B) - 0.72 \times (B-V). \tag{2.1}$$

What do we mean by describing Q as reddening-free? Imagine for the moment a star with no reddening, so that the observed colors are the intrinsic colors $(U-B)_o$ and $(B-V)_o$. Then

$$Q = (U-B)_o - 0.72 \times (B-V)_o. \tag{2.2}$$

Now, imagine instead that the same star has some finite color excess $E(B-V)$. Then the observed colors will be

$$(B-V) = (B-V)_o + E(B-V)$$

and

$$(U-B) = (U-B)_o + E(U-B) = (U-B)_o + 0.72 \times E(B-V)$$

Substitution of these into equation 2.1 will yield 2.2: we get the same Q whether reddening is present or not, as long as the slope in the two-color diagram is know.

We show in the bottom panel of Fig. 9 the CMD using Q as our color, and indeed the cluster members (filled circles) now make a considerably tighter line than in either the $B-V$ or $U-B$ versions.

Let me emphasize that this trick only works for stars hotter than about B5. Cooler than that, lines of constant reddening (lines with slopes of 0.72 in a $U-B$ vs. $B-V$ plane) will intersect the intrinsic colors of much cooler stars: in other words, Q becomes multi-valued with temperature. We give in Table 3 the values of Q computed from the intrinsic colors of FitzGerald (1970) and Schmidt-Kaler (1982), where we have extrapolated in a few cases.

One of the obvious things to do at this point is to determine transformations between the effective temperature and Q by combining the values in Tables 2 and 3. For $Q < -0.4$,

$$\log T_{eff} = 4.055 + 0.041 \times Q + 0.6514 \times Q^2$$

for luminosity class V, while

$$\log T_{eff} = 4.342 + 1.105 \times Q + 1.4793 \times Q^2$$

for luminosity class III, and

$$\log T_{eff} = 4.342 + 1.105 \times Q + 1.4793 \times Q^2$$

for luminosity class I. A complete set of such transformation equations (including conversion from $\log T_{eff}$ to BC) can be found in Table 7 of Massey et al. (1995c), along with a comparison of such fits to the predictions of model atmospheres, but a word of caution is in order: these fits cannot be used with any real reliability at high effective temperatures ($Q < -0.8$) for the reasons emphasized in the previous section. Such transformations are, however, very useful in selecting the stars for which one might want to obtain spectroscopy, but the transformation between Q and effective temperature is just

Type	V			III			I		
	$(U-B)_o$	$(B-V)_o$	Q	$(U-B)_o$	$(B-V)_o$	Q	$(U-B)_o$	$(B-V)_o$	Q
O3	−1.20:	−0.33:	−0.96:	−1.19:	−0.33:	−0.95:	−1.18:	−0.32:	−0.95:
O4	−1.20:	−0.33:	−0.96:	−1.19:	−0.33:	−0.95:	−1.18:	−0.32:	−0.95:
O5	−1.19	−0.32	−0.96	−1.18	−0.32	−0.95	−1.17	−0.31	−0.95
O5.5	−1.19	−0.32	−0.96	−1.18	−0.32	−0.95	−1.17	−0.31	−0.95
O6	−1.18	−0.32	−0.95	−1.17	−0.32	−0.94	−1.16	−0.31	−0.94
O7	−1.17	−0.32	−0.94	−1.14	−0.32	−0.91	−1.14	−0.31	−0.92
O8	−1.14	−0.31	−0.92	−1.13	−0.31	−0.91	−1.14	−0.29	−0.93
O9	−1.13	−0.31	−0.91	−1.13	−0.31	−0.91	−1.14	−0.28	−1.11
O9.5	−1.10	−0.30	−0.88	−1.09	−0.30	−0.87	−1.08	−0.27	−1.11
B0	−1.08	−0.30	−0.86	−1.08	−0.30	−0.86	−1.05	−0.24	−1.11
B0.5	−1.00	−0.28	−0.80	−1.05	−0.28	−0.85	−1.04	−0.22	−0.88
B1	−0.95	−0.26	−0.76	−0.96	−0.26	−0.77	−1.00	−0.19	−0.86
B1.5	−0.88	−0.25	−0.70	−0.94	−0.25	−0.76	−0.96	−0.18	−0.83
B2	−0.81	−0.24	−0.64	−0.93	−0.24	−0.76	−0.93	−0.17	−0.81
B2.5	−0.72	−0.22	−0.56	−0.85	−0.22	−0.69	−0.87:	−0.15	−0.76:
B5	−0.58	−0.16	−0.46	−0.50	−0.16	−0.38	−0.69	−0.09	−0.63
B8	−0.36	−0.11	−0.28	−0.41	−0.10	−0.34	−0.48	−0.02	−0.47
A0	−0.02	−0.01	−0.13	−0.07	−0.03	−0.05	−0.44	+0.01	−0.45

TABLE 3. The intrinsic colors of hot stars as a function of spectral type, taken from FitzGerald (1970) and Schmidt-Kaler (1982), with Q computed assuming a reddening law with $E(U-B)/E(B-V) = 0.72$.

as degenerate as that of the intrinsic colors and effective temperatures—using Q just helps to remove the effects of reddening in the photometric indices.

2.1. Using CCDs to obtain CMDs: a few practical considerations

Although photometry alone can't answer how luminous and massive a particular star is, photometry is a necessary first step in selecting a sample of such stars. Let me provide here a short introduction to CCD photometry.

One of the things to realize is that the telescope and filter and CCD combination one uses is unlikely to be a perfect match to the original UBV system, based as it was on photomultipliers. We observe standard stars with known UBV, and calculate the transformation between our instrumental $u - b$ and $b - v$ system to the standard $U - B$ and $B - V$ system. Such transformations are never linear over the full range of colors, and in practice one tries to observe standards that just bracket the color range of interest. Furthermore, in the case of heavily reddened OB stars (as we are likely to be observing in the Milky Way) we have a real problem: standard stars of similar color are likely to be of very different spectral type as our program objects (they may even be M stars if our O stars are *extremely* reddened!) and so it also behooves us to try to start with a filter system that matches the standard system as closely as possible. In collaboration with Hugh Harris at the USNO, we developed a glass filter set at Kitt Peak some years back that mimics the standard system pretty well with the CCDs now in use. (The recipes for the glass filters may be found by accessing the Direct Imaging Manual for Kitt Peak National Observatory on the World Wide Web: http://www.noao.edu.) The transformation equations from a recent observing run of mine have terms something like this:

$$u = C_u + U + 0.49X + 0.054(U - B)$$

$$b = C_b + B + 0.24X - 0.113(B - V)$$

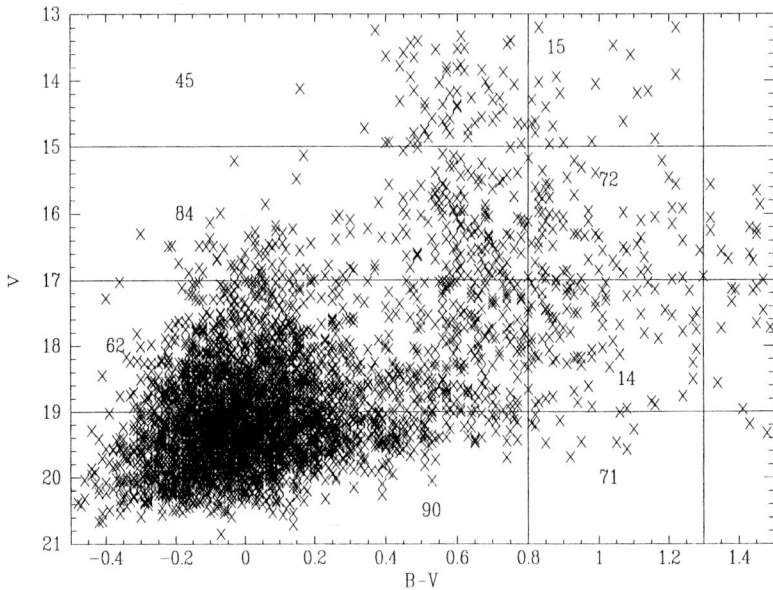

FIGURE 10. CMD of 4000+ stars in M 33 obtained with the KPNO 0.9m telescope; these are the data used to obtain UBV photometry for the UV-bright sources discussed in Massey et al. (1996). The numbers shown in the boxes give the number of expected foreground stars according to the Ratnatunga & Bahcall (1985) model.

$$v = C_v + V + 0.14X + 0.018(B - V)$$
$$r = C_r + R + 0.10X - 0.035(V - R)$$

Here X is the airmass, and the various C's are just zero-points; the point here is that the coefficients on the color-terms are pretty small, suggesting a decent match to the standard system.

Once you've obtained your CCD data and done the standard transformations, you will probably need to do some sort of PSF-fitting if you are studying stars in crowded regions. Mario Mateo has described DoPHOT in some detail during his lectures at this Winter School; I'm a reluctant DAOPHOT enthusiast myself (Stetson 1987), as I *know* it does what I want on specific stars and I am not very interested in statistical results near the crowding limit. On this one point all the experts would agree: you, as a user of such programs, should accept responsibility for testing whether they are giving you valid results or not. Such tests are straight-forward, and you are only irresponsible if you don't bother to make them.

So, let us imagine we were to go out one dark and clear night and take a few images of M 33 using the KPNO 0.9m telescope, Tektronix 2048^2 CCD, and UBV filters, along with plenty of standard stars bracketing the program objects both in airmass and in color. A little DAOPHOTing later, and we might obtain the V vs. $B - V$ diagram of the 4000+ stars shown in Fig. 10. What now?

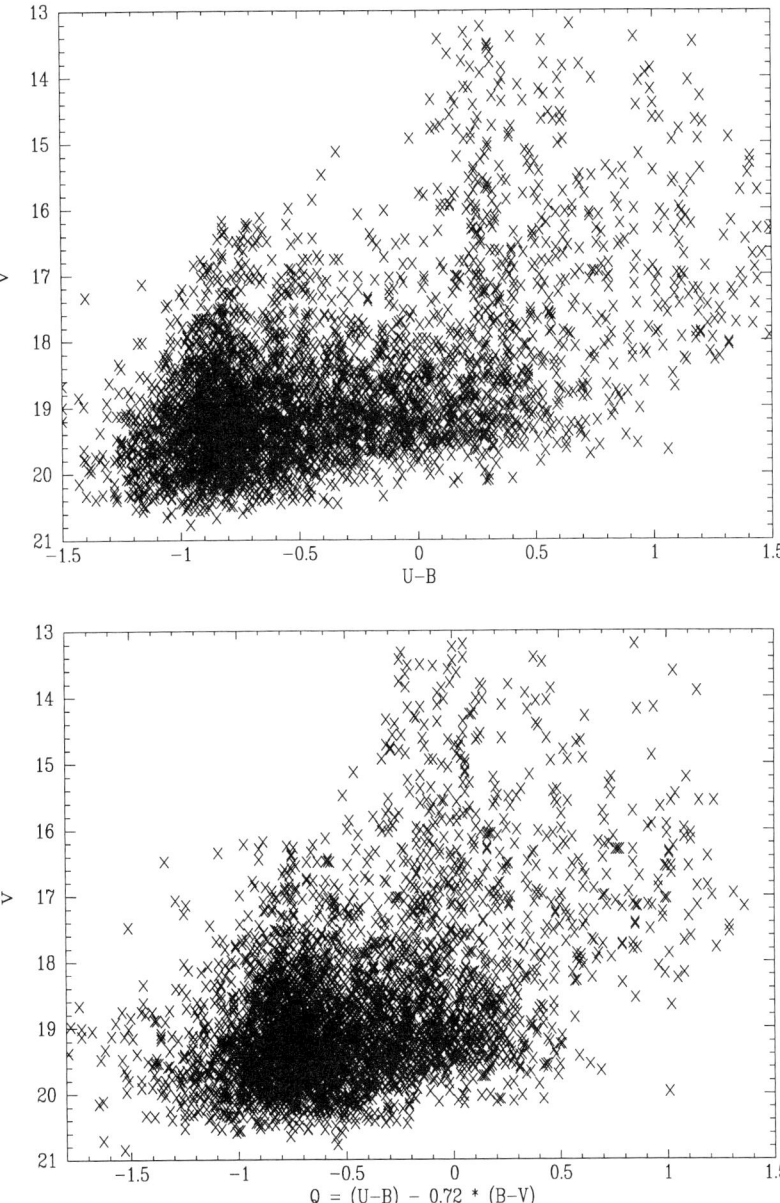

FIGURE 11. The $U-B$ and Q version of the previous figure shows a better defined main-sequence. The distribution still bears little resemblance to that of Fig. 9, as the M 33 data consists of a mix of stars of various ages.

The first question that occurred to me was, which of these stars are expected to be foreground? When we image a galaxy like M 33 we are looking through an appreciable amount of the Milky Way, even though the galaxy is at a reasonably high galactic latitude ($b = -31°$). We can get a crude feeling for how many stars in various parts of this CMD are expected to be foreground interlopers by looking at the Galactic model of Ratnatunga & Bahcall (1985). These numbers are shown in the boxes in the bottom part of Fig. 10.

Galaxy	$100\mathcal{M}_\odot$	$60\mathcal{M}_\odot$	$40\mathcal{M}_\odot$	$25\mathcal{M}_\odot$	$15\mathcal{M}_\odot$	$10\mathcal{M}_\odot$
LMC	13.3	14.3	14.8	15.6	16.8	17.8
SMC	13.3	14.3	14.8	15.6	16.8	17.8
IC 1613	18.4	19.4	19.9	20.7	21.9	22.9
NGC 6822	18.7	19.7	20.2	21.0	22.2	23.2
M 33	18.9	19.9	20.4	21.2	22.4	23.4
M 31	19.1	20.1	20.6	21.4	22.6	23.6
Sex A	19.8	20.8	21.3	22.1	23.3	24.3
IC 10	22.1	23.1	23.6	24.4	25.6	26.6
NGC 1569	23.1	24.1	24.6	25.4	26.6	27.6

TABLE 4. The B apparent brightness of stars as a function of mass on the zero-age main sequence for some nearby galaxies.

(We will see in Sec. 5.4 that foreground contamination is a considerable problem for identifying the red supergiants.) For the stars in the so-called "blue-plume" (really just what extragalactic pundits refer to when they inadvertently come across the "main-sequence") most are indeed going to be M 33 members, but as we get to more neutral and red colors, this isn't true.

We can go ahead and compare the $B-V$ CMD to that obtained by using $U-B$, and that is shown in the top of Fig. 11. The effects of differential reddening are reduced in this figure, and the main-sequence is a little better defined. We do even better by going to the Q vs V plane, shown in the bottom of Fig. 11.

2.2. Spectroscopy

We know from our previous section that if we are to go any further—if we are to actually place stars in the HR diagram—what we need now is spectroscopy. The spectral type will tell us the star's effective temperature, and it will also allow us to determine the star's reddening correction, by comparing the observed $B-V$ to the intrinsic $(B-V)_o$ expected for its spectral class (Table 3). Assuming that the distance modulus (DM) is known, we can then compute $M_V = V - 3.1 \times E(B-V) - DM$. Finally (and I would argue most importantly), the spectral type gives us the bolometric correction as per Table 2, and this now allows us to determine the star's bolometric luminosity $M_{bol} = M_V + BC$. I say "most importantly", because if we look at the evolutionary tracks in Fig. 2, say, we see that the masses that we infer are not so dependent upon the assumed effective temperature as evolution proceeds at nearly constant luminosity. But, paradoxically, the spectral type (which we associate primarily with effective temperature) is actually most crucial for determining the star's luminosity. Note that in this we really don't make that much use of the star's luminosity class, other than that the spectral type and luminosity class together affect the effective temperature and BC that we may wish to adopt (Table 2).

How to best get that spectrum? We give in Table 4 the expected B magnitudes for ZAMS stars of various masses in various galaxies of the Local Group and beyond just to give you an idea of what you might be up against.

Now, Rolf Kudritzki has talked to us about the difference between "optimistic" and "pessimistic" astronomers, and so we need to approach the depressing values in Table 4 in the right frame of mind. On the one hand, I don't think that even the most optimistic of us would consider proposing determining the initial mass function for massive stars in M 31 going down to $10\mathcal{M}_\odot$. On the other hand, a few things do suggest themselves. First, I've given this for zero-age main-sequence stars, and, if there was a scientific goal in simply obtaining spectra of a few stars of 10 $\mathcal{M}_\odot$ in M 31, that actually would be possible: the A-type supergiants in this mass range are expected to be roughly 4 mag or so brighter than the values listed, given the size of the bolometric correction. So, a B value of 23.6 might be out of the question with current ground-based telescopes (yes, even with the mighty Keck, if you wish decent spectra of a large number of stars), $B = 19.6$ is possible with a 4-m class telescope (if you are patient enough). I think that an excellent example of the usefulness of taking spectra of slightly evolved stars is given in the poster by Monteverde—she and her collaborators have been using B-type supergiants (of somewhat higher masses) to derive the oxygen abundance gradient within M 33 (Monteverde et al. 1997). Similarly, the work being done by Kim Venn, Rolf Kudritzki, Sylvia Becker, and others on abundances from A-type supergiants—and potentially using these stars as distance indicators—is another example where obtaining spectra of "a few stars" is scientifically extremely useful, even if a complete sample at that mass range is simply out of the question.

Secondly, I've used the apparent distance moduli at B because this is the spectral region where most of the diagnostic lines are located in the visible. But, that does mean that a little extra extinction makes a huge difference—enough, even, to hide the 0.6 mag difference in the true distance moduli of the two Magellanic Clouds. In some cases the extinction is mainly foreground Galactic (as is the case of NGC 6822 and IC 10), but in others, such as M 31, an "average" value of $E(B-V) = 0.25$ has been assumed, although we know there are regions of considerably less reddening—so this helps by at least half a magnitude right there (Massey et al. 1995a). Or, maybe you can devise a clever scheme that makes use of He I and He II lines in the yellow-red region, where the interstellar extinction is less.

What sort of spectral resolution and signal-to-noise do you need to observe and classify hot stars? There is where I must also distinguish the sort of poor-person's quantitative spectroscopy I do from those of Rolf Kudritzki. There is no question that what he and his collaborators do is a much better procedure, and far more accurate than what I do: by actually doing line fitting, he avoids the uncertainty of assigning spectral types and using the "typical" values for the classification. However, there are of course considerably greater demands required of signal-to-noise. I find I can classify a star with confidence at a SNR of 30 per 3 Å spectral resolution element, and can classify with some increasingly uncertainty down to a SNR of 20. (Below that, I'm just guessing.) Furthermore, I have one additional, subtle advantage: as even better, more improved model atmospheres are calculated, I can very easily determine the transformation from the old scale to the new, as everything is done self-consistently via the spectral types—which don't change.

Why 3 Å resolution? The spectral classification lines for O and B stars are fairly weak (Fig. 6), and so you want to match the resolution with the line widths or you will decrease the contrast between the lines and the continuum. The rotational velocities $v \sin i$ of O-type stars is of order 150-300 km s^{-1}, and hence for spectral resolutions $R < 1000 - 2000$ (>4.4-2.2 Å in the blue) you will degrade the contrast between these (already weak) stellar features and the continuum. By the time you get to the B stars (particularly B-type supergiants), rotational velocities are maybe only 50-100 km s^{-1}, so you really should be using $R \approx 3000$, or 1.5 Å in the blue.

The advent of multiobject fiber spectroscopy provides the means of obtaining the spectra of many stars simultaneously, and in a recent paper my collaborators and I present spectral types for 131 UV-bright sources in M 33 using the new WIYN 3.5-m telescope and multiobject fiber feed Hydra (Massey et al. 1996). These data were obtained during only two nights of observing, and such studies would not have been possible without many more nights of 4-m observing prior to such instrumental innovation. However, a word or two of caution. Greed comes at some price. Although there are 100 fibers with Hydra, in practice one probably devotes 10-20 fibers to sky, and there are maybe 30-40 "prime" targets per configuration. Some significant through-put loses occur compared to a good, high-efficiency spectrograph—I estimate a factor of 3 or so, although others might disagree. Finally, although such work is invaluable for the sort of survey work we were doing, you will certainly achieve better sky-subtraction with a long-slit spectrograph, *and* in cases where nebular contamination is significant, you are never going to achieve good results with fibers as the sky fibers can never be close enough to the object fibers to give a good "local" determination of sky.

3. Finding the evolved descendants of massive stars: LBVs, WRs, and RSGs

In Section 5 we will discuss what we are learning about massive star evolution by comparing the number and distribution of the unevolved massive stars with those of evolved massive stars. We've reviewed in Section 2 some of the methods and pitfalls of finding the unevolved massive stars, and in this section I'd like to do the same thing for their evolved descendents: luminous blue variables (LBVs), Wolf-Rayet stars (WRs) of WC and WN type, and red supergiants (RSGs).

The motivation for this study is described at the onset of these lectures: one would like to know how massive stars evolve as a function of metallicity. Conti (1976) first proposed that massive O-type stars might evolve to WR stars through mass-loss induced by strong stellar winds. It was known that WR stars were massive stars (in some cases the masses could be measured directly from binary systems), and that their strong emission-line spectra revealed the products of CNO H-burning (He and N) for the WN-type Wolf-Rayet stars, or the products of triple-α He-burning (C and O) in the case of WC stars. WR stars showed little or no Hydrogen in their spectra. In the "Conti scenario", a very massive O-type star would strip off its outer most layers, becoming first a WN star and subsequently a WC star. A somewhat less luminous (and less massive) star would have less mass-loss, and during its life might only be able to peel down to a WN star, never making it to the WC stage. An even less luminous star would never become a WR star, but might become a red supergiant.

In the early 1980's I got involved in trying to test this scenario observationally. What was the evidence for or against this? One obvious suggestion was that in a region of higher metallicity, mass-loss rates would be higher, so you would expect the cut-off limit for becoming a WC star would be lower (in mass or luminosity): one would expect the number ratio of WC to WN stars to be a function of metallicity. (The jury has been out on this one until recently, as we'll see in Sec. 5.) Another test was whether RSGs and WRs were found in the same associations, and this was answered early on by Humphreys, Nichols & Massey (1985): no, they weren't. (The foreground contamination in the RSG sample that I will describe shortly does not affect our conclusion, as the comparison was differential: are the number of red stars seen against OB associations different for the associations containing WR stars and those that did not?)

As the community thought and worked on these issues, other questions arose. Stars that were composite in appearance between "extreme" Of stars and late-type WN stars were discovered, the so-called "slash" stars of spectral type Ofpe/WN9 (Walborn 1977; Bohannan 1979; Bohannan & Walborn 1989). Were these actually the missing links between main-sequence O stars and the Wolf-Rayet? And how was it possible for an O star to lose enough mass to become a WR star? The ones with the strongest possible mass-loss rates might manage this in their lifetimes, but a modest little $M_{bol} \approx -9$,

40 $\mathcal{M}_\odot$ star with a mass-loss rate of $10^{-6}\mathcal{M}_\odot$ yr^{-1} just isn't going to make it: in a typical lifetime of 5 Myr (say) you might expect to lose $5\mathcal{M}_\odot$ total, not the $20\mathcal{M}_\odot$ or so needed to strip off the H-rich outer layers. And where did the luminous blue variables fit into the picture? It turned out that these questions might all be related.

3.1. *The LBV content of Local Group galaxies: accidents happen*

A rare class of variable stars are the very bright (*and* luminous!), blue objects once called Hubble-Sandage variables, after the initial discovery of five of these stars in M 31 and M 33 by Hubble & Sandage (1953). As a class, these objects are linked to S Dor in the LMC and η Car in our own Galaxy, with extreme episodic periods of mass loss, during which there is photometric and spectral variability. (For a few years, η Car was the second brightest star in the sky, as has been pointed out by Humphreys & Davidson 1984). These stars are sufficiently luminous that radiation pressure is barely balanced against gravity, the Eddington limit. Sometimes during the evolution to cooler temperatures, an instability sets in (possibly induced by radiation pressure acting on numerous Fe lines in the UV), and the star undergoes an outburst. (An excellent review is given by Humphreys & Davidson [1994].) One possibility is that LBVs provide the mechanism by which O stars can shed enough matter to become WR stars. But, with only a few examples found in nearby galaxies, is this really a possibility?

It may be that detecting LBVs by these flashy displays of photometric variability—when the star changes by several magnitudes visually as the temperature slides around—reveals only the "tip of the iceberg" in terms of their actual numbers (Massey et al. 1995a, 1996). The evidence consists of a number of stars whose spectroscopic properties are identical to those of the known LBVs, as shown in Fig. 12. These stars were observed simply because they were hot and bright; their "UIT" designation comes about as they were discovered on the basis of UV exposures of M 33 (Massey et al. 1996) that were taken with the *Ultraviolet Imaging Telescope* on-board the ASTRO-1 mission. The spectrum of UIT247 was shown by Massey et al. (1995a), who suggested that its spectral similarity to the sanctified LBV Var B made it an excellent candidate for a "quiescent" LBV. This star is also known as B324 from the catalog of Humphreys & Sandage (1980), and three of the participants at this Winter School (one student, one organizer, and one lecturer) had also suggested that that this star was an LBV, based upon spectral variability (Monteverde et al. 1996). Although photometric variability has not yet been established, the fact that the star has changed spectral types requires that M_V must have changed, if M_{bol} has remained constant, as we expect.

Our thinking concerning the actual number of LBVs in nearby galaxies was also changed forever when the Ofpe/WN9 slash star HDE 269858 (R127) in the LMC underwent an S Dor-like eruption in 1981 (Walborn 1982; Stahl et al. 1983). Ten such stars are known in the LMC (Bohannan & Walborn 1989), and six such stars have been spectroscopically confirmed in M 33 as part of our UIT survey, one of which had been previously discovered by Smith, Crowther & Willis (1995). The survey for WR stars described below should be complete for such stars in several selected regions of M 33, but galaxy-wide numbers are not within sight.

In Fig. 13 I show the spectrum of the first of these stars to be spectroscopically identified in M 31. This very nice comparison with the LMC Ofpe/WN9 star HDE 269927c was put together by Paul Crowther when I asked him to comment on a spectrum I obtained at the Multiple Mirror Telescope last September. In case you are curious, the M 31 star is the one known as OB69-WR2 (Massey et al. 1987a); a previous spectrum with worse signal-to-noise and resolution failed to reveal the strong P-Cygni profiles and narrowness of the lines which put this in the same "slash star" category.

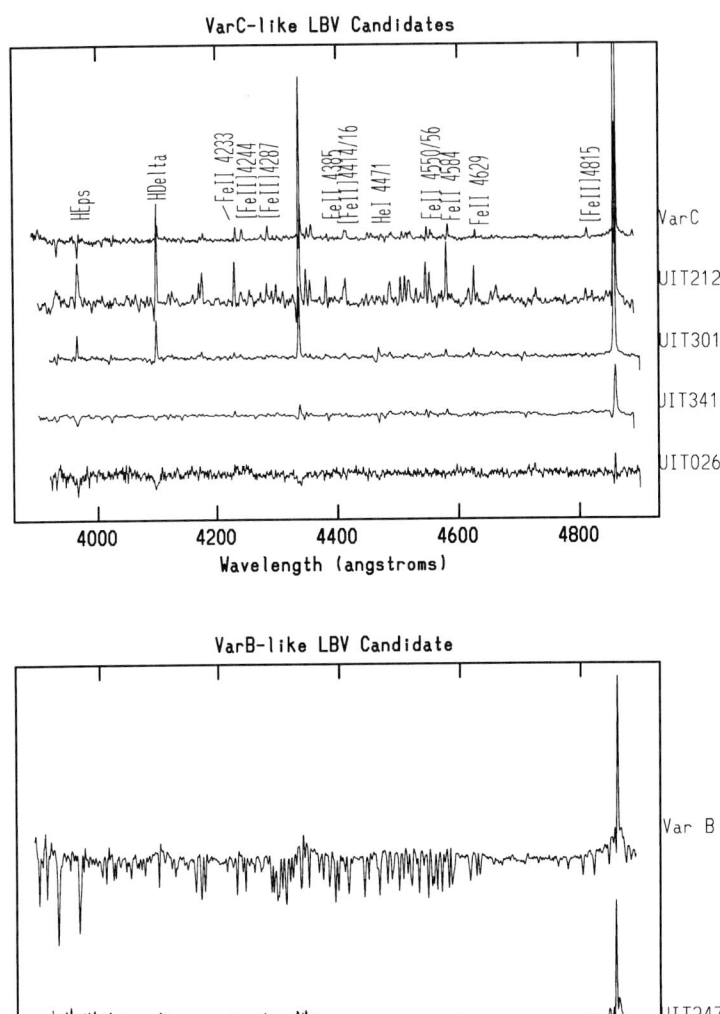

FIGURE 12. Comparisons of the spectra of some stars in M 33 to the known LBVs Var C and Var B. The figures are from Massey et al. 1996. Monteverde et al. 1996 independently demonstrated that the star B324=UIT247 is a spectral variable.

So, in summary: although the stars which are spectroscopically similar to LBVs (B324, for instance) have not been *proven* to be in the same evolutionary stage as bona-fide LBVs, and although we do not know whether all slash stars become LBVs, there is mounting evidence to suggest that the number of stars that undergo episodic periods of high mass-loss may have been very much underestimated. Most of these "LBV candi-

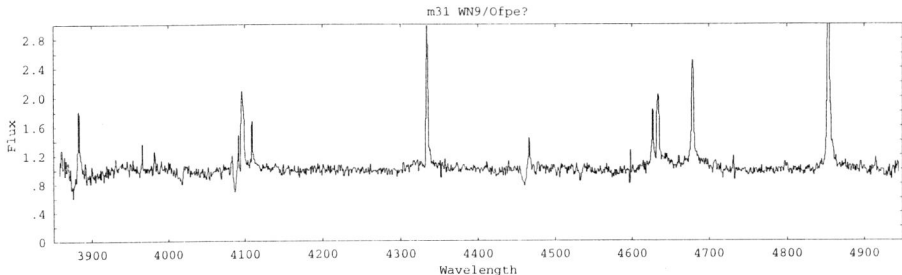
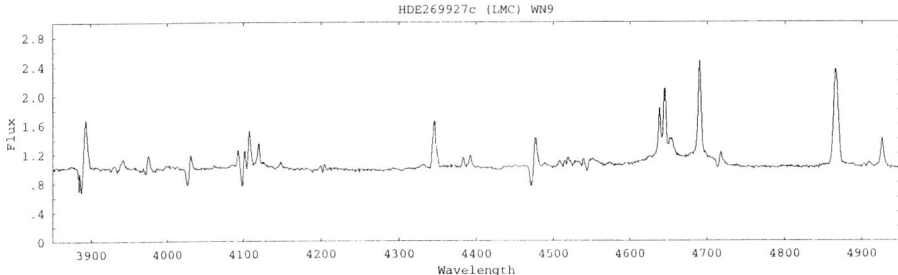

FIGURE 13. A comparison of the M 31 star OB69-WR2 with the slash star HDE 29927c.

dates" have been found pretty much by accident, and a systematic survey is very much needed.

3.2. Wolf-Rayet stars

I show in Fig. 14 the optical spectra of some representative WN and WC Wolf-Rayet stars. For WN stars the spectral subtypes (WN2 through WN9) are based primarily on N V $\lambda 4603, 19$ to N IV $\lambda 4058$ to N III $\lambda 4634, 42$ line ratios, while for WC subtypes (WC4 through WC9) the criteria are based upon the relative ratios of C IV $\lambda 5806$, C III $\lambda 5696$, and O V $\lambda 5592$.

Because of their strong emission lines, it would seem that Wolf-Rayet stars would be relatively easy to find. I remember one of the first 4-m proposals I submitted to do this in M 33; we got the observing time, although I later heard that one of the time allocation committee members had said he was sure he could find these with a 6-inch telescope in his backyard, as the WR emission lines are so strong. Fortunately, he was voted down. In Fig. 15 I show one of the ways that people (including myself) tried to do this at first—it is a low-dispersion objective prism (grism) exposure of NGC 6822. The arrow marks the only Wolf-Rayet star found in this galaxy by this method; it was, in fact, already known. This plate was taken as part of the survey referred to in the introduction; as we shall see, this turns out to be a very poor way of discovering Wolf-Rayet stars.

Why didn't the method work better? With objective prism plates, one is fighting crowding, and, more significantly, the brightness of the night sky.

However, there is a even larger concern that dominates whatever method one uses, if what one is after is some sort of reasonably correct accounting of the relative number

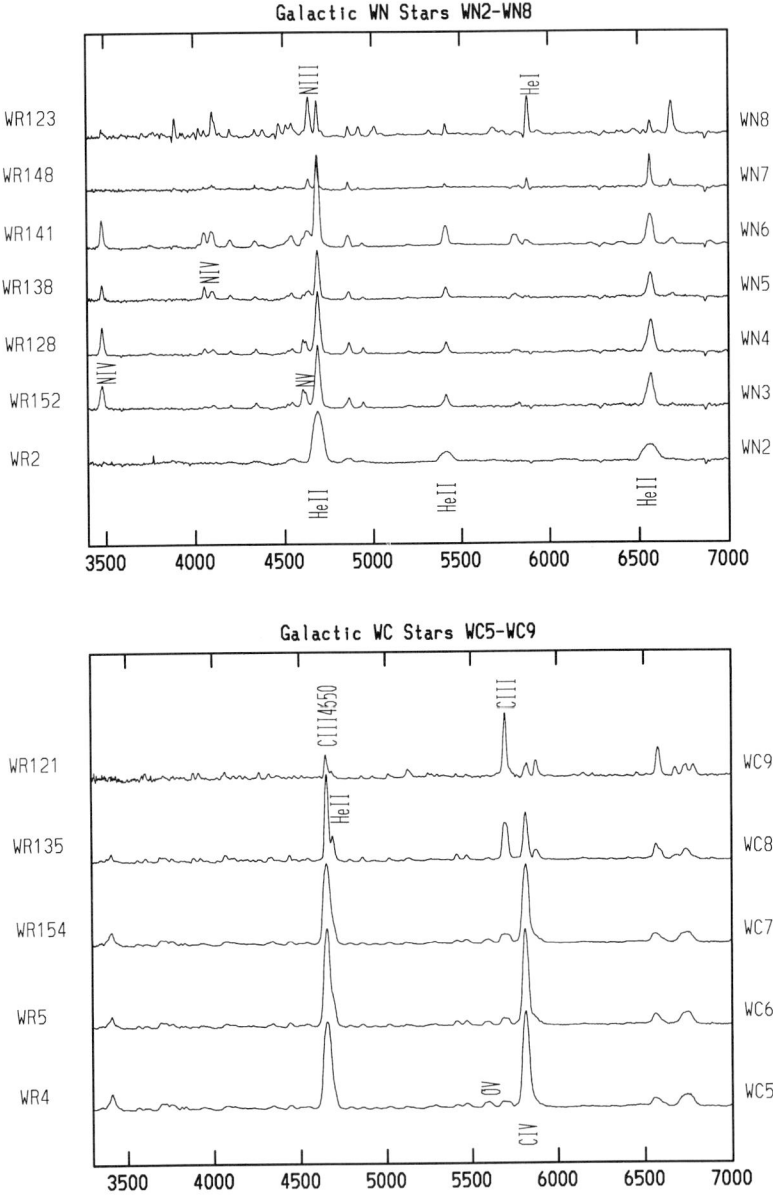

FIGURE 14. The spectra of WN and WC stars. The data were taken with the Intensified Reticon Scanner (KPNO) and SIT-Vidicon (CTIO) back in the early 1980's.

of WC and WN stars. Fig. 16 shows the equivalents widths of the strongest optical line in WC stars (C III $\lambda 4650$) compared to the strongest optical line in WN stars (He II $\lambda 4686$). The data comes from Milky Way and Magellanic Cloud WRs. We see that there is *quite* a difference in the two distributions (note that this is a log plot!): in fact, the median line strength in WC stars is about a factor of 4 stronger than in WN stars. In interpreting this plot, keep in mind that absolute visual magnitudes of many WNs are

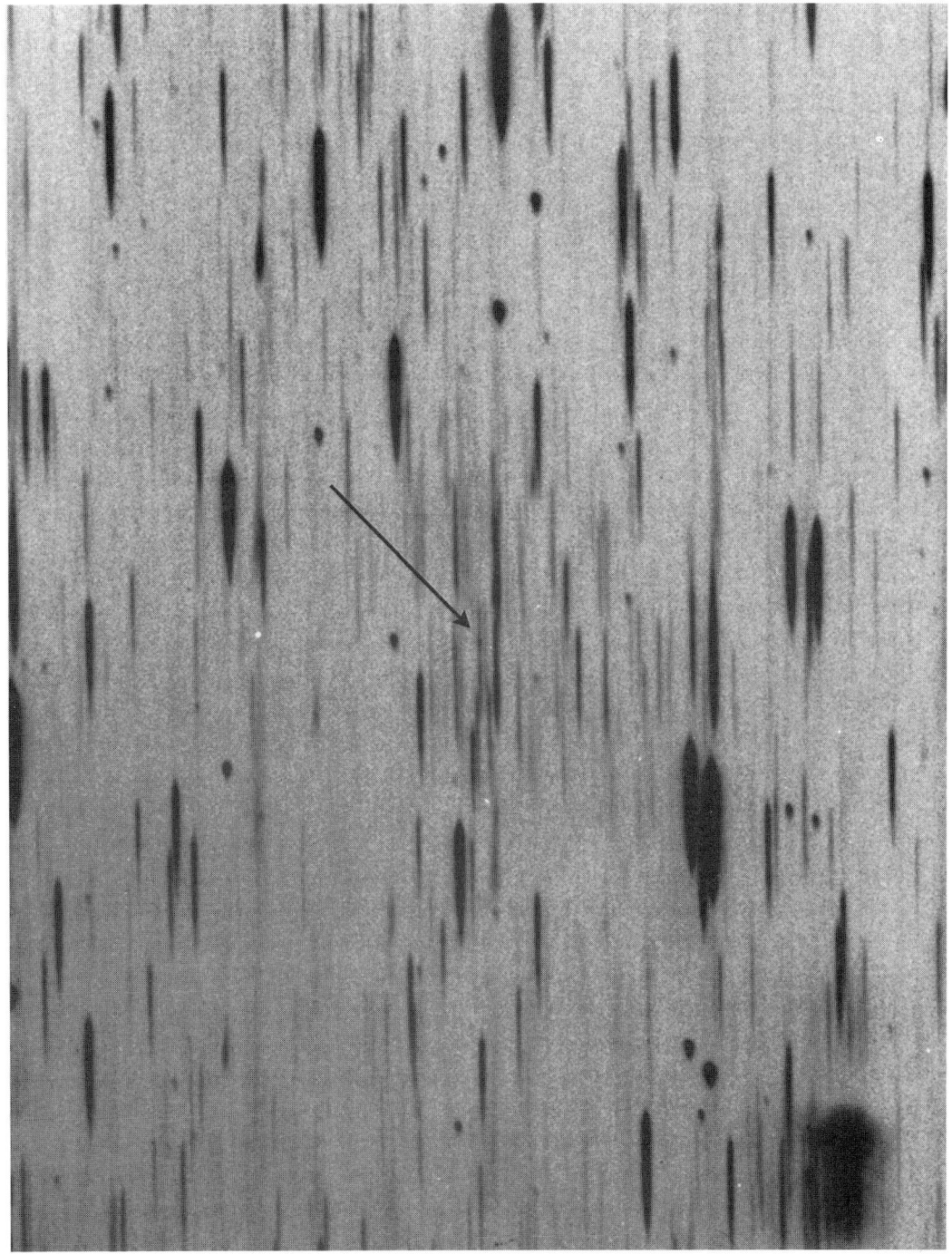

FIGURE 15. A grism exposure obtained of NGC 6822 with the Kitt Peak 4-m telescope. The plate was taken by the author in collaboration with Bruce Bohannan and Peter Conti. The only WR star found on this plate was the one previously known in NGC 6822; subsequent work with interference filter imaging revealed three additional, weaker-lined WR stars.

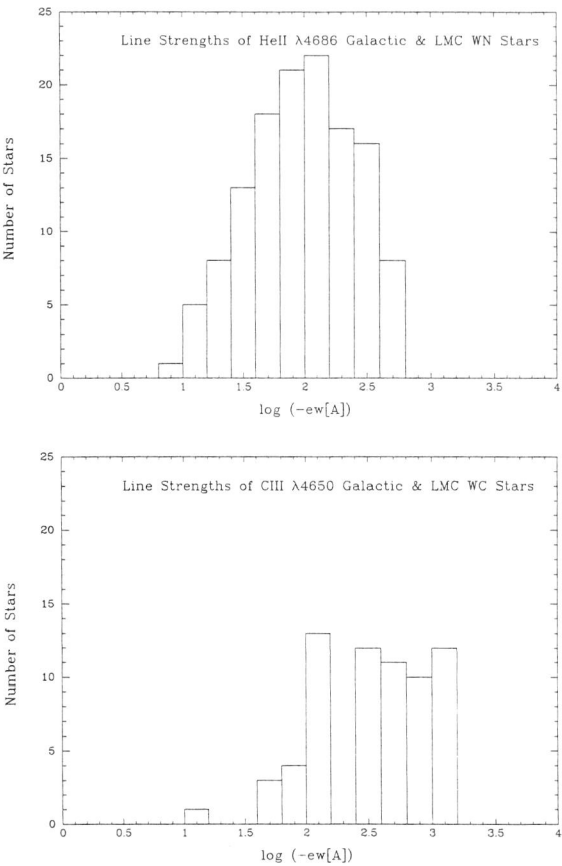

FIGURE 16. The equivalent widths for all the known, non-binary WR stars in the Magellanic Clouds and Milky Way. The data are from Conti & Massey (1989).

comparable to those of the WCs; for instance, $M_V \approx -4.0$ for the WN4 subtype and -3.9 for the WC5 subclass (see Conti & Vacca 1990).

What does this mean for studies of the WR populations in the Local Group? Well, if we are going to compare the number of WC to WN stars, and we want something approaching completeness for WN stars, we are going to have to work something like a factor of 3 or 4 times harder than we would if we just wanted to find most WCs! Although nearly complete samples of Wolf-Rayet stars were identified in the Magellanic Clouds by objective prism surveys (Azzopardi & Breysacher 1979; Breysacher 1981 and references therein), crowding and faintness makes this scheme fail for the more distant members of the Local Group, as witness the fact that only one WR star was detected on the grism plate of NGC 6822 shown in Fig. 15.

According, I designed a set of filters based upon spectrophotometry of WR stars to provide "optimal detection" and separation into WC and WN sequences; these filters were described and used by Armandroff & Massey (1985) to survey NGC 6822, IC 1613, and two regions in M 33 for Wolf-Rayet stars. (In his poster, Pierre Royer describes a beautiful extension of this set that may provide the means to also classify WRs into various subtypes.)

In addition to choosing sensible emission-line and continuum filters, the technique we developed provided a considerable improvement over blinking exposures by eye. For each frame imaged through one of the filters, we performed crowded-field photometry on every star. We obtained then not only the instrumental magnitudes of each object, but just as important, a meaningful estimate of the error on that magnitude. We could then look for stars that were *statistically significantly* brighter in one (or both) of the emission-line filters than in the continuum. Follow-up spectroscopy then allowed us to calibrate this— what fraction of the "3.2σ detections" were real Wolf-Rayets? The results of this, and a new survey, will be discussed in Sect. 5.

3.2.1. Red supergiants

I first got intrigued by the issue of red supergiants (RSGs) in the Local Group when comparing the distribution of blue and red stars in the Humphreys & Sandage (1980) catalog of stars in M 33 (see their Figs. 21, 22, and 23). Basically the distribution of blue stars is pretty clumpy (and indeed this was the basis of defining the OB associations), but the red stars—even the very red stars—show a considerably more uniform distribution. Now, admittedly, one expects the RSGs to be slightly older and hence might have a larger spatial dispersion, but if one considers the fact that the blue stars are dominated by late-B and early-A supergiants of moderate (15-25 $\mathcal{M}_\odot$) high mass (Sec. 1.1), then this discrepancy is surprising: why are the RSGs (which one also expects to be dominated by 15 $\mathcal{M}_\odot$ stars) not found in the same regions? Oddly, the answer is contained implicitly in another paper in the same volume of *Astrophysical Journal Supplements* as the M 33 catalog: the Bahcall & Soneira (1980) model for the Milky Way allows us to estimate the amount of *foreground* contamination by red dwarfs. (The Ratnatunga & Bahcall [1985] analysis previously mentioned does not include sufficient color resolution to be useful for answering this question.) I did a BOTEC ("back-of-the-envelope calculation") and asked the question: how far away would a K or M Galactic dwarf have to be in order to have the same apparent magnitude as a bona-fide RSG in M 33? The results of this simple calculation are shown in Table 5: given that the scale-height for faint M dwarfs is something like 300 pc (see Fig. 2 of Bahcall & Soneira 1980), and that the galactic latitude of M 33 is $-31°$, then you might expect to go about 600 pc before running out of Galactic stars. A 40 $\mathcal{M}_\odot$ star has an $M_{bol} \sim -10$, while a 15$\mathcal{M}_\odot$ star has an $M_{bol} \sim -7$, if we consult the HRDs in Fig. 2, say. At a distance modulus of 24.6, and a bolometric correction ranging from -1.0 (early K) to -3.3 (M5), we might then expect 40 $\mathcal{M}_\odot$ RSGs to be found between $V \sim 15.6 - 18.0$, and 15 $\mathcal{M}_\odot$ RSGs to be found at $V \sim 18.6 - 21.0$, depending upon the spectral class.

What we see immediately in this table, then, is that we should expect foreground contamination to be significant at $V = 15.5$ from K5-M5 foreground stars; at $V = 18.0$ foreground contamination will be significant from M0-M5 stars, but at $V = 20.5$ foreground contamination is significant only for the latest-type M stars. The bottom-line is that we now have a good explanation for why the red stars in Humphreys & Sandage (1980) are distributed the way they are: they are more uniformly distributed than the blue stars because a significant fraction of them may be foreground dwarfs!

So what can be done about this? By playing around with the intrinsic colors of dwarfs vs. supergiants, I eventually came to a conclusion that others had arrived at: there is a pretty good separation in the two-color $B - V$ versus $V - R$ diagram, even with some reddening added to the supergiants. I chose then to go out and image the same fields in NGC 6822, M 31, and M 33 for which I had good UBV data for finding the unevolved massive stars, and for which we had completed surveys for WR stars. The results of this will be described in Sec. 5.4, but for now let me note that the method seems to work

Type	M_V	Distance (pc)		
		V=15.5	V=18.0	V=20.5
K5	8.0	316	1000	3000
M0	9.2	180	575	1800
M2	10.1	120	380	1200
M5	12.3	45	140	440

TABLE 5. The distance at which a foreground red star would have to be for a given V magnitude.

very nicely! A comparison of the distribution of stars in a control field (Selected Area 45 near M 33) compared to stars across the face of M 33 shows that it is indeed possible to separate foreground dwarfs from RSGs via photometry. I am currently in the process of confirming this using the WIYN telescope and Hydra to obtain spectra both of suspected foreground dwarfs and candidate RSGs in NGC 6822, M 31, and M 33.

4. Secrets of star formation as revealed by luminous stars

Garmany, Conti, & Chiosi (1982) began the modern study of the initial mass function (IMF) for massive stars. They constructed a catalog of all known O-type stars, tested it for completeness, computed IMFs both for the regions inwards and outwards of the solar circle, and concluded that there was a galactocentric gradient in the slope of the IMF. As later work showed, the completeness was no where near as good as they thought, and the models did not include a sufficiently wide main-sequence to really allow even differential comparisons, but the paper is really none the worse for having come to the wrong conclusions: it laid out the issues very clearly, got a very interesting problem on the table, and was responsible for inspiring much the further work that followed.

4.1. PDMFs, IMFs, γ, and Γ

Evan Skillman began his lectures on nebular abundances by saying "It's all in Osterbrock"; let me begin here by saying "It's all in Scalo (1986) and Tinsley (1980)." Nevertheless, let me go through this both by words and equations as I think that some of the simplicity can be lost.

First, let me review how we relate the *present day mass function* (what we measure now) to the *initial mass function* (the mass function of newly born stars). Let us consider two extreme regimes:

(a) All the stars in a group are born on Tuesday. We say that the group is then strictly coeval.

(b) Stars are born continuously with a star-formation rate that is constant for at least as long as the life-time of the lowest mass star under consideration.

In each of these two scenarios, how many stars do we see as a function of mass? Let

$$\phi(m)\,\psi(t)\,dm\,dt$$

be the number of stars born per unit area of the galactic disk between time t and $t + dt$ with mass between m and $m + dm$. Then $\phi(m)$ is the initial mass function (basically just a probability function), and $\psi(t)$ is the star formation rate. Of course, in principle, $\phi(m)$ could also be a function of time, but for now let's hope that any dependence on t in ϕ is very, very slow.

In scenario (a) above, in which we have a burst of star formation that begins and ends on Tuesday, then the number of stars we would find on Tuesday night with mass between m and $m + dm$ is just proportional to $\phi(m)$:

$$n(m)\,dm \sim \phi(m)\,dm \tag{4.3}$$

The total number of stars between masses m_1 and m_2 is then

$$n_{m_1}^{m_2} = \int_{m_1}^{m_2} \phi(m)\,dm$$

It is standard practice to assume that the initial function is power law:

$$\phi(m) = Am^\gamma,$$

where $\gamma < 0$; i.e, that fewer high mass stars are born. In that case, the integral becomes easy (maybe this is why it is standard practice!):

$$n_{m_1}^{m_2} \sim m^{\gamma+1} \Big|_{m_1}^{m_2}$$

It's common to refer to $\Gamma = \gamma + 1$ as the slope of the initial mass function, and if we take the log of both sides of equation 4.3, we find

$$\log n = (\gamma + 1) \log m$$

So let's say that we *count* the number of stars as a function of mass in some sample. What we've determined is the *present day* mass function (PDMF). How is this related to the *initial* mass function?

(*a*) In the case of a burst, then at time $t = 0$ (Tuesday night) the PDMF *is* the IMF. At later times $t > 0$, the most massive stars (the ones with main-sequence lifetimes $< t$) will have died. Nevertheless, the slope of the PDMF has the same slope as the IMF for stars whose mass is below this limit:

$$n(m) \sim m^\Gamma \tag{4.4}$$

As long as we're careful to measure the age (or just ignore the highest mass mass bin), then the slope we see today is a true reflection of the IMF. Another way of saying this is that if the mass bins are small enough, then evolution will have no effect on the measured slope of the IMF, but will simply selectively remove the highest mass stars.

(*b*) In the case of *continuous* star formation, stars accumulate. After a time t greater than the hydrogen-burning lifetime of the lowest mass star we are considering, the number of stars we observe as a function of mass (the PDMF) is just the product of the initial mass function with the main-sequence lifetime t_{ms} at a particular mass m:

$$n(m) \sim m^\Gamma \times t_{ms}(m) \tag{4.5}$$

Now, a key point assumed implicitly by the Garmany et al. (1982) study, but which seems often lost on time allocation committees, is that this is really exactly what happens if you average over a *large enough* region in a galaxy: that even though you now have a lot of different star forming regions in your sample, and that most definitely the star formation rate is not constant throughout the region, equation 4.5 still applies, as long as the sample you're dealing with contains enough star-forming regions of different ages.

Finally, let me note that in practice when we measure the PDMF we count the number of stars on the HRD in various mass bins: between $85 \mathcal{M}_\odot$ and $100 \mathcal{M}_\odot$, say, and between $60 \mathcal{M}_\odot$ and $85 \mathcal{M}_\odot$, and so on. We have to make sure to normalize by the bin size; clearly we would get a lot more stars if we simply counted up stars between $60 \mathcal{M}_\odot$ and $100 \mathcal{M}_\odot$ than between $60 \mathcal{M}_\odot$ and $61 \mathcal{M}_\odot$. Scalo (1986) introduces the notation $\xi(\log m)$ to be $\phi(m)$ corrected for the size of the mass bin; i.e., the number of stars born per unit logarithmic (base ten) mass interval per unit area (kpc^2) per unit time (Myr). Taking the log of both sides of either equation 4.4 or 4.5 then leads to

$$\Gamma = d\log\xi(\log m)/d\log m.$$

In the case of an instantaneous burst, ξ is just proportional to n divided by the size of the jth mass bin $[\log m_{j+1}/m_j]$. In the case of continuous star formation, n also needs to be divided by the main-sequence lifetime t_{ms} appropriate to the jth mass bin.

4.2. Doing it in practice

4.2.1. An old war story

So, in the mid-1980's there was this fascinating suggestion made by Garmany et al. (1982) that there was a gradient in the IMF slope Γ with galactocentric distance within the Milky Way, with proportionally more of the highest mass stars being found inwards of the solar circle. I remember coming back to the DAO after one trip to JILA, and telling Sidney van den Bergh about this during morning tea. He expressed some mild skepticism, noting that simple, physical considerations should have it go the other way: radiation pressure on grains should limit the mass of the highest mass star that could form, and so you would expect there to be fewer (or at least less massive upper mass limits) in regions of higher metallicity. (Shields & Tinsley [1976] argue that the upper mass limit should go as $1/\sqrt{z}$.) It got me to wondering about what we would see if we looked at the IMF in the Magellanic Clouds, where the metallicity was considerably more different than in the small range covered by the galactocentric gradient within a few kpc of the sun. Katy Garmany and Peter Conti were beginning to look into this by using the existing OB catalogs of stars in the Magellanic Clouds, and going down to Cerro Tololo to take spectra of these stars, but I found that if I looked at Sanduleak's (1969) catalog for the LMC, or the Azzopardi & Vigneau (1982) catalog for the SMC, it was pretty apparent that stars in regions of strong nebulosity were not included. Given that these H II regions *had* to be excited by O-type stars ("What do you think is ionizing the gas, gumballs?") it seemed to me that some sort of comprehensive program of *UBV* photometry of these regions was needed if we were to have any hopes at all of making progress in measuring the IMF in these two low metallicity systems. CCDs were just then coming on line. (In retrospect, this incompleteness was really only the second biggest problem. The biggest problem was going to be how to get spectra of the hundreds and hundreds of stars we were going to find!) Kathy Eastwood, Katy Garmany, and I proposed to obtain *UBV* images for the stars in the OB associations that had been cataloged by Lucke & Hodge (1970) and Hodge (1985) in the two Clouds, as preparation for determining the initial mass functions. Ten beautiful photometric nights offset the small instrument problem on our first night, when I (single-handedly) destroyed the RCA4 chip.

It is amazing to me to think back on how patient the CTIO Director and TACs were with us, giving us 4-5 nights of 4-m time each year for the follow-up spectroscopy; this went on for most of a decade. I think of this whenever I hear people claim that the National Observatories are unwilling to support long-term studies.

Fig. 17 shows the identification of stars in one of our CCD frames centered on NGC 346 in the SMC, the most striking HII region in that galaxy. Our spectroscopy discovered numerous previously unknown O stars, including new ones as early as type O4; this one study *doubled* the number of known early O-stars in the SMC (see Massey et al. 1989).

Meanwhile, I realized that an analogous program needed to be carried out for the Milky Way OB associations. Again, I was fortunate in the timing: although these regions often covered a degree on a side, a gigantic Tektronix 2048^2 CCD with reasonable U sensitivity was available on the Kitt Peak 0.9m telescope, and I was involved in getting the multi-object fiber-feed Hydra on-line at the 4-m. Extending this project to the

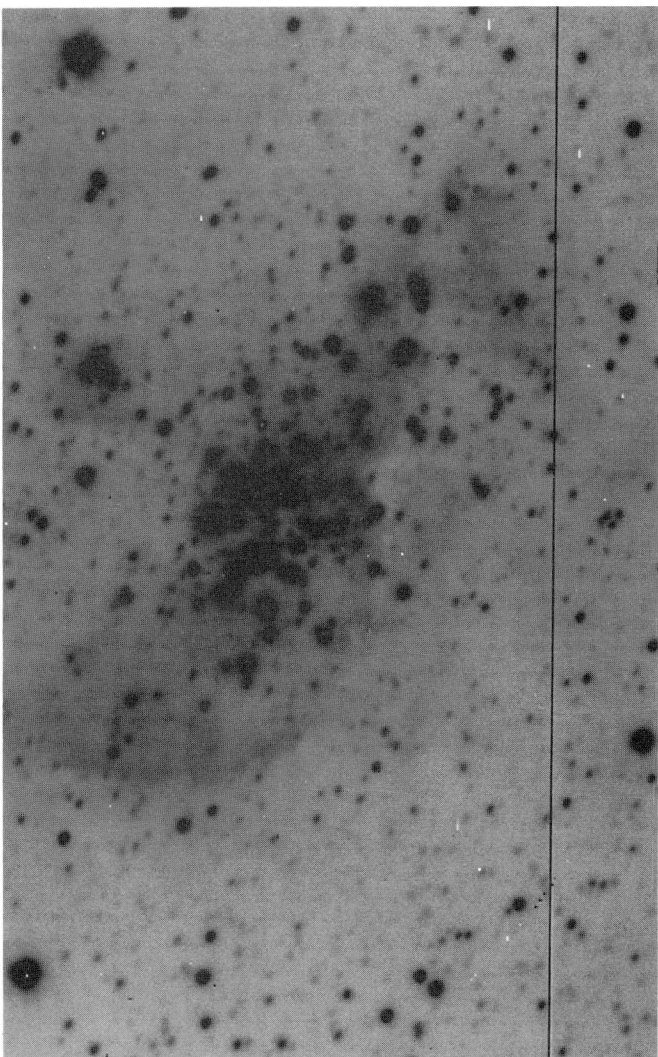

FIGURE 17. The central part of the cluster NGC 346 in the SMC. This CCD image was obtained by the author and Kathy DeGioia-Eastwood in November 1985. Spectroscopy of the blue stars in this one cluster more than doubled the number of known early-type O stars in the SMC.

northern hemisphere Milky Way was quite possible, and only required convincing the time allocation committees that all this wasn't already known.

So, given all of these hundreds and hundreds of spectra, what did we find?

4.3. The initial mass function of the Magellanic Clouds and the Milky Way: the massive star content of OB associations

I list in Table 6 the slopes of the initial mass-functions, the upper-mass limits, and the ages of the various Magellanic Cloud and Milky Way clusters we have studied. The source of the MC data is given in Massey et al. (1995c), while that of the Milky Way is derived in Massey et al. (1995b).

The basic thing to get out of Table 6 is that there really are no differences in the IMF

Association	Age (Myr)	M_{up} ($\mathcal{M}_\odot$)	Γ
	SMC		
NGC 346	2-4	70	-1.3 ± 0.1
	LMC		
LH 9	1-5	55	-1.4 ± 0.2
LH 10	0-3	90	-1.1 ± 0.1
LH 58	2-4	50	-1.4 ± 0.2
LH 117/118	1-3	100	-1.6 ± 0.2
	Milky Way		
NGC 6823	2-7	40	-1.3 ± 0.4
NGC 6871	2-5	40	-0.9 ± 0.4
NGC 6913	4-6	40	-1.1 ± 0.6
Berkeley 86	2-3	40	-1.7 ± 0.4
NGC 7235	(6-11)	15	(-2.0)
NGC 7380	2	65	-1.7 ± 0.3
Cep OB5	2-4	30	-2.1 ± 0.6
IC 1805	1-3	100	-1.3 ± 0.2
NGC 1893	2-3	65	-1.6 ± 0.3
NGC 2244	1-3	70	-0.8 ± 0.3
NGC 6611	1-5	75	-0.7 ± 0.2
Cyg OB2	1-4	110	-0.9 ± 0.2
Tr 14/16	0-3	>120	-1.0 ± 0.2

TABLE 6. The age, highest mass star (M_{up}), and IMF slope Γ for OB associations in the Magellanic Clouds and Milky Way. The data are from Table 5 of Massey et al. (1995b).

slopes between different associations, nor between that of the SMC, the LMC, and the Milky Way. Similarly high mass stars are found in all three systems. This demonstrates that whatever it is that controls the mass of the highest mass star that can form, it isn't radiation pressure acting on grains: there simply is no discernible difference despite the factor of 10 change in metallicity between these three systems.

Now, the alert reader will notice something a little odd about one of the numbers in Table 6, namely that the upper mass star listed for NGC 346 is only 70$\mathcal{M}_\odot$, not the 100-200$\mathcal{M}_\odot$ that Rolf Kudritzki has been telling us about for this star. ("NGC346#1" is also known as Star 3 in Walborn & Blades [1986] or NGC346-355 in Massey et al. [1989].) Remember that for consistency I've simply adopted the bolometric corrections found in Table 2, even for the O3 stars. But, if there is anywhere that this procedure is likely to be in error, it is indeed for the O3 stars. Still, this should not affect the slope of the IMF, any more than selective depletion does. It also perhaps demonstrates how much the very interesting result of Rolf and his colleagues have about these high masses depends upon their having gotten the effective temperature and implied bolometric correction right in dealing with these extremely hot stars.

4.4. Field stars

Let me next turn my attention to the massive-star field population. People are sometimes surprised about this topic as they have in their heads the idea that all stars, and in particular all massive stars, must form as part of large complexes. But what really is the evidence for this?

Garmany, Conti, & Chiosi (1982) show the distribution of O-type stars within 3 kpc of the sun in their Fig. 2. We know now that the statistics for these data were woefully incomplete, particularly in the OB associations. Nevertheless, it is interesting to see the large number of stars (50%) marked as "field stars", i.e., stars that are not (currently) members of OB associations.

I decided to construct an HR diagram for the true "field" population of the Magellanic Clouds, using the Lucke-Hodge OB association boundaries as my guide. I defined a "field star" as one that is so far outside the boundaries of any OB association that the star would not have been able to travel there during its short H-burning lifetime. Churchwell (1991) suggests that velocities of 3 km s^{-1} relative to the parent molecular cloud may be typical, citing the study of Orion by Zuckerman (1973). In 10 Myr we would thus expect a star to travel 1×10^{16} km, or 30 pc. At the distance of the LMC, this corresponds to an angular distance of 2'. (We argue in our paper why this is actually a pretty conservative limit.) Now, at this point someone will invariably ask me about "runaway" O type stars—can't all the O stars found outside of the association boundaries simply be O stars with high peculiar velocities that have been ejected from the associations? In order to explore this possibility, we also defined a population of "extreme" field stars, using as a criterion the separation corresponding to a transverse motion of 100 km s^{-1} in 3 Myr. (None of the six "runaway" stars in the Conti, Leep, & Lorre [1977] study of over 200 O-type stars had peculiar velocities this large.) Further details can be found in Massey et al. (1995c); here let me make just one point: this sample of field stars includes eight O3 and O4 stars. It is pretty hard to imagine the circumstances that could have ejected these vast distances from the nearest OB associations during their very short lifetimes.

We show in Fig. 18 the HRD for these *field* massive stars in the LMC from Massey et al. (1995c). Now, as emphasized in Sect. 1, any V-limited catalog that doesn't go sufficiently deep is going to be incomplete for the hottest stars. Thus, to determine the completeness corrections (as a function of luminosity and effective temperature) we imaged two LMC fields in UBV using the Curtis Schmidt, and subsequently obtained spectra for the bluest members with the CTIO 4-m. The results of this completeness test is found in Fig. 19, and this figure indeed confirms that the hottest and most massive stars have simply been missed. We found that we could reliably correct the HRD for the missing stars down to 25 $\mathcal{M}_\odot$ (Sec. 4).

There are three key results from this field star population:

• The Geneva evolutionary models (Schaller et al. 1992; Schaerer et al. 1993) do an excellent job of reproducing the distribution of stars in the H-R diagram, in the sense that at a given mass range, the same number of stars are found between each 2 Myr isochrone. The main-sequence has the correct width, and there is no evidence of an abundance of "blue loop" stars. This is contrary to the conclusion offered at the Winter School by Cesare Chiosi, who based his comparison on the Fitzpatrick & Garmany (1990) sample. The sample is the source of the "known OB stars" (asterisks) shown in Fig. 19, and one can see just how incomplete that sample was.

• The initial mass function for stars in the field is considerably different than that in the OB associations. We find IMF slopes of $\Gamma = -4.1 \pm 0.2$ for the LMC and $\Gamma = -3.7 \pm 0.5$ for the SMC. Reanalysis of the Garmany, Conti, & Chiosi (1982) field stars using the Geneva tracks (Maeder & Meynet 1988) showed that the IMF slope for Galactic field massive stars is -3.2 ± 1.4, consistent with the values for the Magellanic Clouds.

• Equally high mass stars are found in the field as in the associations, as evidenced by the existence of O3 stars. They simply are not formed in the same proportion to stars of somewhat lower mass. This is just what is expected on theoretical grounds due to the relation between gas temperature and fragment mass (Larson 1985, 1986), if we assume that the field stars are born in lower density (and hence cooler) regions.

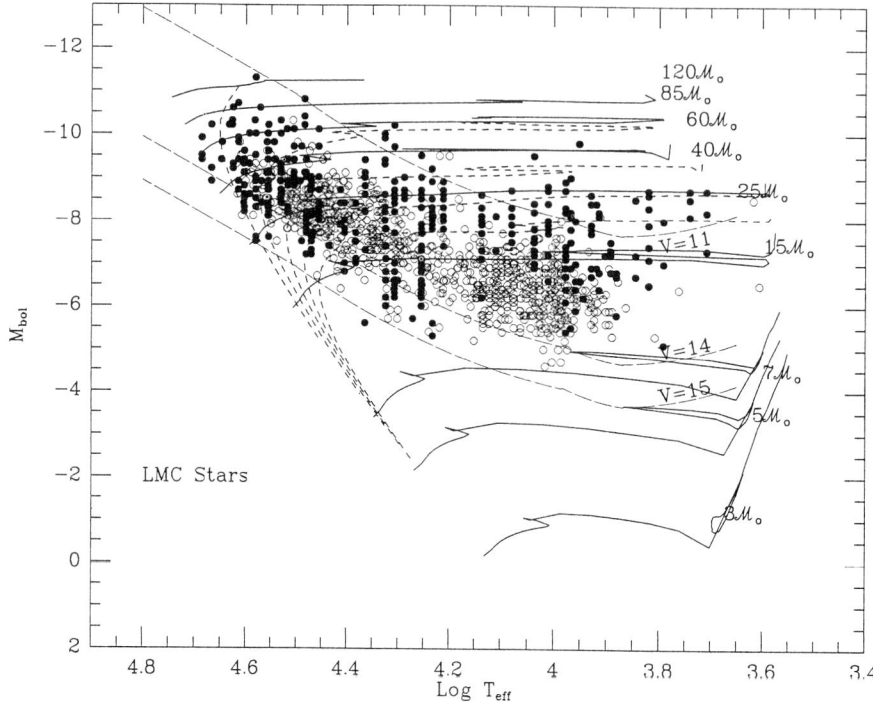

FIGURE 18. The uncorrected sample of field stars in the LMC. This figure is based upon Fig. 4 in Massey et al. (1995c). We have included lines of constant V magnitude in addition to the evolutionary tracks and isochrones.

4.5. Exercise for the student

Given the lines of constant V-magnitude shown in Fig. 18, what exactly could we learn about the IMF by constructing a "luminosity function" for the LMC? (A luminosity function is the cumulative number of stars brighter than a certain V magnitude.) What mass stars would be included in a $V = 11.0$ to $V = 11.5$ magnitude bin, say? Can you see how one could backtrack this reliably to determine the IMF? (Neither can I.)

5. Secrets of stellar evolution revealed by luminous stars

Prior to 1976, the prevailing wisdom was that Wolf-Rayet stars were the product of binary evolution: Roche-lobe induced mass-loss in very massive binaries would result in the (initially) more massive star losing its H-rich outer layers and revealing the products of nuclear burning. Then, at a colloquium held in Liege, Peter Conti (1976) first proposed that maybe WR stars could form from single massive stars via mass-loss simply through their stellar winds, and that in fact, Wolf-Rayet stars might be a normal evolutionary state in the lives of all of the most massive stars. This became known as the "Conti Scenario".

In 1996, another conference was held in Liege to examine what we had learned in the twenty years since this remarkable suggestion. I was invited to review the "Numbers and Distribution of Wolf-Rayet Stars in Local Group Galaxies", and the following section is adapted for this Winter School from that review (Massey 1997).

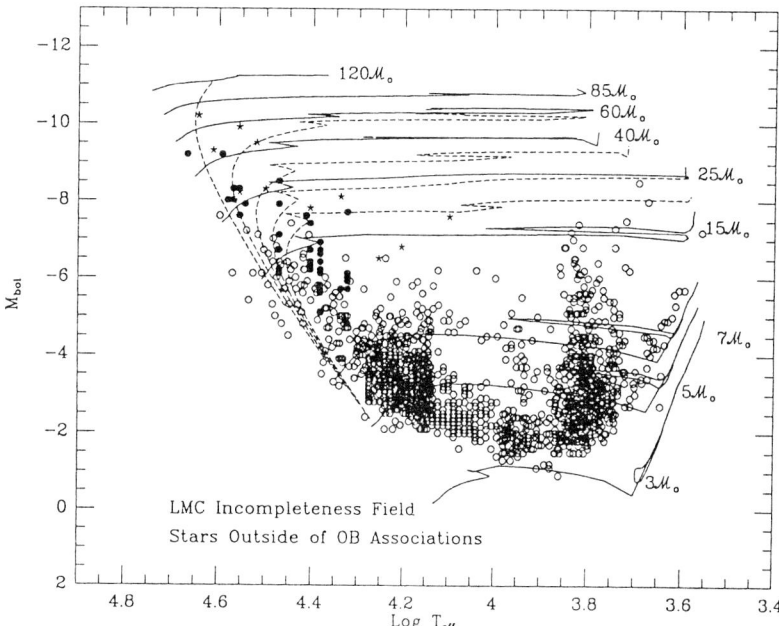

FIGURE 19. The incompleteness correction as a function of mass for the LMC. The asterisks denote the stars already known from catalogs; the other symbols show the stars newly found from the Schmidt field survey described in the text. The filled circles have been placed in the diagram on the basis of spectra; the open circles have been placed by only photometry.

The four things I will cover here are:
- **WC/WN Ratios.** (Sec. 5.1) We have known for years that the relative number of WC to WN stars is significantly lower in the Magellanic Clouds (MCs) than in the Milky Way, and it has been very tempting to interpret this in terms of the lower metallicity of the Clouds. Massey & Conti (1983) established that there is a galactocentric gradient of the WC/WN number ratio in M 33 in the same sense, but comparison of the actual WC/WN numbers have not shown any consistent correlation with metallicity. A new, significantly more complete survey for WR stars in M 33, and some spectroscopic hints that the metallicity in NGC 6822 may actually be lower than in the SMC, have salvaged this relationship: the correlation with metallicity is now very good, although there are some very interesting exceptions (i.e., IC 10) that need to be understood.
- **The Turn-off Masses in Clusters containing WR Stars.** (Sec. 5.2) Recent work has established that most OB associations are extremely coeval (Hillenbrand et al. 1993; Massey et al. 1995b, 1995c). This gives us a *very* powerful tool for directly measuring the mass of the progenitors of WR stars of various subtypes. As an additional benefit, we obtain a direct measure of the bolometric corrections for WR stars. Data on a few Galactic and Magellanic Cloud OB associations will be summarized; further work is in progress.
- **The Relative Number of Massive Unevolved Stars and WRs.** (Sec. 5.3) Recent work among mixed-age populations has established for the first time the actual number of stars with masses $> 40 \mathcal{M}_\odot$ ($N_{>40\mathcal{M}_\odot}$) in the LMC, the Milky Way, and even NGC 6822, M 31, and M 33. The ratio between $N_{>40\mathcal{M}_\odot}$ and the number of WR stars, is not the canonical 10, but rather ≈3. This may indicate that some WRs are

still core-H burning objects (Conti et al. 1995), or it may be that somewhat lower mass stars ($\approx 30 \mathcal{M}_\odot$) evolve to WR stars. The ratio $N_{>40\mathcal{M}_\odot}$/WR is constant to within the uncertainties for all five of these galaxies, suggesting that we are not yet sensitive to metallicity effects.

- **Red Supergiants and WRs**. (Sec. 5.4) A new survey for red supergiants (RSGs) in M 31, M 33, and NGC 6822 is nearing completion with interesting results. I'll summarize what these data are hinting about massive star evolution.

5.1. The WC/WN ratio in Local Group galaxies

In the early 1980's, Peter Conti and I would write observing proposals that began with the statement that the WC to WN ratio was known to show a progression from the solar neighborhood (1:1), to the LMC (1:4.5), to the SMC (1:7), and that this went in the same sense of the metallicities of these same systems ($z = 0.02$, 0.008, and 0.002, respectively, according to Lequeux et al. 1979). This was easy to understand in the context of the "Conti scenario" (Conti 1976; Maeder & Conti 1994): as a massive star evolves, it peels off its outer layers via its stellar wind, revealing first the products of CNO burning (WN stars) and eventually the equilibrium products of He-burning (WC stars). The higher the mass-loss rate (high luminosity and/or high z), the better this works. Thus, stars in galaxies with higher metallicity (or an IMF skewed towards the most massive and luminous stars) should have higher mass-loss rates on average, and hence have an easier time peeling all the way down to a WC star, while stars in lower-metallicity galaxies (or lacking in luminous stars) might never make it to the WC stage at all but end their WR lives as WNs. We proposed to find the WR stars in M 33 and determine their types (looking perhaps for a galactocentric gradient that might be attributable to the known metallicity gradient), and we proposed to do this in M 31, a galaxy whose metallicity gradient wasn't as pronounced but whose overall abundance was comparable or higher to that of the MW. Taft Armandroff and I also struck off on our own and conducted similar surveys in NGC 6822 and IC 1613. (Along the way we also established that the data for the SMC were fairly complete.) It was our hope that by doing this for all of the galaxies of the Local Group that we could do this for—which overlap in metallicity—we could also begin to sort out the effects of z from the effects of different IMF slopes.

Let me cut directly to the chase here: today I think that the WC/WN ratio in *most*—but definitely not *all*—of the Local Group galaxies depends almost solely on metallicity. Now, this differs from what I had to say about this at our last big get-together in Bali (Massey & Armandroff 1991), but improvements in the observational data (both on the direct measurement of IMF slopes in the Milky Way and Magellanic Clouds, and the results of a new CCD search for WRs in M 33) have convinced me that it is the metallicity effects that (usually) dominate. Furthermore, I think that our work on the IMF in OB associations and the field over the past few years have even suggested a way of understanding the notable exception to this trend in the Local Group, namely IC 10.

I'd like to begin by reviewing the observational data on M 33, a galaxy that both educated us and led us astray.

5.1.1. Wolf-Rayet stars in M 33

The first WR stars beyond the Magellanic Clouds were found by Wray & Corso (1972) using on-band, off-band interference-filter photography; 24 (mostly WC) WRs were found in two fields near the center. A grism survey by Bohannan, Conti, & Massey (1985) turned up a few more WRs (Conti & Massey 1981), but the first deep, galaxy-wide survey was obtained by Mike Shara, Tony Moffat, and myself at the CFHT (Massey et al. 1987b) using image tube photography through an interference filter. I spent countless weekends closeted away in Sidney van den Bergh's office at the DAO using his blink comparator to identify stars that might (or might not) blink, and Peter Conti and I subsequently obtained IIDS spectra of many of the candidates with the 4-m Mayall telescope. Observing with Peter is never dull, and as each spectrum began to build up Peter would stare intently at the oscilloscope display, finally declaring the star either "a winner" or "a loser" depending upon whether or not He II λ 4686 and/or C III λ 4650 was present or not. ("Once a loser, always a loser," Peter grumbled upon my attempt to reobserve a couple of stars I was just *sure* had blinked, and he was usually right.) That work resulted in a catalog of the 79 known WR stars in M 33, and established a clear gradient in the ratio of WC to WN stars with galactocentric distance within M 33 (Massey & Conti 1983); the center was dominated by WC stars, while the outer portions were dominated by WN stars. We very cautiously stated that the absolute numbers might be wrong (we were worried even then about selectively missing WN stars) but felt pretty certain about our result that the number ratio changed by about a factor of 4 or 5 going from the center to the outer regions.

A few years later, Taft Armandroff and I started searching for Wolf-Rayet stars using CCD imaging and a set of interference filters designed for "optimal" detection of the WR emission lines, and which also allowed us to distinguish WC from WN photometrically in most cases, although follow-up spectroscopy was needed to confirm what we really had. The CCD removed (most) of the subjective elements, as explained in Sec. 3.2. Our first efforts at this were aimed in surveying the Local Group irregular galaxies NGC 6822 and IC 1613, but we included two "test fields" in M 33 that contained a lot of known WR stars in order to demonstrate that the system worked. (It required a certain amount of talking to convince the CTIO telescope operator that I really wanted to point the telescope to $\delta = +30°$.) The only thing was: when Taft reduced the frames the following summer, he found a lot more WR candidates on the M 33 frames than we had found photographically (Armandroff & Massey 1985=AM85). Follow-up spectroscopy of these new candidates suggested that the photographic survey had been about a factor of two incomplete (Armandroff & Massey 1991=AM91).

Most of the new ones were WNs, and the reason why is easy to understand on the basis of what we saw in Sect. 3.2: the median line strengths of WC stars is about a factor of 4 stronger than for WNs.

The incompleteness issue continued to nag at me for the past decade, and in particular I wondered what if one conducted as sensitive a search for WN stars in the center of M 33 as we had in those two fields? We had claimed that the gradient of WC/WNs with galactocentric distance was real, based upon the argument that our selection effects were probably not much worse in the center than in the outer regions—i.e., that we should be equally bad everywhere. But, I was never comfortable with this since the center regions were a lot more heavily exposed on the photographic plates.

To answer this question, I recently obtained CCD frames of eight $5' \times 5'$ fields using the KPNO 2.1-m telescope ($0.3''$ /pixel) in typically subarcsecond seeing. We included the two AM85 fields to see if there were even more missing WRs, and included a central frame. (At the same time we conducted the search for red supergiants described in Sec. 5.4.) The fields were chosen to include the regions surveyed for O and B stars by Massey et al. (1995a).

What did we find? Summer student Olivia Johnson (Vassar) and I have finished reducing these data, and she, Kathy Eastwood (NAU) and I have obtained spectra of enough of them to have a reasonable handle on what the significance levels really mean. Our survey turned up 129 WR candidates, of which 94 were of sufficiently high significance as to be likely WR stars. Of these 94, 47 had previously been found photographically,

	New			MC83		
ρ	WC	WN	WC/WN	WC	WN	WC/WN
.00—.19	11	14.5	0.76	11	5	2.20
.20—.39	17.5	31	0.57	15	12	1.25
≥ 0.40	4.5	16.5	0.27	11	24	0.46

TABLE 7. The number of WC and WN stars in M 33 as a function of galactocentric distance ρ, normalized to $25'$ (the Holmberg radius), and corrected for projection effects. Fractional stars are listed as we apply a correction for the expected number of "winners" and "losers" based upon the significance level of the detection.

demonstrating that the photographic work had been about 50% complete. Of the new ones, 75% are WNs.

Indeed, we found a slew of new WNs near the center of M 33, although we also found additional ones in the outer regions. The AM85 fields were found to be essentially complete, as we thought—although some of the 3σ detections of AM85 stars were now booming in at 14σ, thanks to the substantial improvements in CCDs! And, there was even a selective bias towards WC stars in the center of M 33, as I had feared—although a substantial gradient (factor of 3) is confirmed, as shown in Table 7 and Fig. 20.

Let me add as a bit of a digression the fact that we have now analyzed the beautiful UV (1500 Å and 2400 Å) images returned by the *Ultraviolet Imaging Telescope* flown on-board the ASTRO-1 shuttle mission (Massey et al. 1996). This survey in fact found virtually *none* of the normal WR stars in M 33—the bolometric corrections even at $\lambda 1500$ render these stars much fainter than the O and B stars. What we did find, though, were the so-called "super WN" stars in M 33's giant H II region NGC 604, objects which we know are very modest analogs of R 136 (Hunter et al. 1996). In addition, the survey detected six Ofpe/WN9 stars, including the one accidentally found by Willis et al. (1992). It is comforting to note that two of these fell into our CCD survey fields, and indeed both were found independently: one as a 13σ detection and the other as a 24σ detection! Also prominent in the UV-selected sample were two B I+WN3 stars, similar to HDE 269546 in the LMC; one is forced to wonder if we understand what is going on with these objects. We also found a number of other interesting objects, including numerous LBV-look-a-likes, but that is a story for another time and place (see Sect. 3.1).

5.1.2. *The relationship between metallicity and WC/WN ratios*

The new M 33 WR CCD survey results let us now compare the absolute WC/WN ratio at different metallicities with those of other Local Group galaxies. The most significant thing to come out of this study is the fact that the WC/WN number ratio now shows a very nice correlation with metallicity for the Local Group galaxies—with one very notable exception that will be discussed shortly. Table 8 gives the actual data, and we see in Fig. 21 what the correlation with metallicity looks like.

NGC 6822, IC 1613, and IC 10 have all been completely surveyed for WR stars via this same technique (deep CCD imaging through interference filters with follow-up spectroscopy), as well as eight regions in M 31. (Of the galaxies beyond the MCs, it was only M 33 that had been studied photographically.) Now, a few additional WR stars have been discovered spectroscopically in the LMC over the past decade or so, but overall the census there is pretty complete; likewise, our knowledge of the WR content of the SMC is probably complete. (Taft and I once obtained CCD images through our WR filters of many of the SMC's OB associations during a night too marginal for our intended project; this survey failed to detect anything new.)

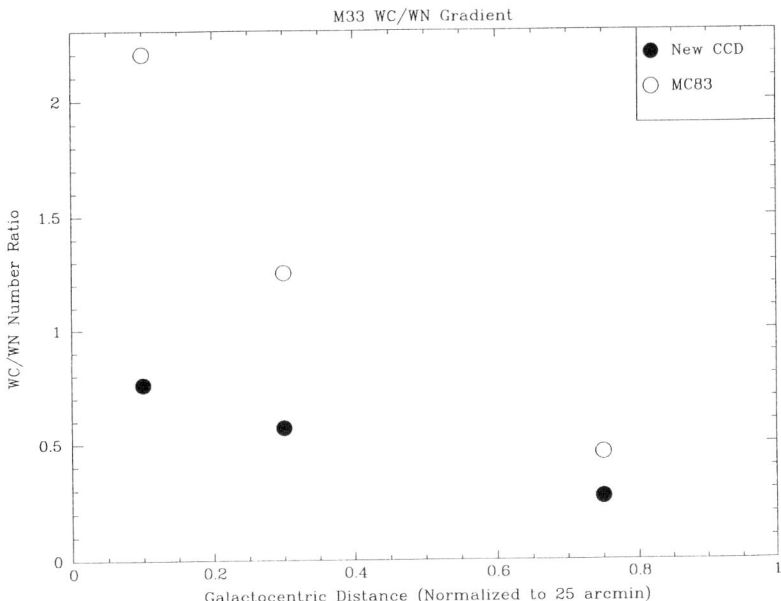

FIGURE 20. The WC/WN number ratio as a function of galactocentric distance ρ in M 33. The data are from Table 7.

Galaxy	WC/WN	WR/kpc^2	Abundance	
			12+log(O/H)	Reference
IC 10 (global)	2.0	5.1	8.2	Lequeux et al. 1979
Milky Way (d <3 kpc)	0.9	2.2	8.7	Shaver et al. 1983
LMC (global)	0.2	2.0	8.4	Russell & Dopita 1990
SMC (global)	0.1	0.9	8.1	Russell & Dopita 1990
M 33 (global)	0.7	2.2	8.4	Vilchez et al. 1988
M 33 CCD fields				
$\rho \sim 0.1$	0.8	≈6.3	8.9	Vilchez et al. 1988
$\rho \sim 0.3$	0.6	≈6.9	8.6	Vilchez et al. 1988
$\rho > 0.4$	0.3	≈1.9	8.4	Vilchez et al. 1988
M 31 (8 CCD fields)	0.8	0.7	9.0	Blair & Kirshner 1985
M 31 (3 "active" fields)	0.9	1.3	9.0	Blair & Kirshner 1985
NGC 6822	0.0	0.6	8.3	Pagel et al. 1980
IC 1613	(1 WC)	0.7	7.9	Talent 1980

TABLE 8. The number ratio of WC to WN stars for Local Group galaxies, along with the WR surface density and the abundances. Data are from Massey & Armandroff (1995), except for M 33, which is based upon the new CCD survey described here.

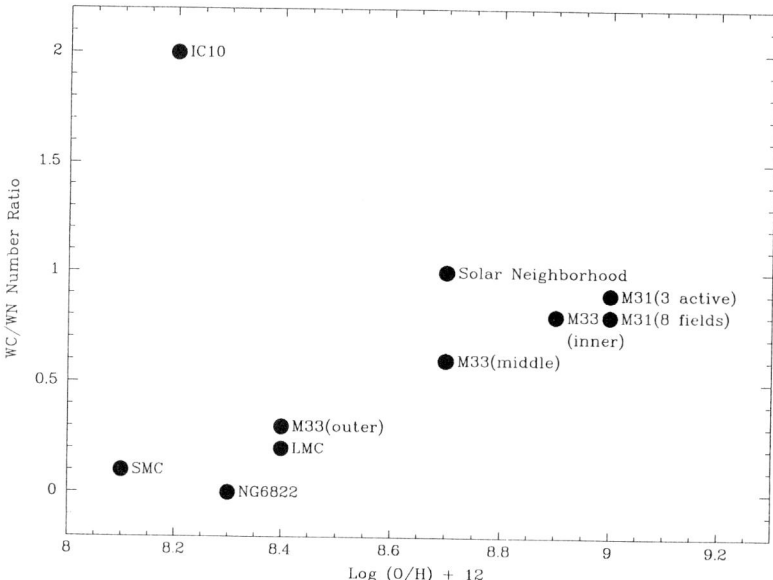

FIGURE 21. The WC/WN number ratio appears to correlate very nicely with metallicity (at least as indicated by the oxygen abundance) with the very glaring exception of IC 10, which appears to be undergoing a starburst.

The correlation is certainly good with the three M 33 points, but it is not perfect. First, let's consider the simple problems: the solar neighborhood, and the NGC 6822/SMC reversal. Given that the WR stars in the MW have been found primarily by objective prism surveys (the Henry Draper catalog), one is forced, I fear, to treat the solar neighborhood (3 kpc of the sun) WC/WN ratio as an upper limit. With *no* extinction, an $M_v = -3.8$ WNE star will be found at $V = 8.5$, within the probable completeness limit for the HD survey. However, the *average* $E(B-V)$ for 10 Galactic OB associations studied by Massey et al. (1995b) (which are found at similar distances) is 0.75 mag, or an implied $A_V = 2.3$ mag. Thus, a "typical" WNE star at 3 kpc is going to be 11th magnitude, and while there may be a few stars that faint in the HD catalog, there aren't many! My point here is merely that if we have to worry so much about selection effects favoring the detection of WC stars (given the 4× greater line strengths!) in M 31, M 33, etc., then I think we have to be equally concerned about the problem locally.

What about the fact that the SMC is more metal-poor than NGC 6822 but sports one WC star to 7 WNs, while NGC 6822 contains four WNs, and no WCs? Well, surely small number statistics play a role here, and I think we have to at least allow for stochastic uncertainties of order $\sqrt{N}$. However, Massey et al. (1995a) also makes the case for NGC 6822 having a lower metallicity than that usually supposed, based upon the weakness of Si lines in the spectra of B supergiants. By comparing stars of the same M_V and spectral type, we concluded that the metallicity appears to be even lower than the SMC.

This tight relation is nice, and is easy then to understand in terms of the Conti scenario; back when it appeared that galaxies of the same metallicity had very different WC/WN ratios, this could not be explained without invoking differing IMFs (see discussion in Massey & Armandroff 1995).

However, there is clearly a conundrum posed by IC 10. Let us consider this interesting galaxy further.

5.1.3. *IC 10: a starburst in the Local Group*

IC 10 is an irregular galaxy located about 1 Mpc out, and it was surveyed for WR stars by Massey, Armandroff & Conti (1992), who were intrigued by Hodge & Lee's (1990) description of this interesting irregular. With the largish distance and high reddening, it occurred to us that we might not find any WRs; imagine our surprise when we found 22 candidates, 16 of which were of high significance. Subsequent spectroscopy confirmed 15, giving IC 10 a galaxy-wide surface-density of WRs higher by a factor of 2 than any other Local Group galaxy: in fact, the global (galaxy-averaged) number of WR/kpc^2 is comparable to that of active star-forming regions in these other galaxies (see Table 8). This is yet another piece of evidence that supports the fact that IC 10 has an extremely high rate of star-formation; data from the Hα luminosity suggests it has one of the highest star-formation rate of any normal irregular galaxy (Hunter, private communication). A comparison of its Hα luminosity with its H I mass led Massey & Armandroff (1995) to call it a classic starburst.

IC 10 has a low metallicity, as measured for its H II regions by Lequeux et al. (1979), but the WC/WN ratio here is 2.0, at least twice that of even the most metal-rich galaxies of the Local Group. How can we explain this? The suggestion we offered in Massey & Armandroff (1995) was that star-formation "vigor" affects the IMF and thus the number of WC to WN stars. Counts of massive stars as a function of mass in the OB associations in the LMC, SMC, and MW (Massey et al. 1995b, 1995c) have found that the IMF is similar in the OB associations of all three galaxies, with a Salpeter slope ($\Gamma = -1.3$). However, a substantial number of massive stars are found in the *field*, so far from OB associations that they must have been *born* in the field during very modest star-forming events, as we discussed in Sec. 4. The IMF of these *field stars* is substantially different, with $\Gamma = -3$ to -4 in all three galaxies. Even though equally massive stars are found in the field than in the OB associations, their numbers (relative to lower-mass stars) are much fewer. This suggests that in regions where the star-formation activity is high, that the IMF may be biased towards higher mass stars. This is just what is expected on theoretical grounds due to the relation between gas temperature and fragment mass (Larson 1985, 1986). Thus, we would not expect metallicity to be the *only* parameter controlling the WC/WN ratio.

I have to acknowledge one defect with this proposed explanation: in the field of the LMC, where we know that the IMF slope is substantially different than in the OB associations (Massey et al. 1995c), we nevertheless find the same WC/WN ratio. So, maybe the explanation we offered for IC 10 is not correct. What is it about the starburst that favors the presence of WC stars relative to WN stars? Do other starbursts show this effect? It seems to me that it is crucial to understand this.

5.2. *The turn-off masses in clusters containing WR stars*

In their study of NGC 6611, Hillenbrand et al. (1993) conclude that their data was consistent with all the highest mass stars being born strictly coevally. As they put it, "the highest mass stars actually could have all been born on a particular Tuesday." This same claim was made by Massey et al. (1995b) in their study of an additional 9 Milky Way

OB associations. Now, the errors on this statement are typically plus or minus a million years, and it is also true that we invariably see *some* indications that a few stars formed earlier than the majority of the cluster, maybe by as much of 10 Myr. But generally, the vast majority of the highest mass stars in the Galactic and LMC associations that have been studied by modern techniques, appear to have been born coevally, with an age spread $\Delta\tau < 1$ Myr.

This provides us then with a very powerful technique for probing the mass range that evolves into various types of WRs (and, for that matter, LBVs and RSGs), namely to determine the "turn-off mass"—the mass of the highest mass star still on the main-sequence—and for clusters and associations containing evolved massive stars.

Fig. 22 shows the HR diagram constructed by Massey & Johnson (1993) for the clusters Tr 14 and Tr 16, which surround and contain the proto-type LBV star η Car. We've included η Car in this diagram by adopting the temperature given by Davidson et al. (1986) and the luminosity from Westphal & Neugebauer (1969), corrected to the revised distance modulus determined by Massey & Johnson. The WR star HD 93162 (WN7+abs) has been placed in this diagram by adopting -4 mags for the bolometric correction (BC), and assigning an effective temperature of 45,000 °K.

Now, there are a couple of things here that I feel are quite interesting concerning the location of the evolved objects. Perhaps the most surprising thing about this diagram is how excellently the location of η Car matches our expectation! This star's location in the HRD is exactly where it should be if this is simply the most massive cluster member still around. (The evolutionary tracks would suggest an age of ≈ 3 Myr.) Given that most of the flux of η Car is in the IR, the determination of its bolometric luminosity using the IR measurement by Westphal & Neugebauer (1969) may be quite good. Personally, I think that this figure is the most compelling observational evidence I've seen that the most massive stars really do turn into LBVs during their evolution. The second point is that the WR star is much lower in luminosity than one would expect. This is identical to what Humphreys, Nichols, & Massey (1985) found when we first tried to play this game for Galactic clusters containing WR stars, namely that invariably the WR stars required a considerably larger bolometric correction to place them on the HR diagram, and I am forced to conclude that the bolometric correction for this star has to be < -5 mag, rather than the -4 assumed in placing it in this diagram.

The mass inferred for the progenitor of the WN7+abs star, and for η Car itself, is >120 $\mathcal{M}_\odot$, where that number comes from (a) adopting the bolometric correction as a function of spectral type determined by Chlebowski & Garmany (1991), and (b) using the evolutionary tracks of the Geneva group (e.g., Schaller et al. 1992).

What about other WR stars? I show in Table 9 the turn-off masses for the well-studied Galactic and Magellanic Cloud clusters.

I am currently in the process of extending this study to additional clusters in the MCs, and working on getting the answers for associations in other galaxies of the Local Group; the goal is to answer directly how the evolution of massive stars depends upon metallicity. In the meanwhile, let me note the following things that we can conclude from Table 9, and the obvious questions that are raised that I am hoping to be able to answer from this on-going study. No obvious differences are found in the progenitor masses of the WC stars in the MW and the Magellanic Clouds. Clearly *some* WC stars come from high mass stars (>70 $\mathcal{M}_\odot$). Do any WC stars come from lower mass? *Some* WN7 stars in the MW come from very high mass stars (>100 $\mathcal{M}_\odot$). Do any come from lower masses? Finally, it would appear that *some* WN4.5-5 stars in the Milky Way come from lower-mass massive stars ($40 \mathcal{M}_\odot$), while *some* early WNs in the LMC seem to have

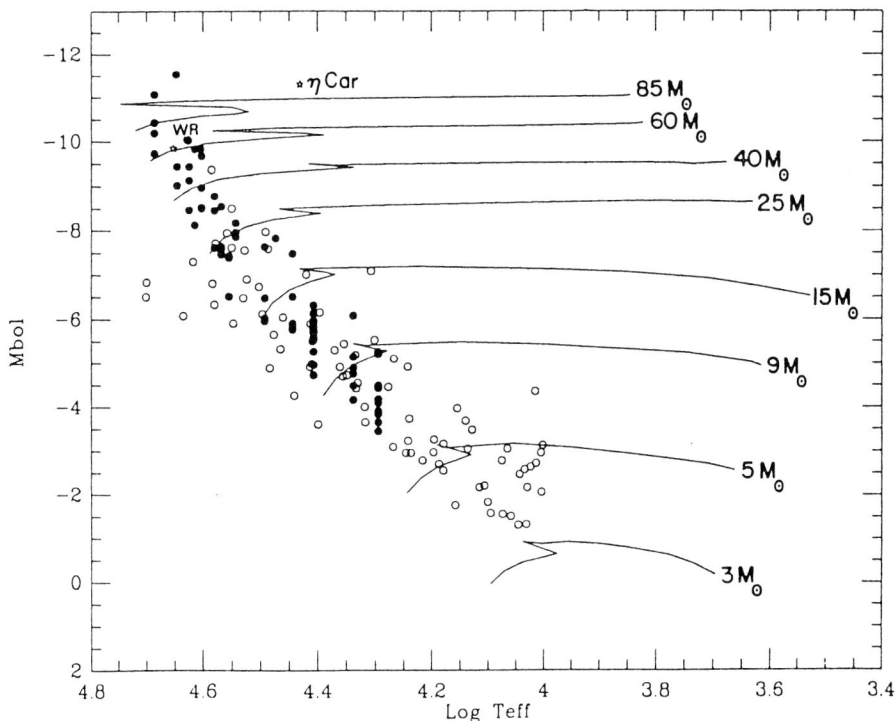

FIGURE 22. The HRD of Tr14/16 from Massey & Johnson (1993). Compare with Fig. 9.

Star	Spectral Type	Cluster Cutoff Mass	M_{bol}	M_V	BC	Cluster
	Milky Way (Massey et al. 1995b)					
HD 190918	WN4.5+O9.5Ia	40	−8.7	−5.8	< −2.9	NGC 6871
V 444 Cyg	WN5+O6	40	−8.7	−5.8	< −2.9	Berkeley 86
WR 144	WC4	110	−10.7	—	—	Cyg OB2
HD 93162	WN7+abs	120	−10.9	−5.9	< −5.0	Tr14/16
	Magellanic Clouds (Massey et al. 1995c)					
HD 5980	OB?+WN3	70	−10.4	−7.5	< −2.9	NGC 346
HDE 269546	B3I+WN3	70	−10.0	−8.6	—	LH 58 (composite)
HD 36521	WC5+OB	70	−10.0	−6.1	< −3.9	LH 58
Br 33	WN3+abs	70	−10.0	−3.6	< 6.4	LH 58
HD 32228	WC5-6+late O	85	−10.5	—	—	LH 9 (blend)

TABLE 9. Turn-off masses of Galactic and MC clusters containing WR stars.

come from higher-mass stars. Are we seeing a metallicity effect? Clearly a lot more data is needed to sort this out, but maybe the answers are within sight.

5.3. The relative number of unevolved massive stars and WRs: the "O/WR" ratio revisited

In Sec. 5.2 we discussed how we are going about using coeval populations to determine the masses of the progenitors of WR stars of differing subtypes, and our plans to extend this to other clusters in order to determine how this depends upon metallicity. However, there is another game that we can play using instead *mixed-age* populations that is almost as much fun, namely trying to compare the number of WR stars with the number of progenitors in order to see what the relative lifetimes are of WR stars as a function of progenitor mass. Let me explain this.

In a mixed-age population (such as a galaxy-wide ensemble, or the population in the field), stars of a given mass are constantly being born, and at the same time, stars of the same given mass are dying. We assume a steady-state condition; this is merely saying that over a few million years, averaged over whatever spatial region we're discussing, that the *average* star-formation rate isn't changing. We then count the number of stars with masses above $40 \mathcal{M}_\odot$ still on the H-burning main-sequence, and compare this to the number of WR stars. If all $40 \mathcal{M}_\odot$ stars turn into WR stars, then this exercise will tell us the relative lifetimes of the two phases. Or, if you prefer, we can adopt a relative lifetime (say, 10%) for WR stars, and look to see what mass unevolved star we have to go down to in order to achieve this. Various authors have attempted to quantify this for Local Group galaxies using the ratio of the "number of O stars" to the number of WR stars; the difficulty is that while the number of WR stars is relatively well known, the "number of O stars" has had to be estimated from such things as luminosity functions and number of H II regions. In any event, it is not the number of O-type stars that we should be concerned about, but rather, the number of core-H burning stars above some mass. (Since the H-burning main-sequence extends further to the right than B0 according to the models, you need to include those stars as well.) Even so, AM91 have argued that in the past such a statistic was essentially meaningless anyway, given the uncertainties in determining the number of unevolved massive stars from the indirect methods then available (i.e., luminosity functions). To underscore how uncertain such techniques are, they noted that Smith (1988) and Maeder (1991) obtained answers for the range of O/WR stars among Local Group galaxies that differed by *two orders of magnitude*.

Massey et al. (1995c) has studied the unevolved massive star content of the *field* of the LMC and SMC, using deep survey fields to determine incompleteness corrections as a function of mass down to 25 $\mathcal{M}_\odot$. Several hundred new spectral types were used to construct an H-R diagram. Similar work has been been carried out by Massey et al. (1995a) in the field and OB associations of NGC 6822, M 31, and M 33; although a number of spectra have been included (an effort which is continuing), most of the information we have on the stellar content of selected regions in these galaxies is based upon *UBV* photometry, and the numbers are considerably less certain. Nevertheless, it is instructive to compare these numbers to the number of WR stars in the same field (Table 10).

The first thing we see is that this number is constant (to within the uncertainties) for all five galaxies in this sample. Thus, if metallicity is playing a role, we are not sensitive to it yet. However, the big surprise is that this number is ≈ 3, not the canonical 10 that we would expect given the relative He-burning and H-burning lifetimes. So, what gives? Could it be that $N_{>40\mathcal{M}_\odot}$ is overly contaminated by He-burning B supergiants

Galaxy	$N_{>40\mathcal{M}_\odot}$	#WRs	$N_{>40\mathcal{M}_\odot}$/WRs
NGC 6822 (6 Fields)	14	4	3.5
M 31 (3 Fields)	30	9	3.3
M 33 (7 Fields)	79	27	2.9
Milky Way (GCC Sample)	161	44	3.7
LMC (Field)	87	38	2.3
LMC (F1+F2)	6	2	3.0

TABLE 10. The number of massive ($> 40\mathcal{M}_\odot$) OB stars ($N_{>40\mathcal{M}_\odot}$) compared to the number of WR stars in mixed-age populations. The data come from Massey et al. (1995a)'s Table 8, except for M 33, which are given new here.

masquerading as H-burning objects? Perhaps—but in their study of the LMC field, Massey et al. (1995c) found excellent agreement in the number of stars between evenly-spaced isochrones, suggesting that most of these stars are just what they appear to be, i.e., stars still on the main-sequence. Still, it is a possibility to keep in mind. Alternatively, it could be that stars of somewhat lower mass evolve to WRs. If we look at the most complete and certain data-set (the LMC), we find that we would get the factor of 10 ratio if the progenitors of WR stars extended down to $30\mathcal{M}_\odot$. (See Table 14 in Massey et al. 1995c.) The third possibility is that some WNs are actually H-burning objects (Conti et al. 1995). Probably some combination of the last two provides the explanation.

5.4. *Red supergiants and WRs*

Although this review is on the "distribution of WR stars" I couldn't resist mentioning some results from our survey for red supergiants (RSGs) among galaxies of the Local Group. First, let me explain the problem. As I argued in Sec. 3.2.1, if we look at a catalog of red stars towards M 33 (e.g., Humphreys & Sandage 1980) we can't tell which of these stars are actually RSGs in M 33, and which of them happen to be foreground Galactic M dwarfs seen in projection against M 33. One can certainly tell by taking spectra, but this entails taking spectra of 600 – 800 stars, and even then the luminosity criterion are subtle, especially if the RSGs have lower metallicity.

It turns out that an excellent discriminant can be found in a two-color $B-V$, $V-R$ plot: the foreground stars have a smaller $B-V$ than RSGs at the same $V-R$ color. Fig. 23 shows the distribution of stars for M 33, NGC 6822, and M 31, where the filled circles are taken from nearby control fields that contain only foreground stars, while the open squares are red stars seen against these galaxies and thus may consist of either foreground or bona-fide RSGs. The solid line shows the expected separation between foreground stars and RSGs, and indeed for the control fields (which contain only foreground M dwarfs), the region below this line contains all the points. For the NGC 6822, M 33, and M 31 two-color plots, I've made cuts at V corresponding to $M_V \approx -6$. And, indeed, in the NGC 6822 and M 33 plots we find that there are a number of stars that are pretty clearly foreground (in that they are down with the control-field points), but also a substantial number of points well above the line. These should be luminous RSGs. However, no such points exist in the M 31 diagram, although the surveyed areas are similar. The brightest RSGs are missing in M 31! (At fainter magnitudes an increasing number of RSG candidates *are* seen in M 31.) Now, this finding differs in two important ways from the survey for RSGs in M 31 by Humphreys et al. (1988), who claim that the most luminous RSGs ($M_V = -8$) *do* exist in M 31, but are just "not found in the relative numbers we would expect," and who attribute this to a low rate of massive-star star formation. First, in the fields we have surveyed, we do *not* find any RSGs that luminous,

142 P. Massey: *Luminous Stars in the Local Group*

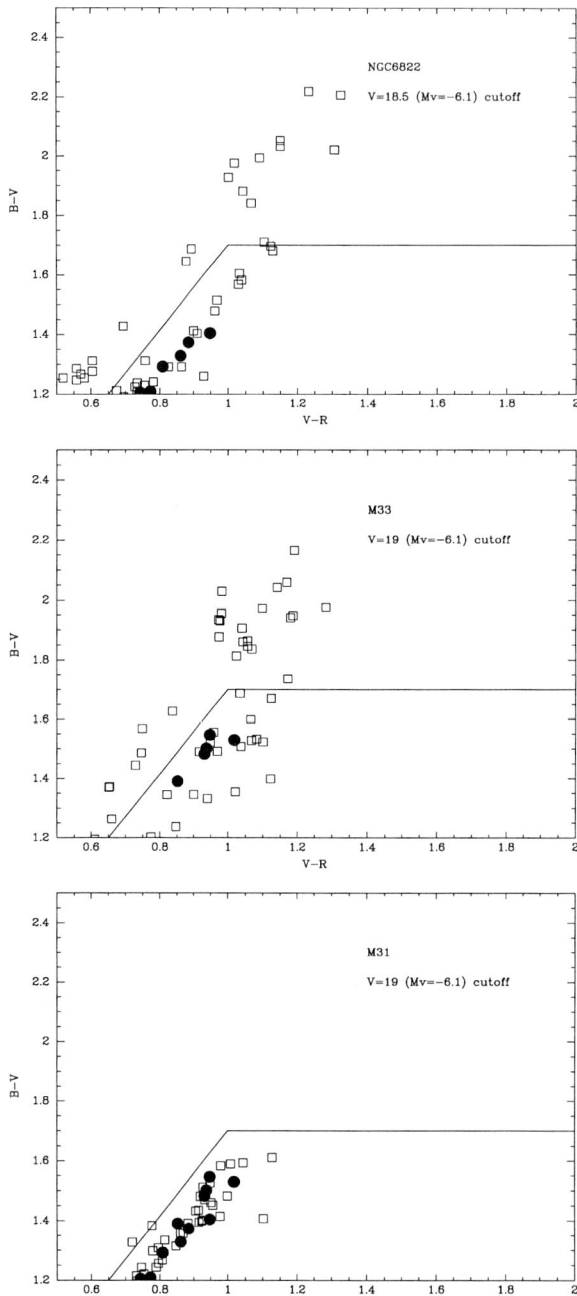

FIGURE 23. Two-color plots for red stars in selected regions of NGC 6822, M 33, and M 31. The data have been truncated at a V magnitude corresponding to $M_V = -6.1$ in each galaxy. Points above the solid line are likely RSGs, while the points below the lines are likely foreground dwarfs. This expectation is confirmed by the solid circles, which are stars observed in neighboring control fields. Note that while there is a preponderance of bright RSGs in M 33 and NGC 6822, these stars are completely missing from the M 31 fields.

although there are certainly some foreground red stars at the corresponding V. Secondly, we cannot claim that the lack of luminous RSGs is due to a low star-formation rate for massive stars, as that would contradict the data that the WR stars themselves provide. We see in Table 8 that the WR surface density is very similar in M 31 and NGC 6822, and so any difference in the RSG content of these two galaxies has to be telling us something about massive star evolution, not star-formation rate (since the number density of WR stars is the same)! We see plenty of $M_V = -6$ RSG candidates in NGC 6822, but none in the fields we've surveyed in M 31 (Fig. 23).

So, what does this mean? The most straightforward interpretation is that in higher metallicity environments more stars are becoming WRs rather than RSGs, as would be expected in the Conti scenario.

5.5. Conclusions

The following are the conclusions I would draw at this time from the numbers and distributions of Wolf-Rayet stars in the galaxies of the Local Group:

- The WC/WN ratio is well correlated with metallicity within individual galaxies of the Local Group, and from galaxy to galaxy, in the sense that there are proportionally more WC stars in regions of higher metallicity. The glaring exception to this is the relatively metal-poor irregular galaxy IC 10, which also appears to be undergoing a starburst. What is it about the high rate of star formation that favors the presence of WC stars over WN stars? It seems to me we must understand the answer to this question.
- The fact that OB associations appear to be highly coeval offers a powerful way of directly answering what mass range progenitors evolve into various objects, including WR stars of different subtype. The data are scant right now, but do show that:
  ○ No obvious differences are found in the progenitor masses of the WC stars in the Milky Way and Magellanic Clouds.
  ○ *Some* WC stars come from high mass stars ($> 70 \mathcal{M}_\odot$).
  ○ *Some* WN7 stars in the MW come from very high mass stars ($> 100 \mathcal{M}_\odot$).
  ○ *Some* WN4-5 stars in the MW come from lower-mass massive stars ($40 \mathcal{M}_\odot$), while *some* early WNs in the Magellanic Clouds have come from higher-mass stars.

Work on this is continuing.

- The relative number of unevolved massive stars and WRs (the "O/WR") ratio) can now be determined with some reliability in the MW, MCs, NGC 6822, M 33, and M 31, at least for unevolved massive stars with masses $> 40 \mathcal{M}_\odot$. The number ratio $N_{>40\mathcal{M}_\odot}/\text{WR}$ is ≈ 3 in all of these galaxies, suggesting that any metallicity effect is still masked by the observational uncertainties. The fact that this number is 3, rather than the canonical 10, suggests that either a large fraction of WR stars are still core-H burning objects (as Conti et al. 1995 have hinted), or that the mass range that turns into WRs is a little lower than usually assumed (30 $\mathcal{M}_\odot$).
- Finally, we have begun to explore the number of RSGs in the same regions of the Local Group we have surveyed for OB stars and for WRs. This work is still in progress, but we can confirm that M 31 is lacking in luminous RSGs compared to NGC 6822 or M 33. This cannot be attributed to a lower star-formation rate, as NGC 6822 and M 31 have the same (global) surface density of WR stars. We conclude that this is an example of metallicity affecting massive star evolution, causing lower-mass stars to turn into WRs in M 31 (rather than into RSGs).

6. Summary: what to take away from all this

When I teach the elementary astronomy classes, I have a pretty definite idea of what I want the students to take away with them. I'm a lot less concerned with whether or not they remember the arguments in the Shapley-Curtis debate than I am with whether or not they understand why the moon's phases are the way they are, and how we know that the universe is expanding. Keeping to the basics, then, I think that the following is what I would most like you to know:

- The *brightest* stars in a nearby galaxy are seldom *the most luminous*. Luminousity is equivalent to mass, in the sense that massive stars evolve at pretty much constant luminosity. It's *difficult* to find the most massive stars in a galaxy. The colors of main-sequence, massive stars don't tell you enough about their temperatures to determine their bolometric corrections with sufficient accuracy to distinguish between two stars with masses that differ by factors of 2-3! To do this right requires spectrsocopy.
- Of the evolved, massive stars, I would maintain that:
  o The LBVs in Local Group galaxies have been found primarily by flashy displays of variability, and may represent only a small fraction of massive stars that pass through an unstable, high-mass-loss period.
  o The Wolf-Rayet stars in Local Group galaxies can be found in a complete sense, but it requires some work to get the majority of the weaker-lined WN stars.
  o The RSG populations of Local Group galaxies are not as well known as many believe, owing to the contamination of foreground K and M dwarfs. It is possible to separate foreground dwarfs from bona-fide RSGs via careful BVR photometry.
- The initial mass function appear to have similar slopes and upper mass limits in the OB associations of the Milky Way, the LMC, and the SMC, with a Salpeter-like slope: $\Gamma = -1.3$—this despite the factor of 10 difference in metallicity between the three systems.
- A significant number of massive stars are born in the *field*, away from any OB association. Studies of this mixed-age population reveals that:
  o The IMF slope of field massive stars is considerably steeper than in OB associations, with proportionally fewer of the highest mass stars being born ($\Gamma = -4$ in the MW, LMC, and SMC). Nevertheless, stars of equally high mass to those found in OB associations *are* found in the field, as witnessed by the presence of O3 and O4 stars far from any OB associations.
  o Stellar evolutionary models give a good match to the distribution of stars in the HRDs of the Magellanic Clouds, suggesting that the isochrones and the (relative) main-sequence lifetimes are pretty well understood. Previous claims to the contrary have been due to a misunderstanding of the completeness of the catalogs of the Magellanic Clouds.
- Studies of Local Group galaxies reveal:
  o The relative number of WC to WN stars is a strong function of metallicity, consistent with the "Conti scenario" in which only the stars with the largest amount of mass-loss (high luminosity and/or high metallicity) can become WC stars.
  o The starburst galaxy IC 10 provides the only exception to this correlation, with a high WC/WN ratio despite low metallicity. Possibly the amount of star-formation "vigor" weights the IMF towards the highest mass stars, consistent with the differences seen between the field and OB association IMFs, and with the predictions of theory.
  o There is an absence of the most luminous RSGs in M 31, despite a high surface density of WR stars, again in accord with the predictions of the Conti scenario.

I want to thank the organizers of this Winter School, Artemio Herrero and Antonio Aparicio, for bringing us together to discuss all of the wonderful topics we've heard about during these two weeks, and I thank the students for their high level of participation and involvement. Before I accepted, I consulted a few of the past lecturers here, and all told me that it was (a) a lot of work, but (b) worth it, and I agree with both of these. I also acknowledge much useful advice from Deidre Hunter on the content and organization of these lectures.

REFERENCES

ARMANDROFF, T. E. & MASSEY, P. 1985 *ApJ* **291**, 685.

ARMANDROFF, T. E. & MASSEY, P. 1991 *AJ* **102**, 927.

AUER, L. H. & MILHALAS, D. 1970 *ApJ* **160**, 233.

AZZOPARDI, M. & BREYSACHER, J. 1979 *A&A* **75**, 120.

AZZOPARDI, M. & VIGNEAU, J. 1982 *A&AS* **50**, 291.

BAHCALL, J. N. & SONEIRA, R. M. 1980 *ApJS* **44**, 73.

BLAIR, W. P. & KIRSHNER, R. P. 1985 *ApJ* **289**, 582.

BOHANNAN, B. 1979 In *Mass Loss and Evolution of O-Type Stars*, ed. P. S. Conti & C. W. H. de Loore (Dordrecht: Reidel), p. 479.

BOHANNAN, B., CONTI, P. S. & MASSEY, P. 1985 *AJ* **90**, 600.

BOHANNAN, B. & WALBORN, N. R. 1989 *PASP* **101**, 520.

BREYSACHER, J. 1981 *A&AS* **43**, 203.

CHLEBOWSKI, T. & GARMANY C. D. 1991 *ApJ* **368**, 241.

CHURCHWELL, E. 1991 In *The Physics of Star Formation and Early Stellar Evolution*, ed. C. J. Lada & N. D. Kylafis (Dordrecht: Kluwer), p. 221.

CONTI, P. S. 1973 *ApJ* **179**, 181.

CONTI, P.S. 1976 *Mem. Soc. Roy. Sci. Liege* 6^e Ser. **9**, 193.

CONTI, P. S. 1986 In *Luminous Stars and Associations in Galaxies*, ed. C. W. H. de Loore, A. J. Willis & P. Laskarides (Dordrecht: Reidel), p. 199.

CONTI, P. S. 1988 In *O Stars and Wolf-Rayet Stars*, ed. P. S. Conti & A. B. Underhill (Washington: NASA SP-497).

CONTI, P. S. & MASSEY, P. 1981 *ApJ* **249**, 471.

CONTI, P. S. & MASSEY, P. 1989 *ApJ* **337**, 251.

CONTI, P. S. & VACCA, W. D. 1990 *AJ* **100**, 431.

CONTI, P. S., LEEP, E. M., & LORRE, J. J. 1977 **214**, 759.

CONTI, P. S., HANSON, M. M., MORRIS, P. W., WILLIS, A. J. & FOSSEY, S. J. 1995 *ApJ* **445**, L35.

DAVIDSON, K., DUFOUR, R. J., WALBORN, N. R. & GULL, T. R. 1986 *ApJ* **305**, 867.

FITZGERALD, M. P. 1970 *A&A* **4**, 234.

FITZPATRICK, E. L. & GARMANY, C. D. 1990 *ApJ* **363**, 119.

GARMANY, C. D., CONTI, P. S. & CHIOSI, C. 1982 *ApJ* **263**, 777.

HANSON, M. M. 1995 PhD Thesis, University of Colorado.

HANSON, M. M., HOWARTH, I. D. & CONTI, P. S. 1997 *ApJ*, in press.

HAYES, D. S. & LATHAM, D. W. 1975 *ApJ* **197**, 593.

HILLENBRAND, L. A., MASSEY, P., STROM, S. E. & MERRILL, K. M. 1993 *AJ* **106**, 1906.

HODGE, P. W. 1977 *ApJS* **33**, 69.

HODGE, P. W. 1978 *ApJS* **37**, 145.

HODGE, P. W. 1979 *AJ* **84**, 744.

HODGE, P. W. 1980 *AJ* **241**, 125.
HODGE, P. W. 1985 *PASP* **97**, 530.
HODGE, P. & LEE, M. G. 1990 *PASP* **102**, 26.
HUBBLE, E & SANDAGE, A. 1953 *ApJ* **118**, 353.
HUMPHREYS, R. M. 1978 *ApJS* **38**, 309.
HUMPHREYS, R. M. 1979a *ApJS* **39**, 389.
HUMPHREYS, R. M. 1979b *ApJ* **234**, 854.
HUMPHREYS, R. M. 1980a *ApJ* **238**, 65.
HUMPHREYS, R. M. 1980b *ApJ* **241**, 587.
HUMPHREYS, R. M. 1980c *ApJ* **241**, 598.
HUMPHREYS, R. M. & DAVIDSON, K. 1979 *ApJ* **232**, 409.
HUMPHREYS, R. M. & DAVIDSON, K. 1984 *Science* **223**, 243.
HUMPHREYS, R. M., NICHOLS, M. & MASSEY, P. 1985 *AJ* **90**, 101.
HUMPHREYS, R. M., PENNINGTON, R. L., JONES, T. J. & GHIGO, F. D. 1988 *AJ* **96**, 1884.
HUMPHREYS, R. M. & SANDAGE, A. 1980 *ApJS* **44**, 319.
HUNTER, D. A., BAUM, W. A., O'NEIL, E. J. & LYNDS, R. 1996 *ApJ* **456**, 174.
LARSON, R. B. 1985 *MNRAS* **214**, 379.
LARSON, R. B. 1986 *MNRAS*, **218**, 409.
LEQUEUX, J., PEIMBERT, M., RAYO, J. F., SERRANO, A. & TORRES-PEIMBERT, S. 1979 *A&A* **80**, 155.
LUCKE, P. B. & HODGE, P. W. 1970 *AJ*, **75**, 171.
MAEDER, A. 1991 *A&A* **242**, 93.
MAEDER, A. & CONTI P. S. 1994 *ARAA* **32**, 227.
MAEDER, A. & MEYNET, G. 1988 *A&AS* **102**, 401.
MASSEY, P. 1997 In *Wolf-Rayet Stars in the Framework of Stellar Evolution, 33rd Liege International Astrophysical Colloquium*, in press.
MASSEY, P. & ARMANDROFF, T. E. 1991 In *Wolf-Rayet Stars and Interrelations with Other Massive Stars in Galaxies*, ed. K. A. van der Hucht & B. Bidayat (Dordrecht: Kluwer), p. 575.
MASSEY, P. & ARMANDROFF, T. E. 1995 *AJ* **109**, 2470.
MASSEY, P. & CONTI, P. S. 1983 *ApJ* **273**, 576 (MC83).
MASSEY, P. & JOHNSON, J. 1993 *AJ* **105**, 980.
MASSEY, P., CONTI, P. S. & ARMANDROFF, T. E. 1987a *AJ* **94**, 1538 (MCA87).
MASSEY, P., CONTI, P. S., MOFFAT, A. F. J. & SHARA, M. M. 1987b *PASP* **99**, 816.
MASSEY, P., PARKER, J. W. & GARMANY, C. D. 1989 *AJ* **98**, 1305.
MASSEY, P., ARMANDROFF, T. E. & CONTI, P. S. 1992 **103**, 1159.
MASSEY P., ARMANDROFF, T. E., PYKE, R., PATEL, K. & WILSON, C. D. 1995a *AJ* **110**, 2715.
MASSEY, P., JOHNSON, K. E. & DEGIOIA-EASTWOOD, K. 1995b *ApJ* **454**, 151.
MASSEY P., LANG, C. C., DEGIOIA-EASTWOOD, K. & GARMANY, C. D. 1995c *ApJ* **438**, 188
MASSEY, P., BIANCHI, L., HUTCHINGS, J. B. & STECHER, T. P. 1996 *ApJ* **469**, 629.
MONTEVERDE, M. I., HERRERO, A., LENNON, D. J., & KUDRITZKI, R. P. 1996 *A&A* **312**, 24.
MONTEVERDE, M. I., HERRERO, A., LENNON, D. J., & KUDRITZKI, R. P. 1997 *A&A* in press.
PAGEL, B. E. J., EDMUNDS, M. G. & SMITH, G. 1980 *MNRAS* **193**, 219.
RATNATUNGA, K. U. & BAHCALL, J. N. 1985 *ApJS* **59**, 63.
RUSSELL, S. C. & DOPITA, M. A. 1990 *ApJS* **74**, 93.

SANDULEAK, N. 1969 *Cerro Tololo Inter-American Observatory Contribution No. 89*.
SCALO, J. M. 1986, *Fund. Cosmic Phys.* **11**, 1.
SCHAERER, D., MEYNET, G., MAEDER, A. & SCHALLER, G. 1993 *A&AS* **98**, 523.
SCHALLER, G., SCHAERER, D., MEYNET, G. & MAEDER, A. 1992 *A&AS* **96**, 269.
SCHMIDT-KALER, TH. 1982 In *Landolt-Bornstein New Series: Group VI, Volume 2b, Astronomy and Astrophysics, Stars and Star Clusters* ed. by K. Schaifers & H. H. Voigt, Springer-Verlag, Berlin, p. 1
SHAVER, P. A., MCGEE, R. X., NEWTON, L. M., DANKS, A. C. & POTTASCH, S. R. 1983 *MNRAS* **204**, 53.
SHIELDS, G. A. & TINSLEY, B. M. 1976 *ApJ* **203**, 66.
SHU, F. H. 1982 *The Physical Universe* (Mill Valley, University Science Books), p. 146.
SMITH, L. F. 1988 *ApJ* **327**, 128.
SMITH, L. J., CROWTHER, P. A. & WILLIS, A. J. 1995 *A&A* **302**, 830.
STAHL, O., WOLF, B., KLARE, G., CASSATELLA, A., KRAUTTER, J., PERSI, P., & FERRARI-TONIOLO, M. 1983 *A&A* **127**, 49.
STETSON, P. B. 1987 *PASP* **99**, 191.
TALENT, D. L. 1980 PhD thesis, Rice University.
TINSLEY, B. M. 1980 *Fund. Cosmic Phys.* **5**, 287.
VACCA, W. D., GARMANY, C. D. & SHULL, J. M. 1996 *ApJ* **460**, 914.
VAN DEN BERGH, S. 1972 *RASCJ* **66**, 237.
VAN DEN BERGH, S. 1975 *ARAA* **13**, 217.
VAN DEN BERGH, S. 1991 *PASP* **103**, 609.
VILCHEZ, J. M., PAGEL, B. E. J., DIAZ, A. I., TERLEVICH, E. & EDMUNDS, M G. 1988 *MNRAS* **235**, 633.
WALBORN, N. R. 1977 *ApJ* **215**, 53.
WALBORN, N. R. 1982 *IAU Circular* No. 3767.
WALBORN, N. R. & BLADES, J. C. 1986 *ApJ* **304**, L17.
WALBORN, N. R. & FITZPATRICK, E. L. 1990 *PASP* **102**, 379.
WESTPHAL, J. A. & NEUGEBAUER, G. 1969 *ApJ* **156**, L45.
WILLIS, A. J., SCHMIDT, H. & SMITH, L. J. 1992 *A&A* **261**, 419.
WRAY, J. D. & CORSO, G. J. 1972 *ApJ* **172**, 577.
ZUCKERMAN, B. 1973 *ApJ* **183**, 863.

Quantitative Spectroscopy of the Brightest Blue Supergiant Stars in Galaxies

By ROLF-PETER KUDRITZKI

Institut für Astronomie und Astrophysik der Universität München, Scheinerstr.1, D-81679 München, Germany

Max-Planck-Institut für Astrophysik, D-85740 Garching bei München, Germany

The spectroscopic analysis of the brightest supergiant stars as individuals provides a unique tool for the determination of the properties of young stellar populations in galaxies. In this series of lectures, the perspectives of future work with the new generation of very large ground-based telescopes are developed and the tools of quantitative spectroscopy that have become available during the past decade are introduced.

It is shown that the physics of the atmospheres of Blue Supergiants are strongly affected by stellar wind outflows contaminating or even dominating the spectra at all wavelengths from X-rays via UV, optical, IR to the radio domain. Consequently, hydrodynamic model atmospheres are used to synthesize and to analyze the spectra. The methods of spectral diagnostics are discussed and it is demonstrated how stellar parameters, abundances and stellar wind properties are determined. Special emphasis is given to very recent results obtained from the analysis of X-ray and IR spectra.

We discuss in detail the applications of the diagnostic techniques on the most luminous and most massive (up to 200 $M_\odot$) stars in the Galaxy and the Magellanic Clouds and on the cluster of massive Supergiants in the central parsec of our Galaxy. Furthermore, it is demonstrated that Blue Supergiants provide an excellent way of determining extragalactic abundances, complementary to that which can be learned from H II - regions. We give examples for the Magellanic Clouds, M33 and M31 and show the first steps out of the Local Group.

Finally, the **Wind Momentum - Luminosity Relationship** is introduced as a new, independent tool for the determination of extragalactic distances. Very recent results obtained for the Galaxy, LMC, SMC, M33 and M31 are presented. The potential of the method is discussed with the conclusion that it **may allow independent distance moduli to be obtained with an accuracy of ten percent out to the Virgo and Fornax clusters of galaxies.**

---

1. Introduction

It has long been a dream of stellar astronomers to study individual stellar objects in distant galaxies by spectroscopy

- to disentangle the kinematics of galaxies and to determine the gravitational potential,
- to determine abundances of the stellar populations and to obtain detailed constraints on the chemodynamical evolution and
- to determine the physical properties of the brightest objects for an estimate of the distances, in this way improving the accuracy of the extragalactic distance scale.

Apart from Supernovae, three classes of stellar objects are ideally suited for such an investigation because of their intrinsic brightness and spectral characteristics: Planetary Nebulae (irradiated by their hot Central Stars), Red Supergiants or Giants and Blue Supergiants.

Planetary Nebulae – although unspectacular point sources at distances beyond the Magellanic Clouds – stick out from the background of faint stars in galaxies by their striking emission line characteristics. Up to 10 percent of the luminosity of the
10^4 $L_\odot$ of the very hot Central Star reprocessed by the surrounding nebula can be

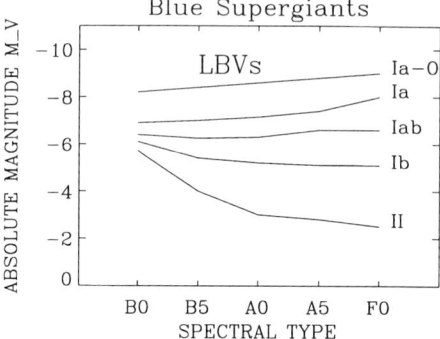

FIGURE 1. Absolute visual magnitudes of galactic A- and B- supergiants as a function of spectral type and luminosity class. Luminous Blue Variables (LBVs) can become brighter than the brightest hypergiants.

reemitted in a single emission line, [OIII] λ 5009. Thus, by on-band/off-band interference filter imaging, hundreds of Planetary Nebulae can be identified per galaxy even at distances beyond the Virgo cluster and can be used to determine distances (see for example Jacoby et al. 1990; Jacoby et al. 1992; Jacoby & Ciardullo 1993; Jacoby 1995; Mendez et al. 1993; Soffner et al. 1996; Soffner & Mendez 1997, Madore 1997). Followup spectroscopy allows the gravitational potential, rotation and dark matter content (Hui et al. 1995; Arnaboldi et al. 1994; Arnaboldi et al. 1996) as well as the chemical evolution history of galaxies (Richer 1993; Richer & McCall 1995) to be investigated. A thorough discussion of the exciting work on extragalactic Planetary Nebulae would require an extra review paper of considerable volume. Here, we can only refer to the papers just cited.

Red Giants and Supergiants are identifiable by their brightness and colour. As demonstrated by Madore 1997, their colour - magnitude diagrams provide an excellent tool to determine extragalactic distances. Very recently with HST, such a stellar population has even been detected in the intergalactic medium of the Virgo cluster (Ferguson 1997, STScI press release) together with a corresponding number of Planetary Nebulae (Arnaboldi et al. 1996). Photometry of these red evolved stars will certainly provide fundamental insight in the population characteristics. However, for quantitative spectroscopy they are too faint, in particular when stepping beyond the Local Group.

On the other hand, Blue Supergiants, in particular those of spectral type A and B are the optically brightest stars in galaxies and are easily found because of their brightness and colour. With absolute magnitudes of about $M_V = -9^m$ (see Fig. 1), the brightest A- and B-Hypergiants in galaxies as distant as the Virgo cluster are expected to be of apparent magnitude $m_V = 22^m$, still accessible for quantitative spectroscopy using 8m-class telescopes. With 4m-class telescopes and efficient spectrographs plus CCD detectors they can be investigated in all Local Group galaxies (Herrero et al. 1994; Monteverde et al. 1996; 1997). Fig. 2 and Fig. 3 give examples of what can be achieved within a 1^h exposure with a 4m-telescope. With HST we can perform quantitative UV spectroscopy of the most UV luminous stars in all Local Group galaxies and we can obtain accurate photometry in crowded fields out to the Virgo cluster and beyond.

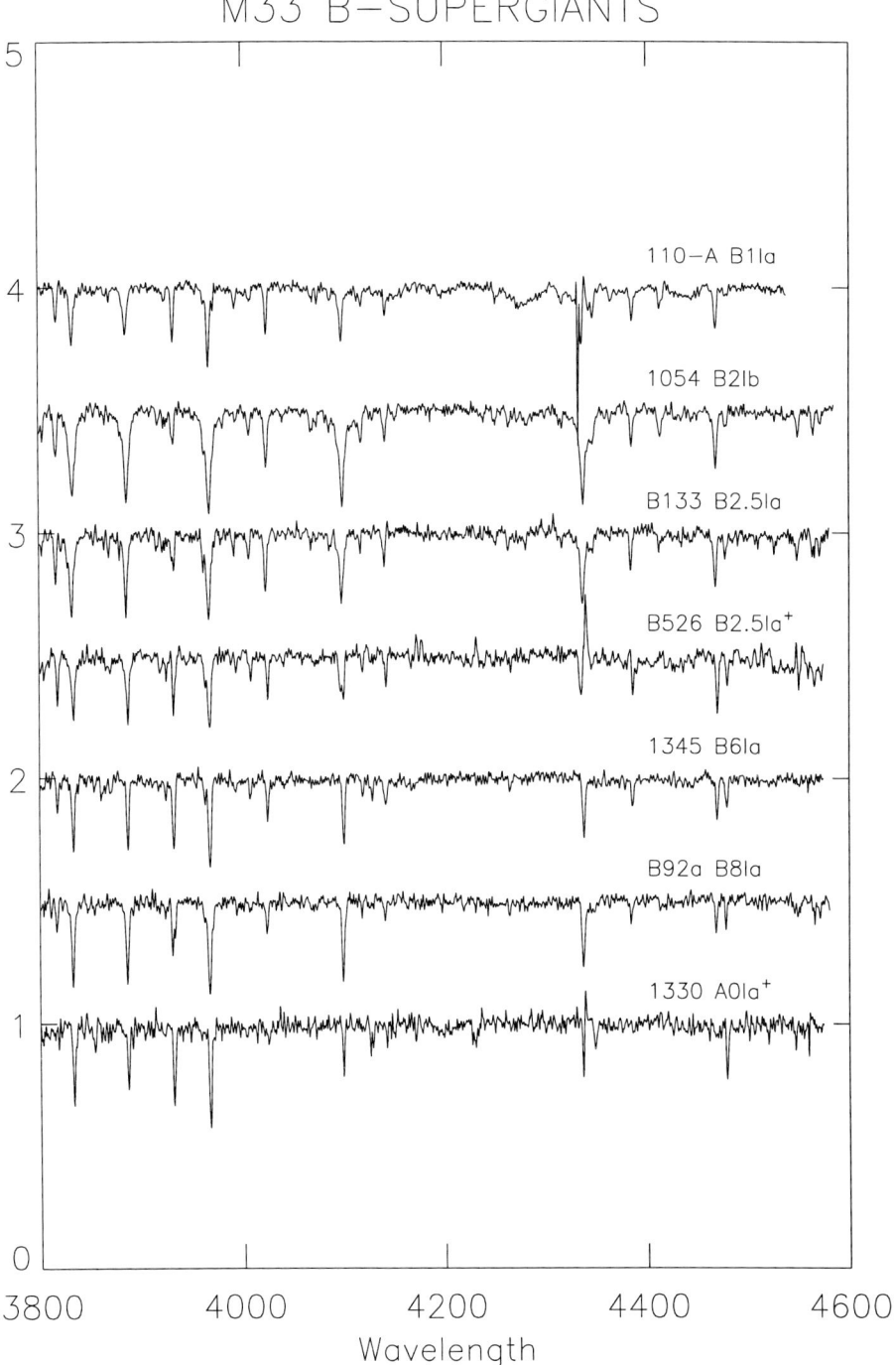

FIGURE 2. A compilation of optical spectra of B-supergiants in M33 as obtained with the William Herschel Telescope (From Monteverde *et al.* 1996).

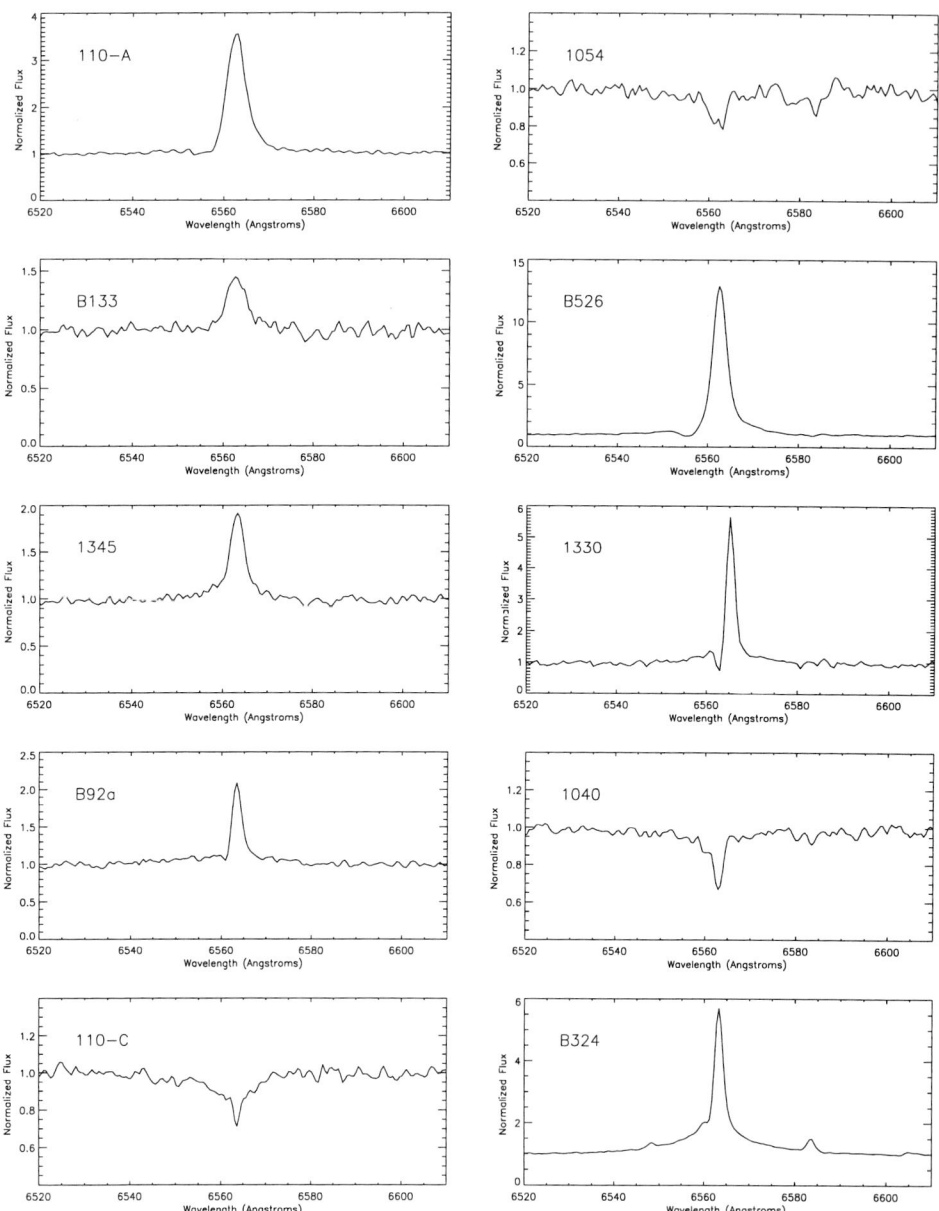

FIGURE 3. Hα line profiles of blue supergiants in M33 as obtained with the William Herschel Telescope indicating strong stellar winds. (From Monteverde et al. 1996).

	A-, B-, O-Supergiants	PN and Central Stars
objects		
masses	20 to 200 M$_\odot$	0.5 to 1.0 M$_\odot$
population	young (starbursts!)	intermediate/old
standard candles	10^6 L$_\odot$	10^4 L$_\odot$
ionization sources	H II - regions, ISM	Planetary Nebulae
nucleosynthesis	stellar winds, SN II	stellar winds, SN I

TABLE 1. Hot stars and their relevance to astrophysics

Therefore, **Blue Supergiants** – or more generally, hot stars – **provide unique tools for investigating the physics of galaxies**. They are ideal tracers of populations with respect to age, abundances and dynamics. They are sources of ionizing radiation thus allowing abundance studies of the ISM emission and absorption lines. They contribute significantly to the cosmic and galactic nucleosynthesis and they can, in principle, be used as standard candles because of their brightness. (Table 1 gives a very short overview of these aspects).

The **Local Group**, on the other hand, **is ideal to study the physics of Blue Supergiants in detail and to provide fundamental calibrations**, because

- the objects are "nearby",
- we encounter a wide range of metallicities,
- we find objects in all stages of evolution.

With these excellent prerequisites, it may be surprising that so little work has been done so far using Blue Supergiants for extragalactic astronomy. Even worse, Blue Supergiants are notorious among extragalactic astronomers as being entirely useless, in particular for the determination of distances. However, step by step, we will show in this paper, this is caused by the fact that, so far, these objects have been used only in the most blunt of possible astronomical ways; looking for strict correlations between simplified spectral types, luminosity classes and absolute magnitudes. These attempts have indeed more or less failed, the physical reason being that **the atmospheres of these objects and their evolution as stars are dominated by strong stellar wind outflows** leading to significant modifications of their observable spectra and the evolutionary scenario (for the latter, see for instance Langer et al. 1994). Signatures of these winds are found everywhere in the spectra, typically strong emission features, originating from ionized metal resonance transitions in the UV or from excited hydrogen, helium and metal lines in the visible and infrared (first examples are given in Fig. 4 and Fig. 5). Thus, from the viewpoint of a classical quantitative spectroscopist, used to working with standard hydrostatic (LTE or NLTE) model atmospheres, these winds are a mess, as they corrupt the spectra with their wind contamination.

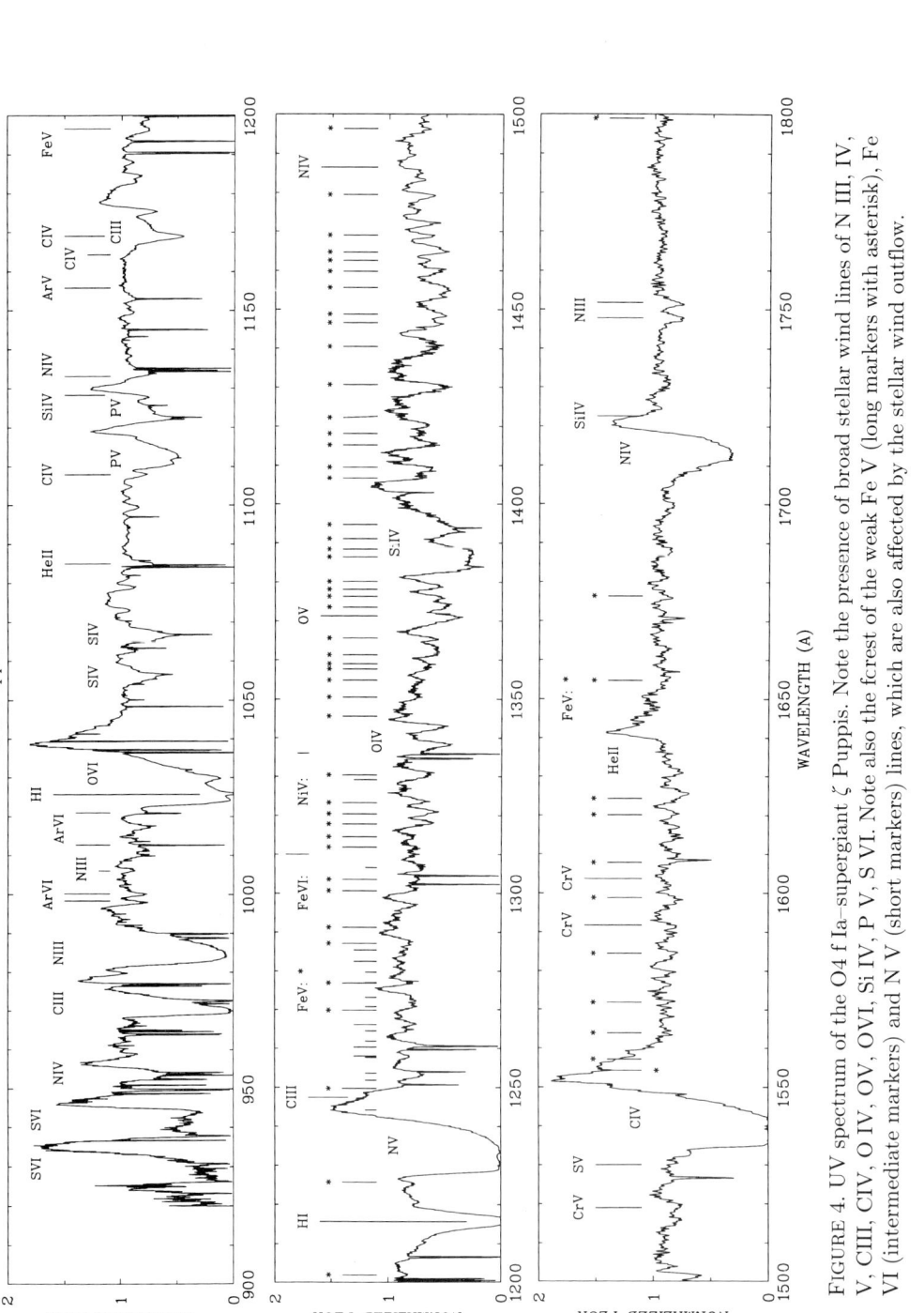

FIGURE 4. UV spectrum of the O4 f Ia–supergiant ζ Puppis. Note the presence of broad stellar wind lines of N III, IV, V, C III, C IV, O IV, O V, O VI, P V, S IV, S VI. Note also the fcrest of the weak Fe V (long markers with asterisk), Fe VI (intermediate markers) and N V (short markers) lines, which are also affected by the stellar wind outflow.

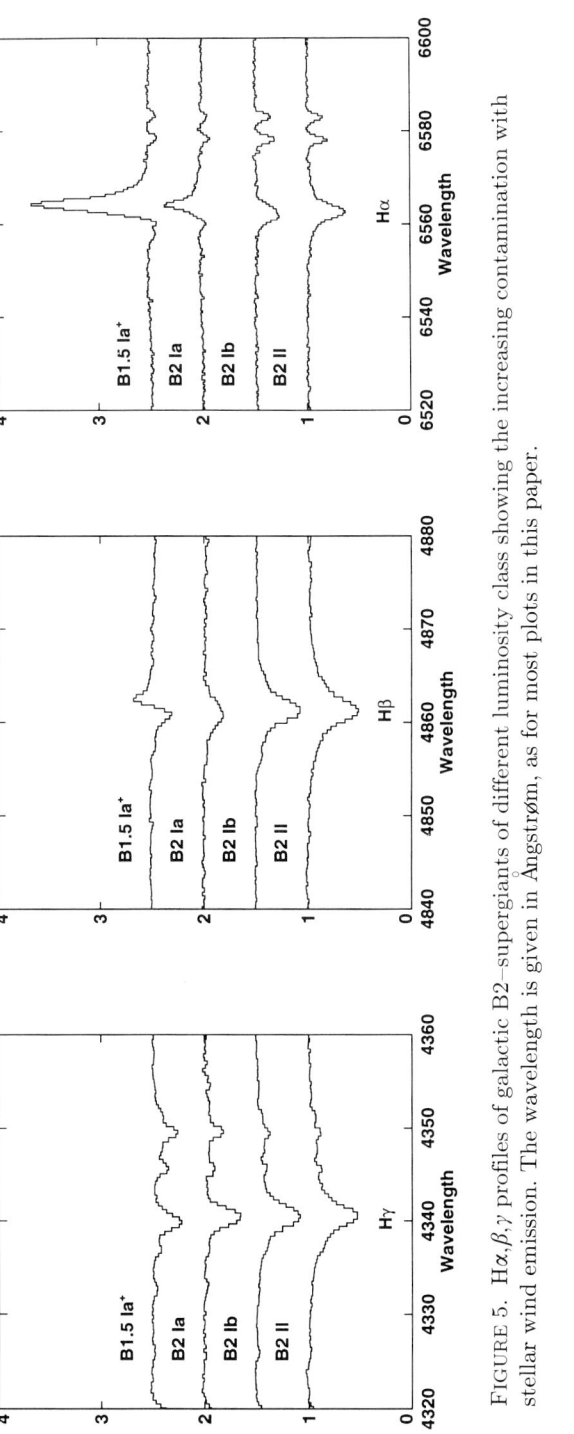

FIGURE 5. Hα,β,γ profiles of galactic B2–supergiants of different luminosity class showing the increasing contamination with stellar wind emission. The wavelength is given in Ångstrøm, as for most plots in this paper.

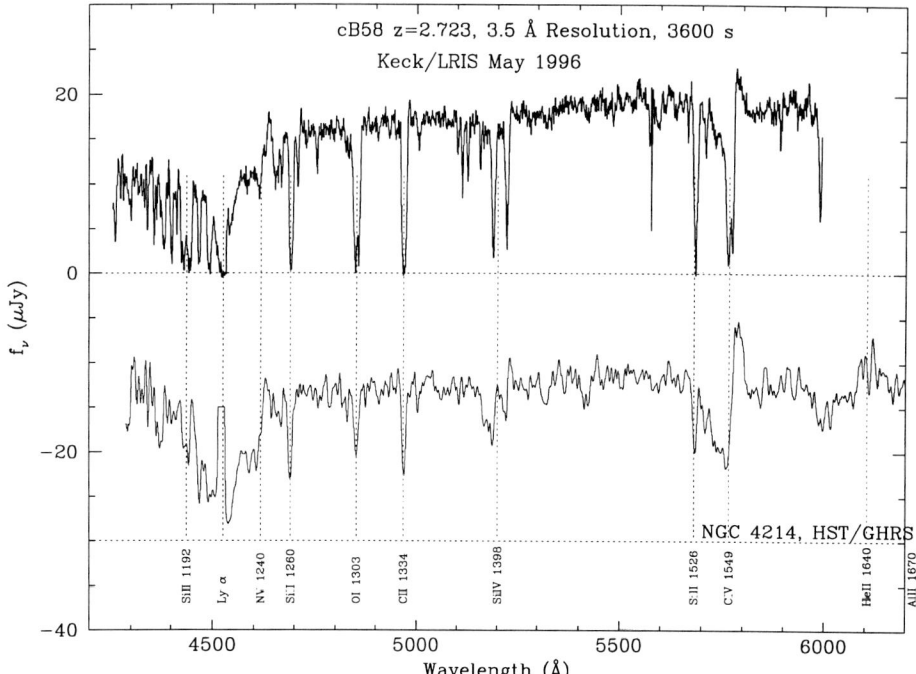

FIGURE 6. Optical spectrum of a galaxy at very high redshift ($z = 2.732$) compared with a "local" starburst. Figure from Steidel and Pettini (priv. comm.), for details see Steidel et al. 1996.

On the other hand, the interpretation of the wind lines, together with the success of radiation driven wind theory, has led to the realization that these **stellar wind features are** also **a gift of nature**. They can be easily identified in medium resolution spectra of Blue Supergiants individually observable out to the Virgo cluster of galaxies or in the integrated spectra of starburst regions in galaxies even much further afield, the most dramatic example being the recent detection of galaxies up to redshifts of $z = 3.5$ by Steidel et al. 1996 (see Fig. 6), which show clear signatures of hot stars and their winds. It is evident that with appropriate atmospheres including the effects of winds, our knowledge of the physics of hot stars will sometimes enable us to interprete these spectra quantitatively and to determine stellar abundances in the early cosmos, as indicated by a few calculated UV model atmosphere spectra redshifted and displayed in Fig. 7.

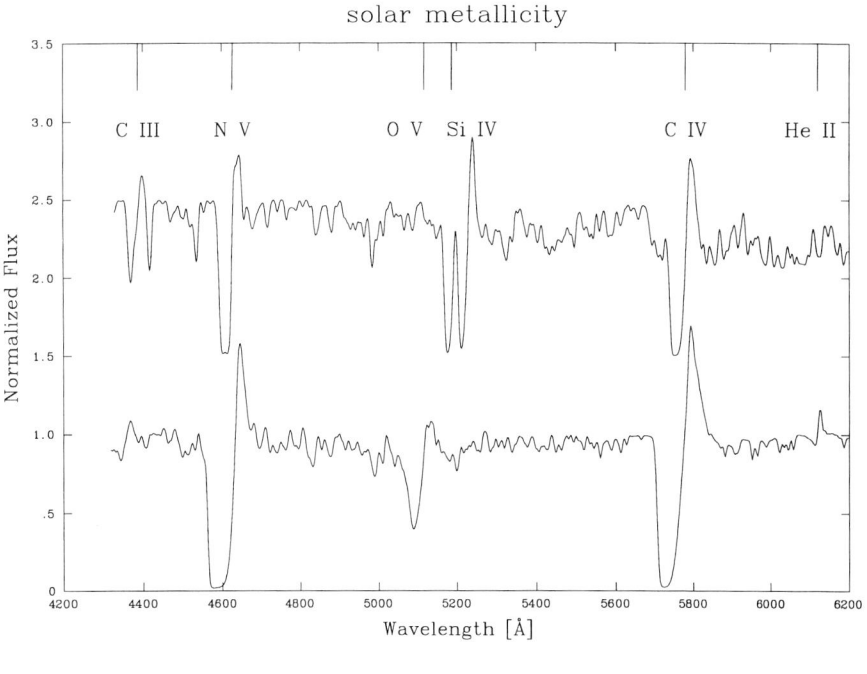
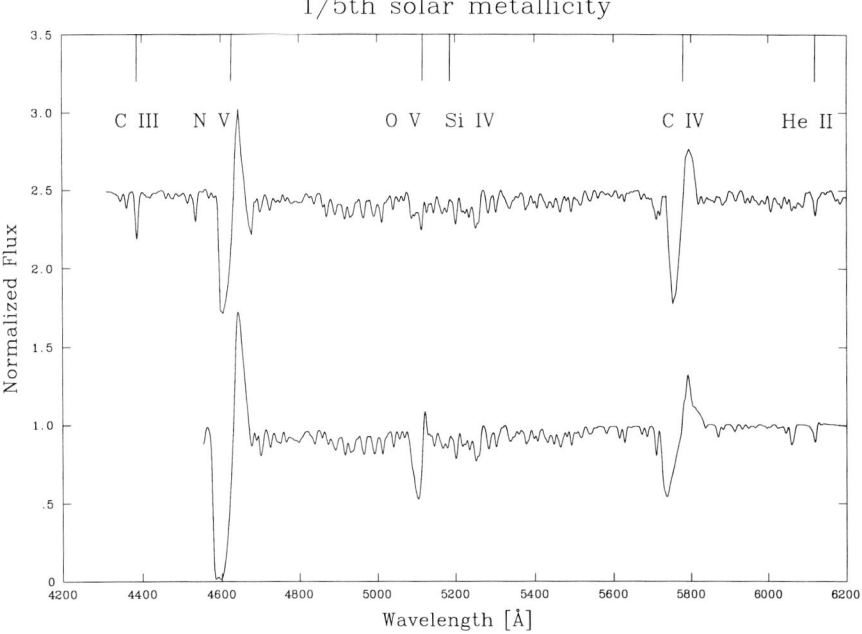

FIGURE 7. Calculated model atmosphere UV spectra of O-stars redshifted to $z = 2.732$ and degraded to 3 Å resolution. **upper panel**: O-supergiants with solar metallicity. Top: spectral type O8, $T_{\rm eff}$=35000K; bottom: spectral type O3, $T_{\rm eff}$=52000K. **lower panel**: O-stars with 1/5 solar metallicity. Top: Main sequence O6 object with $T_{\rm eff}$=43000K; bottom: O3-supergiant with $T_{\rm eff}$=55000K. Note that the simulations do not include the interstellar lines.

Most importantly, **the observation of stellar winds can provide a key to extragalactic distance determinations**. The basic idea is very simple. Since the winds are driven by radiation, the mechanical momentum rate of the stellar wind, $\dot{M}v_\infty$ ($\dot{M}$ is the rate of mass-loss and v_∞ is the terminal velocity of the stellar wind), should be a function of the photon momentum rate provided by the stellar photosphere, which is proportional to the stellar luminosity. Thus, as soon as this **Wind Momentum – Luminosity Relationship (WLR)** is known, one has only to determine the mass-loss rate (from Hα; see the following sections) and terminal velocity (from the P Cygni profiles in the spectrum) to obtain the intrinsic luminosity, and hence the distance.

All that one needs are **hydrodynamic NLTE model atmospheres** incorporating the effects of stellar winds and spherical extension as realistically and consistently as possible. In the following section, we will introduce this new generation of model atmospheres. We will then discuss the methods of spectral diagnostics or – in other words – how we can determine temperatures, gravities, abundances, masses, radii, luminosities, mass-loss rates, stellar wind velocities, distances etc. from the spectra of Blue Supergiants (section 3). We will devote extra sections to spectral diagnostics in new spectral domains, X-rays and the IR (sections 4 and 5). We will discuss the properties of the cluster of very massive supergiants in the central parsec of the Galaxy and we will learn that the most massive stars in the Large Magellanic Clouds have masses clearly in excess of 100 $M_\odot$ (section 6). We will deal with the determination of stellar abundances from blue supergiants in the Local Group and beyond (section 7) and, finally, the Wind momentum - Luminosity Relationship (WLR) will be introduced as a new method for the determination of extragalactic distances (section 8). We will conclude with a discussion of future perspectives (section 9).

2. Atmospheres of luminous hot stars

The goals and objectives for the application of hot star model atmospheres are ambitious. We aim at detailed spectrum synthesis in all spectral domains, X-rays, EUV, UV, optical, IR and radio, and the models should include both, the stellar wind lines formed far out in the atmosphere at high outflow velocities and the "pseudophotospheric" lines formed in deeper atmospheric layers, where the wind velocity is smaller than or of the order of the sound speed. We wish to carry out quantitative spectral analysis , i.e. the spectrum synthesis should allow to determine precise stellar parameters, wind properties and abundances and we should be able to calculate stellar energy distributions and ionizing fluxes that can be used for the interpretation of the emission line spectra of HII-regions and Planetary Nebulae. Combined with stellar evolution calculations, we intend to use our calculated spectra to synthesize spectra of unresolved stellar populations of hot stars in galaxies and to learn about stellar abundances and mass distributions of starbursts at low and high redshifts (see Figs. 6,7).

On the other hand, the physics of the atmospheres of hot luminous stars are complex and very different from standard stellar atmosphere models. They are dominated by the influence of the radiation field, which is characterized by energy densities larger than or of the same order as the energy density of atmospheric matter. This has two important consequences. First, severe departures from Local Thermodynamic Equilibrium of the level populations in the entire atmosphere are induced, because radiative transitions between ionic energy levels become much more important than inelastic collisions. Second, supersonic hydrodynamic outflow of atmospheric matter is initiated by line absorption of photons transferring outwardly directed momentum to the atmospheric plasma. This latter effect is responsible for the existence of the strong stellar winds observed.

In this section, we will concentrate the discussion mostly on the consequences of stellar winds and matter outflow on the atmospheric stratification and the formation of the spectra. A comprehensive and detailed discussion of the crucial effects of the departures from Local Thermodynamic Equlibrium ("NLTE effects") in the atmospheres of hot stars and of the calculation of spectral lines in the presence of outflow velocity fields has already been given by Kudritzki 1988 and will not be repeated here.

2.1. *Hydrodynamic NLTE model atmospheres with spherical extension*

Before discussing the present status of hot star model atmosphere theory we want to briefly describe the history that led to the development of hydrodynamic models. Lucy & Solomon 1970 were the first to find that radiative acceleration caused by line absorption in a few strong UV resonance lines of metals might initiate stellar winds of hot stars, if the stellar luminosity is above a certain threshold. The first consistent hydrodynamical treatment of line driven winds was then introduced in the ingenious paper of Castor, Abbott, & Klein 1975, who gave a beautiful first formulation of the full problem and demonstrated that both, strong and weak lines contribute to the acceleration of stellar winds. Pauldrach, Puls, & Kudritzki 1986 and Friend & Abbott 1986 used for the first time realistic line lists of hundred thousands of metal lines (still in a very approximate NLTE) and an improved treatment of the radiative acceleration (taking into account the fact that the stellar photosphere irradiates its wind as a finite disk rather than as a point source). They achieved reasonable agreement between calculated and observed rates of mass-loss ($\dot{M}$) and terminal velocities (v_∞) of the stellar winds. Kudritzki *et al.* 1987 studied the dependence of the strengths of stellar winds as a function of metallicity. Pauldrach 1987 introduced a full NLTE treatment of the metal lines driving the wind and Puls 1987 investigated the important effect of multiple photon momentum transfer through line overlaps caused by the velocity induced Doppler-shifts.

All these papers mostly concentrated on the hydrodynamics of the stellar wind envelopes around the photospheres, more or less treating the hydrostatic photosphere and the supersonically expanding wind as two different non-interacting entities. Gabler *et al.* 1989 were the first to give up this artificial separation. Driven by the necessity to interpret the optical spectra of hot luminous stars, where many spectral lines such as the Balmer lines, for instance, have contributions from both regions, photosphere and wind, they developed their concept of "Unified Model Atmospheres". Unified Models are spherically extended and yield the entire sub- and supersonic atmospheric structure taking the mass-loss rate, density and velocity structure from the hydrodynamics of radiation driven winds. They can be used to calculate energy distributions simultaneously with "photospheric" and "wind" lines and, most importantly, they can treat the multitude of "mixed cases", where a photospheric line is contaminated by wind contributions. This concept turned out to be extremely fruitful for the interpretation of hot star spectra and has led to the rapid development of significant refinements and improvements in parallel in several groups. Basic papers describing the status of the work in the different groups are Schaerer & Schmutz 1994, Schaerer & de Koter 1997, Pauldrach *et al.* 1994, Taresch *et al.* 1997, Sellmaier *et al.* 1996, Sellmaier 1996, Pauldrach *et al.* 1997, Santolaya-Rey *et al.* 1997, Hillier 1996, Hillier & Miller 1997, Hamann 1996. In the following, we describe the model atmosphere approach developed at the Munich University Observatory.

Fig. 8 gives an overview of the physics. To calculate a stellar atmosphere model the stellar parameters T_{eff} (effective temperature), $\log g$ (logarithm of photospheric gravity), R_* (photospheric radius at a pre-specified optical depth, for instance, $\tau_{ross} = 2/3$; note that T_{eff} and $\log g$ are also defined at this radius) and the abundances Z have to be specified. Then the stationary hydrodynamic equations are solved in spherical symmetry (r is the radial coordinate, ρ the mass density, v the wind velocity, p the gas pressure). The crucial term in the hydrodynamics is the radiative acceleration g_{rad} that has contributions from continuous absorption and scattering (g_{cont}) and line absorption. The calculation of the line acceleration is performed by summing the contributions of lines selected from a list containing more than 2.5 million lines of all important ionization stages (essentially I to VII) from hydrogen to iron group elements. Depending on T_{eff} up to 500000 lines are finally taken into account per model. For each line the oscillator strengths f_{lu}, the statistical weights g_l, g_u and the occupation numbers n_l, n_u of the lower and upper level enter together with the frequency and angle integral over the specific intensity I_ν and the line broadening function ϕ_ν, the latter accounting for Doppler shifts in its argument (ν is the photon frequency in the observers frame and μ is the cosine of the angle between the radial direction and the direction of the light ray for which the intensity $I_\nu(\mu)$ is considered).

Since NLTE effects are extremely important, the occupation numbers are determined by rate equations containing collisional (C_{ij}) and radiative (R_{ij}) transition probabilities. Detailed and very realistic atomic models with a large number of levels and transitions and additional physical processes such as autoionization and dielectronic recombination or absorption by photons emitted in the EUV and X-ray domain by shocks in the wind are taken into account. Fig. 9 gives examples for level diagrams and transition schemes of two individual ionic models. In total, more than 150 such ions with corresponding level diagrams, transition schemes and rate equations are considered (For details, in particular the atomic physics, see Pauldrach et al. 1994 and Taresch et al. 1997). It is important to note that the hydrodynamic equations are coupled directly with the rate equations. The velocity field enters into the radiative transition probabilities while the density is important for the collisional rates and the equation of particle conservation. On the other hand, the occupation numbers are crucial for the hydrodynamics, since the radiative line acceleration dominates the equation of motion.

In addition, the rate equations are coupled to the equations of transfer for the radiation field in each line transition, since the latter enters the radiative transition probabilities. Because of the structure of the transfer equations this coupling is extremely non-linear. Since the absorption coefficient κ_ν depends on occupation numbers (and therefore also on the density) and the velocity field because of the Doppler-shifts in frequency, there is also a coupling of the equations of transfer with the hydrodynamic equations. (S_ν is the source function, i.e. the ratio of the emission to the absorption coefficient).

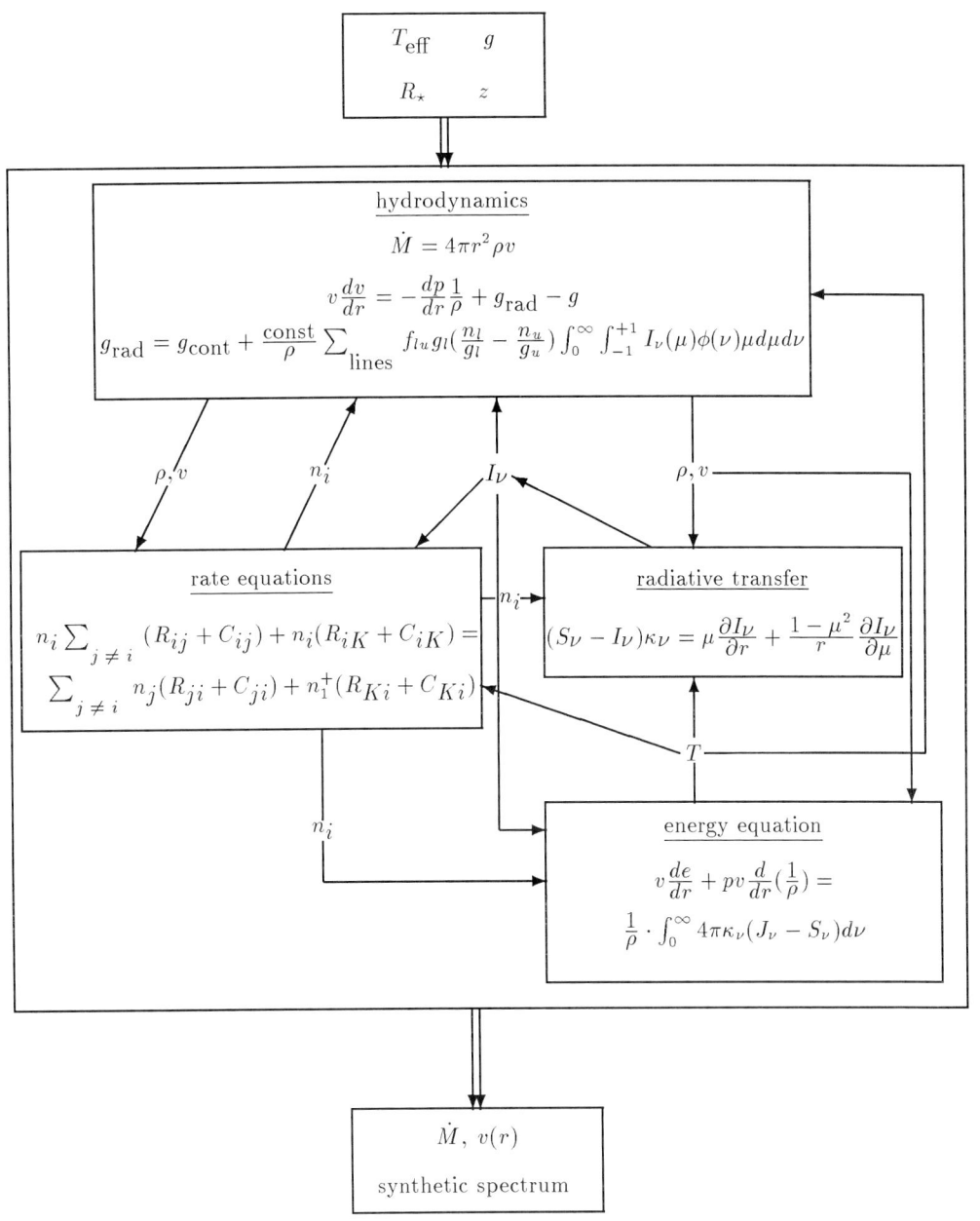

FIGURE 8. The basic equations of stationary hydrodynamic NLTE model atmospheres.

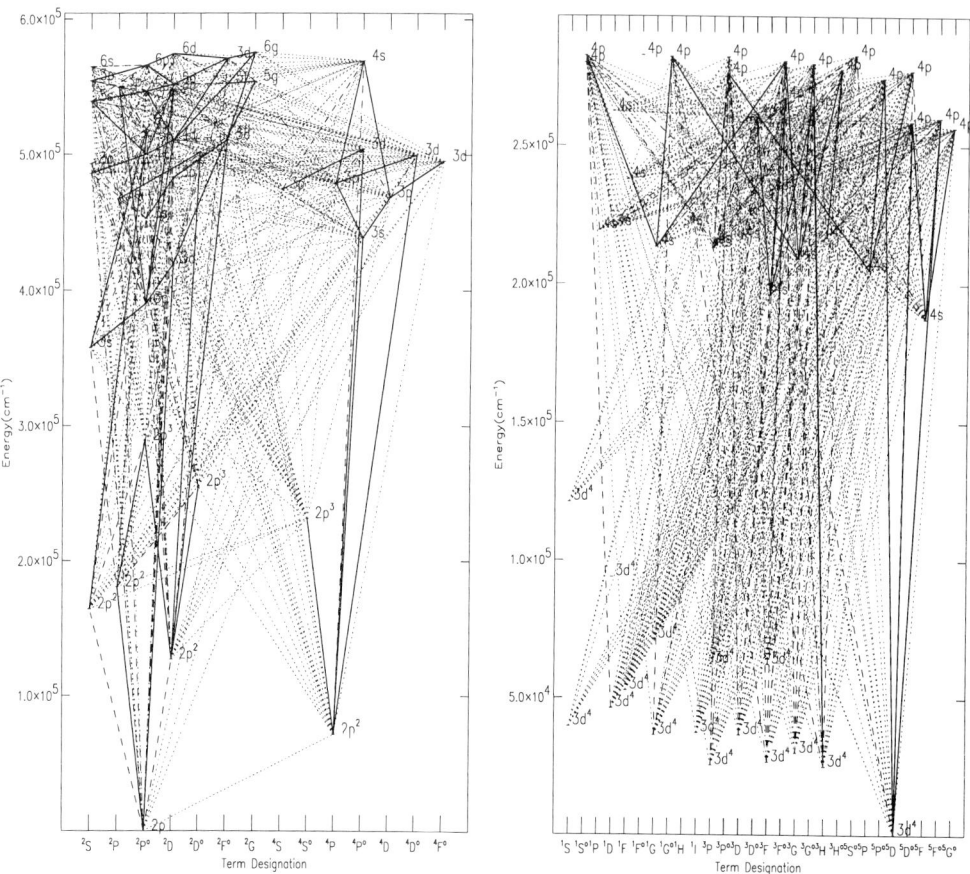

FIGURE 9. Energy level diagram and transition scheme for O IV (left) and Fe V (right) in the Munich model atmosphere code. The transitions are coded by the strengths of their gf-values: solid $gf \geq 1.0$; dashed $1.0 \geq gf \geq 0.1$; dotted $0.1 \geq gf$. In the same way, more than 150 ions are included in the Munich University Observatory stellar wind code developed by A. Pauldrach and collaborators.

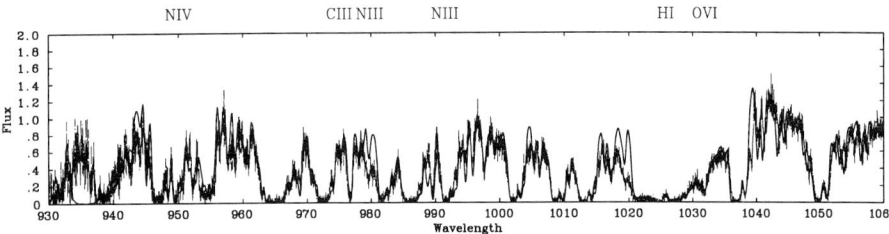
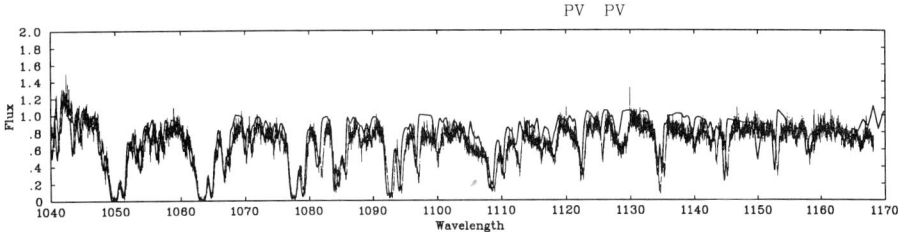
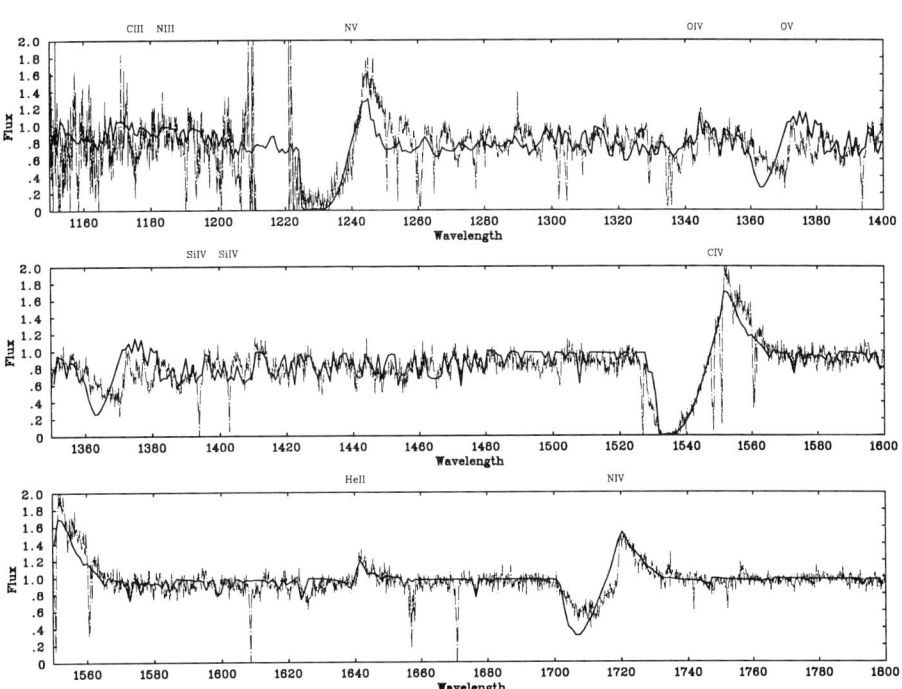

FIGURE 10. Comparison of observed and calculated spectrum of the O3 Iaf+ supergiant HD 93129A. Upper two panels: FUV (ORFEUS), lower three panels: UV (IUE). The ORFEUS spectrum contains many interstellar molecular lines of H_2 and HD which are included in the spectrum fit. The narrow sharp lines in the IUE spectrum are of atomic or ionic interstellar origin and are not included in the spectral fit. (From Taresch et al. 1997).

The local kinetic temperature of the plasma is determined by the energy equation, which contains the specific energy e and the angle averaged intensity J_ν, in addition to the quantities defined previously.

The iterative solution of the total system of equations yields the hydrodynamic and thermodynamic structure of the atmosphere and allows calculations of synthetic spectra to be compared with the observations. Fig. 10 gives a nice example from the recent paper by Taresch et al. 1997.

Two remarks are necessary at the end of this subsection. First, none of the present codes of the different groups (see citations above) solves the set of equations of Fig. 8 without applying substantial approximations. It would be an extra paper to discuss the significance of these approximations in detail. In summary, however, despite these approximations the results obtained with these codes can be regarded as fairly realistic. Second, the theoretical framework introduced so far is based on the assumption of stationarity. This does not mean that we are not aware of the fact that spectral variability of hot stars is observationally well established. But fortunately, the amplitudes of variability in most cases are not very large so that the assumption that the stationary models give a fair description of the time averaged situation and are well suited for the determination of element abundances and stellar parameters appears to justified. Good overviews of observed spectral variability and its interpretation are found in the proceedings of the 1993 Isle aux Coudres Workshop (Moffat et al. 1994), the 1991 STScI Workshop (Drissen et al. 1992) and the 1990 ESO Workshop (Baade 1991). Kaper et al. 1996, Massa et al. 1995, Kaufer et al. 1996a, Kaufer et al. 1996b, Rivinius et al. 1997 are recent papers describing the spectral variability of Blue Supergiants. In section 4, where we discuss the X-ray emission of O-stars, we will again be confronted with this issue.

In the following, we discuss the most important effects caused by the presence of the outflow velocity fields.

2.2. Model stratification

For model atmospheres with matter outflow the location of the **sonic point**, where the velocity field $v(r)$ equals the isothermal sound speed, is crucial. Below this point, the atmospheric structure becomes more and more hydrostatic (see Kudritzki 1988). Around the sonic point and above, however, the stratification is dominated by the velocity field and differs substantially from a hydrostatic model. Above the sonic point, the shape of the velocity field follows the so-called "β - velocity law"

$$v(r) = v_\infty (1 - R_*/r)^\beta, \qquad (2.1)$$

where β depends on the stellar parameters, but is of the order of unity for line driven winds (see Castor, Abbott, & Klein 1975, Pauldrach, Puls, & Kudritzki 1986, Kudritzki et al. 1989, for observational determinations of β, see later sections).

Fig. 11 shows examples of velocity fields in an atmosphere of a typical $O5Ia$ - supergiant for different values of the mass-loss rate $\dot{M}$. With increasing $\dot{M}$ the sonic point moves inward and reaches the depths of the formation of the continuum and weak lines ($-1 \leq \log(\tau) \leq 0$). For $\dot{M} = 5 \times 10^{-5} M_\odot/yr$ the entire atmosphere is dominated by the wind. Note that for radiative transfer of metal lines the **thermal point**, where $v(r)$ becomes equal to the thermal velocity of the corresponding ion, is the crucial location. For metal lines formed around and above this point equivalent widths and line profiles start to differ substantially from hydrostatic calculations (see Kudritzki 1992).

One of the most important consequences of the presence of matter outflow is the change in the density stratification $\rho(r)$, which is determined by the equation of continuity

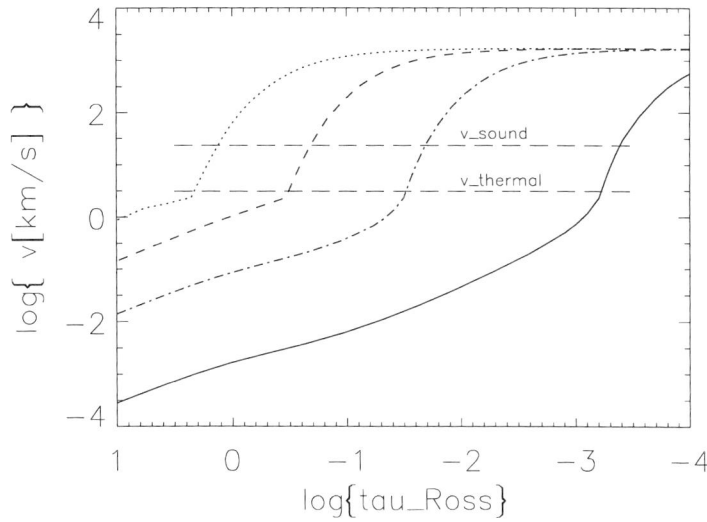

FIGURE 11. Outflow velocity fields as a function of Rosseland optical depth for model atmospheres of an O5 Ia - supergiant with $T_{\text{eff}} = 40000K$, $\log g = 3.5$, $R_*/R_\odot = 16$, $v_\infty = 1700$ km/s and mass-loss rates $\dot{M}$ of 30 (dotted), 5 (dashed), 0.5 (dashed-dotted) and $0.01 \times 10^{-6} M_\odot/\text{yr}$. The horizontal long dashed lines indicate the values of the isothermal sound velocity and the thermal velocity of iron, respectively.

$$\dot{M} = 4\pi r^2 \rho(r) v(r) \qquad (2.2)$$

above the sonic point and obeys a power law with exponent -2, as soon as $v(r)$ approaches the terminal velocity v_∞. As a consequence, **the mass density as a function of optical depth in a model atmosphere with a stellar wind is always smaller than that in a hydrostatic model** (see Fig. 12). This has important consequences for the formation of spectral lines and the ionizing continua.

The change in the density stratification above the sonic point and the corresponding change in scale height is also crucial for the geometrical extension of the atmosphere (see Kudritzki 1988). Above the sonic point the atmospheres become spherically extended and the standard assumption of classic planeparallel model atmospheres becomes invalid (see Fig. 13).

2.3. Energy distributions and ionizing fluxes

Energy distributions are severely affected by the presence of stellar winds, as is demonstrated by Fig. 14. Free-free thermal emission of the spherically extended stellar wind plasma causes a significant IR- and radio- excess at longer wavelengths. But also the ionization edges of bound-free transitions in the UV and EUV are substantially modified. This is caused by two effects: the change in density stratification (see above) and the influence of the velocity induced Doppler-shifts on the optical thickness of resonance lines of ions producing strong absorption edges. A detailed discussion is given by Gabler et al. 1989 and Najarro et al. 1996. Basically, the optical thickness in the resonance transitions is reduced and the detailed balance is modified so that electrons are pumped to the excited level and the ground-states are severely depopulated (see Fig. 15).

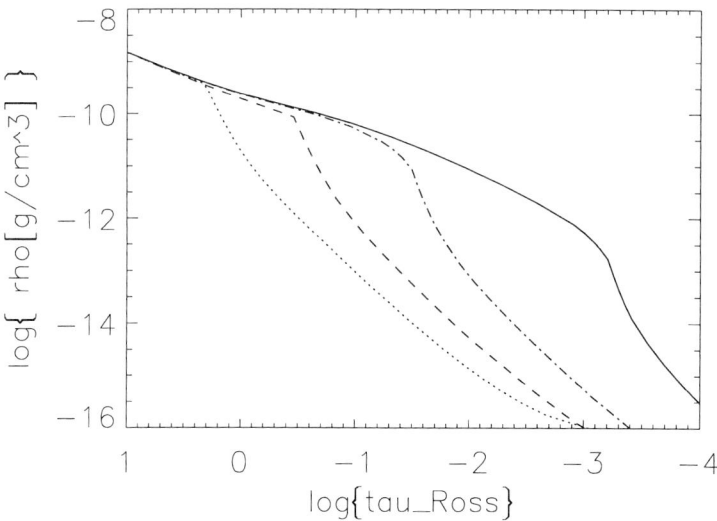

FIGURE 12. Matter density as function of Rosseland optical depth for the same models as in Fig. 11.

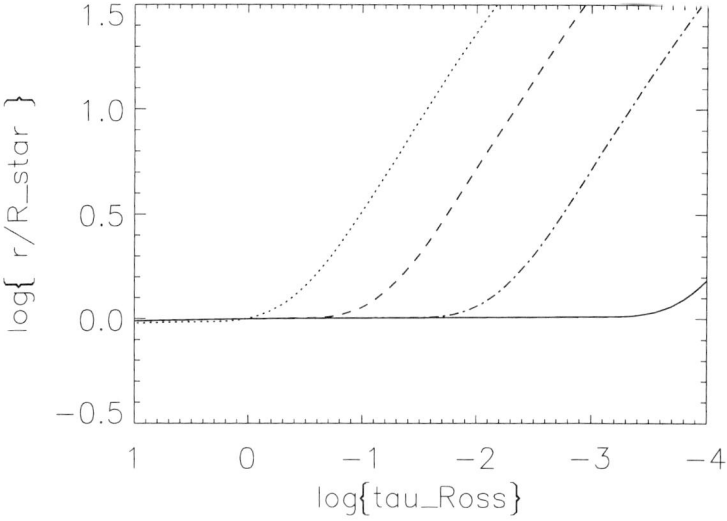

FIGURE 13. Radius as function of Rosseland optical depth for the same models as in Fig. 11.

This has important consequences for the ionization of the ISM by hot stars. Gabler et al. 1989 demonstrated that the effect contributes to the solution of the longstanding problem of the "Zanstra-discrepancy" in Planetary Nebulae. It may also help to understand the ionizing population in Giant Extragalactic H II-regions of high excitation (see Gabler et al. 1992 and Fig. 16). In addition, it is crucial for the ground state continua of H and He I in B-stars, as has been shown recently by Najarro et al. 1996

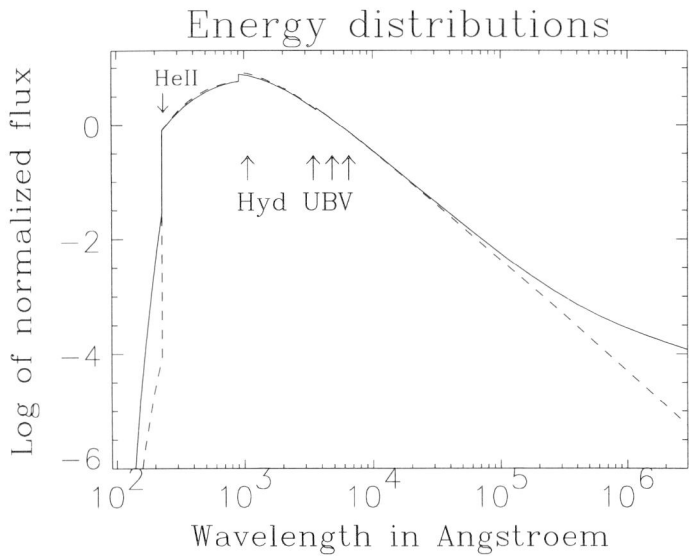

FIGURE 14. Energy distribution of a hot ($T_{\text{eff}} = 47500K$) O-supergiant calculated for a hydrostatic model (dashed) and a hydrodynamic model with a stellar wind (solid). (From Gabler et al., 1989).

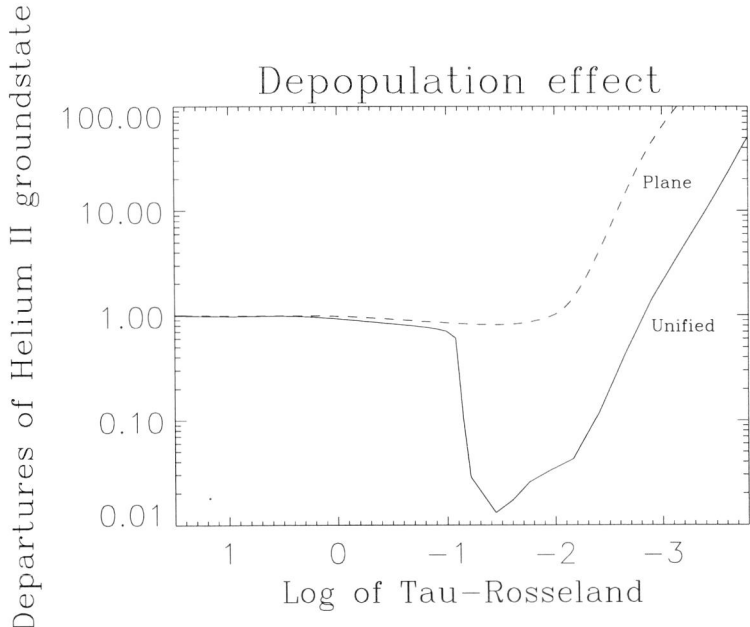

FIGURE 15. NLTE occupation numbers in the ground state of He II (in units of the LTE-value) calculated for a hydrostatic model (dashed) and a hydrodynamic model with a stellar wind (solid). The models are the same as in Fig. 14. (From Gabler *et al.* 1989).

(see Figs. 17, 18). This has important consequences for the ionization of the ISM by a somewhat older population – such as the local ISM – when O-stars are already absent.

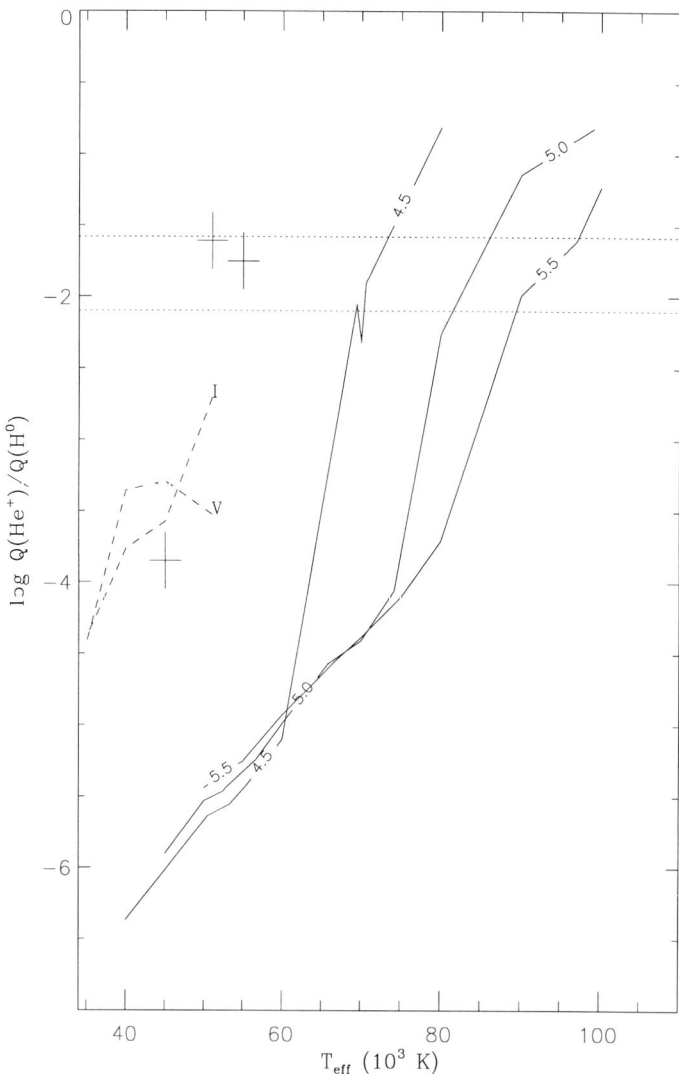

FIGURE 16. Diagram to understand the ionizing population of highly excited Giant Extragalactic H II-regions. The logarithmic ratio of He II ionizing photons to hydrogen ionizing photons is plotted vs. $T_{\rm eff}$. Observed He II and hydrogen nebular recombination lines give ratios between the two horizontal dotted curves and can be compared with the predictions of model atmospheres. Results of three types of model atmosphere are plotted as a function of $T_{\rm eff}$: models without a wind labelled by log g (solid); NLTE model with winds for supergiants (I) and main sequence (V) (dashed); supergiant models with very strong winds (large crosses). Obviously, much smaller effective temperatures are needed as soon as winds are taken into account. (From Gabler *et al.* 1992).

The slope of the ionizing continua can also be changed drastically by the presence of stellar winds. This has been demonstrated by Sellmaier *et al.* 1996, who compared their

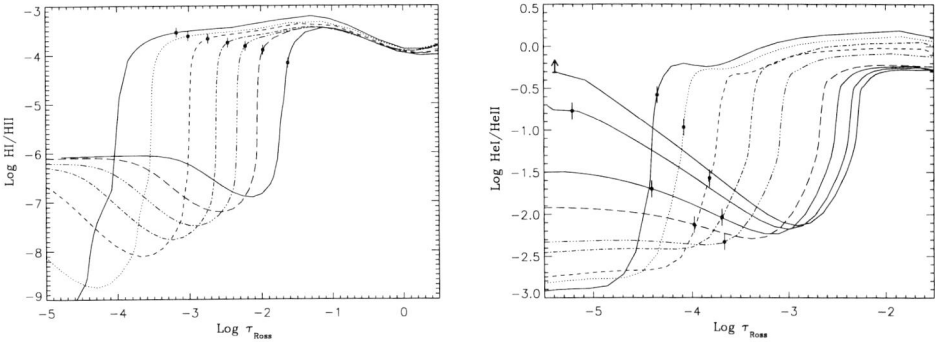

FIGURE 17. Hydrogen (left) and HeI (right) ground state ionization fraction as function of optical depth for a B-giant model atmosphere with moderate mass-loss rates (between 5×10^{-10} to 2.5×10^{-7} $M_\odot/yr$) showing the strong depopulation effects in the ground states. Optical depth unity at the head of the ionizing continuum is indicated for each model. (From Najarro et al., 1996).

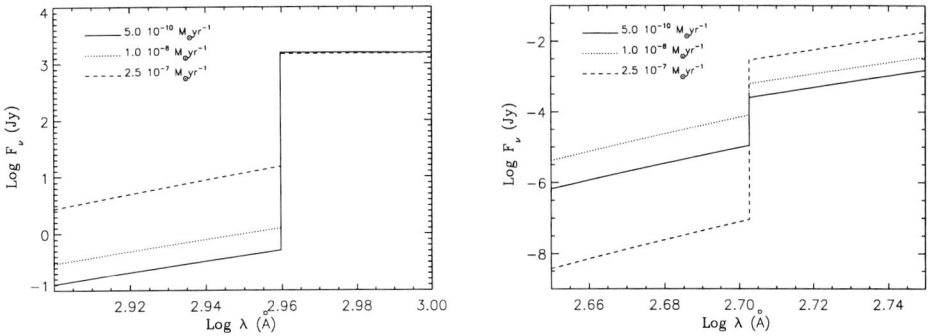

FIGURE 18. Energy distributions around the Lyman-jump (left) and the He I-jump for B-giant models with different mass-loss rates $\dot{M}$. Modifications in $\dot{M}$ lead to enormous changes in the ionizing flux. (From Najarro et al., 1996).

NLTE line blanketed stellar wind models with hydrostatic models (Fig. 19). These new model atmospheres solve the longstanding "$NeIII - problem$" of H II-regions, as they allow the observed Ne III emission lines to be reproduced for the first time with the assumption of reasonable abundances (see Fig. 20).

170 R.P. Kudritzki: *The Brightest Blue Supergiant Stars in Galaxies*

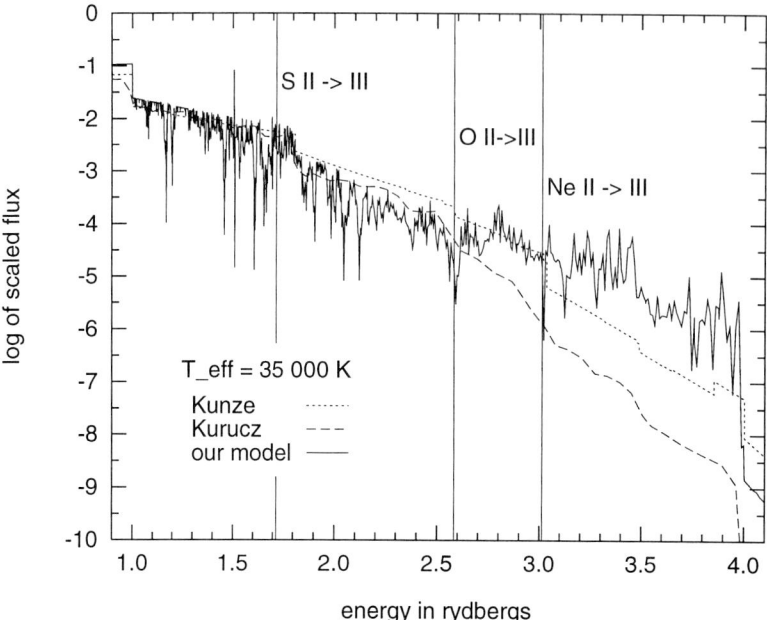

FIGURE 19. Ionizing fluxes at T_{eff}=35000K for a NLTE blanketed model with wind (solid), a NLTE model without wind and metal lines (dotted) and a hydrostatic LTE model with line blanketing (dashed). Vertical lines correspond to energies for the ionization of S II, O II and Ne II, respectively. This shows how the continuum slope is modified by the wind. (From Sellmaier et al. 1996).

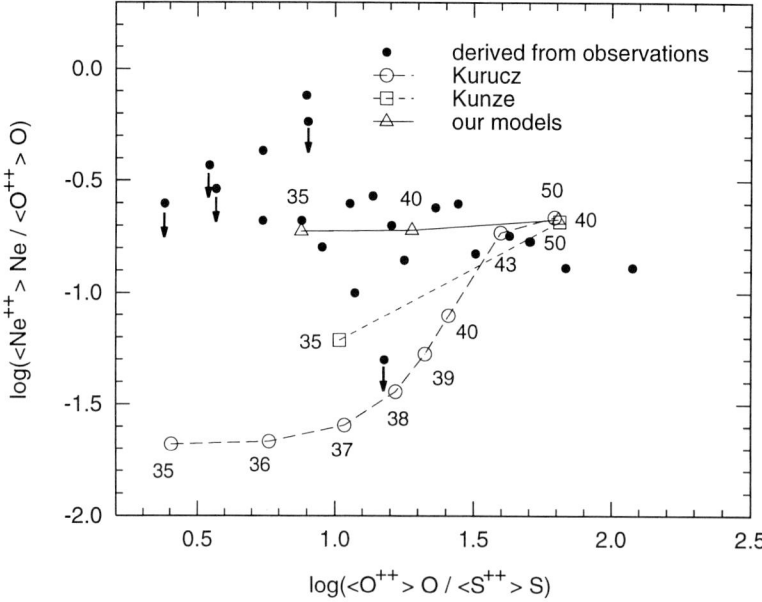

FIGURE 20. Diagnostic diagram of H II-regions showing mean ionization ratios as obtained from the observed emission line spectra (solid dots). Also shown are recombination theory calculations using model atmosphere input fluxes (circles: hydrostatic LTE line blanketed models; squares: hydrostatic NLTE model without metal lines; triangles: NLTE line blanketed models with winds). The numbers refer to T_{eff} in 10^3 K. (From Sellmaier et al. 1996).

2.4. Line diagnostics

The most prominent lines in the optical spectra of hot stars can be severely influenced by the stellar winds. A drastic example is Hα (see Fig. 21). Depending on the strength of the mass-loss rate the line can change from pure absorption to extremely strong emission. This sensitivity of Hα on $\dot{M}$ can be used to determine rates of mass-loss with high precision (see section 3).

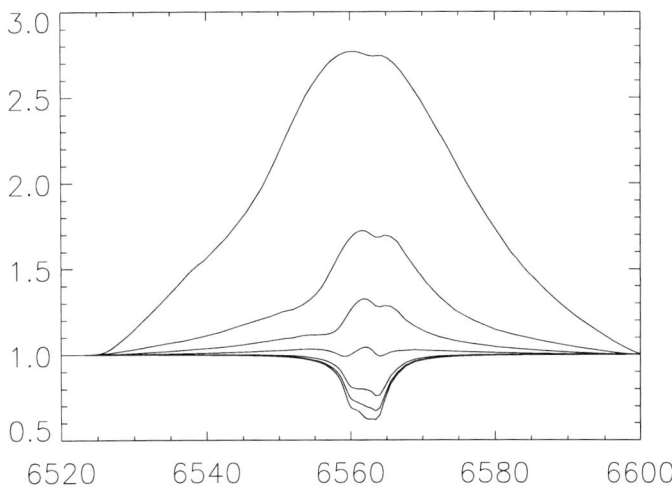

FIGURE 21. Hα profiles for a typical O5 Ia supergiant adopting mass-loss rates of $30, 10, 5, 2.5, 1.0, 0.05$ and 0.01×10^{-6} M$_\odot$/yr.

Higher Balmer lines have smaller oscillator strengths and correspondingly smaller line opacities. As a consequence, the contamination by stellar wind emission is less severe. However, at mass-loss rates of about $10^{-6} M_\odot$/yr a crucial diagnostic line such as $H\gamma$ also starts to be filled in by wind emission thus affecting the absorption profile. Since the strength of the (Stark broadened) absorption wings of $H\gamma$ is a measure of the stellar gravity log g, the neglect of wind contamination will lead to a systematic underestimate of stellar gravities in a quantitative spectral analysis (see Fig. 22).

Helium lines are also affected by the stellar wind. He II λ4686 is a very sensitive stellar wind indicator, as shown by Fig. 23. We note in passing that NLTE model calculations for this line require very careful modelling of the line blocking at the He II resonance line at 304Å and are, therefore, less reliable than calculations for Hα.

The helium ionization equilibrium of He I λ4471 relative to He II λ4542 usually defines the spectral type of O-stars and is the major indicator of effective temperature. Fig. 24 shows that these lines are influenced by the winds as well. The case of He I λ4471 is particularly treacherous. The wind simply weakens the absorption without giving away its presence by a pronounced additional asymmetry. Thus, the use of hydrostatic models makes it all too easy to assign a too high T_{eff} to an O-type supergiant.

Further examples for spectral lines affected by the winds will be given in the following sections.

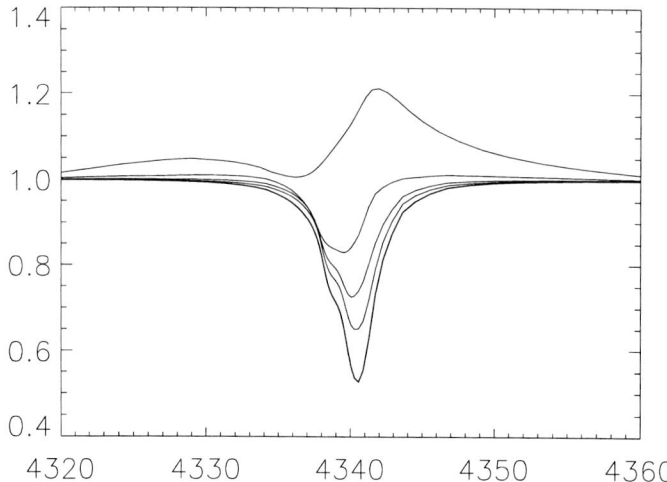

FIGURE 22. Hγ profiles for a typical O5 Ia supergiant adopting mass-loss rates of $30, 10, 5, 2.5$ and 0.01×10^{-6} $M_\odot$/yr.

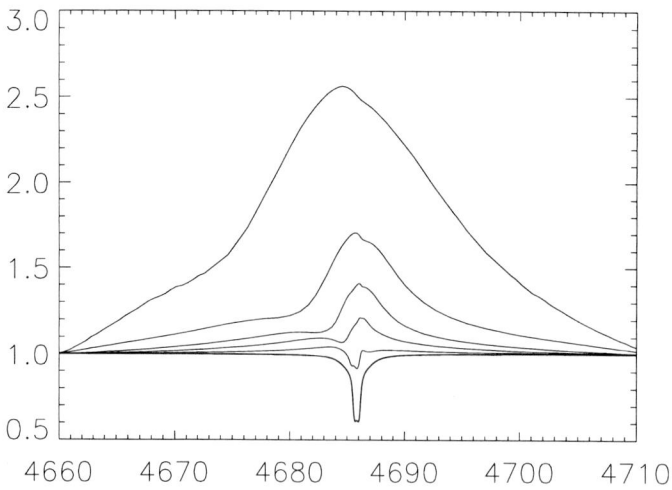

FIGURE 23. He II λ4686 profiles for a typical O5 Ia supergiant adopting mass-loss rates of $30, 10, 5, 2.5, 1.0, 0.01 \times 10^{-6}$ $M_\odot$/yr.

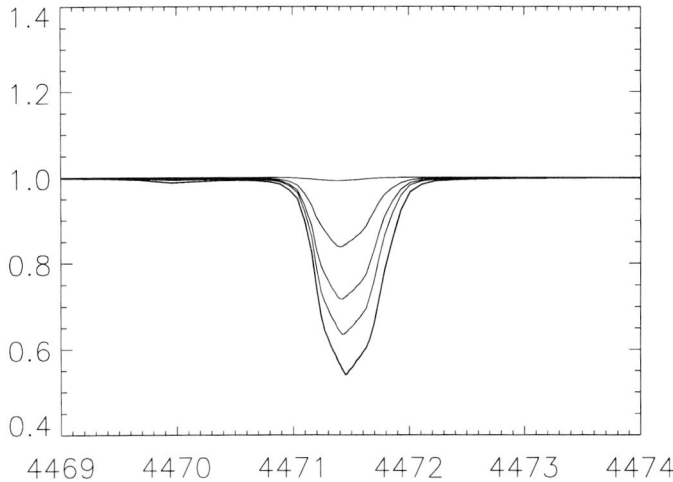

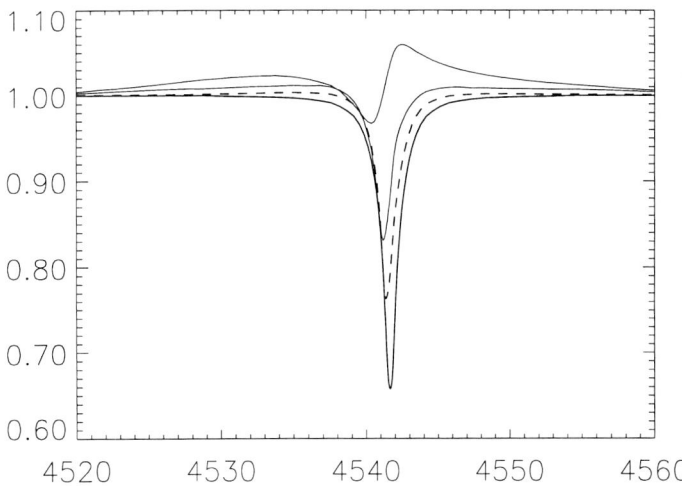

FIGURE 24. He I λ4471 (above) and He II λ4452 (below) profiles for a typical O5 Ia supergiant and mass-loss rates of $30, 10, 5, 2.5$ and 0.01×10^{-6} M$_\odot$/yr.

2.5. NLTE radiative transfer algorithms

For the calculation of weakly wind contaminated lines in the optical spectrum it is crucial that the NLTE radiative transfer in the transition region from the sub- to the supersonic wind is carried out correctly. It turns out that for some ions the so-called "Sobolev approximation" fails badly in the calculation of occupation numbers and that the correct "co-moving frame approach" is needed to obtain reliable results (both methods – with the corresponding references – are discussed in Kudritzki 1988). This was first pointed out by Sellmaier et al. 1993. Very recently, Santolaya-Rey et al. 1997 in their comprehensive model atmosphere paper dealing with supergiants from spectral types A to O addressed this point very carefully. We show three figures from their paper (Fig. 2.5,26,27) demonstrating that the use of the Sobolev-approximation can – in certain cases – lead to strikingly erroneous results.

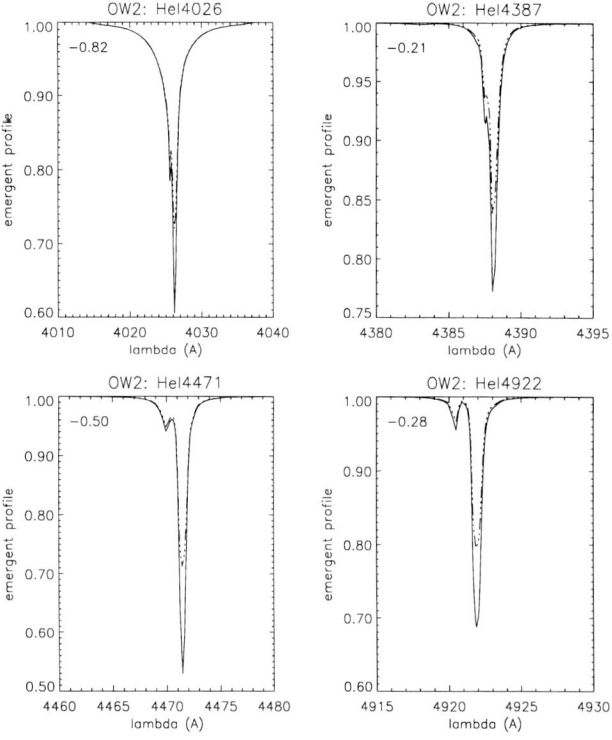

FIGURE 25. He I lines for an O-star model with wind. Dashed: Sobolev-approximation; solid: correct treatment in the comoving frame. (From Santolaya-Rey, Puls, Herrero 1997).

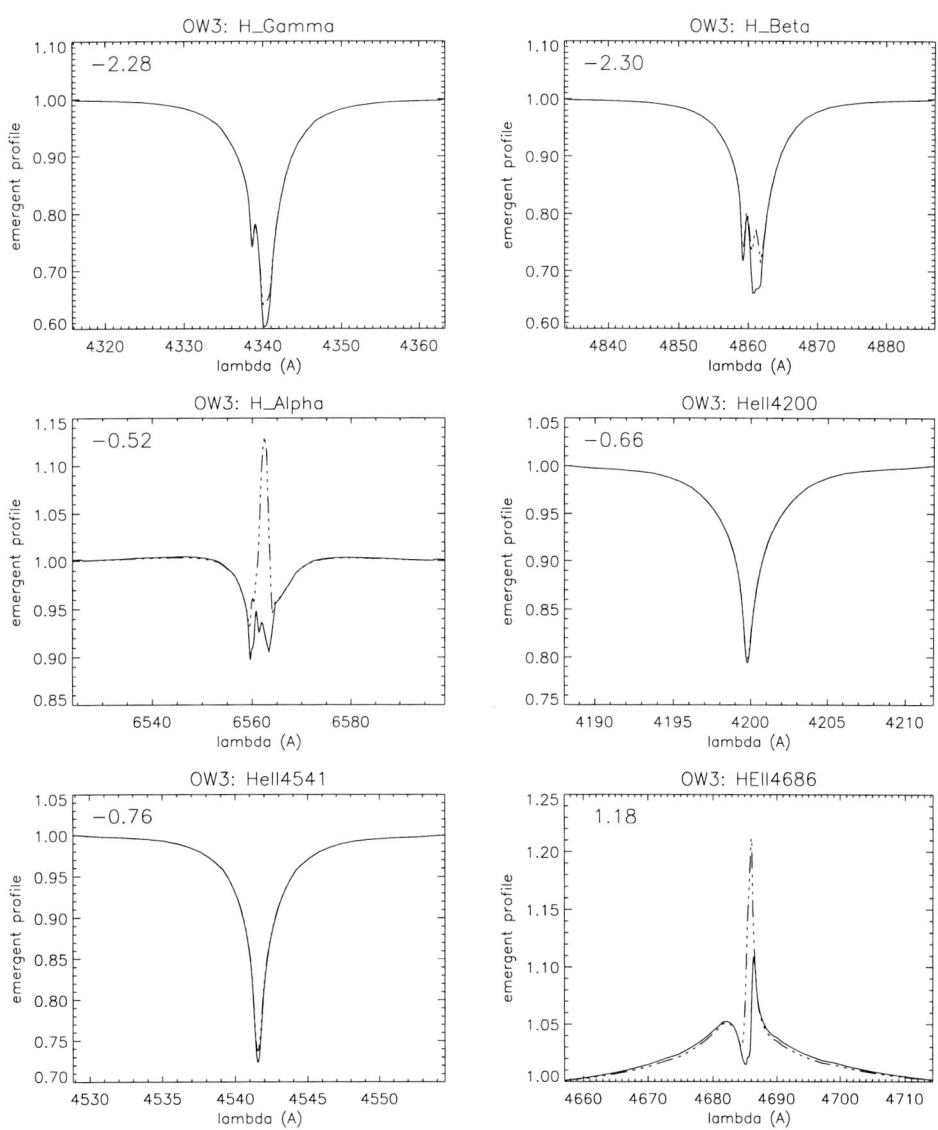

FIGURE 26. Hydrogen and He II lines for an O-star model with wind. Dashed: Sobolev-approximation; solid: correct treatment in the comoving frame. (From Santolaya-Rey, Puls, Herrero 1997).

176   R.P. Kudritzki: *The Brightest Blue Supergiant Stars in Galaxies*

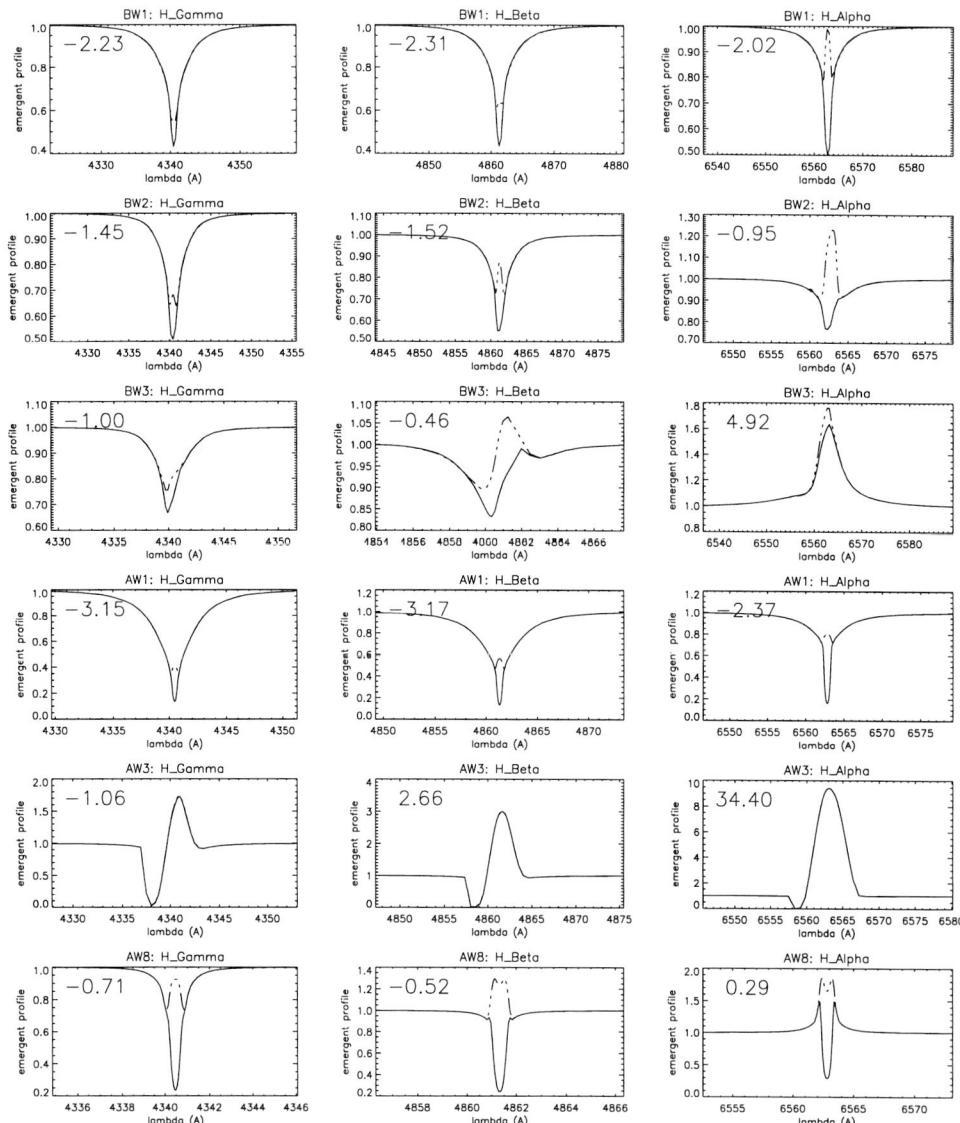

FIGURE 27. Hydrogen Balmer lines for three B-supergiant models (upper three panels) and three A-supergiant models with wind. Dashed: Sobolev-approximation; solid: correct treatment in the comoving frame. (From Santolaya-Rey, Puls, Herrero 1997).

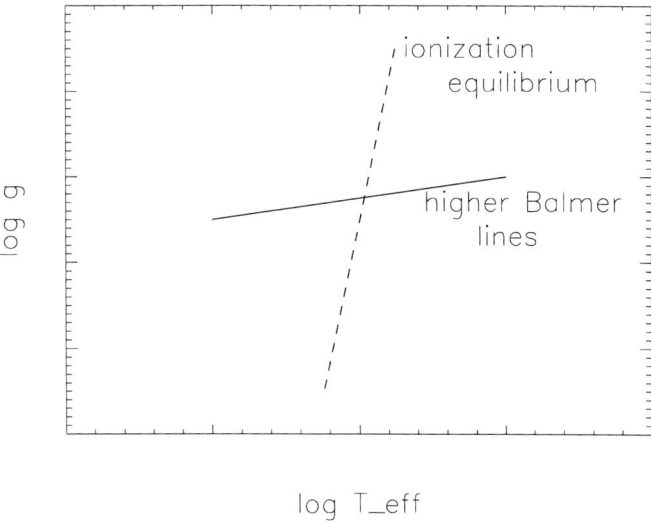

FIGURE 28. Schematic fit-diagram of temperature- and gravity- sensitive spectral lines in the $(\log g, \log T_{\text{eff}})$-plane. Typical ionization equilibria are He I / II for O-supergiants, Si II / III / IV for B-supergiants and Mg I / II for A-supergiants. For further discussion, see text.

3. Methods of spectral diagnostics

In this section we describe the basic techniques to obtain quantitative information about the stellar properties by comparing observed spectra with calculations using the model atmospheres described above. Since the standard methods have been reviewed many times (see Kudritzki 1988; Kudritzki & Hummer 1990), we will be brief and put more emphasis on new methods or new results.

3.1. Effective temperature and gravity

T_{eff} and $\log g$ are the most fundamental atmospheric parameters. They are usually determined by fitting simultaneously two sets of spectral lines, one depending mostly on T_{eff} and the other on $\log g$. Fig. 28 indicates, how this is done, in principle. One constructs fit curves in the $(\log g, \log T_{\text{eff}})$-plane along which the calculated line profiles of different ionization stages ("ionization equilibrium") and the higher (less wind-contaminated) Balmer lines agree with the calculated profiles. The intersection of these fit-curves yields the effective temperature and gravity.

In the fit of the ionization equilibrium spectral lines of two or more ionization stages have to be brought into simultaneous agreement with observations. At different locations in the $(\log g, \log T_{\text{eff}})$-plane this can be achieved only for different abundances. Thus, along the fit curve for the ionization equilibrium in Fig. 28 the abundance of the corresponding element varies and the intersection with the fit curve for the Balmer lines leads to an automatic determination of the abundance. (Note that the old technique of fitting ratios of equivalent widths of lines in different ionization stages and to regard those as being independent of abundance is less reliable, since the lines might be on different parts on the curve of growth). Examples of this technique for O-stars and A-supergiants are given in Figs. 29,30.

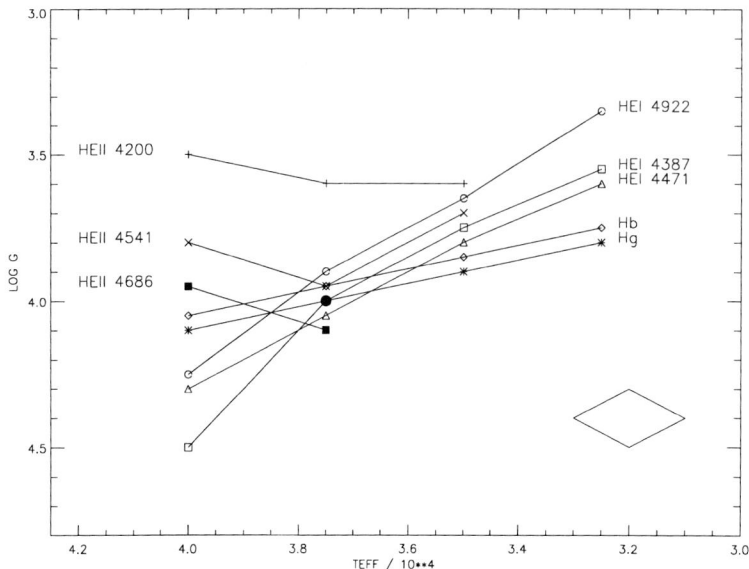

FIGURE 29. Fit diagrams for the Balmer and He I, He II lines of the late O-star 10 Lac adopting solar helium abundance. The diamond indicates the range of uncertainty in T_{eff} and log g. Adapted from Fig.1 of Herrero et al. 1992.

In the past, only ionization those equilibria were used with lines clearly in absorption and as little contaminated by the wind as possible were used. However, this is an unnecessary restriction, in particular in those cases where only one "photospheric" ionization equilibrium is available so that the possible systematic errors of the temperature determination cannot be investigated. One example are O-stars of very early spectral type like O3 or O4. Here only one line, the very weak He I $\lambda 4471$ (relative to He II $\lambda 4542$), was used to determine T_{eff}. Certainly, this is not a good basis for quantitative work. However, with hydrodynamic models it is possible to consider the obviously wind contaminated lines in different ionization stages as well and, thus, to improve the basis for the temperature determination. A first test has been published very recently by Taresch et al. 1997, who used the ionization equilibrium of the wind-affected optical N IV and N V lines in the extreme O3 f- supergiant HD 93129A to determine the effective temperature to $T_{\text{eff}} = 52000K$ (see Fig. 31). This value is in good agreement with the result obtained from the He I line (see Kudritzki et al. 1992).

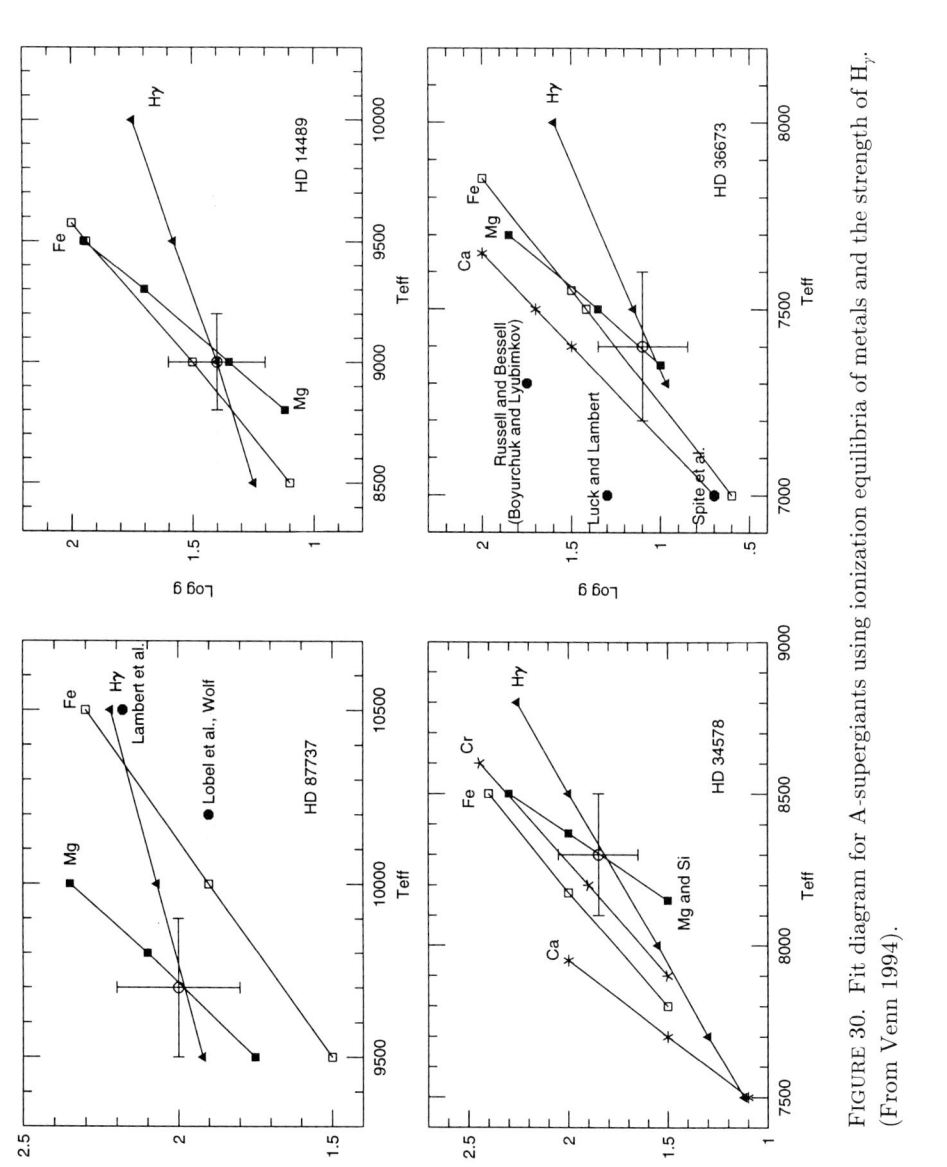

FIGURE 30. Fit diagram for A-supergiants using ionization equilibria of metals and the strength of H_γ. (From Venn 1994).

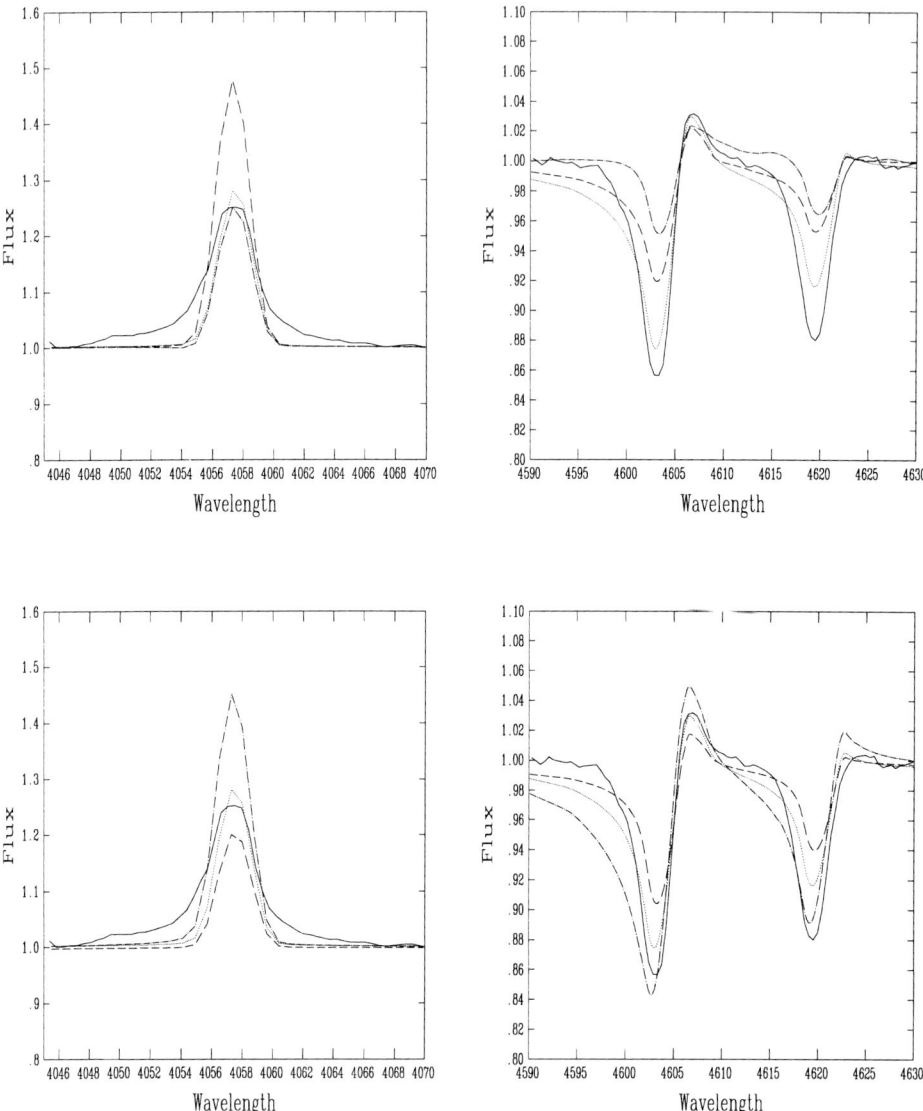

FIGURE 31. Fit of the optical stellar wind nitrogen lines N IV λ4059 (**left**) and N V λ4605 (**right**) of the O3-supergiant HD93129A. **upper panel:** calculations with solar abundance for T_{eff}=50000K (dashed), 52000 K (dotted), 53500 (dashed-dotted). **lower panel:** calculations with T_{eff}=52000K and solar (dashed), factor 2 solar (dotted), factor 5 solar (dashed-dotted) nitrogen abundance. The observed profiles are given by the thick solid curves. (From Taresch et al. 1997).

3.2. Mass-loss rate, terminal velocity and shape of the velocity field

In principle, two types of lines are formed in a stellar wind, *P Cygni profiles* with a blue absorption trough and a red emission peak and *pure emission profiles*. The difference is caused by the re-emission process after the photon has been absorbed within the line transition. If the photon is immediately reemitted by spontaneous emission, then we have the case of line scattering with a source function proportional to the geometrical dilution of the radiation field and a P-Cygni profile will result. If the reemission occurs as a result of a different atomic process, for instance after a recombination of an electron into the upper level or after a spontaneous decay of a higher level into the upper level or after a collision, then the line source function will possibly not dilute and may roughly stay constant as a function of radius so that an emission line results. Typical examples for P-Cygni profiles are resonance transitions connected with the ground level, whereas excited lines of an ionization stage into which frequent recombination from higher ionization stages occurs will produce emission lines. H_α is a typical example for the latter case, as long as the L_α resonance transition is of moderate optical thickness as in O-supergiants and early B-supergiants. For late B- and A-supergiants, when L_α becomes severely optically thick and the corresponding transition is in detailed balance, the first excited level of hydrogen becomes the effective ground state and H_α starts to behave like a resonance line showing also the shape of a P-Cygni profile. (For a more comprehensive discussion of the line formation process in winds, see Kudritzki 1988).

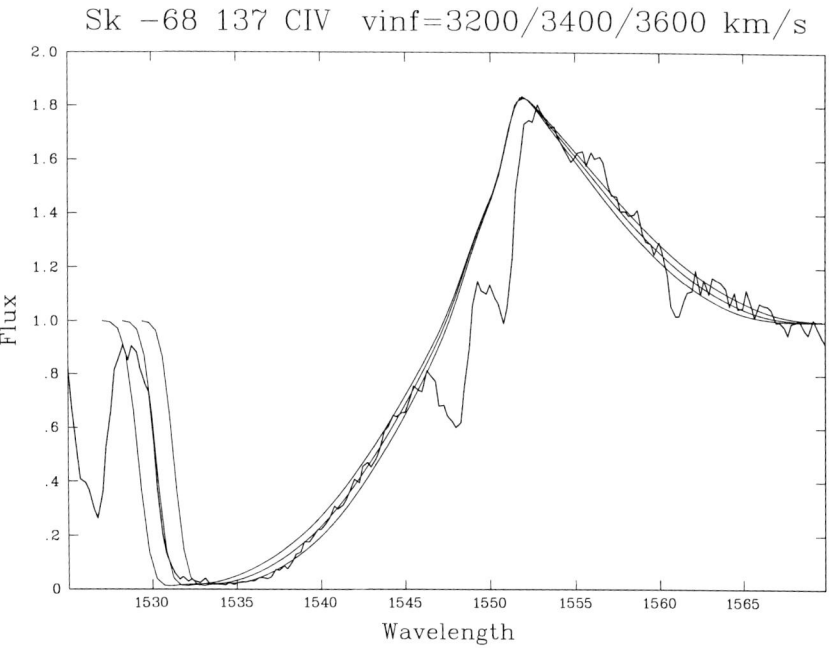

FIGURE 32. Radiative transfer calculations for the UV CIV resonance doublet in the HST FOS spectrum of the LMC O3-star Sk − 67°137 for three different values (3200, 3400, 3600 km/s) of v_∞.

Both types of lines can be used to determine $\dot{M}$, v_∞ and β. Fig. 32 gives an example for the determination of v_∞ from a well saturated UV resonance line. Such lines allow high precision measurements of v_∞ with an accuracy of 5 percent (for details, see Hamann 1981a, Hamann 1981b, Lamers *et al.* 1987, Groenewegen & Lamers 1989, Howarth & Prinja 1989, Lamers *et al.* 1995, Haser *et al.* 1995 and Haser 1995). The shape of the velocity field, i.e. the exponent β can also be determined, mostly from the width of the emission peak. Typical values are $\beta = 0.7$ to 1.5 for OB-stars with supergiants having a clear tendency towards higher β. For A-supergiants values as high as $\beta = 3$ to 4 can be encountered (Stahl *et al.* 1991, McCarthy *et al.* 1997). In addition, the contributions of stochastic motions in the wind measured in terms of "microturbulence velocity" can be determined from the slope of the blue edge of the P-Cygni profile and the wavelength location of the emission peak. Typically v_{turb} is 10 percent of v_∞.

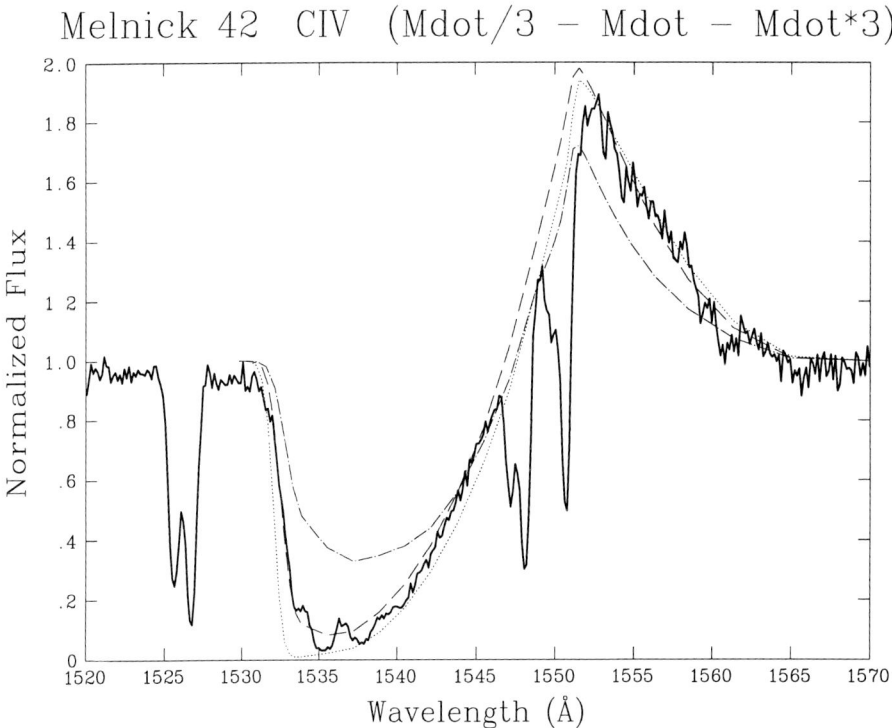

FIGURE 33. Same as Fig. 32, but now using a HST GHRS spectrum of the LMC O3-star Melnick 42 and varying $\dot{M}$ by factors of three. $v_\infty = 3400$ km/s has been adopted.

The strengths of the UV P-Cygni profiles of the metal lines also contain, in principle, the information about the mass-loss rate $\dot{M}$ (see Fig. 33). By means of a radiative transfer calculation it is easy to assign a "line strength" $k(v)$ to each layer in the wind characterized by its value of the outflow velocity $v(r)$, where

$$k(v) \sim \epsilon_{abund} X_{ion} \dot{M} \tag{3.3}$$

and ϵ_{abund} is the element abundance and X_{ion} the degree of ionization of the ionization stage leading to the resonance line observed. In this way beautiful fits of the observed spectra can be obtained (see Fig. 34, 35).

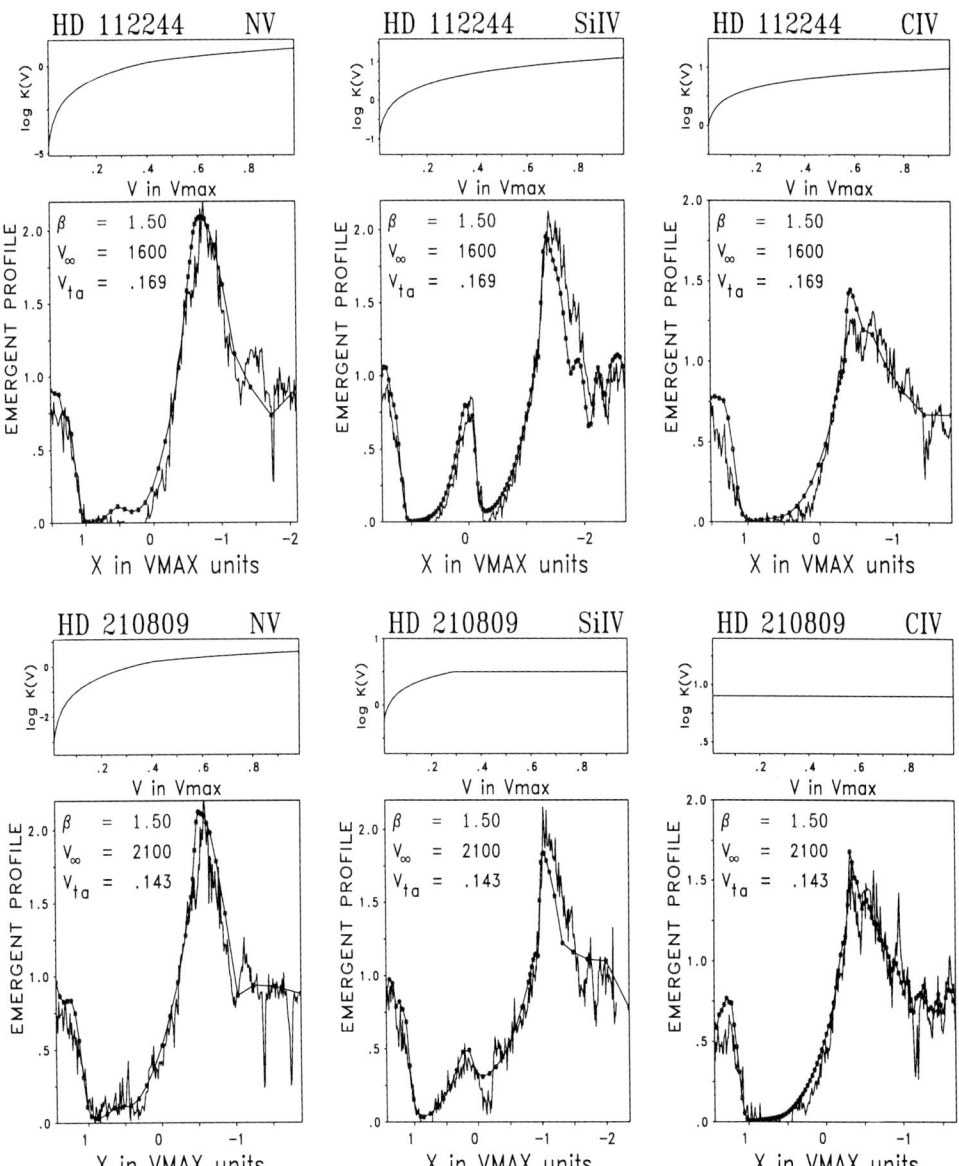

FIGURE 34. Radiative transfer fits of the IUE (high resolution) UV resonance lines of N V, Si IV and C IV of two galactic O-supergiants, HD112244 (O 8.5 Iab (f), upper panel) and HD210809 (O 9 Iab, lower panel). The abscissa is the wavelength displacement from the line centre measured in units of maximum displacement $\lambda_0 v_\infty/c$. Each figure contains the values of β, v_∞ and $v_{ta} = v_{turb}/v_\infty$ obtained as the best fits. Above each profile the corresponding fit of the line strength stratification $k(v)$ is given as a function of $v(r)/v_\infty$. From Haser 1995.

However, despite the impressive fit quality the values of $\dot{M}$ derived are very uncertain. The reason is that the determination of X_{ion} is very uncertain and depends on complicated details of the model computations. In addition, a rather precise determination of the element abundances is needed. This means that mass-loss rates obtained from UV

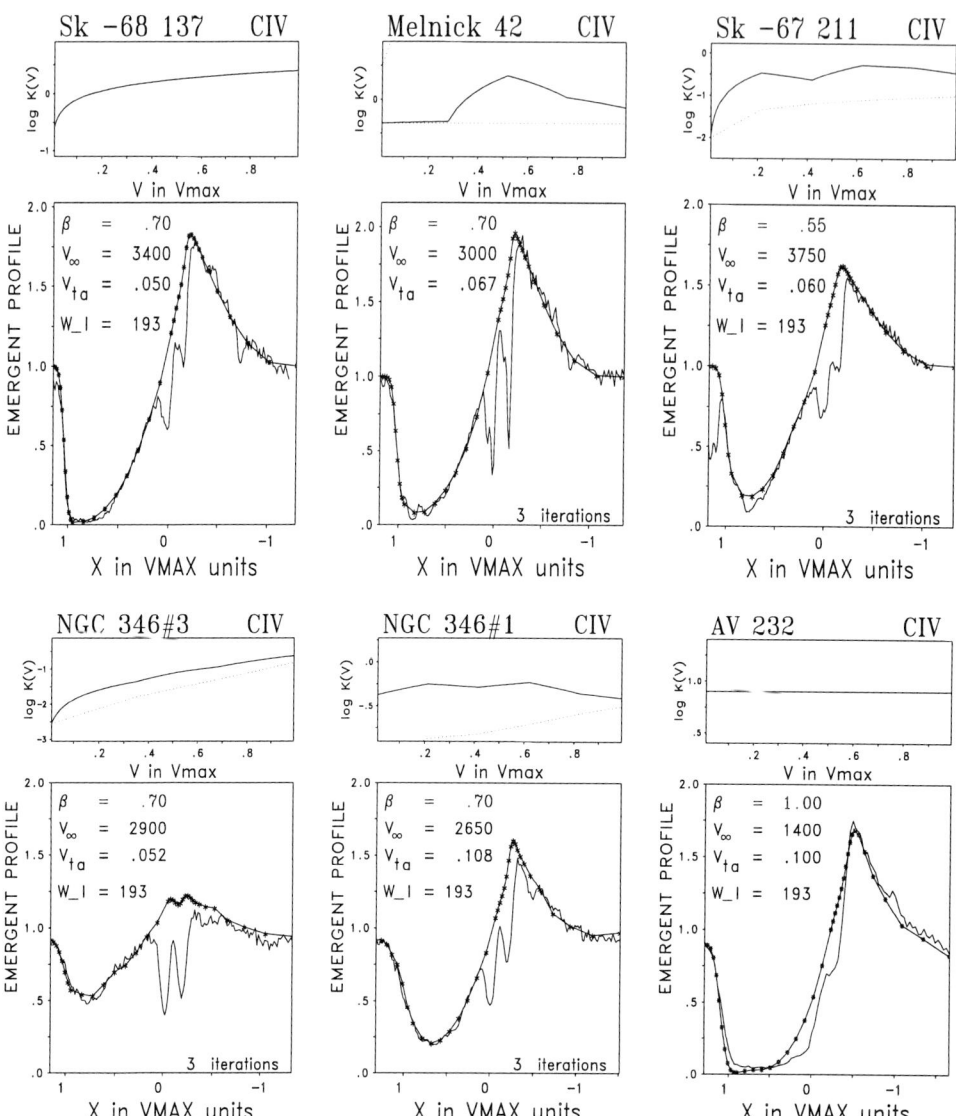

FIGURE 35. Radiative transfer fits of HST C IV wind profiles of three LMC O3-stars (upper panel: Sk -68° 137 (O3 III), Mk 42 (O3 If), Sk -67° 211 (O3 III)) and three SMC O-stars (lower panel: NGC 346 No.3 (O III), No. 1 (O4 III), AV 232 (O7 Iaf^+)) The absorptions near line centre are caused by the interstellar C IV doublet contributions from both the Magellanic Clouds and the Galaxy. The quantity w_I corresponds to the FWHM of the instrumental profile measured in km/s (w_I=77 for Melnick 42). The dotted curves in the log k(v) plot are starting approximations. From Haser 1995.

resonance lines of metals normally have an uncertainty of at least 0.5 dex, much too inaccurate to constrain the influence of mass-loss on stellar evolution.

Fortunately, there is an alternative and much more precise way to determine $\dot{M}$ using the strength of the stellar wind emission in H_α as an indicator of mass-loss rates. The ionization correction for H_α is much simpler and rather precise and the hydrogen abundance is usually much less uncertain. First quantitative NLTE calculations dealing with the wind emission of hydrogen and helium lines go back to Klein & Castor 1978. Then, Leitherer 1988, Drew 1990, Gabler et al. 1989, Gabler et al. 1990, Scuderi et al 1992, Lamers & Leitherer 1993 realized the potential of H_α for the determinations of mass-loss rates. Very recently, Puls et al. 1996 developed an accurate and fast method to obtain mass-loss rates from H_α profiles avoiding the systematic errors inherent in the approach of Leitherer 1988 and Lamers & Leitherer 1993. For a comprehensive discussion of the formation of H_α in hot star winds including detailed profile fits we refer to Puls et al. 1996 and Santolaya-Rey et al. 1997.

Fig. 36 demonstrates how precisely $\dot{M}$ can be determined from H_α in the case of O-supergiants. Fig. 37 gives a few other examples. In optimum cases, when the profile shape also allows a simultaneous accurate determination of β, the error in $\dot{M}$ can be less than 0.1 dex.

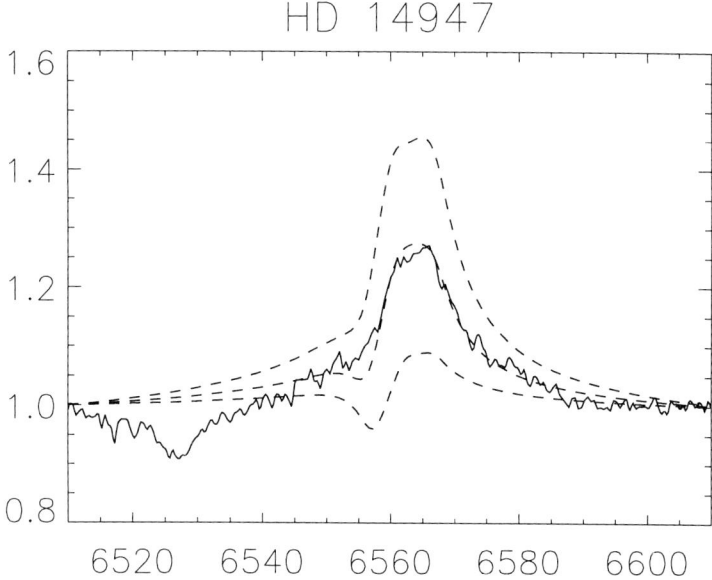

FIGURE 36. H_α line profile of the O 5 Iaf^+ - supergiant HD 14947 compared with unified model calculations adopting 10, 7.5 and 5.0×10^{-6} $M_\odot$/yr, respectively, for the mass-loss rate.

The method works equally well for A- and B-supergiants. Figs. 38, 39 give examples from very recent work. A discussion of observed mass-loss rates and terminal velocities in Local Group Galaxies will be given in section 8.

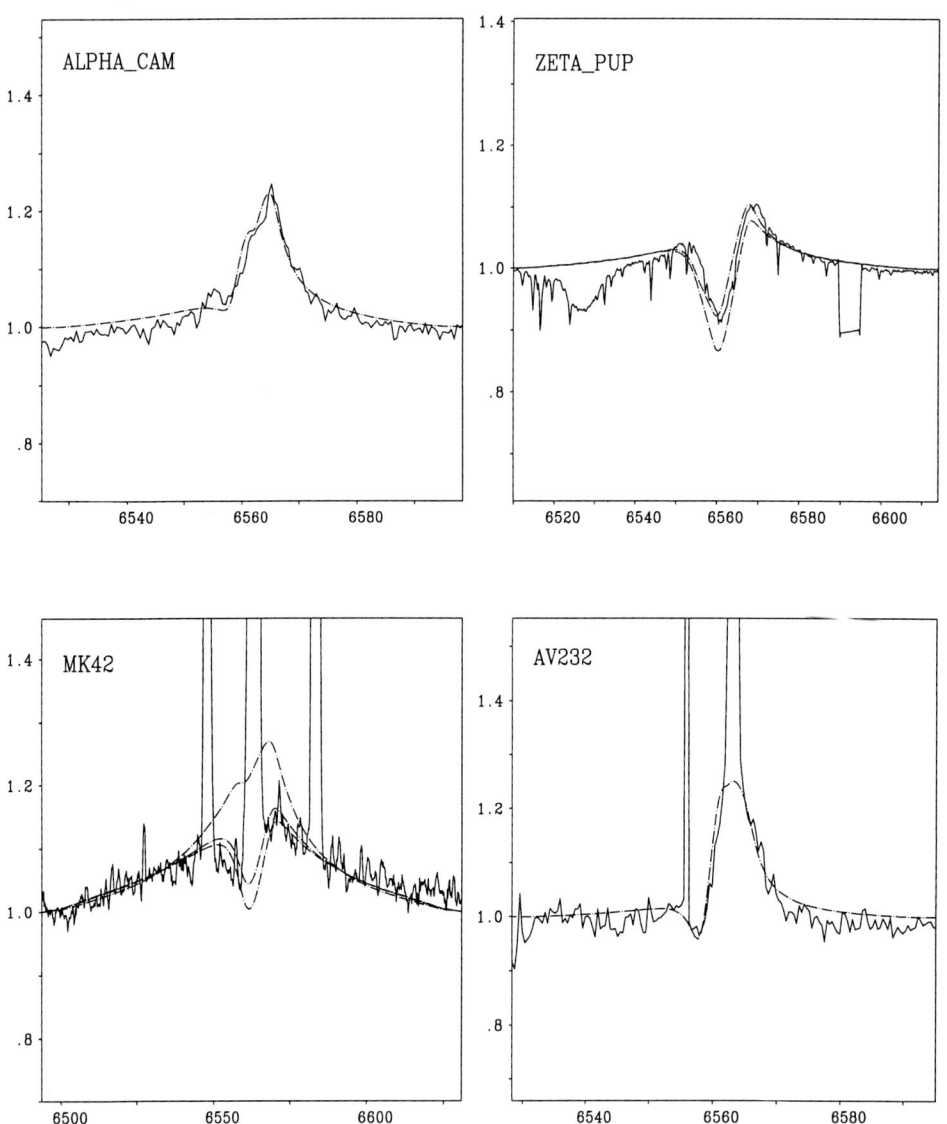

FIGURE 37. H_α profile fits of two galactic (upper panel) and two LMC/SMC (lower panel) O-supergiants. The narrow emission lines originate in the surrounding H II region. See Puls *et al.* 1996.

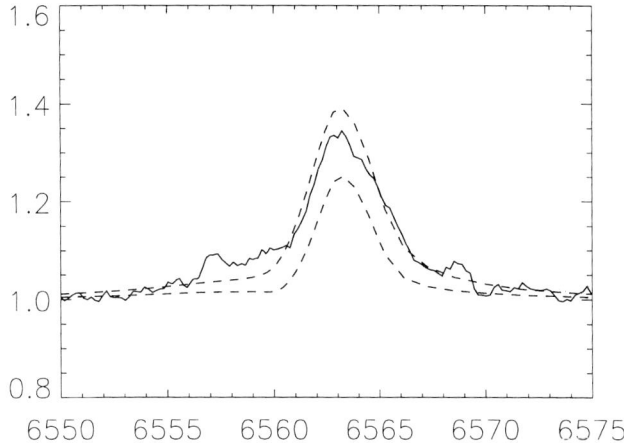

FIGURE 38. H_α line profile of the LMC B0.5 Ia - supergiant Sk-68°41 (taken with the ESO NTT and the EMMI spectrograph) compared with unified model calculations adopting $\beta = 1.5$, $v_\infty = 1050$ km/s (from HST UV profile fits) and $\dot{M} = 1.4$ and 1.8×10^{-6} $M_\odot$/yr.

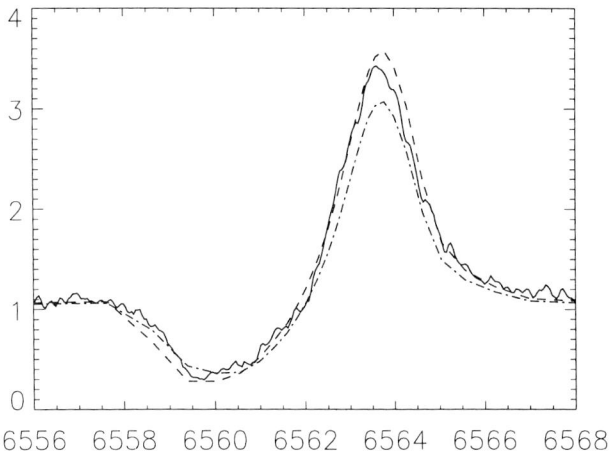

FIGURE 39. H_α line profile of the extreme A-supergiant 41-3654 (A3 Ia-O) in the Andromeda Galaxy M31 taken with the Keck HIRES spectrograph compared with two unified model calculations adopting $\beta = 3$, $v_\infty = 200$ km/s and $\dot{M} = 1.7$ and 2.1×10^{-6} $M_\odot$/yr. Note the H_α P–Cygni profile. From McCarthy et al. 1997.

3.3. Radii, luminosities, interstellar reddening and extinction

With T_{eff}, $\log g$, $\dot{M}$ and $v(r)$ determined from the line profile analysis, the atmospheric stratification of a Blue Supergiant is well known and the emergent flux of a model H_λ^{mod} can be calculated. The connection to the flux F_λ^{obs} observed with a telescope is given by

$$F_\lambda^{obs} = (R_*/d)^2 4\pi H_\lambda^{mod}(T_{\text{eff}}, g) \tag{3.4}$$

If the measured monochromatic fluxes are well calibrated and the correction for interstellar absorption has been applied, this equation yields the stellar angular diameter R_*/d and, if the distance d is known, the stellar radius R_*. The usual way to obtain the interstellar absorption correction is to deredden the observed flux distribution using the standard extinction curves (for instance, Seaton 1979 for the Galaxy and Howarth 1983 for low metallicity ISM as in the SMC) until its slope agrees with the slope of the model atmosphere flux distribution.

In many cases, only wide band photometry is available. Then, one can calculate theoretical intrinsic colours, such as

$$(B-V)_0 = C_{B-V} + 2.5 \log(\frac{\int_0^\infty H_\lambda^{mod} S_B(\lambda) d\lambda}{\int_0^\infty H_\lambda^{mod} S_V(\lambda) d\lambda}) \tag{3.5}$$

and compare them with observed colours to determine the reddening and the extinction. (The constant C_{B-V} has to be calibrated using nearby unreddened objects). If the distance modulus is known, the absolute magnitude M_V follows. The latter is connected to the radius via

$$M_V = C - 5 \log R_* - 2.5 \log(\int_0^\infty H_\lambda^{mod} S_V(\lambda) d\lambda) \tag{3.6}$$

which can then be used to determine the stellar radius (see Kudritzki 1980). Note that **one of the fundamental advantages of quantitative spectroscopy is that interstellar reddening and extinction do not pose a problem, because the intrinsic stellar colours and energy distributions are a bi-product of the spectroscopic analysis**.

Once the radius is known, the stellar luminosity can be calculated from the spectroscopically determined effective temperature.

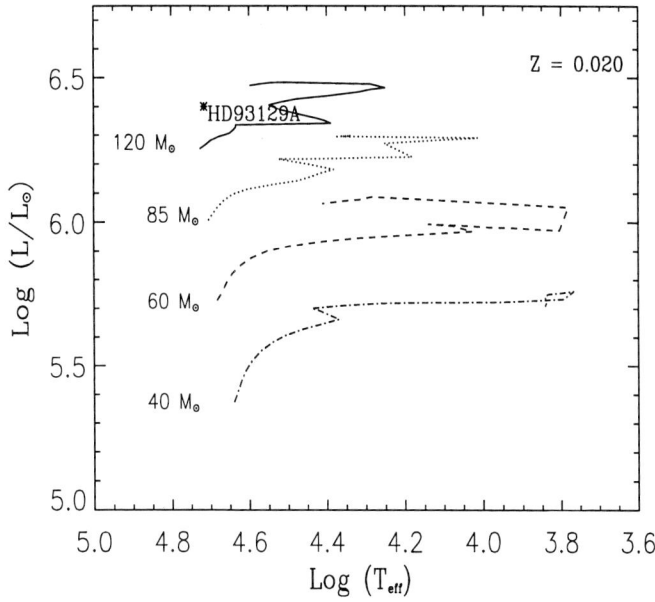

FIGURE 40. Location of the extreme O3-supergiant HD 93129A in the HRD compared with evolutionary tracks from Schaller *et al.* 1992. (From Taresch *et al.* 1997).

3.4. *Masses*

There are three ways to determine masses of Blue Supergiants
- **evolutionary masses** M_{evol} from the location in the HRD and a comparison with evolutionary tracks,
- **spectroscopic masses** M_{spectr} from spectroscopically determined gravity $g = GM_{spectr}/(R_*)^2$, once the radius has been determined and
- **wind masses** M_{wind} from the observed value of the terminal velocity v_∞ via the theory of radiation driven winds, which predicts v_∞

$$\mathbf{v}_\infty = f(T_{eff}, R_*, \mathbf{M}_{wind}) \qquad (3.7)$$

to be a strong function of stellar mass (see Kudritzki *et al.* 1989, Kudritzki *et al.* 1992, also section 8).

As an example we again use the extreme O3-supergiant HD 93129A which was already discussed in section 1 and 2.1. According to Taresch *et al.* 1997 the object has a very high effective temperature T_{eff}=52000K and luminosity $L = 10^{6.4} L_\odot$. The comparison with evolutionary tracks (see Fig. 40) indicates that the mass (at least on the Zero Age Main Sequence) must have been higher than M_{evol}=120 $M_\odot$. This agrees very well $M_{spectr} = 130 M_\odot$ obtained from $\log g = 3.95$ and $R_* = 20 R_\odot$ (see Puls *et al.* 1996).

Fig. 41 shows how the wind mass is determined. Again the value of $M_{wind} = 130 M_\odot$ is in perfect agreement with the other determinations (for a discussion, see Taresch *et al.* 1997).

However, generally the situation is not as good as in the case of HD 93129A. Herrero *et al.* 1992 in their quantitative spectroscopic study of a large sample of galactic O-stars detected a systematic "mass discrepancy". They found that evolved O-stars approaching

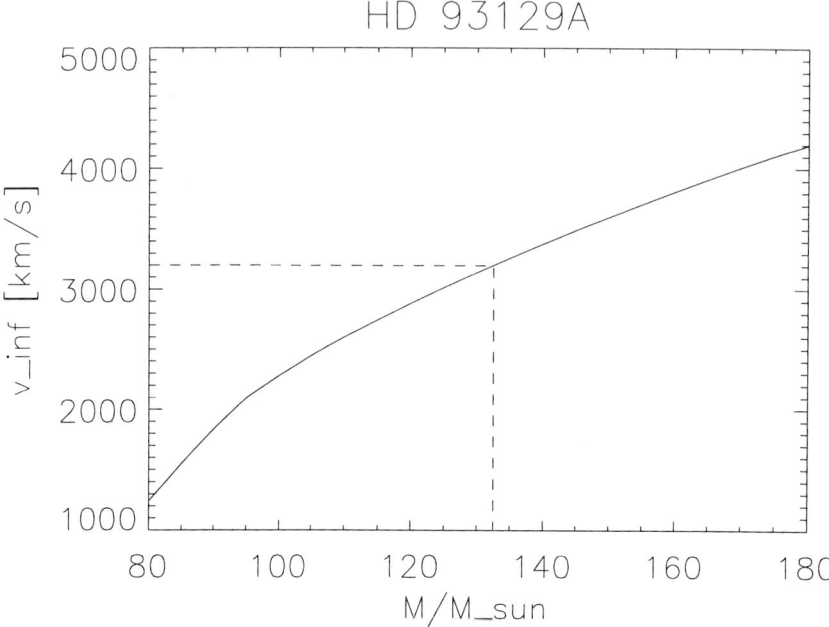

FIGURE 41. The determination of stellar mass of the O3-supergiant HD 93129A using the observed v_∞ (dashed horizontal curve) and a comparison with radiation wind theory, which predicts a strong dependence of v_∞ on stellar mass. From Taresch et al. 1997.

the Eddington-limit have evolutionary masses significantly larger than the spectroscopic and the wind masses (see Fig. 42). Even now it is still a completely open question as to which masses are correct. Langer et al. 1994 claim that stellar evolution including rotation and rotationally induced interior mixing would enhance the luminosity significantly for given mass. In addition, enhanced mass-loss along the evolutionary track would further reduce the mass and bring the evolutionary masses into agreement with the spectroscopic and wind masses. On the other hand, since Herrero et al. 1994 used hydrostatic unblanketed NLTE models in their study (see also Lanz et al. 1996), the stellar gravities and therefore the spectroscopic masses have been systematically underestimated. The question is by how much, and would the use of unified models resolve the discrepancy. A reanalysis is under way.

The older wind masses derived by Herrero et al. 1992 will probably also need revision, since they are based on stellar wind computations using older, less complete and comprehensive atomic physics and line lists than the very recent work by Taresch et al. 1997. Again, a reanalysis is under way.

3.5. Distances

The distance of a normal (non-pulsating, non-eclipsing) star can usually not be determined directly from its spectrum and its apparent brightness only. One always needs additional information either from stellar evolution (telling us how gravities correspond with luminosities) or from cluster membership (using the information about the distances of the other stars in the cluster), if direct parallaxes are not available.

Luminous Blue Supergiants, however, are an exception. Their distances can, in principle, be determined from the properties of their radiation driven

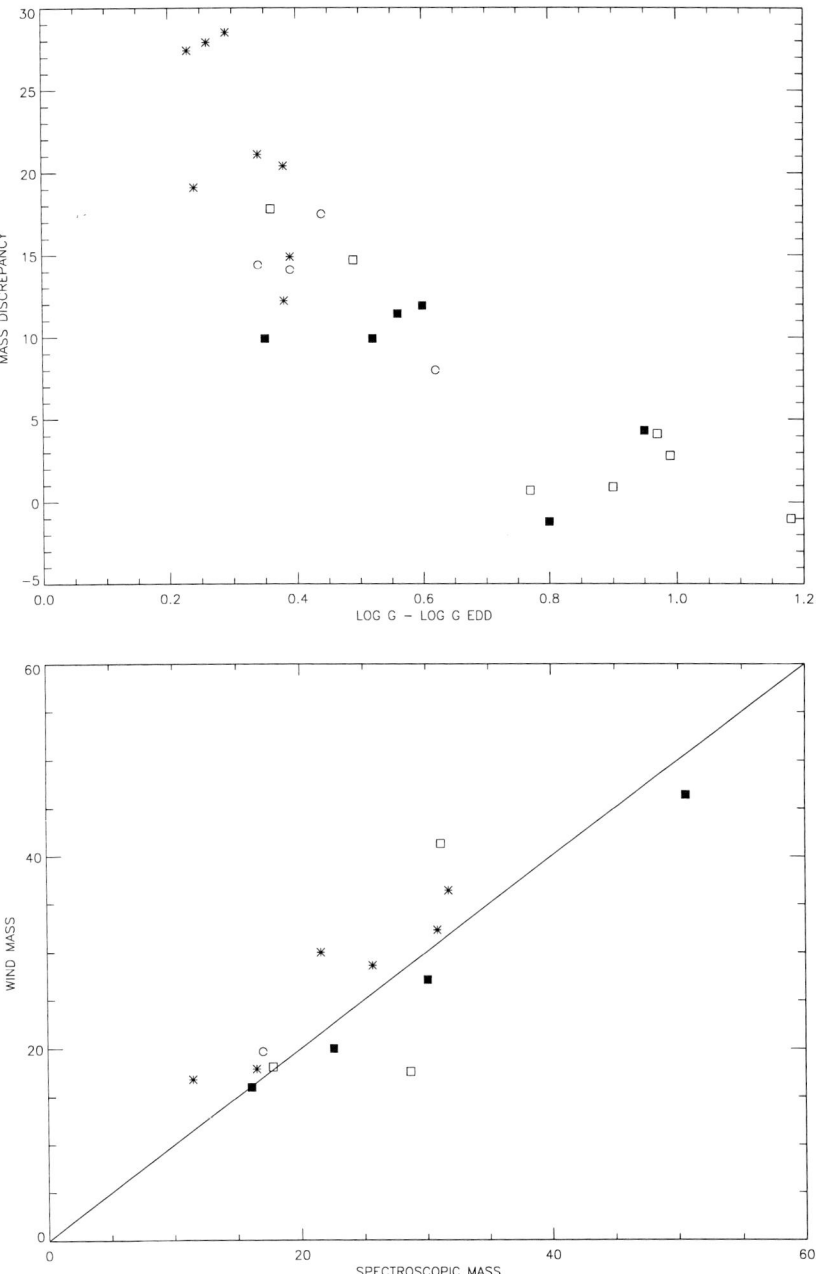

FIGURE 42. The "mass discrepancy" of O-stars. **Upper figure**: The difference $M_{evol} - M_{spectr}$ is plotted against the distance (in log g) from the Eddington limit. **Lower figure**: Comparison of M_{wind} with M_{spectr} showing that these masses agree fairly well. (From Herrero et al. 1992, masses are given in units of $M_\odot$).

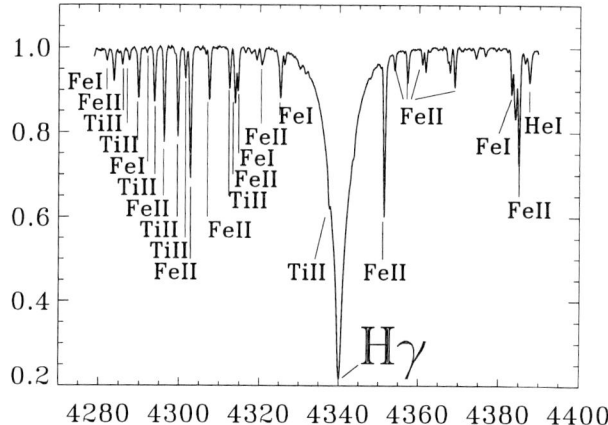

FIGURE 43. Identifications of metal lines in a small spectral interval around Hγ of the A-supergiant HD 87737 (η Leo).

stellar winds. This was pointed out by Kudritzki et al. 1992 who realized that the theory of radiation driven winds predicts that v_∞ and $\dot{M}$ are both independent and almost orthogonal functions of mass and luminosity or gravity and radius

$$\mathbf{v}_\infty = f_1(T_{\text{eff}}, \log \mathbf{g}, \mathbf{R}_*) \tag{3.8}$$

$$\dot{\mathbf{M}} = f_2(T_{\text{eff}}, \log \mathbf{g}, \mathbf{R}_*), \tag{3.9}$$

where f_1 and f_2 are functions that can be expressed analytically (see Kudritzki et al. 1989). Therefore, for a hot luminous supergiant with a spectroscopically well determined effective temperature T_{eff} the characteristic and well measurable quantities of its radiation driven stellar wind, v_∞ and $\dot{M}$, can be used to determine directly radius and gravity and, as a result, distance, luminosity and mass directly from the spectrum. Kudritzki et al. 1992 give several examples to demonstrate how the method would work.

However, the disadvantage of the method is that it relies completely on the theory, which in a quantitative sense has to be perfect. As shown by Puls et al. 1996 this is not the case at the moment. There are certain discrepancies between theory and observations of winds that could affect the method. Therefore, a new method – the **Wind Momentum - Luminosity Relationship** – has been developed recently. This method is again motivated by wind theory but uses a completely observationally calibrated approach. It will be discussed in detail in section 8.

3.6. *Abundances*

The spectra of supergiants show a multitude of metal lines that allow abundance determinations. Optical spectra are ideal for A- and B-supergiants, if the winds are not too strong and the weak metal lines remain in absorption. In this case, classical absorption line work is possible. Fig. 43 gives a quick impression of the amount of information available from a small wavelength interval. The full optical spectra are an abundance treasure, as is indicated by Fig. 44. We will come back to this point when discussing the determination of stellar abundances in other galaxies in section 7 .

For O-stars, a determination of metal abundances from optical spectra is difficult, since

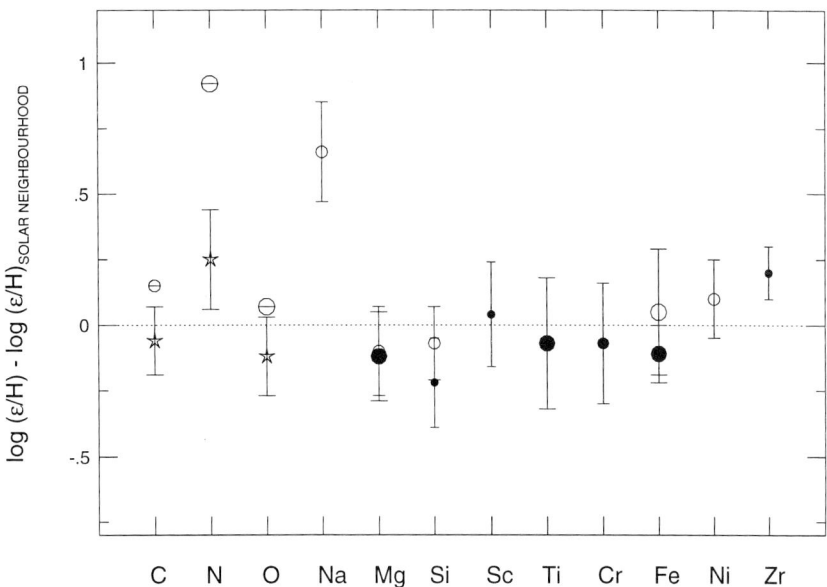

FIGURE 44. Galactic A-supergiant element abundances. Asterisks represent NLTE abundances. Filled/empty circles give LTE abundances for ionic/neutral elements. Symbol sizes are related to the number of targets used in the average value. (From Venn 1994. For further discussion, in particular on N and Na, we refer to this paper)).

only a few lines are present, in particular at earlier spectral types. Thus, UV spectra are needed. Fig. 10 in section 2.1 demonstrates how UV spectrum synthesis can be applied successfully to determine abundances.

With regard to quantitative UV-spectroscopy one point deserves special attention. The UV-spectra of hot stars are usually blended by hundreds of molecular, atomic or ionic interstellar lines (see Fig. 45). In the past, column densities and abundances of these species have been determined while ignoring details of the synthesis of the stellar spectrum. Taresch *et al.* 1997 have shown that this can lead to substantial errors. For accurate work it is absolutely necessary to synthesize the stellar and interstellar spectrum simultaneously to account properly for the blending. An example is given in Fig. 46.

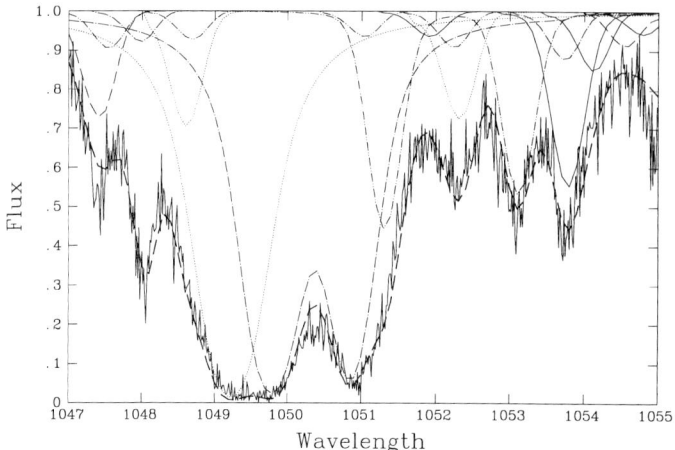

FIGURE 45. FUV spectrum of the O3-supergiant HD 93129A between 1047 and 1055 Å showing mostly molecular lines of H_2 and HD. The dashed curve represents the final interstellar spectrum synthesis. The contributions of individual interstellar line components are also indicated. (From Taresch *et al.* 1997).

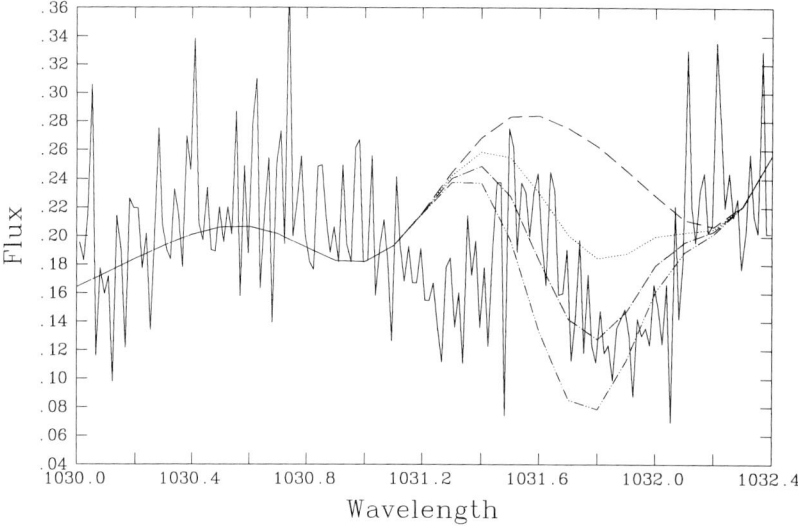

FIGURE 46. Expansion of the profile of the stellar wind OVI resonance line shown in Fig. 10. Calculated stellar contribution (dashed) and the additional contribution of interstellar OVI with log $N(OVI)/cm^2 = 14.1, 14.5, 14.9$, respectively are shown. For the determination of the interstellar column density the shape of the stellar profile is crucial. Note that the spectrum synthesis is incomplete at 1031.3 Å. (From Taresch *et al.* 1997).

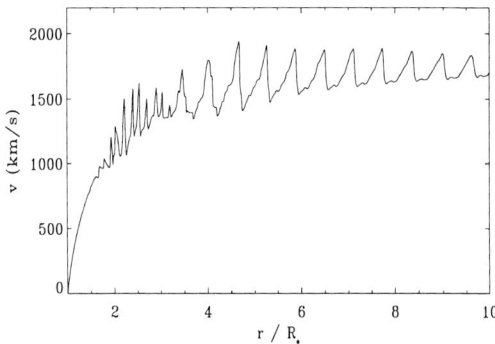

FIGURE 47. Snapshot of the velocity stratification of a time dependent radiation driven wind model calculation. The structures of the shocks and shells are easily seen. (From Feldmeier 1995).

4. The X-ray emission of O-stars

Among the first surprising discoveries of the EINSTEIN observatory was the detection that all O stars are soft X-ray emitters (Harnden et al. 1979; Seward et al. 1979). It was soon found that the X-ray luminosity is roughly correlated with the stellar luminosity: $\log L_x/L_{\rm bol} \approx -7\pm 1$ (Seward et al. 1979; Pallavicini et al. 1981; Chlebowski et al. 1989). The scatter in this relation is amazingly large indicating a dependence on additional parameters. One immediate conclusion from this detection is that the contribution of O-stars to the X-ray emission of galaxies (for instance, the bar of the LMC) can be substantial.

The source of the O-star X-ray emission is widely believed to be shocks propagating through the stellar wind (Lucy and White 1980; Lucy 1982; Cassinelli & Swank 1983; MacFarlane & Cassinelli 1989; Hillier et al. 1993; Cassinelli et al. 1994),
where the shocks may result from a strong hydrodynamic instability of radiation driven winds (Lucy & Solomon 1970; Carlberg 1980; Owocki & Rybicki 1984; Lucy 1984). A corona, as in the sun, can be ruled out as being the source for X-ray emission from O stars on observational grounds (Cassinelli & Swank 1983; Baade & Lucy 1987).

Recent time dependent radiation-hydrodynamic calculations of O-star winds do indeed confirm this picture (Owocki et al. 1988; Owocki 1991; Feldmeier 1995). Strong hot shocks develop and travel through the wind superimposed on the velocity field of the stationary wind (see Fig. 47). On the other hand, the unshocked matter or the cooled post-shock matter, which contains most of the mass in the stellar wind and which produces stellar wind lines observed in the UV, optical and IR, follows the velocity field predicted by a stationary calculation. Thus the X-rays are produced in the hot shocks, whereas the usual stellar wind lines originate from the cool stellar wind material following the quasi-stationary flow.

Unfortunately, these time dependent calculations have not yet reached the the same level of detailed refinement as the stationary calculations of radiation driven winds described in section 2.1 . The consistent coupling of the NLTE treatment and time-dependent hydrodynamics in 3D has still to await another computer generation. (Given the complexity of the problem this is more than understandable). As a consequence, a simplified, empirical approach is necessary for the time being. Such an approach has been developed by Hillier et al. 1993 and Feldmeier et al. 1997 and is described in the following.

For the calculation of X-ray spectra a two-component wind model is adopted:

(i) a "cold" background wind ($T \approx T_{\text{eff}}$) which is described by stationary models, contains most of the mass and therefore produces the UV line spectrum and optical emission lines,

(ii) embedded in the cold wind is an ensemble of randomly distributed, radially directed shocks which heat the gas locally to X-ray temperatures. The solid angle subtended by the shocks is assumed to be $\ll 4\pi$.

The X-ray radiative transfer in this medium is solved numerically by a formal integral in spherical symmetry. The *absorption* coefficient is given by

$$\kappa_\nu = \kappa_\nu^{\text{cold}} + \kappa_\nu^{\text{K-shell}}, \qquad (4.10)$$

where the bound-free and line opacities for the cold wind are taken from the full stationary NLTE wind models as described in section 2.1 , and K-shell absorption is included for the elements C, N, O, Ne, Mg, Si and S (for atomic data of the latter, see Daltabuit and Cox 1972). The abundances that enter into both κ_ν^{cold} and $\kappa_\nu^{\text{K-shell}}$ are determined from the UV and optical spectral analysis. From the total opacity it is found that the winds of typical O supergiants are optically thick at almost all energies observable with ROSAT, except possibly at the highest (Hillier et al. 1993). For the O4f star ζ Puppis, for example, optical depth unity (with $\tau = 0$ at infinity) is reached at $1000 R_*$ for $E = 0.1$ keV, at $100 R_*$ for 0.25 keV, and at $10 R_*$ for 1 keV. This clearly shows the necessity of proper radiative transfer, if one wishes to synthesize the X-ray spectrum of an O-star.

The *emission* coefficient of the embedded hot gas is then given by

$$\epsilon_\nu^{\text{shock}} = \frac{1}{4\pi} f_s n_e n_p \bar{\Lambda}_\nu(T_s, n_e). \qquad (4.11)$$

Here, n_e and n_p are the electron and proton density, respectively, of the stationary cold wind, f_s is the volume filling factor of hot gas, and T_s is the shock temperature. f_s and T_s are the *only* free parameters that enter the model. Finally, the total emission integrated over the shock cooling layer of radial extent l_c is

$$\bar{\Lambda}_\nu = \frac{1}{l_c} \int_0^{l_c} \left(\frac{\rho(l)}{\rho_{\text{cold}}}\right)^2 \Lambda_\nu(T(l)) \mathrm{d}l, \qquad (4.12)$$

where the shock cooling function $\Lambda_\nu(T)$ is calculated directly with the Raymond-Smith code (Raymond & Smith 1977). The following model was adopted for the post-shock density and temperature stratification, $\rho(l)$ and $T(l)$: (i) in the *inner*, dense wind region, where *radiative* losses are the only efficient cooling agent and where the cooling time is short compared with the hydrodynamic timescale, an analytical solution for a stationary cooling layer was used following Chevalier & Imamura 1982; (ii) in the *outer*, thin wind regions, where the gas cools only by *adiabatic expansion* and where the cooling time is long an analytical description was also used following Simon & Axford 1966. The approximate radius at which radiative cooling ceases to be efficient is calculated from the stationary wind structure. Details of this approach are given in Feldmeier et al. 1997. A comparison between observed and synthesized X-ray spectra is shown in Fig. 48 .

This approach has now been applied to the investigation of the X-ray emission of a large sample of O-stars using the ROSAT observatory (first results of this study have been published by Kudritzki et al. 1996b. More details and final results will be given by Palsa 1997 and Palsa et al. 1997). The HRD of the targets studied is shown in Fig. 49. For each of the targets, stellar parameters and stellar wind properties have been de-

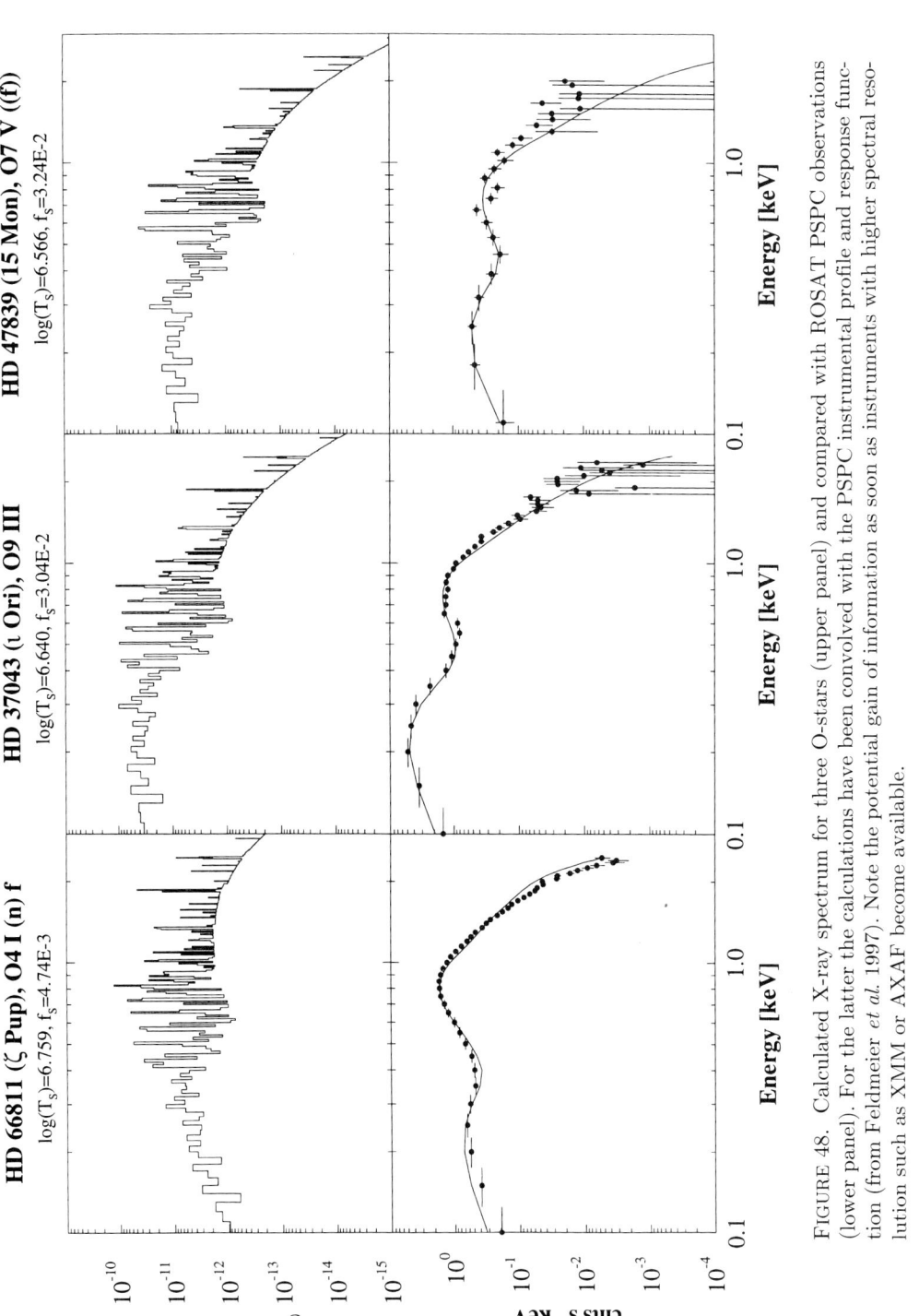

FIGURE 48. Calculated X-ray spectrum for three O-stars (upper panel) and compared with ROSAT PSPC observations (lower panel). For the latter the calculations have been convolved with the PSPC instrumental profile and response function (from Feldmeier et al. 1997). Note the potential gain of information as soon as instruments with higher spectral resolution such as XMM or AXAF become available.

termined from detailed optical and UV spectroscopic studies. In addition, interstellar hydrogen column densities are determined form UV Lyα line, as shown in Fig. 50. This is crucially important to correct for the selective X-ray absorption of the ISM. (Note that procedures involving "X-ray hardness ratios" to determine the interstellar absorption are much less reliable).

The X-ray spectra of the sample allowed the determination of shock temperatures T_s, volume filling factors f_s, and X-ray luminosities L_x for each target. Interesting new correlations were found. The shock temperature depends on the kinetic energy flow in the wind (see Fig. 51)

$$\log T_s \propto \log(\frac{1}{2}\dot{M}v_\infty^2) \qquad (4.13)$$

and the filling factors increase with the average density ρ^{av} in the stellar wind (Fig. 51), which is inversely proportional to the cooling length l_c.

$$\log f_s \propto \log \rho^{av} \propto -\log l_c \qquad (4.14)$$

$$\rho^{av} \propto \dot{M}/(v_\infty R_*^2) \qquad (4.15)$$

In other words, the cooling lengths of the shocks determines how much of the stellar wind volume is filled with hot X-ray emitting gas and the kinetic energy of the radiatively driven wind determines the shock jump velocities or the corresponding shock temperatures. (This purely empirical results awaits further quantitative theoretical explanation).

The X-ray luminosity is then found to depend on both stellar luminosity and filling factor resulting in a new improved correlation (Fig. 52)

$$L_x \propto \left(\frac{\dot{M}}{(R_*^2 v_\infty)}\right)^{-0.47} L_{bol}^{1.15}. \qquad (4.16)$$

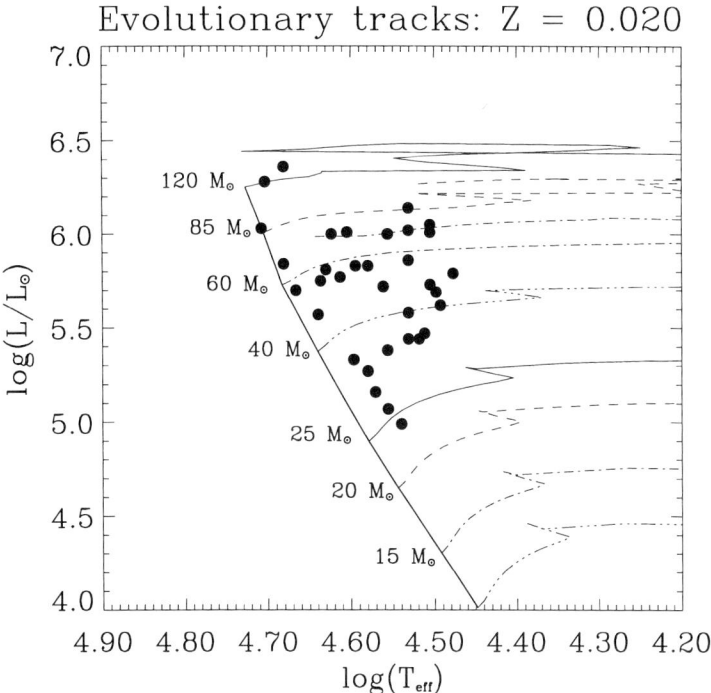

FIGURE 49. The HRD of the O-stars studied with the ROSAT PSPC. Evolutionary tracks are from Schaller *et al.* 1992.

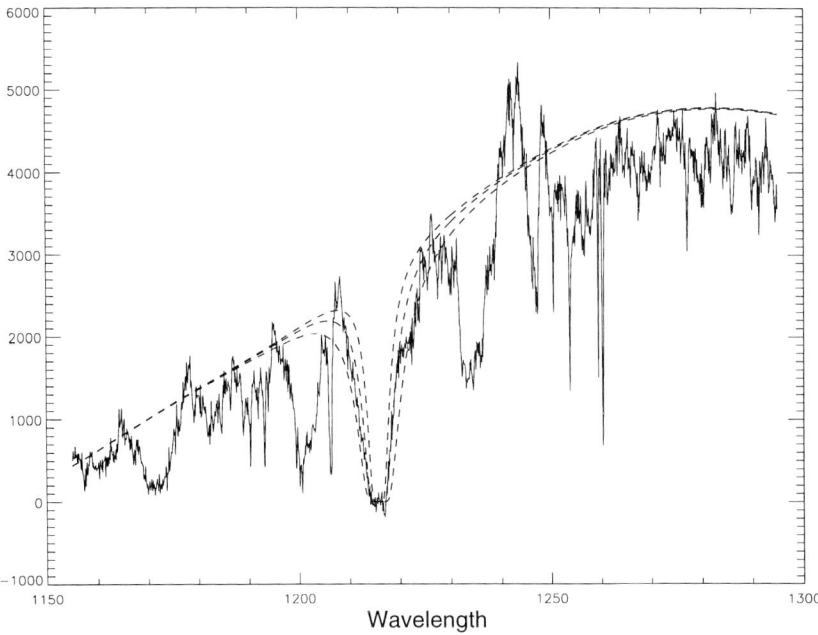

FIGURE 50. The UV spectrum of the O-supergiant ζ Ori around Lyα. Theoretical calculations of interstellar Lyα are shown for three column densities of neutral interstellar hydrogen with $\log(N_H/cm^2) = 20.0, 20.3, 20.5$.

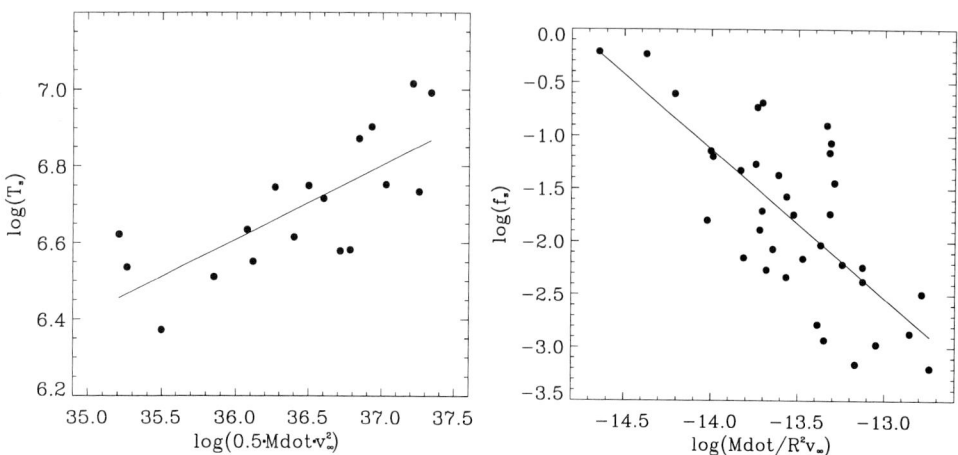

FIGURE 51. **Left:** Shock temperature as a function of kinetic energy of the stellar wind flow. **Right:** Filling factor as function of average stellar wind density.

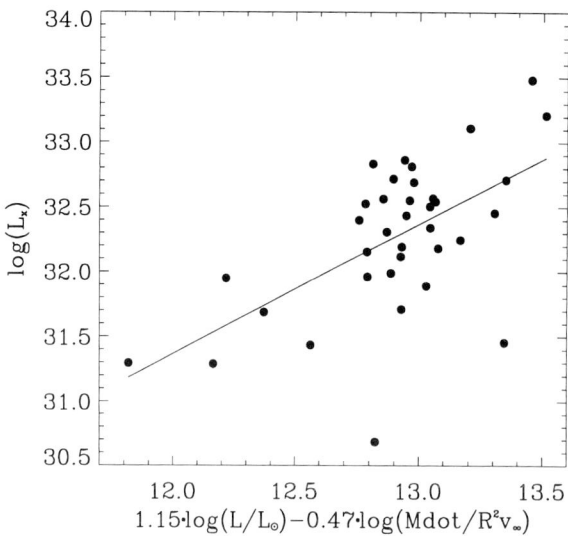

FIGURE 52. The new relation for the X-ray luminosities of O-stars. The two targets falling below the relation are the subject of a separate study.

5. IR - diagnostics of blue supergiants with extreme mass-loss

Gabler et al.1989 in their first paper on unified model atmospheres realized the importance of IR emission and absorption lines for a quantitative understanding of the physics of hot star atmospheres and winds and pointed out the potential of this new spectral range. At the same time – with the revolutionary progress of detector and spectrograph technology – quantitative IR-spectroscopy became a new area of research. Now, the first systematic observational studies of spectra of hot stars in the K-band have been published (Hanson et al. 1996 and Figer et al. 1997) and provide a wealth of material for future quantitative work. They also form a basis for understanding the spectra of starbursting galaxies, in particular in cases where strong absorption by interstellar dust prohibits any kind of optical study.

So far, there are only a few detailed quantitative studies of IR supergiant spectra. In the following we will discuss two of them, which have been published very recently.

5.1. The Luminous Blue Variable P Cygni

P Cygni – the prototype of all Blue Supergiants with extremely strong stellar winds – is a so-called "Luminous Blue Variable" star, i.e. a very luminous star close to the stability limit in the HRD ("Humphreys-Davidson-limit"), where stars become unstable and suffer from strong eruptions accompanied by a remarkable increase in visual brightness. Such objects are sometimes even brighter than the brightest supergiant stars shown in Fig. 1. In other galaxies they are easily identified because of their brightness and variability. It is, therefore, extremely important to test whether the methods of quantitative spectroscopy using hydrodynamical model atmospheres can be applied successfully to these objects as well.

Such a test has been undertaken recently by Najarro 1995, Najarro et al. 1997a, who analyzed the optical and near IR spectrum of P Cygni. The result was very encouraging leading to excellent profile fits and rather precise determinations of the stellar parameters (see Fig. 53). However, after the launch of the **ISO** observatory new IR spectroscopic data taken with the SWS spectrometer became available revealing that despite the very convincing profile fits the best model obtained from the optical analysis was not able to fit all the ISO lines, in particular not those at shorter wavelengths (Fig. 54). Finally, Lamers, Najarro et al. 1996 were able to solve the puzzle. The solution lies in the shape of the velocity field. In the optical analysis a very shallow velocity field with high outflow velocity at the photospheric radius $v_0 = 80$ kms^{-1} and a high "slope exponent" $\beta = 4.5$ (see section 2.2) had been adopted. However, a steeper velocity field ($v_0 = 30$ kms^{-1} and $\beta = 2.5$, see Fig. 55) is necessary to fit the spectra in both wavelength domains (Figs. 57 and 56). The reason is the following. The two velocity fields and the corresponding density stratifications differ substantially in the intermediate part of the wind. Here the continua of the IR lines at shorter wavelengths are formed. (Since the continuum bound-free and free-free opacity increases with wavelength, the optical continuum is formed in deeper layers than the IR continuum). This changes the strengths of the emission lines between 2 and 10 μ.

Fortunately, crucial quantities such as T_{eff}, $\log L$, $\dot{M}$, v_∞ from the quantitative analysis restricted to the optical wavelength range only were not strongly affected by this deficiency. On the other hand, to the understanding the physics of wind acceleration in LBVs and supergiants it is absolutely essential to determine the shape of the velocity field precisely. In this respect, a new spectral range allowing tomography of atmospheric layers previously inaccessible means a breakthrough in the diagnostic techniques. The **ISO Blue Supergiant Project** will, therefore, be continued and we expect a wealth of new information from it.

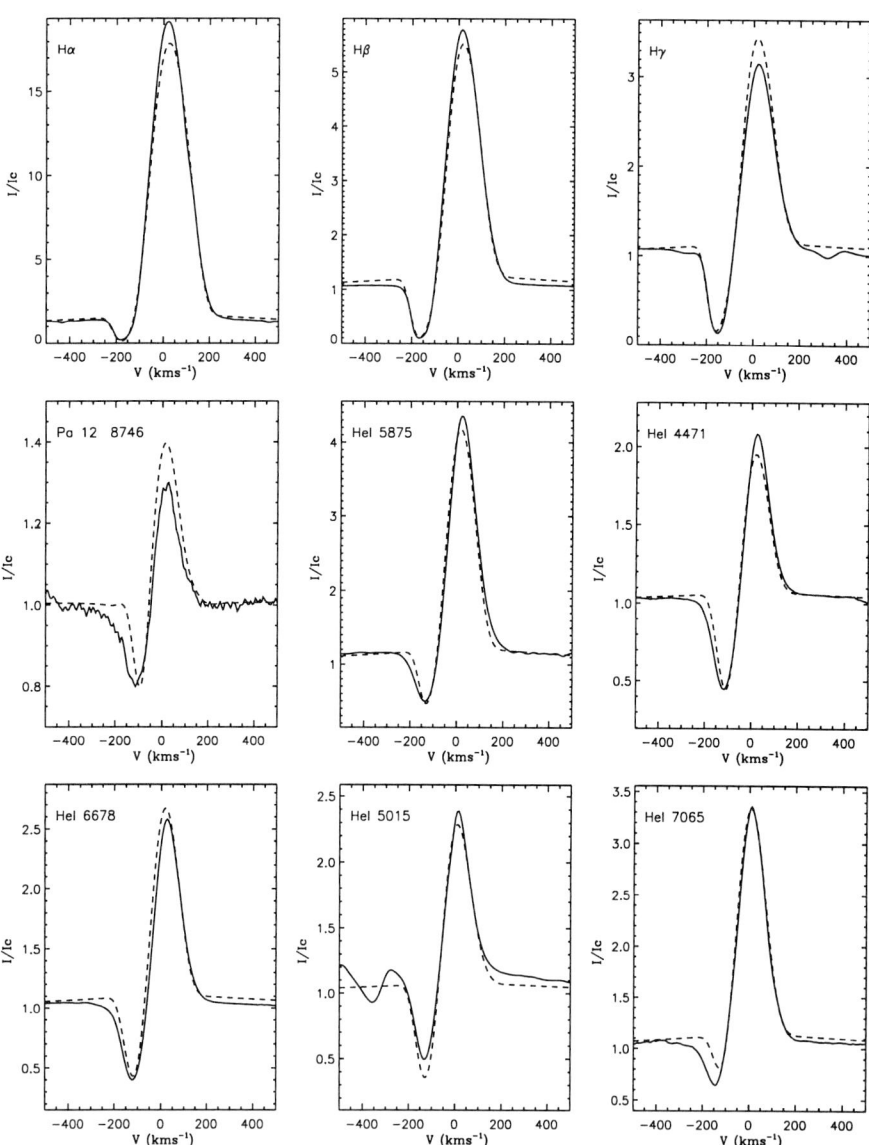

FIGURE 53. Fit of the optical stellar wind lines of P Cygni using information from the optical lines only. From Najarro 1995 and Najarro *et al.* 1997.

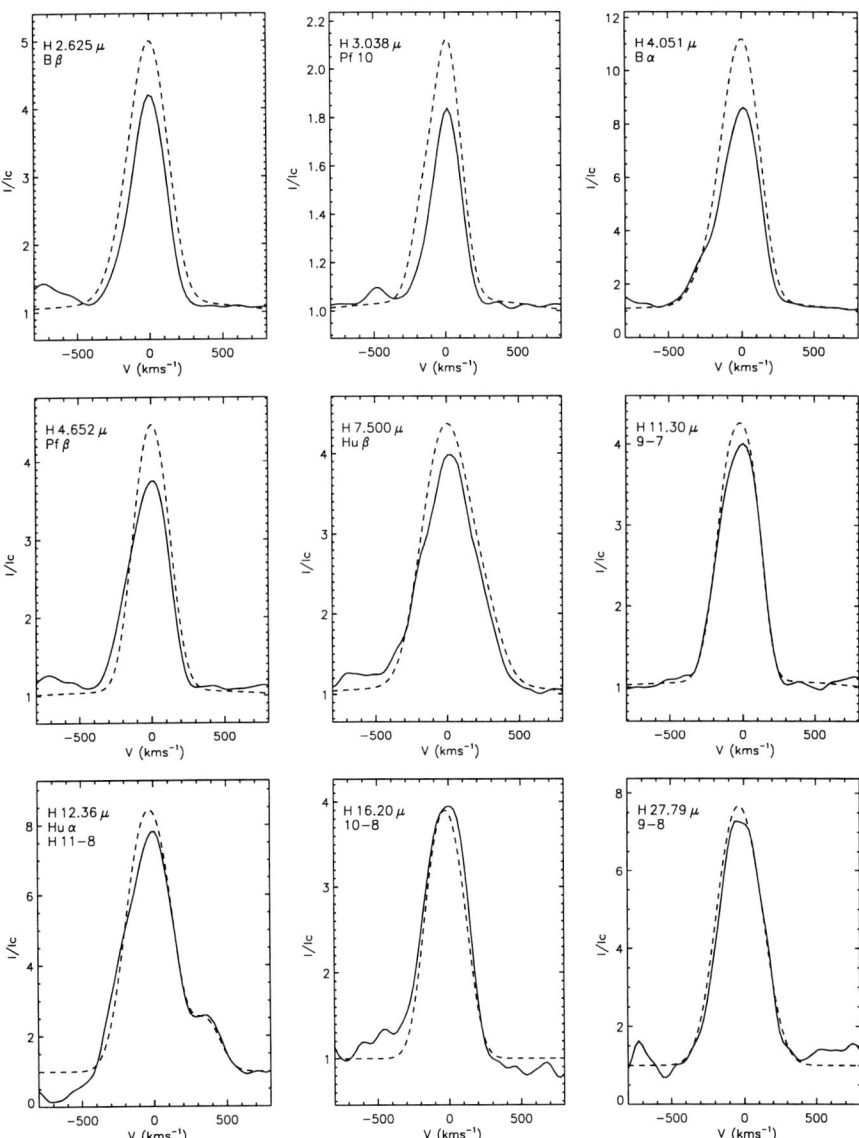

FIGURE 54. Fit of the IR ISO SWS stellar wind lines of P Cygni using the wind model obtained from the analysis of the optical lines. There is a striking discrepancy at shorter wavelengths. From Lamers, Najarro, Kudritzki et al. 1996.

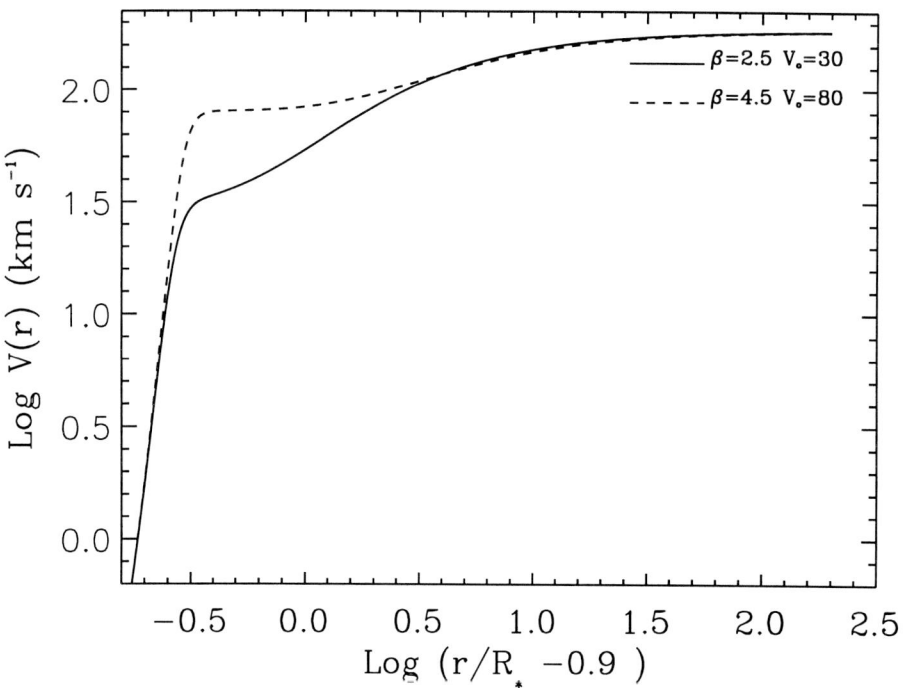

FIGURE 55. Old velocity field as adopted for the analysis of the optical lines (dashed) and new velocity field (solid) fitting both, optical and IR lines. From Lamers, Najarro, Kudritzki *et al.* 1996.

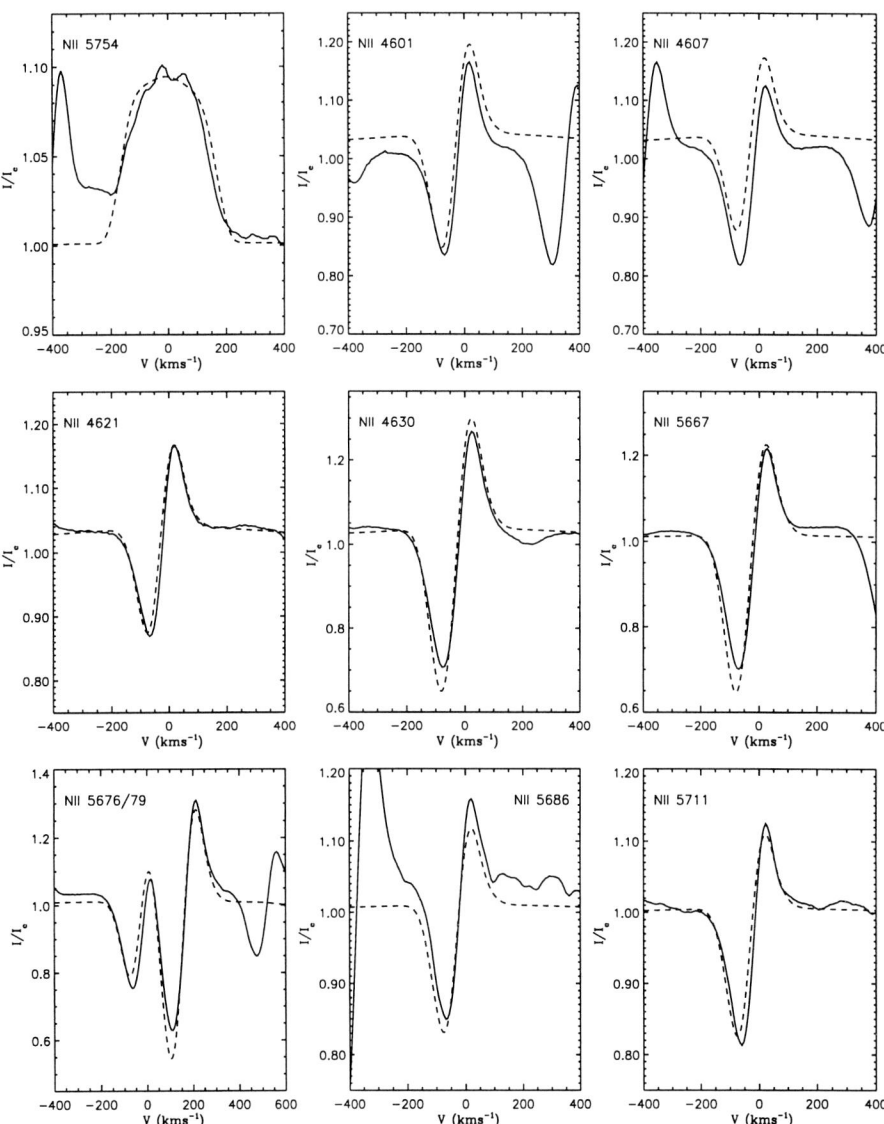

FIGURE 56. Fit of nitrogen lines in the spectrum of P Cygni. The nitrogen abundance is enhanced by a factor of ten relative to solar indicating CNO cycled matter. This agrees well with the high helium abundance ($n_{He}/n_H = 0.3$) found from fitting the helium lines. (From Najarro et al. 1997a).

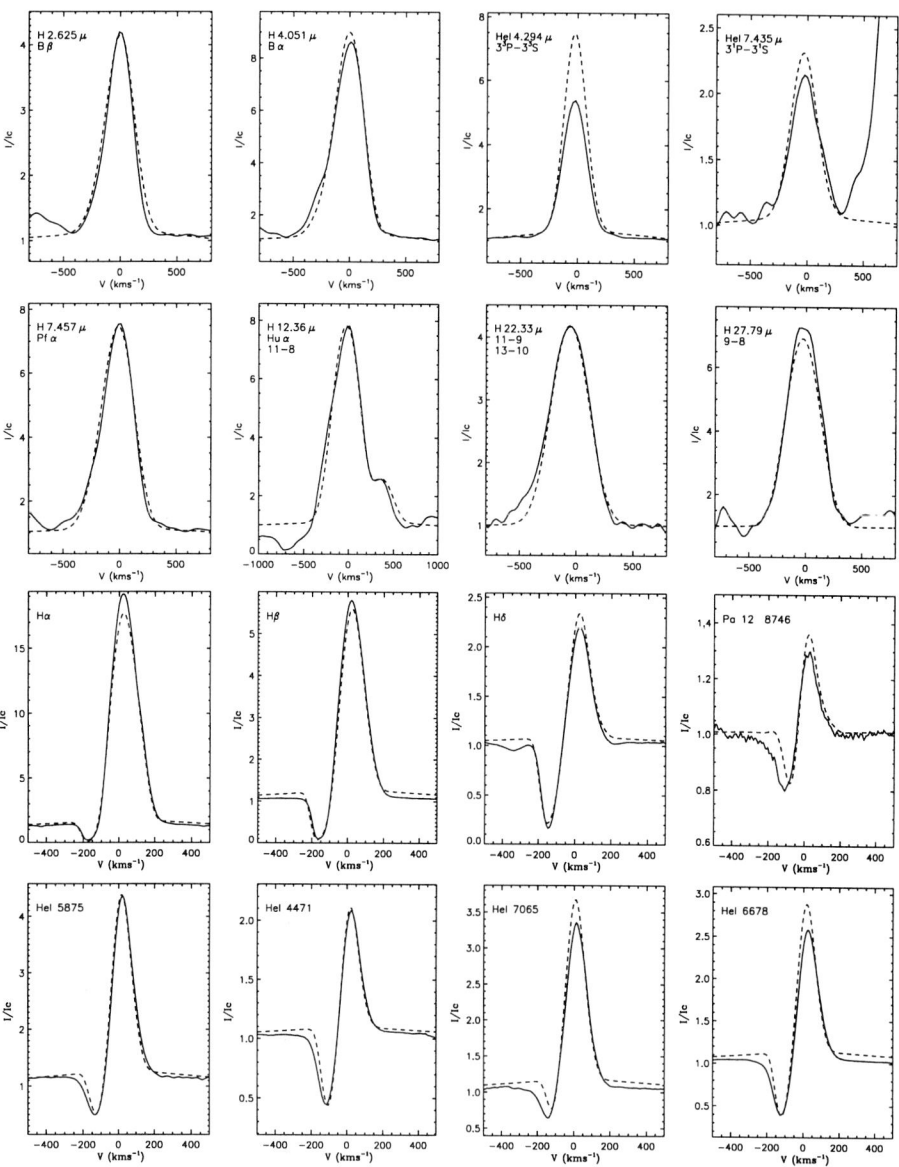

FIGURE 57. The same IR and optical lines as in Fig. 53 and 54, but now calculated with the new velocity field of Fig. 55. (From Lamers, Najarro, Kudritzki et al. 1996).

5.2. The cluster of He I emission line stars in the Galactic Centre

One of the most exciting chapters of ground-based IR astronomy concerns the dense stellar cluster in the Galactic Centre. There is a long-standing debate as to what is the source of luminosity of the central few parsecs of our Galaxy. Is it star formation or a central Black Hole engine ? This debate has gained new momentum with the discovery of a cluster of 2.06μ He I emission line stars in the very centre of the Galaxy using Fabry-Perot spectroscopy techniques (Krabbe *et al.* 1991), 2D-imaging spectroscopy (Krabbe *et al.* 1995) and Adaptive Optics imaging (Eckart *et al.* 1993, Eckart **et al.** 1995). The nature of these emission line objects was unclear in the beginning and stimulated some interesting speculation, for instance of black holes migrating towards the centre via dynamical friction and colliding with red giant stars thus producing He I emission (Morris 1993). However, applying the tools of quantitative stellar spectroscopy in the IR it was possible to prove that we are seeing a cluster of young, massive, luminous Blue Supergiants with very strong stellar winds right in the centre of our Galaxy. They provide the radiation field that ionizes the H II-regions and dominate the energetics via stellar winds and radiation (Najarro *et al.* 1994, Najarro 1995, Krabbe *et al.* 1995, Najarro *et al.* 1997b).

The physics of the 2.06μ He I emission (the exact wavelength is 2.058μ) are rather complex. The line is formed by spontaneous transitions from the 2^1P level to 2^1S (see Fig. 58), where the upper level is populated by recombination processes from above (direct recombination or recombination to higher levels followed by spontaneous cascades into the 2^1P state). However, the strength of the emission depends crucially on the coupling with the ground state 1^1S via the 584 Å resonance transition. The probability of an electron in the level 2^1P decaying via the 2.06μ line is one thousandth of it going through the 584 Å line, because the "branching ratio" of the spontaneous transition probabilities A_{2P-2S}/A_{2P-1S} is roughly 10^3. Hence, every recombination in the 2^1P level will result into a 584 Å photon unless the opacity of this line is large enough to scatter a photon more than thousand times in the volume element until it is degraded into a 2.06μ photon.

Quantitatively, one can describe the processes driving He I 2.06μ into emission by introducing "escape probabilities" β_{line}, which describe the probability of a line photon to escape from the local volume element. Clearly, β_{line} must depend on the optical thickness τ_{line} of the line transition. If τ_{line} is much larger than unity, then β_{line} will approach zero and, conversely, if τ_{line} is very small, β_{line} will become unity. (In the case of supersonically expanding winds ("Sobolev Approximation") the escape probabilities can be expressed analytically as a function of optical thickness (Castor 1970, for an extensive description see Kudritzki 1988)). The statistical equilibrium for the 2^1P level then reads

$$n_e n_{HeII} \alpha_{eff}^{2P} = n_{2P}(A_{2P-2S}\beta_{2P-2S} + A_{2P-1S}\beta_{2P-1S}) \tag{5.17}$$

n_{HeII} is the He II ground state occupation number, n_e the electron density and α_{eff}^{2P} the effective number of recombinations into level 2^1P (for a more detailed discussion see Najarro *et al.* 1994). Rearranging the equation and introducing the branching ratio (see above) one obtains

$$n_{2P} A_{2P-2S}\beta_{2P-2S} = \frac{n_e n_{HeII} \alpha_{eff}^{2P}}{1 + 10^3 \beta_{2P-1S}/\beta_{2P-2S}}. \tag{5.18}$$

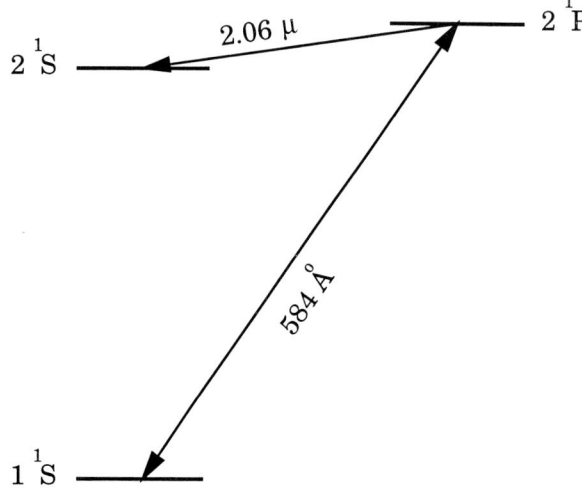

FIGURE 58. The physics of the He I 2.06μ emission line in early type supergiants. For a discussion see text.

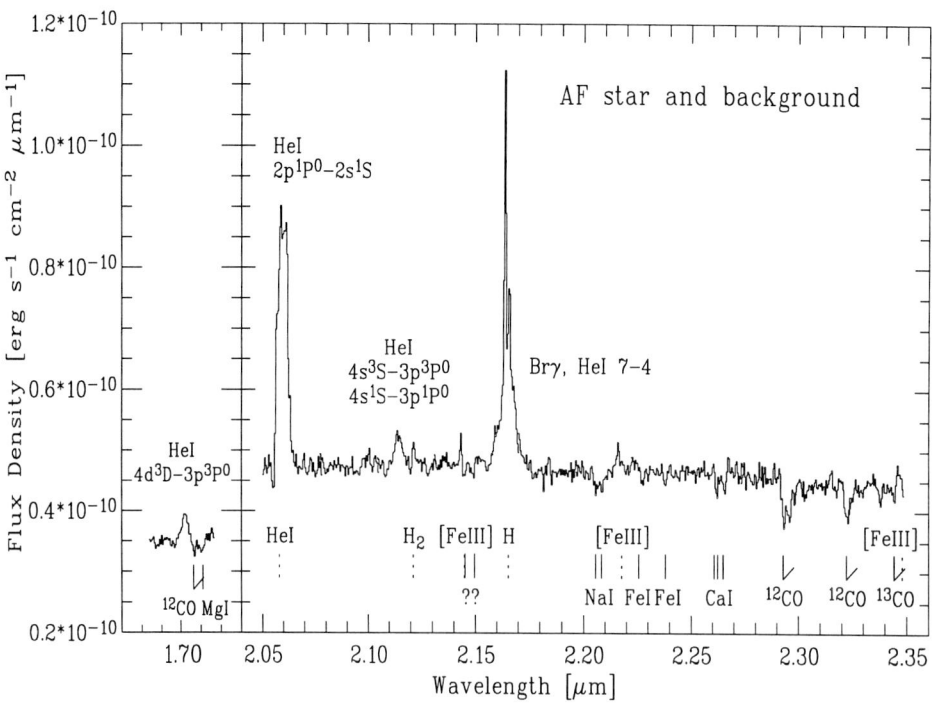

FIGURE 59. ESO NTT IRSPEC spectrum of the brightest Galactic Centre He I emission line star. Note the contaminating narrow emission by surrounding gas (indicated by dashed lines) and absorption features of late type stars (indicated by solid lines). (From Najarro et al. 1994).

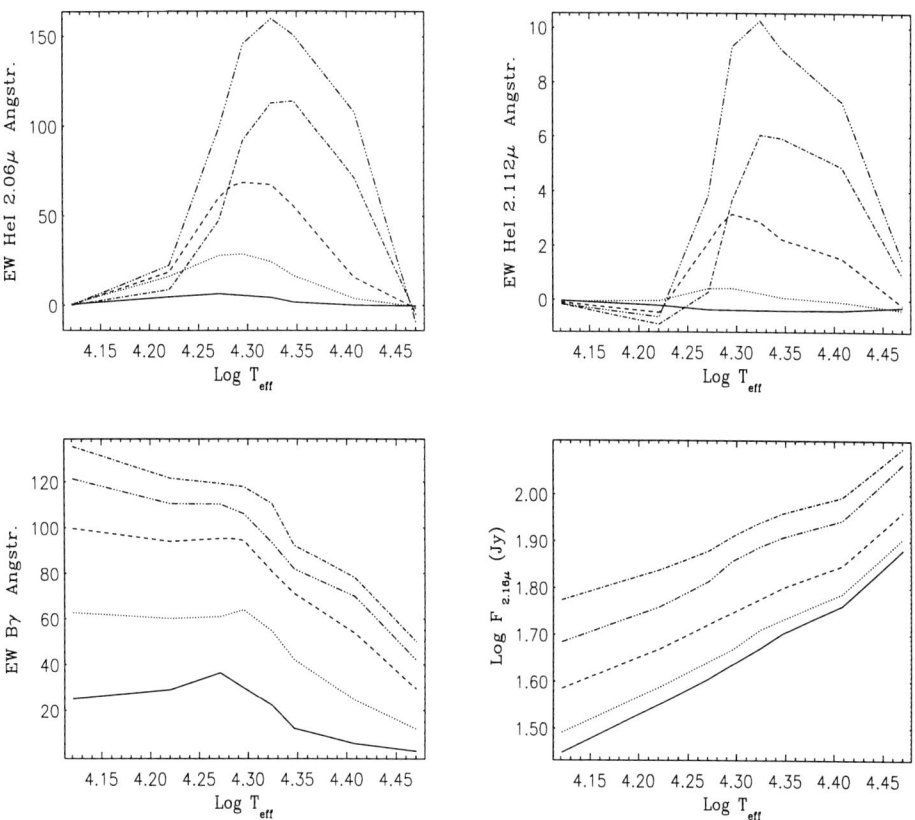

FIGURE 60. Emission line equivalent widths of blue supergiant models with $R_* = 60 R_\odot$, $v_\infty = 500$ km/s and $n(He)/n(H) = 0.125$ (except for curves with three dots between dashes, where $n(He)/n(H) = 0.25$) as a function of stellar effective temperature and mass-loss rate (solid: $\dot{M} = 1$, dotted: $= 2.5$, dashed: $= 5$, dashed-dotted: $= 10 \times 10^{-5} M_\odot/yr$). **upper left**: He I 2.06μ, **upper right**: He I 2.11μ, **lower left**: H B$_\gamma$, **lower right**: continuum K-band flux. (From Najarro 1995).

The quantity on the left side is proportional to the emitted line flux. Obviously, it can only be large if helium is mostly singly ionized, thus restricting the 2.06μ emission to a narrow interval in effective temperature. In addition, β_{2P-1S} has to be very small, requiring a very high optical thickness in the resonance transition. This is the case if the density of the emitting plasma is high (contrary to the case of H II-regions for instance), i.e. when the rate of mass-loss $\dot{M}$ is very high. Thus, **we expect He I 2.06μ in emission in Blue Supergiants with very strong winds in a restricted range of T$_{eff}$ around 30000 K**. This is confirmed by the results of model atmosphere calculations shown in Fig. 60.

Fig. 59 shows a typical K-band spectrum of one of the IR brightest Blue Supergiants in the Galactic Centre. The analysis of these lines yields stellar parameters including helium abundances and the stellar wind parameters. Fig. 61 gives typical line profile fits.

Fits of additional targets are displayed in Fig. 62. The HRD of this spectacular cluster of supergiant stars is shown in Fig. 63 revealing the presence of very massive stars in

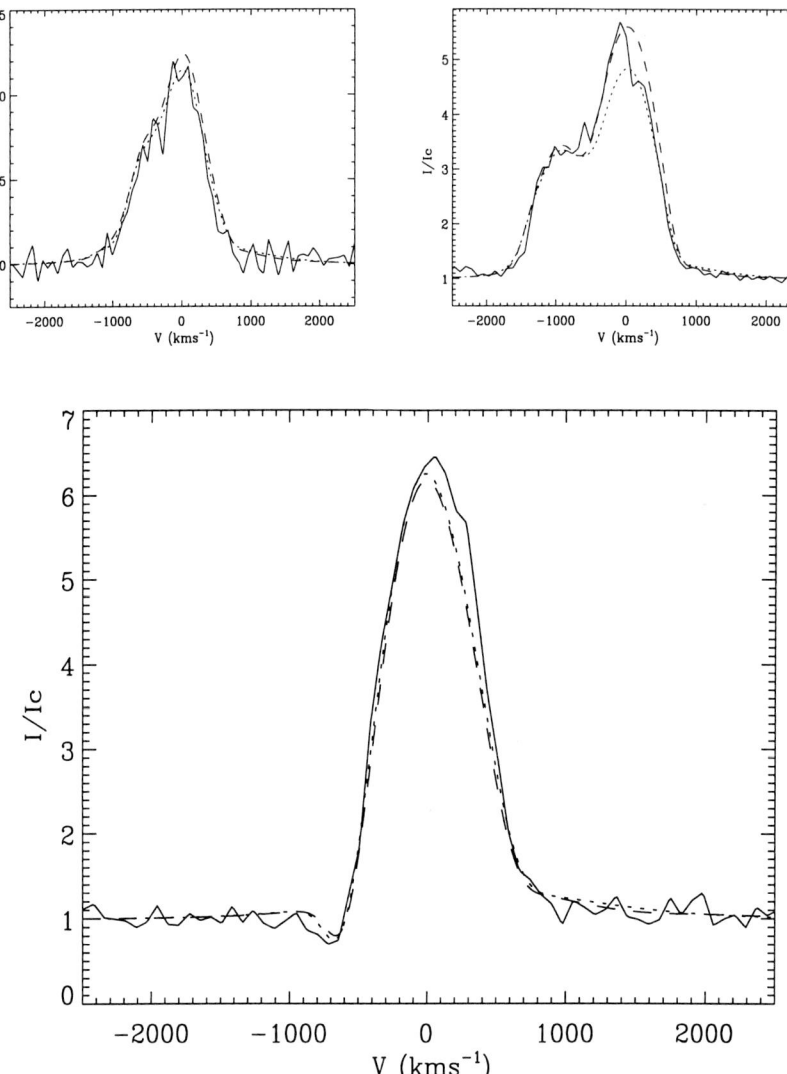

FIGURE 61. Line fits of the brightest Galactic Centre emission line star. **Upper left:** Hydrogen B_γ blended by He I(7-4); **upper right:** Hydrogen P_α blended by He I(4-3); **lower panel:** He I 2.06μ. From Najarro et al. 1994.

an advanced stage of their evolution. All the objects are found to be helium enriched, some of them are already hydrogen poor and almost Wolf-Rayet stars. This is certainly the result of the extremely strong stellar winds with mass-loss rates of the order of $10^{-4} M_\odot/\mathrm{yr}$. The age of the cluster is about five million years. (For more details, see Najarro et al. 1997b).

The individual objects of the cluster can also be used as point masses to probe the gravitational potential of the cluster. Calculating and fitting the IR line spectrum one can also measure the radial velocity of each target precisely (Krabbe et al. 1995). Moreover, proper motions can be measured which can be transformed to tangential velocities. This allows the (anisotropic) velocity dispersion and the rotation in the innermost region of

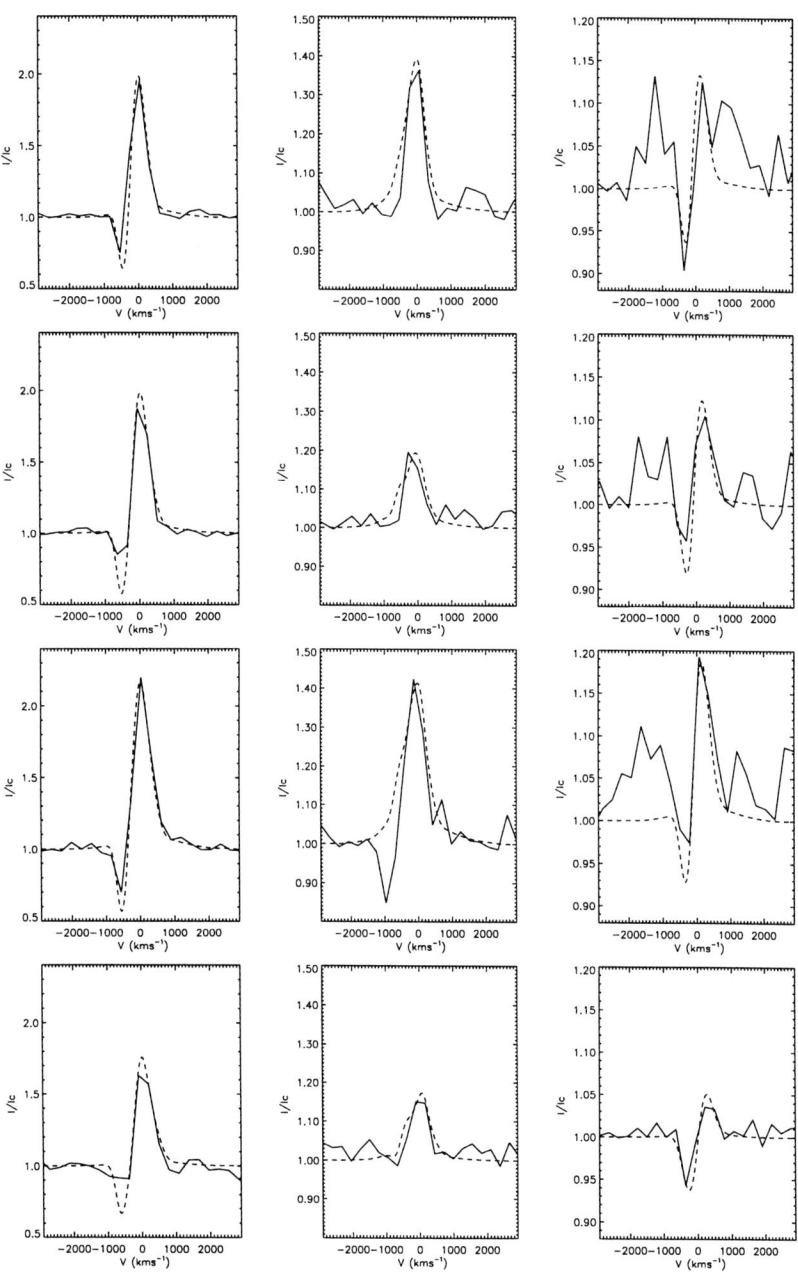

FIGURE 62. K-band line fits of other massive supergiants in the Galactic Centre (IRS 16 sources). **Left**: He I 2.06μ; **middle**: B_γ; **left**: He I 2.22μ. From Najarro et al. 1997b.

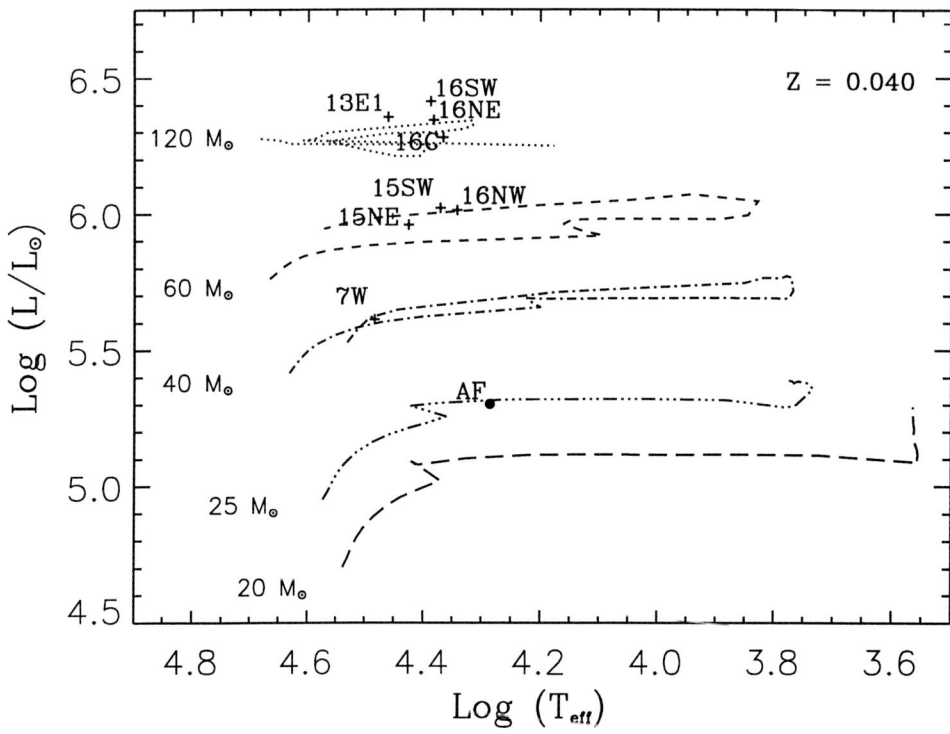

FIGURE 63. The HRD of the cluster of massive stars in the Galactic Centre as obtained from a quantitative analysis of K-band spectra and compared with evolutionary tracks by Schaller *et al.* 1992. From Najarro **et al.** 1997b.

the Galactic Centre to be determined (Eckart **et al.** 1996a, Eckart **et al.** 1996b). The inevitable conclusion of these measurements is that **the Galactic Centre contains a supermassive black hole of** $3 \times 10^6 M_\odot$ **at the position of the non-thermal radio source SgrA***.

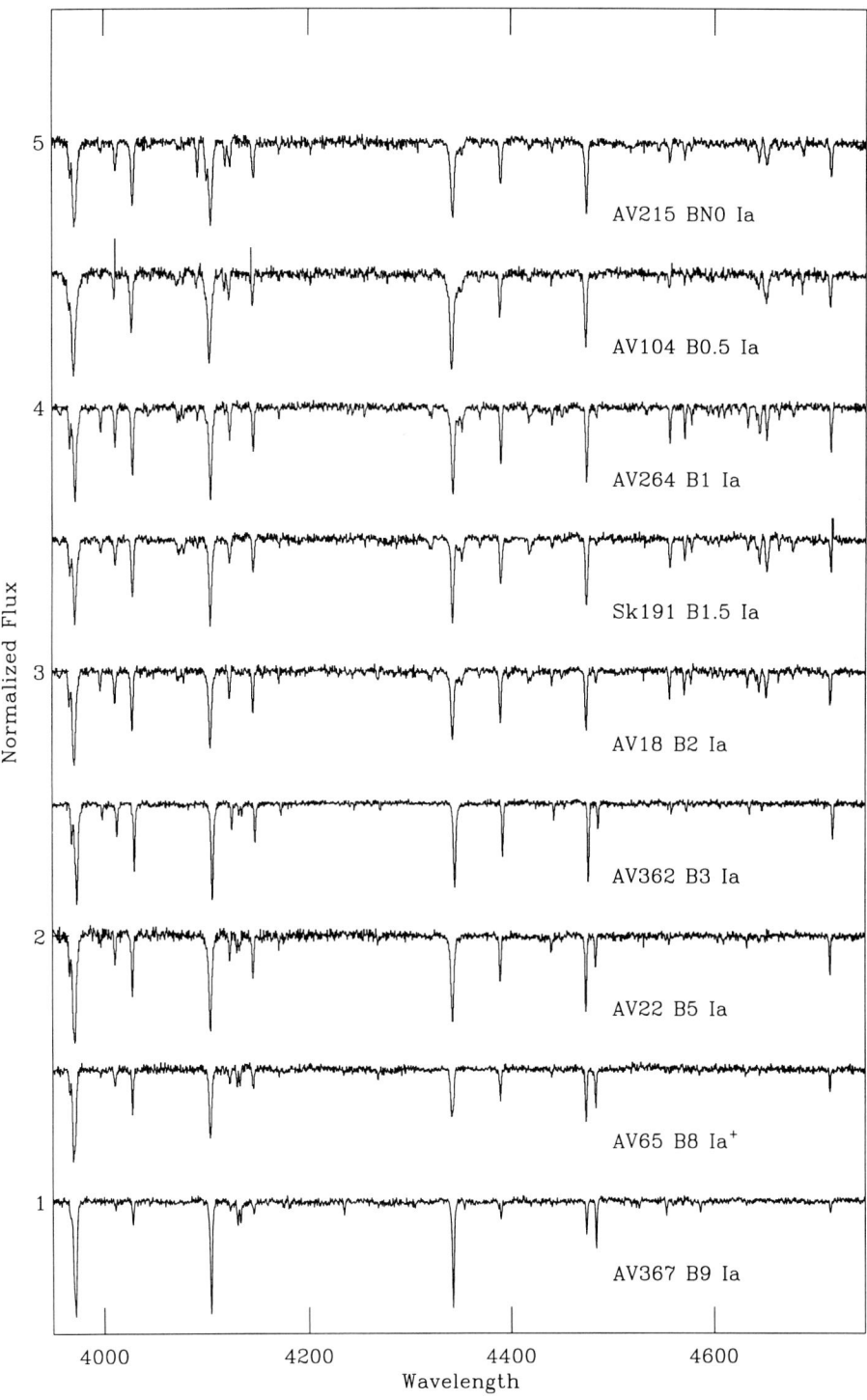

FIGURE 74. The early to late B Ia-supergiants in the SMC (from Lennon, 1996). Comparison with the Galaxy (previous page) shows the striking abundance effects observable.

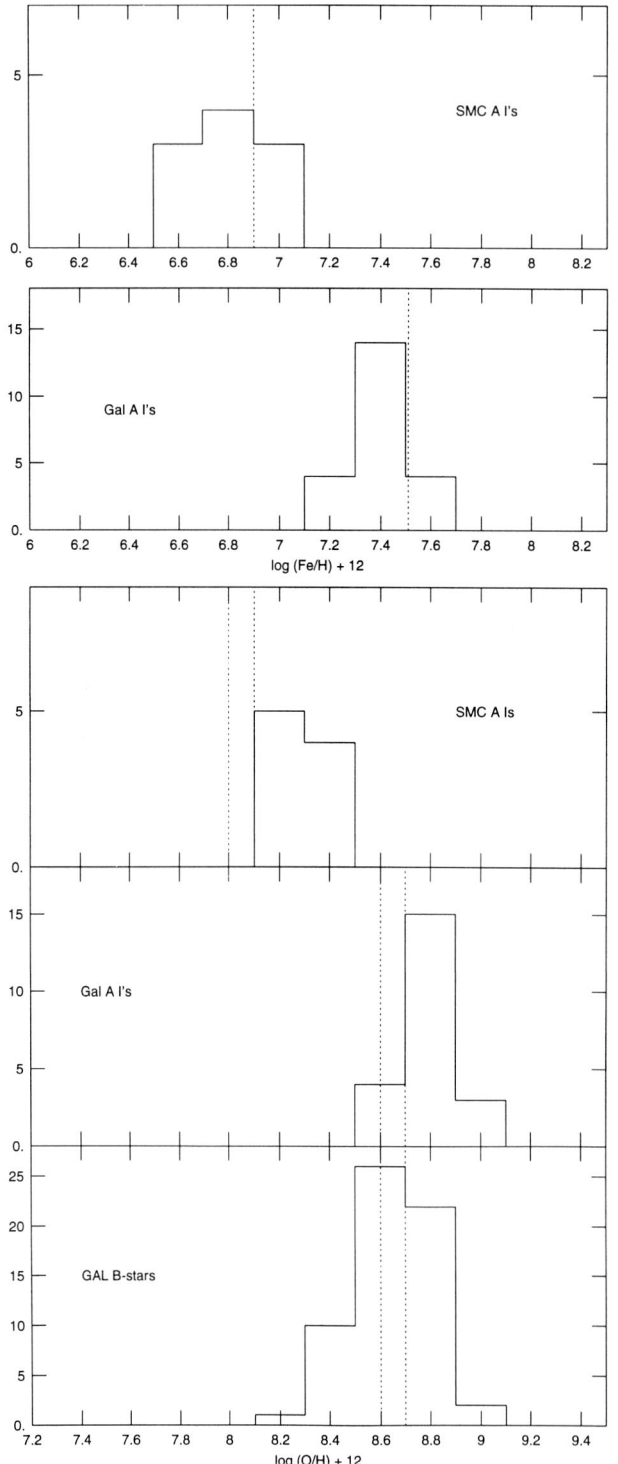

FIGURE 75. Histogram of iron (upper part) and oxygen (lower part) abundances as encountered in the atmospheres of A-supergiants in the Galaxy and SMC. The vertical dotted lines correspond to mean values obtained by other methods (B-stars, H II-region emission lines, F-K supergiants, photometry). (From Venn 1997a,b,c).

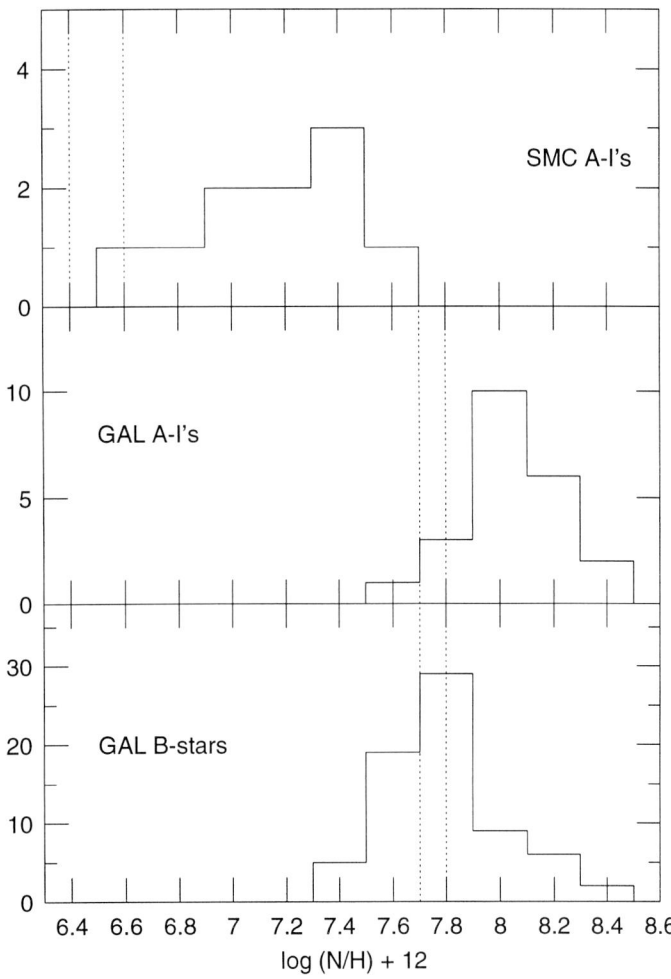

FIGURE 76. Histogram of nitrogen abundances in the SMC and the Galaxy. The vertical dotted lines correspond to mean values obtained from B-stars and H II-regions. (From Venn 1997a,b,c).

7.3. A first step towards stellar abundance gradients in M33 and M31

Radial abundance gradients in the disks of spiral galaxies are predicted to result from the chemodynamical evolution of galaxies. By measuring these gradients one can constrain the models of galaxy evolution. As discussed previously, luminous blue stars provide excellent means for such measurements. Fig. 77 gives an example of very recent work on the abundance gradient in the Galaxy.

Abundance gradients are also crucially important for the determination of extragalactic distances using Cepheids because of the possible dependence of the P-L relation on metallicity. One of the most fundamental test cases are the Local Group spirals M31 and M33. Here, Cepheids observed in fields at different galactocentric distances can be used to investigate the influence of metallicity for the determination of Cepheid distances (Freedman & Madore 1990). However, this requires metal abundances of the corresponding stellar population in the disks of M31 and M33 to be determined as a function of the galactocentric distance. Such a project has been started recently by the "Tenerife-Pasadena-Munich" connection and first results are already available.

McCarthy *et al.*1995 have used the 10m Keck I telescope with the HIRES Echelle spectrograph to study two A-supergiants (designated B-324 and 117-A) at very different galactocentric distances in M33. (Fig. 78 shows a small extract from the spectra compared with spectrum synthesis calculations). It was found that the A0 IaO Hypergiant 117-A in the outskirts of M33 (4.5 kpc away from the centre of M33) is an extremely metal poor object with iron abundances clearly below SMC abundances. On the other hand, B-324, only 1.3 kpc away from the centre, has roughly solar abundances. In the latter case the analysis is difficult, since every single line in the optical spectrum is affected by an extremely strong stellar wind. Therefore, as a first step, the analysis was restricted to the more photospheric absorption components of the line profiles.

In parallel, Herrero *et al.* 1994, Monteverde *et al.* 1996, 1997 have studied optical spectra of B-supergiants in M33 with the WHT on La Palma. The combination of their results with that of McCarthy *et al.* 1995 has led to a first estimate of the stellar abundance gradient in M33 as shown in Fig. 79.

Work on M31 is also in progress. A number of supergiants of spectral types F, A and B has been observed with Keck HIRES and the WHT and the analysis is presently under way. Fig. 80 shows the spectrum of a late A-supergiant in the 10kpc spiral arm of M31. Fig. 81 gives a very first result concerning the oxygen abundance gradient. Fig. 82 shows the the abundance patterns of two objects at distances of 10.4 and 19.7 kpc distance from the centre of M31.

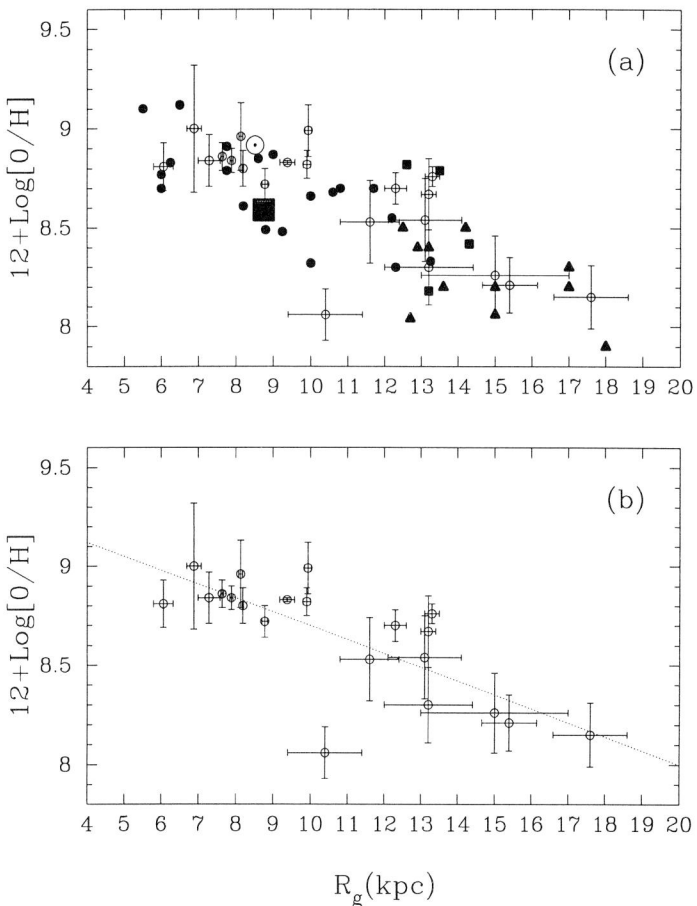

FIGURE 77. The abundance gradient in the galactic disk. **Upper panel**: Oxygen abundances as a function of galactocentric distance for B-type stars (open circles) and H II regions (filled symbols). **Lower panel**: Only B-stars with a least-squares fit through the data points showing a gradient of -0.07 ± 0.01 dex/kpc. (From Smartt and Rolleston, 1997).

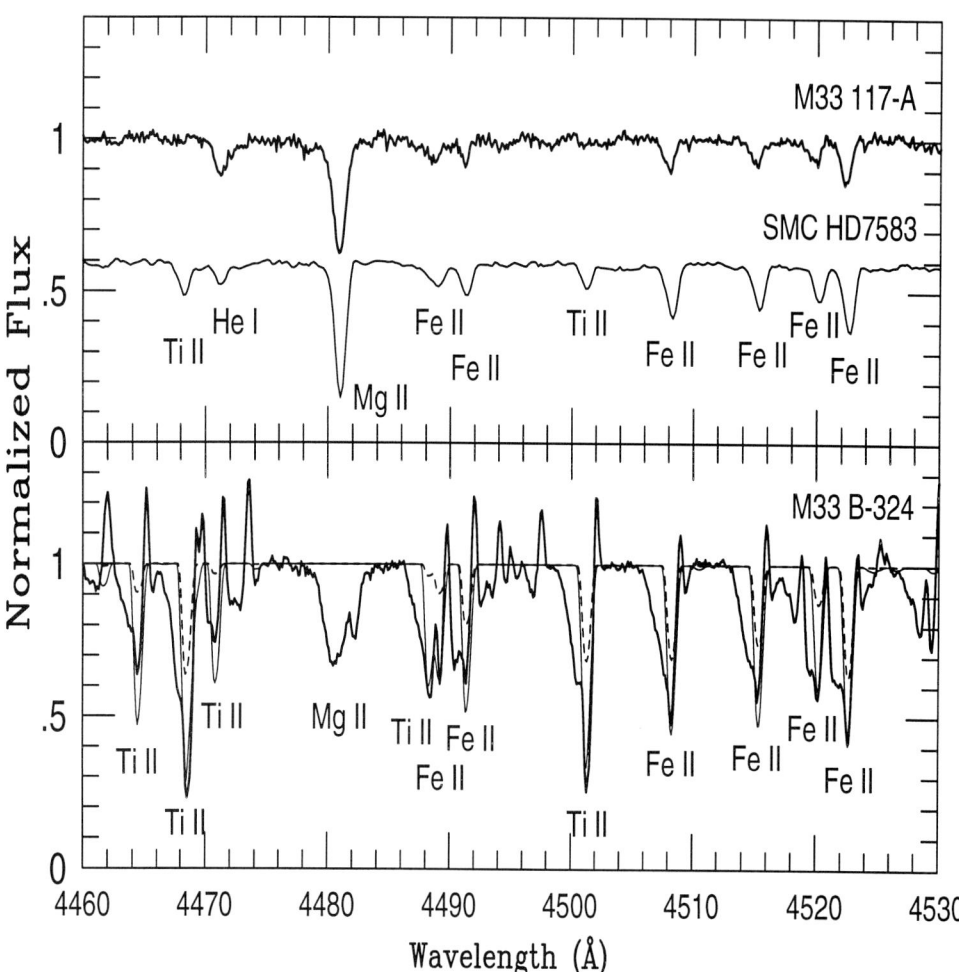

FIGURE 78. Small extract from the Keck HIRES spectra of two A-supergiants in M33. **Upper panel**: Object 117-A is compared to an SMC supergiant of identical spectral type to demonstrate the low metallicity. Quantitative analysis resulted in factor of ten reduction in abundance relative to solar for Fe and Ti, and a factor of five for Mg and Cr. **Lower panel**: Object B-324 showing stellar wind contamination in every line. The spectrum synthesis is restricted to the absorption components only and adopts solar abundance for Fe (three times solar for Ti) for the solid curve and 1/10 solar for the dashed curve. (From McCarthy et al. 1995).

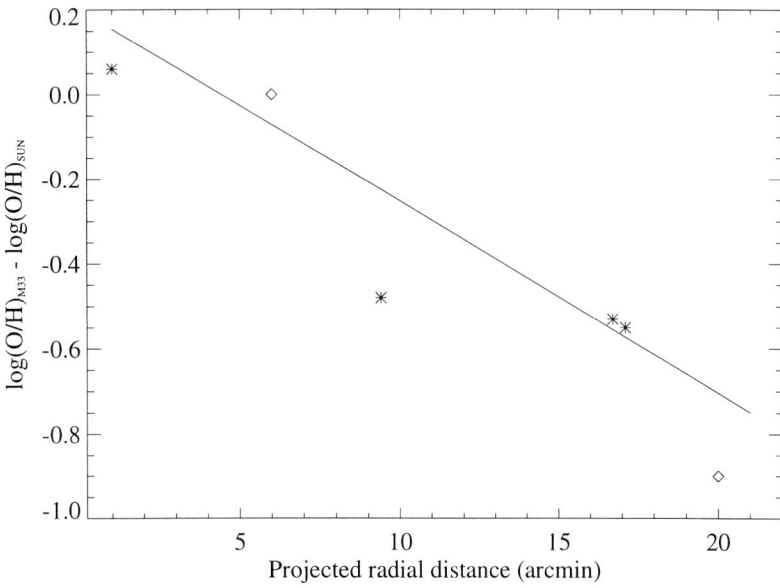

FIGURE 79. The abundance gradient in M33. Asteriks refer to oxygen from B-supergiants, diamonds to iron from A-supergiants, The solid line represents the average result obtained for oxygen from H II-regions. From Monteverde et al. 1997.

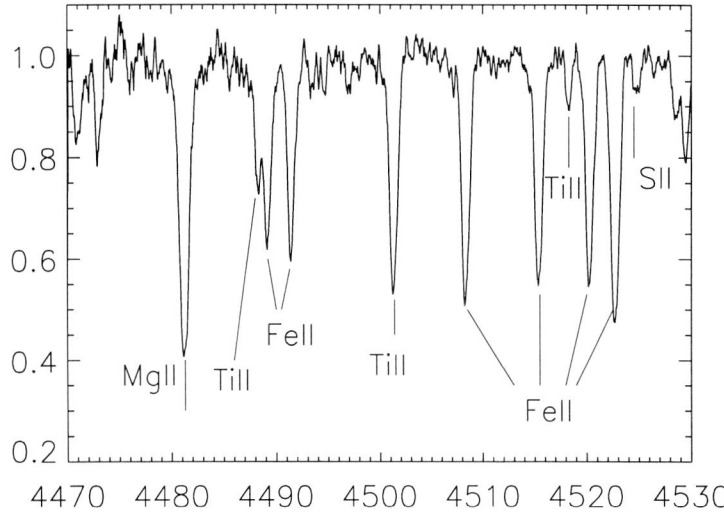

FIGURE 80. Extract from a Keck HIRES spectrum of the A8 Ia supergiant 41-2368 in M31. (From McCarthy, Venn, Lennon, Kudritzki 1997, in prep. for ApJ)

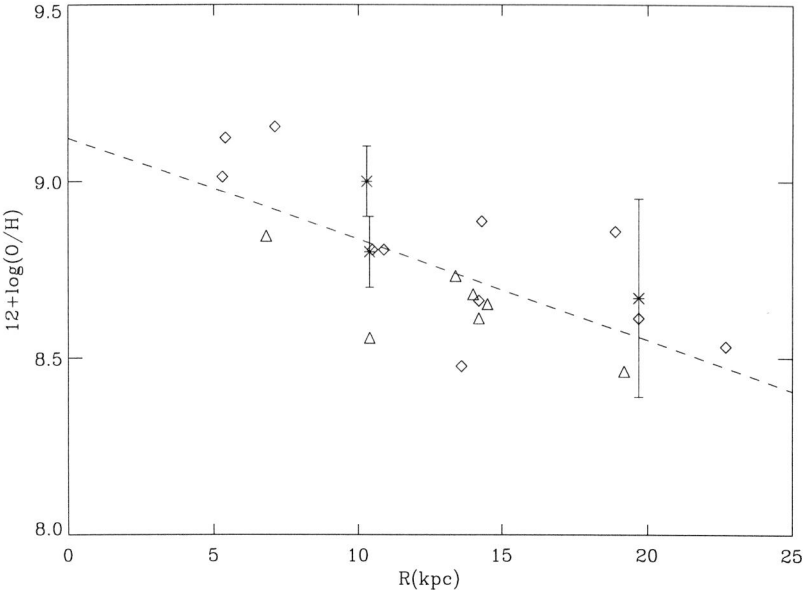

FIGURE 81. The abundance gradient in M31. Asteriks refer to oxygen in A- and F-supergiants, open symbols to nebular oxygen abundances. (From McCarthy, Venn, Lennon, Kudritzki, 1997 in prep. for ApJ)

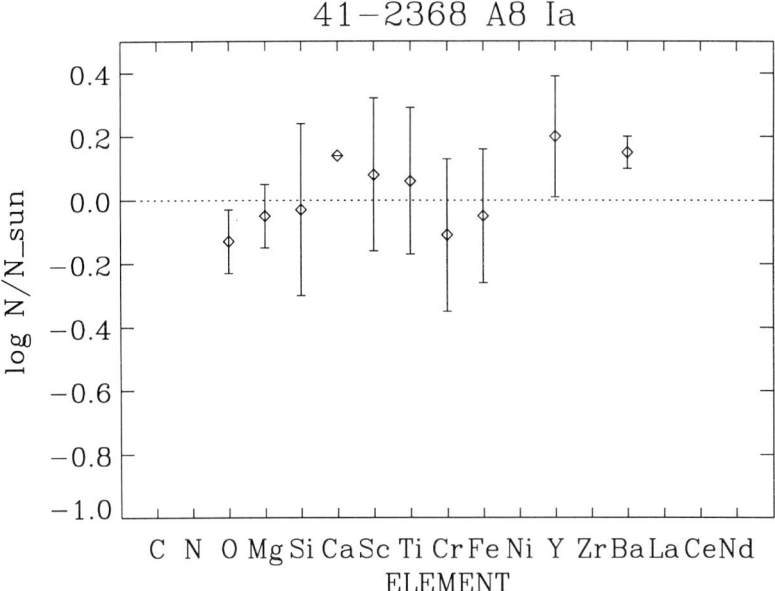

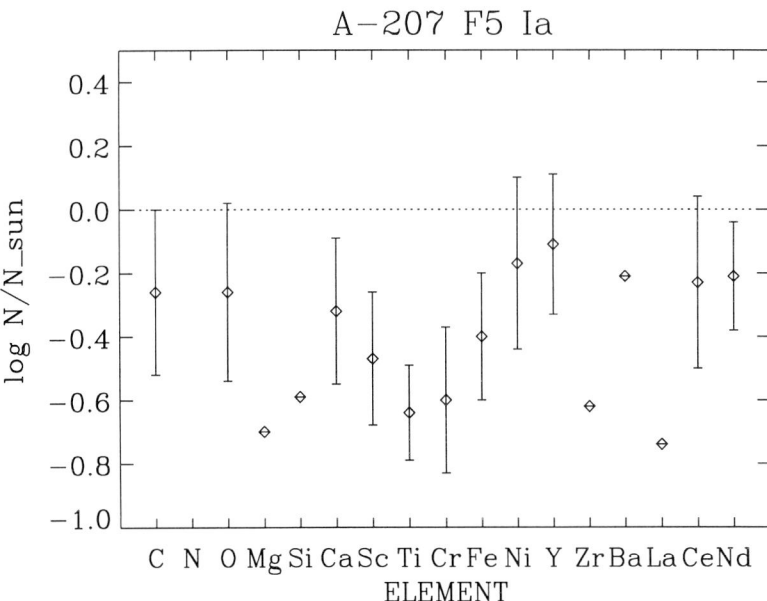

FIGURE 82. Abundance pattern of two supergiants in M31 at 10.4 kpc (upper panel) and 19.7 kpc galactocentric distance as observed with Keck HIRES. Values without error bars are determined from a single line. (From McCarthy, Venn, Lennon, Kudritzki 1997, in prep. for ApJ).

7.4. Stellar abundances beyond the Local Group

With ground-based telescopes of the 8m-class it is definitely possible to extend stellar abundances determinations far beyond the Local Group. As we will discuss in section 9, abundances of the brightest Blue Supergiants in the Virgo and Fornax clusters of galaxies at distances of 15 to 20 Mpc are obtainable. With the Keck telescope, first steps out of the Local Group have already been taken. The first high S/N spectra of 1 Å spectral resolution of supergiants in NGC 2403 – the satellite galaxy of the large spiral M81 – at a distance of 3.1 Mpc have been secured. Fig. 83 demonstrates that the quality of the spectra is excellent and does allow precise abundance work. This means that detailed investigations of spiral and irregular galaxies can be carried out, thereby providing crucial observational constraints on the current chemodynamical models for the evolution of galaxies.

8. The Wind Momentum - Luminosity Relationship and extragalactic distances

Since the detection of stellar winds as a ubiquitous feature across the HRD, the investigation of their properties as function of stellar parameters has been the subject of intense research work. For hot luminous stars, where the winds are driven by radiation, it has become clear that three fundamental correlations exist

- All massive stars with $L \geq 3\ 10^4\ L_\odot$ show stellar winds throughout their lifetime (see Abbott 1979)
- Mass-loss rates increase as function of luminosity with

$$\dot{M} \propto L^x, \tag{8.19}$$

where $x \approx 1.5$ (see, for instance, Garmany & Conti 1984)

- Terminal velocities increase as function of escape velocity from the stellar surface following

$$v_\infty \propto v_{esc}, \tag{8.20}$$

(see Abbott 1982, but also Lamers et al. 1995) with $v_{esc} = (2g_* R_* (1-\Gamma))^{0.5}$. $\Gamma = g_{rad}^{Thom}/g_* \propto L/M_*$ is the ratio of radiative acceleration caused by Thomson electron scattering divided by gravitational acceleration.

These correlations show a large scatter indicating the dependence on additional parameters and the proportionality constants are subject to an intense discussion. This discussion, in particular the one about the strengths of the mass-loss rates $\dot{M}$ is crucially important for stellar evolution, since the evolution of massive stars is affected severely by mass-loss (see Chiosi 1997 or Langer et al. 1994 as examples for the vast amount of literature about this topic). Here, we do not intent to enter into this discussion, since we are interested in a completely different question. We want to investigate what the stellar winds can tell us about the luminosity of stars.

Since the winds of the Blue Supergiants are a result of radiation pressure, we expect that the properties of the stellar winds must somewhere reflect the luminosities and radii of the stars. This is already indicated by the loose correlations of mass-loss rate on luminosity and terminal velocity on gravitational potential mentioned above. With this in mind, we develop the following very simple idea. For radiation driven winds the mechanical momentum flow of a stellar wind $\dot{M}v_\infty$ should be a function of the photon momentum rate L/c provided by the stellar photosphere and interior

$$\dot{M}v_\infty = f(L/c). \tag{8.21}$$

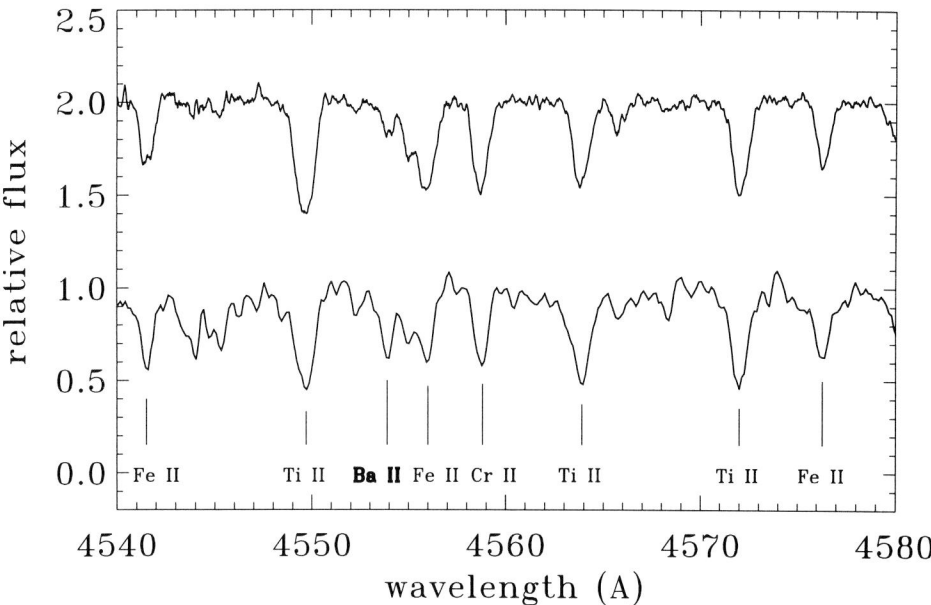

FIGURE 83. **The step out of the Local Group.** Extracted spectral region from Keck HIRES spectra of two late A Ia supergiants. Upper spectrum, M31; **lower spectrum, NGC 2403 at 3.1 Mpc distance.** From work in progress by McCarthy, Kudritzki, Lennon, Venn.

If this is true and if we are able to find this function f, this would enable us to determine directly stellar luminosities from the stellar wind by using the inverted relation

$$L = f^{-1}(\dot{M}v_\infty). \tag{8.22}$$

In other words, by measuring the rate of mass-loss and the terminal velocity directly from the spectrum we would be able to determine the luminosity of a Blue Supergiant. This is an exciting perspective, because it would give us a completely new, purely spectroscopic tool to determine stellar distances. Quantitative spectroscopy would yield T_{eff}, gravity, abundances, intrinsic colours, reddening, extinction, $\dot{M}$ and v_∞. With the luminosity from the above relation one could then compare with the dereddened apparent magnitude to derive a distance.

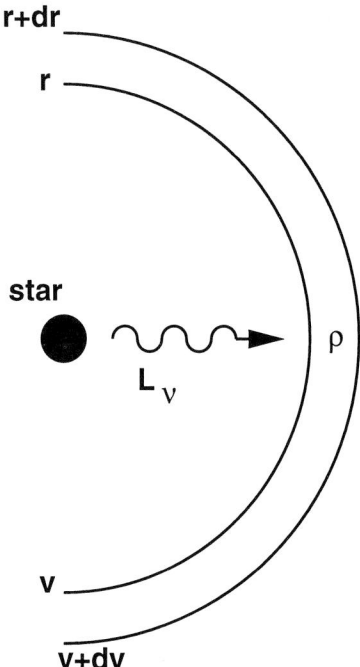

FIGURE 84. Sketch of a Blue Supergiant irradiating its own stellar wind envelope. L_ν is the spectral luminosity at frequency ν. v is the wind outflow velocity at radius r and ρ is the mass density.

For the remaining section we want to follow this idea. First we will use a simplified theoretical description of radiation driven winds to derive the above **Wind Momentum - Luminosity Relationship**. We will then discuss some obvious and necessary refinements of the theory, before we start to compare the theoretical predictions with the observations of Blue Supergiants in the Galaxy. As a next obvious step we show first results of an application to Blue Supergiants in M31. We will finish this section with a discussion of the metallicity dependence of the Wind Momentum - Luminosity Relationship showing results from stars in the Magellanic Clouds and along the metallicity gradient in M33.

8.1. Theory

We start with a very simple description of a line driven stellar wind. We assume that it is stationary and spherical symmetric. From these assumptions the equation of continuity

$$\dot{M} = 4\pi r^2 \rho(r) v(r) \tag{8.23}$$

is derived immediately. Then we consider the star as a point source of photons irradiating and accelerating its own stellar wind (see Fig. 84) and calculate the amount of photon momentum absorbed by one spectral line

$$\frac{L}{c}\frac{L_{\nu_i}(1-e^{-\tau_i})d\nu^{\text{width}}}{L} = \frac{L}{c^2}\frac{\nu_i L_{\nu_i}}{L}(1-e^{-\tau_i})dv. \tag{8.24}$$

The first factor on the left side gives the total photon momentum rate provided by the star, the second describes the fraction absorbed by one spectral line in an outer shell of thickness dr. τ_i is the optical thickness of such an outer shell in the line transition i. $L_{\nu_i} d\nu^{\text{width}}$ is the stellar spectral luminosity at the frequency of line i multiplied by the spectral width of the line. This luminosity can in principle be absorbed by the line if it is entirely optically thick (i.e. $\tau_i \gg 1$). However, depending on the optical thickness only the fraction $(1 - e^{-\tau_i})$ is really absorbed. If we are in the supersonic part of the wind, then the spectral width $d\nu^{\text{width}}$ is not determined by the thermal motion of the ions but rather by the increment of the velocity outflow dv via the Doppler formula

$$d\nu^{\text{width}} = \nu_i \frac{dv}{c}, \qquad (8.25)$$

which leads to the right hand side of equation (8.24).

After calculation of the photon momentum absorbed by a single spectral line we can consider the momentum balance in the stellar wind. The photon momentum absorbed by all lines will just be a sum over all lines i of the expression shown on the right hand side of equation (8.24). Almost all of this absorbed momentum will be transformed into gain of mechanical stellar wind flow momentum $\dot{M} dv$ of the outer shell except the fraction $g(r) dM_r$, which is the momentum required to act against the gravitational force. ($g(r) = GM_*/r^2$ is the local gravitational acceleration and $dM_r = \rho 4\pi r^2 dr$ is the mass within the spherical stellar wind shell)

$$\dot{M} dv = \frac{L}{c^2} \sum_i \frac{\nu_i L_{\nu_i}}{L}(1 - e^{-\tau_i}) dv - G\frac{M_*(1-\Gamma)}{r^2}\rho 4\pi r^2 dr. \qquad (8.26)$$

Note that this momentum balance also includes the photon momentum transfer by Thomson scattering, which leads to the correction factor $(1-\Gamma)$ in the local gravitational acceleration.

Now, the important next step is to deal with the sum over all lines in the above momentum balance. Here, the complication arises from the term in parentheses containing the local optical depth τ_i of each line. τ_i will not only be different for each of the thousands of lines driving the wind, it will also vary through the stellar wind as a function of radius. On the other hand, one of the enormous simplifications in supersonically expanding envelopes around stars is that the optical thickness is well described by (see for instance Castor 1970, Rybicki & Hummer 1978 or Kudritzki 1988)

$$\tau_i = \mathbf{k_i} \kappa_{\text{Thom}} \frac{v_{\text{therm}}}{dv/dr}, \qquad (8.27)$$

where v_{therm} is the thermal velocity of the ion and $\mathbf{k_i}$ is the (dimensionless) **line strength** defined as

$$\mathbf{k_i} = \frac{\kappa_i}{\kappa_{\text{Thom}}}, \qquad (8.28)$$

i.e. the opacity of line i

$$\kappa_i = \frac{1}{\Delta\nu_{\text{Dopp}}} \frac{\pi e^2}{m_e c} n_l f_{lu}(1 - \frac{n_u}{n_l}\frac{g_l}{g_u}) \qquad (8.29)$$

in units of the continuous Thomson scattering opacity or in units of the local density

$$\kappa_{\text{Thom}} = n_e \sigma_e = \rho s_e. \qquad (8.30)$$

σ_e is the cross section for Thomson scattering of photons on free electrons, n_e the local number density of free electrons. In a hot plasma with hydrogen as the main constituent mostly ionized and with $Y_{He} = n_{He}/n_H$ and I_{He} the number of electrons provided per helium nucleus we have (m_H is the mass of the hydrogen atom)

$$s_e = \frac{1 + I_{He}Y_{He}}{1 + 4Y_{He}} \frac{\sigma_e}{m_H}. \tag{8.31}$$

Thus, if the degree of ionization of helium is roughly constant as a function of radius r in the wind, s_e is also constant and we have

$$\tau_i = \mathbf{k_i} s_e \rho(r) \frac{v_{therm}}{dv/dr}, \tag{8.32}$$

with the line strength $\mathbf{k}_i$ proportional to oscillator strength f_{lu}, wavelength λ_i and the occupation number n_l of the lower level divided by the mass density ρ

$$\mathbf{k_i} \propto \frac{n_l}{\rho} f_{lu} \lambda_i. \tag{8.33}$$

Thus, the line strength is roughly independent of the depth in the atmosphere and is determined by atomic physics (f_{lu}, λ_i) and atmospheric thermodynamics (n_l/ρ). Since ρ and dv/dr vary strongly through the wind a line can be optically thick in deeper layers

$$\tau_i \gg 1 \Longrightarrow (1 - e^{-\tau_i})dv \approx dv \tag{8.34}$$

and can become optically thin further out

$$\tau_i \ll 1 \Longrightarrow (1 - e^{-\tau_i})dv \approx k_i \rho dr. \tag{8.35}$$

For the calculation of the momentum transfer in equation (8.26) this means that line contributions can have an entirely different functional form depending on the optical thickness of the lines.

This problem can be solved in a very elegant way by introducing a **line strength distribution function**

$n(k,\nu)d\nu dk$ = number of lines with ν_i from $(\nu, \nu + d\nu)$ and with k from (k, k+dk).

Since the hydrodynamic model atmosphere codes (see section 2) contain atomic data and occupation numbers for millions of lines, we can investigate the physics of the line strength distribution function. As it turns out (Castor, Abbott, & Klein 1975, Puls 1987, Kudritzki et al. 1988, Puls 1993, Springmann 1997, Springmann et al. 1997) the distribution in line strengths – to a very good approximation – obeys a power law

$$n(k,\nu)d\nu dk = g(\nu)d\nu k^{\alpha-2}dk, 1 \le k \le \infty \tag{8.36}$$

independent (to first order) of the frequency. The exponent α depends weakly on T_{eff} and varies (in the temperature range of OB-stars) between

$$\alpha = 0.6 \ldots 0.7. \tag{8.37}$$

α is mostly determined by the atomic physics and basically reflects the distribution function of the oscillator strengths. One can, for instance, show that for the hydrogen atom the distribution of the Lyman-series oscillator strengths is a power law with exponent $\alpha = 2/3$ (Puls 1993, see also Appendix A). It is important to realize that α is not

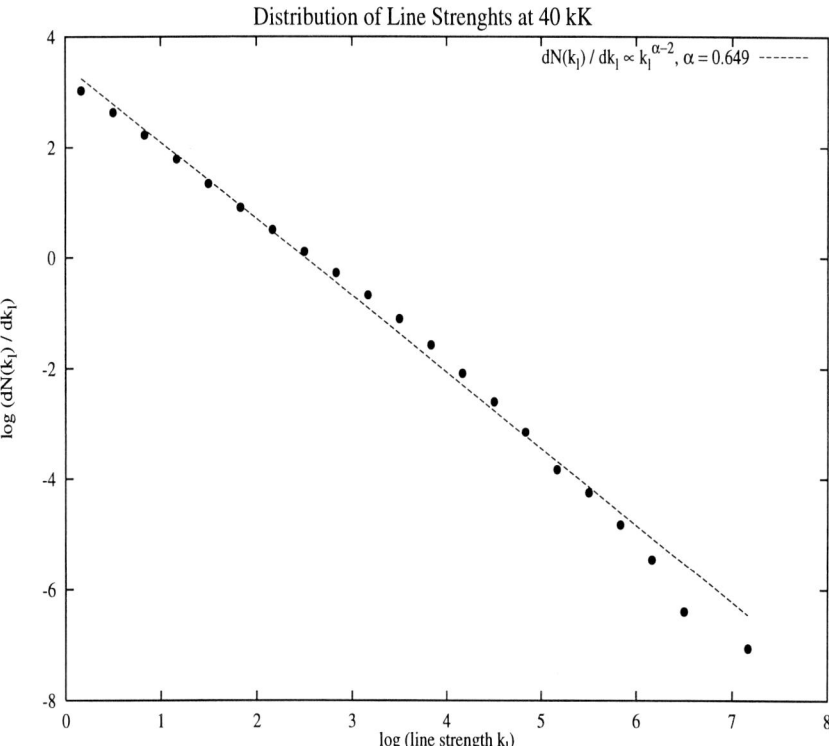

FIGURE 85. Logarithmic plot of the line strength distribution function for a stellar wind model with $T_{eff} = 40000K$. Dots are numbers from the line list in the model atmosphere code. The dashed curve is the power law fit. For this fit the last three points at the largest k-values, which correspond to a total of only six lines, were ignored. From Springmann et al. 1997.

a free parameter but, instead, well determined from the thousands of lines taken into account in the model atmosphere calculations. An example for the power law dependence of line strengths is given in Fig. 85.

With the line strength distribution function of equation (8.36) we can replace the sum in the momentum balance of equation (8.26) by a double integral, which is then solved taking equation (8.32) into account (see also Kudritzki 1988).

$$\sum_i \frac{\nu_i L_{\nu_i}}{L}(1 - e^{-\tau_i}) \longrightarrow \int_0^\infty \int_0^\infty (1 - e^{\tau(k)})\frac{\nu L_\nu}{L}n(k,\nu)d\nu dk \longrightarrow N_o\{\frac{dv/dr}{\rho}\}^{\alpha-1} \quad (8.38)$$

This means that the momentum transfer from photons to the stellar wind plasma depends non-linearly on the gradient of the velocity field. The degree of the non-linearity is determined by the steepness of the line strength distribution function α. N_o is proportional to the number of lines in the line strength interval $1 \leq k \leq \infty$.

With equation (8.38) we can return to the total momentum balance equation (8.26) to obtain

$$\dot{M}dv = \frac{L}{c^2}N_o\{\frac{dv/dr}{\rho}\}^{\alpha-1}dv - G\frac{M_*(1-\Gamma)}{r^2}4\pi r^2 \rho dr. \quad (8.39)$$

Inserting

$$4\pi\rho = \frac{\dot{M}}{r^2 v} \tag{8.40}$$

(from the equation of continuity 8.23) we then arrive at the non-linear differential equation for the velocity field

$$r^2 v \frac{dv}{dr} = \frac{L}{\dot{M}^\alpha} \frac{N_o}{c^2} (4\pi)^{\alpha-1} \{r^2 v \frac{dv}{dr}\}^\alpha - GM_*(1-\Gamma). \tag{8.41}$$

which looks much more complicated than it really is. The solution is easy (Kudritzki et el. 1989). From the different mathematically possible solutions we select the only one having a velocity field with a smooth transition from being subsonic at small radii to supersonic at larger radii. We obtain $\dot{M}$ as the uniquely determined eigenvalue of the problem

$$\dot{M} \propto L^{1/\alpha} \{M_*(1-\Gamma)\}^{1-1/\alpha}. \tag{8.42}$$

and a velocity gradient

$$r^2 v \frac{dv}{dr} \propto GM_*(1-\Gamma) \tag{8.43}$$

leading to a "$\beta - type$" velocity field with $\beta = 0.5$ (see section 2)

$$v(r) \propto v_{esc} \{1 - R_*/r\}^{0.5}, v_{esc} \propto \{\frac{GM_*(1-\Gamma)}{R_*}\}^{0.5} \tag{8.44}$$

and a terminal velocity proportional to the escape velocity v_{esc}.

$$v_\infty \propto v_{esc}. \tag{8.45}$$

Combining equations 8.42 and 8.45 we obtain the wind momentum rate

$$\dot{M} v_\infty \propto \frac{1}{R_*^{0.5}} L^{1/\alpha} \{M_*(1-\Gamma)\}^{3/2-1/\alpha}, \tag{8.46}$$

which – as expected – depends strongly on the luminosity but also on the photospheric radius and the expression in the parentheses, which contains the stellar mass and distance from the Eddington limit. It is this expression which can vary significantly for different Blue Supergiants and, therefore, causes the large scatter in the observed correlations (8.19, 8.20) of mass-loss rates with luminosity and terminal velocity with escape velocity as discussed previously (see equations 8.42 and 8.45). However, for the product of mass-loss rate and terminal velocity, the stellar wind momentum rate, the exponent of the term in brackets should be – thanks to the laws of atomic physics – very close to zero, since $\alpha \approx 2/3$. This means that to first order the wind momentum rate should be determined by

$$\dot{\mathbf{M}} \mathbf{v}_\infty \propto \frac{1}{R_*^{0.5}} \mathbf{L}^{1/\alpha} \tag{8.47}$$

This is the Wind Momentum - Luminosity Relationship. It predicts a strong dependence of wind momentum rate on the stellar luminosity with an exponent determined by the statistics of the strengths of the tens of thousands of lines driving the wind. It also contains a weak dependence on stellar radius which comes from the fact that the stellar wind has to work against the gravitational potential when accelerated by photospheric photons.

8.2. Refinements

In deriving the Wind Momentum - Luminosity Relationship analytically in the last subsection we made several assumptions that require justification and refinement. First, we assumed that the power law exponent for the distribution function of line strengths in equation (8.36) is independent of frequency. A detailed comparison with the hundred thousands of lines calculated in NLTE reveals that this is not completely true and that α can vary slightly as a function of wavelength. However, this effect is not dramatic, since for the sum or the integral in equation (8.38) the lines are weighted with the relative flux distribution and, therefore, it is sufficient to determine α properly only for those lines in the spectral region corresponding to the maximum of νL_ν and to use the resulting value as being a representative for the total distribution of line strengths. Consequently, the analytical result is not affected.

More serious is the effect that the ionization and excitation of metals may change through the wind. Since ionization depends mostly on the ionizing radiation field, which dilutes with geometrical dilution factor

$$W(r) = \frac{1}{2}\{1 - [1 - (\frac{R_*}{r})^2]^{0.5}\}, \qquad (8.48)$$

and since recombination follows the electron density $n_e(r)$, the change of ionization through the wind is mostly a function of $n_e(r)/W(r)$. This change of ionization also causes changes in the line strength distribution function through the wind, which have to be considered. It turns out that this can be done by evaluating first the line strength distribution function and the integral in equation (8.38) at a prespecified depth in the wind, for instance where $n_e(r)/W(r) = 10^{11} \text{cm}^{-3}$, and then by accounting for relative changes in the ionization as a function of radius by multiplication by a factor $(n_e/W)^\delta$ (see Abbott 1982, Pauldrach 1987, Pauldrach et al. 1994)

$$N_o\{\frac{dv/dr}{\rho}\}^{\alpha-1} \longrightarrow N_o\{\frac{dv/dr}{\rho}\}^{\alpha-1}[n_e/W]^\delta. \qquad (8.49)$$

In this way an excellent analytical representation of the numerical calculation of the sum in equation (8.38) as a function of depth in the wind is obtained. On the other hand, the functional form of the differential equation for the velocity field (8.41) is changed. Typical values for δ are $\delta = 0.05\ldots 0.1$.

Even more crucial than the influence of changes of the ionization through the wind are the deviations from the point source geometry of the stellar radiation irradiating the stellar wind. A stellar photosphere for a wind at, for instance, a radius of $r = 2R_*$ is not a point source but provides photons from a stellar surface of finite angular diameter. Castor, Abbott, & Klein 1975 showed that this corresponds to an additional correction factor to be applied to equation (8.49) of the form

$$CF(r, v, \frac{dv}{dr}) = \frac{1}{\alpha+1}\frac{1}{1-u}\frac{r^2}{R_*^2}[1 - \{1 - \frac{R_*^2}{r^2}(1-u)\}^{\alpha+1}] \qquad (8.50)$$

where

$$\frac{1}{u} = \frac{r\,dv}{v\,dr}. \tag{8.51}$$

Pauldrach, Puls & Kudritzki 1986 and Friend & Abbott 1986 were able to show that this "finite cone angle correction factor" changes the dynamics of radiation driven winds substantially. It is smaller than unity close to the photosphere, namely $CF \approx \frac{1}{\alpha+1}$ for $r \approx R_*$, but becomes larger than unity at larger radii and approaches unity from above at infinity (for an analytical discussion, see Kudritzki et al. 1989).

With the corrections accounting for changes in the ionization through the wind and the finite cone angle of the stellar surface, the former differential equation (8.41) for the velocity field now reads

$$r^2 v \frac{dv}{dr} = \frac{L}{\dot{M}^\alpha} \frac{N_o}{c^2} (4\pi)^{\alpha-1} \{r^2 v \frac{dv}{dr}\}^\alpha [\frac{n_e}{W}]^\delta CF(r,v,\frac{dv}{dr}) - GM_*(1-\Gamma). \tag{8.52}$$

This equation is by far more complicated than the previous one and was therefore not treated by Castor, Abbott, & Klein 1975. However, Pauldrach, Puls, & Kudritzki 1986 were later able to obtain numerical solutions showing that the shape of the velocity field changes the exponent $\beta = 0.5$ to $\beta = 0.7\ldots 1.0$ and that the terminal velocities and mass-loss rates are modified significantly. Kudritzki et al. 1989 developed analytical solutions and scaling relations for $\dot{M}$ and v_∞ of the form

$$\dot{M} \propto L^{1/\alpha'} \{M_*(1-\Gamma)\}^{1-1/\alpha'} [\frac{g_*}{R_*}]^{\delta/2\alpha'} \tag{8.53}$$

and

$$v_\infty \approx 2.24 \frac{\alpha}{1-\alpha} v_{\text{esc}}, \tag{8.54}$$

where α' is given by

$$\alpha' = \alpha - \delta. \tag{8.55}$$

This leads to a modified formula for the wind momentum rate

$$\dot{M} v_\infty \propto \frac{1}{R_*^{0.5}} L^{1/\alpha'} \{M_*(1-\Gamma)\}^{3/2-1/\alpha'} [\frac{g_*}{R_*}]^{\delta/2\alpha'}. \tag{8.56}$$

However, since both exponents $3/2 - 1/\alpha'$ and $\delta/2\alpha'$ are close to zero the **Wind Momentum - Luminosity Relationship of equation (8.47) remains valid.**

8.3. *The observed Wind Momentum - Luminosity Relationship in the Galaxy*

Using the spectroscopic techniques described in section 3 a first sample of galactic supergiants of spectral types A, B and O has been analyzed with regard to stellar parameters, mass-loss rates and terminal velocities (Puls et al. 1996, Kudritzki et al. 1997). As an example a typical fit of an H_α-profile of a galactic A-supergiant is shown in Fig. 86. Except for the small bump in the P-Cygni absorption trough (which we attribute to time dependent perturbations as frequently encountered in A-supergiant winds, see Kaufer et al. 1996a) the observed profile is very well represented by the unified model atmosphere calculation allowing a very precise determination of $\dot{M}$.

Similar fits as displayed in Fig. 86 for a total of 19 galactic targets in a luminosity

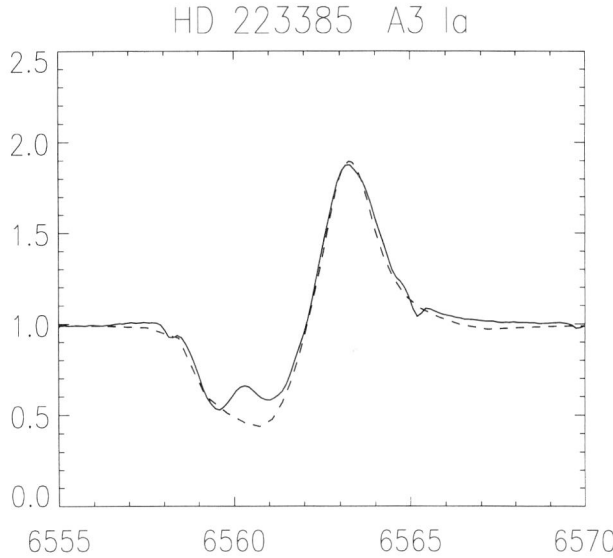

FIGURE 86. H_α-profile fit of the galactic A3 Iae supergiant HD 223385 with a unified stellar wind model adopting $T_{\rm eff} = 8400K$, $\log g = 0.8$, $R/R_\odot = 180$ and $\dot{M} = 5.5 \times 10^{-7} M_\odot/yr$, $v_\infty = 200$ km/s.

range between 1×10^5 to $3 \times 10^6 L_\odot$ then lead to the **empirical Wind Momentum - Luminosity Relationship shown in Fig. 87.** The relationship is tight with relatively small scatter. It confirms – in principle – the theoretical predictions of sections 8.1 and 8.2. However, there is an indication that the slope for A- and B- supergiants is steeper than that for O- supergiants. We obtain ≈ 2.5 for the former and ≈ 1.5 for the latter corresponding to $\alpha' = 2/5$ and $2/3$, respectively. We are presently investigating whether the atomic physics and thermodynamics in the cooler winds of later spectral types can explain these smaller values of α'. In any case, the definite existence of an observed Wind Momentum - Luminosity Relationship for the Blue Supergiants has encouraged us (the Tenerife - Pasadena - Munich connection) to continue with the project and to start the first crucial tests beyond our own galaxy.

8.4. A first investigation in M31

The obvious next step to test the reliability of the Wind Momentum - Luminosity Relationship for extragalactic distance determinations is to investigate Blue Supergiants in the Andromeda Galaxy M31. For this purpose, we are using the WHT and the HIRES spectrograph (Vogt et al. 1994) on the 10m Keck I telescope. With Keck, it is possible to obtain spectra of similar quality with respect to spectral resolution and signal-to-noise in M31 (or M33) as was previously only possible in the Galaxy and the Magellanic Clouds (see also the discussion in section 7). We have, therefore, undertaken a systematic spectroscopic study of A-supergiants in M31 and M33 to determine wind momenta and to test the WLR method. The first results for two targets in M31 have been published very recently by McCarthy et al. 1997 and are presented here in the following.

The two A-supergiants 41–3654 and 41–3712 were selected as initial targets on the basis of an earlier, medium resolution survey by Herrero et al. 1994, who confirmed

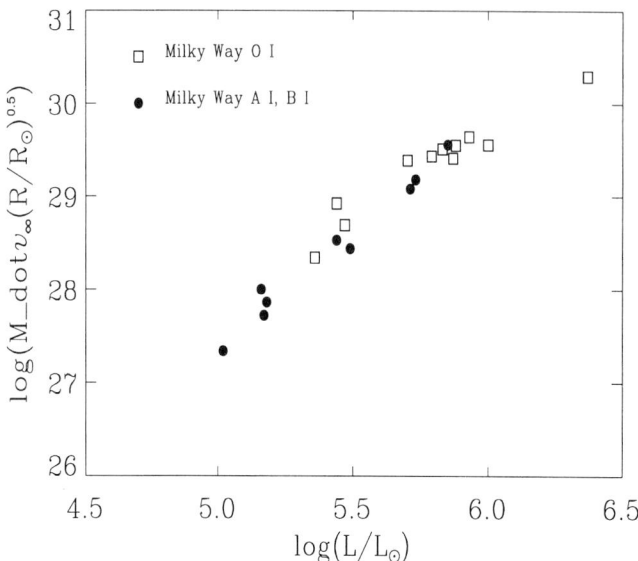

FIGURE 87. The observed Wind Momentum - Luminosity Relationship of A-, B- and O-supergiants in the Galaxy. All mass-loss rates have been determined from H_α. The terminal velocities are derived from the blue edges of UV P-Cygni profiles for the O- and B-supergiants and from H_α for the A-supergiants (data from Puls et al. 1996 and Kudritzki et al. 1997).

that they are very likely isolated stars because all lines present are consistent with a single stellar temperature. They are located 10.2 kpc away from the centre of the galaxy, where analyses of the emission line spectra of H II regions and supernovae remnants in M31 indicate that at this galactocentric distance abundances are solar (Blair, Kirshner, & Chevalier 1982; Blair & Kirshner 1985). The photometric parameters and spectral types are given in Table 2.

The sequence of H_α profile fits of 41-3654 shown in the next figures gives an impression as to how precisely the wind momentum can be determined. Fig. 88 demonstrates the influence of $\dot{M}$ on the $H\alpha$ profile. It affects mostly the strength of the emission, whereas the effect on the absorption is weaker. If all other parameters (v_∞, β, v_{turb}, see section 3) were known, $\dot{M}$ could be determined to an accuracy better than 10%.

A very important physical process when fitting spectral line profiles of A-supergiants very close to the Eddington limit is incoherent scattering of line photons by free electrons. As we all know from our first lectures in physics, this scattering process – normally referred to as "Compton scattering" – is incoherent in a coordinate frame comoving with the electron. On the other hand, for optical photons with energies $h\nu$ very small compared with the rest energy of the electrons $m_e c^2$ the photon is scattered elastically with energy and frequency remaining constant in the electron's frame ("Thomson scattering"). However, since the electrons have thermal velocities, they move relative to the observer and this can lead to Doppler shifts of the line photons

object	41–3654	41–3712
V	16.47	16.19
B − V	+0.29	+0.29
Sp. Type	A2 Ia-O	A3 Ia-O
T_{eff} (K)	8600 ± 200	8500 ± 200
$\log g$	0.9 ± 0.15	0.9 ± 0.15
$E(B-V)$	0.26	0.24
V_0	15.66	15.44
M_V	−8.59	−8.8
$R/R_\odot$	230	265
$\log L/L_\odot$	5.41	5.51
$v \sin i$ (km/s)	37 ± 5	33 ± 6
v_{rad} (km/s)	-90.34 ± 0.86	-116.43 ± 0.48

TABLE 2. Basic stellar properties of the two A-supergiants in M31 studied by McCarthy *et al.* 1997.

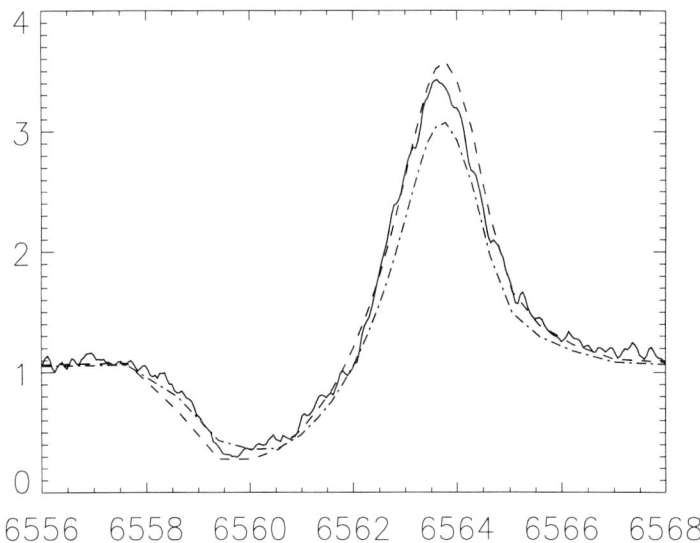

FIGURE 88. The influence of $\dot{M}$ on the Hα profile. Two models with $\dot{M} = 1.65$ and $2.15 \times 10^{-6} M_\odot$/year (dashed-dotted, dashed) and otherwise identical parameters are shown superimposed on the observed profile of 41–3654. As in all the other plots theoretical profiles are rotationally broadened by $35\,\mathrm{km\,s^{-1}}$. From McCarthy *et al.* 1997.

$$\frac{\Delta \nu}{\nu} = \frac{v_e}{c} \cos\theta, \quad (8.57)$$

where θ is the angle between the direction of the scattered photon and the velocity vector of the electron. Since the electron mass is very small, the thermal velocities $\langle v_e^{therm} \rangle_{av} = \{\frac{2kT}{m_e}\}^{1/2}$ are high and the redistribution of line photons can occur over a large range in wavelength (≈ 20Å).

The importance of incoherent electron scattering is seen in Fig. 89. There are emission wings observed far out in the profile, which can only be reproduced if incoherent electron scattering is taken into account. However, it is essential to realize that this scattering process does not take place in the outer wind, where the peak of the H_α emission occurs. It is a photospheric effect caused by the extremely low gravity, which enhances the influence of Thomson scattering relative to bound-free and free-free absorption, and by the overpopulation of the higher levels relative to the n=2 level of hydrogen. The latter leads to a local photospheric emission line core, which is then scattered by the free photospheric electrons into the far wings. The observed central emission lobe (and the blue absorption) are formed far out in the wind and are unaffected by the incoherent scattering process, because the electron optical depth is much too small. This means that the widely used approximation of obtaining the scattering wings by folding the strong P-Cygni profile emerging from the wind through an additional purely electron scattering layer of significant optical depth (Bernat & Lambert 1978, Stahl & Wolf 1980, Wolf et al. 1981, Wolf & Stahl 1982) is entirely incorrect when applied to A-supergiants, since no significant electron scattering layer is left above the wind layers that contribute to the emission and the blue part of the absorption. Unfortunately, the correct numerical radiative transfer treatment of incoherent electron scattering is rather complex and requires some computational effort. On the other, with present day workstations and clever programming it is possibly to include this important effect routinely in the quantitative spectroscopic analysis. Santolaya-Rey et al. 1997 give a short review in their paper.

As discussed in sections 3 and 4, signatures of chaotic motion on mesoscopic (optically thin) scales superimposed on the stellar wind outflow can be identified from the shapes of the P-Cygni profiles. This process is accounted for by introducing a microturbulence velocity v_{turb} in addition to the thermal velocity in the broadening function of the local line absorption coefficient. The significance of v_{turb} is demonstrated by Fig. 90. Its effect on the wavelength position of the emission peak was first detected and discussed by Hamann (Hamann 1980, Hamann 1981a, Hamann 1981b) in his investigation of UV metal resonance line P-Cygni profiles of O-stars. This is the first study using $H\alpha$ profiles to detect microturbulence in A-star winds.

The parameter β, which describes the slope of the stellar wind velocity field (see section 3) also affects the $H\alpha$ profile because it changes the density stratification according to the equation of continuity (section 8.1). A velocity field with a higher β is shallower, resulting in a higher density closer to the star and thus producing a higher emission peak. To determine β one has to calculate, for every β, a grid of models with different $\dot{M}$ and to adjust $\dot{M}$ such that the height of the emission profile is reproduced. Then the width of the emission and the shape of the absorption allow the determination of β, as demonstrated by Fig. 91.

Of course, the terminal velocity v_∞ also affects the width of the absorption and emission. To determine v_∞ one has to again adjust $\dot{M}$ for every selected value of v_∞ such that the height of the emission peak is always reproduced. In this way the width and shape of the calculated lines allows v_∞ to be found as illustrated by Fig. 92. Fortunately, although changes in v_∞ and β both affect the width and shape of the profile, their influence on the absorption shape is different so that both parameters can be uniquely determined.

In this way, all the parameters can be obtained quite accurately. Table 3 summarizes the results obtained by McCarthy et al. 1997. Fig. 93 gives an impression of the final H_α fit obtained for 41-3654. The results are extremely encouraging. They demonstrate that diagnostics of Blue Supergiant winds are possible with high precision in rather distant galaxies. In particular, the wind momenta are well determined and allow a comparison with A- and B-supergiants in the Galaxy (Fig. 94). It is very satisfying that the first

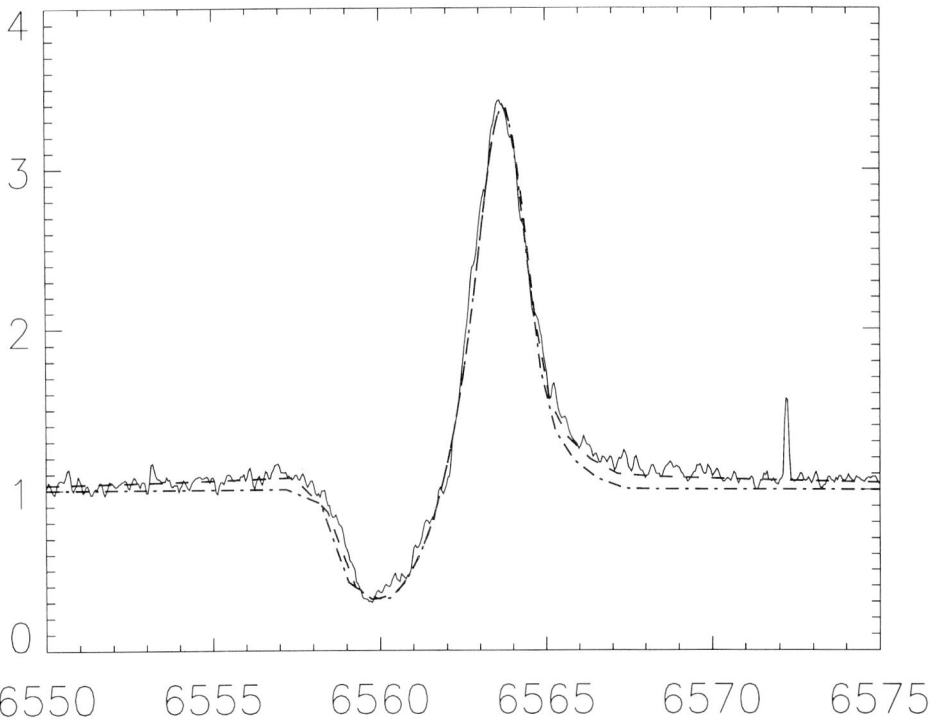

FIGURE 89. The influence of incoherent electron scattering on the Hα profile. The dashed profile includes incoherent electron scattering, whereas in the dashed-dotted profile it was neglected. Note that all other calculated profiles shown in the other figures of this subsection take incoherent electron scattering into account. (From McCarthy et al. 1997).

object	41–3654	41–3712
$\dot{M}$ $(10^{-6}\,M_\odot/\mathrm{y})$	1.90 ± 0.25	1.10 ± 0.20
v_∞ (km s^{-1})	200 ± 25	200 ± 25
β	$3.0^{+1.0}_{-0.5}$	2.0 ± 0.5
v_{turb} (km s^{-1})	20 ± 5	20 ± 5
$\log[\dot{M} v_\infty (R/R_\odot)^{0.5}]$ (cgs)	28.56 ± 0.11	28.35 ± 0.12

TABLE 3. Stellar wind parameters of the two A-supergiants in M31 studied by McCarthy et al. 1997.

two targets studied carefully in M31 lie just above and below the corresponding galactic WLR with deviations from the linear regression comparable to those of the galactic stars.

McCarthy et al. used the result displayed in Fig. 94 to investigate the accuracy in distance determination possible with the WLR method. For this purpose they trans-

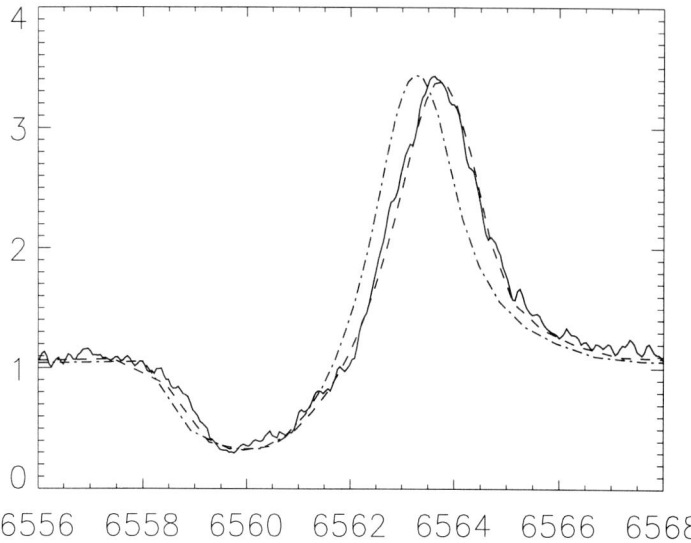

FIGURE 90. The influence of v_{turb} on the Hα profile. Two models with $v_{turb} = 0$ and $20\,\mathrm{km\,s^{-1}}$ (dashed-dotted, dashed) and otherwise identical parameters are shown superimposed on the observed profile of 41–3654. (From McCarthy et al. 1997).

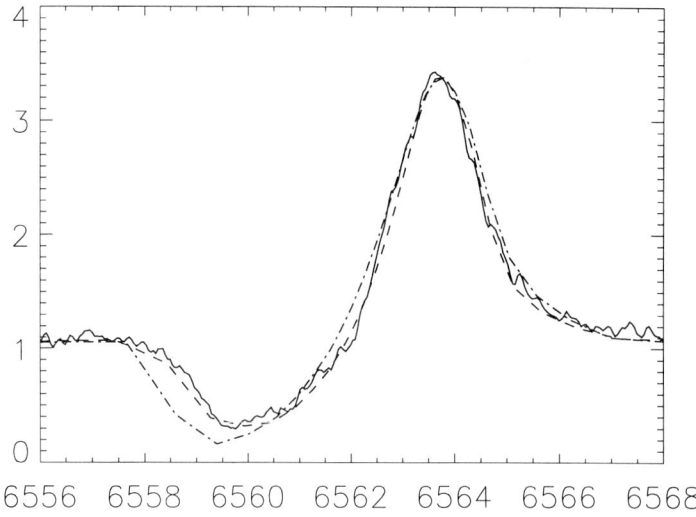

FIGURE 91. The determination of β. Two models with $\beta = 2.0$ and 3.0 (dashed-dotted, dashed) and $\dot{M}$ adopted to fit the height of the emission peak are shown superimposed on the observed profile of 41–3654. All other parameters are identical. (From McCarthy et al. 1997).

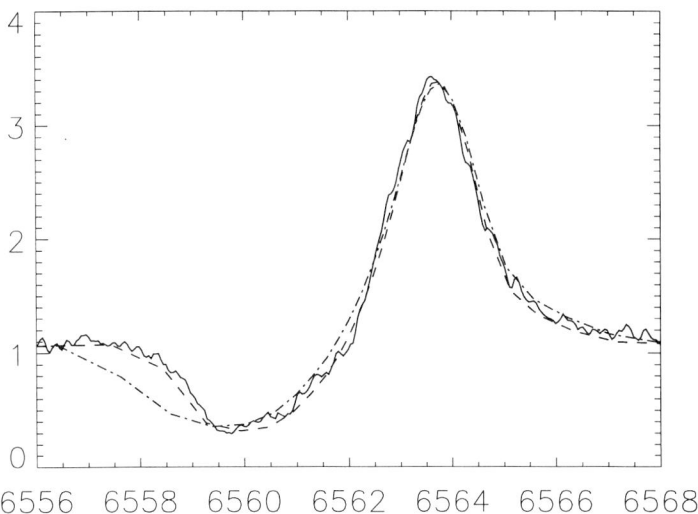

FIGURE 92. The determination of v_∞. Two models with $v_\infty = 200$ and $250\,\mathrm{km\,s^{-1}}$ (dashed, dashed-dotted) and $\dot{M}$ adopted to fit the height of the emission peak are shown superimposed to the observed profile of 41–3654. All other parameters are identical. (From McCarthy *et al.* 1997).

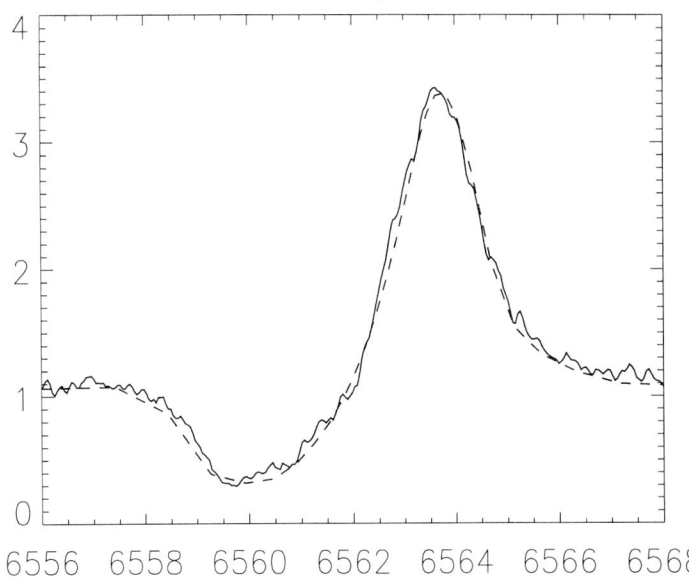

FIGURE 93. Hα profile of the final model of 41–3654 (dashed) compared with the observation. (From McCarthy *et al.* 1997).

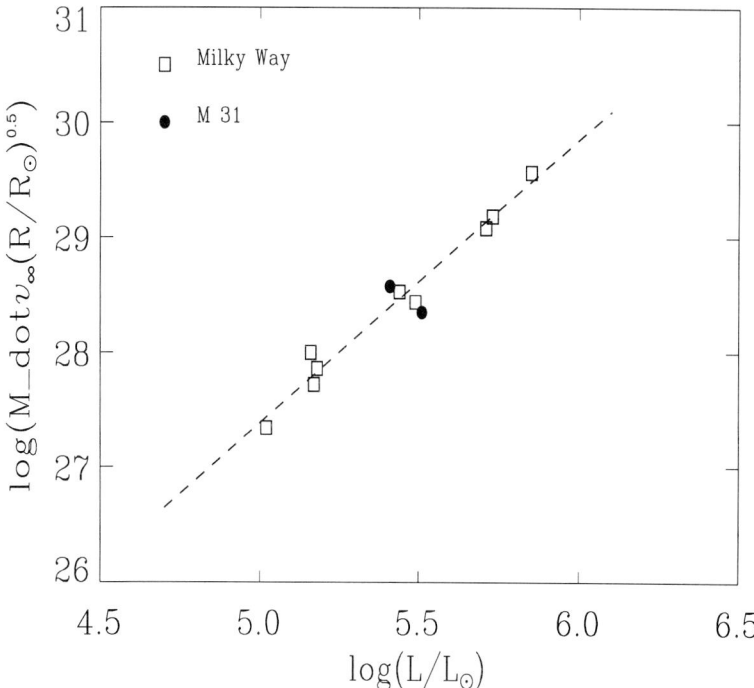

FIGURE 94. Wind momentum (in cgs units) and luminosity of the two A-supergiants in M31 compared with galactic A- and B- supergiants. The dashed curve is the linear regression for the galactic supergiants. (From McCarthy et al. 1997).

formed the deviations of 41–3654 and 41–3712 from the galactic regression curve into uncertainties in distance modulus or, in other words, they determined how the distance modulus of each target would need to be modified to shift it exactly onto the galactic WLR regression line in the Wind momentum – Luminosity plane. If the logarithm of the wind momentum is y and the logarithm of luminosity is x, then the linear regression of Fig. 94 is

$$y = ax + b \tag{8.58}$$

with $a = 2.467$ and $b = 15.050$.

Assuming a distance d_0 we have determined the values y_0 and x_0 for every target, thereby giving a residual in y of

$$\epsilon_0 = y_0 - ax_0 - b \tag{8.59}$$

Assuming a different distance $d_1 = \alpha \, d_0$ would change y_0 and x_0 to the new values:

$$x_1 = x_0 + 2\log(\alpha) \tag{8.60}$$
$$y_1 = y_0 + 2\log(\alpha) \tag{8.61}$$

since R changes proportionally to α and $\dot{M}$ changes proportionally to $R^{1.5}$ (see Puls et al. 1996). Requiring $\epsilon_1 = 0$ for the new distance, we derive from eqs. (8.58), (8.59), (8.60) and (8.61)

$$\log \alpha = \frac{\epsilon_0}{2(a-1)} \tag{8.62}$$

Bearing in mind that the change in distance modulus is $\Delta\mu = 5\log\alpha$, we obtain $\Delta\mu = +0.31$ and -0.50 mag for 41–3654 and 41–3712, respectively.

These are uncertainties in distance modulus that are comparable to those for individual Cepheids in galaxies if the Period – Luminosity Relation is applied. **Enlarging the number of observed supergiants with well determined wind momenta up to 20 would give a determination of distance modulus with an error of ± 0.1 mag or smaller**. The advantage of the supergiant method is that an individual reddening (and therefore extinction) can be derived for every object by comparing its observed colors with the model predictions.

8.5. Metallicity dependence. The Magellanic Clouds and the abundance gradient in M33

It is very obvious that the Wind Momentum - Luminosity Relationship must depend on metallicity. Since the winds are driven by photon momentum transfer through metal line absorption, the wind momentum rate must be a function of stellar metallicity. In the theoretical treatment of line driven winds, the chemical composition enters via equation (8.33) in section 8.1. This means that the line strength of an individual line becomes larger (or smaller) when the abundance becomes larger (or smaller). In consequence, when the integral in equation (8.38) over the line strength distribution function is carried out, the quantity N_o proportional to the number of lines with line strengths $k \geq 1$ becomes larger. As has been shown by Abbott 1982, Kudritzki et al. 1987 and Puls et al. 1996, this results in a metallicity dependence of $\dot{M}$ of the form

$$\dot{M} \propto [\frac{\epsilon}{\epsilon_\odot}]^{\frac{1-\alpha}{\alpha'}}, \qquad (8.63)$$

where ϵ is the average metallicity. The analysis of O-stars in the LMC and SMC by Puls et al. 1996 has indeed confirmed this metallicity dependence. Recent results for the WLR in the LMC and SMC also including wind momenta from A- and B-supergiants are shown in Figs. 95, 96. A small shift in wind momentum is found for the LMC, corresponding to an abundance difference of 0.3 dex (adopting $\alpha' = 0.56$ and $\delta = 0.07$, see Puls et al. 1996). For the SMC the shift is more significant indicating an abundance lower by 0.8 dex. The differences in abundance inferred from shifts in wind momentum are in rough agreement with the abundances obtained from the analysis of the spectra, as discussed in sections 7.1 and 7.2. However, there are still not enough objects studied in the Magellanic Clouds to allow a firm conclusion. An open question – only to be answered by an investigation of a larger sample – is, whether the exponents α and α' change with abundance. This work is presently under way.

For the applicability of the WLR method for extragalactic distance determinations the metallicity dependence does not pose a problem. We can use the Magellanic Clouds as an empirical metallicity calibrator of the relationship. Since we will, anyhow, have to take spectra to apply the method, we will always be able to determine abundances directly from the spectrum (see section 7). And since we will have an empirically calibrated metallicity dependence of the relationship, we will always be able – quite precisely – to determine the luminosity. In this sense, the method is intrinsically superior to the Cepheid-method, as the metallicity will be determined directly from the spectra of the objects used for the determination of distances.

For the determination of extragalactic distances using either Cepheids or Blue Supergiants it is important to realize that abundances may vary not only from galaxy to galaxy but also within one galaxy. The abundance gradients discussed in sections 7.2 and 7.3 are a typical example. Blue Supergiants in the outskirts of a spiral galaxy will

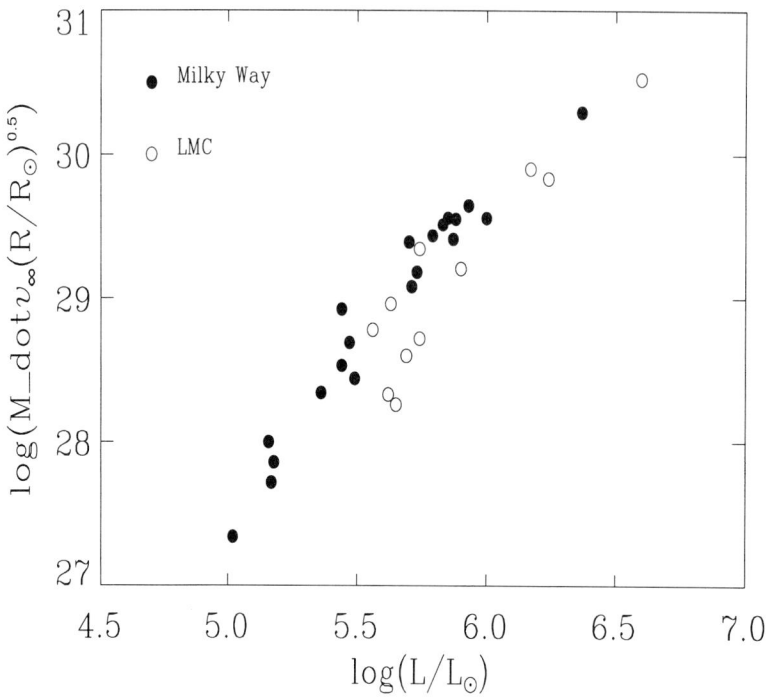

FIGURE 95. Wind momentum (in cgs units) and luminosity of A-, B- and O-supergiants in the Galaxy and in the LMC.

therefore have weaker winds than objects in the inner regions. The pulsation properties of Cepheids might differ as well.

To investigate the influence of abundance gradients on the WLR method we have started a project on Blue Supergiants in M33, a Local Group galaxy with a well pronounced abundance gradient identified in stars as well as in H II-regions (see section 7.3). The first results have been presented by McCarthy et al. 1995, where two A-supergiants have been studied, B 324 and M 117-A. In section 7.3 the determination of abundances for these two objects has already been discussed. B 324 located only 1.3 kpc away from the centre of M33 has roughly solar abundance, whereas M 117-A at 4.5 kpc galactocentric distance has a metallicity clearly below the SMC. The winds of these two objects are strikingly different. B 324 has a very strong H_α emission peak leading to $\dot{M} = 1.2 \times 10^{-5} M_\odot/yr$. M 117-A, on the other hand, has a strange H_α profile (see Fig. 97) with symmetric emission, but central absorption. As the calculations by McCarthy et al. 1995 show, the emission is purely photospheric, a NLTE effect caused by the fact that the object is close to the Eddington-limit. The depth of the central absorption, however, is influenced by the stellar wind and leads to a very low mass-loss rate of $\dot{M} = 2.5 \times 10^{-8} M_\odot/yr$. The corresponding wind momenta are shown in Fig. 98 and compared with the Galaxy and the SMC. As to be expected from the abundance analysis, the solar composition object coincides roughly with the galactic WLR. On the

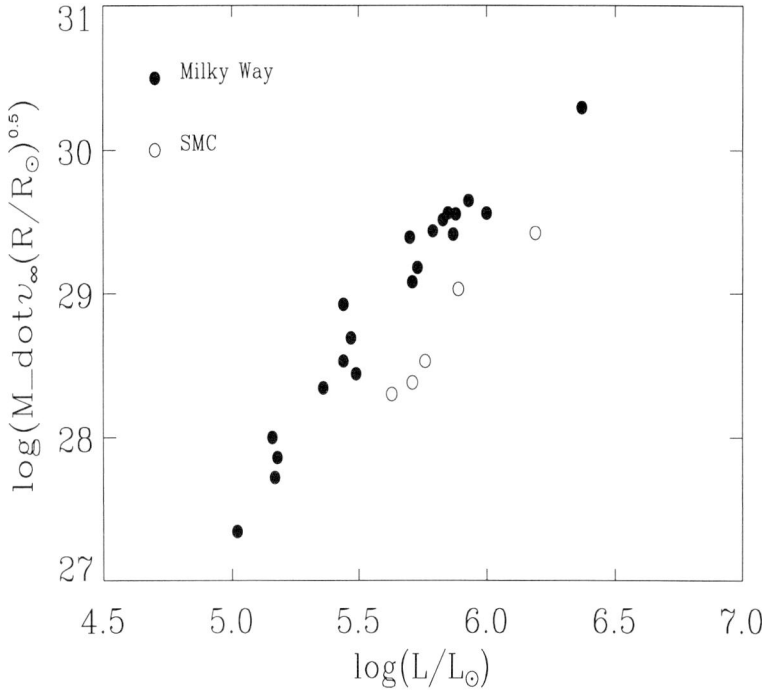

FIGURE 96. Wind momentum (in cgs units) and luminosity of A-, B- and O-supergiants in the Galaxy and in the SMC.

other hand, the extremely metal poor object falls below the SMC relationship, again in agreement with the expectation. **We conclude, that with an empirical calibration of wind momentum as a function of metallicity it will also be possible to apply the WLR method for distance determinations to spiral galaxies with a strong metallicity gradient.**

8.6. *The potential of the WLR-method*

From the previous subsections we have learned that the uncertainties in WLR distance moduli might be comparable to those obtainable for Cepheids in galaxies. The advantage of the WLR-method is that individual reddening (and therefore extinction) as well as metallicity can be derived directly from the spectrum of every object. In addition, it is a new independent primary method for distance determinations and can, therefore, contribute to the investigation of systematic errors of extragalactic distances. However, the crucial question is to what distances can the method be applied.

Objects like 41–3654 and 41–3712 in M31 would be of apparent magnitude 20 in galaxies 6 Mpc away, certainly not a problem for medium (2 Å) resolution spectroscopy with 8 m class telescopes. Even in a galaxy like M100 at 16 Mpc distance (Freedman *et al.* 1994b; Ferrarese *et al.* 1996) these objects would still be accessible at magnitudes around 22.5 and would yield wind momentum distances if medium resolution spectroscopy on

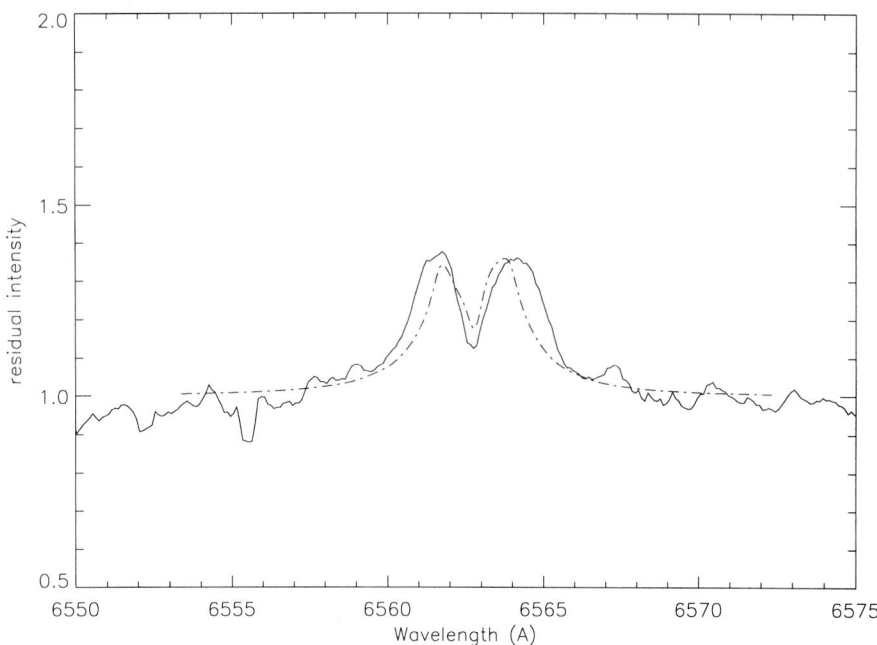

FIGURE 97. H_α profile of the extremely metal poor M33 A0 IaO-Hypergiant 117-A fitted with a unified model calculation assuming a very low mass-loss rate of $\dot{M} = 2.5 \times 10^{-8} M_\odot/\text{yr}$. (From McCarthy et al. 1995).

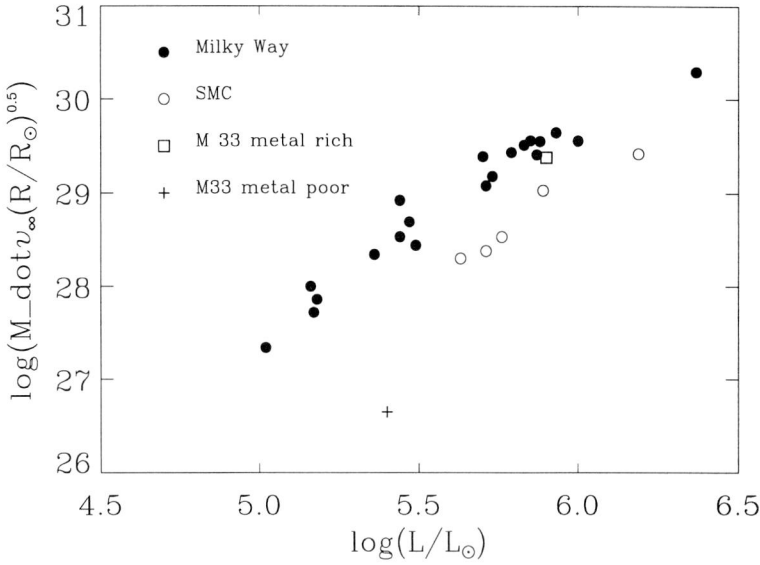

FIGURE 98. Wind momentum (in cgs units) and luminosity of the two A-supergiants in M33 compared with galactic and SMC supergiants. Open square: M33 A-supergiant with galactic metallicity. Cross: Extremely metal poor A-supergiant in the outskirts of M33. (From McCarthy et al. 1995).

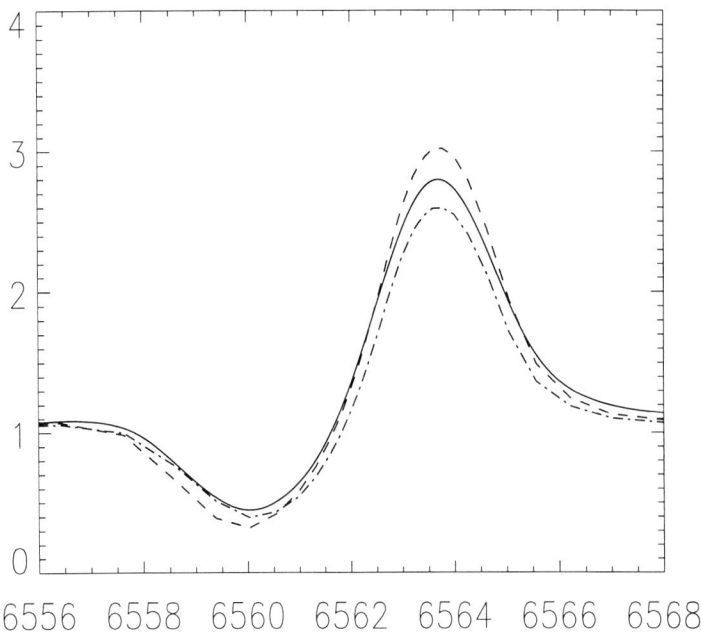

FIGURE 99. H$_\alpha$ fit of the A-supergiant 41-3654 as in Fig.87 but degraded to a spectral resolution of 3000 to demonstrate that also at medium resolution a determination of wind momenta is possible. (From McCarthy *et al.* 1997).

a 8 m class telescope were combined with HST photometry. Indeed, the HST colour magnitude diagram published by Freedman *et al.* 1994a may show the presence of such objects in M100.

One might ask, of course, whether a medium resolution of 2 Å would still allow a sufficiently accurate determination of effective temperature, surface gravity, abundances, and wind momentum. We note that the rotational velocities of A-supergiants are typically of the order of $40\,\mathrm{km\,s^{-1}}$, a broadening which is matched by a spectral resolution of roughly 1.4 Å at Hα. Therefore we expect that the medium resolution situation will not be dramatically worse, provided sufficient S/N can be achieved and accurate sky-, galaxy-, and H II region-background subtraction can be performed at very faint magnitudes. We present a simple illustration of this point in Fig. 99, which is a degradation of Fig. 88 to a resolution of 2 Å, from which we conclude that a determination of wind momentum will not be severely affected by the lower resolution.

In summary, the WLR results obtained so far in the Local Group galaxies M33 and M31 with the Keck I telescope are very encouraging. We are optimistic that after further tests and calibration steps, the WLR method can make an independent contribution to the determination of extragalactic distances. The obvious next steps for future work are:

- the addition of more objects with reliable distances to the galactic WLR,
- the investigation of the impact of spectral variability,

- further calibration of the WLR metallicity dependence using LMC and SMC supergiants,
- the addition of more targets in M31 and M33 including HST photometry and spectroscopy (the latter for B-supergiants to measure v_∞ from UV resonance lines) to obtain independent distances to these galaxies and to study their abundance gradients, and finally
- the investigation of blue supergiants beyond the Local Group out to the Virgo cluster of galaxies

Work in all of these directions is now under way. **We are confident that, in the end, we will have a new distance determination method capable of reaching out to the Virgo and Fornax clusters of galaxies.**

9. Extragalactic stellar astronomy – a vision

Presently, we are witnessing a revolution in observational astronomy. The refurbished Hubble Space Telescope is producing images of impressive quality with regard to spatial resolution and photometric accuracy. With its new imaging spectrograph STIS the HST will provide deep spectroscopy in crowded fields including the extremely important UV spectral range. It will be the most efficient spectroscopic UV telescope in the history of astronomy. The ROSAT observatory has opened new horizons in X-ray astronomy and following missions such as XMM and AXAF will enable us to do detailed quantitative X-ray spectroscopy. ISO is providing us with beautiful spectra and images in the far-IR, a new spectral window. The follow-up mission FIRST is close to being decided upon and the fantastic project of a Next Generation Space Telescope (NGST) – an infrared 8m-telescope in space – has entered a very serious phase of discussion.

Ground-based astronomy is also in a revolutionary phase. Adaptive Optics and Interferometry have been or are becoming reality. Imaging spectroscopy in the K-band with high spatial and medium spectral resolution allowing precise multi-object spectrophotometry has been introduced as a new observational tool. Optical multi-object spectrographs have become standard. The new generation of 8 to 10m class telescopes is in operation (Keck I and II, HET) or will soon be completed (ESO VLT, Subaru, Gemini, LBT). These telescopes are or will be equipped with extremely efficient faint object/multi-object spectrographs of low to medium resolution (and slitlets long enough to enable accurate subtraction of night sky lines) allowing quantitative spectroscopy down to $m_V = 23^m$.

There are many branches in astrophysics, where these new observational facilities will have an enormous impact. Among these, the extragalactic application of stellar physics or, as I call it, **"extragalactic stellar astronomy"** will certainly be one of the important directions of the future. With the new diagnostic tools from hydrodynamic model atmospheres in our hands, the use of individual Blue Supergiants to investigate the physics of other galaxies as far out as the Virgo or Fornax cluster looks like an astrophysical goldmine. The general strategy, how to exploit this goldmine, has been developed in the previous sections. However, as always, a general strategy is not enough. It is the overwhelming amount of concrete hard work on the details which in the end leads to a success of such a project or not. Here, the hard astronomical work at the beginning is the identification and selection of appropriate target candidates. This will be based on survey-like work using ground-based and HST (the extragalactic distance scale Key Project) photometry as well as ground-based low resolution multi-object spectroscopy. Many of the relatively brightest targets per galaxy will have to be abandoned, as they will turn out to be multiple either from their images or from the low resolution spectra. Our experience from the work in the Local Group spirals M31 and M33 (for instance, Herrero et al. 1994) is that, from a sample selected on the basis of ground-based (mostly aperture) photometry and low resolution spectra, one half of the targets showed traces of multiplicity in good quality medium resolution spectra and could not be used, whereas the remaining half was suitable for quantitative work. At the moment, we are beginning to gain further experience with preselection from HST WFPC2 images and low resolution spectroscopy using the Palomar 5m, the WHT and the ESO NTT. This will be the critical part of the project. Later on, quantitative multi-object spectroscopy with exposures of several hours using 8m-class telescopes of a well selected sample of non-multiple Blue Supergiants per galaxy will be straightforward and yield invaluable information about young stellar populations, chemodynamical evolution and distances. This kind of work will not be restricted to a small number of galaxies. We will be in a position to study a large number and a large variety of normal spirals of different type, metal poor irregulars, actively star forming or star bursting galaxies, etc. This study will yield – besides accurate distances – very detailed information about abundances ($\alpha-$, iron-group, s-process elements) in different locations of these galaxies. It will, therefore, allow us to investigate whether the old concept of chemical evolution of galaxies undergoing mixing on large scales is reliable. It will be fun to learn that it is not.

Acknowledgements

It is a pleasure to thank the organizers of the IAC Winter School, my co-lecturers and – last, but not least – the students for a wonderful week. This School has become famous among astronomers and I have now learned why. It is the unique mixture of perfect organization, hospitality, scientific and intellectual challenge, soccer matches and Cuba Libre, which is hard to forget.

The work presented in my lectures is the result of a collaboration between many colleagues and friends over many years. For their contributions I wish to thank Artemio Herrero and Ilu Monteverde from the IAC, Jim McCarthy and Barry Madore (Caltech), Wendy Freedman (Carnegie), Kim Venn (Munich, now Minnesota), Stephen Smartt (Belfast, now La Palma) and the Munich crowd (in alphabetical order) Sylvia Becker, Keith Butler, Achim Feldmeier, Alex Fullerton, Rudi Gabler, Stefan Haser, Margie and Danny Lennon, Paco Najarro, Ralf Palsa, Adi Pauldrach, Joachim Puls, Florian Sellmaier, Uwe Springmann and Gudrun Taresch. Special thanks go to Uli Hopp for identifying hundreds of inconsistencies in the draft manuscript with his sharp extragalactic eyes.

Appendix A. Oscillator strengths distribution function of hydrogen

In section 8.1 we have argued that in stellar winds driven by line absorption the distribution function of line strengths is a power law reflecting mostly the atomic physics of oscillator strengths. To make this more plausible we use the hydrogen atom as an example (closely following the derivation given by Puls 1993).

The oscillator strengths for transitions to the hydrogen ground-state are approximately given by ("Kramers formula")

$$f_{1j} = \frac{32}{33^{0.5}\pi}(1-\frac{1}{j^2})\frac{1}{j^3} \approx \frac{A}{j^3}, j \gg 1. \tag{A 64}$$

Considering a highest main quantum number j_{max}, the corresponding oscillator strength is

$$f_{1j_{max}} = \frac{A}{j_{max}^3} \tag{A 65}$$

and is smaller than all the other f-values corresponding to lower j

$$f_{1j_{max}} \leq f_{1j}, j \leq j_{max} \tag{A 66}$$

Taking into account the selection rules we compute the total number of transitions into the ground-state from all j with $2 \leq j \leq j_{max}$ as

$$n = j_{max} - 1 \tag{A 67}$$

which means that the number $n(f \geq f_{1j_{max}})$ of oscillator strengths larger than the value corresponding to $f_{1j_{max}}$ is (we use equation A67 and replace j_{max} by equation A65)

$$n(f \geq f_{1j_{max}}) = A^{1/3}f_{1j_{max}}^{-1/3} - 1 = \int_{f_{1j_{max}}}^{\infty} dn(f). \tag{A 68}$$

By differentiating this expression we obtain

$$\frac{dn}{df} \propto f^{-4/3} = f^{\alpha-2}, \alpha = \frac{2}{3} \tag{A 69}$$

This means that the hydrogen oscillator strengths obey a power law with $\alpha = 2/3$. It is not completely surprising that the statistics of millions of other lines of other ions at the end lead to comparable exponents.

REFERENCES

ABBOTT D.C. 1979, Proc. IAU Symp. **83**, eds. P.S. Conti & C.W.H. de Loore, p.237

ABBOTT D.C. 1980, ApJ **242**, 1183

ABBOTT, D.C. 1982, ApJ **259**, 282

ARNABOLDI, M., FREEMAN, K.C., HUI, X., CAPACCIOLI, M., FORD, H. 1994, ESO Messenger **76**, 40

ARNABOLDI, M., FREEMAN, K.C., MENDEZ, R.H., CAPACCIOLI, M., CIARDULLO, R., FORD, H., GERHARD, O., HUI, X., JACOBY, G.H., KUDRITZKI, R.P., QUINN, P.J. 1996, ApJ **472**, 145

BAADE D., LUCY L.B. 1987, A&A **178**, 213

BAADE, D. (ED.) 1991, "Rapid Variability of OB-Stars: Nature and Diagnostic Value", ESO Conference and Workshop Proceedings No. 36

BERKHUIJSEN, E.M., HUMPHREYS, R.H., GHIGO, F.D., & ZUMACH, W. 1988, A&AS **76**, 65

BERNAT, A. P. & LAMBERT, D.L. 1978, PASP **90**, 520

BIANCHI, L., LAMERS, H., HUTCHINGS, J.B., MASSEY, P., KUDRITZKI, R.P., HERRERO, A., & LENNON, D.J. 1994, A&A **292**, 213

BLAIR, W.P., KIRSHNER, R.P., & CHEVALIER, R.A. 1982, ApJ **254**, 50

BLAIR, W.P., & KIRSHNER, R.P. 1985, ApJ **289**, 582

CARLBERG R.G. 1980, ApJ **241**, 1131

CASSINELLI J.P., SWANK J.H. 1983, ApJ **271**, 681

CASSINELLI J.P., COHEN D.H., MACFARLANE J.J., SANDERS W.T., WELSH B.Y. 1994, ApJ **421**, 705

CASTOR, J.I. 1970, MNRAS **149**, 111

CASTOR, J.I., ABBOTT, D.C., & KLEIN, R.I. 1975, ApJ **195**, 157

CHIOSI, C. 1997, this volume

CHEVALIER R.A., IMAMURA J.N. 1982, ApJ **261**, 543

CHLEBOWSKI T., HARNDEN F.R., SCIORTINO S. 1989, ApJ **341**, 427

CONTI., P.S., BURNICHON, M.L. 1975, A&A **38**, 467

DALTABUIT E., COX D. 1972, ApJ **173**, L13

DREW, J.E. 1989, ApJ Suppl. **71**, 267

DRISSEN, L., LEITHERER, C., NOTA, A. (EDS.) 1992, "Nonisotropic and Variable Outflows from Stars", ASP Conf. Series, Vol. 22

DRISSEN, L., MOFFAT, A.F.J., WALBORN, N.R., SHARA, M.M. 1995, AJ **110**, 2235

ECKART., A., GENZEL, R., HOFMANN, R. et al. 1993, ApJ Letters **407**, L77

ECKART, A., GENZEL, R., HOFMANN, R., SAMS, B.J., TACCONI-GARMAN, L.E. 1995, ApJ Letters **445**, L23

ECKART, A., GENZEL, R. 1996a, NATURE, 383, 415

ECKART, A., GENZEL, R. 1996b, MNRAS (in press)

FELDMEIER A. 1995, A&A **299**, 523

FELDMEIER A., KUDRITZKI R.P., PALSA R. et al., 1997, A&A **320**, 899

FERRARESE, L. et al. 1996, ApJ **464**, 568

FIGER, D.F., MCLEAN, I.S., NAJARRO, F. 1997, ApJS, in press

FREEDMAN, W., & MADORE, B. 1990, ApJ **365**, 186

FREEDMAN, W., et al. 1994a, ApJ **435**, L31

FREEDMAN, W. et al. 1994b, Nature **371**, 757

FRIEND, D.B., ABBOTT, D.C. 1986, ApJ **311**, 701

GABLER, R., GABLER, A., KUDRITZKI, R.P., PAULDRACH, A.W.A., & PULS, J. 1989, A&A **226**, 162

GABLER, A., GABLER, R., KUDRITZKI, R.P., PULS, J., PAULDRACH, A. 1990 Proc. 1st Boulder-Munich Workshop, PASPC **7**, p. 218

GABLER, R., KUDRITZKI, R.P. & MENDEZ, R.H. 1991, A&A **245**, 587

GABLER, R., GABLER, A., KUDRITZKI, R.P. & MENDEZ, R.H. 1992, A&A **265**, 656

GARMANY, C.D., CONTI, P.S. 1984, ApJ **284**, 705

GOULD, A. 1994, ApJ **426**, 524

GROENEWEGEN, M.A.T., & LAMERS, H.J.G.L.M. 1989, A&A **79**, 359

HAMMAN, W.R. 1980, A&A **84**, 342

———. 1981a, A&A **93**, 353

———. 1981b, A&A **100**, 169

HAMANN, W.R. & KOESTERKE, L. 1996, Proc. 33rd Liege International Astrophysical Colloquium 'WR stars in the framework of stellar evolution", ed. J.M. Vreux, invited paper, p. 491

HANSON, M.M., CONTI, P.S., RIEKE, M.J. 1996, ApJS **107**, 281

HARNDEN F.R., BRANDUARDI G., ELVIS M. et al., 1979, ApJ **234**, L51

HASER, S.M., LENNON, D.J., KUDRITZKI, R.P., PULS, J., PAULDRACH, A., BIANCHI, L., & HUTCHINGS, J.B. 1995a, A&A **295**, 136

HASER, S.M. 1995, thesis University of Munich

HASER, S.M., PAULDRACH, A., LENNON, D.J., KUDRITZKI, R.P., LENNON, M., PULS, J., VOELS, S.A. 1997, A&A submitted

HERRERO, A., KUDRITZKI, R.P., VILCHEZ, J.M., KUNZE, D., BUTLER, K. & HASER, S.M. 1992, A&A **261**, 209

HERRERO, A., LENNON, D.J., VILCHEZ, J.M., KUDRITZKI, R.P., & HUMPHREYS, R.H. 1994, A&A **287**, 885

HILLIER D.J., KUDRITZKI R.P., PAULDRACH A.W. et al., 1993, A&A 276, 117

HILLIER, D.J. 1996, Proc. 33rd Liege International Astrophysical Colloquium 'WR stars in the framework of stellar evolution", ed. J.M. Vreux, invited paper, p. 509

HILLIER, D.J., MILLER, D.L. 1997, ApJ, submitted

HOWARTH, I. 1983, MNRAS **203**, 301

HOWARTH, I.D., PRINJA, R.K. 1989, ApJ Suppl. **69**, 527

HUI, X., FORD, H.C., FREEMAN, K., & DOPITA,M. 1995, ApJ **449**, 592

HUMPHREYS, R.M., MASSEY, P., & FREEDMAN, W.L. 1990, AJ **99**, 84

JACOBY, G.H., CIARDULLO, R., FORD, H.C., 1990, ApJ **356**, 332

JACOBY, G.H., BRANCH, D., CIARDULLO, R., ET AL. 1992, PASP **104**, 533

JACOBY, G.H., CIARDULLO, R. 1993, Proc. *IAU Symp. 155 'Planetary Nebulae"*, Kluwer, eds. Weinberger, R., Acker, A., page 503, invited paper

JACOBY, G.H. 1995, Proc. 'Science with the VLT", eds. J.R. Walsh and I.J. Danziger, Springer Verlag, page 267, invited paper

KAPER, L., HENRICHS, H. F., NICHOLS, J. S., SNOEK, L. C., VOLTEN, H., & ZWARTHOED, G. A. A. 1996, A&AS **116**, 257

KAUFER, A., STAHL, O., WOLF, B., GÄNG, TH., GUMMERSBACH, C.A., KOVÁCS, J., MANDEL, H., & SZEIFERT, TH. 1996a, A&A **305**, 887

KAUFER, A. ET AL. 1996, A&A **314**, 599

KLEIN, R.I, CASTOR, J.I. 1978, ApJ **220**, 902

KRABBE, A., GENZEL, R., DRAPATZ, S., ROTACIUC, V. 1991, ApJ Letters **382**, L19

KRABBE, A., GENZEL, R., ECKART, A., NAJARRO, F., LUTZ, D., CAMERON, M., KROKER, H., TACCONI-GARMAN, L.E., THATTE, N., WEITZEL, L., DRAPATZ, S., GEBALLE, T., STERNBERG, A., KUDRITZKI, R. 1995, ApJ Letters **447**, L95

KUDRITZKI, R.P. 1980, A&A **85**, 174

KUDRITZKI, R.P., PAULDRACH, A.W.A., PULS, J. 1987, A&A **173**, 293

KUDRITZKI, R.P. 1988, in 18th Advanced Course of the Swiss Society of Astrophysics and Astronomy (Saas-Fee Courses) Radiation in Moving Gaseous Media", eds. Y. Chmielewski and T. Lanz, published and sold by Geneva Observatory, pages 1 - 192, invited paper

KUDRITZKI, R.P., PAULDRACH, A., PULS, J. 1988, "Radiation driven winds of hot luminous stars" in; *O-stars and Wolf-Rayet stars* eds. P.S. Conti & A.B. Underhill, NASA Sp-497, p. 173-198

KUDRITZKI, R.P., PAULDRACH, A.W.A.,, PULS,J., & ABBOTT, D.C. 1989, A&A **219**, 205

KUDRITZKI, R.P., HUMMER, D.G. 1990, Annu. Rev. Astron. Astrophys. **28**, 303 [A

KUDRITZKI, R.P. 1992, A&A **266**, 395

KUDRITZKI, R.P., HUMMER, D.G., PAULDRACH, A.W.A., PULS, J., NAJARRO, J., & IMHOFF, J. 1992, A&A **257**, 655

KUDRITZKI, R.P., LENNON, D.J., HASER, S.M., PULS, J., PAULDRACH, A., VENN, K., & VOELS, S.A. 1996a, in *Science with the Hubble Space Telescope II* eds. P. Benvenuti et al., 285–296

KUDRITZKI, R.P., PALSA, R., FELDMEIER, A., PULS, J., PAULDRACH, A. 1996, Proc. of "Röntgenstrahlung from the Universe", eds. H.U. Zimmermann et al., MPE Report 263, page 9

KUDRITZKI, R.P., PULS, J., LENNON, D.J., VENN, K.A., MCCARTHY, J.K., & HERRERO, A. 1997, A&A, in preparation.

LAMERS, H., CERRUTI-SOLA, M., PERINOTTO, M. 1987, ApJ **314**, 726

LAMERS, H., LEITHERER, C. 1993, ApJ **412**, 771

LAMERS, H., SNOW, T.P., LINDHOLM, D. ApJ **455**, 269

Lamers, H., Najarro, F., Kudritzki, R.P. et al. 1996, A&A Letters **315**, L229

LANGER, N., HAMANN, W.R., LENNON,D.J. ET AL. 1994, A&A 290, 819

LANZ, T., DE KOTER, A., HUBENY, I., HEAP, S.R. 1996, ApJ **465**, 359

LEITHERER, C. 1988, ApJ **326**, 356

LENNON, D.J., KUDRITZKI, R.P., BECKER, S.T., BUTLER, K. EBER, F., GROTH, H.G., & KUNZE, D. 1991, A&A **252**, 498

LENNON, D.J., DUFTON, P.L., FITZSIMMONS, A. 1992, A&AS **94**, 569

———. 1993, A&AS **97**, 559

LENNON, D.J. 1997, A&A **317**, 871

LUCY, L.B., SOLOMON, P. 1970, ApJ **159**, 879

LUCY L.B. 1982, ApJ **255**, 286

LUCY L.B. 1984, ApJ **284** 351

LUCY L.B., WHITE R.L. 1980, ApJ **241**, 300

MACFARLANE J.J., CASSINELLI J.P. 1989, ApJ **347**, 1090

MADORE, B. 1997, this volume

MASSA, D. et al. 1995, ApJ **452**, L53

MCCARTHY, J.K. 1990. In 2nd *ESO/ST-ECF Data Analysis Workshop*, ed. D. Baade & P.J. Grosbøl (Garching: European Southern Observatory), 119.

MCCARTHY, J.K., LENNON, D.J., VENN, K.A., KUDRITZKI, R.P., PULS, J., & NAJARRO, F. 1995, ApJ **455**, L35

MCCARTHY, J.K., & NEMEC, J.M. 1996, ApJ, in press

MCCARTHY, J.K., KUDRITZKI, R.P., LENNON, D.J., VENN, K.A., PULS, J. 1997, ApJ in press

MENDEZ, R.H., KUDRITZKI, R.P., JACOBY, G.H., CIARDULLO, R. 1993, A&A **275**, 534

MOFFAT, A. F. J., OWOCKI, S. P., FULLERTON, A. W., ST-LOUIS, N. (EDS.) 1994, "Instability and Variability of Hot-Star Winds", Dordrecht: Kluwer, also Ap&SS, Vol. **221**

MONTEVERDE, I., HERRERO, A., LENNON, D.J., & KUDRITZKI, R.P. 1996, A&A **312**, 24

———. 1997, ApJ (Letters), in press

MORRIS, M. 1993, ApJ **408**, 496

NAJARRO, F., HILLIER, D.J., KUDRITZKI, R.P. et al., 1994, A&A **285**, 573

Najarro, F. 1995, PhD thesis, University of Munich

NAJARRO., F., KUDRITZKI, R.P., CASSINELLI, J.P., STAHL, O., HILLIER, J.D. 1996, A&A **306**, 892

NAJARRO, F., HILLIER, D.J., STAHL, O. 1997a, A&A, submitted

NAJARRO, F., HILLIER, D.J., KUDRITZKI, R.P. et al. 1997b, A&A, submitted

OWOCKI S.P. 1991, in: Crivellari L., Hubeny I., Hummer D.G. (eds.) Stellar atmospheres: beyond classical models. Kluwer, Dordrecht, p. 235

OWOCKI S.P., RYBICKI G.B. 1984, ApJ **284**, 337

OWOCKI S.P., CASTOR J.I., RYBICKI G.B. 1988, ApJ **335**, 914

PALLAVICINI R., GOLUB L., ROSNER R. et al., 1981, ApJ 248, 279

PALSA, R. 1997, PhD thesis, University of Munich, in preparation

PALSA, R., KUDRITZKI, R.P., FELDMEIER, A., PULS, J., PAULDRACH, A., HILLIER, J.D. 1997, in prep. for A $ A

PAULDRACH, A.W.A., PULS, J., & KUDRITZKI, R.P. 1986, A&A **164**, 86

PAULDRACH, A.W.A. 1987, A&A **183**, 295

PAULDRACH, A.W.A., KUDRITZKI, R.P., PULS, J., BUTLER, K., & HUNSINGER, J. 1994, A&A **283**, 525

PAULDRACH, A.W.A., SELLMAIER,F., ET AL. 1997, A&A in preparation

PULS, J. 1987, A&A **184**, 227

PULS, J. 1993, Habilitationsschrift, University of Munich

PULS, J., KUDRITZKI, R.P., HERRERO, A., PAULDRACH, A., HASER, S.M., LENNON, D.J., GABLER, R., VOELS, S.A., VILCHEZ, J.M., WACHTER, S., & FELDMEIER, A. 1996, A&A **305**, 171

RAYMOND J.C., SMITH B.W. 1977, ApJS 35, 419

RICHER, M.G. 1993, ApJ **415**, 240

RICHER, M.G., MCCALL, M.L. 1995, ApJ **445**, 642

ROZANSKI, R., & ROWAN-ROBINSON, M. 1994, MNRAS **271**, 530

RYBICKI, G.B., HUMMER, D.G. 1978, ApJ **219** 654

RIVINIUS, TH. ET AL. 1997, A&A **318**, 819

SAHA, A., SANDAGE, A., LABHARDT, L., SCHWENGELER, H., TAMMANN, G.A., PANAGIA, N., & MACCHETTO, F.D. 1995, ApJ **438**, 8

SANTOLAYA-REY, E., PULS, J., & HERRERO, A. 1997, A&A, in press

SCHAERER,D., SCHMUTZ, W. 1994, A&A **288**, 231

SCHAERER,D., DE KOTER, A. 1997, A&A in press

SCHALLER, G., SCHAERER, D., MEYNET, G., MAEDER, A. 1992, A&A, Suppl. **96**, 269

SCUDERI, S., BONONNO, G. ET AL. 1992, ApJ **392**, 201

SEATON, M. 1979, MNRAS **187**, 73p

SELLMAIER, F., PULS,J., KUDRITZKI, R.P., GABLER, A., GABLER, R., VOELS, S.A. 1993, A&A **273**, 533

SELLMAIER, F., YAMAMOTO, T., PAULDRACH, A.W.A., RUBIN, R.H. 1996, A&A Letters **305**, L37

SELLMAIER, F. 1996, thesis, University of Munich

SEWARD F.D., FORMAN W.R., GIACCONI R. et al., 1979, ApJ **234**, L55

SIMON M., AXFORD W.I. 1966, Planet. Space Sci. **14**, 901

SIMON, K.P., JONAS, G., KUDRITZKI, R.P., RAHE, J. 1983, A&A **125**, 34

SMARTT, S.J., ROLLESTON, W.R. 1997, ApJ Letters, submitted

STEIDEL, C.C., GIAVALISCO, M., PETTINI, M., DICKINSON, M., ADELBERGER, K.L. 1996, ApJ., Letters, in press

SOFFNER, T., MENDEZ, R.H., JACOBY, G.H., CIARDULLO, R., ROTH, M.M., KUDRITZKI, R.P. 1996, A&A **306**, 9

SOFFNER, T., MENDEZ, R.H. 1997, A&A, in press

SPRINGMANN, U. 1997, thesis, Munich University

SPRINGMANN,U., PULS, J., PAULDRACH, A., KUDRITZKI, R.P. 1997, A&A, submitted

STAHL, O., WOLF, B. 1980, A&A **90**, 338

STAHL, O., AAB, O., SMOLINSKI, J.,& WOLF, B. 1991, A&A **252**, 693

STIFT,M.J. 1995, A&A **301**, 776

TARESCH, G., KUDRITZKI, R.P., HURWITZ, M. ET AL. 1997, A&A in press

WALBORN, N.R., NICHOLS-BOHLIN, J., PANEK, R.J. 1985, "IUE Atlas of O-type Spectra", NASA Ref. Publ. 1155

WALBORN, N.R., LENNON, D.J., HASER, S.M., KUDRITZKI, R.P., VOELS, S.A. 1995a, PASP **707**, 104

WALBORN, N.R., PARKER, J., NICHOLS, J.S. 1995b, "IUE Atlas of B-type Spectra", NASA Ref. Publ. 1363

WOLF, B., STAHL, O., DE GROOT, M., STERKEN, C. 1981, A&A **99**, 351

WOLF, B., STAHL, O. 1982, A&A **112**, 111

VAN DEN BERGH, S. 1964, ApJS **9**, 65

VENN, K.A. 1994, ApJS **99**, 659

VENN, K.A. 1995, ApJ **449**, 839

VENN, K.A. 1997a, in Advances in Stellar Evolution, eds. R.T. Rood & A. Renzini, Cambridge University Press

VENN, K.A. 1997b, in Luminous Blue Variables: Massive Stars in Transition, eds. A. Nota & H. Lamers, Astronomical Society of the Pacific

VENN, K.A. 1997c, A&A, in prep.

VOGT, S.S., et al. 1994, Proc. SPIE, 2198, 362

Calibration of the Extragalactic Distance Scale

By BARRY F. MADORE[1], WENDY L. FREEDMAN[2]

[1]NASA/IPAC Extragalactic Database, Infrared Processing & Analysis Center, California Institute of Technology, Jet Propulsion Laboratory, Pasadena, CA 91125, USA

[2]Observatories, Carnegie Institution of Washington, 813 Santa Barbara St., Pasadena CA 91101, USA

The calibration and use of Cepheids as primary distance indicators is reviewed in the context of the extragalactic distance scale. Comparison is made with the independently calibrated Population II distance scale and found to be consistent at the 10% level. The combined use of ground-based facilities and the *Hubble Space Telescope* now allow for the application of the Cepheid Period-Luminosity relation out to distances in excess of 20 Mpc. Calibration of secondary distance indicators and the direct determination of distances to galaxies in the field as well as in the Virgo and Fornax clusters allows for multiple paths to the determination of the absolute rate of the expansion of the Universe parameterized by the Hubble constant. At this point in the reduction and analysis of Key Project galaxies $H_o = 72$ km/sec/Mpc ± 2 (random) ± 12 [systematic].

1. Introduction to the lectures

Even some of the earliest tentative steps out of our Milky Way galaxy and into the local Universe were guided by Cepheid variables. Indeed, the very first convincing demonstration of the size of our Universe rested on the identification by Edwin Hubble (1929) of Cepheid variables in the nearest galaxies NGC 6822, M31 and M33. Armed with empirical knowledge of the relationship between luminosity and period, discovered for Cepheids by Henrietta Leavitt only a few years earlier, Hubble used the Cepheids to set the scale size for galaxies themselves, and for the Universe in which they are expanding away from each other.

This series of lectures will detail the role played by Cepheids in the calibration of the extragalactic distance scale. In preparation for that application the physics of Cepheids will be outlined and their empirical calibration will be presented. The properties of individual Cepheids varying through their pulsational cycle will be detailed, and the structure of the instability strip for the time-averaged properties of Cepheids will be discussed. Concerns about the metallicity sensitivity of the Period-Luminosity relation will be dealt with in a number of ways, from theoretical considerations acting as a guide, to direct tests and finally intercomparisons of independent distance determinations to serve as external checks. Means of determining distances freed from the effects of interstellar reddening and extinction will also be outlined.

Secure with the foundations provided by Cepheid variables and an application of their period-luminosity relation, we move on the recent use of the *Hubble Space Telescope* in the rapid determination of distances to 'nearby' galaxies useful for calibrating so-called 'secondary distance indicators' needed to extend the distance scle out past the perturbing effects of the Virgo cluster and well out to cosmologically significant distances unaffected by peculiar motions and large-scale flows.

We digress slightly en route to highlight a new distance indicator that is fundamentally different from and yet absolutely complementary to the Cepeids as a distance indicator, the tip of the red giant branch (TRGB) method. The TRGB is well understood to mark

2. Cepheids

These lectures will first concentrate on our developing understanding of the very basic principles characterizing the variability seen in individual Cepheids as well as the systematics and trends observed for Cepheids as an ensemble. The emphasis will be on Cepheids as accurate extragalactic distance indicators, and will only touch upon the debate concerning the exact value of the zero point later when the recent results from the Hipparcos satellite are discussed. Previous work on the zero point of the Cepheid period-luminosity relation is adequately covered in recent commentaries by Feast & Walker (1987), Walker (1988) and Schmidt (1991) all of which suggest that there is a convergence of opinion at the level of about ±0.10 mag. As optimistic as these reports are, readers are still referred to the cautioning remarks by Turner (1990).

We begin by presenting a physical basis from which to view Cepheids, as distance indicators. (However, for dissenting views on this whole process interested readers are referred to Clube & Dawe 1980, and to Stift 1982, 1990). We then go on to discuss the difficulties and uncertainties caused by reddening and metallicity effects. Following sections put into perspective the explicit determination of reddening made possible by employing panoramic and long-wavelength detectors. We then review the status of the Cepheid-based distances to Local Group galaxies and then link in to those beyond the Local Group being observed by the *Hubble Space Telescope*. The impact of these distances on a determination of the Hubble constant concludes the lectures. Later sections look at the differences between the PLC and the PL relation, discuss an implicit method for dealing with reddening, and finally, examine the prospects for determining reddenings to individual Cepheids.

For historical notes on Cepheids as distance indicators the interested reader is referred to the monographs by Fernie (1969), Sandage (1972, 1988a), Stothers (1983), Walker (1988), Madore (1986) and Tanvir (1997) and references therein. Other more recent reviews of Cepheids in the context of the extragalactic distance scale can be found in Madore (1985), Freedman (1988), Feast & Walker (1987), Madore & Freedman (1991), Jacoby et al. (1992), and Freedman & Madore (1996). A review by Hodge (1981) on the extragalactic distance scale is itself especially relevant in as much as many of the major topics of concern raised by him regarding the Cepheid PL relation have now largely been addressed at least by direct observations. The evolutionary status of Cepheids is most recently reviewed by Chiosi (1990), while Simon (1990) gives a detailed look at the convergence of techniques used to calibrate the Galactic PL relation, and the confrontation of these observations with pulsation theory. And finally, Chiosi, Wood & Capitanio (1993) have published theoretical PL relations which they then map to both optical and near-infrared wavelengths in a modern attempt to illucidate the effects of metallicity on those zero points.

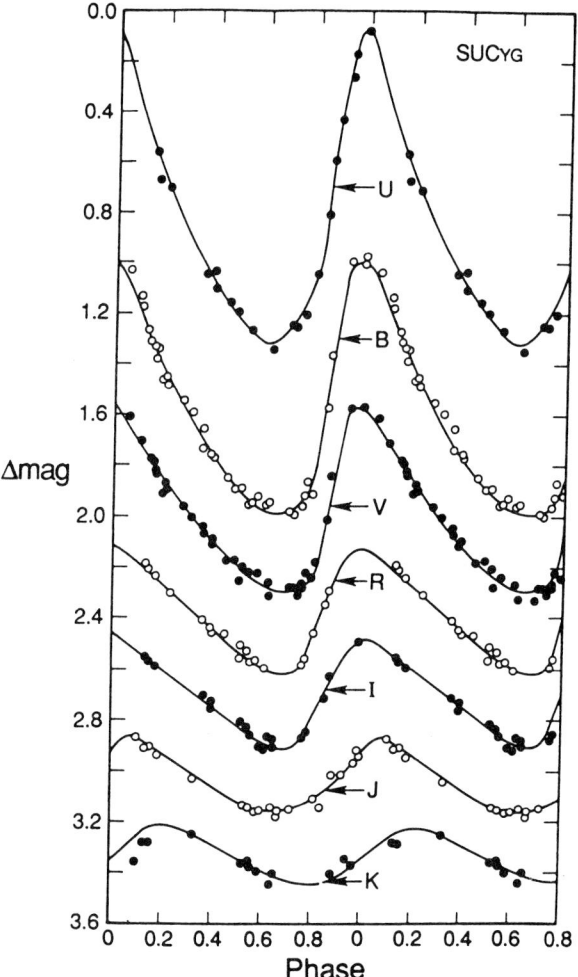

FIGURE 1. Variations of amplitude and phase for a typical Galactic Cepheid as a function of increasing wavelength. Note the monotonic drop in amplitude, the progression toward more symmetric light variation, and the phase shift of maximum toward later phases, all with increasing wavelength.

3. Brief summary of the observed properties of cepheid variables

At the time of this writing, Cepheids have now been surveyed for and discovered in over two dozen galaxies. Including surveys of the Magellanic Clouds, light curves for approximately 3,000 Cepheids have been obtained. Over 1,100 Cepheids are known in each of the Magellanic Clouds. These numbers are rapidly increasing as additional galaxies are being surveyed with the *HST*, and as the MACHO survey for microlensing in fields toward the Large Magellanic Cloud (Cook 1996) progresses.

Cepheids are high-luminosity, radially-pulsating, variable stars. Their spectral types range from early F to late K (see Code 1947; Kraft 1960); while their intrinsic brightnesses at visual wavelengths range from $-2 > M_V > -7$ mag making them ideal candidates for distance indicators on Galactic and extragalactic scales. The periods of Population I classical Cepheids range from approximately 2 to over 100 days. Cepheids are characterized by their distinctive light curves having a rapid rise to maximum light, followed by a slower linear decline to minimum light. Due to their variability they are easily isolated, identified and classified. Below a period of about 9 days (*e.g.*, Bohm-Vitense 1994) Cepheids are observed to pulsate in both the fundamental, in addition to the first harmonic (overtone) mode. But perhaps most importantly for the distance determinations, detailed stellar pulsation models of these objects indicate that we understand the basic physics underlying their luminosities, colors and periods. In this regard, Cepheids appear to stand alone in the extragalactic distance scale; however we shall argue in this series of lectures that there is a second, fully complementary path using the brightest Population II red giant stars that has similar accuracy and precision to the Population I Cepheid route.

Although Cepheids exhibit strong correlations between their periods, luminosities and colors, the amplitudes of Cepheids do not appear to correlate with other observables. Rather, examples of Cepheids having amplitudes from 0.4 to about 1.5 mag in the V band can be found at almost all periods, with the greatest upper bound on the amplitudes occurring preferentially around 25–30 days (*e.g.*, see Schaltenbrand & Tammann 1970). On the other hand, for an individual Cepheid the monochromatic amplitude is observed to decrease toward redder wavelengths (see Figures 1 and 2). As a practical matter, the large amplitudes at bluer wavelengths suggest that Cepheid searches should be carried out at shorter wavelengths (Freedman, Grieve & Madore 1985; Madore & Freedman 1991).

Finally, in searching for Cepheid variables, it is useful to know their frequency relative to other types of variables of comparable luminosity. Averaging over the discovery statistics for complete surveys in fields published before the launch of HST, for the SMC, LMC, M31 and NGC 300, Madore & Freedman (1985) found that more than 70% of all variables reported are classical Cepheids, 16% are irregular variables, 5% are Mira-type long-period variables, 7% are eclipsing variables, and the remaining 2% are W Virginis stars.

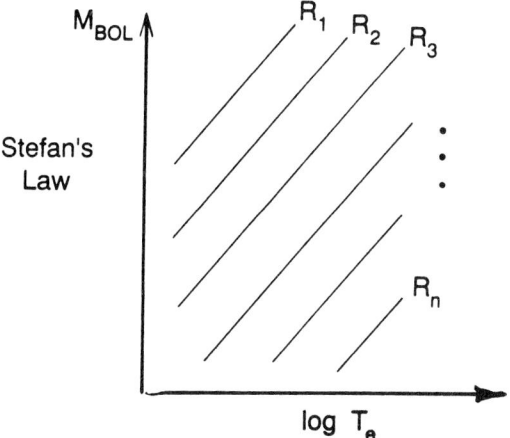

FIGURE 2. Stefan's Law expressed in graphical form projected onto the theoretical $M_{BOL} - logT_e$ plane where loci of constant radius are indicated by upward sloping lines.

4. Simple physical considerations

The basic physics connecting the luminosity and color of a Cepheid to its period is well understood. Using Stephan's law
$$L = 4\pi R^2 \sigma T_e^4$$
the bolometric luminosities L of all stars, (including Cepheids), can be derived. The radius R is a geometric term, parameterizing the total emitting surface area $4\pi R^2$ and the effective temperature T_e is a thermal term, used to parameterize the areal surface brightness, given by σT_e^4. Expressed in magnitudes, Stephan's Law becomes
$$M_{BOL} = -5\, logR - 10\, logT_e + C$$
and it is schematically shown in Figure 2. It should be noted that *the entire $M_{BOL} - log\, T_e$ plane is mapped by* Equation (2), and that within the context of this relation alone, for any value of $log\, T_e$, an unbounded range of radius R is independently possible. However, once R and T_e are each specified, M_{BOL} is uniquely defined. Additional constraints, outside of Stephan's Law itself, must be involved if bounds on permitted values of the independent geometric and thermal variables are to be considered. An important constraint is provided by stellar evolution. In terms of timescales and allowed equilibrium configurations, the core-hydrogen-burning main sequence is one of the most striking and well known examples of such a 'constraint' on populating the HR diagram. Hydrostatic equilibrium can be achieved for long periods of time along the hydrogen-burning main sequence, and as a result we are constrained to observe most of the stars there most of the time. There are of course other constraints.

For mechanical systems it is well known that $P\rho^{1/2} = Q$, where Q is a structural constant, and P is the natural free pulsation period, determined by gravity through ρ, the mean density of the system, in turn defined by $M = 4/3\pi R^3 \rho$, where M is the total

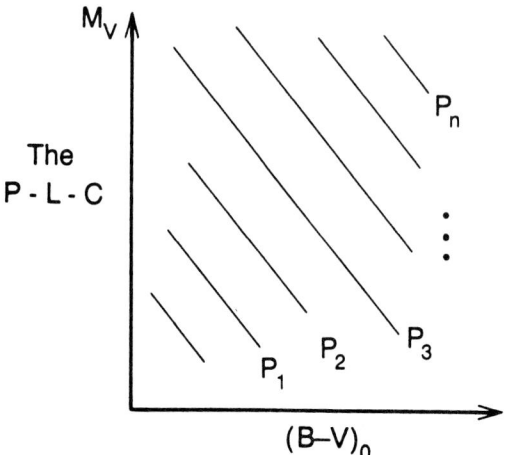

FIGURE 3. The PLC relation expressed in graphical form as projected onto the observational $M_V - (B-V)$ plane, where loci of constant period are downward sloping lines.

mass of the system. If it is assumed that mass is predominantly a function of R and T_{eff}, then the pulsation period can be used as the second observable parameter instead of requiring the radius to be observed directly.

If we then linearly map $\log(T_e)$ into an observable intrinsic color (*i.e.*, $B-V$), and map radius into an observable period, we thereby predict a new two-parameter description of the luminosity of (pulsating) stars. This is precisely the physical basis for the period-luminosity-color (PLC) relation for Cepheids, as was so elegantly introduced and explained over a quarter of a century ago (Sandage 1958, Sandage & Gratton 1963, Sandage & Tammann 1968). In its linearized form for pulsating variables, Stefan's law takes on the following form of the PLC: $M_V = \alpha \log P + \beta (B-V)_o + \gamma$. In analogy to plotting Stephan's Law in the theoretical $M_{BOL} - \log T_e$ above, Figure 4 shows the PLC mapped into the observational $M_0 - (B-V)_0$ color-magnitude diagram. Again the entire plane is mapped. To see how the PLC relates to the more commonly referred to relations (the PL and PC relations) the reader is referred to a more detailed discussion in the sections ahead, the caption to Figure 4, and the Cepheid section in Jacoby et al. (1992) where an empirical analog of Figure 4 is given.

Stars can be found evolving across many parts of the color-magnitude diagram. However, only in very narrowly defined regions do they become pulsationally unstable. Cepheid pulsation in particular occurs because of the changing atmospheric opacity with temperature in the helium ionization zone. This zone acts like a heat engine and valve mechanism, alternately trapping and then releasing energy, thereby periodically forcing the outer layers of the star into motion against the restoring force of gravity. Not all stars are unstable to this mechanism. The cool (red) edge of the Cepheid instability strip is thought to be controlled by the onset of convection, which then prevents the helium ionization zone from driving the pulsation (see Baker & Kippenhahn 1965; and Deupree 1977 for references). For hotter temperatures a blue edge is encountered when the helium ionization zone is found too far out in the atmosphere for significant pulsations to

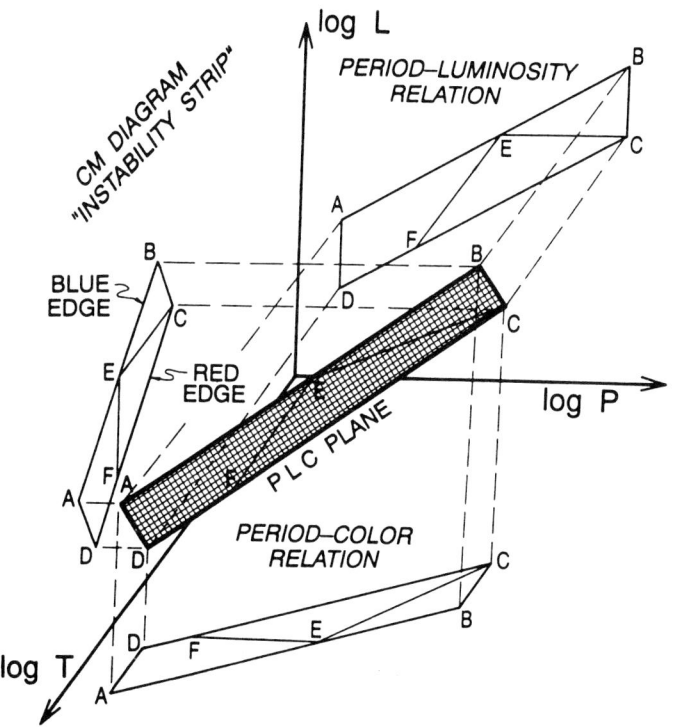

FIGURE 4. The Cepheid Manifold: Projections of the PLC plane (shown shaded) onto the three principal co-ordinate systems (luminosity [L], increasing up, period [log P], increasing to the right and color [(B–V)] becoming bluer to the lower left). The backward projection onto the L-P plane gives the period-luminosity relation. Projecting to the left gives the position of instability strip within the color-magnitude diagram. Projecting down gives the period-color relation.

occur. Further details and extensive references can be found in the monograph on stellar pulsation by Cox (1980).

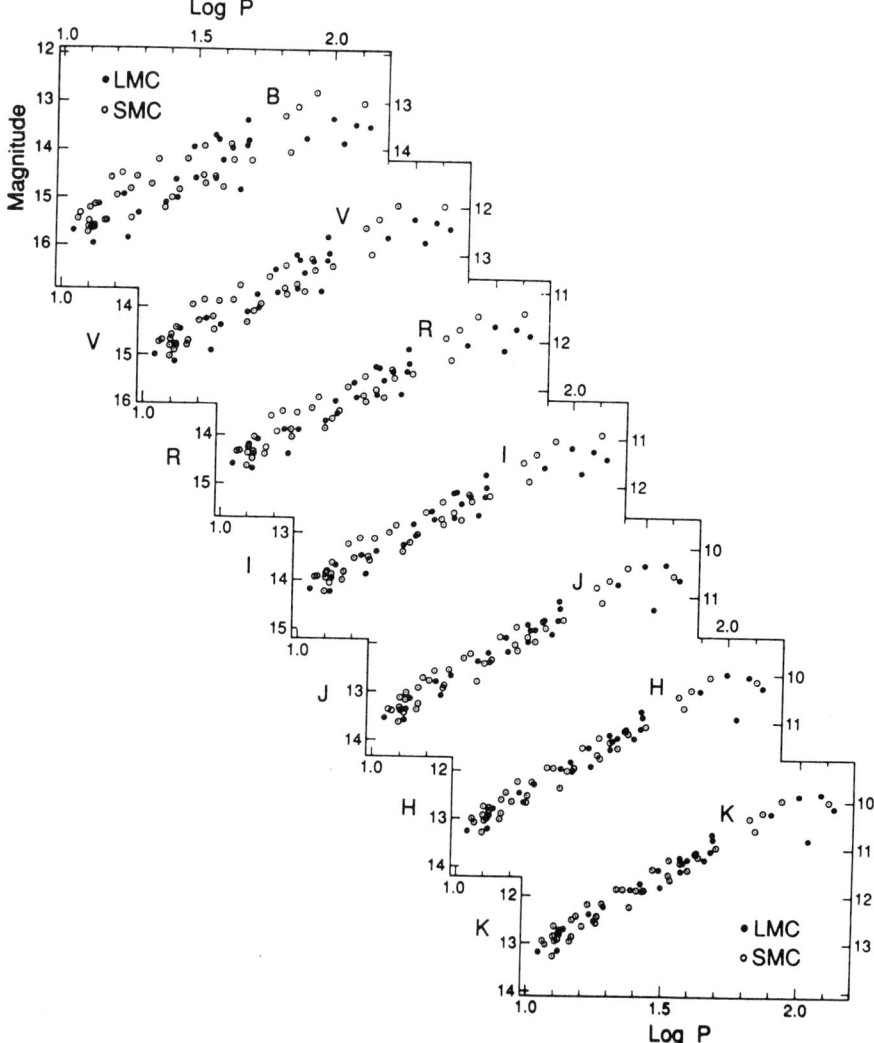

FIGURE 5. Magellanic Cloud Cepheid period-luminosity relations at seven wavelengths, from the blue to the infrared, constructed from a self-consistent data set (Madore & Freedman 1991). LMC Cepheids are shown as filled circles; SMC data, shifted to the LMC modulus, are shown as open circles. Note the decreased width and the increased slope of the relations as longer and longer wavelengths are considered.

5. Observational considerations

5.1. General Issues

By the 1960's the instability strip had been observed to have the following general properties: periods for Cepheids ranged from several days to a few hundred days; at constant period, the B magnitude total width of the PL relation was about 1.2 mag; the V magnitude width was measured to be about 0.9 mag; and the $(B-V)$ color width was found to be about 0.4 mag, with the reddest Cepheids being the faintest at any given period. In a practical sense, this meant that in estimating distances, any individual Cepheid could deviate from the statistical ridge line by up to ± 0.6 mag in B; and such an error (if applied to one Cepheid) would translate into an equivalent error of about 30% in distance. Large samples can decrease the error on the apparent modulus inversely with the square root of the number; a formal error of only 10% being possible with a sample containing as few as a dozen Cepheids.

The discussion in the preceding section concerned an idealized PLC relation, expressed in its linearized form. Some of the difficulties encountered in the empirical calibration of this relation will be discussed in the present section. We concentrate on extragalactic studies. And so we do not discuss the issue of how best to determine independent distances and independent reddenings for the Galactic population of classical Cepheids: as plentiful as they are Galactic Cepheids in the field are problematic, while only a handful of (short-period) Cepheids are contained in open clusters, which are generally sparsely populated, and often heavily obscured. The interested reader is however referred to the recent papers by Fernie (1990), Fernie & McGonegal (1983), Feast & Walker (1987) and Jacoby et al. (1992) for a pathway into the literature on this Galactic approach to the calibration.

Because they are nearby, and because of the large numbers of Cepheids cataloged in them (Payne-Gaposchkin 1971, Payne-Gaposchkin & Gaposchkin 1966), the Magellanic Clouds have long been the testbed for calibrations of the period-luminosity and period-luminosity-color relations. Indeed, the original discovery of these relations was made in the Magellanic Clouds (Leavitt 1906). These same large samples allowed the first estimates of the slope of the PL relation and first approximations to the period and color dependences of the PLC. Now with a number of analyses of the geometric expansion parallax to the LMC via the remnant of Supernova 1987A there are geometric distance modulus estimates of 18.50 ± 0.13 mag (Panagia et al. (1991), 18.38 ± 0.07 mag (Schmidt-Kaler 1992), 18.52 ± 0.13 mag (McCall 1993), 18.61 ± 0.11 mag (Crotts, Kunkel & Heathcote 1995), $<18.37 \pm 0.04$ mag (Gould & Uza 1997) for the LMC. While the precision of this measurement might be expected to improve with time as the expansion continues to be monitored the systematics (assumptions about ring geometry and placement of the SN with respect to the main body of the LMC, etc) will soon dominate the solution. Nevertheless, it is reassuring that the majority of the determinations agree extremely well with alternative estimates of the LMC distance modulus, and that an independent check of the zero point (at the 10-15% level) is provided by the measurement of RR Lyrae distances (e.g., Reid & Freedman 1994), which also are in good agreement with the Cepheids (for more recent reviews see Westerlund 1990, 1997 and references therein).

For work on the distance scale, the existence of a statistical relation between period and luminosity is of such great utility by itself, that it is of little wonder that concern about second-order effects in the calibration (i.e., those aspects above and beyond establishing the slope and zero point) were **not** of immediate import in the earliest studies. Some of these issues are: the origin of the scatter in the PL relation; the systematic effects of reddening; the systematic effects of metallicity (*e.g.*, Gascoigne 1974; Stothers 1983; Freedman & Madore 1990), companions (*e.g.*, Madore 1977; Coulson, Caldwell & Gieren 1986); CNO abundance, helium abundance and mass loss (*e.g.*, Lauterborn, Refsdal & Weigert 1971; Lauterborn & Siquig 1974; Cox, Michaud & Hodson 1978; Becker & Cox 1982); magnetic fields (*e.g.*, Stothers 1982); the possibility of curvature in the PL relation (*e.g.*, Fernie 1967, Sandage & Tammann 1968, 1969); the relative disposition and the slopes of the red and blue edges of the instability strip (Fernie 1990), and the physical origin of these constraints (*e.g.*, Iben & Tuggle 1972a,b; 1975; Chiosi, Wood, Bertelli & Bressan 1991), *etc.* Unfortunately, several of the corrections to be considered are probably manifesting themselves simultaneously, and at the same level of numerical significance.

Before we can approach an empirical determination of the coefficients in the PLC (or any determination of their variation with metallicity) we must solve the reddening problem. While theory predicts a finite width to the instability strip (with temperature/color being the controlling parameter), and while metallicity is a quantity that is known to be different from galaxy to galaxy, (and it is known to vary systematically within individual galaxies) only when reddening has been accounted for, can we go on to look for meaningful correlations of luminosity residuals with intrinsic color and/or metallicity, for instance. To decouple and solve for the effects of metallicity, reddening, and intrinsic temperature variations, high precision photometry is a prerequisite, and at least as many independently measured quantities are needed as there are parameters to be determined.

5.2. *Reddening*

Interstellar grains, within our Galaxy, along the line of sight to a nearby galaxy, or within the galaxy being studied, will each result in light being selectively scattered and absorbed. If any one of these components of extinction is not accounted for, a Cepheid in an external galaxy will appear fainter and more distant than it actually is, and at the same time it will appear redder and cooler than it truly is. Systematic errors will thereby creep into the distance scale.

Accounting for the Galactic foreground component associated with dust in the plane of our own Milky Way is relatively straightforward and will not be discussed here in any detail. The use of foreground stars and/or reddening maps, generated from galaxy counts in combination with neutral hydrogen studies (see Burstein& Heiles 1984) appear to be quite reliable and are widely used. Since most external galaxies subtend an angular size small in comparison to expected variations of extinction across the line of sight, and since most extragalactic studies are also done at fairly high Galactic latitudes, these foreground Galactic extinction corrections are relatively small and of low variance.

Dealing with the reddening internal to the parent galaxy itself is more problematic. In the earliest studies it was simply ignored. Even if this simplification had proven true for the first few specific cases, there is no reason to believe that it would have obtained in general.

To illustrate the systematic effects of reddening on the observed PL and PLC relations, we consider the following example. Suppose for the moment that the instability strip is intrinsically very narrow both in color and in magnitude at fixed period. Now consider a sample of Cepheids drawn from this strip in a nearby galaxy, where on average the reddening is $E(B-V) = 0.2$ mag, with a standard deviation of $\pm$ 0.1 mag. These stars, differentially obscured, would be observed to have a period-color relation with a full ($\pm$ two-sigma) color width of $\sim$0.4 mag, a B period-luminosity relation with a magnitude width of 1.7 mag and a V period-luminosity relation with a width of 1.3 mag (for a ratio of $A_V / E(B-V) = 3.3$). As "predicted" from a general consideration of the theoretical PLC, the deviations in the period-luminosity relation would be found to be correlated, with residuals from the period-color relation, apparently confirming the theory. But none of this correlation would be intrinsic, of course, despite it being very well defined. Unfortunately too, solutions for distances would be systematically in error since the ridge line of the data (defining the mean PL relations) would be displaced from the intrinsic strip by $A_B = 4.3\ E(B-V)$, always towards larger apparent distances.

The above example is extreme, but it illustrates the point that any attempt to disentangle the effects of differential reddening and true color deviations within the instability strip must rely first on a precise and *thoroughly independent* determination of the intrinsic structure of the Period-Luminosity-Color relation. In order to achieve that calibration high-quality, independent reddenings and distances to individual calibrator Cepheids must be available. The uncertainty involved in undertaking this first step will affect all future results based on those assumptions. Below we discuss old methods that have been adopted to deal with the reddening problem, and emphasize new methods that have been brought to bear with the introduction of panoramic, digital detectors operating at optical and now at near-infrared wavelengths.

5.3. Metallicity Sensitivity of the PL Relation

The chemical composition of a star plays a role in setting the rate of energy generation, it affects the evolution off of the main sequence, and it determines the wavelength distribution of the emergent flux. The role of metallicity in the evolution of individual Cepheids in specific, and for the forms of PL, PC and PLC, in general, has been a matter of conjecture and debate for several decades now. Stothers (1988) reviewed the theoretical aspects of this problem in great detail, concluding that at blue wavelengths the effect could be of concern to the distance scale. While the effects are generally conceded to be smaller at longer wavelengths, only recently has strict attention been paid to quantifying the effects of metallicity for near-infrared bandpasses.

Two independent physical mechanisms contribute to the effect of metallicity on the mean color of Cepheids. First of all it is expected that the observed colors of Cepheids should vary (in a wavelength-dependent way) as a function of differing amounts of atmospheric metal-line blanketing. In addition, changes in the mean metallicity of the star as a whole are predicted to affect the interior opacities, resulting in equilibrium radius changes and different mean surface brightnesses (effective temperatures) for the same nuclear-generated luminosity. Detailed studies suggest however, that the effect of metallicity on the observed colors is largest for atmospheric line blanketing, compared to changes forced upon the interior structure of the star.

The most recent self-consistent modeling effort is that of Chiosi, Wood & Capitanio (1993). These authors produced linear nonadiabatic pulsation calculations for a grid of Cepheid models with a range of masses, various effective temperatures, and chemical compositions ranging from 1/4 to solar metallicity, for a variety of mass-luminosity relations. The theoretical fluxes were then convolved with the $UBVR_cI_c$ passbands for comparison with observations. These authors conclude that the agreement between theory and ob-

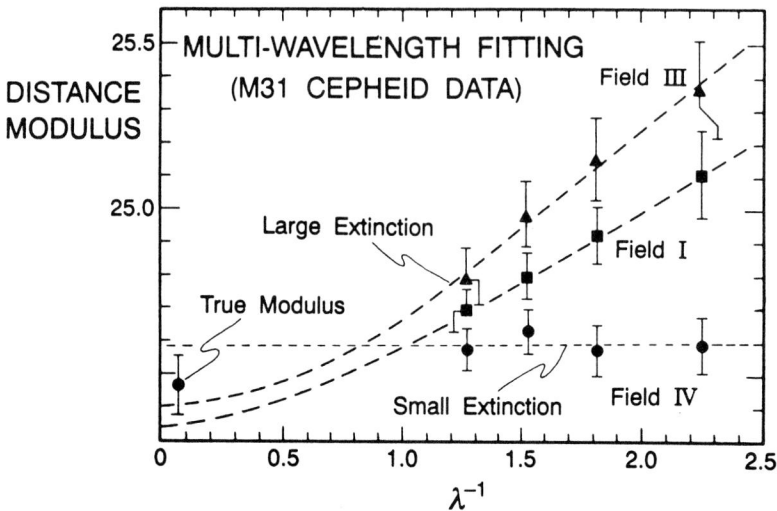

FIGURE 6. Multiwavelength fits of normal reddening lines to *BVRI* apparent distance moduli for M31 Cepheids (Freedman & Madore 1990). The three fields in M31 have distinctly different average reddenings associated with them, however the true modulus, as given by the intercept at $1/\lambda = 0$ is quite stable, indicating little residual sensitivity of the Cepheid PL relation to metallicity.

servation is best at longer wavelengths, noting a discrepancy between the observed and predicted (B-V, Log P) relations between the SMC. the LMC, and the Galaxy. Their predicted uncertainty of the true distance modulus, after correcting for reddening, is $\delta(m-M) = -1.7\delta Z$ (at log P = 1.0). Hence, for the entire range of chemical compositions represented by the low-metallicity SMC (Z = 0.004) and the solar-metallicity Galactic Cepheids (Z = 0.016), a very small abundance effect (amounting to only 0.02 mag, full range) is predicted.

The predictions of Chiosi *et al.* (1993) are consistent with the results of a test of the metallicity sensitivity of the Cepheid PL relation by Freedman & Madore (1990). These authors undertook the first empirical test of the metallicity sensitivity of the PL relation by observing samples of Cepheids in three fields at differing distances from the nucleus of M31. The observational test is differential, and thus independent of any absolute metallicity calibration. From multiwavelength (BVRI) observations, true distance moduli and reddenings were determined for each of the three fields. The range in metallicity over the three fields is approximately a factor of 5. Large (0.80 mag) differences were found between the *apparent blue* moduli, whereas the maximum differences in the *apparent I-band* moduli amounted to only 0.3 mag. After correcting for reddening, a small (0.3 mag peak-to-peak) range in the true distance moduli was found, having low overall statistical significance, but still in the same sense as predicted by theory.

Despite the good agreement with the predictions of Chiosi *et al.* (1993), the results of Freedman & Madore (1990), are smaller (by about a factor of 3) than earlier predictions by Stothers (1988) for B and V-band photometry. Subsequently, the Freedman & Madore

(1990) data have been reanalyzed by Gould (1994), who concludes that the M31 data are consistent with a larger metallicity dependence. However, see Chiosi et al. (1993) for their cautionary remarks about the predictions from (B-V) photometry. Furthermore, a large metallicity dependence for the Cepheid PL relation does not appear to be consistent with the very good agreement of the (lower metallicity) Population II distance indicators (e.g. RR Lyraes and first ascent red giant branch (TRGB) stars) to be discussed later.

While the importance of metallicity effects is still not totally resolved, all theoretical models to date predict that at progressively longer wavelengths, the effects of metallicity differences should be very small. We are currently completing a five-year program in which we hve obtained JHK near-infrared array photometry of the Freedman & Madore sample of M31 Cepheids (Freedman, Madore & Sakai 1998, in preparation.) The longer wavelength baseline added to the optical data will offer much tighter constraints on this empirical metallicity calibration. As a further check of the metallicity sensitivity of the PL relation, the M31 test has been carried over to 2 fields in M101 (Kennicutt et al. 1998) as part of the Key Project on the Extragalactic Distance Scale. Once again the test indicates the as far as the methodology of using V and I-band data to determine extinction-corrected moduli, the metallicity effect is small, on the order of $\delta(m-M)_o/\delta[O/H] = -0.24 \pm 0.16$ mag/dex. Even at this level of dependence it is to be emphasized that the resulting effect on the distance scale will only amount to a few percent. This is so because the LMC metallicity (–0.4 dex) is very close to the mean metallicity of the galaxies being observed in the Key Project (–0.3 dex). Obviously individual distances of galaxies with metallicities significantly larger or smaller than the metallicity of the LMC will systematically deviate (if not corrected for metallicity) but no large bias is expected in the mean. Obviously NICMOS data would be extremely valuable in dealing with this issue empirically.

6. Advances driven by new technology

6.1. Application of Near-Infrared Techniques

Although the first near-infrared observations of Galactic Cepheids were made nearly 30 years ago (Wisnewski & Johnson 1968) their applicability to the distance scale was not appreciated until quite recently. In the first of a series of papers McGonegal et al. (1982) unambiguously demonstrated that once periods were adopted from optical data, near-infrared observations of Cepheids provided a number of distinct advantages.

It was, of course, anticipated that by going to the infrared concerns about total and/or differential reddening would be significantly reduced. For instance, a blue extinction of $A_B = 0.32$ mag (typical for Cepheids in the LMC, for instance) would translate to a total correction of $A_K = 0.05$ mag at 2.2μm; the full correction in the near infrared being comparable to the uncertainty alone associated with most optical extinctions!

But it was also immediately clear from the outset (see for example Figure 5) that the infrared had other advantages directly applicable to the establishment of the Cepheid distance scale. First of all, the decrease in observed width of the PL relation was dramatic. Even single observations of the Magellanic Cloud Cepheids, (uncorrected to mean light) produced a PL relation with remarkably small scatter (±0.2 mag). This narrowing of the width is due to two effects: a decreased sensitivity to differential reddening, but more significantly, due to a much decreased sensitivity of the infrared surface brightness to the temperature width of the instability strip. For exactly the same reason, the amplitudes of individual Cepheids (shown in Figure 1, as plotted earlier by Wisnewski&

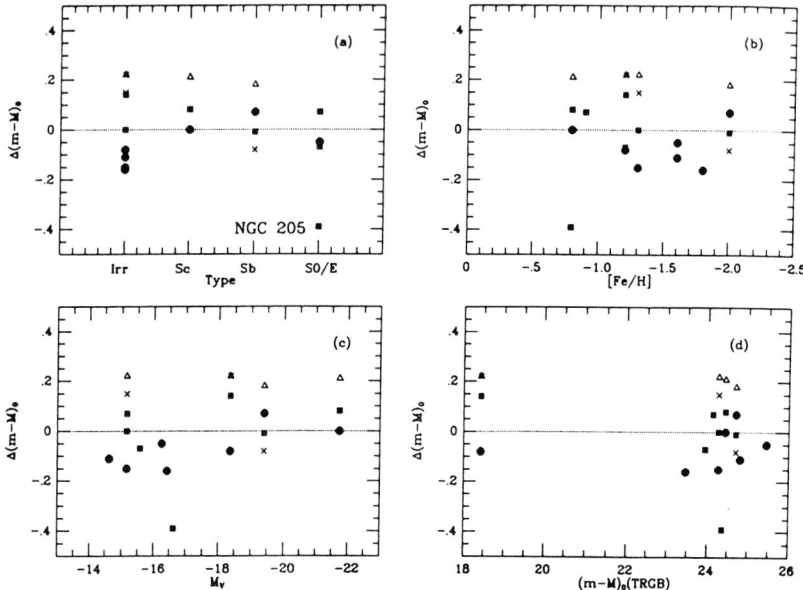

FIGURE 7. A differential comparison of distance moduli determined from the Population I Cepheid PL relation and the Population II TRGB magnitude as a function of Hubble type of the parent galaxy (upper left) absolute magnitude of the parent galaxy (lower left), the TRGB modulus (lower right) and finally the metallicity of the parent galaxy (upper right). Within the scatter no trends are apparent, especially with metallicity.
(adapted from Lee, Freedman & Madore 1993)

Johnson 1968) also decrease with the wavelength of the observations. Thus Cepheids observed at long wavelengths and at random points in their cycle are (1) closer to their time-averaged mean magnitudes than the equivalent observation in the blue, and (2) the mean magnitudes themselves are in fact coming from a narrower projection of the instability strip into the infrared plane than the equivalent blue PL relation. From B to K, a typical Cepheid amplitude drops from 1.0 mag to 0.4 mag, while the width of the PL relation decreases from 1.2 mag to 0.5 mag. As a result, for distance determinations even single, random-phase observations of known Cepheids, when made in the near-infrared, are comparable in accuracy to complete time-averaged magnitudes (derived from a dozen or more observations) in the blue.

The periods for many extragalactic Cepheids had already been determined and published from photographic studies made at blue wavelengths in many long-term studies by Hubble, Baade, Sandage, Swope, Payne-Gaposchkin, Gaposchkin and others. As a result, in only a matter of nights, it was possible to reobserve the entire galaxy sample in the near infrared and thereby provide new accurate distances, almost unaffected by absorption effects.

However, until recently, with the advent of near infrared arrays, there have been limitations in the application of the IR to the Cepheid distance scale, even for nearby Local Group galaxies. Single-channel infrared detectors, available at the outset, were aperture devices which had to be 'chopped', 'on' and 'off' the source in order to continuously monitor the intense and fluctuating terrestrial sky contribution to the signal. For Galactic Cepheids, and even for those in the Magellanic Clouds, a typical aperture of 5 arcsec could be placed over a star and chopped to a nearby reference region a few arcseconds away with relatively small uncertainty. However, for Cepheids at larger distances, the near-infrared observations of Cepheids in M31 (Welch, McAlary, McLaren & Madore 1986) and M33 (Madore *et al.* 1985) show much more noise than could be attributed to photon statistics alone. The most likely source of error is crowding and confusion; that is, contamination in one of the two comparison fields where 'skies' were being measured, or contamination in the object aperture itself. Although no systematic error is expected from this contamination, the random errors are appreciable. New observations are being obtained with near infrared InSb and HgCdTe arrays and two-dimensional image analysis is allowing more accurate infrared magnitudes to be measured in M31 and M33.

7. CCDs and multiwavelength coverage

Much was learned from the early near-infrared observations that could be applied to optical observations, given accurate data gathered over a sufficiently broad wavelength range. CCDs provided just that opportunity, bridging the development gap slowly being filled by the relatively small-area IR arrays. Given a wavelength sensitivity running from the B band (0.45μm) to the I band (0.7μm), CCDs afford the opportunity to gather seeing-limited, panoramic, digital data, which can subsequently be reduced using local sky subtraction and point-spread-function fitting techniques, to derive accurate magnitudes and colors. Crowding and confusion errors can be dealt with at the one square-arcsecond level. Use of CCDs and near-IR arrays decrease the areal confusion by about an order of magnitude over the near-infrared apertures.

CCDs also offer the advantage of a large wavelength coverage, thereby allowing an explicit determination of the reddening from the optical data itself. In this case, it is not necessary to rely purely on foreground estimates and/or on assuming that additional reddening is of negligible importance. Given a knowledge of the interstellar extinction law as a function of wavelength it is possible to fit all of the data simultaneously. Freedman (1988b) introduced this new approach to determining true moduli for extragalactic Cepheids using multiwavelength data, applying it first to single-epoch observations of Cepheids in IC 1613, and later refining it and expanding its application to data obtained for M31, M33, and NGC 300 as cited below. For a detailed discussion of the technique and its implementation the interested reader is referred to those papers. Briefly stated, one determines differential apparent moduli, scaled against the corresponding LMC PL relations. By assuming that all of the difference as a function of inverse wavelength can be attributed to selective absorption, fitting an interstellar extinction law to the data, simultaneously estimates the total (foreground plus internal) absorption and the true distance modulus, relative to the LMC.

In the following we briefly review the systematics of the PL relation as observed at wavelengths ranging from the blue to the near-infrared. To do this we concentrate on self-consistent data sets assembled for this purpose by Madore & Freedman (1991) and plotted in Figure 5. The stars included in this compilation are Cepheids in the LMC and SMC for which there are time-averaged mean magnitudes at all seven wavelengths (excluding R for the LMC sample which largely lacks this bandpass in published Cepheid observations). Note that the slope and the dispersion of the PL relation change systematically as a function of the effective wavelength of the filter bandpass. These impressions are quantified in the equations presented at the end of Section 8. In any case, as the longer-wavelength data are considered, it is clear that both quantities (slope and dispersion) have already begun to converge on an asymptotic value (set by the period radius relation, which because of its geometrical nature is largely wavelength independent). From this point of view it makes no practical sense to observe Cepheids at wavelengths much beyond $2\mu m$ if the aim is to decrease the dispersion in the observed PL relation. Fortunately this regime is still accessible from the ground and is not far into the thermal IR where background effects become extremely large.

For comparison we show the rate of fall-off in the monochromatic extinction as a function of wavelength scaled to the blue extinction. As is well appreciated, the extinction does continue to decrease with increasing wavelength, making it sensible to extend the observations as far into the infrared as is practical. Of course a decrease in extinction by a factor of over 5 is realized at K in comparison to B, so for most purposes this too is a reasonable wavelength at which to stop the effort.

The impact of panoramic linear detectors, such as the CCD, on the study of extragalactic Cepheids has been significant. It is now possible to obtain from the ground, high-quality light curves of Cepheids in galaxies 2 Mpc away, ranging from blue and visual wavelengths, as well as reaching out to nearly one micron with CCD detectors, and then extending to $2.2\mu m$ with the newly available HgCdTe and InSb infrared areal detectors. With HST the distances reached have now exceeded 20 Mpc, and NICMOS is being applied to a number of these galaxies in follow-up studies in the near infrared.

8. Obtaining accurate cepheid distances

Once new extragalactic Cepheids are found at least four issues need to be adequately addressed, all of which are tightly coupled to common sets of observations: (1) Periods have to be determined, (2) Complete light curves have to be delineated, (3) Mean magnitudes must be derived, and finally, (4) Accurate colors are required for reddening determinations. Needless to say it would be hard to derive (1) the periods (or prove that a star is in fact a Cepheid) without (2) the light curves; and vice versa. Similarly, (3) and (4), accurate mean magnitudes and colors generally depend on correct period-phasing and proper lightcurve fitting. But the requirements for accurate periods are in fact quite different from the requirements for accurate magnitudes. The number of data points required to yield a time-averaged magnitude (of specified precision) increases as the square of the lightcurve amplitude. This makes colors and mean magnitudes based on short wavelength observations more costly in observing time than their longer-wavelength counterparts. On the other hand, for fixed photometric uncertainties, periods increase in accuracy almost linearly with the time interval over which the observations are spaced. Furthermore periods good to a few percent can be obtained using only moderately accurate photometry after only a dozen or so cycles, thereby making the time constraint a minimal one.

Finally, one must contend with the intrinsic width of the instability strip as projected into the PL relation. Increasing N, the numbers of Cepheids is the most obvious solution here. For the B-band the equivalent dispersion in magnitude in the Cepheid PL relation is ± 0.35 mag. The error in the mean apparent distance modulus decreases like $\sqrt{N}$. In the absence of reddening then it would appear that for *apparent* distance moduli alone, a dozen Cepheids will give the requisite accuracy in the mean. But of course the real problem, once again, comes when trying to deal with reddening. And an example using two band-passes only illustrates this graphically. In such a case the ensemble-averaged extinction essentially comes from differencing the mean apparent moduli found at two different wavelengths. Multiplying this difference by the ratio of total-to-selective extinction appropriate to those two wavelengths and subtracting the product from the mean apparent modulus gives the final true modulus. If 10% in distance is the goal (0.2 mag in true modulus), then for the filter combination VI, simple arithmetic shows that two to three dozen Cepheids are required to establish the mean moduli to such a degree that the reddening corrected modulus has a final error of less than 0.2 mag. Of course either increasing the number of wavelengths and/or increasing the wavelength baseline will each reduce the final error on the mean without demanding an increase in sample size.

In closing this section we present our adopted fiducial multiwavelength PL relations. We emphasize that these relations differ slightly from those published by other workers up to this point, because they are derived from sets of data which are now totally self-consistent. Specifically, all of the PL relations are based on the same stars in order to eliminate sample-dependent variations in the solutions. Furthermore, Cepheids with $\log P > 1.8$ are excluded from the least-squares fits due to uncertainties in their reddenings and their evolutionary status. The LMC data set (scaled and dereddened as outlined in the next section) has been chosen as fiducial because of its large sample size, large wavelength coverage and because the LMC is very close to being face-on, thereby minimizing the effects of back-to-front geometry on the solutions. The relations are centered on $\log P = 1.0$, the mid-point of the range of periods considered here. Errors on the quoted coefficients are given after each of the values. Following each of the PL relations, the quantity in square brackets is the rms dispersion about the mean for that relation.

$$M_B = -2.43(\pm 0.14)\,(\log P - 1.00) - 3.50(\pm 0.06)\ [\pm 0.36]$$
$$M_V = -2.76(\pm 0.11)\,(\log P - 1.00) - 4.16(\pm 0.05)\ [\pm 0.27]$$
$$M_R = -2.94(\pm 0.09)\,(\log P - 1.00) - 4.52(\pm 0.04)\ [\pm 0.22]$$
$$M_I = -3.06(\pm 0.07)\,(\log P - 1.00) - 4.87(\pm 0.03)\ [\pm 0.18]$$

[Note that the RI magnitudes are on the Cousins system, while our JHK magnitudes are on the CIT/CTIO system. There are 32 LMC Cepheids for which BVI photoelectric photometry is available in the range $0.2 < \log P < 1.8$. R photometry is not available for many of the stars used above; however, R magnitudes were derived using the methodology set out in Freedman (1988b).]

Finally, we give below consistent PL solutions, based on a smaller set consisting of only 25 LMC stars, each of which has $BVRIJHK$ photometry available (given the same conditions outlined above):

$$M_B = -2.53(\pm 0.28)\,(\log P - 1.00) - 3.46(\pm 0.12) \quad [\pm 0.40]$$
$$M_V = -2.88(\pm 0.20)\,(\log P - 1.00) - 4.12(\pm 0.09) \quad [\pm 0.29]$$
$$M_R = -3.04(\pm 0.17)\,(\log P - 1.00) - 4.48(\pm 0.08) \quad [\pm 0.25]$$
$$M_I = -3.14(\pm 0.17)\,(\log P - 1.00) - 4.84(\pm 0.06) \quad [\pm 0.21]$$
$$M_J = -3.31(\pm 0.11)\,(\log P - 1.00) - 5.29(\pm 0.05) \quad [\pm 0.16]$$
$$M_H = -3.37(\pm 0.10)\,(\log P - 1.00) - 5.65(\pm 0.04) \quad [\pm 0.14]$$
$$M_K = -3.42(\pm 0.09)\,(\log P - 1.00) - 5.70(\pm 0.04) \quad [\pm 0.13]$$

The effective wavelengths for each of the seven bandpasses were chose to be appropriate for a G-star spectrum (see for example, Bessell 1979) where, for future reference, we have adopted: $B(0.444\ \mu m)$, $V(0.550\ \mu m)$, $R\ (0.653\ \mu m)$, $I\ (0.789\ \mu m)$, $J\ (1.25\ \mu m)$, $H\ (1.60\ \mu m)$, $K\ (2.17\ \mu m)$.

Finally, it should be noted that there are external checks on the Cepheid calibration and distance scale derived from it that have been applied to galaxies within the Local Group. An extensive review of these methods (including the use of RR Lyrae stars, red giant luminosity functions, novae and long-period variables, to name just a few) confirms (conservatively at the ± 0.2 mag level) the basic solidity of the Cepheid calibration (van den Bergh 1989, de Vaucouleurs 1991). For details the interested reader is referred to those reviews and the many references cited therein. Because many of these independent methods that provide checks on the Cepheid distance scale use intrinsically fainter stars it is unlikely that galaxies significantly beyond the Local Group will be of much use in further refining the agreement (or disagreement) between the various estimators. However, more extensive and more precise observations of those same (faint) distance indicators within these and other Local Group galaxies will be crucial for fine tuning the calibration and may be especially helpful in establishing the level at which metallicity corrections are needed in Population I and Population II distance indicators alike.

9. Local Group galaxies

We now quickly review the status of the individual galaxies in the Local Group for which (non-photographic) digital data on known Cepheids have been obtained in the last few years. This discussion and that in the next section update and supersede similar overviews published earlier by Madore (1985) and Walker (1987).

For the LMC we adopt hereafter 18.5 mag for its true modulus and 0.10 mag for $E(B-V)$, and scale all other Cepheid-based distances assuming a value for the total-to-selective absorption of $R_V = A_V/E(B-V) = 3.3$ (appropriate for the later spectral types of Cepheids). In a later section we discuss the impact of recent parallax observations of Galactic Cepheids by the Hipparcos satellite, and conclude that these calibrations are consistent at the 10% level. All period-luminosity fits are done over the range $0.2 < \log P < 1.8$. Furthermore, as discussed in Freedman (1988b), all fits are carried out for a self-consistent set of stars in the LMC defined by the simultaneous availability of photometry at B, V and I wavelengths. Although the sample of Cepheids with B and V photometry alone is about a factor of two larger, the comparison of inconsistent data sets can lead to erroneous results. It should be noted that although we quote the sources of original data throughout the next two sections, *the distances and reddenings given for each galaxy here are based on a new and homogeneous application of the multiwavelength fitting procedures* discussed earlier in this review (and described in detail in Freedman, Wilson & Madore 1991). In this procedure we have adopted the reddening law as determined by Cardelli, Clayton & Mathis (1989). By definition, our multiwavelength fitting procedure does not yield information about the distance and/or reddening to the LMC, as they are adopted *ab initio*. For completeness, we also review data for Cepheids in the Magellanic Clouds.

9.1. The Magellanic Clouds

For a relatively complete bibliography of modern photographic and photoelectric observations of Magellanic Cloud Cepheids the reader is referred to Tables 2 and 4 in Madore (1985); they will not be repeated here. Moreover, Caldwell & Laney (1991) have reviewed progress in the southern hemisphere on calibrating the Cepheid PL relations. They use data available in Madore (1985), in addition to more recently published near-infrared *JHK* data available for many dozens of Cepheids observed by Welch *et al.* (1987) and also Laney & Stobie (1986a,b). Similar to Welch & Madore (1984) these latter papers concentrate, among other things, on using the near-infrared data to determine the back-to-front geometry of the two Magellanic Clouds based on the ability of near-infrared observations of Cepheids to give extremely precise distances to individual stars. In this regard, it should be noted that Visvanathan (1989) has calibrated the Cepheid PL relation at 1.05μm and applied it to Cepheids in the SMC (Mathewson, Ford & Visvanathan 1986, 1988), also aiming to probe the structure of the SMC in the context of a tidal encounter/disruption model.

9.2. IC 1613

Sandage (1971) published photographic light curves and periods for 25 of the confirmed Cepheids discovered, but never published, by Baade. Both single-phase, near-infrared *H*-band observations of 10 of those Cepheids (McAlary, Madore & Davis 1984) and single-phase, multiwavelength *BVRI* observations (Freedman 1988b) of 11 of them have now been published. Freedman's work incorporated the near-infrared data. Using the new fitting procedure (rather than an earlier adopted linear extinction approximation), and excluding the lower-accuracy *H* band data, yields a total mean reddening of $E(B-V) = 0.02$ mag, and a true modulus of 24.42 ± 0.13 mag, corresponding to a distance of 765 kpc. To maintain homogeneity for comparison of relative distances, this value supersedes the distance modulus 0.12 mag lower, quoted by Freedman (1988b).

It is also worth noting at this point that Freedman's (1988b) conclusions concerning the universality of the slope of the PL relation (once brought in to some degree of doubt because of the uncertainties in the faint-magnitude calibration of the photographic data on the IC 1613 Cepheids) has subsequently been bolstered by the additional analysis of 16 newly confirmed Cepheids in IC 1613, as discussed in Carlson & Sandage (1990).

In comparison to any other member of the Local Group, IC 1613 offers the best opportunity for a ground-based telescope to provide a definitive calibration of the multiwavelength period-luminosity relation for Classical Cepheid variables. It is nearby and therefore offers the opportunity for accurate photometry; it has very low foreground (and internal) extinction and yet it is distant enough that back-to-front effects are negligible.

Finally, IC 1613 (and its Cepheid population) is metal poor, both in an absolute sense and with respect to the Galactic Cepheids in clusters. Comparison of a low-metallicity calibration through IC 1613 with the high-metallicity calibrations through the LMC and Milky way, respectively, may shed additional light on the metallicity sensitivity of the Cepheid PL relation zero point.

9.3. NGC 6822

One of the most extensive study of the dwarf irregular Local Group galaxy NGC 6822 is the photographic work of Kayser (1967). This study built on the original work of Hubble (1925) who found several Cepheids and a number of bright irregular variables in this galaxy. Unfortunately NGC 6822 is fairly close to the Galactic plane ($b = -18$ degrees), resulting in large, and still somewhat uncertain, foreground reddening estimates. Published estimates for the reddening to the Cepheids in NGC 6822 range from $E(B-V) = 0.19$ to 0.42 mag.

Results on the Cepheid distance to NGC 6822 have been appearing slowly. Both Hodge (1977), and then van den Bergh & Humphreys (1979) have reported photoelectric BV observations of the 65-day Cepheid, V7 in NGC 6822. Multiwavelength $BVRI$ CCD data have been obtained by the authors, but are as yet unpublished; while Schmidt, Spear & Simon (1986), Schmidt & Spear (1987), Schmidt & Simon (1987), and Schmidt & Spear (1989) have published some CCD observations of Cepheids in NGC 6822, which were obtained for other reasons. Of late, the only directed study of the Cepheid distance to NGC 6822 is the paper by McAlary, Madore, McGonegal, McLaren & Welch (1983) on near-infrared H-band aperture photometry of 9 Cepheids. Unfortunately, an independent determination of the foreground/internal reddening was not attempted because only one wavelength was involved in the new study. Visvanathan (1989) observed three Cepheids in NGC 6822 once each at 1.05 μm and derived a true modulus of 23.26 mag (scaled to an LMC true modulus of 18.5 mag). On the other hand, random-phase I-band CCD data (Lee, Freedman & Madore 1992) yield a true modulus of 23.59 mag.

Pending complete publication of the new CCD data, we can use the photographic observations of Kayser (1967) in combination with the H-band observations of McAlary, Madore, McGonegal, McLaren & Welch (1983) to provide a multiwavelength fit and solve for the true distance modulus and reddening to NGC 6822. In this application we have rederived all apparent moduli with respect to our internally self-consistent set of LMC Cepheid data. Excluding Kayser's variable No. 30 (which falls many sigma above the mean B PL relation), we find $(m-M)_B = 24.66 \pm 0.06$ mag, $(m-M)_V = 24.50 \pm 0.08$ mag and $(m-M)_H = 23.77 \pm 0.17$ mag, resulting in $(m-M)_O = 23.66$ mag with $E(B-V) = 0.26$ mag. The reddening is well within the range of previous estimates quoted above and is remarkably close to Kayser's original estimate of 0.27 mag. Despite this formal solution, it was clear that the Cepheids in NGC 6822 could profit from a modern investigation at several wavelengths. Accordingly Gallart, Aparicio & Vilchez (1996) using single-phase (multiwavelength) CCD data of the Cepheids in NGC 6822 have derived a new estimate of distance and reddening. They have also compared there result with the the distance calculated using the tip of the red giant branch as a distance indicator. A distance modulus of $(m-M)_O = 23.49 \pm 0.08$ mag and a reddening of $E(B-V) = 0.24 \pm 0.03$ mag are derived.

9.4. M33

Hubble (1926) discovered 35 Cepheids in the inner regions of M33 and determined their periods. These photographic data were later recalibrated by Sandage (1983). Sandage & Carlson (1983a) added identifications, periods and photographic mean magnitudes for 13 new Cepheids in an outer region of this galaxy, while Kinman, Mould & Wood (1987) then surveyed the main body of M33 using photographic plates taken at the prime focus of the KPNO 4m in search of long-period variables, and in the process discovered 54 new Cepheids. All four of these publications include finder charts for their variables.

Building on these discovery papers, single phase H-band observations (Madore, McAlary, McLaren, Welch, Neugebauer & Mathews 1985) were made of 15 Cepheids in M33. These were then augmented by and incorporated into a multiwavelength study of 19 Cepheids (Freedman, Wilson & Madore 1991) using CCD observations to obtain light curves and therefore time-averaged magnitudes and colors. It was the H-band study of the M33 Cepheids that first noted the basic limitation due to crowding and confusion on the application of aperture techniques to the infrared Cepheid distance scale. JHK array imaging of individual Cepheids in M33 and M31 is being done by the authors using the Palomar 200-inch, so some of those early limitations should soon be lifted. Until those data are available the CCD observations alone indicate that for M33, $A_B = 0.41$ mag and the true modulus is 24.63 ± 0.09 mag, corresponding to a linear distance of 840 kpc.

5.5 M31

Baade & Swope (1963, 1965) and Gaposchkin (1962) used 200-inch plate material to catalog Cepheid variables in M31. With the exception of the outermost Field IV (for which there is both B and V data) only blue photographic magnitudes are available, making reddening estimates for the Cepheids both circumstantial and rather unreliable. Nearly a quarter of a century later Welch, McAlary, McLaren & Madore (1986) obtained single-phase H-band observations of 22 Cepheids in M31. More recently Freedman & Madore (1990) have published multiwavelength PL relations for 38 Cepheids in three of the Baade & Swope M31 fields resulting in independent reddenings and true distance moduli for each of the regions. A consistent true distance modulus of 24.44 ± 0.10 mag (corresponding to 770 kpc) is determined here with mean reddenings ranging from $E(B-V) = 0.00$ mag in Field IV to 0.25 mag in Field III. Those data are shown plotted in Figure 6.

10. Beyond the Local Group

To discover Cepheids in galaxies, and then confidently determine their periods, light curves, mean magnitudes and colors is an extremely demanding task. So it is perhaps not too surprising that although Cepheids in galaxies as far away as M81 were known to Baade in the early 1960's, up until the 1980's the only major galaxy outside of the Local Group for which a definitive study had been published of its variable star content was the solitary, late-type spiral galaxy, NGC 2403 (Tammann & Sandage 1968). Only in the last decade have CCD surveys begun to come to the fore, as the last of the photographic surveys for Cepheids in galaxies easily resolved from the ground, are now being published. CCDs have small fields of view, but they have significantly higher quantum efficiency than photographic emulsions, especially at longer wavelengths. Accordingly, they are in fact allowing significant progress to be made in searching for Cepheids in the few remaining galaxies near enough to be resolved from the ground. Progress on those fronts is now reviewed.

10.1. NGC 2403

Morphologically very similar to M33 (and its southern counterpart NGC 300) NGC 2403 is a highly resolved late-type spiral galaxy often associated with M81 and a handful of lower luminosity spiral and irregular galaxies (*e.g.*, de Vaucouleurs 1975). Many dozens of photographic plates taken by Mount Wilson and Palomar observers were searched for variables by Tammann & Sandage (1968) resulting in the discovery of 17 Cepheids. Due to plate limitations, only the brightest Cepheids in the outer parts of the galaxy were found. Even so only the brightest parts of the blue light curves were securely calibrated and only fragmentary information was available in the visual bandpass. Nevertheless, good periods were determined by these authors and (assuming no extinction within NGC 2403 itself) an apparent modulus was estimated from the brightness of the apparently least reddened Cepheid seen at maximum light. These data have been re-analyzed by Madore (1976) and then by Hanes (1982) and again by Rowan-Robinson (1985) with different sets of assumptions, and consequently diverging conclusions. New data were obviously required before significant progress could be made beyond arcane arguments concerning methodology.

I-band CCD observations of 8 of the known Cepheids in NGC 2403 are discussed by Freedman & Madore (1988) who derive a tentative true modulus of 27.51 mag. Their adopted error of ± 0.24 mag includes the uncertainty for the still undetermined amount of reddening internal to NGC 2403 itself. Given what we now know about the difficulties inherent in doing near-infrared aperture photometry in confusion-limited cases, the modified J-band observations reported by McAlary & Madore (1984) and the conclusions drawn therein, are thus superseded.

10.2. M81

Baade (1963) was aware that Cepheids had been detected in M81 but it was another 20 years before any results were published. Because of the high surface brightness of the disk of M81, photometry of individual stars is very difficult, making the detection of variable stars such as Cepheids extremely taxing, especially near the plate limit. Nevertheless Sandage (1984) succeeded in determining the periods for two Cepheids in M81 using B-band photographic plate material. Freedman & Madore (1988) obtained I-band CCD photometry of these stars (based on finder charts and periods made available by Sandage (1987, private communication), and derived a true modulus of 27.59 ± 0.31 mag adopting a foreground Galactic extinction of $A_B = 0.15$ (Burstein & Heiles 1984). This modulus corresponds to a distance of 3.30 Mpc which is still uncertain at the 15% level in as much as the sample is small and no corrections for reddening internal to M81 itself were included. HST observations by Freedman *et al.* (1994) yield a reddening-corrected distance of 3.6 ± 0.3 Mpc, based on 30 Cepheids discovered in two fields (one along the major axis of M81, the other positioned along the minor axis).

10.3. *M101*

Sandage & Tammann (1974) attempted to find Cepheids in their 200-inch photographic plate material on M101. Their failure led them to conclude that M101 has a distance modulus in excess of 29.3 mag. Later, using CCD detectors Cook, Aaronson & Illingworth (1986) did succeed in detecting at least two Cepheids at $R \sim 23$ mag (with preliminary periods of 37 and 47 days) in a relatively uncrowded outer region of this galaxy. Given the uncertainties in the absolute zero point of the calibration of the Cepheid PL relation and the unknown extinction internal to M101 itself these data give an apparent R-band modulus of 29.38 mag. Other fields in M101 also observed by this group (Cook, Aaronson & Illingworth 1989), and their results based on 4 Cepheids (Alves & Cook 1995) give a true distance modulus of 29.08 ± 0.13 mag (6.4 ± 0.4 Mpc).

More recently Kelson *et al.* (1996) have used *HST* to image an outer field in M101 and discovered 29 Cepheids, from which they derive $E(B-V) = 0.03$ mag and a true distance modulus of 29.34 ± 0.17 mag, corresponding to a distance of 7.4 ± 0.6 Mpc. This latter distance compares very favorably with the planetary nebula luminosity function distance 7.7 ± 0.5 Mpc derived by Feldmeier, Ciardullo & Jacoby (1996).

10.4. *NGC 300*

Graham (1984) conducted a photographic study of NGC 300 using the Tololo 4m reflector and was able to discover 18 Cepheids, determine their periods, and estimate mean B and V magnitudes. Madore, Welch, McAlary, & McLaren (1987) subsequently observed 2 of these Cepheids at H. Walker (1988) later criticized Graham's photoelectric calibration sequence and suggested a correction to the photographic photometry of the Cepheids. Visvanathan (1987) observed three Cepheids in NGC 300 at 1.05 μm, and derived a true modulus of 25.80 maag (scaled to our adopted LMC modulus). Finally, Freedman *et al.* (1991) present *BVRI* CCD observations of 16 Cepheids in NGC 300 giving revised periods and time-averaged magnitudes and colors. A true distance modulus of 26.67 mag, corresponding to a distance of 2.16 Mpc, is derived from a multiwavelength analysis of the CCD data of the 10 Cepheids with complete light curves.

10.5. *NGC 247 and NGC 7793*

Catanzarite, Freedman, Madore & Horowitz (1998, in preparation) have found 9 Cepheids in NGC 247, which turns out to be located at a distance that is nearly a factor of two larger than that of NGC 300. The significantly larger distance, with the ensuing crowding problems and brighter absolute magnitude cut-off in the observational data made this object far harder to determine a Cepheid-based distance than was originally anticipated. The requisite data for NGC 7793 have also been obtained and are currently being reduced.

10.6. *WLM*

A photographic study of the Wolf-Lundmark-Melotte galaxy has been published by Sandage & Carlson (1985a) where 15 Cepheids are identified and periods determined (ranging from 3.3 to 9.6 days). The CCD study of WLM published by Ferraro, Fusi Pecci, Tosi & Buonnano (1989) unfortunately included none of the Sandage & Carlson Cepheids, but the CCD photometry does indicate that the photographic photometry is too bright by +0.4 mag in both B and V (as derived from the intercomparison of the two studies given in their Table 2.

Applying the B offset to the photographic data for the Cepheids in WLM and rederiving the apparent modulus as discussed above we find $(m - M)_B = 25.19$. Adopting $A_B = 0.09$ mag (Burstein & Heiles 1984), gives a true modulus of 25.10 mag. From preliminary reductions of I-band CCD observations of 5 WLM Cepheids, Lee, Freedman & Madore (1992) find a true modulus of 24.89 ± 0.15 mag, corresponding to a linear distance of 0.95 Mpc.

10.7. IC 10

Saha et al. (1996) reported the discovery 4 Cepheids in this galaxy and derive a true distance modulus of 24.59 ± 0.30 mag (corresponding to a distance of 830 Mpc) after correcting for $E(B-V) = 0.94$mag. Wilson et al. (1996) then obtained JHK observations of these same Cepheids and derived a slightly revised true modulus of 24.57 ± 0.21 mag, giving a distance of 820 kpc. The largest uncertainty at this point must be the adopted foreground reddening correction which is considerable. A TRGB/Cepheid study of IC 10 (Sakai, Madore & Freedman 1998) is nearing completion, and should throw some light on this problematic object.

10.8. NGC 3109

¿From photographic material Demers, Kunkel & Irwin (1985) report discovering 5 Cepheids with periods in the range 10 to 23 days in the nearly edge-on, Magellanic-type irregular galaxy NGC 3109. Sandage & Carlson (1988) also using photographic material, but with better plate scale, were able to observe deeper into the luminosity function and determine the periods for 29 Cepheids. The periods of their Cepheids ranged from 6 to 31 days, confirming only 2 out of the 5 earlier-reported periods. Using the data from Sandage & Carlson, and adopting a foreground extinction of $A_B = 0.14$ mag from Burstein & Heiles (1984), we derive a true modulus of 25.94 mag for NGC 3109, corresponding to 1.54 Mpc. No correction for extinction internal to NGC 3109 itself has been applied.

10.9. Pegasus = DDO 216

Hoessel it et al. (1990) report the discovery of 31 variable stars in Pegasus, with 7 being likely Cepheids. Based on these observations they derive a distance modulus of 26.22 ±0.20 mag, corresponding to a geometric distance of 1.75 Mpc.

10.10. Leo A = DDO 069

Hoessel et al. (1994) report the discover of 14 variable stars in Leo A of which only 5 had sufficient phase coverage for them to be classified as Cepheids. Based these stars (corrected for foreground Galactic extinction) the authors derive a true distance modulus of 26.74 mag, corresponding to a distance of 2.2 Mpc, placing Leo A at the outermost edge of the Local Group.

10.11. GR8 = DDO 155

Tolstoy et al. (1995) found a total of six variables in the dwarf irregular galaxy GR8, only one of which was classified as a Cepheid. After correcting for Galactic extinction the single star gave a distance modulus of $(m - M)_0 = 26.75 \pm 0.35$ mag, corresponding to a linear distance of 2.24 Mpc.

10.12. Sextans A and Sextans B

In a photographic survey of Sextans A, Sandage & Carlson (1983b) discovered 5 Cepheids whose periods ranged from 15 to 25 days. In a later paper Sandage & Carlson (1985b) recalibrated the Sextans A photometry, and added data on the adjacent dwarf-irregular galaxy Sextans B, in which they discovered 4 Cepheids having periods ranging from 7 to 28 days. However, CCD data for stars in Sextans A (Hoessel, Schommer & Danielson 1983) after transforming from the original Gunn system to the BV system give a zero-point difference between the two data sets amounting to 0.2 mag. Walker(1987) then used a CCD with standard Johnson filters to study this problem further, and concluded that the Sandage & Carlson B-band data are too faint by 0.16 mag in the color range appropriate to Cepheids. With the original Sextans A data transformed to the Sextans B photometric scale as described by Sandage & Carlson (1985b), and then further corrected using Walker's offset, we derive true moduli of 25.59 and 25.64 mag for Sextans A and B, respectively, after adopting foreground extinctions of $A_B = 0.06$ and 0.05 mag (Burstein & Heiles 1984). The above true moduli correspond to 1.31 and 1.34 Mpc, respectively. Finally, it should be noted that for Sextans A, Visvanathan (1987) reports a true modulus of only 25.35 mag, based on single-phase 1.05 μm observations of three Cepheids.

10.13. NGC 2366 = DDO 042

Tolstoy et al. (1996) found 6 Cepheids out of a total of 13 variables stars in this M81 Group galaxy, NGC 2366. After correcting for foreground Galactic extinction they derive $(m - M)_O = 27.68 \pm 0.20$ mag, corresponding to a distance of 3.44 Mpc.

10.14. M83 (= NGC 5236), IC 5152, and the Phoenix Dwarf

Caldwell & Schommer (1990) have been monitoring the luminous southern spiral M83 for Cepheids, while Caldwell, Schommer & Graham (1988) report actually having discovered Cepheids in two southern dwarf irregular galaxies, IC 5152 and the Phoenix dwarf, as well as in M83. No details have yet been published.

10.15. The Centaurus Group

Finally, it should be reported that at least two groups are attempting to discover Cepheids in the Centaurus Group. Walker (private communication) has been obtaining CTIO prime-focus CCD frames of galaxies in this cluster, but his coverage is presently insufficient to securely identify variables, and therefore no periods have yet been reported. On the other hand, Tammann et al. (1991) have a Key Project underway at ESO which has nine half nights used in early 1991 to begin a search for Cepheids in the Centaurus Group late-type galaxies NGC 5236, NGC 5264 and AM 1324-411.

11. The Hubble constant

Of course, while distances to individual galaxies are of value in their own right, there is a more compelling reason for undertaking this extremely time-consuming task of Cepheid discovery and calibration. That goal is the Hubble constant, a measure of the size-scale and indeed the time-scale of the Universe. This goal was deemed sufficiently important that the design of the *Hubble Space Telescope* was in large measure constrained to meet the minimum requirements needed to undertake this task from space: to discover Cepheids in galaxies well beyond the Local Group, out to and including the Virgo and Fornax clusters.

Even so, one or more secondary distance indicators are needed to get to 'cosmologically significant' distances where the pure Hubble flow can be reliably measured. And it is generally accepted that these secondary distance indicators are best calibrated by the Cepheid PL relation. At the moment secondary methods with small measured internal dispersions are the best contenders for confidently extending the distance scale, with the intent of determining the value of the Hubble constant and placing a credible uncertainty. They are: the Tully-Fisher relation, Type II SN, Type Ia SN, the surface brightness fluctuation technique and the Faber-Jackson relation. The planetary nebula luminosity function, although showing great promise as a precise distance indicator, has not been pressed to distances any further than those already directly probed by Cepheids themselves. The latter three, because they apply primarily to early-type spirals, S0 galaxies or ellipticals might seem to be outside of the sphere of influence offered by the Cepheid calibration, but this is not strictly true. With the availability of *HST* a volume of space at least one thousand times larger than is regularly available to ground-based telescopes is now accessible. This volume includes several groups of galaxies that are known to contain both early-type systems and late-type spirals. Confirmed group membership then allows Cepheid distances to be brought into the calibration of galaxies that do not themselves contain Cepheids. This has in fact been the case in several attempts to calibrate these same relations from the ground. Specifically, M31 and M81 (which have Cepheid-based distances) with their significant bulge luminosity have been used to calibrate the PNLF (Ciardullo, Jacoby, Ford & Neill 1989) and an application of the Faber-Jackson relation (Dressler 1987). For the surface-brightness-fluctuation technique the apparent association of M32 again with M31 has been used as a calibrating path. For the future, obvious groups with a mix of Hubble types, worthy of detailed distance determinations are the Centaurus (NGC 5128) Group, and the Leo I (M96 = NGC 3368) Group; with the NGC 2841 Group, the NGC 1023 Group and the NGC 2997 Group offering some interesting potential as well.

12. The future

Any future ground-based attempts to detect Cepheids in external galaxies must carefully consider the question of how the periods are going to be unambiguously determined. The rapid success enjoyed by the recalibration of the extragalactic distance scale using infrared detectors was in no small part due to the fact that enormous amounts of telescope time had already been invested in discovering those Cepheids and determining their periods. Unlike all previous investigations we are now confronted by a new problem in the application of Cepheids to the extragalactic distance scale. When the light curves are sparsely sampled over time intervals that are only a few cycles in duration there is a tight coupling between the accuracy of the period determinations and the photometric accuracy of the observations. Co-ordinated ground-based observations are difficult to schedule, and even more difficult (if not impossible to guarantee). With space observations there is however the immense advantage that the exposures can be optimally scheduled to minimize aliasing effects and also provide for uniform coverage of the phased light curve. Details of the methodology employed by the Key Project will soon be given in Madore & Freedman (1998).

13. Contrasting aspects of the PL and PLC

A point of some confusion which needs clarification when discussing Cepheids as distance indicators resides in the difference between the PLC and the PL relation. As discussed earlier the PLC is simply a restatement of Stefan's Law, and it is therefore applicable to all stars on an individual basis. The PL and PC relations, on the other hand, are statistical relations for ensembles of stars; and they are in turn the result of constraints on Stefan's Law. That constraint is manifest in nature as a strip in which stars are unstable to pulsation. Here we expand on and illustrate this statement of differences.

A PLC relation exists for all stars. That is, for all combinations of temperature and luminosity one can calculate a period, since the PLC embodies exactly the same (universally applicable) principles as does Stephan's Law, from which it is derived. To use the PLC formalization, one must of course observe a color, correct it to an intrinsic color, and then independently determine a fundamental period in order to calculate a luminosity. But while all stars may have mathematically and physically well-defined fundamental-mode periods, not all stars are unstable to these oscillations, and so their pulsational periods often are not manifest directly.

Much as Nature has provided us with a useful and powerful constraint on Stephan's Law, through the hydrogen burning main sequence, Nature has also provided a different constraint on Stephan's Law (*i.e.*, a constraint on the ubiquitous PLC relation) by defining a narrow zone in which stable pulsation can and does occur. This alternate 'constraint' manifests itself as the Cepheid instability strip. *The Cepheid instability strip itself, should not be confused with the period-luminosity-color relation.* The Cepheid instability strip defines a range of luminosities, colors, and periods over which pulsation is a stable mode for the star and is therefore an observable. But this constraint does not control the detailed correlations between period, temperature and luminosity, nor does it control the interdependence of the observed parameters for individual stars. In much the same way, the main sequence is a strip of stable hydrogen burning, where stars are constrained to spend a large fraction of their luminous lifetimes. While the gross details of the main sequence are controlled by a single parameter (the mass of the star) the individual stars in this narrowly defined range (the main sequence) still obey the two-parameter Stephan's Law.

The Cepheid constraint, in the form of an instability strip, controls the *statistical* properties of the *ensemble* of Cepheid variable stars. As such, physical laws *external to Stephan's Law* are responsible for the now famous group statistical trends of the period-luminosity and period-color relations. But these trends are incomplete (and even sometimes misleading) descriptors for individual stars: that is to say *the properties of individual stars can never be accurately defined by the constraints on the PLC relation, but only by the PLC relation itself.*

To illustrate how fundamentally different these two concepts (of the underlying equations and the overlaying constraints) are, the interested reader is referred to an earlier paper on the subject (Madore & Freedman 1991). However, the flavor of the argument

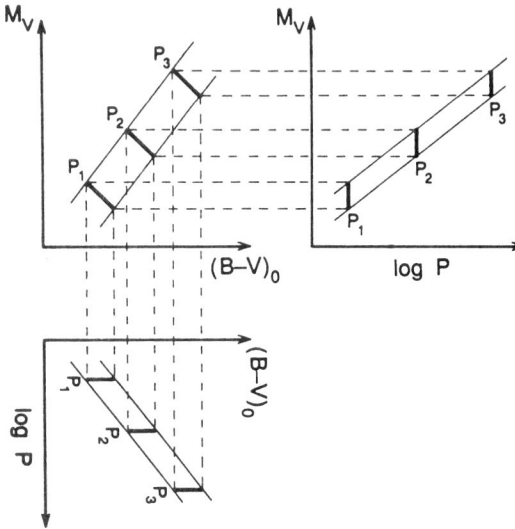

FIGURE 8. Projections of the Instability Strip. The Cepheid instability strip is shown plotted in the upper left-hand panel. Lines of constant period (P_1, P_2, P_3, etc) are shown as thick lines crossing between the red and blue edges of the instability strip, which act as constraints on the underlying PLC relation which would otherwise fill the plane, as in Figure 3. The broken horizontal lines show how the instability strip and the lines of constant period map over into the period-luminosity plane, while the vertical lines show the mapping down into the period-color plane (which can be viewed in its normal orientation by rotating the diagram 90 degrees counterclockwise.

can be had from considering Figure 8. The solid slanting lines in each figure represent the lines of constant period for the same underlying PLC relation. The heavy pair of lines cutting the constant-period lines represent two extreme examples of hypothetical boundaries to the instability strip inside of which stars are allowed to pulsate. If we now project each of these diagrams into alternate representations of the data, using the period as the abscissa, we get the corresponding period-luminosity and period-color relations as given in Figure 9 and 10. In the period-luminosity representation of the data contained in Figure 9, (where the instability strip is essentially vertical in the CM diagram), there is a strong *statistical* trend of luminosity with period. However in the corresponding period-color plot (Figure 10), there is no correlation at all, in a statistical sense; yet for individual stars the period, luminosity and color are in any case perfectly correlated through the period-luminosity-color relation. Conversely, in Figure 9, there is no statistical Period-Luminosity relation; there is a very strong Period-Color relation, and yet again precisely the same PLC relation is generating the data. Only the constraints have changed.

The distribution of observed data points (drawn from the underlying PLC) may be constrained both by observational selection effects and/or by different physics, without necessarily violating the two-parameter period-luminosity-color relation. Accordingly, differences in the statistical correlations (PL, PC, *etc.*) do not, of necessity, signal deeper-seated differences in the properties of the individual stars. On the other hand differences

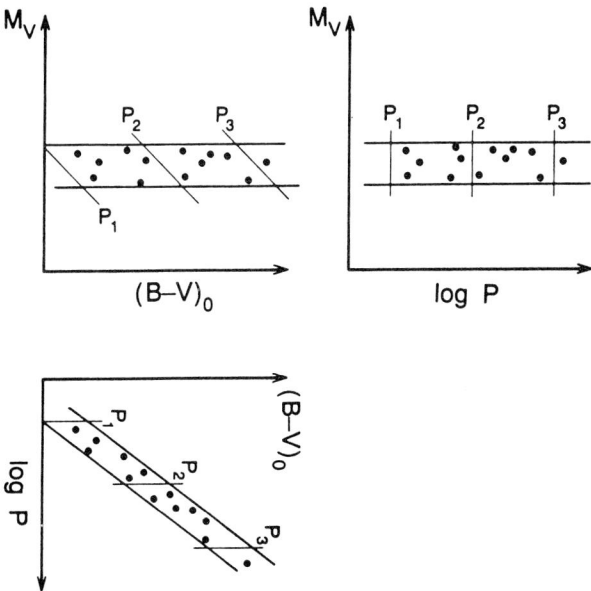

FIGURE 9. The same as Figure 8 except now a different instability strip is imposed on the underlying universal PLC. This instability strip is horizontal in the CM diagram resulting in a definite statistical correlation of color with period (lower panel, rotated by 90 degrees), but *no trend of luminosity with period* (right-hand panel).

in the bulk properties of an ensemble may be taken as fair warning that the detailed physics is changing too. In fact it is difficult to understand why and how the two properties could be decoupled.

It is of course historically very important that the instability strip is naturally oriented in such a way as to give not only a statistical PL, but also a statistical PC relation, for it was in attempts to utilize both of these relations (and to understand the "scatter" about their means) that ultimately led to the empirical formulation of a PLC (Sandage 1958).

14. A reddening-free formulation of the PL relation

The absolute calibration of colors and/or magnitudes of Cepheids is obviously wrought with many traps and uncertainties in methodology and procedure. One very appealing way to proceed is in an *implicit* formulation and calibration of the problem. Rather than explicitly solving for the reddening/extinction star by star, one can begin quite differently:

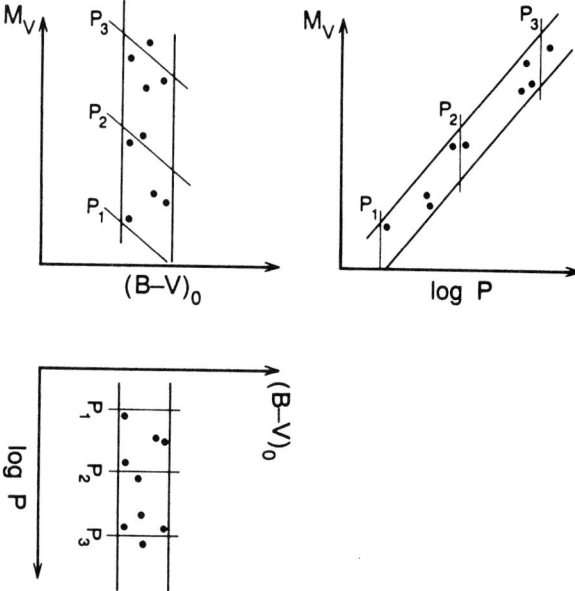

FIGURE 10. The same as Figure 8 except now a second instability strip is imposed on the underlying universal PLC. This instability strip is vertical in the CM diagram resulting in a definite statistical correlation of luminosity with period (right-hand panel), but showing *no trend of color with period* (lower panel, rotated by 90 degrees).

First adopt a standard reddening line for the color system being employed, then form linear combinations of the magnitudes and colors so that the numerical quantity formed cancels the reddening (star by star) without ever explicitly determining it. Then this formulation can be used both for the calibration objects and for the target Cepheids. Forming this reddening-free magnitude directly parallels the more common reddening-free application to colors, such as $Q = (U-B) - X(B-V)$ as defined by Johnson (1963). An example follows.

Suppose that the observed magnitude of the Cepheid is V and its apparent color is $(B-V)$. As in the standard procedure, where individual reddening corrections are made, we assume that the ratio of total-to-selective absorption is known from independent determinations, and that its value is $R = A_V/E(B-V)$. Then it naturally follows that a reddening free quantity W (called the Wesenheit function by Madore 1982) can be formed where $W \equiv V - R(B-V)$. A simple expansion of the relevant terms shows that for small amounts of reddening the numerical value of W is independent of extinction, and equal to the value that would be calculated if the intrinsic magnitudes and colors were known (which they now need not be). That is, by definition,

$$W = V_0 + A_V - R\,[(B-V)_0 + E(B-V)]$$
$$W = V_0 + R\,E(B-V) - R\,(B-V)_0 - R\,E(B-V)$$
$$W = V_0 - R\,(B-V)_0 \equiv W_0$$

There exists some confusion in the literature as to the real motivation and physical justification for creating W for Cepheids. W in fact is a reddening-free quantity for stars, not exclusively for Cepheids. It is strictly defined by the properties of the interstellar medium, not by any properties of the stars to which it is to be applied. One is not at liberty, for example, to change the value of R from its normal value of 3.3, unless that correction derives from an independent study of the extinction law itself. Accordingly, with the exception of the small known dependence of R on $(B-V)$ and $E(B-V)$ itself, (manifest as non-linear curvature terms at large values of these quantities; true for all applications of reddening corrections, not just W) one cannot manipulate the form of W at all.

The reason for some of the confusion dates back to a similar realization of the PLC introduced by van den Bergh (1975), also called W. In that first appearance of a linear combination of V and $(B-V)$ the complete reddening independence of the function was not fully implemented, as will become clear now. In the following discussion we will use W_{BFM} and $W_{vdB} = V - \beta(B-V)$ to distinguish between the differing definitions used by Madore (1982) and van den Bergh (1975), respectively.

For Cepheids one can begin to see what W_{BFM} is, both in terms of its parent relationship, the period-luminosity-color relation, and in terms of its own definition. Adopting a linearized form of the period-luminosity-color relation we have,

$$V = V_0 + A_V$$
$$V = M_V + mod_0 + A_V$$
$$V = \alpha \log P + \beta\,(B-V)_0 + \gamma + mod_0 + A_V$$

where mod_0 is the true distance modulus. By definition,

$$W_{BFM} \equiv V - R\,(B-V)$$

so, by substituting the period-luminosity-color relation into the definition of W_{BFM} we get

$$W_{BFM} = \alpha \log P + \beta\,(B-V)_0 + \gamma + mod_0 + A_V - R\,(B-V)$$

or, expanding the reddening terms, regrouping and cancelling, we get

$$W_{BFM} = \alpha \log P + [\beta - R](B-V)_0 + \gamma + mod_0$$

since, by definition, A_V equals $R\,E(B-V)$. For Cepheids, the W_{BFM} can be reformulated as a period-luminosity-color relation. Here it must be emphasized that the zero point γ and the slope of the period dependence α in the W_{BFM} formulation of the period-luminosity-color relation are identical to their counterparts in the V formulation; only the coefficient multiplying the intrinsic color changes from β to $[\beta - R]$.

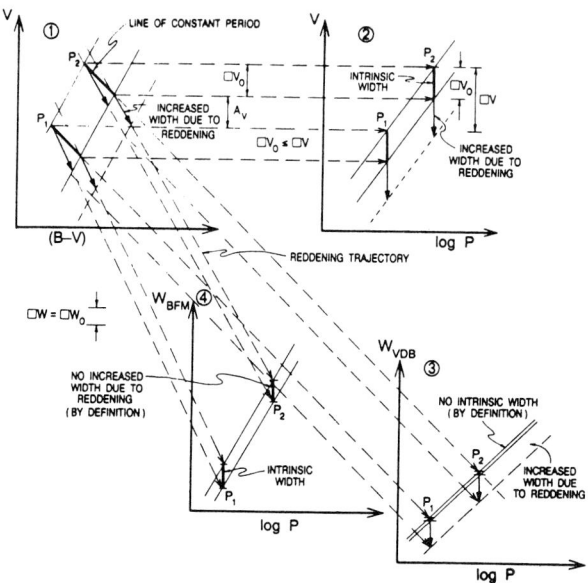

FIGURE 11. Projecting the observed instability strip, in the presence of reddening. (1) In the upper left panel the instability strip superimposed on the PLC is shown projected into the color-magnitude plane. Lines of constant period (P_1 and P_2) are shown as thick solid lines slanting down to the right. The blue and red edges of the instability strip are shown as thinner solid lines sloping down to the left. Arrows indicated the magnitude and direction of reddening which acts to increase the apparent width of the distribution by systematically scattering points to redder colors and to fainter magnitudes. A dashed line parallel to but fainter than the red edge of the instability strip illustrates the bounds of this effect. (2) The upper right panel is constructed from the first panel by projecting the instability strip and the reddening vectors into the Period-Luminosity plane. There is a systematic increase in the width of the apparent PL relation due to extinction, as in the CM diagram. (3) If the slopes of the lines of constant period are known *a priori* then the projection of the instability strip into the $W_{vdB} - \log P$ plane can be performed. By definition the intrinsic width goes to zero in this projection, but because the slope of the lines of constant period are not exactly parallel to the reddening trajectory, extinction does project into this plane and will widen the relation, and systematically shift the ridge line. (4) Since reddening trajectories are, in general, already determined *a priori* it is possible to project the instability strip into the $W_{BFM} - \log P$ plane where (by definition) extinction effects are eliminated. In this case, however the intrinsic width does still contribute to the dispersion, but the disposition of the individual stars within this strip is not effected by reddening.

Here now is the point of confusion. van den Bergh (1969) noted the numerical similarity of $R = 3.2$ to the then espoused value of $\beta = 2.7$, and at that point he chose his definition of $W_{vdB} \equiv V - \beta(B-V)$ which is quite a different approach. W_{BFM} has the advantage of being a strictly-defined, reddening-free magnitude adopting a well-determined quantity R, which does not attempt to anticipate the *a priori* unknown value of β. Furthermore residual scatter in the W_{BFM}-$\log P$ relation is also thereby distinctly of interest in its own right. Unlike scatter in the observed PL or PC relations, scatter in the W_{BFM}-log P relation cannot be due to reddening effects, be they differential or total. One is not at liberty to adjust the coefficient R for various samples of Cepheids (to minimize the scatter in the residual, for example) unless there is *independent* evidence that the ratio of total-to-selective absorption is different for that particular galaxy. The remaining scatter in a $W_{BFM} - logP$ plot will be due to a combination of intrinsic (color?) correlations and photometric errors. As is so often the case it will be an understanding of the photometry and its quality that will ultimately limit our understanding of the intrinsic interrelations.

In order to illuminate the differences between the W_{BFM} and W_{vdB} Figure 11 portrays the mapping of the PLC from its projection onto the traditional color-magnitude diagram (as an instability strip) and then into the three "period-luminosity" planes.

The upper left-hand panel shows a portion of the Cepheid instability strip in a color-magnitude diagram. The upwardly slanting solid lines give the red and blue edges of the strip, while the heavy downward-slanting lines line labeled P_1 and P_2 are representative lines of constant period. Arrows indicate the slope of a reddening/extinction line. The upwardly sloping broken line indicates the apparent red edge of the instability strip as defined by stars reddened away from the intrinsic line.

The horizontal broken lines leading from Panel 1 to Panel 2 show the first mapping of the instability strip into the $V - logP$ plane. The sloping lines of constant period in Panel 1 now become vertical, with downward extensions due to reddening increasing the apparent dispersion at fixed period.

Panel 3 to the lower right shows the effect of forming W_{vdB} given *a priori* knowledge of the value of β the slope of the lines of constant period in the PLC. As can be seen the the projected width in the W_{vdB}-log P plane is non-zero in the presence of reddening. While the intrinsic width of the instability strip does project to zero (by definition, once β is known from independent sources) there is residual widening due to reddening

Panel 4 shows the effect of forming W_{BFM}. Since the slope of a reddening line is well known from independent studies, differential reddening has no effect on the width at constant period. The only factor contributing to the width in W_{BFM} is the intrinsic width of the PLC (and as mentioned earlier photometric errors in the magnitudes and colors, which broaden all of the above relationships and projections).

Given Figure 12 it is perhaps worth noting that in the presence of differential reddening it is clearly inappropriate to determine the value of β by minimization of residuals, as would be the case in an unrestricted application of a multi-linear regression fit to PLC data, for instance. For the example at hand, one can see immediately by inspection that a slope somewhere between the true value of β and the reddening slope R would give the minimum residuals plotted as a function of period. The numerical value of that parameter for any given set of Cepheids is however of no universal significance.

15. Comments on reddening determinations

It should, by now, go without saying that the removal of reddening is essential to any study that aspires to a calibration of the intrinsic PLC relation for Cepheids. What is all too often forgotten in the process is that the systematics of this first step will not only effect the calibration, they may well predetermine it. Some authors have claimed that color-color plots for Cepheids can be accurately calibrated and confidently used to determine reddenings to Galactic and extragalactic Cepheids. We examine in some detail now one such case, that of Feast and his collaborators (hereafter referred to as the South

African Group). Their calibration is as widespread as its implications, and therefore we discuss it in some detail.

The South African calibration of the reddening determination begins with Galactic Cepheids (Dean, Warren & Cousins (1978). It has then been applied to Magellanic Cloud Cepheids (Martin, Warren & Feast 1979), where they derive a form of the PLC. And then they have further generalized the dereddening procedure for all extragalactic Cepheids of arbitrary metal abundance (Caldwell & Coulson 1985). Beginning with the specific case of the Galactic calibration the South African Group accept at least four crucial assumptions. Assumption (1): There exists a relation between $(B-V)_0$ and $(V-I)_0$ that is *dispersionless*. Assumption (2): A shape for this relation is obtained by following single stars through their pulsation cycle. Assumption (3) Since no single star cycles through all of the intrinsic colors expected for Cepheids, in practice a variety of trajectories need to be superposed by moving the narrow color-color loops formed by individual stars (of differing periods) back along reddening lines until they overlap and form an *empirically non-linear* but continuous *(imposed minimum dispersion)* sequence. Assumption (4): A zero point can be fixed by placing stars of independently known reddening (*e.g.,* Galactic cluster Cepheids and very near-by field objects) into this relation. And finally, Assumption (5): Metallicity corrections can be found that consist of simple zero-point shifts on the colors as deduced by the authors from published theory. This formalism and calibration has been adopted in one form or another by the South African Group for all of their subsequent analysis of Magellanic Cloud data and their calibration of the PLC as they have applied it to the extragalactic distance scale (including most recently Laney & Stobie 1994).

We now illustrate the dangers inherent in each of these assumptions. We start with the generally accepted, linear form of the PLC. It is written twice below: once with the color $(B-V)_0$, and then again with the color $(V-I)_0$ as the temperature indicator:

$$M_V = \alpha \log P + \beta (B-V)_0 + \gamma$$
$$M_V = \delta \log P + \epsilon (V-I)_0 + \eta$$

We then equate these two expressions and regroup like terms, eliminating M_V, thereby leading to the following equation, applicable to the color-color plane:

$$(B-V)_0 = \left[\frac{\delta-\alpha}{\beta}\right] \log P + \left[\frac{\epsilon}{\beta}\right](V-I)_0 + \left[\frac{\eta-\gamma}{\beta}\right]$$

By claiming that for Cepheids there is a unique and dispersionless relation between any two intrinsic colors (despite empirical evidence to the contrary, Cousins 1978a,b) the South African Group is explicitly ignoring the period dependence in the above equations. The tacit assumption then is that both colors are insensitive to gravity effects, and therefore the combination of the equations given above are forced by them to be degenerate in period (*i.e.,* $\alpha = \delta$). The consequences of such an assumption are two-fold: not only are the derived reddenings wrong at the level that surface-gravity difference effects from star to star are manifest in the time-averaged colors, but the inferred intrinsic colors are deceiving. If the South African calibration tracks the ridge line of the true instability strip in the color-color plane then the following statements can be made: Obviously the stars residing on the constant period lines intrinsically to the blue of the mean will systematically have their reddenings underestimated by this method (while stars whose intrinsic colors have them on the red side of the instability strip will have their reddenings overestimated). But moreover, as Figure 13 illustrates, should the relative slope of the constant-period line be less than the slope of the reddening line in the color-color plot,

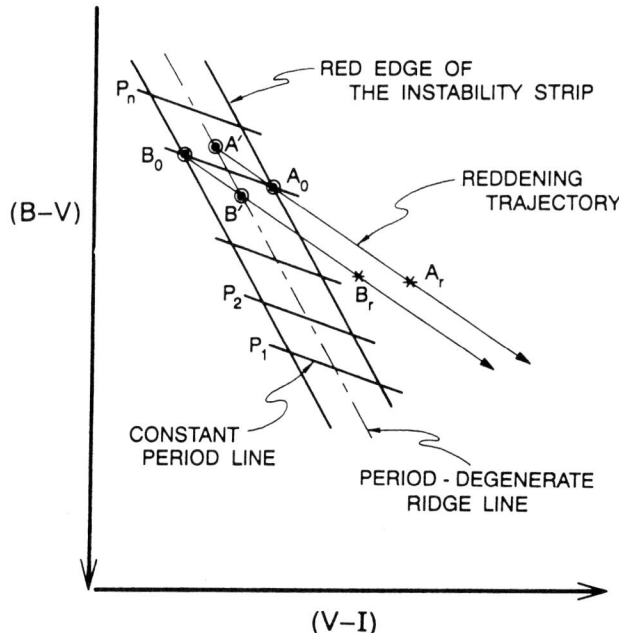

FIGURE 12. Projection of the PLC into the $(B - V) - (V - I)$ color-color plane. As shown by foregoing equation involving (B-V),(V-I) and log(P) in Section 15, the projection of the Cepheid PLC into the color-color plane results in a finite region crossed by lines of constant period and bounded by the blue and red edges of the instability strip. Representative reddening trajectories are shown passing through the ends of a typical constant-period line such that their projection back onto the central ridge line illustrates the reddening errors made if one erroneously adopts a dispersionless relation between these two colors. Notice how an a Cepheid that intrinsically resides on one (red/blue) edge of the instability strip gets forced to the opposite (blue/red) edge by this incorrect procedure.

then the inferred intrinsic position of the stars within the instability strip will be totally inverted from their true mapping!

While ignoring the period dependence in a second PLC relation may be a convenient and simplifying first approximation to a difficult problem, continuing on to add higher-order non-linear (curvature) terms, and also use theoretical displacements of the origin (attributed to metallicity effects) seems to be, at the very least, somewhat premature.

Furthermore, validity of Assumption (2) is questionable. It explicitly equates the instantaneous, time-dependent behavior (of luminosity and color) of an *individual* Cepheid of fixed mass, period, and dynamically evolving surface gravity (as it cycles through radius, luminosity and temperature with phase) with the time-averaged properties of *an ensemble of* stars each having different masses, different mean surface gravities, temperatures and periods. However, there is no *a priori* justification for assuming that the behavior of a single star during its cycle is anything more than qualitatively indicative of the way in which the (time-averaged properties of the) instability strip are mapped in luminosity and color when changing from star to star. For example, no one seriously considers determining the slope of the color term β, in the PLC by looking at how an individual star changes its luminosity with color (as a function of phase). Likewise no one should seriously use the color-color trajectory of individual Cepheid to calibrate the complex mapping of a variety of stars and their PLC into the color-color plane.

Despite the small formal uncertainties in the resulting PLC fit determined by this methodology (see Feast 1991 for a recent example) it should be emphasized that almost all methods which "correct" for differential absorption within the PL relation will give rise to a PLC of smaller (apparently intrinsic) scatter. Furthermore, the determination of the color term β in the PLC depends sensitively and explicitly on there being accurate individual reddenings for each of the stars entering the calibration. For the reasons outlined above, we do not believe that an accurate value of β has yet been determined precisely because the systematics of determining reddenings have yet to be fully appreciated or adequately addressed.

Because of its inherent complexity we are not in a position to solve the reddening problem at this time; however its impact on the distance scale can be minimized by accepting that reddening is a systematic problem, and realizing that it can be dealt with effectively by at least three available means, which involve (1) moving as far to the infrared as is practical, so as to reduce the extinction problem to the level of other systematic and random errors, (2) combining multiwavelength (visual to near-infrared) data for significant numbers of Cepheids in a given galaxy, and determining the ensemble-averaged extinction (and true modulus) using an independently calibrated wavelength-dependent extinction law, or (3) using reddening-free formulations which are designed to cancel out any and all extinction on a star-by-star basis, without ever attempting to determine the amount explicitly. Each of these alternatives are dealt with in some detail in this review and its other appendices.

It is clear however that if an *explicit* solution for individual reddenings is ever to be found, it will most likely come through an investigation of extragalactic, not Galactic, Cepheids. The LMC has acted as the focal-point for calibration purposes for some years now, primarily because it has a large population of Cepheids with known periods, and because it is sufficiently close that accurate photometry can be obtained for its stars with relative ease. However, the LMC Cepheids do individually suffer from some degree of extinction internal to the LMC itself, and also varying amounts of foreground Galactic extinction. The SMC has a similarly large population of Cepheids. Although it is somewhat further away than the LMC, it is not excessively so. And while it is generally accepted that in comparison to the LMC the extinction is less internal to and in front of the SMC, it is now known (primarily from studies of the Cepheids!) that the back-to-front geometry of the SMC is such that appreciable differential modulus residuals are affecting the magnitudes. Any empirical correlations between reddening-corrected colors and extinction-corrected magnitudes would have this (geometric) noise to contend with.

We suggest then that the best place for future work on the intrinsic calibration problem is not the Magellanic Cloud system, but the Local Group galaxy IC 1613. The foreground reddening to IC 1613 is, by all estimates, very low and probably quite uniform, considering the high Galactic latitude and small angular size of this galaxy as compared to the either of the Magellanic Clouds. In a like way to the SMC the extinction internal to IC 1613 also appears to be quite small. With the surveys of Sandage (1971) and Carlson & Sandage (1990) now complete, there is also a sizable population of Cepheids in IC 1613 to work with. Of course crowding is more of a problem for photometry of individual Cepheids in IC 1613 as compared to the LMC, for example; but considering

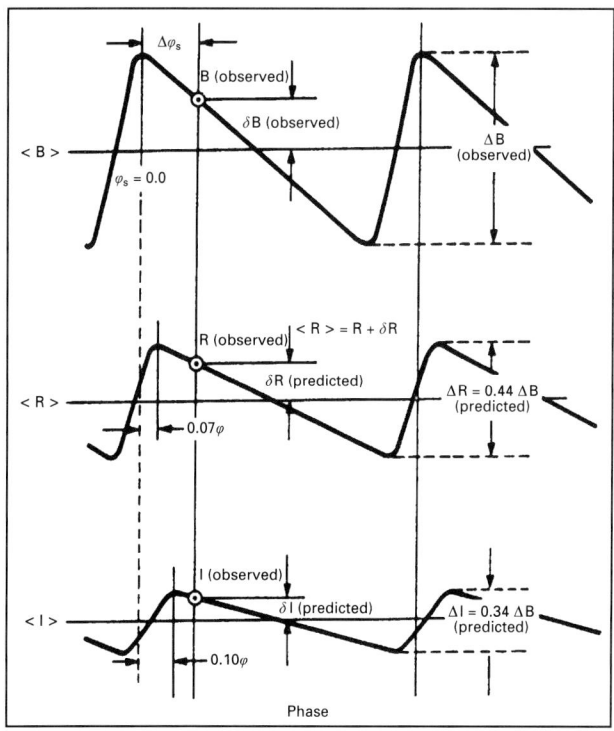

FIGURE 13. Cepheid light curves as a function of observed wavelength. A schematic illustration of the relative decrease in amplitude and shift in phase associated with observing Cepheids at progressively longer wavelengths (adapted from Freedman 1988a).

the success had with photoelectric photometers using aperture sizes in excess of 10 arcsec when working on LMC/SMC Cepheids it is realistic to expect that point-spread-function fitting routines (effectively working on one-arcsec scales) will be able to do at least as well. And the quality of $BVRI$ light curves obtained for Cepheids in NGC 300 (Freedman et al. 1992), at a distance about three times further than IC 1613, seems to bear out this expectation. Of course IC 1613 is only one galaxy, representing a single cut of metallicity, still the Cepheid calibration is at that stage where even one solid observational study can contribute a great deal to the effort.

16. Comparisons with other distance indicators

As recently as a decade ago, the range of published Cepheid distances to nearby galaxies was sufficiently large that by adopting one or another calibration could result in differences in the Hubble constant of almost a factor of two. Fortunately, a decade later, distance determinations to nearby galaxies have been obtained with new CCD data using a number of independent techniques (Cepheids, RR Lyraes, TRGB, and even type II supernovae in the case of SN 1987 in the LMC). Now that photometry with linear detectors is available for a variety of methods, and corrections for reddening can be applied, the

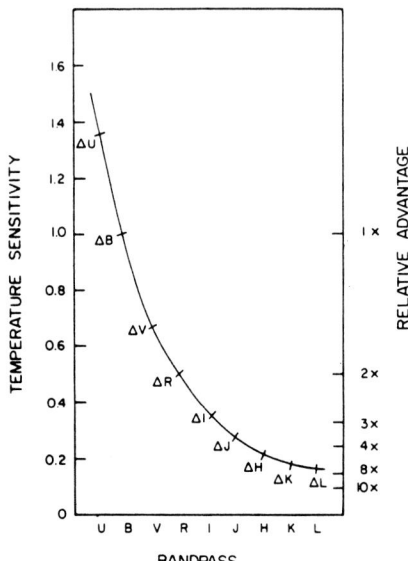

FIGURE 14. Amplitudes and/or instability strip width for Cepheids as a function of increasing wavelength (adapted from Freedman, Grieve & Madore 1985). The large relative amplitudes (large strip widths) at blue and ultraviolet wavelengths are due to the higher sensitivity of monochromatic surface brightness to temperature variations at short wavelengths; low amplitudes (small instability strip widths) in the red and near-infrared are dominated by the small percentage changes in surface area, driven by the radial pulsation.

distances to nearby galaxies have converged to *full range* differences of less than 0.3 mag (*i.e.*, 15% in distance, Freedman & Madore 1993). Moreover, the excellent agreement of individual distances gives no indication of large remaining systematic errors.

However, as noted by Freedman & Madore (1993), although the Cepheid and RR Lyrae distances agree to within their stated errors, the differences are systematic (in the sense that the RR Lyrae distances are smaller than the Cepheid distances). This effect has been noted and discussed by Walker (1992), Hazen & Nemec (1992) and Reid & Freedman (1994) in the case of the LMC, and Saha et al. (1992) in the case of IC 1613. Most recently, this effect has been discussed by van den Bergh (1995). As yet unresolved are the slope and zero points of the relation between absolute magnitude and the metallicity for RR Lyrae stars, as well as the metallicity sensitivity of the Cepheid PL relations as a function of wavelength. However, the fact that the Population I Cepheid distances now agree as well as they do (to within 0.15 to 0.30 mag) with the Population II RR Lyrae and TRGB distances, is consistent with the predictions of a small Cepheid PL dependence on metallicity by Chiosi et al. (1993), and the observational result of Freedman & Madore (1990).

17. The key project

The Cepheid distance scale provides the zero-point calibration for most of the relative distance indicators in use today: the Tully-Fisher relation for spiral galaxies, the surface-brightness fluctuation method for elliptical galaxies, the planetary nebula luminosity

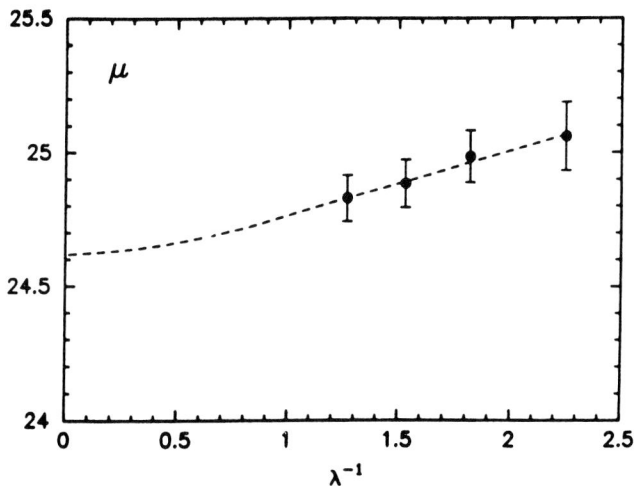

FIGURE 15. BVRI apparent distance moduli for Cepheids in M33, plotted as a function of inverse wavelength ($1/\lambda$ in inverse microns) from Freedman, Wilson & Madore (1990). The broken line is a fit of a Galactic extinction law to the data, with E(B-V) = 0.10 mag.

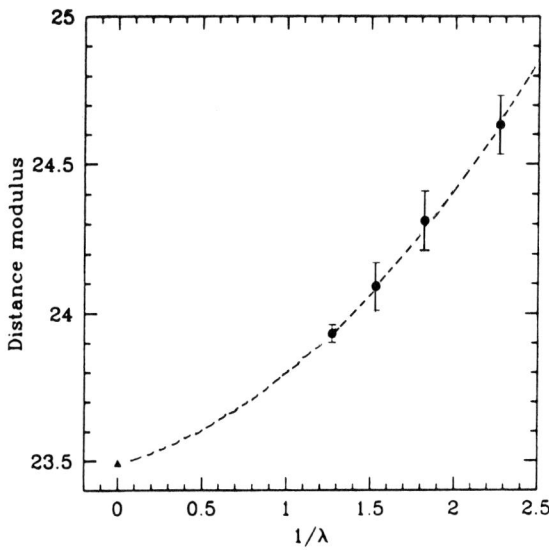

FIGURE 16. BVRI apparent distance moduli plotted as a function of inverse wavelength for Cepheids in NGC 6822. The filled triangle marks the true modulus = intercept of the fit at the origin $1/\lambda = 0.0$ for E(B-V) = 0.21 ± 0.03 mag (from Gallart, Apparicio & Vilchez 1996).

function, Types I and II supernovae, brightest stars in galaxies, and the globular cluster luminosity function (for a recent overview see Jacoby et al. 1992). The Cepheid distance scale also lies at the heart of the *HST Key Project on the Extragalactic Distance Scale* (Freedman 1994a,b,c; Kennicutt, Freedman & Mould 1995) and in several other *HST* distance scale programs (*e.g.*, Sandage et al. 1994; Saha et al. 1994, 1995; and Tanvir et al. 1995).

The Key Project on the Extragalactic Distance Scale has been designed to use Cepheid variables to determine (Population I) primary distances to a representative sample of galaxies in the field, in small groups, and in major clusters. The galaxies were chosen so that each of the secondary distance indicators with measured high internal precisions could be accurately calibrated in zero point and intercompared on an absolute basis. These data will then be used as secondary calibrations and applied to independent galaxy samples at cosmologically significant distances. Cepheid distances to the Virgo and Fornax clusters provide an alternative route to the secondary calibrations. The intention is to derive a value for the expansion rate of the Universe, the Hubble constant, to an accuracy of 10% (for additional details see Kennicutt, Freedman & Mould 1995, Freedman, Madore & Kennicutt 1997).

17.1. *Goals*

A measurement of the Hubble constant to 10% provides an immense challenge given the history of systematic errors in the extragalactic distance scale. For this reason, the Key Project has been designed to allow many independent cross-checks of both the primary and secondary distance scales. The goals of the Key Project are described in more detail in Kennicutt, Freedman & Mould (1995) and Freedman et al. (1994a). Briefly, there are four primary goals: (1) To discover Cepheids, and thereby measure accurate distances to spiral galaxies located in the field and in small groups that are suitable for the calibration of several independent secondary methods. (2) To make direct Cepheid measurements of distances to three spiral galaxies in each of the Virgo and Fornax clusters. (3) To provide a check on potential systematic errors in the Cepheid distance scale via independent distance estimates to the nearby galaxies, M31, M33 and the Large Magellanic Cloud (LMC). And (4) to undertake an empirical test of the sensitivity of the zero point of the Cepheid PL relation to metallicity as described previously.

17.2. *First Results*

Prior to the 1994 repair mission, the Key Project team was still able to undertake a search for new Cepheids in the nearby galaxy M81. These observations were undertaken to provide a test of the new discovery algorithms (Madore & Freedman 1998) which were designed to optimally detect Cepheids with a range of (unknown) periods, using a minimum of spacecraft time, and further restricted by the small (60-day) observing windows available only once in any given year. 30 Cepheids were discovered in two fields searched in M81 and a reddening-corrected distance modulus of 27.80 ± 0.20 mag was derived (Freedman et al. 1994b). Previous ground-based attempts to discover variables in this galaxy yielded only two confirmed Cepheids, one of which was intentionally targeted by the Key Project as a test of the search procedure. This Cepheid was recovered and confirmed to have a period in agreement with the more extensive ground-based determination derived from decades worth of data.

Immediately following the December 1993 repair mission, BVR images of the Virgo spiral galaxy M100 were obtained as part of a collaboration between the WFPC2 IDT and the H0 Key Project teams. ALLFRAME photometry was obtained for over 30,000 stars. By overlaying the position of the mean Cepheid instability strip on the resulting color-magnitude diagrams, it was possible to demonstrate that stars were present with the magnitudes and colors expected for Cepheid variables at the distance of the Virgo cluster. Given this success, a sequence of 12 V and 4 I exposures was begun in April 1994. Twenty high signal-to-noise Cepheid variables were found, from which a reddening corrected distance of 17.1 ± 1.8 Mpc (or a distance modulus of 31.16 ± 0.20 mag) was determined to M100 (Freedman et al. 1994a). Allowing for the uncertainty in the position of M100 with respect to the Virgo cluster core, in addition to the uncertainty in the Virgo cluster recession velocity, a preliminary value of the Hubble constant of $H_0 = 80 \pm 17$ km/sec/Mpc was determined. A discussion of the random and systematic errors is found in Freedman et al. (1994a). Recently, a new determination of the distance to M100 has been made based on a larger sample of over 50 Cepheids and an improved calibration (Ferrarese et al. 1996). A value of 15.8 ± 1.5 Mpc is obtained, in good agreement, to within the measurement uncertainties, with the earlier value.

Other galaxies currently being analyzed as part of the HST Key Project include NGC 925 (Silbermann et al. 1996) in the NGC 1023 Group; NGC 3351 in the Leo I group (Graham et al. 1997); two fields in M101, discussed above in the context of extending the metallicity test (Kelson et al. 1996; Stetson et al. 1998, in preparation; Kennicutt et al. 1998, in preparation); NGC 7331 (Hughes et al. 1998, in preparation), a Tully-Fisher calibrator in the field; NGC 4414 (Turner et al. 1998, in preparation), a distant and fairly inclined early-type spiral useful for calibrating a number of secondary methods including type Ia supernovae. Recently, NGC 1365, a galaxy in the Fornax cluster (Silbermann et al. 1998; Madore et al. 1998, in preparation) has been observed for Cepheids, and the results of that survey are discussed in the closing sections of this lecture series. Other galaxies completed at the time of writing include NGC 3621 (Rawson et al. 1998) and NGC 2090 (Phelps et al. 1998).

18. Other ground-based work

18.1. *Galactic Cluster and LMC Calibration*

While currently the most commonly adopted route for calibrating the zero point of the Cepheid PL relation is through the Large Magellanic Cloud, a more traditional approach is to use individual Cepheids in Galactic clusters. Following this route, main sequence fitting and B-star reddenings can provide (in principle) individual moduli, absolute magnitudes and intrinsic colors for the member Cepheid(s). In practice this approach has turned out to be limited in a number of ways: very few clusters have known Cepheids as bona fide members, those examples generally only cover periods up to about ten days, and many of the clusters are either very sparse or behind considerable amounts of (perhaps variable) extinction, or both.

While sample size and the period distribution of that sample is not likely to be changed considerably, the treatment of the reddening and measurement of the true distance moduli can be re-examined with new technology. In a long-term collaboration Persson et al. (1996) have embarked on a project at Las Campanas and Palomar of infrared imaging of the main sequences of all Galactic clusters containing classical Cepheids. These clusters include NGC 7790, NGC 129, NGC 6649, NGC 6664 and Lynga 6. A parallel study of many of these clusters is also underway by Fry & Carney (1997, private communication).

Finally, in addition to the recalibration of Galactic clusters, multiple epochs of JHK photometry are being obtained for a sample of over 90 Cepheids in the LMC. This sample will serve as a fiducial calibrating sample for anticipated NICMOS observations of Cepheids. Also at Las Campanas we have been obtaining JHK observations of the known Cepheids in IC 1613 and NGC 300.

18.2. Recent Ground-Based Searches for Cepheids

Easily accessed only from the southern hemisphere, the South Polar (Sculptor) Group contains the last of the relatively nearby spiral galaxies to be searched for Cepheids from the ground. Three of its members (NGC 247, NGC 300 and NGC 7793) are of the proper type and inclination to become Tully-Fisher calibrators, given independent distance measurements. Using the CTIO 4m and the Las Campanas 2.5m we have been following up the Cepheids discovered in NGC 300 (Graham 1984, Madore et al. 1987) and monitoring NGC 247, NGC 7793 (and NGC 253) for new variables. Based on *BVRI* CCD data of 10 Cepheids Freedman et al. (1992) determined a distance to NGC 300 of 2.1 ± 0.1 Mpc.

19. Helium core flash and the tip of the red giant branch as a primary distance indicator

19.1. Introduction

In his 1930 *Harvard Observatory Monograph:* "Star Clusters," Harlow Shapley outlined various methods for the estimation of distances to Galactic globular clusters. Noteworthy among these methods is the one that assumes that the apparently brightest (giant) stars are all of the same absolute magnitude. Tested for consistency in those cases where more than one method of distance determination could be made (with RR Lyrae stars, for instance), and roughly corrected for foreground contamination (choosing the magnitude of the sixth-brightest star as a robust estimator), Shapley went on to map out the size scale of the Galaxy as defined by the old Population II globular cluster system. Although the exact details of his approach are now known to be flawed, the method as applied was of sufficient precision that a revolutionary view of the size of our Galaxy was obtained, and the Milky Way as an island universe started to take on a tangible reality of its own.

Before pursuing the historical path relating to the successful application of brightest red giant stars as *extragalactic* distance indicators we digress slightly in the next section to present a set of criteria that any extragalactic distance indicator might be judged by. It will rapidly become clear that many of these criteria are probably mutually exclusive in a practical sense (ultra-luminous, locally calibrated and theoretically understood), but ideals are seldom realized in full measure in the real world. Nevertheless we can establish metrics for relative performance.

20. The ideal distance indicator

In an extragalactic context the requirements for an ideal distance indicator are indeed demanding:

(1) High luminosity is a pre-requisite, given the enormous distances that must be bridged in passing out of the Milky Way, beyond the Local Group and eventually into regions of pure and undisturbed Hubble flow.

(2) The ideal distance indicator (be it derived from a luminosity or a measure of size) needs to have the lowest possible intrinsic dispersion. Low-precision distance indicators all eventually run afoul of bias at large distances where only the largest and/or intrinsically brightest objects fall into limiting-case samples. Blindly calibrated against the mean these exceptional objects always end up underestimating distances, if not properly corrected for selection effects.

(3) A secure empirical calibration requires that examples of the distance indicator be found locally so that detailed and repeated tests can be made at high signal-to-noise, before applications are made at the limits of detectability.

(4) In order to anticipate problems, predict trends, and understand exceptions, a firm theoretical basis for the distance indicator is required. In the consideration of many distance indicators a theoretical understanding is all too often lacking.

(5) Lacking a detailed theoretical understanding of the distance indicator, one might require that a demonstrable empirical proof that the luminosity (or size) being used is

CATALOGUE OF GLOBULAR CLUSTERS.—

N. G. C.	Ellipticity	Orientation	Photographic Magnitude				Adopted Modulus
			Var.	Bright	6th	30th	
104	8	−55	.....	13.09	12.4	13.4	14.17
288	9	.....	.....	14.80	14.5	15.1	15.81
362	8	+65:	15.5	14.12	13.5	14.8	15.55
1261	9.5	.....	.....	.....	.....	.....	16.72
1851	9	−75	.....	.....	.....	.....	15.78
1904	9	+ 5	.....	15.29	15.01	15.72	16.54
2298	8	+39:	.....	.....	.....	.....	17.12
2419	9	−56	.....	.....	.....	.....	17.41
2808	8	+84	.....	14.9	14.3	15.4	16.05
3201	9	.....	14.52	13.52	13.3	13.8	14.81

FIGURE 17. A reproduction of part of Shapley's original 1930 table showing the first use of the brightest red giants in globular clusters to determine distances.

insensitive to (or at least can be easily corrected for) known effects. Age differences, chemical composition variations, population size, environment differences and, of course, interstellar reddening are just a few of the known outstanding variables that should be considered.

(6) The ideal distance indicator should also be universally available. It should be found in spiral galaxies, ellipticals and irregulars, if clusters, groups and the field are to be equally sampled without bias. This universality constraint obviously weighs heavily against Population I distance indicators, because of the lack of any significant star formation in elliptical and S0 galaxies at the present time.

(7) Singular events are to be avoided. The hope is to find a distance indicator that has more than an occasional or unpredictably fleeting presence in a galaxy, but rather that it is abundant and always found whenever and wherever it is needed. Follow-up observations or applying new technology is problematic for one-time, historical events. Supernovae of all types suffer from this problem.

(8) Finally, the identification of the ideal distance indicator should be unambiguous. Two, three or more types and sub-types of the distance indicator, each having subtle differences in their properties inevitably require additional observations to resolve differences. Cepheids (Classical Population I versus Population II W Virginis stars) suffered from this flaw early in the history of their calibration; supernovae suffer from the (growing) diversity of types to this very day (compare the observations required, as the singular event is unfolding, to definitively distinguish between a supernovae of SN Type II as opposed to a supernova of SN Type Ia [Branch-Normal], for example).

21. Some history concerning the red giant branch

Following Shapley's brave application of brightest red giants to the Galactic distance scale, continued applications were ironically being made at the same time as new obser-

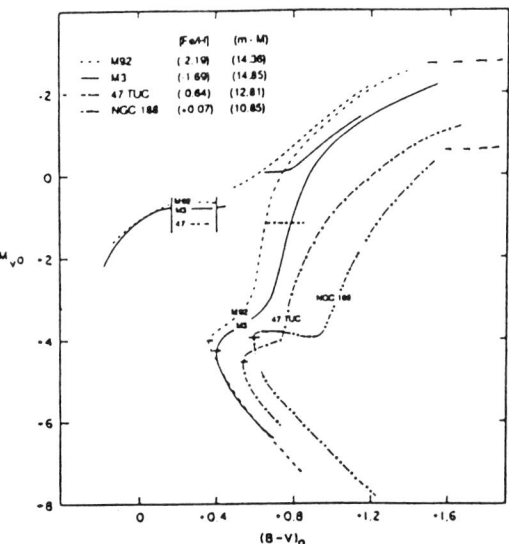

FIGURE 18. A schematic representation of the effect of metallicity on the morphology of the red giant branch when observed at optical (UBV) wavelengths

vational data were being assimilated (sometimes by the very same people) which were showing that the method was fatally flawed. Flawed in the optical, that is. Late into the 1960's and 70's Arp in his review and even Harris (1974) in his doctoral thesis used the apparent B-band luminosities of the brightest (few) red giants as statistical indicators of distances to globular clusters. And this proceeded in the face of compelling evidence (and theory) that the giant branch morphology was a sensitive function of age and most especially a function of metallicity (see Figure 19 for examples). Bolometric corrections, driven primarily by atmospheric line-blanketing effects in the optical resulted in differences of up to 2 mag between the absolute magnitudes of the brightest red giant stars in metal-rich (line-blanketed) systems as compared to those in metal-poor globular clusters. It was not until the detailed and calibrated studies of the color-magnitude diagrams (CMDs) of Galactic globular clusters by Arp, Sandage, Wildey and others, using more modern photometric systems, that the absolute UBV magnitudes of the brightest giants in globular clusters were found to be less than ideal as distance indicators. The detailed morphology of the giant branch, (especially its height above the horizontal branch) and its terminal color were found to depend critically on the mean metallicity of the system.

The application of the then new CCD detectors to unexplored wavelengths led Mould & Kristian (1986) to reconsider the tip of the red giant branch as a potential distance indicator. With admittedly sparse statistics, they noted in a series of papers that the dominant feature in the CMDs of Local Group galaxy halos was the presence of a giant branch population of stars, showing a wide range of colors, but generally terminating at high-luminosity at a well-defined and fairly constant magnitude. Of the four galaxies that they initially imaged (M31, M33, NGC 147 and NGC 205) only M31 had an independent and reasonably secure distance published (based on Cepheids), while the distance to M33 was still under considerable debate. At the conclusion of their 1986 paper, Mould &

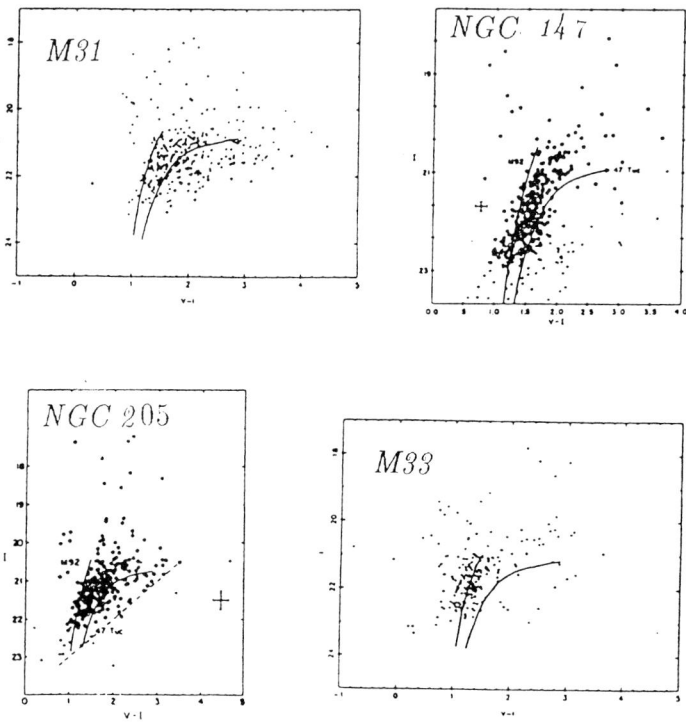

FIGURE 19. A sampling of early color-magnitude diagrams for Local Group galaxies and their halos.

Kristian reasonably called for more work before the TRGB method could be considered a mature distance indicator. Their requests included: (1) The examination of a larger sample of Local Group galaxies, and (2) an absolute calibration through Galactic globular clusters based directly in the Cousins I-band system.

Answering the call for a direct calibration of the method using Galactic globular clusters Da Costa & Armandroff (1990) published a grid of standard globular cluster giant branches observed in the Kron-Cousins VI system. And they found 'acceptable agreement' between the bolometric magnitudes of the brightest giants and the theoretical predictions for the luminosities of helium core flash. At the same time considerably more observational work was being undertaken which supported the use of the TRGB as a distance indicator in the I band. Freedman (1988) made the first combined CCD-based determination of a TRGB distance and a Cepheid-based distance for IC 1613. While confusion over the disparate Cepheid distance moduli to M33 finally resolved itself with the application of multiwavelength CCD observations of the variable stars in that galaxy (Freedman, Wilson & Madore 1991), the result again confirmed the TRGB calibration. A comprehensive review and survey of published results pertaining to the comparison of

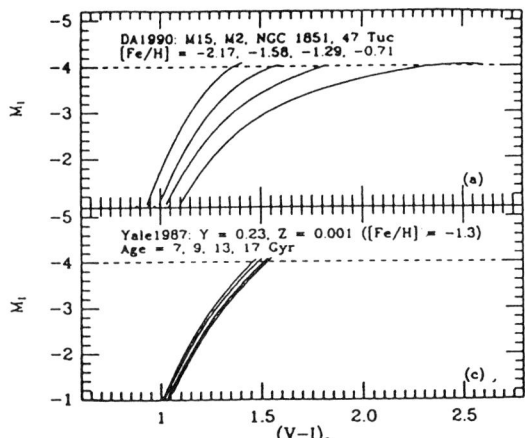

FIGURE 20. Sensitivity of the TRGB to age and metallicity effects (adapted from Lee, Freedman & Madore 1993). The upper panel shows fiducial giant-branch lines for a series of Galactic globular clusters having a range of metallicities. Note how the terminal I-band magnitude is a very weak function of chemical composition. The lower panel shows the (negligible) effect of age (7-17 Gyr) on the RGB peak luminosity, based on Revised Yale isochrones calculated for [Fe/H] = −1.3 dex.

tip distances in comparison to RR Lyrae and Cepheid distances was given by Lee, Freedman & Madore (1993), and the interested reader is referred to that paper for additional details and references.

The TRGB method, as applied by Lee, Freedman & Madore (1993) is based on a calibration established for the metallicity range of $-2.2 < [Fe/H] < -0.7$ dex, derived by using empirical loci of the red giant branches for four Galactic globular clusters given in the paper by Da Costa & Armandroff (1990). Da Costa & Armandroff estimated distances by adopting the metallicity–M_V magnitude relation for RR Lyraes based on the theoretical horizontal branch models for $Y_{MS} = 0.23$ of Lee, Demarque & Zinn (1990). At present, the exact calibration at the higher end of metallicity range (*i.e.*, metallicities exceeding the most metal-rich Galactic clusters) and its practical effects on TRGB distances to composite systems still remains uncertain.

22. Concerns and technical issues

There are of course a number of procedural concerns that must be attended to in both calibrating and then applying the TRGB Method to extragalactic distance determinations. We briefly discuss the most important of these in this section. Some of these problems can be anticipated and dealt with in the design phases of the observing program (for example, crowding, signal-to-noise, *etc.*); some require careful post-processing (the removal of cosmic rays, background galaxies, *etc.*), or require complementary observations (to distinguish AGB stars, and deal with high-metallicity effects). Others will hopefully diminish in importance with additional time and effort (such as zero point uncertainties).

22.1. Detecting the Tip

Until recently, the apparent magnitude of the TRGB was simply estimated by visual inspection of the color magnitude diagram. Lee, Freedman & Madore (1993) introduced a quantitative edge-detection method (the Sobel filter) to both identify the position and estimate the uncertainty of the TRGB observed luminosity. When this filter is convolved with the luminosity function, the output response function peaks where the discontinuity is the largest. Further refinements to this method have subsequently been introduced by Sakai, Madore & Freedman (1996a,b).

22.2. Asymptotic Giant Branch Stars

Normal globular clusters have asymptotic giant branch stars. These evolutionarily-advanced stars are found at magnitudes above and below the TRGB, loosely paralleling the red giant branch to the blue and exceeding the first ascent red giant branch stars at higher luminosities. In mixed-age populations intermediate-mass stars can also evolve up the AGB sequence populating those luminosity intervals above the TRGB with even brighter and often very much redder stars than normally found in old, metal-poor, pure Population II systems.

AGB stars can be a source of additive noise in the luminosity function near the TRGB, as they can be slightly brighter than $M_I \sim -4.0$ mag. Fortunately, the AGB luminosity function is known to be flat and/or only rather slowly rising, thus making it extremely unlikely to be misidentified with the much more abrupt TRGB signature (when the edge–detection filter is passed through the luminosity function, it would respond more prominently to the pronounced discontinuity of the RGB tip). The AGB problem can also be minimized by observing the red giant branch population preferentially in the outermost halos of the galaxies where intermediate–age populations are less likely to be present. Furthermore, the added advantage of working in the lower-surface density halos is that crowding effects are also minimized.

22.3. Red Supergiants

Depending on the mix of Pop I and II, very massive stars can add contamination to the reddest portions of the CMD by contributing evolved supergiant stars. These objects are in proportion to the recent star formation rate and are spatially co-located with their immediate progenitors, the blue (plume) main sequence stars. Avoiding regions of easily identified active star formation reduces this contaminant effectively to zero. Nevertheless these red supergiants are both rare (in comparison to TRGB stars) and show no characteristic discontinuity in their gradually increasing apparent luminosity function. They can add noise to the TRGB detection, but they are not likely to be mis-identified with it. This is especially true given that the blue population effectively predicts their presence and rough numbers.

22.4. High-Metallicity Populations

At high metallicity (beyond [Fe/H] = -0.7 dex) bright stars suffer noticeable line blanketing even in the I-band (*e.g.*, Bica, Ortolani & Barbuy 1994). However the effect is to make these high-metallicity tip stars fainter than their low-metallicity counterparts. This would be a problem (which could be calibrated in principle) for the determination of distances to pure, high-metallicity systems, but whether such 'pure' systems do exist or even could exist is unlikely. In terms of detection thresholds, mixed-metallicity populations (that is, any system that has low-metallicity stars as the precursor population to the next generation of higher-metallicity red giants) would first reveal their (bluer, brighter) low-metallicity stars, and thereby define the jump in TRGB luminosity function at a magnitude corresponding to the low-metallicity calibration independent of the fainter (high-metallicity population) tip stars mixed in below.

Although the magnitude of the TRGB has been shown both observationally and theoretically to be extremely stable at $M_I \sim -4.0$ mag, this stability has only been solidly demonstrated and calibrated in the metallicity range defined by Galactic globular clusters (i.e., $-2.1 < [Fe/H] < -0.7$ dex). At the higher metallicity end of this range, little data have been obtained to demonstrate whether the constancy of the TRGB magnitude prevails, or if not, how it changes in detail. Of course, one can reduce this uncertainty by restricting TRGB observations to the halos of galaxies where color gradients suggest lower metallicities. And, in practice, for most irregular galaxies, and the outer regions of spiral galaxies, the available lower-metallicity calibration will be sufficient.

22.5. Crowding

Crowding will limit the discovery and the photometry in all TRGB applications. However, given even a rough estimate of the expected distance the local surface brightness (in the halo) will predict the expected crowding at the magnitude corresponding to the tip. Using computer simulations, Madore & Freedman (1995) showed that the TRGB method could be applied out to 3 Mpc $[(m - M) \sim 27.5$ mag$]$ from the ground, and from space at least a factor of four further in distance $[(m - M) \sim 30.5$ mag$]$, being limited primarily by integration time rather than crowding in the latter case.

22.6. Sufficient Signal

While this may seem to be a simple matter of combining aperture and integration time to reach the requisite signal-to-noise ratio, other factors preempt extended integration. Variable seeing is a critical issue for ground-based attempts to go to the limit of this method; while the limiting case of using HST encounters a variety of subtle but important issues relating to charge-transfer efficiency, fixed pattern noise, extensive cosmic ray removal, etc. when extremely long intergations are called for. Given these limitations, with its present detectors, HST may be considered to have an operating range of ~ 10 Mpc for applying the TRGB method; beyond that, extreme care should be taken in the acquisition, processing, and interpretation of the data.

22.7. Background Galaxies and Quasars

Background galaxies as a source of noise can be dealt with in a variety of ways. First they can be resolved. In most cosmological models, galaxies are not expected to be much less than an arcsec for the magnitudes of interest in our applications, where I < 25 mag. Based on simple profile-fitting, galaxies can therefore be discriminated from stars and eliminated early in the analysis process Very distant, point-like objects are not expected to have any abrupt discontinuity in their apparent luminosity function and could at worst contribute a 'background' noise component which can be further reduced by color selection, eliminating the very bluest contaminants (certain quasars) and the very reddest (background ellipticals).

22.8. Cosmic Rays

Careful planning and post-processing can deal with cosmic-ray events in the CCD images used to detect the TRGB. Median-filtering of multiple exposures, combined with image-profile fitting selection, set to reject 'sharp' events can reduce cosmic-ray noise to an acceptable level of contamination.

22.9. RR Lyrae Distance Scale

The absolute magnitude of the TRGB rests on the globular cluster distance scale, which in turn depends on the calibration of the absolute magnitudes of the horizontal branch RR Lyrae stars. And that calibration, in form and zero point, is still controversial (see, for example, Sandage & Cacciari 1990, Renzini 1991, and Carney, Storm & Jones 1992, for the full range of opinions). We have adopted the calibration of Da Costa & Armandroff (1990) which is $M_V(RR) = 0.17[Fe/H] + 0.82$ mag. As discussed in Walker (1992), Saha, Freedman, Hoessel & Mossman (1990), and Freedman & Madore (1993), the fainter ($M_V(RR) \sim 1.0$ mag) alternative calibration disagrees with the Cepheid distance scale to overlapping galaxies at the 0.2-0.3 mag level. Our currently adopted zero point is $\sim$0.8 mag, and falls between the aforementioned extremes.

23. An overview of the theoretical underpinnings: core helium ignition

The evolution of a post-main-sequence low-mass star up the red giant branch is one of the best-understood phases of stellar evolution (*e.g.*, Iben & Renzini 1983). For the stars of interest to us in the context of the TRGB, a helium core forms at the center of the star, supported by degenerate electron pressure. Surrounding the core, and providing the entire luminosity of the star is a hydrogen-burning shell. The 'helium ash' from the shell rains down on the core increasing its mass systematically with time. The temperature of the (basically isothermal) core and therefore the luminosity generation rate at its surface are simple functions of core mass and core radius. In analogy with the white dwarf equation of state and the consequent scaling relations that interrelate core mass (M_c,) and core radius (R_c) for degenerate electron support, the core (= shell) temperature (T_c) and the resulting shell luminosity are simple functions of M_c and R_c:

$$T_c \sim M_c/R_c$$

$$L_c \sim M_c^7/R_c^5$$

meaning that as the core mass increases (and the radius shrinks) the luminosity increases due to both effects. That is, the star secularly ascends the red giant branch to ever-increasing luminosities, *and higher core temperatures*. And it is the latter increase that stops the progression in luminosity: When T_c exceeds a physically well-defined temperature (set by nuclear physics), helium in the core will ignite. While this does provide a new and additionally powerful source of energy, helium core ignition does not make the star brighter, but rather it all but eliminates the shell source by explosively heating and thereby lifting the electron degeneracy within the core. This dramatic change in the equation of state is such that the entire internal structure of the star is rearranged so quickly that the core flash (which generates the equivalent instantaneous luminosity of the entire galaxy) is internally quenched in a matter of seconds, inflating the core and settling down to a helium core-burning main sequence, far removed in luminosity and effective temperature from the RGB. This phase change marks the TRGB. And nuclear physics fundamentally controls the stellar luminosity at which the RGB is truncated, essentially independent of the chemical composition and/or residual mass of the envelope sitting above the core. This is the underlying power of the TRGB: it is a physically based and theoretically well understood distance indicator.

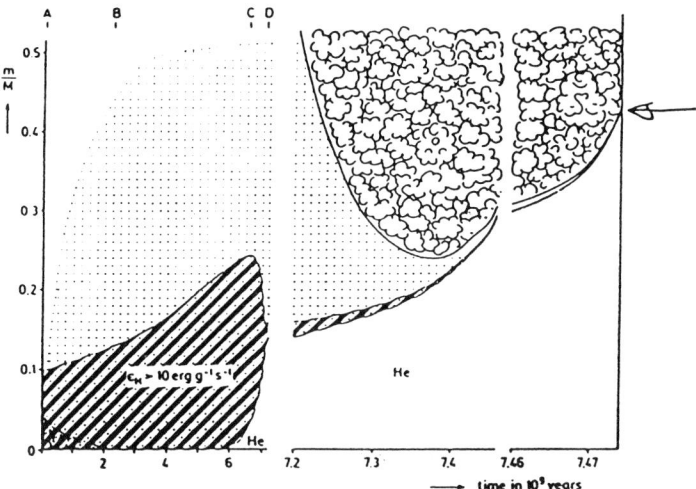

FIGURE 21. A schematic representation of the time evolution of the interior of a low-mass star from core hydrogen burning (A-B-C) through shell-hydrogen burning (D), ending with abrupt core Helium ignition at 7.47×10^9 years (arrow). Adapted from Thomas (1967).

24. Recent applications of the TRGB method

24.1. Ground-based Studies: The Local Group

A comparison of 10 galaxies with estimated TRGB distances with those having primary distances estimated using Cepheids or RR Lyrae stars was presented by Lee, Freedman & Madore (1993). These galaxies covered a wide range of morphological types (dwarf ellipticals: NGC 147, NGC 185 and NGC 205, an early–type spiral: M31, a late–type spiral: M33, and several irregulars: NGC 3109, NGC 6822, IC 1613 and WLM), covering a range of luminosity, metallicity, and distances up to $(m-M) = 25.5$ mag. The results of this comparison with Cepheid and RR Lyrae distances were extremely encouraging: the difference in the *relative* distances amounts to less than $\pm 10\%$ (rms), or 5% in distance.

Motivated by this result, we undertook a number of follow-up studies including simulations (Madore & Freedman 1995) to investigate the sources of error and factors limiting the application of this method. Then, using the combined facilities of the Palomar 5m and Las Campanas 2.5m we have begun a complete survey of TRGB distances to all galaxies in the Local Group and as far beyond it as aperture and seeing conditions will permit. To date the results for two galaxies slightly peripheral to the Local Group, Sextans A and Sextans B have been submitted for publication. These objects were chosen to allow a further comparison with the Cepheid distance scale, given that Cepheids have been observed in both of these galaxies and the derived distances indicated that the TRGB would be readily detected from the ground. And indeed it was, with both galaxies showing the tip at I ∼ 21.7 mag. Other galaxies currently being under study are NGC 3109, WLM, M33, and the various dwarf elliptical companions to M31.

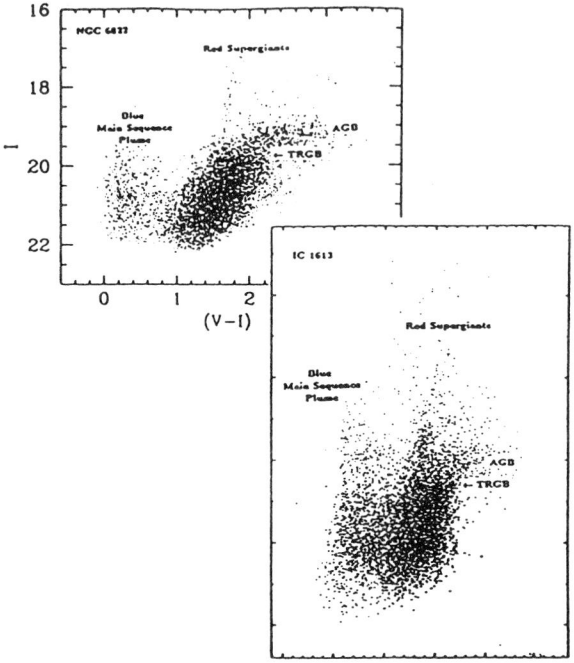

FIGURE 22. A montage of modern I-(V–I) color-magnitude diagrams for the two Local Group galaxies IC 1613 (lower right) and NGC 6822 (upper left) While red and blue supergiants are marked along with the AGB population, it is clear that the dominant contributor is the RGB with its broad and distinctive upper luminosity plateau marking the TRGB. The NGC 6822 data are adapted from the thesis of Gallart (1996).

24.2. HST Applications: Inside 10 Mpc

A number of researchers have already begun to use HST to apply the TRGB method to determining distances beyond the seeing/resolution limit of ground-based telescopes. For instance the halo of the peculiar elliptical radio galaxy NGC 5128 (= Cen A) has been resolved by Soria et al. (1996) and their I-(V-I) color-magnitude diagram shows the TRGB at I = 24.1 mag, thereby providing a distance of 3.6 ±0.2 Mpc. Similarly, Elson (private communication) reports the detection of the TRGB in the southern hemisphere lenticular galaxy NGC 3115, and quotes a preliminary distance of ∼10 Mpc. And most recently Sakai et al. (1997) report the detection of the TRGB in the giant elliptical galaxy NGC 3379 (= M105) at a distance of 11.5 ±1.6 Mpc. This latter application is of special interest given that NGC 3379 is a member of a group (the Leo I Group) which has two spiral galaxy members which have had independent Cepheid-based distances derived. Tanvir et al. (1995) find a distance of 11.9 ±0.9 Mpc for NGC 3368 (= M96), while

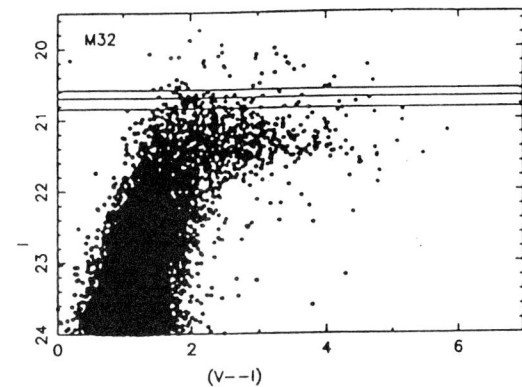

FIGURE 23. The I-(V–I) color-magnitude diagram for stars in M32 as derived from WFPC-2 HST observations by Grillmar et al. (private communication). Note that while the reddest (metal-rich) bright giant stars are found progressively at fainter magnitudes the bluest (metal-poor) giants still define the brightest tip. The horizontal lines mark the expected magnitude for the TRGB based on three apparent Cepheid moduli derived for various fields in M31, which is thought to be the parent galaxy of M32, and therefore at the same distance.

Graham et al. (1997) derive a slightly lower modulus for their target spiral NGC 3351, corresponding to a distance of 10.0 ±0.3 Mpc. While the differences in these Population I distances may indicate back-to-front geometry, they agree, individually and in the mean, to within ~10% of the Population II TRGB distance.

At the time of writing the authors were aware of two HST proposals to pursue applications of the TRGB method in the Virgo cluster. One proposal by Harris and collaborators was to image two low-metallicity dwarf galaxies directly; those exposures are still to be taken. On the other hand, early reports indicate that Ferguson et al. apparently have detected the red giant population *between* galaxies in the Virgo cluster core. However, at these large distances, very long integration times are essential to achieve the required signal-to-noise needed to detect stars at $I > 27$ mag, and consequently no color information is yet in hand.

25. The scorecard

As a concluding summary we return to the list proffered at the beginning of the article as criteria that a successful extragalactic distance indicator might aspire to, and try to assess the TRGB in that context.

Luminous: Indeed the luminosity generated by the helium core flash is enormous, being comparable to the instantaneous output of the rest of galaxy. Sadly the duration of this burst is only a few seconds, and even more tragically, virtually all of that energy is absorbed internally within the star as it restructures itself en route to quiescent helium core burning on the horizontal branch. Nevertheless the luminosity at which the upward evolution of the red giant star is terminated by core ignition is still quite luminous by

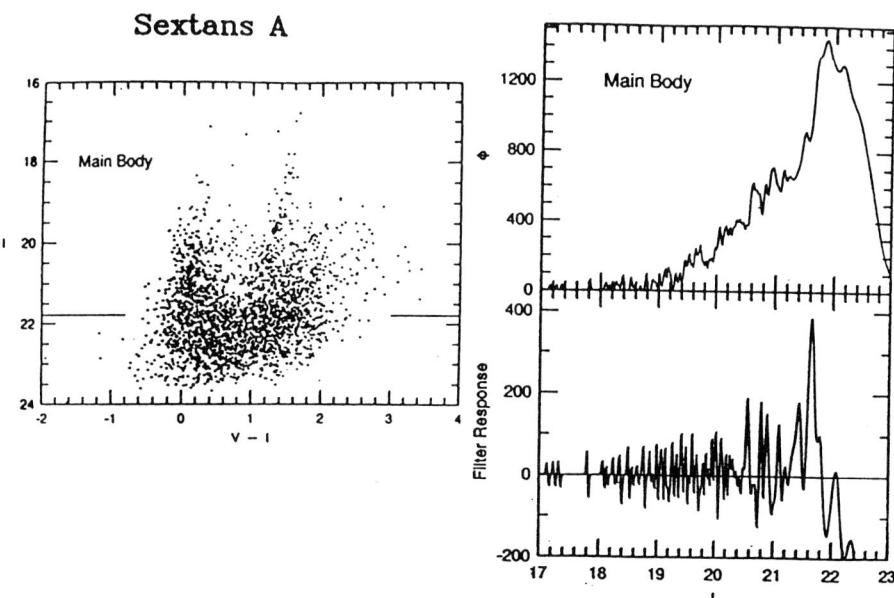

FIGURE 24. Sextans A: The left panel shows the I-(V–I) color-magnitude diagram for the main body of the dwarf irregular galaxy Setxans A. The horizontal lines at I ~ 21.7 mag mark the derived apparent magnitude level of the TRGB. The upper right panel shows the apparent I-band luminosity function for the stars in Sextans A; the lower right panel shows the Sobel response function identifying the position of the discontinuity in the luminosity function (Sakai, Madore & Freedman (1996a).

stellar standards. In the I band the TRGB magnitude is as bright as a 5-day Cepheid variable.

Dispersionless: By extragalactic standards, it is fair to say that any distance indicator that can claim a precision of ±0.20 mag (or 10% in distance) qualifies as being an excellent distance indicator. The TRGB method boasts an externally defined dispersion of less than ±0.1 mag, comparable to Cepheid distance moduli.

Calibrated: For the metallicity range $-2.2 < [Fe/H] < -0.7$ dex, the I-band color-magnitude diagrams for Galactic globular clusters calibrate the zero point of the TRGB method to the same level that the RR Lyrae distance scale is now known. As the RR Lyrae zero point is improved, the TRGB zero point will also get better with time.

Understood: To first order nuclear physics determines the bolometric luminosity at which all low-mass stars terminate.

Insensitive: Theory predicts (and observations confirm) that the bolometric (I-band) magnitude is insensitive to chemical composition variations in the range $-2.2 < [Fe/H] < -0.7$ dex, and for ages in the calculated range 7–17 Gyr.

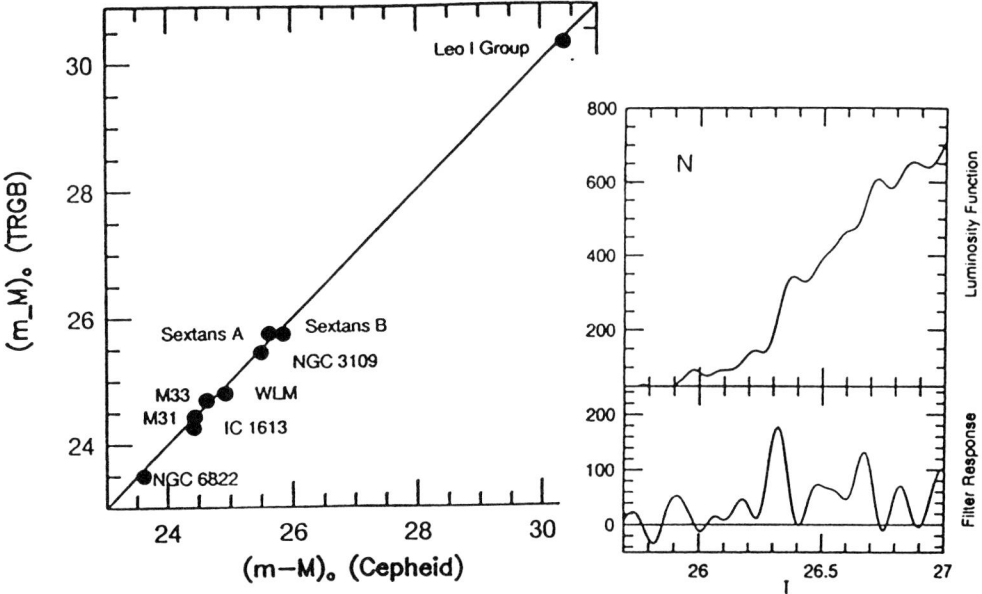

FIGURE 25. NGC 3379 and the Comparison of Cepheid Distances with the TRGB. The right panel shows both the I-band luminosity function (upper) and the tip response function (lower) for the HST observation of stars in the giant elliptical galaxy NGC 3379 (Sakai et al. (1997). The left panel shows a comparison of the NGC 3379 TRGB distance (marked by the Leo I Group) with the corresponding Cepheid-based distance to the group. Other Cepheid-TRGB comparisons are also shown for closer galaxies, Sextans A, NGC 3109, M33, etc.

Abundant: The halos of all galaxies are synonymous with Population II stars. If there is a measurable surface brightness defining these halos then the red giant branch stars must be present in abundance.

Ubiquitous: Every galaxy (elliptical, spiral or irregular) must have had a first generation of stars. The low-mass and low-metallicity members of that generation of stars must now be populating the red giant branch, and defining the tip.

Unambiguous: At the color and approximate magnitude level expected for TRGB stars other populations of stars do little more than add noise to the distinctive discontinuity in the red-giant-branch luminosity function. However, care should be taken that the TRGB is truly triggering the edge detector by having a fair sample of luminosity function well sampled below the tip before incompleteness in the photometry sets in. All too often that may prove to be a luxury in limiting applications, but the warning still stands.

26. Discussion

The tip of the red giant branch (TRGB) method is as precise and as accurate as the Cepheid PL relation. Moreover, an extremely useful feature of the TRGB method is that it is applicable to any galaxy containing a detectable population of old, metal-poor, low-mass stars. In practice, this means that nearly all nearby galaxies, regardless of Hubble type and/or inclination, can now be placed on a common distance scale out to the crowding and the flux limits of CCD detectors and telescopes. These distances will allow, for the first time, a complete sample of accurate, uniformly-measured distances to all galaxies within the Local Group and out to distances of ~ 10 Mpc.

27. Implications of the Hipparcos observations of galactic cepheids

Hipparcos parallaxes have recently become available for a sample of Galactic Cepheids, and we have used these new distances to calibrate the Cepheid period-luminosity (PL) relation at six wavelengths ($BVIJHK$). Comparing these calibrations with previously published multiwavelength PL relations we find agreement to within 0.07 ± 0.14 mag, or 4 ±7% in distance. Unfortunately, the current parallax errors for the fundamental pulsators (ranging in signal-to-noise = π/σ_π from 0.3 to 5.3, at best) preclude an unambiguous interpretation of the observed differences, which may arise from a combination of true distance modulus, reddening and/or metallicity effects. We explore these effects and discuss their implications for the distance to the Large Magellanic Cloud (LMC) and the Cepheid-based extragalactic distance scale. These results suggest a range of LMC moduli between 18.44 ±0.35 and 18.57 ±0.11 mag; however, other effects on the Cepheid PL relation (*e.g.*, extinction, metallicity, statistical errors) are still as significant as any such reassessment of its zero point.

27.1. *Introduction*

Feast and Catchpole (1997 = FC97 hereafter) have recently published the first results on parallaxes to Galactic Cepheids based on measurements from the Hipparcos satellite. They list data for the 26 highest signal-to-noise Cepheid parallaxes; and after an extensive series of reductions (see their Table 2) they conclude that the best fit PL relation for the visual bandpass is $M_V = -2.81\ log(P) - 1.43$, with a standard error on the Hipparcos zero point of ±0.10 mag, adopting the slope from prior work on LMC Cepheids. The authors go on to apply this V-band solution to determining the distance modulus of the LMC corrected for $E(B-V) = 0.074$ mag. Adding a metallicity correction of +0.042 mag and adopting $<V>_o - log(P)$ from Caldwell & Laney (1991) gives $(m - M_V)_o^{LMC} = 18.70 \pm 0.10$ mag. In this *Letter* we go beyond the V-band PL relation and explore the implications of the Hipparcos data for the multiwavelength calibrations of the Cepheid PL relation from the blue (B-band) out into the near infrared (2.2μm K-band).

28. Comparison with V-band period-luminosity relations

In Figure 27 we compare differentially four calibrations (heavy dotted lines) of the V-band Cepheid PL relation with the FC97 Hipparcos-based relation (solid horizontal lines). The first two comparisons (in the upper two panels) are with the relations given by Madore & Freedman (1991; hereafter MF91), derived from self-consistent sets of LMC Cepheid data whose stars either had complete $BVRI$ observations (MF91.1 containing 32 Cepheids) or complete $BVRIJHK$ observations (MF91.2 containing 25 stars)[†] These first two solutions indicate the sensitivity of slopes and zero points to sample selection,

---

[†] Tanvir (1997) has suggested that there may be small corrections (ranging from 0.02 to 0.09 mag) to the published I-band magnitudes of these LMC Cepheids arising from the originally sparse sampling and consequent averaging of their light and color curves. For the past five years we have been obtaining new VI CCD observations of the LMC calibrators at Las Campanas and now also at Siding Springs Observatories. These new data are designed to address those concerns.

which are considerable, but within the quoted statistical uncertainties: ±0.11 and ±0.20, respectively for the slopes, and ±0.05 and ±0.09 mag, for the zero points. So as to make the subsequent comparisons consistent, the original Sandage & Tammann (1968) calibration (ST68.1 in the lower left panel) has been placed on the modern Hyades/Pleiades Galactic cluster distance scale by applying a single offset of +0.13 mag derived from the average difference between the absolute magnitudes of the Cepheids used in the 1968 calibration updated to Feast and Walker (1987), their Table 2. This distance scale corresponds to a Hyades modulus of 3.27 (see Pel 1985) and uses the Pleiades main sequence, at a modulus of 5.57 (van Leeuwen, 1983) to effectively correct for the over-metallicity of the Hyades with respect to the older Galactic clusters in which the Cepheid calibrators are found.‡ Finally, the FW87 calibration itself is plotted in the lower left panel. In all panels the dashed horizontal lines represent the fiducial Hipparcos calibration flanked by thin parallel lines at ±0.10 mag.

The error bars of all of the plotted previously published relations overlap with errors quoted for the Hipparcos solution (a formal uncertainty was not given by ST68, so we have arbitrarily assigned them an error of ±0.05 mag). However, the offsets are not randomly distributed, with each of the solutions appearing to be systematically fainter in V with respect to the Hipparcos calibration by about 0.1 mag. We discuss the significance and possible implications of this difference in the following sections.

29. Multiwavelength period-luminosity relations

In Madore & Freedman (1991) we published fiducial PL relations in seven bandpasses: $BVRIJHK$. These were all based on selecting self-consistent sets of previously published LMC Cepheid data, scaled to an LMC true distance modulus of 18.50 mag and applying a single line-of-sight reddening correction using $E(B-V) = 0.10$ mag. Thirty-two stars were available for a calibration of $BVRI$ PL relations; 25 stars were used for an alternative set of $BVRIJHK$ calibrations. In the following we compare those multiwavelength PL relations with the Hipparcos sample of Galactic Cepheids, individually corrected for foreground reddening and scaled to their geometric parallax distances.

We have collected from the literature multiwavelength ($BVIJHK$) mean magnitudes for as many of the Hipparcos-calibrating Cepheids as have been published (notably for the infrared Wisniewski & Johnson 1968, Welch et al. 1984, Laney & Stobie 1992 and reference therein). These form rather disjoint subsets. After eliminating the suspected overtone pulsators listed by FC97, the total available sample with parallaxes drops from 26 to 20. Of these only 7 have mean magnitudes published at all six wavelengths, while 10 and 13 Cepheids, respectively have either $BVIJK$ or $BVJK$ magnitudes in common. We have analyzed these four groups of stars independently, but self-consistently, in the following way.

Using the Hipparcos parallaxes and Galactic reddenings adopted by FC97 from Fernie, Kamper & Seager (1993) scaled to the various wavelengths using the extinction law

‡ At the February 14, 1997 meeting of the Royal Astronomical Society in London on February 14, 1997 F. van Leeuwen and C.S. Hansen Ruiz reported a true distance modulus of 5.29 ± 0.06 mag for the Pleiades cluster, based on Hipparcos trigonometric parallaxes. Following the Venice Meeting in June 1997 the value had changed only slightly to 5.33 ±0.06 mag (C. Turon, private communication) If adopted, this Pleiades modulus would make the Galactic-cluster-based calibrations approximately 0.3 mag fainter than the FC97 solution plotted in Figure 27. At this point in time the Galactic cluster zero point appears to be in a state of flux, and we will not comment on it further, except to note that the Hipparcos calibration will undoubtedly converge on a more accurate zero point than we have access to at this precise moment.

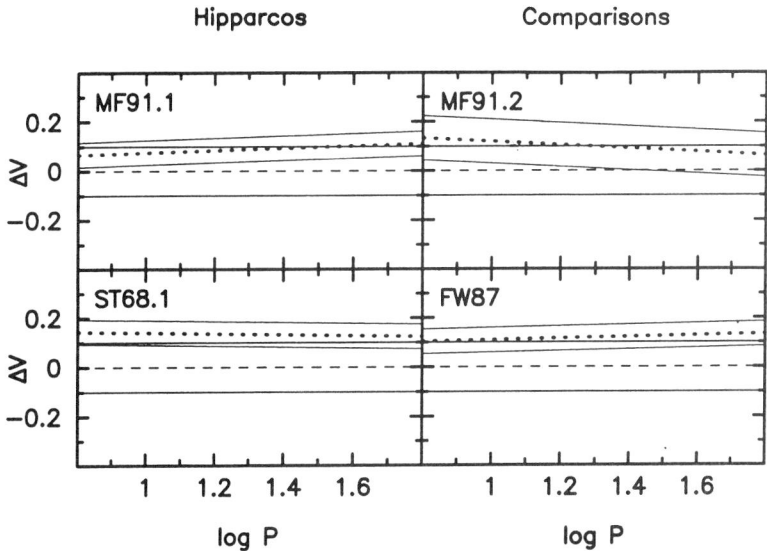

FIGURE 26. Differential comparison of recently published V-band PL relations (heavy lines) relative to the Hipparcos calibration (thin lines). Plotted is the difference [V - V(Hipparcos)] versus log P, in the sense that if Hipparcos is brighter the difference shown is positive.

of Cardelli, Clayton & Mathis (1989), we derived absolute magnitudes for each of the Cepheids in each of the observed wavelengths. (We note that these corrections for interstellar extinction are not inconsiderable, ranging up to 2 mag in B for several stars). The resulting PL relations are shown in the six panels of Figure 28. Error bars are one-sigma uncertainties from the quoted parallaxes. Note the highly correlated nature of the individual data points about the fiducial lines. And too, we remind the reader that the computation of distances and their related errors from observed parallaxes is non-trivial (Brown et al. 1997), as distances are not linearly related to parallaxes, and parallax errors can subtly bias samples. A full treatment of this issue is beyond the scope of this paper, but we note that selection biases at least are minimized for stars having the smallest reported errors. As discussed by Brown et al., given the true parallax distribution the expected biases follow naturally; however, corrections to the *observed* parallaxes require assumptions about the true distribution, and detailed modeling. Fortunately for this application the Cepheid sample is not parallax-selected; the objects being chosen in advance based on their optical variability, periods and apparent magnitudes.

The differences between these individual (trigonometric) absolute magnitudes and the predicted $BVIJHK$ magnitudes derived from the mean PL relations of MF91 (solid lines in Figure 28) are each plotted in Figure 29 against the corresponding B-band residual. The $(B-V)$ intrinsic color residuals are plotted against the B-band residuals in the upper

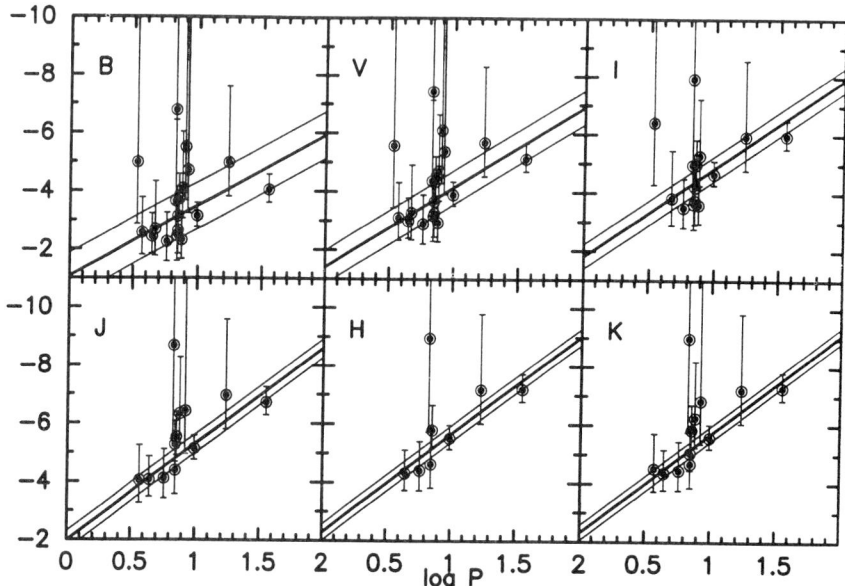

FIGURE 27. Multiwavelength Period-Luminosity relations for Cepheids with Hipparcos parallaxes, plotting all stars that have data available at the particular wavelength, as noted in the upper left corner of each panel. In each panel the solid sloping line is not a fit to the data, but rather it is the published calibration of Madore & Freedman (1991) flanked by thin parallel lines representing the 2-σ limits quoted by them as being the intrinsic width of the instability strip at each wavelength.

right panel. The individual residuals at a given wavelength contain random contributions from the parallax uncertainties, reddening errors, and finally the intrinsic (temperature-induced) magnitude residuals which reflect the finite width of the Cepheid instability strip. The observed residuals are however extremely large (nearly 5 mag peak-to-peak) and are almost certainly dominated by the (achromatic) errors in the parallaxes, given the strict unit-slope correlation of the mag-mag residuals, and the total lack of any correlation between the magnitude-color residuals (Figure 29).

Wavelength-dependent offsets between the six mean solutions independently will reflect (1) errors in the adopted true distance to the LMC (which set all of the zero points in the MF91 multiwavelength PL relation calibrations), (2) reddening errors in the adopted extinction to the LMC sample of calibrating Cepheids, and finally (3) intrinsic differences between the LMC and Galactic Cepheids, for example, due to metallicity.

Our first solution considers the largest data set (in terms of parallaxes) but the one that is most restricted in terms of wavelength coverage: it consists of 19 Cepheids observed in B and V. Weighted by the square of the signal-to-noise ratio in the Hipparcos parallax, the residuals were summed and averaged at each of the two wavelengths giving mean

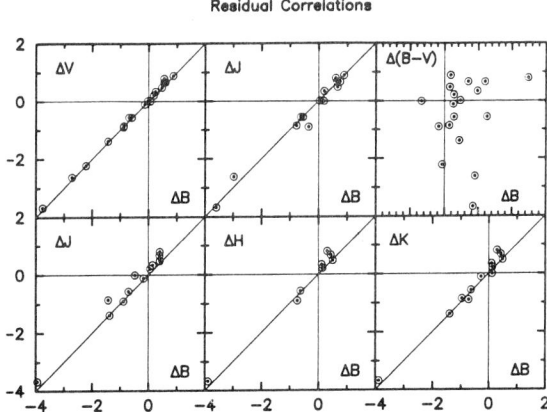

FIGURE 28. B-band residuals from the multiwavelength Period-Luminosity relations in Figure 28 are sequentially plotted as a function of residuals from each of the other five PL relations and (upper right panel) against the (B–V) color residuals. The total lack of correlation in the latter instance is unexpected except in the limit where the residuals are dominated by distance errors in the derived parallaxes. This latter situation is apparently the case given the strong (unit-slope) correlations of the residuals in each of the other panels, regardless of wavelength.

offsets between the LMC calibration and the Galactic Cepheids. The variance in each mean offset was then calculated from the average of the squares of these same residuals again inversely weighted by the variance in the individually quoted parallaxes. The differences are $\Delta B = +0.23 \pm 0.35$ mag and $\Delta V = +0.16 \pm 0.28$ mag, in the sense that the LMC Cepheid calibration appears to be too faint with respect to the Galactic calibration. (Further restricting the sample to only those 12 stars with $\pi/\sigma_\pi > 2.0$ changes ΔB to $+0.22 \pm 0.24$ mag and ΔV to $+0.15 \pm 0.17$ mag.)

If the (statistically marginal, but apparently systematic) differences in the B and V solutions were to be ascribed to reddening alone, then the Galactic data and the LMC calibration can be reconciled by invoking an increase of $\Delta E(B-V) = 0.07$ mag in the adopted mean reddening to the LMC Cepheid sample. This is consistent with a similar suggestion regarding the LMC Cepheid calibration made recently by Bohm-Vitense (1997) based on different data. This reddening solution has the consequence that it would also require the distance modulus of the LMC to be revised *downwards* by −0.06 mag to 18.44 mag; the uncertainty on this offset being at least as large as the uncertainty in the individual moduli (±0.3 mag), depending on the degree of correlation in those cumulative uncertainties. This particular path, of a reddening solution, cannot be considered definitive. Other possibilities are: (1) the LMC true modulus should be increased by $(0.23 + 0.16)/2 = +0.20$ mag, without any change to the foreground reddening, or (2) that there are differential metallicity corrections amounting to −0.23 and −0.16 mag that need to be applied at the B and V wavelengths, respectively. Of course any suitably contrived linear combination of the above three effects could also be invoked. More constraints on the problem are obviously needed.

An alternative possibility is that some of the wavelength-dependent effects seen in the comparison of Galactic (high metallicity) data with the LMC (lower metallicity) data could be due to chemical composition differences between the two samples. Taken at face value the dependence of the apparent V modulus on metallicity would be very large, $\Delta V/\Delta[Fe/H] = 0.16/0.15 = 1.1(\pm 1.9)$ mag/dex, assuming that the full offset in V noted in the above comparison is due to metallicity, and adopting a metallicity underabundance of 1.4× between the LMC and the Solar neighborhood (see, for example, FW87). However, we note that this effect is basically indistinguishable from reddening in its form (as evidenced by our first set of solutions), and that the offset (whatever its origin) when treated as reddening leads to a true distance modulus for the LMC that is unchanged, from previous assumptions, at 18.50 mag. Given this apparent degeneracy between reddening and metallicity, and the current large uncertainties in the parallaxes, assessing the dependence on metallicity from these data alone will remain problematic.

To obtain added leverage on the solution, moving to the infrared has numerous well known advantages, as first articulated in McGonegal, McLaren, McAlary & Madore (1982): reddening effects are known to decrease with wavelength, in a well defined and calibrated manner; and simultaneously, metallicity effects are also expected to decrease in amplitude with increased wavelength.

Our second solution is based on 13 Cepheids each having $BVHK$ data in common. This four-color solution gives a derived reddening *increase* for the LMC Cepheid sample of $+0.04 \pm 0.08$ mag, with no formal offset in the derived 18.50 ± 0.13 mag true modulus for the LMC. Our next approximation employs 10 Cepheids each now having $BVIJK$ mean magnitudes. Here the formal solution for the true modulus for the LMC is 18.53 ± 0.14 mag, with a corresponding increase in the mean reddening of $+0.06 \pm 0.07$ mag. Finally, we have analyzed a sample of 7 Galactic Cepheids, each having $BVIJHK$ photometry, to obtain one last solution: $\Delta E(B-V) = 0.07 \pm 0.07$ mag with $(m-M)_{LMC} = 18.57 \pm 0.11$ mag. The fit to this final set of observations is shown in Figure 30; the χ^2 weighted residual fitting surface being shown as an inset. The individual apparent moduli discussed here, and their errors, are summarized in Table 1.

Finally, if we now adopt the metallicity correction of $\Delta V = 0.04$ mag advocated by FC97 and assume that the effects at JHK are negligible, (and eliminate B and I from the solution given that metallicity corrections for these filters are not well defined at this time) we find for this 4-color solution $\Delta E(B-V) = 0.06 \pm 0.11$ mag with $(m-M)_{LMC} = 18.57 \pm 0.11$ mag. This is virtually indistinguishable from the full $BVIJHK$ solution given above.

30. Discussion

We have used the Hipparcos parallaxes of nearby Galactic Cepheids to explore corrections to the multiwavelength Period-Luminosity relations for LMC Cepheids. The latter are based on an LMC data set scaled to a true distance modulus of 18.50 mag and an adopted foreground reddening of $E(B-V) = 0.10$ mag. Although the current uncertainties in the parallaxes are large and still dependent upon the specific subsets of the Cepheids chosen for the comparison, the agreement is good, indicating that to within ± 0.14 mag (or, 7% in distance) the previously adopted zero point is substantially correct. Based on different subsamples of data either having BV, $BVJK$, $BVIJK$ or $BVIJHK$ photometry, LMC moduli, ranging from 18.44 to 18.57 mag are derived. These results, summarized in Table 2, differ from the value of 18.70 mag of FC97, which are based

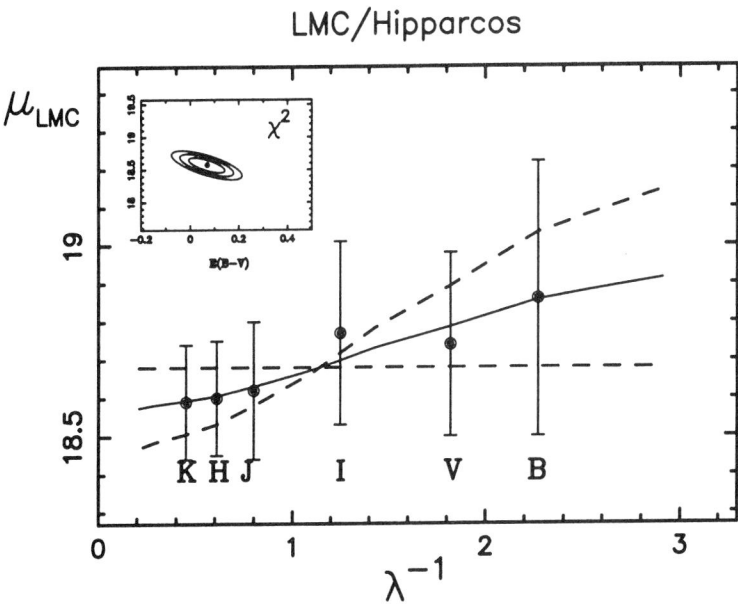

FIGURE 29. Apparent modulus plots for LMC Cepheids observed at $BVIJHK$ scaled to the Hipparcos zero point and using the published multiwavelength PL solutions of Madore & Freedman (1991). The solid line is a weighted χ^2 fit of a reddening line to the data; the broken line indicates the one-sigma limits on that solution. Inset (top left) shows the χ^2 surface indicating the minimization solution for the modulus and reddening and the interdependence of their associated errors.

Table 1. Multiwavelength Reddening Solutions

Filters	No. Stars	$E(B-V) \pm \sigma$	$\mu_{LMC} \pm \sigma$
BV	19	$0.17 \pm \cdots$	18.44 ± 0.35
BVJK	13	0.14 ± 0.08	18.50 ± 0.13
BVIJK	10	0.16 ± 0.07	18.53 ± 0.14
BVIJHK	7	0.17 ± 0.07	18.57 ± 0.11
V_cJHK	8	0.16 ± 0.11	18.57 ± 0.11

solely on the reddening-corrected V photometry of Caldwell & Laney (1991), externally adjusted for metallicity. The Hipparcos data alone do not allow us to discriminate between metallicity effects and the physically distinct possibility of added reddening to the LMC.

To alleviate the ambiguity posed by the need to simultaneously solve both for reddening and metallicity effects on the Cepheid distances we are currently deriving OB-star reddenings along the individual lines of sight to several dozen LMC Cepheids. This will allow us to decouple the reddening determinations from metallicity effects, and go beyond the use of a single mean (foreground + internal) reddening for the LMC calibrating Cepheid sample. Preliminary reductions indicate that the variance from field to field is large (ranging from $E(B-V) = 0.00$ up to 0.40 mag) while still indicating that an average value of $< E(B-V) > = 0.10$ mag is appropriate for the LMC calibrating Cepheids. Details will be presented in Madore, Freedman & Pevunova (1998 in preparation).

We close by noting that at least three other very recent determinations of the true modulus to the LMC fall on either side of the value 18.50 mag adopted by MF91 in setting a zero point for the Cepheid distance scale. Both Reid (1997) and Gratton et al. (1997) derive large LMC moduli (18.65 ± 0.10, and 18.63 ± 0.06 mag, respectively) using Hipparcos-based calibrations of the Galactic globular cluster and RR Lyrae distance scale. On the other hand, Gould & Uza (1997) have re-analyzed the SN 1987A supernova "light echo" and derive *an upper limit* of $\mu_{LMC} <$ 18.37 ±0.04 mag for the LMC true distance modulus; although they note that if the ring is slightly elliptical ($b/a \sim 0.95$) this upper limit increases to $<$ 18.44 ± 0.05 mag. A value of 18.56 ± 0.05 mag has been derived by Panagia et al. (1996) from the same data. Until these differences are fully understood and resolved, and given the remaining uncertainties in the Hipparcos Cepheid parallax data we prefer to adopt a true distance modulus of 18.50 mag for the LMC, but now bounded by an uncertainty of ± 0.15 mag, defined to fully encompass the above range of recently published values. This value is consistent with other estimated distances to the LMC based on a wide variety of methods (for a comprehensive modern review see Westerlund 1997). Viewed in that perspective the Hipparcos data confirm the Cepheid distance scale at better than the ±10% level (95% confidence).

31. Implications of a cepheid distance to the Fornax cluster

Thirty-seven long-period Cepheid variables have been discovered in the Fornax cluster spiral galaxy, using the Hubble Space Telescope. The resulting V and I period-luminosity relations give a true modulus of $\mu_o = 31.28\pm0.07$ mag, corresponding to a distance of 18.0±0.6 Mpc. A Cepheid distance to the Fornax cluster offers several means of estimating the Hubble constant. First, associating this distance with the Fornax cluster as a whole gives a local Hubble constant of $H_o = 73$ $(\pm7)_{random}$ $[\pm18]_{systematic}$ km/sec/Mpc. Second, the Fornax cluster provides a means of calibrating a wide variety of secondary distance indicators. Recalibrating the Tully-Fisher relation using NGC 1365 and 6 nearby spiral galaxies, applied to 15 clusters out to 100 Mpc gives $H_o = 76$ $(\pm2)_r$ $[\pm8]_s$ km/sec/Mpc. A broad-based set of differential moduli established from Fornax out to Abell 2147, nearly a factor of ten in distance further, gives $H_o = 72$ $(\pm1)_r$ $[\pm7]_s$ km/sec/Mpc. With the addition of two Type Ia supernova calibrators in Fornax and correcting the supernova peak luminosities for decline rate, gives $H_o = 68$ $(\pm5)_r$ $[\pm8]_s$ km/sec/Mpc, out to a distance in excess of 500 Mpc. Seven Cepheid-based distances to groups of galaxies out to and including the Virgo and Fornax clusters yield $H_o = 70$ $(\pm3)_r$ $[\pm16]_s$ km/sec/Mpc. These major distance determination methods agree to within their statistical errors. The resulting value of the Hubble constant, encompassing all those determinations which are based directly on Cepheids or tied to secondary distance indicators, is found to be $H_o = 72$ $(\pm3)_r$ $[\pm12]_s$ km/sec/Mpc, out to cosmologically significant distances.

31.1. Introduction

Hubble (1929) announced his discovery of the expansion of the Universe nearly 70 years ago. Despite decades of effort, and continued improvements in the actual measurement of extragalactic distances, convergence on a consistent value for the absolute expansion rate, as parameterized by the Hubble constant, H_o, was not forthcoming. However, progress in the last few years has been rapid and dramatic (see, for instance, Freedman, Madore & Kennicutt 1997; Mould Sakai, Hughes & Han 1997; Tammann & Federspiel 1997). This accelerated pace has occurred primarily as a result of the improved resolution of the Hubble Space Telescope (and its consequent ability to discover classical Cepheid variables at distances a factor of ten further than can routinely be achieved from the ground), giving accurate zero points to a number of recently refined methods which can measure precise relative distances beyond the realm of the Cepheids. These combined efforts are providing a more accurate distance scale for local galaxies, and are indicating a convergence among various secondary distance indicators in establishing an absolute calibration of the far-field Hubble flow.

Soon after the December 1993 HST servicing mission it was clear that the measurement of Cepheids in the Virgo cluster (part of the original design specifications for the telescope) was feasible (Freedman, et al. 1994a). And although the subsequent discovery of Cepheids in the Virgo galaxy M100 (Freedman, et al. 1994b) and subsequent refinements (Ferrarese et al. 1995) were important steps in resolving outstanding differences in the extragalactic distance scale (Mould et al. 1995), the Virgo cluster is complex both in its geometric and its kinematic structure, and there still remain large uncertainties in both the velocity and distance to this cluster. Virgo clearly was, and still is, not an ideal test site for an unambiguous determination of the cosmological expansion rate or the

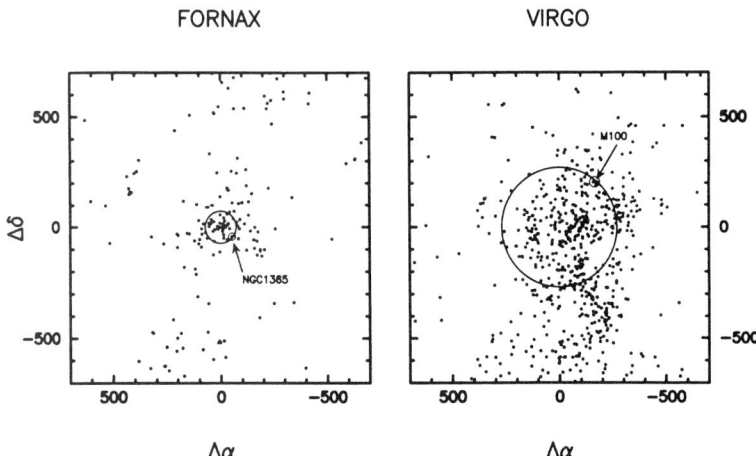

FIGURE 30. A comparison of the distribution of galaxies as projected on the sky for the Virgo cluster (right panel) and the Fornax cluster (left panel). M100 and NGC 1365 are each individually marked by arrows showing their relative disposition with respect to the main body and cores of their respective clusters. Units are arcmin.

calibration of secondary distance indicators. In this paper we discuss the implications of a Cepheid distance to the next major clustering of galaxies, the Fornax cluster, which is a much less complicated system than Virgo.

32. NGC 1365 and the Fornax cluster

The Fornax cluster is comparable in distance to the Virgo cluster (de Vaucouleurs, 1975), but it is found almost opposite to Virgo in the skies of the southern hemisphere. The Fornax cluster is less rich in galaxies than Virgo (Ferguson & Sandage 1988), but it is also substantially more compact than its northern counterpart (Figure 31). As a result of its lower mass, the influence of Fornax on the local velocity field is less dramatic than that of the Virgo cluster. And because of its compact nature, questions concerning the membership of individual galaxies in Fornax are less problematic, while the back-to-front geometry is far less controversial than any of these same points raised in the context of the Virgo cluster complex. Clearly, Fornax is a much more interesting site for a test of the local expansion rate.
Although the goals of the Key Project on the Extragalactic Distance Scale (Kennicutt, Freedman & Mould 1995) are far broader than simply investigating the distances to a few nearby clusters, there are several important reasons to secure a distance to the Fornax cluster. It is both a probe of the local expansion velocity field, and Fornax is a major jumping-off point for a variety of secondary distance indicators which can be used to probe a volume of space at least 1,000 times larger. To obtain a distance to the Fornax

cluster, the Key Project is configured to monitor three member galaxies; the first of these, discussed here, is the Seyfert 1 galaxy NGC 1365, a strikingly picturesque, two-armed, barred-spiral galaxy with an active galactic nucleus. In the coming year, two additional galaxies, NGC 1425 and NGC 1326A, are slated for imaging with HST.

At least three lines of evidence independently suggest that the first galaxy to be observed, NGC 1365 is a physical member of the Fornax cluster. First, NGC 1365 is almost directly along our line of sight to Fornax. The galaxy is projected only ~70 arcmin from the geometric center of the cluster, whereas the diameter of the cluster is ~200 arcmin (Ferguson 1989; see Figure 31). In addition, NGC 1365 is also coincident with the Fornax cluster in velocity space. The systemic (heliocentric) velocity and velocity dispersion of the main population of galaxies in Fornax are well defined: 30 spirals/irregular galaxies give V = 1,415 km/sec and $\sigma = \pm 347$ km/sec, 70 E/SO galaxies give V = 1,473 km/sec with $\sigma = \pm 335$ km/sec, and the combined sample gives $\sigma = \pm 340$ km/sec. The observed velocity of NGC 1365 (+1,636 km/sec) is only +181 km/sec larger than the mean velocity of the Fornax cluster as a whole, which based on 100 galaxies is found to be 1,455±34 km/sec (Schroder 1995; Schroder & Richter 1997; Han & Mould 1990; NED); with the mean velocity of the spirals agreeing with the mean for the ellipticals to within 60 km/sec). The velocity off-set of NGC 1365 is only half of the cluster velocity dispersion. Finally, we note that for its rotational velocity, NGC 1365 sits only 0.02 mag from the central ridge line of the *apparent* Tully-Fisher relation relative to other cluster members defined by recent studies of the Fornax cluster (Bureau, Mould & Staveley-Smith 1996; Schroder 1995).

On the other hand, it is often noted that NGC 1365 is impressively large in its angular size, and that it is very bright in apparent luminosity as compared to any other galaxy in the immediate vicinity of the Fornax cluster. However, corrected for an inclination of 44 degrees, the 21cm neutral hydrogen line width of NGC 1365 is found to be ~575 km/sec (Bureau, Mould & Staveley-Smith 1996; Mathewson, Ford, & Buchhorn 1992). Using the Tully-Fisher relation as a *relative* guide to intrinsic size and luminosity, this rotation rate places NGC 1365 among the most luminous galaxies in the local Universe; brighter than M31 or M81, and comparable to NGC 4501 in the Virgo cluster or NGC 3992 in the Ursa Major cluster. We therefore conclude that NGC 1365 is in all respects apparently normal, (albeit large and luminous) and that its distance is indicative of the ensemble distance to the other spiral and elliptical galaxies constituting the Fornax cluster proper.

33. HST Observations

Using the *Wide Field and Planetary Camera 2* on HST, we have obtained a set of 12-epoch observations of NGC 1365. These observations were begun on August 6, and continued until September 24, 1995. The observing window of 44 days was selected to maximize target visibility, without necessitating any roll of the targeted field of view. Sampling within the window was prescribed by a power-law distribution, tailored to optimally cover the light and color curves of Cepheids with anticipated periods in the range 10 to 60 days (see Freedman *et al.* 1994a for additional details). Contiguous with 4 of the 12 V-band epochs, I-band exposures were also obtained so as to allow reddening corrections for the Cepheids to be determined. Each V-band epoch made use of the F555W filter and consisted of two exposures split between orbits (and allowing for cosmic ray rejection); a total of 5,100 sec of V-band data were obtained at each epoch in the course of the monitoring program. The I-band exposures (F814W) totaled 5,400 sec each, again cosmic-ray split and accumulated over two orbits.

All frames were pipeline pre-processed at the *Space Telescope Science Institute* in Baltimore and subsequently analyzed in Pasadena using ALLFRAME (a suite of special-purpose stellar photometry packages (Stetson 1994)). A second independent reduction is being performed using the DoPhot photometry package. At this juncture we are still allowing for the possibility of a 10% systematic offset in the true modulus derived from the two packages (Table 3); the results discussed here are based solely on the ALLFRAME analysis. Zero-point calibrations for the photometry were adopted (Holtzmann, J., *et al.* 1995; Hill, *et al.* 1997), which agree to 0.05 mag on average. Details on the DoPHOT and ALLFRAME reduction and analysis of this data set are presented elsewhere (Silbermann, *et al.* 1997). We are also currently undertaking artificial star tests on these frames to quantify the uncertainty due to crowding.

ERROR BUDGET THE CEPHEID DISTANCE TO NGC1365

Source of Uncertainty on the Mean	Description of Uncertainty	Percentage Error
LMC	**CEPHEID PL CALIBRATION**	
[A] LMC True Modulus	Independent Estimates = 18.50 ±0.10 mag	5%
[B] V PL Zero Point	LMC PL $\sigma_V = (0.27)/\sqrt{31} = \pm 0.05$ mag	3%
[C] I PL Zero Point	LMC PL $\sigma_I = (0.18)/\sqrt{31} = \pm 0.03$ mag	2%
[SC] Systematic Uncertainty	[A] + [B] + [C] combined in quadrature	6%
NGC 1365	**CEPHEID TRUE DISTANCE MODULUS**	
[D] HST V-Band Zero Point	On-Orbit Calibration: ±0.05 mag	3%
[E] HST I-Band Zero Point	On-Orbit Calibration: ±0.05 mag	3%
[M1] Cepheid True Modulus	[D][E] are uncorrelated, but coupled by reddening law: $\sigma_{\mu_o} = \pm 0.15$ mag	7%
[F] Cepheid V Modulus	NGC 1365 PL $\sigma_V = (0.32)/\sqrt{37} = \pm 0.05$ mag	3%
[G] Cepheid I Modulus	NGC 1365 PL $\sigma_I = (0.31)/\sqrt{37} = \pm 0.05$ mag	3%
[M2] Cepheid True Modulus	[F] and [G] are partially correlated, giving $\sigma_{\mu_o} = \pm 0.06$ mag	3%
[Z] Metallicity	M31 metallicity gradient test gives $\sigma_{\mu_o} = \pm 0.08$ mag	4%
	M101 calibration gives +0.14 mag of $\Delta\mu_o/\Delta[Fe/H] = -0.25$ mag/dex	7%
[H] Reduction Package	Systematic differences in aperture corrections	8%
[J] Random Errors	[M1]+ [M2] combined in quadrature	8%
[K] Systematic Errors	[SC] + [Z] + [H] combined in quadrature	12%
	D = 18.0 Mpc ± 1.4 (random) ± 2.2 [systematic]	

Note: There are 32 Cepheids in the LMC with published VI photometry [18 = M3]. The measured dispersions in the period-luminosity relations at V and I are 0.27 and 0.18 mag, respectively.

34. Cepheids in NGC 1365

Representative light curves for 9 of the 52 Cepheid candidates discovered in NGC 1365 are given in Figure 32. The phase coverage in all cases is sufficiently dense and uniform that the form of the light curves is clearly delineated. This allowed 37 of these variables to be unambiguously classified as high-quality Cepheids with their distinctively rapid brightening, followed by a long linear decline phase. Periods, obtained using modified Lafler-Kinman algorithm (Lafler, & Kinman, 1965), are are judged to be statistically good to a few percent, although in some cases ambiguities larger than this do exist as a consequence of the narrow observing window and the restricted number of cycles (between 1 and 5) covered within the 44-day window.

The resulting V and I period-luminosity relations for the complete set of 37 Cepheids

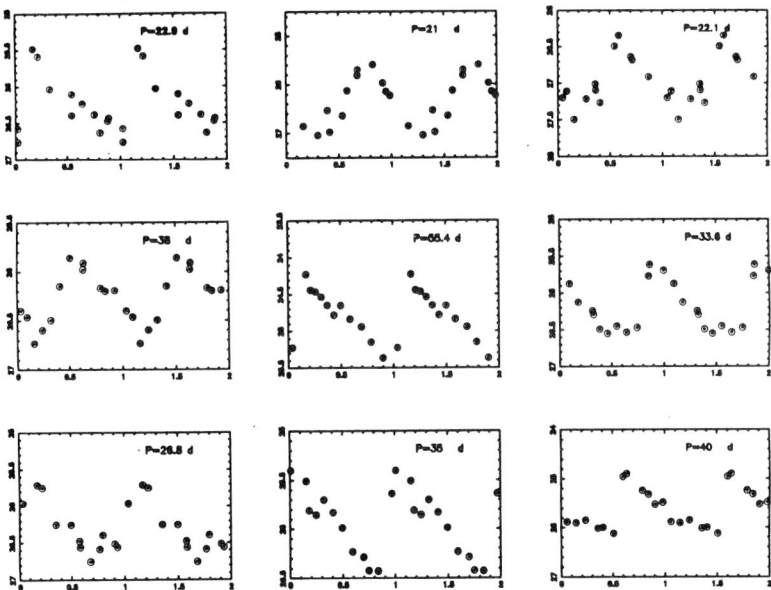

FIGURE 31. Representative V-band light curves for 9 of the 37 Cepheid variables found in the Fornax cluster galaxy, NGC 1365. Magnitudes are plotted along the vertical axis, Two repeated cycles in phase are plotted along the horizontal axis.

are shown in the upper and lower panels of Figure 33, respectively. The solid line is a minimum χ^2 fit to the fiducial PL relation for LMC Cepheids (Madore & Freedman 1991), corrected for $E(B-V)_{LMC} = 0.10$ mag, scaled to an LMC true distance modulus of $\mu_o = 18.50$ mag, and shifted into registration with the Fornax data. [Recent results from the Hipparcos satellite bearing on the Galactic calibration of the Cepheid zero point (Feast & Catchpole 1997; Madore & Freedman 1997) indicate that the LMC calibration is confirmed at the level of uncertainty indicated in Table 4, with the possibility that a small (upward) correction to the LMC reddening is in order.] The derived apparent moduli are $\mu_V = 31.67 \pm 0.05$ mag and $\mu_I = 31.57 \pm 0.04$ mag. Correcting for a derived total line-of-sight reddening of $E(V-I)_{N1365} = 0.10$ mag (based on the Cepheids themselves) gives a true distance modulus of $\mu_0 = 31.43 \pm 0.06$ mag. This corresponds to a distance to NGC 1365 of 18.0±0.6 Mpc. The quoted error *at this step in the analysis* quantifies only the statistical uncertainty generated by photometric errors in the ALLFRAME data combined with the intrinsic magnitude and color width of the Cepheid instability strip.

35. The Hubble constant

We now discuss the impact of a Cepheid distance to the Fornax cluster in estimating the general expansion rate of the Universe. Below we present and discuss three independent estimates, where the analysis is based both on the new Fornax distance and the distances to other Key Project galaxies. At the end we intercompare the results for convergence and consistency. The first estimate is based solely on the Fornax cluster, its velocity and its Cepheid-based distance. This scrutinizes the flow sampled in one particular direction at a distance of ~20 Mpc. We then examine the inner volume of space, leading up to and

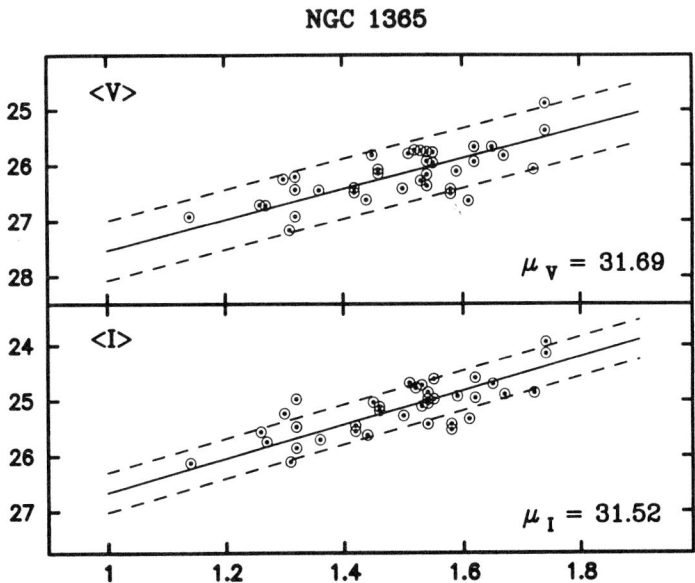

FIGURE 32. V and I-band Period-Luminosity relations for the full set of 37 Cepheids monitored in NGC 1365. The fits are to the fiducial relations given by Madore & Freedman (1991), shifted to the apparent distance modulus of NGC 1365. Dashed lines indicate the expected intrinsic (2-sigma) width of the relationship due to the finite temperature width of the Cepheid instability strip.

including both the Virgo and Fornax clusters. This has the added advantage of averaging over different samples and a variety of directions, but it is still limited in volume (to an average distance of ~ 10 Mpc), and it is subject to the usual caveats concerning bulk flows and the adopted Virgocentric flow model (Table 4). The third estimate comes from using the Cepheid distance to Fornax to lock into secondary distance indicators, thereby allowing us to step out to cosmologically significant velocities (10,000 km/sec and beyond) corresponding to distances greater than 100 Mpc. Local flow uncertainties then are replaced by largescale flow uncertainties; while the systematically secure Cepheid distances are replaced by currently more controversial secondary distance indicators. This is done in order to increase volume and the sample. Averaging over the sky, and working at large redshifts, alleviates the flow problems. Examining consistency between independent the secondary distance estimates, and then averaging over their far-field estimates should provide a systematically secure value of H_o and, more importantly, a measure of its external error. Comparison of the three 'regional' estimates (Fornax, local and far-field) then can be used to provide a check on the systematics resulting from the various assumptions made independently at each step.

36. The Hubble constant at Fornax

36.1. Uncertainties in the Fornax Cluster Distance and Velocity:

(1) Distance — The two panels of Figure 31 show a comparison of the Virgo and Fornax clusters of galaxies drawn to scale, as seen projected on the sky. The comparison of apparent sizes is appropriate given that the two clusters are at approximately the same distance from us. In the extensive Virgo cluster (right panel), the galaxy M100 can be seen marked ~4 degrees to the north-west of the elliptical-galaxy-rich core; this corresponds to an impact parameter of 1.3 Mpc, or 8% of the distance from the LG to the Virgo cluster. The Fornax cluster (left panel) is more centrally concentrated than Virgo, so that the back-to-front uncertainty associated with its three-dimensional spatial extent is reduced for any randomly selected member. Roughly speaking, converting the total angular extent of the cluster on the sky (~3 degrees in diameter (7)) into a back-to-front extent, the error associated with any randomly chosen galaxy in the Fornax cluster, translates into a few percent uncertainty in distance; and that uncertainty in distance will soon be reduced when the two additional Fornax spirals are observed with HST in the coming year.

(2) Velocity — Here, we note that the infall-velocity correction for the Local Group motion with respect to the Virgo cluster (and its associated uncertainty) becomes a minor issue for the Fornax cluster. This is the result of a fortuitous combination of geometry and physics. We now have Cepheid distances from the Local Group to both Fornax and Virgo. Combined with their angular separation on the sky this immediately leads to the physical separation between the two clusters proper. Under the assumption that the Virgo cluster dominates the local velocity perturbation field at the Local Group **and** at Fornax, we can calculate the velocity perturbation at Fornax (assuming that the flow field amplitude scales with $1/R_{Virgo}$ and characterized by a R^{-2} density distribution (Schechter 1980)). From this we then derive the flow contribution to the measured line-of-sight radial velocity, as seen from the Local Group. Figure 34 shows the distance scale structure (left panel) and the velocity-field geometry (right panel) of the Local Group–Virgo–Fornax system. Adopting an infall velocity of the Local Group toward Virgo of +200 km/sec (10) with an uncertainty of ±100 km/sec, the flow correction for Fornax is only –44 ±22 km/sec.

(3) H_o at Fornax, and its Uncertainties — Correcting to the barycentre of the Local Group (−90 km/sec) and compensating for the −44 km/sec component of the Virgocentric flow, derived above, we calculate that the cosmological expansion rate of Fornax is 1,321 km/sec. Using our Cepheid distance of 18.0 Mpc for Fornax gives $H_o = 73$ (±7)$_r$ [±20]$_s$ km/sec/Mpc. The first uncertainty (in parentheses) includes random errors in the distance derived from the PL fit to the Cepheid data, as well as random velocity errors in the adopted Virgocentric flow, combined with the distance uncertainties to Virgo propagated through the flow model. The second uncertainty (in square brackets) quantifies the currently identifiable systematic errors associated with the adopted mean velocity of Fornax, and the adopted zero point of the PL relation (combining in quadrature the LMC distance error, a measure of the metallicity uncertainty, and a generous estimate of the possible differences in the true modulus that might be generated from adopting different stellar photometry packages). Finally, we note that according to the Han-Mould model (Han & Mould 1990), the so-called "Local Anomaly" gives the Local Group an extra velocity component of approximately +73 km/sec towards Fornax. If we were to add that correction our local estimate, the Hubble constant would increase to $H_o = 77$ km/sec/Mpc.

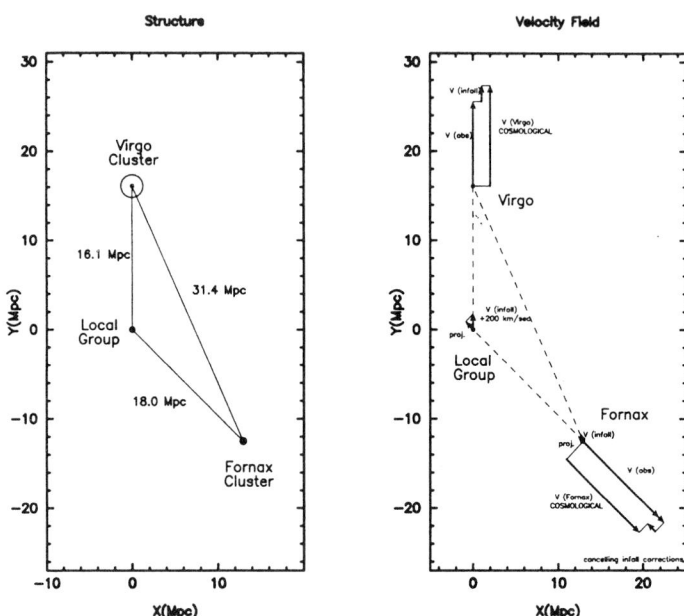

FIGURE 33. Relative geometry (left panel), and the corresponding velocity vectors (right panel) for the disposition and flow of Fornax and the Local Group with respect to the Virgo cluster. The circles plotted at the positions of the Virgo and Fornax clusters have the same angular size as the circles minimally enclosing M100 and NGC 1365 in the two panels of Figure 31.

Given the highly clumped nature of the local universe and the existence of large-scale streaming velocities, there is still a lingering uncertainty about the total peculiar motion of the Fornax cluster with respect to the cosmic microwave background restframe. Observations of flows, and the determination of the absolute motion of the Milky Way with respect to the background radiation suggest that line-of sight velocities $\sim$300 km/sec are not uncommon (Coles & Lucchin 1995). The uncertainty in absolute motion of Fornax with respect to the Local Group then becomes the largest outstanding uncertainty at this point in our error analysis: a 300 km/sec flow velocity for Fornax would result in a systematic error in the Hubble constant of $\sim$20%. We shall however be able to look from afar, and revisit this issue, following an analysis of more distant galaxies made later in this section.

37. The nearby flow field

We now step back somewhat and investigate the Hubble flow between us and Fornax, derived from galaxies and groups of galaxies inside 20 Mpc, each having Cepheid-based distances and expansion velocities individually corrected for a Virgocentric flow model (Kraan-Korteweg 1986). Figure 35 captures those results in graphical form. At

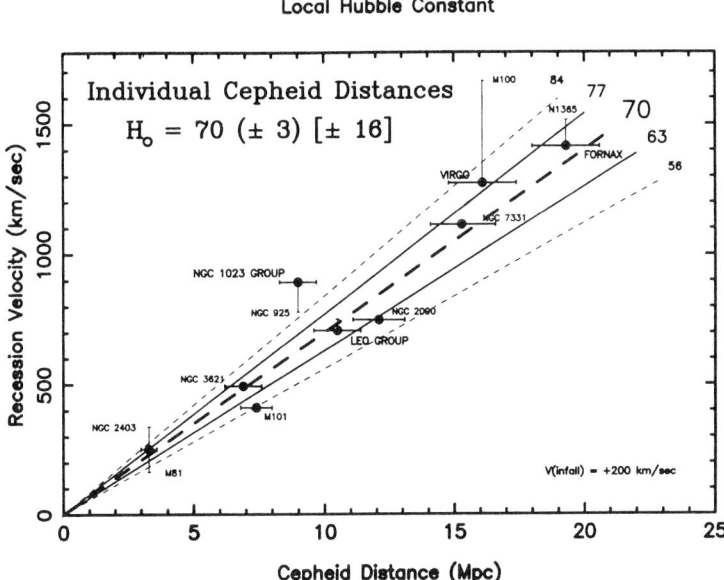

FIGURE 34. The velocity-distance relation for local galaxies having Cepheid-based distances. Circled dots mark the velocities and distances of the parent groups or clusters. The one-sided "error" bars with galaxy names attached mark the velocities associated with the individual galaxies having direct Cepheid distances. The broken line represents a fit to the data giving $H_o = 70 \pm 3$ km/sec/Mpc. The 95% confidence interval on the observed scatter is ± 16 km/sec/Mpc, and is shown by the thin diverging broken lines; the solid lines indicate one-sigma limits.

3 Mpc the M81-NGC 2403 Group (for which both galaxies of this pair have Cepheid distance determinations) gives $H_o = 75$ km/sec/Mpc after averaging their two velocities. Working further out to M101, the NGC 1023 Group and the Leo Group, the calculated values of H_o range from 62 to 99 km/sec/Mpc. An average of these independent determinations including Virgo and Fornax, gives $H_o = 70(\pm 3)_r$ km/sec/Mpc. This determination, as before, uses a Virgocentric flow model with a $1/R_{Virgo}$ infall velocity fall-off, scaled to a Local Group infall velocity of +200 km/sec, which was determined *ab initio* by minimizing the velocity residuals for the galaxies with Cepheid-based distances. The foregoing determination of H_o is again predicated on the assumption that the inflow-corrected velocities of both Fornax and Virgo are not further perturbed by other mass concentrations or large-scale flows, and that the 25,000 Mpc^3 volume of space delineated by them is at rest with respect to the distant galaxy frame. To avoid these local uncertainties we now step out from Fornax to the distant flow field. There we explore three applications: (i) Use of the Tully-Fisher relation calibrated by Cepheids locally, and now including NGC 1365 and about two dozen additional galaxies in the Fornax cluster. Ultimately these calibrators are tied into the distant flow field at 10,000 km/sec defined by

the the Tully-Fisher sample of galaxies in clusters (Aaronson, Huchra, Mould, Sullivan, Schommer & Bothun 1980). (ii) Using the distance to Fornax to tie into averages over previously published differential moduli for independently selected distant-field clusters, (iii) Recalibrating the Type Ia supernova luminosities at maximum light, and applying that calibration to events as distant as 30,000 km/sec.

Table 4

ERROR BUDGET ON THE HUBBLE CONSTANT

Source of Uncertainty on the Mean	Description of Uncertainty	Percentage Error
FORNAX CLUSTER	**EXPANSION VELOCITY AND INFERRED DISTANCE**	
(L) Velocity Dispersion	± 34 km/sec = $\pm 340/\sqrt{N-1}$ (No. of galaxies = 100) at $< V > = 1,321$ km/sec	3%
(M) Geometry of Cluster	± 0.4 Mpc at 18.0 Mpc	2%
(N) Virgocentric Flow	± 22 km/sec on -44 km/sec along the Local Group line of sight	2%
(O1) Bulk Flow	± 300 km/sec	23%
Random Errors	(J) + (L) + (M) combined in quadrature	10%
Systematic Errors	(K) + (N) + (O1) combined in quadrature	28%
	$H_o = 73$ km/sec/Mpc ± 7 (random) ± 19 (systematic)	
LOCAL FLOW	M81, M101, N2090, N3621, N7331, VIRGO, FORNAX	
(P) Random Motions	± 2 km/sec/Mpc = $\pm 5/\sqrt{N-1}$ (No. of galaxies = 7)	4%
(O2) Bulk Flow	± 300 km/sec at $V(max) = +1,400$ km/sec	21%
Random Errors	(P) = total observed scatter	4%
Systematic Errors	[SC] + [Z] + [O2] combined in quadrature	23%
	$H_o = 70$ km/sec/Mpc ± 3 (random) ± 16 (systematic)	
DISTANT FLOW	I. TULLY-FISHER: 16 CLUSTERS TO 10,000 km/sec	
(S) Observed Scatter	± 0.04 mag = $\pm 0.16/\sqrt{N-1}$ (No. of clusters = 16)	2%
(R) TF Zero Point	$\sigma(mean) = \pm 0.13$ mag = $\pm 0.40/\sqrt{N-1}$ (No. of calibrators = 11)	6%
(O3) Bulk Flow	± 300 km/sec evaluated at 10,000 km/sec	3%
Random Errors	(S)	2%
Systematic Errors	[SC] + [Z] + [R] + [O3] combined in quadrature	11%
	$H_o = 76$ km/sec/Mpc ± 2 (random) ± 8 (systematic)	
DISTANT FLOW	II. HYBRID METHODS: 17 CLUSTERS TO 11,000 km/sec	
(U) Observed Scatter	$\pm 0.02 = \pm 0.08/\sqrt{N-1}$ (No. of clusters = 17)	2%
(O4) Bulk Flow	± 300 km/sec evaluated at 11,000 km/sec	3%
(T) Fornax Distance	[SC] + [Z] combined in quadrature	10%
Random Errors	(U)	2%
Systematic Errors	[T] = [SC] + [O4] + [Z]	10%
	$H_o = 72$ km/sec/Mpc ± 1 (random) ± 7 (systematic)	
DISTANT FLOW	III. Type Ia SN: 20 EVENTS OUT TO 20,000 km/sec	
(T1) Peak Luminosity	± 0.11 mag = $\pm 0.45/\sqrt{N-1}$ (No. of SNIa = 16)	6%
(V1) Random Motions	± 300 km/sec at 5,000 km/sec	6%
(O5) Bulk Flow	± 300 km/sec at 20,000 km/sec	2%
(Q1) SNIa Zero Point	$\sigma(mean) = \pm 0.18$ mag = $\pm 0.45/\sqrt{N-1}$ (No. of calibrators = 7)	9%
Random Errors	(T1) + (V1) combined in quadrature	8%
Systematic Errors	[SC] + [O5] + [Q1] combined in quadrature	12%
	$H_o = 68$ km/sec/Mpc ± 5 (random) ± 8 (systematic)	

38. Beyond Fornax: the Tully-Fisher relation

Quite independent of its association with the Fornax cluster as a whole, NGC 1365 provides an important calibration point for the Tully-Fisher relation which links the (distance-independent) peak rotation rate of a galaxy to its intrinsic luminosity. In the left panel of Figure 36 we show NGC 1365 (in addition to NGC 925 (Silbermann 1996), NGC 4536 (Saha, Sandage, Labhardt, Tammann, Macchetto & Panagia 1996) and NGC 4639 (Sandage, Saha, Tammann, Labhardt, Panagia & Macchetto 1996)) added to the ensemble of calibrators having published Cepheid distances (Freedman 1990), and I-band magnitudes (Pierce 1994 and references therein). As mentioned earlier NGC 1365 does now provide the brightest data point in the relation; additional galaxies soon to be added include NGC 2090 (Phelps 1997), NGC 3351 (Graham 1997) and NGC 3621 (Rawson 1997).

Table 5

SUMMARY

Method	Hubble Constant	(Random)	[Systematic]
Fornax Cluster	73 km/sec/Mpc	±7 (random)	±19 [systematic]
Local Flow	70 km/sec/Mpc	±3 (random)	±16 [systematic]
Tully-Fisher	76 km/sec/Mpc	±2 (random)	± 8 [systematic]
Hybrid Methods	72 km/sec/Mpc	±1 (random)	± 7 [systematic]
Type Ia SNe	68 km/sec/Mpc	±5 (random)	± 8 [systematic]
Modal Average:	72 km/sec/Mpc	±2 (random)	±12 [systematic]
Major Systematics:	±10% [FLOWS]	±8% [LMC]	±4% [Fe/H]

Notes: (1) The measured scatter of the $N = 5$ tabulated values of the Hubble constant about the derived mean of 72 km/sec/Mpc is ±3 km/sec/Mpc; the formal error on the mean (due to random errors) is then $3/\sqrt{N-1} = 1.5$ km/sec/Mpc.
(2) The systematic error due to large-scale flows is the average of the ±300 km/sec term on each of the five methods. (23%, 21%, 3%, 3% and 2%, respectively, as given in Table 2)
(3) Calculated for differences in the five Hubble constants with respect to the mean, and scaled to their externally quoted errors, the reduced $\chi^2 = 0.78$. This is only slightly smaller than expected by chance, and suggests that the random errors on the individually determined values of the Hubble constant are realistic.
(4) The concordance between the local and far-field values of the the Hubble constant argue that there is no large flow of the local supercluster with respect to the 20,000 km/sec volume probed by the SNe. At face value the differences in Hubble constants admit a local flow of ~100 km/sec. If so, the (averaged) systematic error due to large scale flow perturbations drops from a dominant 10% down to <3-4%, leaving the LMC distance as the leading source of systematic error on H_o.
(5) The systematic uncertainty due to metallicity is derived from the work on M31 (Freedman & Madore 1990) which reports a marginal detection of a dependence of Cepheid luminosities on metallicity, at the level of ~0.2 mag/dex. A similar test using Cepheids at various radii in M101 is being undertaken by the Key Project Team, and that analysis is currently underway (Kennicutt, et al. 1998).

Although we have only the Fornax cluster for comparison at the present time, it is interesting to note that there is no obvious discrepancy in the Tully-Fisher relation between galaxies in the (low-density) field and galaxies in this (high-density) cluster environment. The NGC 1365 data point is consistent with the data for other Cepheid calibrators. Adding in all of the other Fornax galaxies for which there are published I-band magnitudes and inclination-corrected HI line widths provides us with another comparison of field and cluster spirals. In the right panel of Figure 36 we see that the 21 Fornax galaxies (shifted by the true modulus of NGC 1365) agree extremely well with the 9 brightest Cepheid-based calibrators. The slope of the relation is virtually unchanged by this augmentation; with the scatter about the fitted line increasing somewhat to ±0.35 mag. (Nevertheless the small intrinsic scatter in the relation greatly diminishes the impact of Malmquist-bias.) In following applications we adopt $M_I = -8.80\ log(\Delta V - 2.445) + 20.47$ as the best-fitting least squares solution for the calibrating galaxies.

Han (1992) has presented I-band photometry and neutral-hydrogen line widths for the

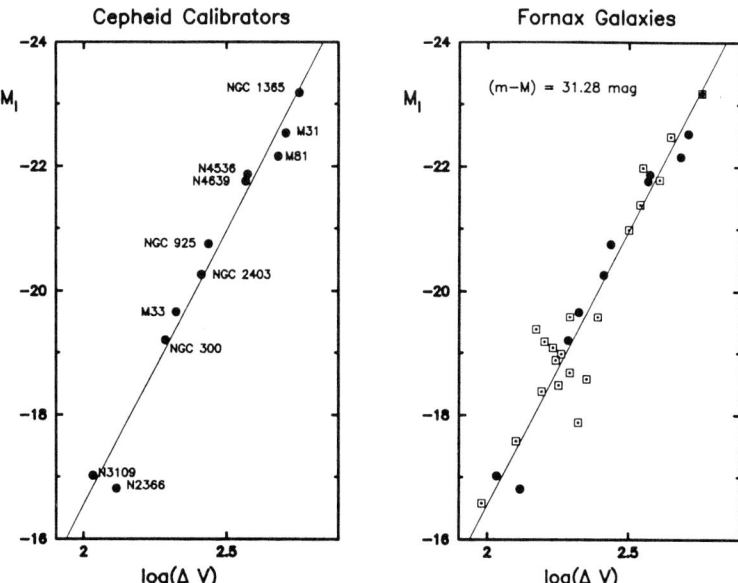

FIGURE 35. Tully-Fisher relations. The left panel shows the absolute I-band magnitude, M_I versus the inclination-corrected 21-cm line widths for galaxies having individually determined Cepheid distances. NGC 1365 is the brightest object in this sample; the position of this *cluster spiral* is fully consistent with an extrapolation of the relation defined by the lower luminosity *field galaxy* sample. The right panel shows the calibrating sample (filled circles) superimposed on the entire population of Fornax spiral galaxies for which I-band observations and line widths are available; the latter being shifted to absolute magnitudes by the Cepheid distance to NGC 1365.

determination of Tully-Fisher distances to individual galaxies in 16 clusters out to redshifts exceeding 10,000 km/sec. We have rederived distances and uncertainties to each of these clusters using the above-calibrated expression for the Tully-Fisher relation. The results are contained in Figure 37, where a linear fit to the data gives a Hubble constant of $H_o = 76$ km/sec/Mpc with a total observed scatter giving a formal (random) uncertainty on the mean of only ±2 km/sec/Mpc. It is significant that neither Fornax nor Virgo deviate to any significant degree from an inward extrapolation of this far-field solution. At face value, these results provide evidence for both of these clusters having only small motions with respect to their local Hubble flow.

39. Beyond Fornax: other relative distance determinations

In addition to the relative distances using the Tully-Fisher relation discussed above, a set of relative distance moduli based on a number of independent secondary distance indicators, including brightest cluster galaxies, Tully-Fisher and supernovae is also avail-

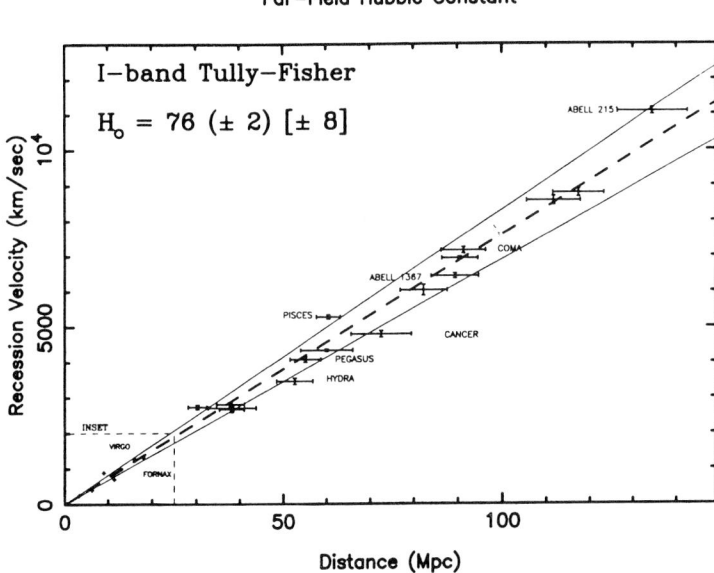

FIGURE 36. The velocity–distance relation for 16 clusters of galaxies out to 11,000 km/sec, having distance moduli determined from the I-band Tully-Fisher relation. A fit to the data gives a Hubble constant of $H_o = 76 \pm 2$ km/sec/Mpc. The solid lines mark one-sigma bounds on the observed internal scatter.

able (Jerjen & Tammann 1993). We adopt, without modification, their differential distance scale and tie into the Cepheid distance to the Fornax cluster, which was part of their cluster sample. The results are shown in Figure 38 which extends the velocity-distance relation out to more than 160 Mpc. No error bars are given in the published compilation but it is clear from the plot that the observed scatter is fully contained by 10% errors in distance or velocity. This sample is now sufficiently distant to average over the potentially biasing effects of large-scale flows, and yields a value of $H_o = 72(\pm 1)_r$ km/sec (random), with a systematic error of 10% being associated with the distance (but not the velocity) of the Fornax cluster. Again the coincidence of H_o at Fornax with that for the far field, argues for Fornax being relatively at rest with respect to the microwave background.

40. Beyond Fornax: type Ia supernovae

In a separate paper (Freedman, et al. 1997) details are reported on the impact of a Cepheid distance to Fornax specifically on the calibration and application of Type Ia supernovae to the extragalactic distance scale. Various calibrations dealing with interstellar extinction and/or decline-rate correlations are presented. Application to the distant Type Ia supernovae (Hamuy et al. 1995) gives $H_o = 68\,(\pm 8)_r$ km/sec/Mpc.

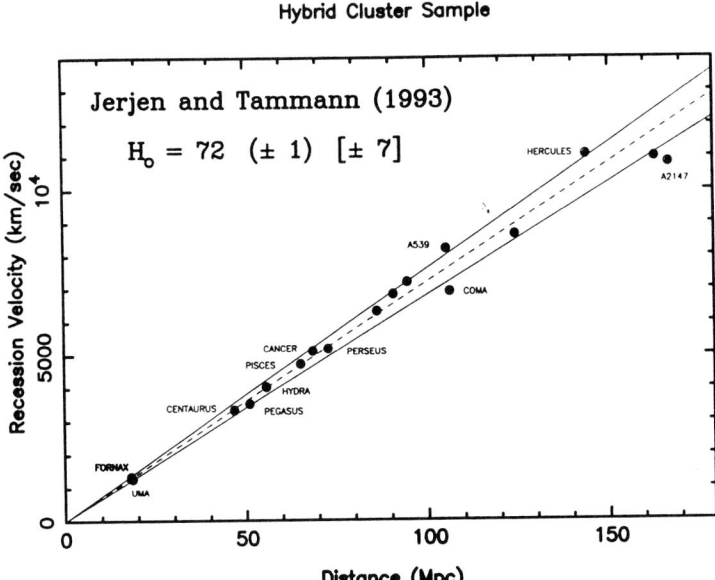

FIGURE 37. The velocity–distance relation for 17 clusters of galaxies with published differential distance moduli (Jerjen & Tammann 1993) scaled to the Fornax cluster. A fit to the data gives a Hubble constant of $H_o = 72 \pm 1$ km/sec/Mpc. As in Figure 37, the solid lines mark one-sigma bounds on the observed internal scatter.

41. Cosmological implications

Given the consistency of Hubble constants derived, both locally and at large recessional velocities, then we can state that H_o falls within the full-range extremes of 75±1 and 68±5 km/sec/Mpc, giving formally $H_o = 72(\pm 2)_r \, [\pm 12]_s$ km/sec/Mpc out to a velocity-distance 0.1c (30,000 km/sec.) These results are summarized graphically in Figure 39 and numerically in Table 5.

A value of the Hubble constant, in combination with an independent estimate of the average density of the Universe, can be used to estimate a dynamical age for the Universe (e.g., see Figure 40). For a value of of $H_o = 72(\pm 2)_r$ km/sec/Mpc, the age ranges from a high of ~12 Gyr for a low-density ($\Omega = 0.2$) Universe, to a young age of ~9 Gyr for a critical-density ($\Omega = 1.0$) Universe. These ages change to 15 and 7.5 Gyr, respectively allowing for a systematic error of ±10 km/sec/Mpc.

Other, independent constraints on the age of the Universe exist; most notably the ages of the oldest stars, as typified by Galactic globular clusters. These ages traditionally are thought to fall in the range of 14±2 Gyr (Chaboyer, Demarque, Kernan, & Krauss 1996), however the subdwarf parallaxes obtained by the Hipparcos satellite (Reid 1997) may reduce these ages considerably. For $\tau = 14$ Gyr and $\Omega = 1.0$, H_o would have to be ~45 km/sec/Mpc. If constrained by the stellar ages, and interpreted within the context of the

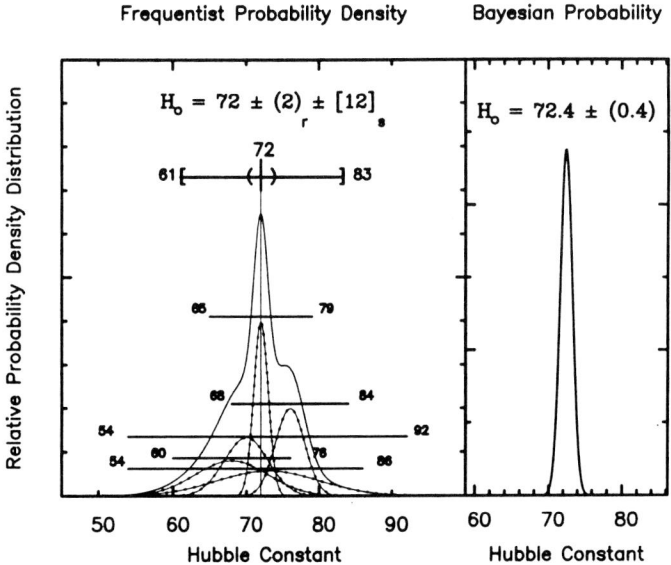

FIGURE 38. A graphical representation of Table 5 showing the various determinations of the Hubble constant, and the adopted mean. Each value of H_o and its statistical uncertainty is represented by a Gaussian of unit area (linked dotted line) centred on its determined value and having a dispersion equal to the quoted *random* error. Superposed immediately above each Gaussian is a horizontal bar representing the one sigma limits of the calculated *systematic errors* derived for that determination. The adopted average value and its probability distribution function (continuous solid line) is the arithmetic sum of the individual Gaussians. (This simple representation treats each determination as independent, assuming no *a priori* reason to prefer one solution over another.)

standard Einstein-de Sitter model, our value of $H_o = 72$ km/sec/Mpc, is incompatible with a high-density ($\Omega = 1.0$) model universe without a cosmological constant (at the 2.5-sigma level defined by the identified systematic errors).

42. Conclusions

In the past decade considerable progress has been made in understanding the systematics of the Cepheid distance scale. The amplitudes of individual Cepheids decrease at longer wavelengths. In addition, the width of the instability strip decreases as a function of increasing wavelength. By making use of these two observed properties, it has been possible (1) to reobserve known Cepheids efficiently and optimally, and (2) to optimize the methods by which new Cepheids are being discovered. Moreover, the ability to obtain multiwavelength data has made it possible to undertake corrections for reddening, and to begin to explore the residual effects of metallicity on the PL relation zero point.

Also in the past decade, a number of completely independent external checks of the

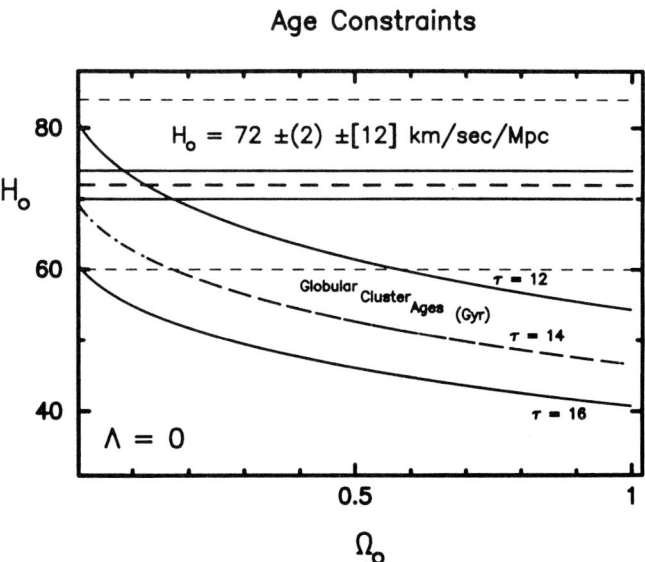

FIGURE 39. Lines of fixed time representing the theoretical ages of the oldest globular cluster stars are shown for 12, 14 and 16 Gyr, plotted as a function of the expansion rate H_o and density parameter Ω_o, for an Einstein-de Sitter universe with the cosmological constant $\Lambda = 0$. The thick dashed horizontal line at $H = 72$ $(\pm 2)_r$ $[\pm 12]_s$ km/sec/Mpc is the average value of the Hubble constant given in Table 5. The parallel (solid) lines on either side of that solution represent the one-sigma random errors on that solution. Systematic errors on the solution for H_0 are represented by thin dashed lines at 61 and 83 km/sec/Mpc. The only region of (marginal) overlap between these two constraints is in the low density ($\Omega < 0.2$) regime, unless $\Lambda \neq 0$. If the globular cluster ages are assumed to place *a lower bound* on the age of the Universe, the region of plausible overlap between the two solutions is more severely restricted to even lower density models.

Cepheid distance scale within the Local Group have been made using RR Lyrae stars, and also using the Pop II TRGB method. These nearby distances now agree to within $\pm 10\%$ rms. Several programs are currently underway to increase the accuracy of the calibration of the Cepheid period-luminosity relation, as well as to test for potential remaining systematic effects in the zero point.

Cepheids have now been discovered out as far as the Virgo cluster and observations have just been completed for their discovery in the Fornax cluster galaxy NGC 1365; their implications are discussed below. As part of the H_0 Key Project, Cepheid distances to approximately 20 galaxies will be obtained which will allow the calibration of 5-6 independent secondary methods, several of which can be applied at velocity distances out to or beyond 10,000 km/sec where the effects of peculiar flows are minimized. Independent, direct Cepheid distances will be measured to three galaxies in each of the Virgo and Fornax clusters. At the end of this program, there will be a solid basis for the intercomparison of secondary distance indicators and therefore for an accurate application to the problem of determining the size scale of the Universe and its absolute expansion rate, the Hubble constant.

We thank the various time allocation committees and observatories that have over the years supported our observational program into the systematic calibration and subsequent application of Cepheids to the extragalactic distance scale; these include Las Campanas, Palomar, KPNO, CTIO and the CFHT, and most recently, the HST. WLF acknowledges support from the NSF in the form of grants AST 87-13889 and 91-16496. Support for this work was also provided by National Aeronuatics and Space Administration (NASA) through grant number GO-2227 from the Space Telescope Science Institute (STScI), which is operated by the Association of Universities for Research in Astronomy, Inc., under NASA contract NAS 5-26555. BFM thanks NASA for support through STScI, JPL, LTSA, and the NASA/IPAC Extragalactic Database (NED). We both also wish to thank our collaborators on the distance scale for their contributions over the past decade: R. Bernstein, J. Catanzarite, I. Horowitz, M.G. Lee, D. Murphy, E. Persson, D. Welch, C. Wilson, and the members of the HST Key Project team: R. Kennicutt, J. Mould, F. Bresolin, L. Ferrarese, H. Ford, B. Gibson, J. Graham, M. Han, P. Harding, J. Hoessel, R. Hill, J. Huchra, S. Hughes, G. Illingworth, D. Kelson, L. Macri, R. Phelps, S. Sakai, A. Saha, N. Silbermann, P. Stetson, and A. Turner. We acknowledge having had illuminating correspondence with Drs. M. Feast, A. Gould, N. Reid and F. van Leeuwen in the course of preparing these lecture notes.

REFERENCES

AARONSON, M., MOULD, J., HUCHRA, J., SULLIVAN, W.T., SCHOMMER, R.A., & BOTHUN, G.D. 1980, *Astrophys. J.*, **239**, 12

ALVES, B.R. & COOK, K.H. 1995, *Astron. J.*, **110**, 192

BAADE, W. 1963 in Evolution of Stars and Galaxies, ed. C. Payne-Gaposchkin, (Harvard, MIT Press).

BAADE, W. & SWOPE, H.H. 1963, *Astron. J.*, **68**, 435

BAADE, W.1 & SWOPE, H.H. 1965, *Astron. J.*, **70**, 212

BAKER, N.H. & KIPPENHAHN, R. 1965, *Astrophys. J.*, **142**, 868

BECKER, S.A. & COX, A.N. 1982, *Astrophys. J.*, **260**, 707

BUREAU, M., MOULD, J.R., & STAVELEY-SMITH, L. 1996, *Astrophys.J.*, **463**, 60

BURSTEIN, D. & HEILES, C. 1984, *Astrophys. J. Suppl.*, **54**, 33

BICA, E., ORTOLANI, S. & BARBUY, B. 1994, *Astron. Ap.*, **106**, 161

BOHM-VITENSE, E. 1994, *Astron. J.*, **107**, 673

BOHM-VITENSE, E. 1997, *Astron. J.*, **113**, 13

CALDWELL, J. A.R. & COULSON, I.M. 1985, *Mon. Not. R. Astron. Soc.*, **212**, 879

BROWN, A.G.A., ARENOU, F., VAN LEEUWEN, F., LINDGREN, L., & LURI, X. 1997, preprint

CALDWELL, J. A. 1991, in *The Magellanic Clouds and Their Dynamical Interactions with the Milky Way*, IAU No. 148, eds. R. F. Haynes & D. K. Milne, (Reidel: Dordrecht)

CALDWELL, N., SCHOMMER, R.A. & GRAHAM, J. 1988, PASP, **100**, 1217

CALDWELL, N., & SCHOMMER, R.A. 1988, *ASP Conf. Series*, **4**, 77.

CALDWELL, J.A.R., & LANEY, C.D. IN IAU SYMP. 148, The Magellanic Clouds, ED. R. HAYNES, & D. MILNE (DORDRECHT: KLUWER), 249

CALDWELL, N., SCHOMMER, R.A. & GRAHAM, J.A. 1988, *Bull. Amer. Astron. Soc.*, 20, 1084

CARDELLI, J.A., CLAYTON, G.G. & MATHIS, J.S. 1989, *Astrophys. J.*, **345**, 245

CARLSON, G. & SANDAGE, A.R. 1990, *Astrophys. J.*, **352**, 587

CARNEY, B.W., STORM, J., & JONES, R.V. 1992, *Astrophys. J.*, **386**, 663

CHABOYER, B., DEMARQUE, P., KERNAN, P.J., & KRAUSS, L.M. 1996, *Science*, **271**, 957

CHIOSI, C., 1990, ASP CONF. SER. **11**, *Confrontation Between Stellar Pulsation and Evolution*, EDS. C. CACCIARI & G. CLEMENTINI (ASP: SAN FRANSISCO), P. 158

CHIOSI, C., WOOD, P. & CAPITANIO, N. 1993, *Astrophys. J. Suppl.*, **86**, 541

CIARDIULO, R., JACOBY, G.H., FORD, H.C. & NEILL, J.D. 1989, *Astrophys. J.*, **339**, 53

CLUBE, S.V.M., & DAWE, J.A. 1980, *Mon. Not. R. Astron. Soc.*, BF 190, 591

CODE, A.D. 1947, *Astrophys. J.*, **106**, 309

COLES, P., & LUCCHIN, F. 1995, IN *Cosmology*, WILEY, 399

COOK, K., AARONSON, M. & ILLINGWORTH, G. 1986, *Astrophys. J.*, **301**, L45

COOK, K. 1996 IN *Astropysical Applications of Stellar Pulsation*, ED. R.S. STOBIE, IAU COLL. 155, (CAMBRIDGE UNIV. PRESS: CAMBRIDGE), IN PRESS

COULSON, I.M., CALDWELL, J.A.R. & GIEREN, W.P. 1986, *Astrophys. J.*, **303**, 273

COUSINS, A.W.J. 1978A, *Observatory*, **98**, 54

COUSINS, A.W.J. 1978B, *Mon. Not. Astron. Soc. So. Africa*, **37**, 62

COX, A.N., MICHAUD, G. & HODSON, S.W. 1978, *Astrophys. J.*, **222**, 621

COX, J.P. 1980, IN *Theory of Stellar Pulsation*, (PRINCETON UNIVERSITY PRESS, PRINCETON)

DA COSTA, G.S., & ARMANDROFF, T.E. 1990, *Astron. J.*, **100**, 162

CROTTS, A.P.S., KUNKEL, W.E., & HEATHCOTE, S.R. 1995 *Astrophys. J.*, **483**, 724

DEAN, J.F., WARREN, P.R. & COUSINS, A.W.J. 1978, *Mon. Not. R. Astron. Soc.*, **183**, 569

DEMERS, S., KUNKEL, W.E., & IRWIN, M.J. 1985, *Astron. J.*, **90**, 1967

DEUPREE, R.G. 1977, *Astrophys. J.*, **215**, 620

DE VAUCOULEURS, G., 1975, IN *Stars and Stellar Systems*, 9, (A.R. SANDAGE, M. SANDAGE, J. KRISTIAN, EDS.), UNIV CHICAGO PRESS, 557

DRESSLER, A., 1987, *Astrophys. J.*, **317**, 1

FEAST, M.W., & CATCHPOLE, R.M. 1997, *Mon. Not R. Astron. Soc.*, **000**, 000

FEAST, M.W. & WALKER, A.R. 1987, *Ann. Rev. Astron. Ap.*, **25**, 345

FELDMEIER, J.J., CIARDULLO, R., & JACOBY, G.H. 1996, *Astrophys. J. (Lett.)*, **461**, 25

FERNIE, J.D. 1969, *Publ. Astron. Soc. Pacific*, **81**, 707

FERNIE, J.D. 1990, *Astrophys. J.*, **354**, 295

FERNIE, J.D. & MCGONEGAL, R. 1983, *Astrophys. J.*, **275**, 783

FERNIE, J.D., KAMPER, K.W., & SEAGER, S. 1993, *Astrophys. J.*, **416**, 820

FERRARESE, L. et al. 1996, *Astrophys. J.*, IN PRESS

FERRARO, F.R., FUSI PECCI, F., TOSI, M. & BUONNANO, R. 1989, *Mon. Not. R. Astron. Soc.*, **241**, 433.

FEAST, M.W., & CATCHPOLE, R.M. 1997, *Mon. Not. R. Astron. Soc.*, IN PRESS

FERGUSON, H.C., & SANDAGE, A.R. 1988, *Astr. J.*, **96**, 1520

FERGUSON, H.C. 1989, *Astr. J.*, **98**, 367

FREEDMAN, W.L. 1988A, *Astrophys. J.*, **326**, 691

FREEDMAN, W.L. 1988B, IN ASP CONFERENCE SERIES, **4**, ED. S. VAN DEN BERGH & C. PRITCHET (PROVO, BRIGHAM YOUNG UNIVERSITY PRESS) P. 24

FREEDMAN, W.L. 1988C, *Astron. J.* **96**, 1248

FREEDMAN, W.L. 1990, *Astrophys. J. (Lett.)*, **355**, L35

FREEDMAN, W.L., GRIEVE, G.R. & MADORE, B.F. 1985, *Astrophys. J. Suppl.*, **59**, 311

FREEDMAN, W.L., HOROWITZ, I., MADORE, B.F., MOULD, J. & GRAHAM, J. 1988A, IN ASP CONFERENCE SERIES, **4**, ED. S. VAN DEN BERGH & C. PRITCHET (PROVO, BRIGHAM

Young University Press) p. 207

Freedman, W.L., Horowitz, I., Madore, B.F. & Mould, J. 1988b, PASP, **100**, 1220

Freedman, W.L. & Madore, B.F. 1988, *Astrophys. J.*, **332**, L63

Freedman, W.L., & Madore, B.F. 1990, *Astrophys. J.*, **365**, 186

Freedman, W.L., & Madore, B.F. 1996, in *Clusters, Lensing and the Future of the Universe*, ASP Conf. Series, eds. V. Trimble & A. Reisenegger, p. 9

Freedman, W.L., Madore, B.F. & Kennicutt, R.C. 1997, in *The Extragalactic Distance Scale*, eds. M. Livio, M. Donahue & N. Panagia, (Cambridge, Cambridge Univ. Press), p. 171

Freedman, W.L., Madore, B.F., Stetson, P.B., et al. 1994, *Astrophys. J.*, **435**, L31

Freedman, W.L. & Madore, B.F. 1990, *Astrophys. J.*, **365**, 186

Freedman, W.L. & Madore, B.F. 1993, in *New Perspectives on Stellar Pulsation and Pulsating Variable Stars*, eds. J.M. Nemec & J.M. Matthews, (Cambridge Univ. Press: Cambridge)

Freedman, W.L., Wilson, C.D. & Madore, B.F. 1991, *Astrophys. J.*, **372**, 455

Freedman, W.L. et al. 1992, *Astrophys. J.*, **396**, 80

Freedman, W.L. et al. 1994a, *Nature*, **371**, 757

Freedman, W.L. et al. 1994b, *Astrophys. J.*, **427**, 628

Freedman, W.L. et al. 1994c, *Astrophys. J.*, **435**, L31

Freedman, W.L., et al. 1997, in preparation

Gallart. C., Aparicio, A. & Vilchez, J.M. 1996, *Astron. J.*, **112**, 1928

Gaposchkin, S. 1962, *Astron. J.*, **67**, 334

Gould, A. & Uza, O. 1997, *Astrophys. J.*, , submitted (=astro-ph/9705051)

Graham, J.A. 1984, *Astron. J.*, **89**, 1332

Graham, J.A., et al. 1997, *Astrophys. J.*, **477**, 535

Gratton, R.G., Fusi Pecci, F., Carretta, E., Clenentini, G., Corsi, C.E., & Lattanzi, M. 1997, *Astrophys. J.*, , submitted (=astro-ph/9704150)

Gould, A. 1994, *Astrophys. J.*, **426**, 542

Hamuy, M. 1995, et al. *A. J.*, **109**, 1

Han, M.S. 1992, *Astrophys. J. Suppl.*, **81**, 35

Han M., & Mould J.R. 1990, *Astrophys. J.*, **360**, 448

Hanes, D.A. 1982, *Mon. Not. R. Astron. Soc.*, **201**, 145

Hill, R., et al. 1997, *Astrophys. J. Suppl.*, in press

Hodge, P.W. 1977, *Astrophys. J. Suppl.*, **33**, 69

Hodge, P.W. 1981 *Ann. Rev. Astron. Ap.*, **19**, 357

Holtzmann, J., et al. 1995, *Publs Astron. Soc. Pacif.*, **107**, 156

Hoessel, J.G., Abbott, M.J., Saha, A., Mossman, A.E.,& Danielson, G.E. 1990, *Astron. J.*, **100**, 1151

Hoessel, J.G., Schommer, R.A. & Danielson, G.E. 1983, *Astrophys. J.*, **274**, 577

Hoessel, J.G. et al. 1990, *Astron. J.*, **100**, 1151

(Hoessel, J.G., Saha, A., Krist, J., & Danielson, G.E. 1994, *Astron. J.*, **108**, 645

Hubble, E.P. 1925, *Astrophys. J.*, **62**, 409

Hubble, E.P. 1926, *Astrophys. J.*, **63**, 236

Hubble, E.P. 1929, *Proc. Nat. Acad. Sci.*, **15**, 168,

Iben, I., 1974, *Ann. Rev. Astron. Astrophys.*, **12**, 215

Iben, I. & Renzini, A. 1983, *Ann. Rev. Astron. Ap.*, **21**, 271

Jacoby et al. 1992, *Publ. Astron. Soc. Pacific*, **104**, 599

JERJEN, H., & TAMMANN, G.A. 1993, *Astr. Ap.*, **276**, 1

JOHNSON, H. L. 1963, STARS & STELLAR SYSTEMS, VOL. III, *Basic Astronomical Data*, ED. K.A. STRAND, (UNIV. CHICAGO PRESS, CHICAGO), P. 204

KAYSER, S.E. 1967, *Astron. J.*, **72**, 134

KENNICUTT, R.C., FREEDMAN, W.L., & MOULD, J.R. 1995, *Astron. J.*, **110**, 1476

KELSON, D., et al. 1996, *Astrophys. J.*, **463**, 26

KINMAN, T., MOULD, J. & WOOD, P.R. 1987, *Astron. J.*, **93**, 833

KENNICUTT, R.C., FREEDMAN, W.L., & MOULD, J.R. 1995, *Astron. J.*, **110**, 1476

KRAFT, R.P. 1960 *Astrophys. J.*, **131**, 330

KRAAN-KORTEWEG, R. 1986, *Astron. Astrophys. Suppl*, **66**, 255

LAFLER, J., & KINMAN, T.D. 1965, *Astrophys. J. Suppl.*, **11**, 216

LANEY, C.D. & STOBIE, R.S. 1986A, *So. African Astron. Obs. Circ.*, **10**, 51

LANEY, C.D. & STOBIE, R. S. 1986B, *Mon. Not. R. Astron. Soc.*, **222**, 449

LANEY, C.D., & STOBIE, R.S. 1992, *Astron. Astrophys. Suppl.*, **93**, 93

LANEY, C.D., & STOBIE, R.S. 1994, *Astron. Astrophys.*, **266**, 441

LAUTERBORN, D., REFSDAL, S. & WEIGERT, A. 1971, *Astrophys. J.*, **10**, 97

LAUTERBORN, D. & SIQUIG, R.A. 1974, *Astrophys. J.*, **191**, 589

LEAVITT, H. 1906, *Ann. Harvard Coll. Obs.*, **60**, 87

LEE, M.G. 1993, *Astrophys. J.*, **408**, 409

LEE, Y.-W. DEMARQUE, P. & ZINN, R. 1993, *Astrophys. J.*, **350**, 155

LEE, M.G., FREEDMAN, W.L. & MADORE, B.F. 1993, *Astrophys. J.*, **417**, 553

LEE, M.G., FREEDMAN, W.L. & MADORE, B.F. 1993, *Astrophys. J.*, **417**, 553

MADORE, B.F. 1976, *Mon. Not R. Astron. Soc.*, **177**, 157

MADORE, B.F. 1982, *Astrophys. J.*, **253**, 575

MADORE, B.F. 1985, *Cepheids: Theory and Observations*, ED: B.F. MADORE, CAMBRIDGE UNIV. PRESS: CAMBRIDGE, P. 166

MADORE, B. 1986 IN *Galaxy Distances and Deviations from Universal Expansion*, ED. B. F. MADORE & R. B. TULLY (DORDRECHT, REIDEL), P. 29

MADORE, B.F. et al. 1987, *Astrophys. J.*, **320**, 26

MADORE, B.F. et al. 1997, *Nature*, SUBMITTED

MADORE, B.F. & FREEDMAN, W.L. 1985, *Astron. J.*, **90**, 1104

MADORE, B.F., & FREEDMAN, W.L. 1991, *Publ. Astron. Soc. Pacific*, **103**, 933 (MF91)

MADORE, B.F., & FREEDMAN, W.L. 1995, *Astron. J.*, **109**, 1645

MADORE, B.F., & FREEDMAN, W.L. 1997, *Astrophys. J. (Lett.)*, IN PRESS.

MADORE, B.F., FREEDMAN, W.L. & LEE, M.G. 1993, *Astron. J.*, **106**, 2243

MADORE, B.F., FREEDMAN, W.L., & PEVUNOVA, O. 1998, *Astron. J.*, (IN PREPARATION)

MADORE, B.F., MCALARY, C.W., MCLAREN, R.A., WELCH, D.L., NEUGEBAUER, G. & MATTHEWS, K. 1985, *Astrophys. J.*, **295**, 560

MADORE, B.F., WELCH, D.L., MCALARY, C.W. & MCLAREN, R.A. 1987, *Astrophys. J.*, **320**, 26

MARTIN, P.R., WARREN, P.R. & FEAST, M.W. 1979, *Mon. Not. R. Astron. Soc.*, **188**, 139

MATHEWSON, D.S., FORD, V.L. & VISVANATHAN, N. 1986, *Astrophys. J.*, **301**, 664

MATHEWSON, D.S., FORD, V.L. & VISVANATHAN, N., 1988, *Astrophys. J.*, **333**, 617

MATHEWSON, D.S., FORD, V.L., & BUCHHORN, M. 1992, *Astrophys. J. Suppl.* **81**, 413

MCALARY, C.W. & MADORE, B.F. 1984, *Astrophys. J.*, **282**, 101

MCALARY, C.W., MADORE, B.F. & DAVIS, L.E. 1984, *Astrophys. J.*, **276**, 487

MCALARY, C.W., MADORE, B.F., MCGONEGAL, R., MCLAREN, R.A. & WELCH D.L. 1983,

Astrophys. J., **273**, 539

MCCALL, M.L., 1993, *Astrophys. J. (Lett.)*, **417**, L75

MCGONEGAL, R., MCLAREN, R.A., MCALARY, C.W. & MADORE, B.F., 1982, *Astrophys. J. (Lett.)*, **257**, L33

MOULD, J.R. & KRISTIAN, J. 1990, *Astrophys. J.*, **354**, 438

MOULD, J.R. & KRISTIAN, J. 1986, *Astrophys. J.*, **305**, 591

MOULD, J.R., KRISTIAN, J., & DA COSTA, G.S. 1983, *Astrophys. J.*, **270**, 471

MOULD, J.R., KRISTIAN, J., & DA COSTA, G.S. 1984, *Astrophys. J.*, **278**, 575

PANAGIA, N., et al. 1996, UNPUBLISHED POSTER PAPERS FROM THE STScI SYMPOSIUM, *The Extragalactic Distance Scale*

PANAGIA, N., GILMOZZI, R., MACCHETTO, F., ADORF, H.-M., & KIRSCHNER, R.P. 1991, *Astrophys. J.*, **380**, L23

PAYNE-GAPOSCHKIN, C. 1971, *Smithsonian Contr. Ap.*, **13**, (SMITHSONIAN INST. PRESS, WASHINGTON D.C.)

PAYNE-GAPOSCHKIN, C. & GAPOSCHKIN, S. 1966, *Smithsonian Contr. Ap.*, **9**, (SMITHSONIAN INST. PRESS, WASHINGTON D.C.)

PEL, J.W. 1985, IN IAU COLLOQ. 82, *Cepheids: Theory and Observations*, ED. B.F. MADORE (CAMBRIDGE: CAMBRIDGE UNIV. PRESS), 1

PHELPS, R., et al. 1998, *Astrophys. J.*, IN PREPARATION

PIERCE, M. 1994, *Astrophys. J.*, **430**, 53

RAWSON, D.M., et al. 1998, *Astrophys. J.*, IN PREPARATION

REID, I.N. & FREEDMAN, W.L. 1994, *Mon. Not. R. Astron. Soc.*, **267**, 821

REID, I.N. 1997, *Astron. J.*, IN PRESS (=ASTRO-PH/9704078)

ROWAN-ROBINSON, M. 1985, IN *The Cosmological Distance Ladder*, (FREEMAN, SAN FRANSISCO)

SAHA, A. et al. 1992, *Astron. J.*, **104**, 1072

SAHA, A., HOESSEL, J.G., KRIST, J., & DANIELSON, G.E. 1996 *Astron. J.*, **111**, 197

SAHA, A. et al. 1995, *Astrophys. J.*, **438**, 8

SAHA, A., FREEDMAN, W.F., HOESSEL, J.G., & MOSSMAN, A.E. 1992, *Astron. J.*, **104**, 1072

SAHA, A., SANDAGE, A.R., LABHARDT, L., TAMMANN, G.A., MACCHETTO, F.D., & PANAGIA, N. 1996, *Astrophys. J.*, **466**, 55

SAHA, A. et al. 1994, *Astrophys. J.*, **425**, 14

SAKAI, S., MADORE B.F., & FREEDMAN, W.L. 1996A, *Astrophys. J.*, **461**, 713

SAKAI, S., MADORE, B. F., & FREEDMAN, W. L. 1996B, *Astrophys. J.*, **480**, 589.

SAKAI, S., MADORE B.F., FREEDMAN, W.L. LAUER, T.R., AJHAR, E.A., & BAUM, W.A. 1997, *Astrophys. J.*, (IN PRESS).

SANDAGE, A.R. 1958, *Astrophys. J.*, **127**, 513

SANDAGE, A.R. 1972, *Q.J.R.A.S*, **13**, 202

SANDAGE, A.R. 1983, *Astron. J.*, **88**, 1108

SANDAGE, A.R. 1984, *Astron. J.*, **89**, 621

SANDAGE, A.R. 1988A, *Astrophys. J.*, **331**, 605

SANDAGE, A.R. 1988B, *Publ. Astron. Soc. Pacific*, **100**, 935

SANDAGE, A.R., & CACCIARI, C. 1990, *Astrophys. J.*, **350**, 645

SANDAGE, A.R. & CARLSON, G. 1983B, *Astrophys. J.*, **258**, 439

SANDAGE, A.R. & CARLSON, G. 1983A, *Astrophys. J.*, **267**, L25

SANDAGE, A.R. & CARLSON, G. 1985A, *Astron. J.*, **90**, 1464

SANDAGE, A.R. & CARLSON, G. 1985B, *Astron. J.*, **90**, 1019

SANDAGE, A.R. & CARLSON, G. 1988, *Astron. J.*, **96**, 1599

SANDAGE, A.R. & GRATTON, L. 1963, IN *Star Evolution*, (NEW YORK: ACADEMIC PRESS), p. 11
SANDAGE, A.R., SAHA, A., TAMMANN, G.A., LABHARDT, L., PANAGIA, N., & MACCHETTO, F.D. 1996, *Astrophys. J. (Lett.)*, **460**, L15
SANDAGE, A.R. & TAMMANN, G.A. 1968, *Astrophys. J.*, **151**, 531
SANDAGE, A.R. & TAMMANN, G.A. 1969, *Astrophys. J.*, **157**, 683
SANDAGE, A.R. & TAMMANN, G.A. 1971, *Astrophys. J.*, **167**, 293
SANDAGE, A.R. & TAMMANN, G.A. 1974, *Astrophys. J.*, **194**, 223
SANDAGE, A.R, & TAMMANN, G.A. 1997, *Mon. Not. R. Astron. Soc.*, (SUBMITTED)
SANDAGE, A.R. et al. 1994, *Astrophys. J.*, **423**, L13
SCHALTENBRAND, R. & TAMMANN, G.A. 1970, *Astron. Astrophys.*, **7**, 289
SCHECHTER, P. 1980, *Astron. J.*, **85**, 801
SCHMIDT, E. 1991, OBSERVATORY,
SCHMIDT, E. & SIMON, N.R. 1987, IN *Stellar Pulsation*, EDS. A.N. COX, W.M. SPARKS, AND S.G. STARRFIELD, (SPRINGER-VERLAG, BERLIN), P. 180
SCHMIDT, E. & SPEAR, G.G. 1987, *Bull. Amer. Astron. Soc.*, **19**, 1036
SCHMIDT, E. & SPEAR, G.G. 1989, *Mon. Not. R. Astron. Soc.*, 236, 567
SCHMIDT, E., SPEAR, G.G. & SIMON, N.N. 1986, *Bull. Amer. Astron. Soc.*, **18**, 964
SCHMIDT-KALER, TH. 1992, *ASP Conf. Ser.*, **30**, 195
SCHRODER, A. 1995, DOCTORAL THESIS, UNIVERSITY OF BASEL
SCHRODER, A., & RICHTER, O.-G., 1997, IN PREPARATION
SILBERMANN, N.A., et al. 1996, *Astrophys. J.*, **470**, 1
SILBERMANN, N.A., et al. 1998, *Astrophys. J.*, IN PREPARATION
SIMON, N., 1990, ASP CONF. SER., **11**, *Confrontation Between Stellar Pulsation and Evolution*, ED. C. CACCIARI (ASP: SAN FRANSISCO), P. 193
SORIA, R., et al. 1996, *Astrophys. J.*, **456**, 79
STELLINGWERF, R.F. 1986, *Astrophys. J.*, **303**, 119
STETSON, P.B. 1994, *Publs Astron. Soc. Pacif.*, **106**, 250
STIFT, M.J. 1982, *Astron. Astrophys.*, **112**, 149
STIFT, M.J. 1990, *Astron. Astrophys.*, **229**, 143
STOTHERS, R. 1982, *Astrophys. J.*, **255**, 227
STOTHERS, R. 1983, *Astrophys. J.*, **274**, 20
STOTHERS, R. 1988, *Astrophys. J.*, **329**, 712
TAMMANN, G.A. & SANDAGE, A.R. 1968, *Astrophys. J.*, **151**, 825
TANVIR, N.R. et al. 1995, *Nature*, **377**, 27
TANVIR, N.R. 1997, IN *The Extragalactic Distance Scale*, EDS. M. LIVIO, M. DONAHUE & N. PANAGIA, (CAMBRIDGE, CAMBRIDGE UNIV. PRESS), P. 91
THOMAS, H.-C. 1967 *Zeit. Ap.*, **67**, 420.
TOLSTOY, E., SAHA, A., HOESSEL, J.G., & DANIELSON, G.E. 1995, *Astron. J.*, **109**, 579
TOLSTOY, E., SAHA, A., HOESSEL, J.G., & MCQUADE, K. 1996, *Astron. J.*, **110**, 1640
TURNER, D. G. 1990, *Publ. Astron. Soc. Pacific*, **102**, 1331
VAN DEN BERGH, S. 1995, *Astrophys. J.*, **446**, 39
VAN DEN BERGH, S. 1975, STARS & STELLAR SYSTEMS, VOL. IX, *Galaxies and the Universe*, ED. A.R. SANDAGE, M. SANDAGE & J. KRISTIAN, (UNIV. CHICAGO PRESS, CHICAGO), P. 509
VAN DEN BERGH, S. 1979, *Astron. J.*, **84**, 604
VAN LEEUWEN, F. 1983, PH.D. THESIS, LEIDEN UNIV., NETHERLANDS

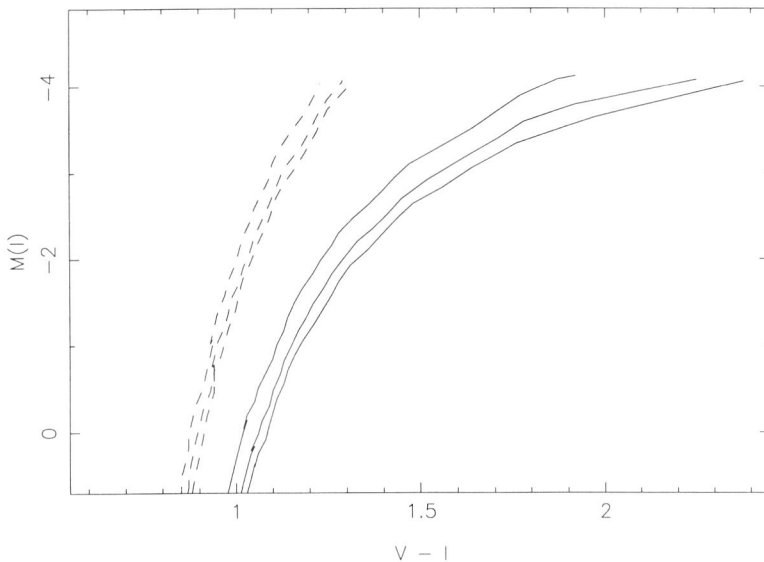

FIGURE 1. Red giant branches from the Bertelli et al. (1994) isochrone set for Z=0.0004 (log Z/Z_{sun} = -1.7), dashed lines, and Z=0.004 (log Z/Z_{sun} = -0.7), solid lines, for ages of 5, 10 and 15 Gyr. At fixed abundance, younger ages are bluer in colour.

a stellar population to which this abundance estimation technique can be applied. This age limit results simply from the fact that younger, more massive stars do not develop significant electron-degenerate cores as they leave the main sequence in the same way as do older, lower mass stars. Consequently, the younger stars do not undergo a full RGB evolution up to the helium flash luminosity. Instead, core helium burning begins under non-degenerate conditions at luminosities well below that of the helium flash. Hence in a stellar population with an age less than ~1 – 2 Gyr (the actual limit depends somewhat on abundance, see Bertelli et al. 1994, for example), a full red giant branch equivalent to that seen in an older population would not be observed.

(b) The bolometric luminosity of the helium flash (the tip of the Red Giant Branch) is largely independent of both Age and Abundance.
This is illustrated in Fig. 2 where red giant branches from Bertelli et al. (1994) are shown for three compositions (Z = 0.0004, 0.004 and 0.008) and three ages (5, 10 and 15 Gyr). While there is a small increase in M_{bol}(RGB tip) with Z, this variation is clearly small enough that M_{bol}(RGB tip) can serve as a distance indicator (i.e. a "standard candle"). The reason for this relative constancy of the He core flash luminosity is that the temperature of the electron-degenerate, approximately-isothermal core in which the flash occurs, is set by the characteristics of the thin H-burning shell on the outside of the core, and these are largely independent of both the amount of material above the shell and the abundance (e.g. Eggleton 1968).

Generally, measurements in the infrared are required to determine m_{bol}(RGB tip) directly, but if the bolometric correction to, for example, the RGB tip magnitude in the I-band is known, then the corresponding m_{bol} can be determined. This is the basis of the so-called "Tip of the Red Giant Branch" (TRGB) distance indicator (e.g. Da Costa & Armandroff 1990, Lee et al. 1993a, Madore, these lectures). The TRGB scale can be set by either directly adopting the theoretical calculations of the helium flash luminosity, or by calibrating observed values of I(TRGB) or m_{bol}(TRGB) on a chosen distance scale, such as a globular cluster horizontal branch luminosity scale. For relatively metal-poor

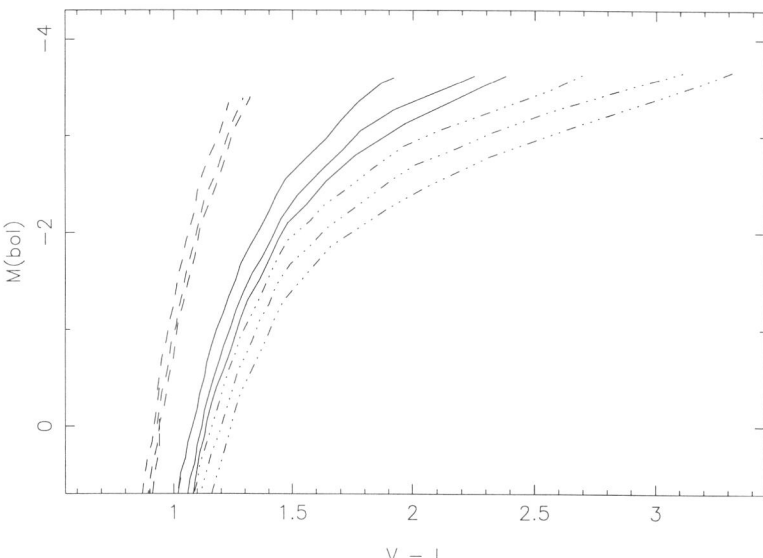

FIGURE 2. Red giant branches from the Bertelli et al. (1994) isochrone set for Z=0.0004 (log Z/Z_{sun} = −1.7), dashed lines, Z=0.004 (log Z/Z_{sun} = −0.7), solid lines, and Z=0.008 (log Z/Z_{sun} = −0.4), dash-dotted lines, for ages of 5, 10 and 15 Gyr. At fixed abundance, younger ages are bluer in colour. Note that the luminosity of the red giant branch tip is almost independent of age and abundance.

systems ([Fe/H] ≤ −0.7), the TRGB in I, rather than in m_{bol}, is used as a distance indicator since the value of M_I(TRGB) does not vary strongly with [Fe/H] or age. For more metal-rich systems, however, blanketing of the red giant atmospheres by molecules such as TiO, for example, depresses the I-band magnitude of the RGB tip. Consequently, the TRGB technique cannot be used in this situation unless the abundance and age are known independently. This effect is illustrated in Fig. 3 which is identical to Fig. 2 except that M_I is plotted instead of M_{bol}. The depression of I(TRGB) at the cooler temperatures reached by the metal-rich red giants is evident. Nevertheless, it is also evident from Fig. 1 and/or Fig. 3 that with observations that cover the brightest two or three magnitudes or so of the red giant branch, we can determine estimates for *both* the distance and abundance of a stellar population (provided it is not too metal-rich).

(c) The temperature or colour of core helium burning stars of low mass is dependent on both Age and Abundance.
Fundamentally, age information for a stellar population is contained in the luminosity of the main sequence turnoff. But even with HST, observations of an old (age ≥ 10 Gyr) main sequence turnoff (at $M_V \approx +4$) are practical only for the Galaxy's companions. On the other hand, it is possible with HST to reach, in all Local Group galaxies, magnitudes that are somewhat fainter than the magnitude ($M_V \approx +0.6$) of the "horizontal branch", the locus in a c-m diagram formed by core helium burning stars of low mass. Thus it seems reasonable to assess what kind of age information can be gained from observations that reach this magnitude level. Obviously, for those stellar populations whose main sequence turnoff is brighter than the horizontal branch luminosity, the age information is directly accessible. This situation corresponds to ages younger than 1 – 2 Gyr; older than this the main sequence turnoff is fainter than the horizontal branch luminosity. In such circumstances, age information can then sometimes be gleaned from the properties of the core helium burning stars that make up the horizontal branch.

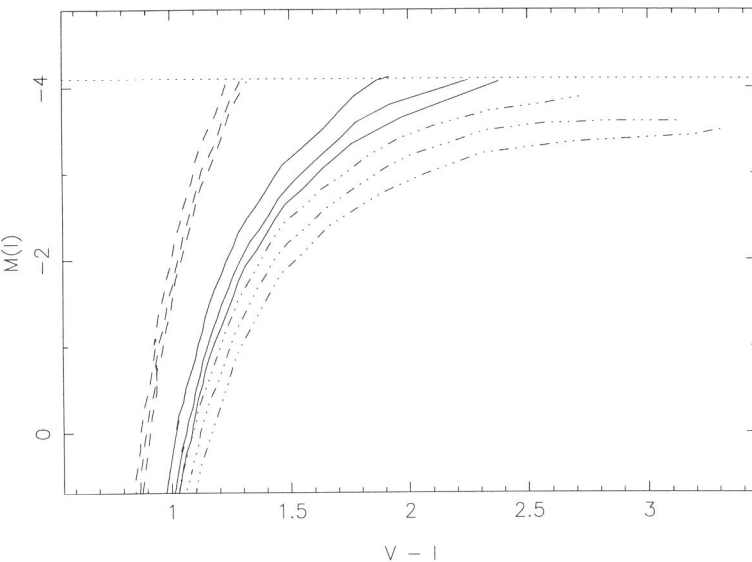

FIGURE 3. Red giant branches in the I-band from the Bertelli et al. (1994) isochrone set for Z=0.0004 (log Z/Z$_{sun}$ = −1.7), dashed lines, Z=0.004 (log Z/Z$_{sun}$ = −0.7), solid lines, and Z=0.008 (log Z/Z$_{sun}$ = −0.4), dash-dotted lines, for ages of 5, 10 and 15 Gyr. At fixed abundance, younger ages are bluer in colour. The dotted horizontal line is for M$_I$ = −4.1. Note that the M$_I$ values for the red giant branch tip are depressed below this value in the metal-rich case.

The location of a low mass (meaning a star that undergoes a full RGB evolution) core helium burning star in a c-m diagram is determined primarily by the star's envelope mass, the difference between the total mass and the mass of the star's core. The sense of the effect is that lower envelope masses imply hotter temperatures or bluer colours on the horizontal branch. Consequently, since younger stars leave the main sequence with larger total masses than older stars, and since the core mass does not change much with total mass, at fixed abundance younger stars will have redder colours than older stars during the core helium burning phase of evolution. Younger core helium burning stars also tend to be somewhat brighter, particularly for ages less than ∼5 Gyr. On the other hand, at fixed age, lower abundance stars have smaller total masses at the turnoff and so will lie to the blue of higher abundance stars during core helium burning evolution. Thus the location of a low mass core helium burning star in the c-m diagram is a function of both age and abundance. But if we have independent information on the abundance, for example from the RGB colour, then the morphology of the core helium burning stars in the c-m diagram can yield age information (e.g. Lee et al. 1994, Sarajedini et al. 1995).

These concepts are illustrated in Fig. 4 where the locations of initial horizontal branches from the isochrone set of Bertelli et al. (1994) are shown for a number of ages and abundances. In this figure it can be seen that at fixed abundance, decreasing the age first shifts the location of the initial horizontal branch to the red and then raises the luminosity somewhat. It is also clear that this effect is more marked at older ages and lower abundances since in these situations the envelope mass is smaller. In particular, it can be seen that blue horizontal branch stars, and RR Lyrae variables (which are intermediate colour horizontal branch stars whose surface temperatures place them in the instability strip), occur only for the oldest ages. Thus the occurrence of such stars in a stellar population is strong evidence for the existence of old ($\geq$ ∼10 Gyr) stars in that population (note though that the reverse is not true — since not all red giants generate

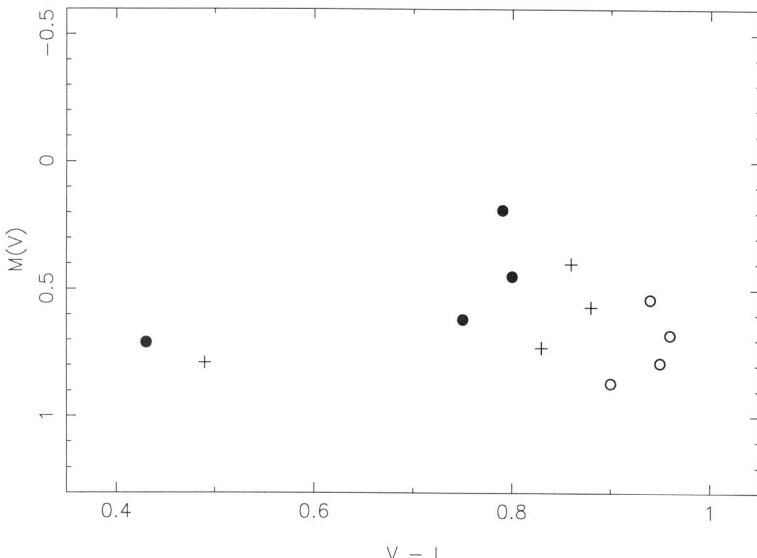

FIGURE 4. Initial horizontal branch locations as a function of age and abundance from the Bertelli et al. (1994) isochrone set. Abundances are Z=0.0004 (filled circles), Z=0.001 (plus signs) and Z=0.004 (open circles), and for each abundance, ages of 2, 5, 10 and 15 Gyr are shown. The oldest age in each case has the bluest colour and the faintest luminosity.

RR Lyrae progeny, the absence of RR Lyrae variables does not imply the absence of an old population). On the other hand, at younger ages and/or higher metallicities, since the envelope masses are higher, the core helium burning stars are confined to relatively red colours. This is the origin of the term "red clump" which is often used to describe the appearance of the horizontal branch in the c-m diagrams of younger/more metal-rich systems. A full discussion of the details of core helium burning evolution are beyond the scope of these lectures but see, for example, Lee et al. (1994) and the references therein. It is necessary to keep in mind only that evolution away from the initial location, and the range in envelope mass that results from varying amounts of mass loss on the RGB, both contribute to a sizeable colour spread in observed horizontal branches, even at constant age and abundance.

(d) Asymptotic Giant Branch (AGB) stars with sufficient envelope mass can rise to luminosities above the RGB tip.
When helium is exhausted in the core of a horizontal branch star, then provided the star has sufficient envelope mass, it evolves back towards the red giant branch becoming an asymptotic giant branch star. The luminosity that the star then reaches before it sheds its remaining envelope as a planetary nebula, with the core becoming a white dwarf, is dependent on the envelope mass. Observationally, it appears that in galactic globular clusters with [Fe/H] $\leq$ −1.0, the AGB evolution terminates at a luminosity comparable to the red giant branch tip (e.g. Frogel & Elias 1988). But in younger (or in more metal-rich) populations, the envelope mass is larger and the star can then evolve to higher luminosities. Thus, to some extent, the luminosity reached by upper-AGB stars can serve as an age indicator. In particular, the occurrence of upper-AGB stars with luminosities significantly above the RGB tip is an unambiguous sign of the presence of a population whose age is younger than that of the galactic globular clusters (provided the abundance is not too high).

These concepts are illustrated in Fig. 5 where upper AGB loci are shown for four dif-

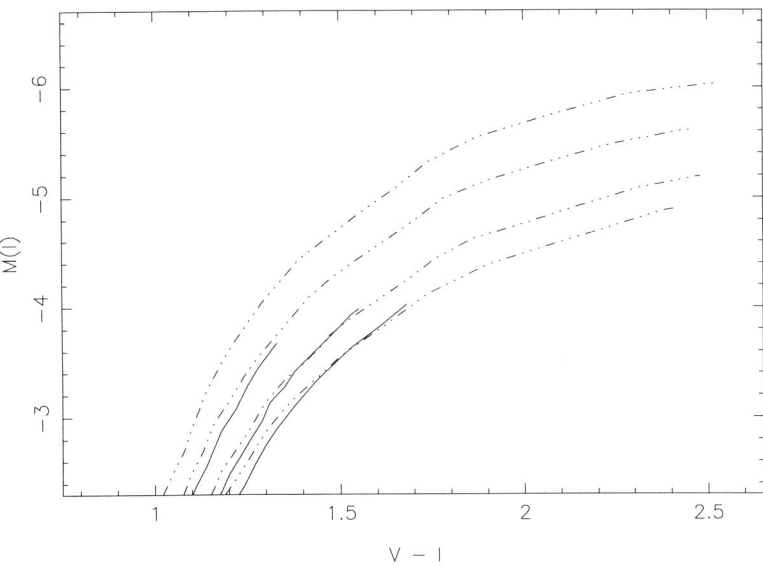

FIGURE 5. Upper asymptotic giant branches in the I-band (from the Bertelli et al. (1994) isochrone set) are shown as dash-dotted lines for Z=0.001 (log Z/Z$_{sun}$ = −1.3) and ages of 1, 2, 5 and 10 Gyr. The upper-AGBs for younger ages are initially bluer and rise to higher luminosities. The corresponding red giant branches, which exist only for the three older ages, are shown as solid lines.

ferent ages at an abundance of Z=0.001 (log Z/Z$_{sun}$ = −1.3). The luminosities which the upper-AGB stars reach are clearly brighter for the younger ages. Figure 5 also illustrates that the upper-AGB stars reach quite red colours or low temperatures. A detailed discussion of upper-AGB evolution is beyond the scope of these lectures (see Chiosi et al. 1992 and Marigo et al. 1996 together with the references therein, for example) but three points should be kept in mind. First, given that the AGB termination luminosity is a function of envelope mass, then anything that changes the envelope mass will affect the luminosity evolution. In particular, these low surface gravity, high luminosity stars will be undergoing significant mass loss on the AGB. Theoretically, this mass loss can be parametrized in a number of ways (e.g. Vassiliadis & Wood 1993) but the details of the evolution will be sensitive to the adopted mass loss law (see, for example, Appendix A of Gallart et al. 1996b). Second, upper-AGB evolution is rapid, so that a large sample of progenitors is required in order to observe a substantial number of stars in the upper-AGB phase. Therefore, for example, the luminosity of a single upper-AGB star is not likely to be a good indicator of the AGB termination luminosity in an underlying stellar population. Third, during upper-AGB evolution, the star undergoes thermal pulses which can bring freshly synthesized carbon into the surface layers. In most cool giants, the abundance of oxygen in the atmosphere exceeds that of carbon which is mostly locked up in CO molecules. For that reason M stars show oxide bands, such as those of TiO, in their spectra. But if the abundance of carbon in the atmosphere exceeds that of oxygen, a carbon star is formed. These are stars whose spectra are dominated by molecular bands from C_2, CH and CN, for example, rather than TiO. Thus the occurrence of carbon stars in a stellar population can be taken as another indicator of the presence of a population whose age is significantly less than the age of the galactic globular clusters. Such carbon stars are common, for example, in the 1 – 3 Gyr old star clusters in the LMC (e.g. Frogel et al. 1990).

2.1. An example

In the preceding section we have assembled the tools for investigating a resolved stellar population (that is not too metal-rich), particularly as regards its population older than 1 or 2 Gyr. The distance can be derived from the luminosity of the RGB tip and the mean colour of the giant branch gives an estimate of the mean abundance. Age information is provided by both the morphology of the horizontal branch and by the luminosity of any upper-AGB stars. Let us now apply this information to a particular example. The example I've chosen is a historical one: Baade's (1944a,b) papers that introduced the subject of stellar populations and which fundamentally influenced decades of subsequent work. Baade had at his disposal the following observations: (i) on red sensitive photographic plates he could resolve the brightest stars in the bulge of M31 and in its dwarf elliptical galaxy companions M32, NGC 205, NGC 147 and NGC 185. (ii) With blue sensitive plates he could not detect or resolve these same stars. He therefore concluded that the brightest stars in these systems were red. (iii) He further noted that in the Sculptor dwarf spheroidal galaxy, which had been discovered some years earlier, the brightest stars were red and of similar luminosity to the brightest red giants in globular clusters. The Sculptor system was also known to contain RR Lyrae variables which are common in globular clusters. From these data alone, Baade then concluded that the stellar populations of the bulge of M31 and its companions were similar to those of galactic globular clusters. Indeed, he went further and proposed that the stellar populations of all early-type galaxies (E and S0, both dwarfs and giants) were like those of galactic globular clusters. Over the years this proposition has come to be interpreted as indicating that early-type galaxies are dominated by populations as old the galactic globular clusters, though Baade, lacking our knowledge of stellar ages, did not himself draw this conclusion. But this interpretation has led, for example, to the Local Group dwarf spheroidal galaxies being regarded merely as "puffed-up globular clusters", an inference that, as we shall see in a subsequent lecture, is often far from the truth.

Given what we have discussed above concerning stellar evolution, should Baade have drawn the conclusions that he did? Probably not. Certainly the observations of a globular-cluster-like giant branch and the presence of RR Lyrae variables in the Sculptor dwarf spheroidal do suggest that this dwarf does contain an old population, but as for the bulge of M31 and its dwarf elliptical companions, no such conclusion can be drawn without further information. The bright red stars might, for example, be upper-AGB stars considerably younger than globular cluster stars. Certainly, as we have seen, the occurrence of a red giant branch on its own signifies nothing more than the presence of a population whose age is greater than 1 – 2 Gyr. Thus it is fair to say that the interpretation which led to the generally accepted notion that the stellar populations of E galaxies are basically old, is fundamentally flawed, though the notion itself may well be correct (see Freedman 1994, 1995 for further discussion of this issue). However, the real purpose of highlighting this example to you is to illustrate that many of the so-called "classic concepts" in Astronomy, though they may well be correct, are often based on invalid or out-dated interpretations: you should never take them for granted even if they have the status of "accepted wisdom".

3. "Old" populations in the Magellanic Clouds

In this section we consider the basic question *"Do the LMC and the SMC contain an old population and is it coeval with the oldest globular clusters in the Milky Way?"* Clearly the answer to this question is of considerable significance for theories of galaxy formation. Before addressing this question, however, it is important to define some terminology that will be used throughout these lectures. First, the term *Old* will be used to mean a stellar population that is older than approximately 10 Gyr. Second, the term *Intermediate-Age* will signify a population that has an age between $\sim 1-2$ Gyr and ~ 10 Gyr. Here the lower limit is set at the age beyond which stars evolving away from the main sequence develop significant electron-degenerate cores. As discussed above, such stars evolve as red giants until they ignite He under degenerate conditions (the helium flash) at a luminosity that is largely independent of age and abundance. The actual age at which a full red giant branch appears is somewhat composition dependent (e.g. Bertelli et al. 1994) but $\sim 1-2$ Gyr is a reasonable limit to adopt. The adopted upper age limit for an intermediate-age population of ~ 10 Gyr is again approximate, being defined as the age beyond which upper-AGB stars with luminosities significantly above the red giant branch tip do not occur. Again as noted above, from a theoretical point of view, the upper-AGB luminosity reached by a star corresponding to a particular turnoff mass (age) is strongly dependent on the assumed mass loss rate, and is, as a result, somewhat uncertain. On the other hand, upper-AGB stars evolve rapidly and thus observationally, the probability of an upper-AGB star being found near the AGB tip is strongly dependent on the size of the underlying progenitor sample. Nevertheless, it is generally accepted that AGB stars in metal-poor galactic globular clusters, which have ages of perhaps $13-15$ Gyr, do not reach luminosities brighter than the red giant branch tip. In the SMC, however, upper-AGB (carbon) stars are found in the cluster Kron 3 (age ≈ 8 Gyr, Rich et al. 1984) with luminosities at least 0.5 mag in M_{bol} above the RGB tip, while in the SMC cluster NGC 121 (age ≈ 12 Gyr, Stryker et al. 1985) the brightest AGB candidate (star V8) has a bolometric magnitude comparable to that of the red giant branch tip. Consequently, adopting ~ 10 Gyr as the upper limit for the age of an intermediate-age population does not seem unreasonable. Finally, the term *young* or *younger* will be used for stellar populations that have ages less than $\sim 1-2$ Gyr. The evolution of stars in this age range is qualitatively different from that for older stars.

3.1. *The Large Magellanic Cloud*

In the Galaxy the oldest identified objects are the globular clusters and so the first way to answer the question posed in the beginning of this section is to look for "globular clusters" in the LMC. Using a variety of criteria, such as low metal abundance, presence of RR Lyrae and/or blue horizontal branch stars in c-m diagrams, integrated colours, absolute magnitudes, etc, Suntzeff et al. (1992) conclude that there are 13 star clusters in the LMC that can be classified as "old" objects. But what direct evidence do we have concerning the ages of these clusters? From the ground it is possible to observe the main sequence turnoff in those LMC globular clusters that lie away from the crowded central regions of this galaxy. The data (e.g. Walker 1992) clearly establish that these old LMC star clusters are not very different in age from their galactic counterparts. However, the precision with which this statement can be made is limited to $\sim \pm 2-3$ Gyr. Unfortunately, an uncertainty of this size is significant: for example, it is comparable to the total *range* in age claimed to exist among the globular clusters of the outer galactic halo (e.g. Lee et al. 1994). The required faint limiting magnitude and the effects of image crowding on the photometry mean that it is unlikely these limits can be significantly improved using only ground-based data. It should, however, be possible to measure relative ages for LMC globular clusters to a precision of ~ 1 Gyr with HST imaging. Consequently, the results from a number of approved HST/WFPC2 imaging programs for LMC globular clusters are eagerly awaited.

In the meantime, an indirect approach based on horizontal branch (HB) morphology considerations may provide additional insight. As discussed in the second section, the morphology of the HB is a function of both abundance and age. In particular, for any (fixed) abundance [Fe/H] $\leq$ –1.0, the HB morphology becomes redder with decreasing age. Lee et al. (1990) introduced the HB morphology index $i = (B-R)/(B+V+R)$ where B is the number of blue HB stars, V is the number of RR Lyrae variables (which are also found on the HB) and R is the number of red HB stars. Clearly the index i varies from $i = +1$ for a pure blue HB to $i = -1$ for a pure red HB, and at fixed abundance, younger populations have smaller i values. Now if we look at this HB morphology index as a function of abundance for galactic globular clusters that lie within $\sim$8 kpc of the galactic center, then there is very little scatter about a mean relation (cf. Fig. 6). This suggests that these inner galactic halo globular clusters have closely similar ages (e.g. Lee et al. 1994). On the other hand, in a similar plot for the galactic globular clusters that lie beyond $\sim$20 kpc from the galactic center, the HB morphology – abundance relation is not as well defined (cf. Fig. 6). Further, the relation is shifted towards redder morphologies and the scatter is notably larger. In this situation there is a need for a "second parameter" (the first parameter is abundance) to explain the HB morphologies. In a galaxy wide context, this "second parameter" is usually interpreted as age — the outer galactic halo globular clusters are somewhat younger on average (by perhaps 2 – 3 Gyr) and there is a larger age scatter (of perhaps 3 – 4 Gyr, e.g. Lee et al. 1994). This age variation interpretation remains controversial (e.g. Richer et al. 1996) but nevertheless it is one of the cornerstones of the Searle-Zinn model for the formation of the outer galactic halo. In this model (Searle & Zinn 1978), the outer galactic halo is formed principally from the disruption of small satellite galaxies which, prior to their disruption, were independently evolving objects.

How then do the LMC old star clusters fit into this context? With ground-based data it is relatively straightforward to determine both abundances and HB morphology indices for a number of LMC globular clusters. The results of this are shown in Fig. 6 (see Da Costa 1993 for details). It is evident that with the possible exception of Hodge 11 (the LMC cluster with the bluest HB morphology) the LMC clusters occupy a similar location to the outer galactic halo globular clusters. If we follow Lee et al. (1994) and interpret this diagram in terms of the "age is the second parameter" hypothesis, then it is apparent that the LMC globular clusters are, on average, some 2 to 3 Gyr younger than the inner galactic halo globular clusters. This similarity to the outer galactic halo globular clusters is emphasized in Suntzeff et al. (1992, see also Da Costa 1993) who point out that the mean cluster metallicity, the absolute magnitude distribution, and the relative number of RR Lyraes per unit cluster luminosity for the LMC old clusters are also very similar to the equivalent quantities for the outer galactic halo globular clusters. Indeed Suntzeff et al. (1992) argue that the old cluster population of the LMC can serve as a model for the cluster systems of the "satellite galaxies" accreted by the Galaxy during the halo formation process outlined by Searle & Zinn (1978).

So, in answer to our initial question, the properties of the LMC globular clusters imply the existence of at least some old population in the LMC, though with an age perhaps 2 to 3 Gyr younger than the oldest globular clusters in the Galaxy. Is this also the case for the field population? Field RR Lyraes are known to occur throughout the LMC and this is good evidence for the existence of an old field population. However, the detailed study of a very large sample of LMC field RR Lyrae variables in Alcock et al. (1996) suggests that, as for the old LMC star clusters, the age of the field RR Lyrae population in the LMC is less than that of the oldest galactic globular clusters. Alcock et al. (1996) argue from the surfeit of low amplitude Bailey type b variables in their sample that the underlying morphology of the LMC field horizontal branch is predominantly red.

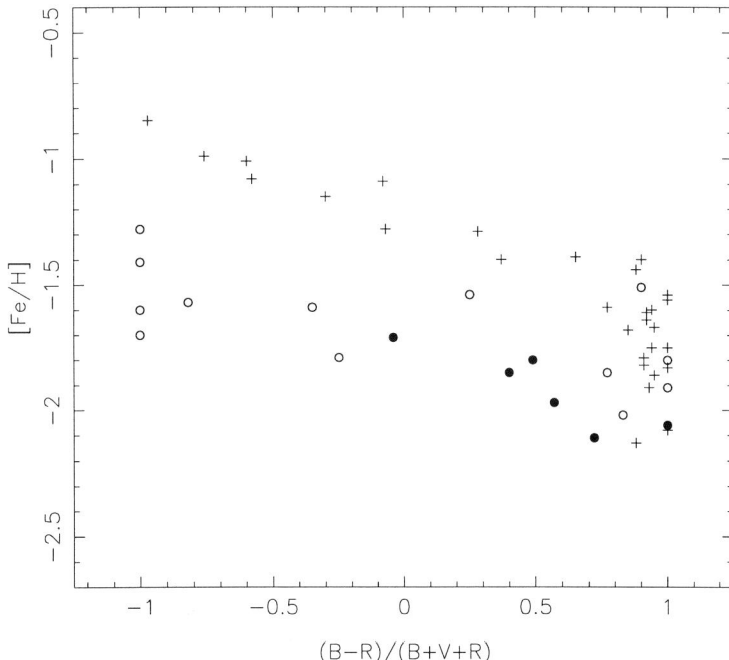

FIGURE 6. The relations between horizontal branch morphology index (B–R)/(B+V+R) and abundance [Fe/H] for inner ($R_{gc} \leq 8$ kpc) galactic halo globular clusters (+ signs), outer ($R_{gc} \geq 20$ kpc) galactic halo globular clusters (open circles), and LMC globular clusters (filled circles). At fixed abundance, younger objects have lower values of the morphology index (redder horizontal branches).

Further, spectra of a small number of variables indicate a modal abundance value of [Fe/H] ~ -1.7 which, when coupled with a red HB morphology, implies the younger age (cf. Fig. 6). Indeed Alcock et al. (1996) suggest that the "old" field population in the LMC may even be a Gyr or so younger than the old LMC cluster population.

In this context though, it is important to remember that the old component in the LMC is a relatively minor one. As discussed in detail in Mateo's lectures, the field (and cluster) population of the LMC in regions away from sites of current or recent star formation is dominated by a component whose age, determined from main sequence turnoff observations, is only a few Gyr. Thus LMC field region c-m diagrams reveal relatively few blue HB and RR Lyrae stars and the "horizontal branch" region of the c-m diagram is dominated by the strong red clump of core-helium burning intermediate-age stars (e.g. Stryker 1984, Bertelli et al. 1992, Vallenari et al. 1996a,b). For this reason also, the majority of the red giants that are found in outlying LMC field regions are predominantly intermediate-age stars and not old objects. The significance of this result will become apparent in later sections.

3.2. *The Small Magellanic Cloud*

If the oldest population in the LMC is represented by the old star clusters, then it is logical to expect that this will also be the case for the SMC. But we first have to consider if we should expect any such star clusters in the SMC, given that there only 13 or so in the more luminous LMC. If we scale the LMC old star cluster number by the ~ 2 magnitude difference in present day luminosity between the LMC and the SMC (which may or may not be the correct scaling factor), then perhaps 2 old star clusters are expected in the

SMC†. The best candidate for such a cluster is NGC 121. With an absolute magnitude of $M_V \approx -8$ and an abundance of [Fe/H] ≈ -1.4 (Stryker et al. 1985), NGC 121 has many of the characteristics of an old globular cluster. In particular, it possesses a small number of RR Lyrae variables (e.g. Walker & Mack 1988) though c-m diagrams for this cluster (e.g. Tifft 1963) show that the HB morphology is quite red, the RR Lyrae variables being on the extreme blue side of the cluster horizontal branch. Thus on the basis of this HB morphology (and abundance) we would expect the cluster to be relatively young (cf. Fig. 6). This is confirmed by the work of Stryker et al. (1985) who suggested from their main sequence turnoff photometry that the age of NGC 121 is 12 ± 2 Gyr, or some 3 Gyr or so younger than the age of galactic globular clusters for the same isochrone set. Alternatively, Da Costa (1993) has compared the NGC 121 data of Stryker et al. (1985) with that of Green & Norris (1990) for the galactic globular cluster NGC 362, which has a similar abundance to NGC 121. In this comparison NGC 121 is clearly seen to be younger by perhaps ∼1 – 2 Gyr) relative to NGC 362, which is itself younger by perhaps ∼2 –3 Gyr relative to most galactic globular clusters (e.g. Green & Norris 1990). Thus, as for the LMC, it appears that the oldest star cluster in the SMC is notably younger than its counterparts in the inner galactic halo. In other words, it appears that the epoch of old star cluster formation in both the LMC and the SMC occurred significantly after the commencement of globular cluster formation in the inner halo of the Galaxy.

As regards the field population of the SMC, there is not as much information available as there is for the LMC. Field RR Lyrae stars are found throughout the SMC (e.g. Graham 1975, Smith et al. 1992) and, as for the LMC, this is indicates the presence of an old population. But the SMC lacks large area studies reaching faint limiting magnitudes equivalent to those of Bertelli et al. (1992) and Vallenari et al. (1996a,b) for the LMC. Nevertheless, we can draw on the work of Suntzeff et al. (1986) who have conducted a spectroscopic and photometric study of a sample of SMC field red giants in a region near NGC 121. They find a mean abundance of <[Fe/H]> $\approx -1.6 \pm 0.1$ with an intrinsic abundance range from [Fe/H] ≈ -1.0 to [Fe/H] ≈ -2.0 (Suntzeff et al. 1986). Of course we know from the discussion in Section 2 that the ages of these red giants are not well determined — they could lie between ∼1 – 2 Gyr and ∼15 Gyr. If, however, we assume that these red giants are predominantly as old as the inner galactic halo globular clusters, then we would expect for this mean abundance that the horizontal branch morphology in the corresponding c-m diagram would contain a large number of blue HB stars (cf. Fig. 6). But in fact the existing c-m diagrams for SMC field regions (e.g. Gardiner & Hatzidimitriou 1992) show a dominant red HB (and/or a red clump) with few blue HB stars present. Consequently, as for the LMC, there does not appear to be any sizeable field population in the SMC with an age comparable to that of the inner galactic halo globular clusters. Interestingly, however, the c-m diagram for the field region near NGC 121 studied by Stryker et al. (1985) does not show the relatively bright main sequence turnoff, corresponding to an age of a few Gyr, seen in the LMC field. Instead, Stryker et al. (1985) suggest an age of at least 8 Gyr for the field stars in this region. Thus while larger area studies are required, it is possible that the bulk of the SMC field population is generally older than it is in the LMC. In particular, it appears that the SMC field is not dominated by a few Gyr old population as is the case

---

† Globular-cluster-like star clusters are not common among the dwarf irregular galaxies of the Local Group. Neither NGC 6822 (which is comparable to the SMC in luminosity) nor IC 1613 (which is ∼1.3 mag fainter) appear to possess any luminous centrally concentrated star clusters. However, WLM, which has $M_V \approx -14$, does possess a bright ($M_V \approx -8.5$) compact globular-cluster-like star cluster (Ables & Ables 1977), whose integrated $(B-V)_0$ is consistent with that expected for an old metal-poor globular cluster.

for the LMC field; a dominant ~10 Gyr SMC field population would be consistent with all existing observations.

4. Local Group dE and dSph galaxies

4.1. Overview

The Local Group contains (at least) 16 dwarf elliptical (dE) and dwarf spheroidal (dSph) galaxies: the 9 dSph companions to the Galaxy, the 3 dSph and 3 dE companions to M31 and the relatively isolated dE system Tucana. These dwarf galaxies are fundamentally characterized by, first, a generally smooth, symmetrical appearance (i.e. a lack of obvious internal structure, though occasionally a distinct nucleus is seen), and second, by an overall lack of gas, dust and young stars, though all of these quantities are found in small amounts in some systems. For example, relatively young blue stars are found in the central regions of the M31 dE companions NGC 185 and NGC 205, as are dust clouds and small amounts of HI and CO. The HI content (or lack thereof) of all these systems, however, can be characterized by noting that $M_{HI}/L_B < \sim 10^{-2}$, even when HI is detected.

Prior to discussing the overall properties of these galaxies, however, we must first deal with the question of terminology. For example, why is the term "dwarf spheroidal" being used in addition to "dwarf elliptical", and why is M32 not included in the list of M31 dE companions? The term "dwarf spheroidal" was originally applied to the low luminosity companions of the Galaxy whose c-m diagrams were basically similar to those of galactic globular clusters. When van den Bergh (1972a) discovered three similar low luminosity companions to M31, these objects (And I, And II and And III) also gained the dSph classification. But strictly speaking, there is no real reason to distinguish these "dwarf spheroidals" from dwarf ellipticals such as the more luminous M31 dE companions NGC 147, NGC 185 and NGC 205. Nevertheless, both terms will continue to be used here.

But exactly what is a dwarf elliptical? One of the more significant advances in the understanding of galaxy morphology came about in the mid-80's with the recognition that ellipticals and dwarf ellipticals follow different relations between their structural parameters (Wirth & Gallagher 1984, Kormendy 1985). For example, in the surface brightness, absolute magnitude plane, elliptical galaxies define a sequence of increasing surface brightness with decreasing luminosity, whereas the dwarf ellipticals have fainter surface brightnesses at fainter luminosities (e.g. Fig. 1 of Binggeli 1994). At intermediate luminosities (representing bright dwarf elliptical systems and faint ellipticals, both of which are intrinsically rare objects) there may be a degree of overlap between the two sequences and exact classification is ambiguous (e.g. Prugneil 1994). However, it is generally accepted that M32, with its centrally concentrated high surface brightness structure, falls within the elliptical sequence and thus it is not a dwarf elliptical in the sense we are using the term here. Instead it is often referred to as a "compact elliptical" (cE). We should keep in mind though that this terminology is not universally accepted. For example, while Kormendy (see Kormendy & Bender 1994) also refers to the sequence showing decreasing surface brightness with increasing luminosity as the elliptical galaxy sequence, he designates a low luminosity member of this sequence as a "dwarf elliptical"(!). On the other hand, he refers to the galaxies in the sequence showing increasing surface brightness with increasing luminosity as "spheroidal galaxies" and to low luminosity members of that class as "dwarf spheroidal galaxies"(!). So, in his terminology, M32 is a "dwarf elliptical", NGC 205 is a "spheroidal galaxy" and Tucana

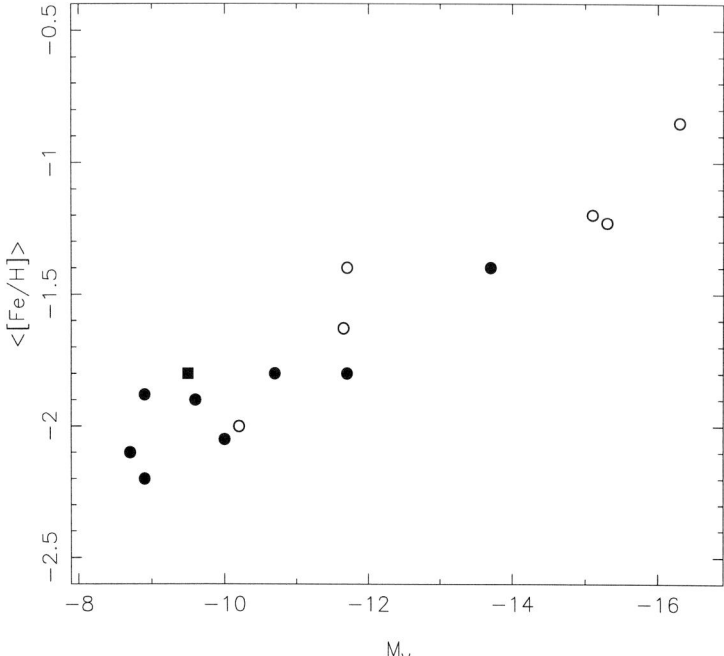

FIGURE 7. The relationship between mean abundance <[Fe/H]> and absolute visual magnitude for the dE and dSph galaxies in the Local Group. Companions to the Galaxy are shown as filled circles, companions to M31 as open circles, and the isolated Local Group dE Tucana as a filled square, respectively. Typical error bars are ±0.2 dex in <[Fe/H]> and ±0.5 mag in M_V.

is a "dwarf spheroidal galaxy". This terminology will not be used here. Nevertheless, regardless of whether M32 is a compact elliptical (and therefore worthy of inclusion in Mateo's lectures) or a dwarf elliptical, we will discuss its stellar population below, so that this important galaxy is not overlooked.

In addition to photometric quantities such as surface brightness parameters, length scales (e.g. exponential law scale lengths or core radii) and total magnitudes, it is also possible, for the Local Group dE and dSph systems, to determine mean abundances. These mean abundances can be derived either from spectroscopy of individual red giants for the closest systems, or photometrically from giant branch colours. It is then possible to investigate correlations between these quantities. For example, as illustrated in Fig. 7, mean abundance is well correlated with total magnitude over a factor of ∼25 in abundance and ∼600 in luminosity. Note that the Sagittarius dSph is not plotted in Fig. 7, nor in any of the other figures, because its total magnitude, for example, is poorly known at best. Further, the degree to which its length scale and surface brightness parameters have been altered by its interaction with the Galaxy is also poorly constrained at present.

The Local Group dE and dSph galaxies also follow a well-defined surface brightness, absolute magnitude relation, illustrated in Fig. 8, and a surface brightness, length scale relation. Since these quantities are all correlated, there are also surface brightness, mean abundance relations, etc, and it is not obvious, at least from the Local Group systems, which of these relations are the more fundamental. Nevertheless, there is no indication that any of these relations are any different for the M31 companions than for the dSph companions to the Galaxy. Thus, it is tempting to assume that these relations represent characteristics of the dE formation process rather than the influence of a nearby large

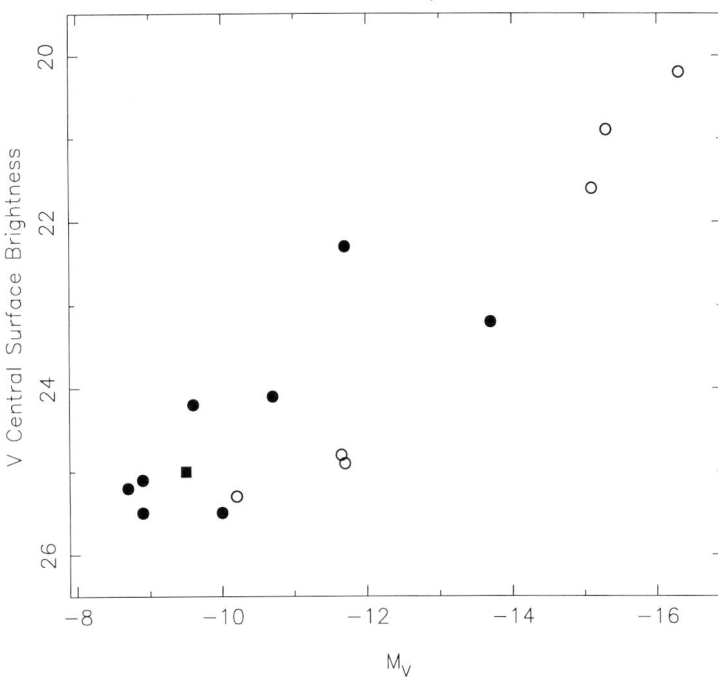

FIGURE 8. The relationship between central surface brightness in the V band and absolute visual magnitude for the dE and dSph galaxies in the Local Group. Companions to the Galaxy are shown as filled circles, companions to M31 as open circles, and the isolated Local Group dE Tucana as a filled square, respectively. Typical error bars are ±0.1 mag in central surface brightness (larger for the Galaxy's companions) and ±0.5 mag in M_V.

galaxy. However, we need to establish the validity of these relations beyond the Local Group before succumbing to this temptation.

The major question that we will focus on, however, is the star formation histories of these dwarf galaxies. For many years it was thought that the star formation histories of Local Group dSph and dE galaxies were basically globular cluster like — essentially no star formation subsequent to an initial formation episode. We know now that while this is perhaps true for some objects, it is certainly not the case for others, indeed there is a rich and surprising variety of star formation histories among the Local Group dSph and dE galaxies. Before considering the data for individual objects, however, we begin with a discussion of the globular clusters systems of the Local Group dSph and dE galaxies.

Observationally, it appears that only the more luminous dSph and dE galaxies possess globular cluster systems. Fornax and Sagittarius, the most luminous of the galactic dSphs, possess 5 and 4 globular clusters, respectively (see, for example, Beauchamp et al. 1995 and the references therein; Da Costa & Armandroff 1995 and the references therein) while the M31 dE companions NGC 147, NGC 185 and NGC 205 all possess globular cluster systems containing up to half dozen or more objects each (e.g. Da Costa & Mould 1988). But the less luminous Local Group dEs and dSphs lack globular clusters. In most instances the globular clusters associated with the dE and dSph galaxies appear to be analogues of the galactic halo globular clusters. In Sagittarius, for example, which is sufficiently nearby that old main sequence turnoffs can be reached from the ground, two of the clusters (M54 and Ter 8) appear to be as old as the inner galactic halo globular clusters (Sarajedini & Layden 1995, Ortolani & Gratton 1990) though for M54 this result is inferred from the horizontal branch morphology rather than from the main

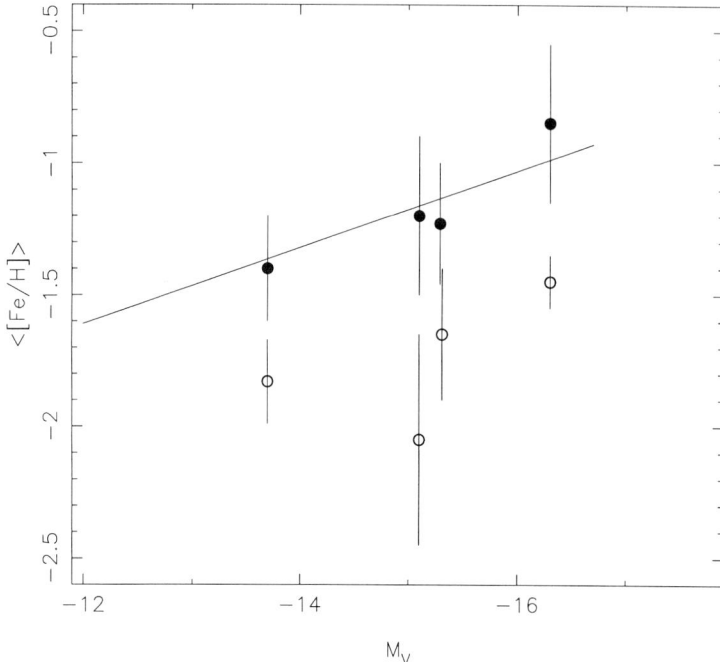

FIGURE 9. The relationship between mean abundance for the field population (filled symbols) and mean abundance for the corresponding globular cluster system (open symbols) for the Local Group dSph and dE galaxies (in order of increasing luminosity) Fornax, NGC 147, NGC 185 and NGC 205. The mean abundance for the globular clusters associated with the Sagittarius dSph is also lower than that for the Sgr field stars (e.g. Da Costa & Armandroff 1995) but no point is plotted since the Sgr total luminosity is poorly known. The straight line represents the (luminosity, abundance) relation defined by all Local Group dSph and dE galaxies (cf. Fig. 7).

sequence turnoff luminosity. The other two clusters (Arp 2 and Ter 7), however, are both somewhat younger (Buonanno et al. 1995a,b). It is also possible that there is a range of age among the Fornax globular clusters. While three of the four clusters studied in detail† show relatively blue horizontal branch (HB) morphologies and steep red giant branches consistent with their low metallicities, cluster 1 has a redder HB than expected for its metallicity (see Beauchamp et al. 1995, Smith et al. 1996 and references therein). In other words, the "second parameter effect" is also found in Fornax. Interpreting the second parameter as age would then require cluster 1, which lies in the outskirts of Fornax, to be somewhat younger than the other clusters. Further, if we again interpret HB morphology as an age indicator, then the Fornax cluster system as a whole may be significantly younger (by perhaps 3 – 4 Gyr) than the globular clusters of the inner halo of the Galaxy (e.g. Zinn 1993). Confirmation of these results, however, must await HST/WFPC2 observations of the main sequence turnoff in the Fornax globular clusters, a difficult but feasible observation.

As regards the globular clusters associated with the M31 dE companions, no cluster c-m diagrams are available as yet. But integrated spectra of these clusters reveal (with the exception of NGC 205 – Hubble V which is apparently a cluster with an age of ∼0.5 – 1 Gyr, Da Costa & Mould 1988) line strengths that are consistent with those of galactic

† Little reliable information is available for the c-m diagram morphology of Fornax cluster 4 since this compact cluster, which is the most metal-rich of the Fornax system, is projected on a very crowded field close to the centre of the galaxy (e.g. Beauchamp et al. 1995).

globular clusters (Da Costa & Mould 1988), i.e. the dE globular clusters appear old and metal-poor. As is the case for the Fornax and Sgr cluster systems, each dE globular cluster system shows a substantial abundance range — from abundances as low as any seen for the galactic globular clusters up to [Fe/H] $\approx$ –1.3, which is not much less than the upper abundance limit for globular clusters in the "halo" (as distinct from the "disk") system of galactic globular clusters (e.g. Da Costa & Armandroff 1995 and references therein).

In all cases (including Fornax and Sgr), however, the mean abundance of the globular cluster system is lower than that of the field stars in the same dwarf galaxy (the lack of sizeable abundance gradients in dE and dSph systems facilitates this comparison). This result is illustrated in Fig. 9 where data from Da Costa & Mould (1988) are shown. The globular cluster system mean abundances are offset from those for the field stars by $\sim$0.5 dex, though it seems that the cluster system mean abundances increase with increasing luminosity in much the same way as do the field star mean abundances. Similar results are found for more luminous galaxies (e.g. Harris 1991) though in this case the extent to which the mean abundance of the globular cluster system increases with increasing galaxy luminosity is a controversial subject (see, for example, Ashman & Bird 1993; also Fig. 2 of Da Costa 1994). Conventionally, the difference in mean abundance between the field stars and the globular cluster system in a galaxy is taken to indicate that the globular clusters represent a population that formed before the bulk of the field stars. While this is certainly true in the LMC where the bulk of the field stars are considerably younger than the LMC globular clusters, and probably true in the Fornax and Sagittarius dSph systems where the bulk the field population is also comparatively young (see sections 4.2.1 and 4.2.3), the hypothesis remains to be verified in other situations. The Fornax and Sagittarius dSph systems can play an important role here, since for both these systems it is possible to establish via direct observation the age of all the globular clusters (already done to some extent for Sgr) *and* the "age" of the field population.

We now turn to consideration of the observations relevant to the field star formation histories of the Local Group dSph and dE galaxies. The dSph companions to the Galaxy will be discussed first since, either from the ground, or with HST/WFPC2, it is possible to reach in these systems luminosities that are fainter than that of the main sequence turnoff for an old population. Next we will consider the M31 dSph and dE companions (as well as M32) where with HST/WFPC2 it is possible to reach magnitudes somewhat fainter than that of the horizontal branch. Finally, we will describe recent results for the Tucana dE system and then end this section with a brief discussion of the overall results. Much of the material to be described can also be found in Da Costa (1997).

4.2. *Milky Way dSph galaxies*

If you are ever asked to provide an example of a dwarf elliptical galaxy that corresponds to the classic "single old age" population concept, then the dSph companion to the Galaxy Ursa Minor should be your answer. The c-m diagram for the brighter stars in Ursa Minor of Cudworth et al. (1986) shows the strong blue horizontal branch with a small number of RR Lyrae variables characteristic of an old metal-poor population. The assignment of an old age to Ursa Minor is confirmed by the CCD main sequence turnoff photometry of Olszewski & Aaronson (1985). They indicate that in their data there is no evidence for any significant age range within the Ursa Minor stellar population. Further, they demonstrate that their results are consistent with this dSph being the same age as the Galaxy's old metal-poor halo globular clusters (e.g. M92). A small number of blue stragglers are seen in the Olszewski & Aaronson (1985) c-m diagram, but such stars are also seen in, for example, the c-m diagram of the low central concentration metal-poor

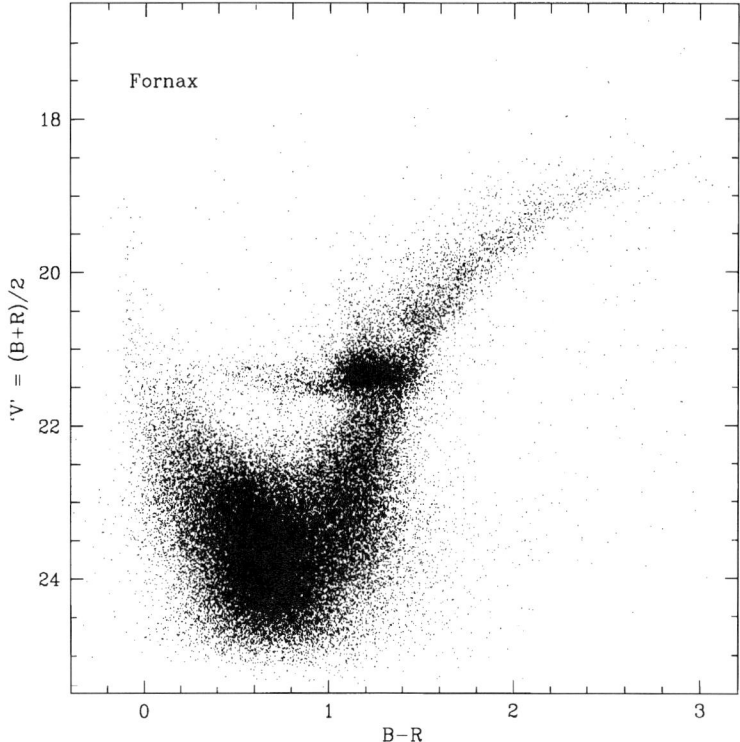

FIGURE 10. A colour-magnitude diagram for the Fornax dSph derived by Stetson and collaborators from CTIO 4m prime focus CCD observations.

globular cluster NGC 5053 (Nemec & Cohen 1989). Thus there is no reason to assume that these blue straggler stars are any different in age from the rest of the Ursa Minor population. But the other dwarf spheroidal companions to the Galaxy do not appear to be as "simple" as Ursa Minor...

4.2.1. *Fornax*

Figure 10 shows a new c-m diagram (Stetson 1997) for the Fornax dwarf spheroidal derived by Stetson and his collaborators from observations made at the Cerro Tololo Inter-American Observatory in Chile. A number of noteworthy features are visible in this c-m diagram which contains over 80,000 stars. In particular, at V ≈ 21.4, there is an obvious "red clump" of core helium burning stars. As discussed in Section 2, this clump consists of relatively massive stars from an intermediate-age population. On the other hand, at slightly fainter luminosities and bluer colours, a second group of core helium burning stars are seen. These are the less massive horizontal branch stars characteristic of an old population. Thus from just this observation of a complex morphology for the core helium burning stars in the Fornax c-m diagram, it is already evident that this dSph has had a vastly different star formation history from Ursa Minor. The existence of upper-AGB stars, visible in Fig. 10 as red (B–R > ~2.6) stars beyond the RGB tip, is also an indication of the presence of an intermediate-age population. Many of these stars have been spectroscopically confirmed as upper-AGB carbon stars and their luminosities, determined from infrared photometry, reach as bright as $M_{bol} \approx -5.5$ (see Azzopardi 1994 and references therein). However, the proximity of Fornax allows imaging to reach considerably fainter stars, stars that are on the main sequence, and this provides more direct information on the star formation history.

The main sequence in Fig. 10 is obviously strongly populated up to V ≈ 23, a luminosity which corresponds to an age of a few Gyr at the distance of Fornax. Of particular interest is the roughly even distribution with luminosity, for V ≥ ~23.0, of the "subgiant" stars, i.e. the stars between the main sequence and the giant branch. This relatively even distribution indicates that star formation in Fornax was reasonably continuous from an initial epoch perhaps ~13 – 15 Gyr ago until ~3 – 4 Gyr ago at which time the rate of star formation seems to have decreased. It is apparent from Fig. 10 though that star formation in Fornax did not cease at this epoch — the main sequence is still populated brighter than V ≈ 23 and there is a further break in the main sequence luminosity function at V ≈ 21.5 corresponding to an age of ~1 Gyr. Remarkably, there is a sparse continuation of the main sequence up to V ≈ 19 and these luminous blue stars would appear to have ages as young as perhaps a few hundred Myr or less! This type of star formation history is one we might more readily associate with a dwarf irregular galaxy (see Section 5) but, at least in its central regions (which are those that have been searched), Fornax does not contain any detectable amounts of HI (Knapp et al. 1978). A sensitive HI survey of the entire area of Fornax, however, is needed before we can be certain this dSph contains no gas. Nevertheless, at least with current data, it would appear we cannot classify Fornax as a dIrr in quiescence though clearly this dSph would have been classified as a dIrr a Gyr or more ago, a relatively small fraction of the Hubble time. We have then the following picture of the history of star formation in Fornax: stars formed in Fornax approximately continuously until a few Gyr ago, after which it continued at a lower rate until perhaps ~200 Myr ago, though there is no current star formation. This history certainly stands in stark contrast to the classic "single old population" concept!

4.2.2. *Carina*

The Carina dSph has been known for some time (e.g. Mould & Aaronson 1983, Mighell 1990, Smecker-Hane et al. 1994) to be a dSph that contains a substantial intermediate-age population. Figure 11 shows a new deep c-m diagram for Carina derived from ground-based imaging at CTIO (Smecker-Hane et al. 1996). As is the case for Fornax (cf. Fig. 10), in Fig. 11 the red clump of intermediate-age core helium burning stars is readily distinguishable from the bluer and fainter horizontal branch of old core helium burning stars. In particular, the horizontal branch extends to blue colours and RR Lyrae variables are present (Saha et al. 1986). Thus an old population definitely occurs in this galaxy.

However, unlike the situation in Fornax (cf. Fig. 10), the subgiant portion of Fig. 11 is *not* evenly populated. In particular, there is a notable deficiency of subgiant stars between R ≈ 22.3 and R ≈ 23.0. This deficiency necessarily implies that there must have been a marked decrease in the star formation rate at the corresponding epoch in the dwarf's history. In fact interpreting Fig. 11 with theoretical models (Smecker-Hane et al. 1996) suggests that this large reduction in the star formation rate occurred between an initial episode of star formation some ~11 to 13 Gyr ago and a major episode that took place from ~3.5 to 6 Gyr ago (Smecker-Hane et al. 1996). This situation contrasts rather strongly with that in Fornax where the star formation was apparently reasonably continuous over a similar period of time. Smecker-Hane et al. (1996) go on to conclude that there have been further low intensity star formation episodes in Carina at epochs of ~2 Gyr and ~1 Gyr ago. But there has been little, if any, subsequent star formation and Carina does not now contain detectable amounts of HI (Mould et al. 1990).

There are two further points to note concerning these Carina results. First, it appears

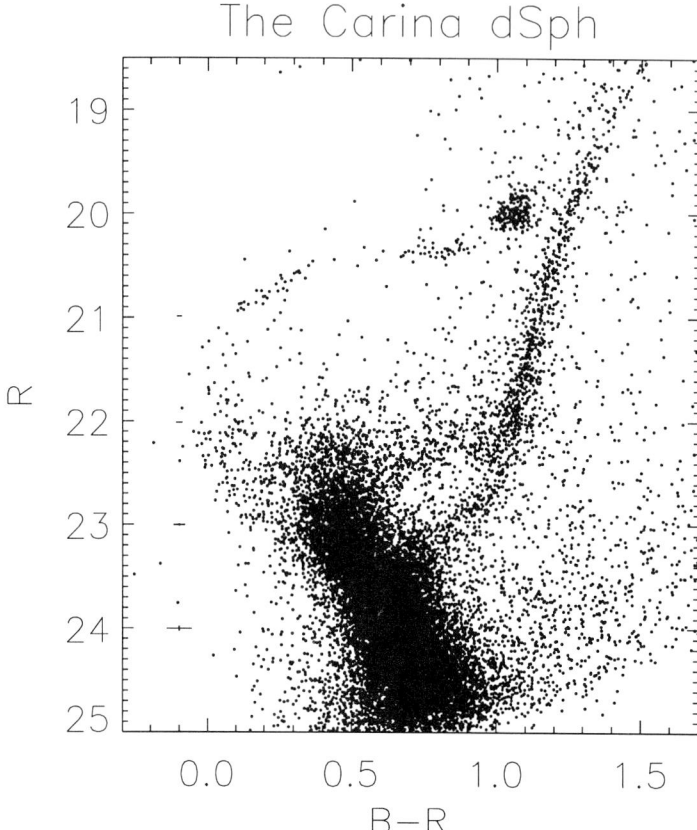

FIGURE 11. The colour-magnitude diagram for the Carina dSph of Smecker-Hane et al. (1996).

that the star formation events were relatively long lived (timescale ~1 Gyr or longer), so that certainly the term "burst of star formation" is not appropriate; "episode of star formation" is preferable. Further, during these star-forming episodes, Carina must have resembled a low luminosity dIrr. For example, assuming a normal IMF, the star formation rate inferred for the major 3.5 – 6 Gyr episode is about 10^{-3} to 10^{-4} solar masses per year. This rate is quite comparable to observed present-day star formation rates in low luminosity dIrrs; e.g. a current star formation rate of 8×10^{-4} solar masses per year is listed by Hunter & Gallagher (1985) for the dIrr DDO 53 which has $M_B \approx -13$.

Second, the giant branch in Fig. 11, which is fed by all the populations, is apparently quite narrow (cf. the wide Fornax giant branch in Fig. 10). Although detailed modelling is required for confirmation, this result suggests that the abundance range in Carina may be relatively small, despite the numerous enrichment events (Smecker-Hane et al. 1994). Spectroscopy of a small sample of Carina red giants (see Da Costa 1994) supports this view. Unless the IMF is unusual, this apparent lack of any significant abundance range can only be understood by insisting that the enrichment products from supernovae are not retained by Carina, but are instead lost in, for example, a metal-enhanced galactic wind (e.g. Vader 1986). A new spectroscopic study of a large sample of Carina red giants could therefore be used to explore in some detail the enrichment processes in this dwarf galaxy.

4.2.3. Other galactic dSph companions

From either ground-based or HST/WFPC2 observations, c-m diagrams that reach fainter than the magnitude expected for an old main sequence turnoff are now available for all the galactic dSph companions. And while the data for some of these dSphs lack the quality of the Fornax and Carina observations discussed above, the results nevertheless offer further support for the notion that the galactic dSph companions possess a wide variety of star formation histories.

For example, the Leo I system contains a large intermediate-age population (Lee et al. 1993c, Demers et al. 1994) and there are hints (Mateo et al. 1994) that the star formation in this dSph may have been episodic in the same manner as Carina. Similarly, Mighell & Rich (1996) have used HST/WFPC2 imaging data to conclude that the typical member of Leo II has an age of $\sim$9 Gyr, considerably younger than the age of most galactic halo globular clusters. Their observations in fact suggest that star formation in Leo II continued for approximately 7 Gyr, from $\sim$14 Gyr to $\sim$7 Gyr ago, and then terminated, as there is no indication of any subsequent star formation. The bulk of the Leo II stars apparently formed over a $\sim$4 Gyr interval but at present there is no real evidence to support or refute the occurrence of distinct star formation episodes such as are seen in Carina (Mighell & Rich 1996).

The recent HST/WFPC2 observations of Draco discussed by Grillmair et al. (1997) suggest that this dSph is similar to Ursa Minor in that there is little evidence for any extended or on-going star formation. Further, the observations appear to indicate that Draco is as old as the galactic halo globular clusters M68 and M92, to within 1 or 2 Gyr. This is a somewhat surprising result given the large difference in horizontal branch morphology between Ursa Minor and Draco (these two dSphs have similar mean abundances). It will be particularly interesting to see if theoretical horizontal branch models can reproduce the observed HB morphology difference without violating the apparently tight limit on any possible age difference. If this is not possible, then the prevailing view that "age is the (global) second parameter" may need to be reassessed. As regards whether or not there has been any extension of the star formation epoch in Draco, it must be kept in mind that the region of Draco imaged by Grillmair et al. (1997) is only a very small fraction of the entire galaxy. Consequently, the possibility of additional minor amounts of star formation in Draco subsequent to the initial episode cannot be ruled out on the basis of the Grillmair et al. (1997) observations alone. Indeed Stetson (1997) hints that such additional star formation has occurred.

One final example worth discussing is the case of the Sagittarius dSph. This dSph is known to contain RR Lyrae variables (e.g. Mateo et al. 1995b, Alard 1996, Alcock et al. 1997) so an old population is again present. As noted in Section 4.1, there are four globular clusters associated with Sgr and results for three of these clusters on their own suggest that star formation has been on-going in Sgr. In particular, the Sgr clusters Arp 2 (Buonanno et al. 1995a) and Ter 7 (Buonanno et al. 1995b) are both approximately 3 – 4 Gyr younger than galactic globular clusters of similar abundance. On the other hand, the Sgr cluster Ter 8 (Ortolani & Gratton 1990) is indistinguishable in age from the old halo globular clusters such as M92. As for the field population in Sgr, the extended nature of this dSph and its location behind the galactic bulge make generating "clean" c-m diagrams difficult. Nevertheless, Fahlman et al. (1996) conclude, as do Mateo et al. (1995a), from observations that reach below the main sequence turnoff in Sgr, that the age of the bulk of the population in this dSph is somewhat younger than that of the galactic halo globular clusters, and is comparable to that inferred for the clusters Arp 2 and Ter 7. When considered with the existence of upper-AGB carbon stars in Sgr

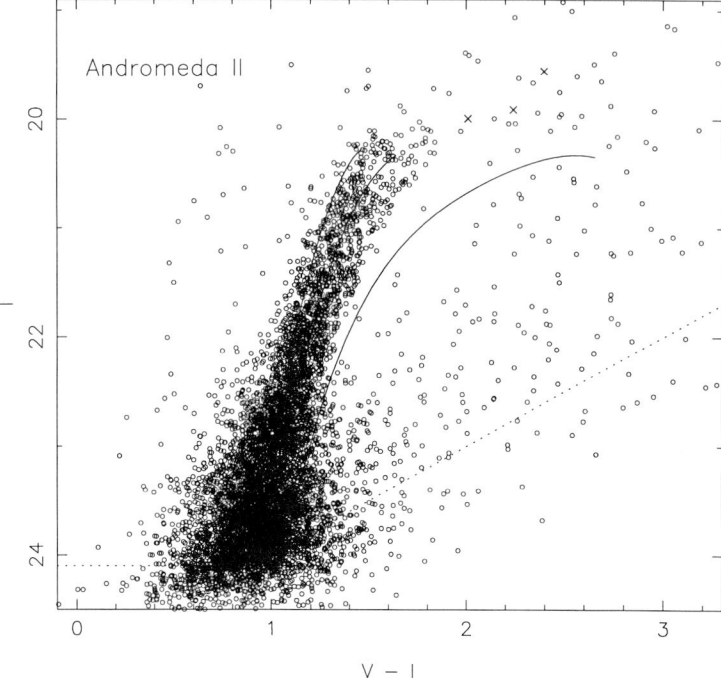

FIGURE 12. A colour-magnitude diagram for the M31 dSph companion And II derived from KPNO 4m observations (see Armandroff 1994). Giant branches for the galactic globular clusters M15, M2 and 47 Tuc are superposed for an And II distance modulus of $(m-M)_0 = 24.30$ and a reddening $E(B-V) = 0.035$ mag. The three stars classified spectroscopically as upper-AGB stars by Aaronson et al. (1985) are plotted as ×-symbols. Dashed lines show where the data become seriously incomplete.

(Ibata et al. 1995), stars that are a signature of an intermediate-age population, these data together make a good case for an extended star formation history in this dSph also.

4.3. M31 dSph galaxies

The dwarf spheroidal companions to M31, known as And I, And II and And III, were discovered by van den Bergh (1972a) from a photographic survey of 350 square degrees of sky surrounding M31†. Subsequent imaging (van den Bergh 1972b, 1974) showed that these objects resolved into stars at about the same magnitude as the M31 dE companion NGC 185, thus establishing their association with M31. These dSphs lie (in projection) approximately 45, 130 and 60 kpc from M31, respectively, a range of distances that are similar to the galactocentric distances of the Galaxy's dSph companions.

Recent work on these M31 dSph companions has shown that they are very similar in their overall properties to their galactic counterparts (e.g. Armandroff 1994). In particular, the surface photometry study of Caldwell et al. (1992) has indicated that the M31 dSphs have total magnitudes, surface brightnesses and length scales that are similar to those of the galactic dSphs (cf. Fig. 8). This similarity extends also to flattenings: the galactic dSphs range from $b/a = 0.9$ for Leo II through $b/a = 0.45$ for Ursa Minor,

† van den Bergh also discovered a fourth object, And IV, which because of its higher surface brightness and more compact nature, he suggested was either a "star cloud" in M31's outer disk or a background dwarf (irregular) galaxy. Subsequent work by Jones (1993) has shown that this system is indeed more probably a very large open cluster or a densely populated stellar association in M31 rather than a background galaxy.

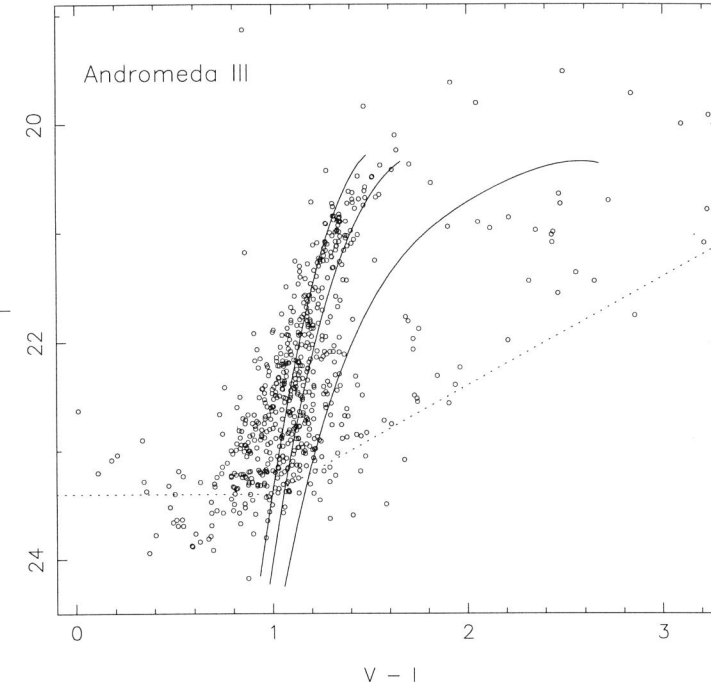

FIGURE 13. The colour-magnitude diagram for the M31 dSph companion And III from Armandroff et al. (1993). Giant branches for the galactic globular clusters M15, M2 and 47 Tuc are superposed for an And III distance modulus of $(m-M)_0 = 24.30$ and a reddening $E(B-V) = 0.05$ mag. Dashed lines show where the data become seriously incomplete.

while the M31 dSphs range from $b/a = 1.0$ for And I though $b/a = 0.4$ for And III. Ground-based c-m diagrams, which reveal the upper two or three magnitudes of the giant branch, are also available for all three systems: And I (Mould & Kristian 1990), And II (König et al. 1993, Armandroff 1994) and And III (Armandroff et al. 1993). As for the galactic dSphs, these c-m diagrams reveal well populated red giant branches whose tip magnitudes provide distance information and whose mean colours provide abundance information. Figures 12 and 13 show c-m diagrams for And II and And III.

The mean abundances derived from the giant branch photometry are consistent with those expected from the relations between mean abundance, absolute magnitude and surface brightness exhibited by the galactic dSphs (cf. Fig. 7). Further, again like the galactic dSphs, there is evidence for internal abundance ranges in each And dSph system. As regards the occurrence of intermediate-age populations, Armandroff et al. (1993) use the number of stars above the red giant branch tip to argue that a 3 – 10 Gyr old population contributes approximately $10 \pm 10\%$ of the visual luminosity in And I and in And III. This is a small value when compared to galactic dSphs such as Carina or Fornax, but is not dissimilar to Sculptor, for example. The largest observational contribution to the uncertainty in these figures comes from the small numbers of stars involved and from the uncertainty in the number of field stars (both M31 and galactic foreground) that must be subtracted from the total number of upper-AGB candidates. For And II no quantitative estimate of the intermediate-age population fraction is available as yet, but there is unambiguous evidence that such a population exists since three And II stars have been spectroscopically classified as upper-AGB stars (1 carbon star, 1 marginal carbon star and 1 M giant) by Aaronson et al. (1985). Armandroff (1994) in fact indicates

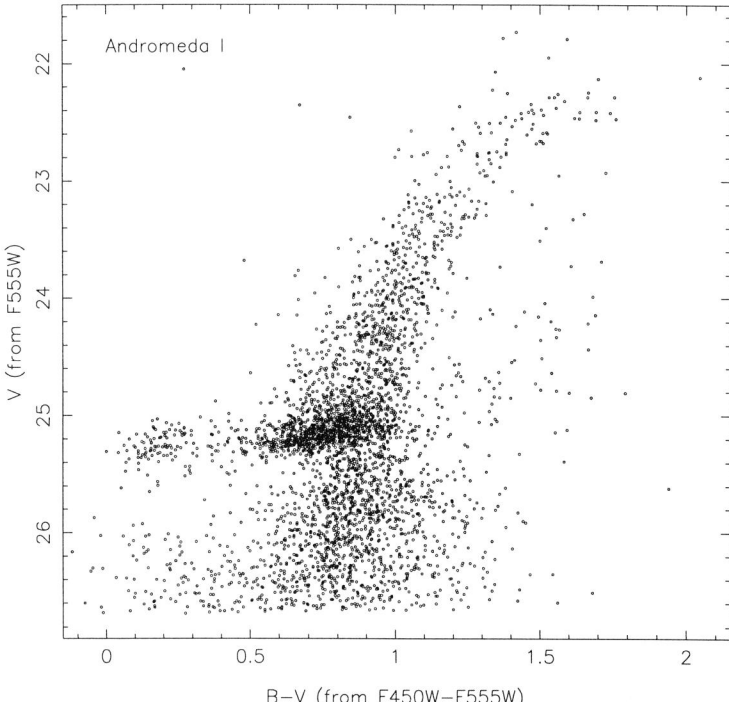

FIGURE 14. A colour-magnitude diagram for the M31 dSph companion And I from the HST/WFPC2 observations of Da Costa et al. (1996).

that Cook, Olszewski & Suntzeff have identified 8 carbon star candidates (including the known object) in And II via narrow band imaging. It is thus likely that this M31 dSph companion has a substantial intermediate-age population. No carbon star candidates were found in either And I or And III, however, which is consistent with the broad-band imaging results.

Thus it appears likely that the variety of star formation histories found among the galactic dSphs also occurs to some extent in the M31 dSph companions. We now turn to a discussion of the recent HST/WFPC2 data for And I of Da Costa et al. (1996), which reveals for the first time the morphology of the horizontal branch in this system.

4.3.1. *And I*

Figure 14 shows a c-m diagram for And I based on the HST/WFPC2 observations described in Da Costa et al. (1996). A number of results are immediately apparent from this c-m diagram. First, in accord with the ground-based studies, there is no sign of any significant upper-AGB population, so that And I does apparently lack any substantial intermediate-age component. Second, the core helium burning stars are found principally as red horizontal branch stars, though blue horizontal branch stars are also clearly present. Further, analysis of the individual data frames reveals the presence of RR Lyrae variables. Light curves for 4 variables found in this way from the PC frames are shown in Fig. 15; a total of almost 50 similar variables have been discovered in these data. Together with the blue horizontal branch stars, the presence of these RR Lyrae variables again argues the case for at least some old population in this dSph galaxy.

However, given the mean metal abundance of this dSph ($<$[Fe/H]$> = -1.45 \pm 0.2$, Da Costa et al. 1996), the dominance of red horizontal branch stars suggests that the bulk

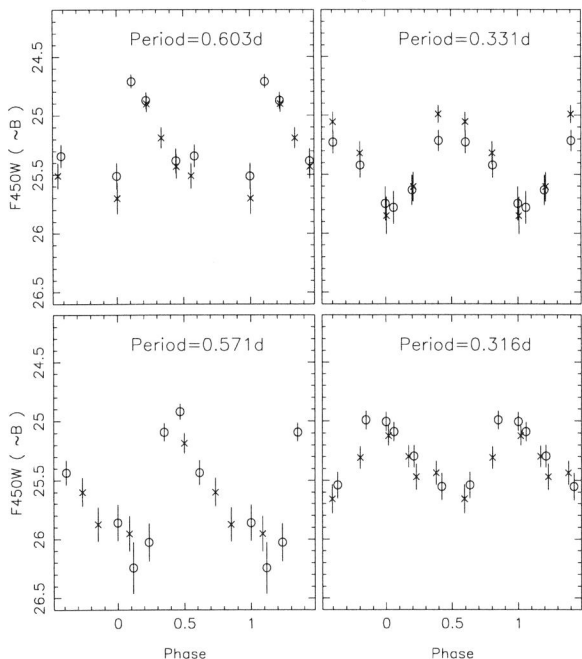

FIGURE 15. Light curves for 4 RR Lyrae variables found on the PC images of And I (from Da Costa et al. (1996).

of the And I stellar population is somewhat younger than the galactic globular clusters, especially those of the inner galactic halo (cf. Fig. 6). For example, the galactic globular cluster M5 has a very similar abundance to And I but its c-m diagram shows many blue horizontal branch stars and only a few red ones (e.g. Buonanno et al. 1981), unlike And I. Based on theoretical models of horizontal branch morphology as a function of age and abundance (e.g. Sarajedini et al. 1995), the age of the bulk of the stellar population of And I is estimated as ~10 Gyr (Da Costa et al. 1996). This interpretation is supported by the Mighell & Rich (1996) results for Leo II, a galactic dSph with a mean abundance and horizontal branch morphology very similar to those of And I. In this case the age inferred from the horizontal branch morphology is confirmed by direct observations of the main sequence turnoff (Mighell & Rich 1996). Thus, since the blue horizontal branch and RR Lyrae stars in And I are likely to be at least 2 to 3 Gyr older than the bulk of the population, we again have evidence supporting the existence of an extended epoch of star formation in a dSph galaxy.

There is one further intriguing result for And I: the horizontal branch morphology shows a *radial gradient* with the blue horizontal branch (and RR Lyrae) stars being relatively more common outside the dSph's core radius (Da Costa et al. 1996). In other words, the blue horizontal branch and RR Lyrae star population is less centrally concentrated than the dominant red horizontal branch population. This morphology gradient could reflect either a metallicity gradient (in the sense of more metal-rich stars at smaller radii) or an age gradient (younger stars at smaller radii). There is, however, no change in the mean colour of the giant branch with radius and Da Costa et al. (1996) use this to set an upper limit of ~0.2 dex for any change in the mean abundance over the observed radius interval, which extends from the centre of And I to ~1.2 core radii. Thus it seems we are seeing an age gradient — And I was apparently more extended when the minor (10 – 15%) old component formed than when the majority of its stars formed. There is some evidence (cf. Da Costa et al. 1996) that this effect is also seen in the Leo II and Sculptor dSphs, though it is apparently not present in Carina (Smecker-Hane et al. 1994). The existence of this radial gradient in And I, regardless of its origin, has a significant implication: it suggests strongly that this dSph (at least) formed and evolved as an independant system rather than as debris from the disruption of a larger system (as is sometimes argued for some, or indeed all, of the galactic dSphs, e.g. Lynden-Bell & Lynden-Bell 1995 and the references therein).

4.4. M31 dE galaxies

Ever since their resolution into stars (Baade 1944a,b), the dE companions to M31 NGC 147, NGC 185 and NGC 205, together with their attendant globular cluster systems, have been extensively studied objects. Colour-magnitude diagrams derived from ground-based imaging are now available for all three systems (e.g. NGC 147, Mould et al. 1983, Davidge 1994; NGC 185, Lee et al. 1993b, Martinez-Delgado et al. 1997; NGC 205, Mould et al. 1984, Lee 1996). These c-m diagrams reveal the upper two or three magnitudes of the red giant branch and have been used, *inter alia*, to determine mean abundances and abundance dispersions. Deep ground-based imaging has also revealed the existence of RR Lyrae variables in each of these dEs (Saha et al. 1990, Saha & Hoessel 1990, Saha et al. 1992, respectively) from which we can immediately conclude that each dwarf contains at least some old stars. All three also possess their own globular cluster systems whose properties were discussed in Sect. 4.1. NGC 185 and NGC 205 both contain small amounts of atomic and molecular gas, dust, and young stars in their central regions, but NGC 147 appears to lack all of these quantities (e.g. Young & Lo 1997 and references therein). We shall return to a discussion of this aspect of these dEs below.

The major issue concerning these dwarfs at present however, is whether or not they contain any significant intermediate-age population. This controversy can be viewed in some sense as a dialogue between those who hold to the classic view, that E galaxies (and in this case dE galaxies) possess only old stellar populations, and those who regard the classic view as likely to be overly simplistic. For NGC 185 and NGC 205 this issue has added importance in that the results are relevant to the origin of the young stars found in the centres of these dEs: do these young stars in the galaxy centres represent the first occurrence of star formation since early epochs or are they just the most recent event in an ongoing or cyclical process? We review here the arguments for and against the existence of intermediate-age stars in these dE galaxies, considering each in turn.

For NGC 147, in their pioneering study of the outer regions of this dE, Mould et al. (1983) found no evidence to support the existence of stars above the red giant branch tip. They set an upper limit of ∼10% for the fraction of stars in their field that have ages between ∼1 and 12 Gyr, thus supporting the "classic" picture. On the other hand, Davidge (1994) in his study of the central regions of NGC 147, reached the conclusion that a minor intermediate-age component, with upper-AGB stars as bright as $M_{bol} \approx -5$, was present. Such a population might have an age of perhaps ∼5 Gyr. It is worth keeping in mind here that in a population with stars whose abundances exceed [Fe/H] ≈ -1.0, as is the case for some fraction of the NGC 147 population, AGB stars as bright as $M_{bol} \approx -4.5$ can occur without the need to invoke the presence of an intermediate-age population (e.g. Frogel & Elias 1988). Thus, given that Davidge (1994) did not monitor the effects of image crowding on his data (cf. Martinez-Delgado & Aparicio 1997), and that the number of stars brighter than $M_{bol} \approx -4.5$ in the luminosity function is small, the case for an intermediate-age population in NGC 147 from Davidge's results could be perhaps best described as "not proven". However, the recent HST/WFPC2 data for

NGC 147 of Han et al. (1997), which has much better resolution and is thus much less susceptible to image crowding effects, does confirm the presence of an intermdiate-age population in this dE. Further, Han et al. (1997) find that the NGC 147 upper-AGB stars are more centrally concentrated than the general population, thus providing consistency with the Mould et al. (1983) results for the outskirts of this dE galaxy. The Han et al. (1997) results will be discussed in more detail in Sect. 4.4.2 below.

For NGC 185, Lee et al. (1993b) have produced c-m diagrams for the central regions from ground-based imaging obtained in $\sim$0.7″ seeing. They report that three distinct populations can be identified in their c-m diagram data. These are: a strong red giant branch population, an upper-AGB population of stars with luminosities above that of the red giant branch tip, and a small centrally concentrated population of "blue/yellow" stars which they interpret as young objects. The upper-AGB stars are claimed to reach a luminosity of $M_{bol} \approx -5.0$ corresponding, in their interpretation, to an age of $\sim3 \pm 3$ Gyr (Lee et al. 1993b). They suggest that "most of the AGB stars above the tip of the RGB are likely to belong to an intermediate-age population". Lee et al. (1993b), however, did not evaluate the effects of image crowding on their photometry.

Martinez-Delgado et al. (1997), on the other hand, disagree with the Lee et al. (1993b) results as regards the presence of an intermediate-age population in NGC 185. In particular, Martinez-Delgado et al. (1997) argue that in the outer parts of NGC 185, where image crowding is not a serious concern, contamination from foreground stars prevents a definitive answer regarding the status of the few stars found above the red giant branch tip. But for the inner regions, they claim that the stars found above the RGB tip are principally the result of image crowding effects and *not* an indication of the presence of an intermediate-age population (Martinez-Delgado et al. 1997). Their results are supported by the numerical simulations presented in Martinez-Delgado & Aparicio (1997). Here, a synthetic c-m diagram, generated using the Bertelli et al. (1994) isochrones and containing only stars older than 12 Gyr (i.e. a pure "old" population), is subjected to varying degrees of image crowding. Martinez-Delgado & Aparicio (1997) show that in cases where the image crowding is "severe", which in their interpretation corresponds to the entire region of NGC 185 studied by Lee et al. (1993b), the effect is to scatter some stars to magnitudes as much as half a magnitude brighter than their true magnitudes. This effect then generates a luminosity function which extends to brighter limits than it would in the absence of image crowding. Consequently, Martinez-Delgado & Aparicio (1997) maintain that the Lee et al. (1993b) claim for an intermediate-age population in NGC 185 is at least questionable, if not spurious.

There are three points that need to be made here. First, the Martinez-Delgado & Aparicio (1997) results undoubtedly show that in regions where image crowding is non-negligible, its effects must be fully determined before valid astrophysical conclusions can be drawn from observed c-m diagrams. Second, in generating model c-m diagrams, one must always keep in mind the inherent characteristics of the input data. For example, the Bertelli et al. (1994) isochrones used by Martinez-Delgado & Aparicio (1997) contain, even for pure old metal-poor populations, AGB stars with luminosities up to $\sim$0.5 to 0.6 mag (in M_{bol}) above the RGB tip (see, for example, Martinez-Delgado & Aparicio 1997, Fig. 1, panel a). Such upper-AGB stars are not found in metal-poor galactic globular clusters (e.g. Frogel & Elias 1988, Renzini 1992). One could then argue from the Martinez-Delgado & Aparicio (1997) input luminosity function and image crowding results, together with the Martinez-Delgado et al. (1997) observations, that NGC 185 does possess stars with luminosities up to $\sim$0.5 mag in M_{bol} above the RGB tip. It is just the age of these stars (i.e. intermediate-age, $\leq \sim$10 Gyr, versus old, $\geq \sim$10 Gyr) that is controversial. Third, this debate will not actually be settled without further observations. These could be, for example, ground-based narrow band imaging, perhaps with follow-up low resolution spectroscopy, to identify a population of upper-AGB carbon stars. Such stars are an unambiguous signature of the presence of an intermediate-age population. Alternatively, imaging in the infrared JHK bands, where image crowding in less significant for the brightest stars, can establish bolometric magnitudes directly. Finally, broad band optical imaging with HST/WFPC2 could also play a significant role since the degree of image crowding is much reduced in such high resolution images.

For NGC 205 a similar situation to NGC 185 exists. Mould et al. (1984) have presented a c-m diagram for a region in the outskirts of this dE which shows the first two magnitudes or so of the red giant branch. Like NGC 185, no evidence is found in this c-m diagram for the presence of any significant number of stars with luminosities above that of the RGB tip. However, Mould et al. (1984) do report the existence of a small population of NGC 205 stars whose magnitudes are up to $\sim$0.5 mag brighter than the RGB tip. The interpretation of these stars is not clearcut: it is possible that they do represent an older intermediate-age population (age $\geq$ $\sim$8 Gyr) but other interpretations are equally viable. For example, the statistics are such that a large fraction could be M31, or foreground, stars (Mould et al. 1984). Mould et al. (1984) do, however, use the lack of brighter stars to set an upper limit of $\sim$10% for any ($\sim$1 – 8 Gyr) intermediate-age population in this NGC 205 field.

Colour-magnitude diagrams for the inner regions of NGC 205 have been produced from ground-based imaging (0.6″ – 1.0″ seeing) by Davidge (1992) and Lee (1996). Presumably because of the crowded nature of the fields, neither data set unambiguously reveals the tip of the red giant branch. But both reveal apparently significant numbers of upper-AGB stars (the Lee (1996) data also contain information on the young population in NGC 205's central regions. This will be discussed further below). The candidate upper-AGB stars in these c-m diagrams are more numerous and reach brighter magnitudes than is the case for NGC 185. In particular, the brightest stars have I $\approx$ 19 and $\sim$1.7 $\leq$ V–I $\leq$ $\sim$3.0 and such stars reach M_{bol} $\approx$ –5.5 or about 0.5 mag brighter than the equivalent limit for NGC 185 (Lee et al. 1993b). Given the results of Martinez-Delgado & Aparicio (1997), is it reasonable to claim that these NGC 205 stars are genuine intermediate-age objects, rather than being artifacts of image crowding? If so, then the corresponding age could perhaps be as young as $\sim$1 Gyr (Davidge 1992, Lee 1996). In this particular case the answer to the question is probably "yes", though of course a full analysis along the lines suggested by Martinez-Delgado & Aparicio (1997) should be carried out to confirm this assertion. The reasons for this response are twofold. First, the candidate stars are up to $\sim$0.5 mag brighter than is the case for NGC 185. Martinez-Delgado & Aparicio (1997) have shown that image crowding can scatter stars as much as 0.5 mag above their true magnitude, but promoting significant numbers of stars by $\sim$1 mag seems less likely. After all, an unresolved blend of two identical RGB-tip stars is brighter only by 0.75 mag. Second, Richer et al. (1984) have used narrow band imaging to argue for the presence of upper-AGB carbon stars in these central regions of NGC 205. Richer et al. (1984) employ three filters with $\sim$100Å widths: the 71 filter is centered on a TiO band near 7100Å, the 78 filter is predominantly continuum and the 81 filter covers a strong CN band. Consequently, carbon stars have large (81-78) indices at a given V–I colour while M stars with the same V–I colour have small or negative values of (81-78). In their photometry Richer et al. (1984) claim to have rejected all stars which were "obviously confused" (i.e. crowded) but still found seven carbon star candidates from 68 stars redder than V–I = 1.6. Further, the V and I photometry for these stars (Richer et al. 1984) places them among the upper-AGB candidates in the Davidge (1992) or Lee (1996) c-m diagrams, and the brightest have M_{bol} $\approx$ –5.0 corresponding to an age of a few Gyr. As a result, there seems little reason to doubt that NGC 205 does contain a sizeable intermediate-age population in its inner regions. This inner region also contains 3 globular clusters, two of which are apparently old, metal-poor objects while the third, Hubble V, has an age of perhaps 500 Myr to 1 Gyr (Da Costa & Mould 1988, Lee 1996). NGC 205 also has a nuclear star cluster with hydrogen line strengths that are comparable to, or perhaps somewhat stronger than, the H-line strengths in the spectrum of Hubble V (Da Costa & Mould 1988). Bica et al. (1990) suggest an age of 100 to 500 Myr for its dominant stellar population. All these results then suggest that star formation in NGC 205 has been an ongoing process. We now consider in more detail the young populations in the central regions of NGC 185 and NGC 205.

4.4.1. *The central regions of NGC 185 and NGC 205*

The central regions of NGC 185 and NGC 205, but not NGC 147, have been known for many years to contain small populations of luminous blue stars as well as distinct clouds of dust (e.g. Baade 1951; Hodge 1963, 1973). These same regions are now also known to contain modest amounts of both HI and, through the detection of CO emission, molecular gas (see Young & Lo 1997 and the references therein). Young & Lo (1997) give HI masses of $\sim 4 \times 10^5$ M$_{sun}$ and $\sim 1 \times 10^5$ M$_{sun}$ for NGC 205 and NGC 185, respectively, and report a 3σ upper limit of 3×10^3 M$_{sun}$ for the mass of HI in NGC 147. The corresponding M$_{HI}$/L$_B$ values are 2×10^{-3}, 1×10^{-3} and $<4 \times 10^{-4}$, respectively, in solar units.

There are two fundamental and related issues that need to be addressed here. The first concerns the star formation: do the young stars represent a current episode in an ongoing or cyclical process, for example? The second relates to the origin of the gas. Here there would seem to be two alternatives: that the gas has been acquired from sources external to the dwarfs, perhaps from M31, or that it has an internal origin in the mass lost by evolving stars (it would seem unlikely, given the low HI masses, that the gas could be left over from the formation epoch). Any consistent explanation of these issues must also endeavour to explain why NGC 147, which is very similar to NGC 185 in most of its properties, lacks *any* indication of gas, dust and young stars in its central regions. Further, in considering these issues it must also be kept in mind that in both NGC 185 and NGC 205, the gas, dust and young stars are all confined to a central region that is much smaller than the overall size of the galaxies. Indeed, in both dEs the "Population I" characteristics are restricted to regions with radii of $\sim 1 - 2'$, or less than ~ 500 pc, which corresponds to 1 or 2 exponential scale lengths for these galaxies (e.g. Caldwell et al. 1992). Consequently, we are dealing with essentially a central, as distinct from a global, phenomenon that has little effect on the overall properties of these dEs. Further, at the distance of Virgo, for example, the regions containing the young stars, gas and dust would be only a few arcseconds in size at most, and therefore perhaps easily overlooked.

As regards the star formation, here is definitely an area where the high resolution of HST/WFPC2 could make a dramatic impact. But a number of results are available, in any case, from existing ground-based observations. First, the absence of HII regions indicates, assuming a normal IMF, that neither dE is currently forming stars (the extended Hα emission in NGC 185 (e.g. Young & Lo 1997) is believed to be from a supernova remnant, Gallagher et al. 1984). Second, the c-m diagram studies of Lee et al. (1993b) and Lee (1996) do yield some information on the ages of the young stars, though the crowded nature of the fields makes definitive answers difficult to obtain. It does appear, for example, that the NGC 205 blue star population is younger than that in NGC 185 — the brightest blue stars are up to ~ 1 mag brighter in NGC 205. Lee (1996) estimates an age of ~ 20 Myr for the brightest NGC 205 blue stars, so star formation has apparently occurred in the recent past. However, the number of stars is small and the errors are large, so this age could easily be a substantial underestimate. Lee (1996) also suggests the presence of core helium burning stars with ages of perhaps 80 to 200 Myr, but whether this population blends with the ∼1 Gyr population inferred from the brightest upper-AGB stars in an approximately continuous star formation history remains unclear. For NGC 185 the status of the blue/yellow stars is also unclear (i.e. are they main sequence or core helium burning objects?) but an age exceeding ∼100 Myr for the youngest stars seems a reasonable estimate. The bulk of the blue/yellow stars in the c-m diagram, however, are perhaps ∼500 Myr old (Lee et al. 1993b). Again the crowded nature of the field means the data are not of sufficient quality to thoroughly investigate the underlying star formation history.

As regards the distribution and kinematics of the gas, Young & Lo (1997) report that in neither dE is the gas found as a rotating disk or ring. Rather it has a clumpy structure and there is little indication of any net rotation (Young & Lo 1997). For NGC 205 Young & Lo (1997) suggest that the gas and the (dominant old) stars in the central region are dynamically distinct systems. For example, the HI velocity centroid is ∼20 kms^{-1} different from the central stellar velocity and moreover, the HI shows a velocity gradient which the stars do not (Young & Lo 1997 and references therein). This suggests that the gas in NGC 205 is not yet in a stable long-lived configuration. For example, it may have recently fallen in and has not yet had time to spread into a complete disk or ring (Young & Lo 1997). In this respect it is interesting to note that the outer isophotes of NGC 205 are distorted (e.g. Hodge 1973) perhaps indicating an interaction with M31 and Bender et al. (1991) offer kinematic evidence to support this possibility. It is therefore conceivable that gas from M31 has been drawn into the central regions of NGC 205 as a result of a tidal interaction.

For NGC 185, the HI does not show any systematic velocity gradients nor velocity offsets and thus the gas and stars could be part of the same dynamical system (Young & Lo 1997). This is consistent with an internal origin for the gas as mass lost by evolving stars, a suggestion originally made by Gallagher & Hunter (1981). Welch et al. (1996) estimate that only ∼1 Gyr or so is required for the observed gas mass in NGC 185 to accumulate from stellar mass loss. While this may seem a plausible scenario, the lack of gas, and indeed of stars younger than ∼7 – 8 Gyr even its central regions (Han et al. 1997, see following section), in the otherwise very similar system NGC 147, argues against it, though in that case the fate the material presumably lost by evolving stars remains unexplained. We now turn to a discussion of the HST/WFPC2 results of Han et al. (1997) for NGC 147.

4.4.2. *NGC 147*

In a recent paper Han et al. (1997) have presented results from an HST/WFPC2 study of NGC 147. Two fields were imaged, the first near the centre of the galaxy and the second, an outer region ∼4′ south of the centre. The total exposure times for the central field were considerably shorter (by a factor of ∼10) than for the outer region, and thus information at the faintest levels in the inner region c-m diagram is less reliable than it is for the outer parts. The inner regions are also somewhat more crowded but the difference in limiting magnitude between the two fields means that the difference in degree of image crowding is much less than one might otherwise have expected. The inner and outer region c-m diagrams contain roughly 78,000 and 38,000 stars, respectively! Both show strong red giant branches whose mean colours do not provide any compelling evidence for the existence of a radial abundance gradient: <[Fe/H]> ≈ –0.9 for the inner field and <[Fe/H]> ≈ –1.0 for the outer field. There does, however, appear to be a decrease in the *abundance dispersion* with increasing radius. Han et al. (1997, Fig. 9) indicate that $\sigma([\text{Fe/H}])$ declines from $\sim$0.6 in the region closest to the galaxy's center to $\sigma([\text{Fe/H}]) \approx 0.4$ in the outer parts. Interestingly, the intrinsic colour dispersion at $M_I \approx -3.0$ found by Han et al. (1997) in their outer region, $\sigma(\text{V--I}) = 0.13$, is identical to that found from the ground-based data of Mould et al. (1983), which applies to a field somewhat further from the galaxy's centre. Thus the change in abundance dispersion is apparently restricted solely to the inner parts ($r \leq \sim 1.5' - 2'$) of this galaxy.

As regards the core helium burning stars, in the central region c-m diagram they are apparently restricted to a "red clump" and there is no clear indication of any horizontal branch component. However, at these magnitudes the inner data have very large errors and, as a result, it is not possible to definitely rule out the presence of horizontal branch stars. Indeed it is not clear that even the term "red clump" should be used, at least in the sense that this term has been used in these lectures, where it carries the connotation of a younger age. For example, in a globular cluster such as 47 Tuc, which has an abundance only slightly higher than the mean for the NGC 147 inner field given by Han et al. (1997), the red horizontal branch stars are separated from the red giant branch by only $\sim$0.25 mag in V--I (e.g. Da Costa & Armandroff 1990) which, at this luminosity, is also approximately the difference in colour between the 47 Tuc giant branch and that of a metal-poor system like M15. Consequently, given the colour errors in the inner field at this magnitude, $\sigma(\text{V--I}) \approx \pm 0.12$ mag, and given the presence of a sizeable abundance spread ($\sigma([\text{Fe/H}]) \approx 0.6$), it is clear that the inner region data do not permit any strong inferences to be drawn from the morphology of the core helium burning stars in this particular c-m diagram.

For the outer region, however, where the errors at this luminosity are considerably smaller, the morphology exhibited by the core helium burning stars in the c-m diagram is more certain. There is a clear dominance of "red clump" (red horizontal branch?) stars, together with an obvious horizontal branch structure that extends to moderately blue colours ($(\text{V--I})_0 \approx 0$). This horizontal branch morphology is consistent with the detection of RR Lyrae variables in this galaxy (Saha et al. 1990). As for other systems, the presence of blue horizontal branch stars and RR Lyrae variables immediately testifies to the existence of at least some old population in this dE. Han et al. (1997) also show that the bluer horizontal branch stars (those with $(\text{V--I})_0 \leq 0.5$) have a more extended radial distribution than, for example, the "red clump" stars. This situation is reminiscent of the situation in And I described above. Like Da Costa et al. (1996) for that dSph, Han et al. (1997) conclude that since there is no obvious abundance gradient, the effect represents an age gradient with relatively more older stars at larger radii in NGC 147. As for the dominant "red clump", Han et al. (1997) conclude that this is an indicator of a "large and perhaps even dominant intermediate-age population". While this may be the case (especially if the intermediate-age stars are relatively "old"; i.e. age $\sim$7 -- 8 Gyr or so), it is by no means proven, especially given the relatively high mean abundance and large abundance dispersion in this region of NGC 147. As Han et al. (1997) themselves admit, a full modelling of the c-m diagram is required to quantify more clearly this dE's star formation history.

There are, however, a few additional inferences that can be drawn from these c-m diagrams (Han et al. 1997). First, in the outer field c-m diagram there is no indication of the presence of main sequence stars to a level at least a magnitude fainter the horizontal branch. This indicates that star formation in the outer parts of NGC 147 ceased more than 1 -- 2 Gyr ago. Second, both the inner and outer c-m diagrams show a population of stars that lie above the red giant branch tip and which, at least in the inner region, extend to quite red colours ($(\text{V--I})_0 \geq 3.0$). There are also clearly relatively more of these stars in the inner field. Indeed Han et al. (1997, Fig. 10) demonstrate quantitatively that these stars are more centrally concentrated than the majority of the population. They interpret these stars as extended (upper) AGB stars and thus as evidence for the existence of an intermediate-age population in NGC 147. Further, they note that the higher central concentration of these stars is consistent with the age gradient interpretation of the more diffuse distribution of the bluer horizontal branch stars: both suggest younger stars are found more frequently closer to the dE's centre.

The radial gradient in the relative distribution of the upper-AGB stars is in fact the strongest argument in favour of their being intermediate-age stars. If they were simply the progeny of the older population, which cannot be ruled out *a priori* since the majority of the stars have $-3.5 \geq M_{bol} \geq -4.0$ (Han et al. 1997) and stars with these luminosities are known in pure old populations with metallicities similar to that of NGC 147 (e.g. Frogel & Elias 1988), then one would not expect to see a change with radius in the ratio of upper-AGB stars to all stars. Fig. 10 of Han et al. (1997) shows, however, that this ratio does change with radius, especially within the inner $\sim 1.5'$ or so, the same region where the increase in abundance dispersion is most marked. Thus the central regions of NGC 147, while not as dramatically different from the rest of the galaxy as are the central regions of NGC 185 and NGC 205, is still in some sense a special place. Indeed the size of this region, at $\sim 1.5'$ or ~ 300 pc, is quite comparable in size to the region in NGC 205 that contains the young stars, gas and dust.

The relatively low luminosity of the NGC 147 upper-AGB stars suggests that these stars are from a relatively "old" (age $> \sim$few Gyr) intermediate-age population. But when considered together with the existence of relatively blue horizontal branch stars and RR Lyrae variables in this dE, the case is quite strong that star formation in NGC 147 was an extended process that took place over at least several Gyr. Given this result and those for the other M31 dE and dSph companions, it certainly seems quite likely that the diversity of star formation histories seen among the galactic dSphs is also present in M31's companions, the objects that provided the basis for the "classic" picture of the stellar populations of dE (and E) galaxies. Clearly, there is now little reason to suggest that this "entirely old population" scenario is at all appropriate for the majority of dE systems, whether within the Local Group or beyond it.

4.4.3. *M32*

As the nearest example of an elliptical (as distinct from a dwarf elliptical) galaxy, M32 plays a pivotal role in the study of stellar populations. On the one hand the galaxy has a high surface brightness centrally concentrated light profile which, together with a relatively low velocity dispersion, facilitates the acquisition of high quality integrated spectra. On the other hand, it is sufficiently close that individual stars can be resolved and studied to form colour-magnitude diagrams. In principle, this dichotomy then provides an ideal opportunity to check the validity of population synthesis models constrained by the integrated spectrum. As is well known (e.g. Worthey 1994), population synthesis models often have difficulty separating the effects on integrated spectra of age from those of metallicity. Thus the verification procedure that M32 potentially offers could play a vital role in underpinning the acceptance of population synthesis models in those many situations when only integrated spectra are available.

In reality, however, the situation is not so straightforward. For example, the steep density profile and the high surface brightness that are a boon for integrated light spectroscopy mean that reliable c-m diagrams for the central regions of M32 are very difficult to generate — the degree of image crowding is extremely high even with HST/WFPC2. Thus available M32 c-m diagram studies are necessarily for regions well beyond where the integrated spectra apply. Consequently, the size and extent of any radial (abundance and/or age) gradients become important issues. Further, M32 lies close to the outer disk of M31 (in projection) so that contamination of c-m diagrams by M31 stars is an additional non-negligible complication. Indeed, the steep density profile results in a relatively narrow radius interval over which it is possible to minimize both crowding and contamination from M31.

But what do the observations of M32 reveal? As regards integrated spectroscopy, for many years now population synthesis models of M32's integrated spectrum have consistently argued for the presence of a dominant intermediate-age ($\sim$6 – 8 Gyr) population in (the central regions of) this galaxy (e.g. O'Connell 1980, Rose 1994, Hardy et al. 1994 and the references therein). This result is in direct conflict with the much older age "expected" from the classical picture of E galaxies, in which they are regarded as passively evolving old stellar systems.

As for optical studies of individual stars in M32, c-m diagrams have been determined by, for example, Freedman (1989), Davidge & Neito (1992) and Davidge & Jones (1992). In particular, Freedman (1989) noted that the brightest stars in her field (centered $\sim$2′ from the nucleus) had I $\approx$ 19.5. Further, the number of stars above the red giant branch tip suggested the presence of a significant intermediate-age population with an age perhaps as young as 5 Gyr if M32 is at a comparable distance to M31 (as is now generally believed to be the case). Freedman (1989) also noted the presence of a large metallicity spread in the c-m diagram and gave a lower limit to the mean abundance as <[Fe/H]> $\geq$ –0.5 dex. Similarly, Davidge & Neito (1992) used observations taken in excellent seeing to suggest the presence of AGB stars as bright as I $\approx$ 19.0 within 30″ of M32's centre. Such stars would be even younger than the AGB candidates in the Freedman (1989) c-m diagram. Finally, Davidge & Jones (1992) find only a few stars brighter than I $\approx$ 20 in their c-m diagram study of a field adjacent to that of Freedman (1989). They concluded, in contrast to Freedman (1989), that the number and luminosity of the stars above the red giant branch tip was consistent with that expected for an old population. Nevertheless, they use the difference between their results and those of Davidge & Neito (1992) to suggest that the outer regions of M32 may be different from the central regions.

In interpreting these c-m diagrams two things must be kept in mind. First, M32 is a metal-rich system. As noted in Sect. 2 (cf. Fig. 3), at higher metallicities the giant branch in (I, V–I) c-m diagrams becomes almost horizontal. Further, it does not reach I magnitudes as bright as any metal-poor population and extends to very red colours. Consequently, very deep limiting magnitudes in V are required to ensure that metal-rich red stars are not underrepresented in the resulting c-m diagrams. Second, as noted above, contamination from M31 is significant, the fields are extremely crowded and the number of brighter stars is small. Thus the interpretation of optical c-m diagrams is necessarily subject to large uncertainties.

These interpretation uncertainties can be alleviated to some extent, however, by imaging in the JHK infrared bands rather than in the optical. This approach has the distinct advantage that the resulting photometry is a much more direct measure of the bolometric luminosities for metal-rich stars than is I band photometry. Such JHK imaging studies have been carried out for M32 by Freedman (1992), who imaged a $\sim$100″ × 40″ region within the area of her earlier optical study, and by Elston & Silva (1992) who imaged a $\sim$4′ × 4′ region centered $\sim$3′ east of the centre of M32. Both studies find a significant population of stars with luminosities well in excess of the first giant branch tip. The brightest of these stars have $M_{bol} \approx$ –5.5, brighter by $\sim$1 mag than the brightest stars seen in equivalent studies of the galactic bulge (e.g. Frogel & Whitford 1987). These IR data seem to provide a strong case for the presence of a distinct intermediate-age population in the outskirts of M32. However, one must be a little careful when pursuing this interpretation and again take note of the relatively high metallicity of M32. For example, while in metal-poor (old) globular clusters AGB stars are not found at luminosities above the RGB tip, this is not the case for more metal-rich (old) globular clusters. In these clusters AGB stars can be found as bright as $M_{bol} \approx -4.6$ (Frogel & Elias 1988, see also Guarnieri et al. 1997). Thus since M32 has a similarly high abundance, one should not be surprised by the presence of stars as bright as this luminosity even if the population is purely old. Indeed, the luminosity function of Elston & Silva (1992) shows a distinct drop (by a factor of ~ 2) at approximately this magnitude. However, in this same luminosity function, there remains an excess of stars beyond $M_{bol} \approx -4.8$ reaching as bright as $M_{bol} \approx -5.5$; these stars are almost certain to be from an intermediate-age population. Elston & Silva (1992) estimate the age of these stars as 2 – 4 Gyr while Freedman (1992) suggests 4 ± 3 Gyr. Unfortunately, while these observations do seem to indicate the probable presence of an intermediate-age population in M32, they do not tell us whether the corresponding intermediate-age population *in the central regions* is sufficient to allow concordance with the integrated spectra results. Imaging of the centre of M32 with the new NICMOS infrared camera on HST should permit investigation of this possibility.

The recent HST/WFPC2 observations of Grillmair et al. (1996), however, do provide some further insight relevant to this issue. Although their observations are for a field $\sim 1.5'$ from the galaxy's centre, they reach sufficiently faint to reveal, for the first time, the full extent and morphology of the M32 giant branch in the V and I bands. The colour distribution for the wide giant branch seen in their c-m diagram is consistent with a smooth metallicity distribution skewed towards metal-rich stars. The peak of the abundance distribution is slightly below solar abundance (the actual value depends somewhat on the assumed age of the stars) and there is a low metallicity tail that extends to $[Fe/H] \approx -1.5$. Interestingly, most of the candidate intermediate-age upper-AGB stars seen in the ground-based optical c-m diagrams appear to have been artifacts of image crowding, since they are largely absent in the better resolution HST/WFPC2 data which shows almost no stars brighter than $I \approx 20$. On the other hand, these observations yield no information on the status of the infrared luminous upper-AGB candidates since they cannot be uniquely identified in the V and I data. However, the HST/WFPC2 observations do go faint enough to detect core helium burning stars in M32. These are found to occur as a strong red clump, but given the high mean metallicity inferred from the giant branch, this result does not constrain the age of the stars significantly (Grillmair et al. 1996).

While establishing the existence of a wide abundance range in M32 is the principle result of their paper, Grillmair et al. (1996) go on to conclude that at least half of the stars in the field studied must be older than 8 Gyr. As to whether this conflicts with the younger ages inferred for the central regions from population synthesis models, the answer is not clear cut. Grillmair et al. (1996) quote the unpublished results of González (1993, UCSC PhD thesis) whose integrated spectra reveal a substantial difference in the strength of the Hβ feature between the nuclear region and the regions beyond it. This difference in Hβ line strength between the centre and more distant regions is also evident in the data of Hardy et al. (1994). According to Grillmair et al. (1996), when combined with their c-m diagram results, these integrated spectra observations imply a somewhat more metal-rich and younger population in the inner regions, in agreement with the results from the population synthesis modelling. But it is probably fair to say

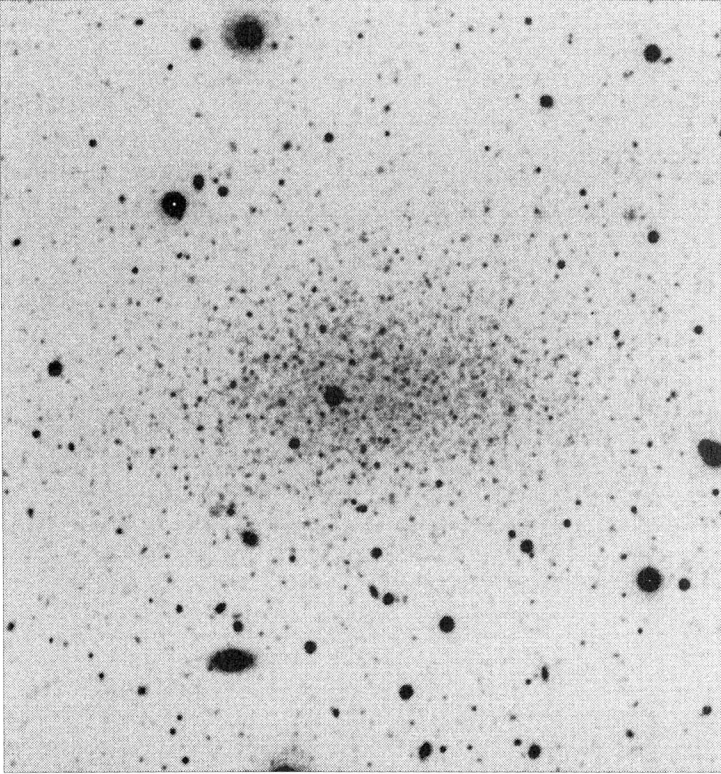

FIGURE 16. The Tucana Dwarf Elliptical Galaxy. This B-band image is from a 1000 sec exposure obtained by the author with the AAT prime focus CCD camera. North is at the top and East is to the left. The horizontal width of the area depicted is 270 arcseconds in size.

that the issue of the age of the bulk of the population in the central regions of M32 remains unresolved at the present time.

4.5. The Tucana dE

The Local Group dE known as Tucana was accidentally "discovered" by Lavery in 1990 on the SERC J Sky Survey films (though it is listed in the Southern Galaxy Catalog of Corwin et al. 1985). Described as an "extremely faint smudge", follow-up CCD imaging resolved the system into individual faint stars indicating that it was relatively nearby and therefore probably a Local Group member (Lavery & Mighell 1992). The appearance of this dwarf galaxy, which Lavery & Mighell (1992) classify as dE5, is illustrated in Fig. 16. Subsequent ground-based studies (e.g. Da Costa 1994, Saviane et al. 1996) have confirmed that Tucana lies at the fringe of the Local Group ∼900 kpc from the Galaxy. Intriguingly, given that all other Local Group dE and dSph galaxies have been found in the vicinity of M31 or the Galaxy, Tucana lies in a relatively isolated location far from any of the large luminous galaxies of the Local Group. It therefore provides us with a further opportunity to investigate the extent to which the properties of dE and dSph systems are influenced by their environment.

The surface photometry data of Da Costa (1994) and Saviane et al. (1996) show that, like most dE and dSph galaxies, the light profile of Tucana can be well fit by either an exponential law ($I(r) \approx I_0 e^{-r/r_{scale}}$) or by the profile of a low central concentration King (1966) model. More significantly, the length scales, surface brightnesses and total magnitude that result from these fits are entirely consistent with those for the other Local Group dSph and dE galaxies. This is illustrated in, for example, Fig. 8 where it is evident that Tucana is indistinguishable from other Local Group dSphs of similar luminosity. The large flattening of Tucana (b/a ≈ 0.5) is also not unusual — the galactic companion dSph Ursa Minor and the M31 dSph companion And III have similar axis ratios.

Ground-based c-m diagrams (e.g. Da Costa 1994, Saviane et al. 1996), which reach 2 or 3 magnitudes fainter than the RGB tip, reveal a dominant population of apparently old, metal-poor red giants. Based on the colour of the red giant branch, the mean abundance of Tucana is <[Fe/H]> ≈ –1.8 ± 0.2 which, when combined with the dwarf's absolute magnitude, places Tucana with the dSph companions to the Galaxy and to M31 in the mean abundance – absolute magnitude relation (see Fig. 7). The c-m diagrams also show no indication of any current or recent star formation, in contrast to the situation seen in the Phoenix and LGS 3 dIrr galaxies, which have similar luminosities and which also have dominant apparently old metal-poor populations (see Section 5.3). Further, there is no evidence in the Tucana c-m diagrams for any intermediate-age population in the form of upper-AGB stars similar to those found in some of the other Local Group dSph and dE galaxies. Thus there is little evidence from these data to suggest that Tucana has had any significant star formation after its initial episode.

Oosterloo et al. (1996) have conducted a sensitive search with the Australia Telescope Compact Array for HI gas in Tucana, but none was detected coincident with the optical image†. The 5σ upper limit corresponds to a HI mass of 1.5×10^4 solar masses within Tucana's effective radius, and this yields an HI mass to blue luminosity ratio limit of $M_{HI}/L_B < 3 \times 10^{-2}$. This lack of gas is entirely consistent with the HI content limits for other Local Group dE and dSph galaxies (Oosterloo et al. 1996). Again, though, this result contrasts with the situation for Phoenix and LGS 3 which both contain small but detectable amounts of HI (see Section 5.3).

Overall these results suggest that Tucana, despite its isolated location, is in every way consistent in its properties with the galactic and M31 dSph and dE companions, and this in turn suggests that environment does not play a major role in determining the gross properties of dSph and dE galaxies, at least within the Local Group.

The ground-based results for Tucana's stellar population are reinforced by the HST/WFPC2 imaging study of Seitzer et al. (1997) whose data are shown in the c-m diagram of Fig. 17. There is no indication of any significant population younger than ∼2 Gyr (the main sequence of such a population would be visible in the c-m diagram if it were present) nor are there any bright red giants that could be intermediate-age upper-AGB stars in Tucana (the stars brighter than V ≈ 22 in the c-m diagram are most likely galactic foreground stars). The colour of the red giant branch is consistent with a mean abundance of <[Fe/H]> ≈ –1.8 and there is evidence for a modest abundance spread. The horizontal branch is relatively evenly populated (cf. Fig. 14, the equivalent c-m diagram for And I) and, as might be expected, a number of RR Lyrae variables have been detected among the blue HB stars in Fig. 17. The presence of these variables and the blue HB stars indicates there is a definite component in Tucana that is comparable in age to the galactic globular clusters. On the other hand, given the mean abundance of Tucana and the lack of any sizeable internal abundance range, the substantial red HB population seen in Fig. 17 must be drawn from a somewhat younger population. The

---

† An HI cloud was detected some 15′ away from Tucana, but this cloud is interpreted as a foreground high velocity HI cloud, probably associated with the Magellanic Stream, unrelated to Tucana (Oosterloo et al. 1996).

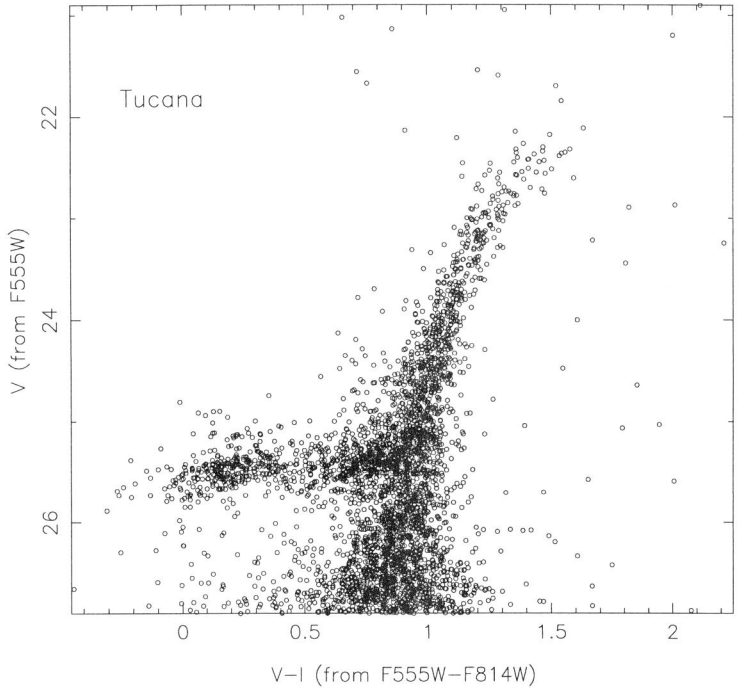

FIGURE 17. A colour-magnitude diagram for the Tucana Local Group dE from the HST/WFPC2 observations of Seitzer et al. (1997).

lack of upper-AGB stars, however, constrains this population to be no more than 3 or 4 Gyrs younger than the older stars. Thus, unlike many of the galactic and M31 dSph and dE companions, we have in Tucana a dE that does not reveal any compelling evidence for significant star formation much beyond an initial episode that lasted perhaps a few Gyr at most. This lack of any evidence for significant on-going star formation in Tucana is somewhat unexpected, given that to some extent (e.g. van den Bergh 1994b) it is the dSphs that lie furthest from their "parent" galaxy (i.e. the Galaxy or M31) which reveal the strongest evidence for star formation beyond that of an initial episode. Thus it appears that environment, at least within the Local Group, is also not a strong influence on long term star formation history.

4.6. *Discussion*

Within the Local Group then, we have examples of dE and dSph galaxies whose star formation histories range from *exclusively old* (e.g. Ursa Minor) through *mainly old* (e.g. Tucana, Leo II, NGC 147, And I) to *mainly intermediate-age, but "episodic"* (e.g. Carina, Leo I) and *mainly intermediate-age, but "continuous"* (e.g. Fornax). Is there an explanation for this remarkable diversity? At the moment, unfortunately, the answer to this question is "no" because we lack an understanding of the fundamental physical mechanism(s) that is, or are, responsible for driving star formation in these galaxies. Is it purely stochastic? Hopefully not. What role does the presence of a large parent galaxy play? The results for Tucana suggest that perhaps it is not large but, for example, is there a connection with the orbit of a satellite around a parent? Such a connection might be expected if, for example, star formation is "triggered" at perigalacticon. On this point at least we will soon have new observational input as proper motions for the nearer galactic dSphs are measured. Further, what role do the dark matter halos, in which all these systems are presumably embedded, play? These are all questions to which there are no clear answers at the present time, but they are questions that must be addressed if we are to make progress in understanding the origin and evolution of these Local Group dwarf galaxies and their more distant cousins.

5. Local Group dIrr galaxies

5.1. Overview

In addition to the dE and dSph galaxies, the Local Group also contains a number of dwarf Irregular (dIrr) galaxies. Once again the term "dwarf" implies relatively low luminosities, typically fainter than $M_V \approx -16$ or -17. So strictly the SMC is a dwarf, while the LMC is not. The term "irregular" arises from the optical appearance of the galaxies: unlike the smooth, structureless (or almost structureless) appearance of the dE and dSph galaxies, dIrr galaxies show structure on many size scales since their current star formation is not, in general, uniformly distributed over the face of the galaxy. This structure can range in size from individual HII regions and associations of young stars up to large star forming complexes that may cover an appreciable fraction of the dwarf's surface area. But such structure is rarely ordered in the way it is in a spiral galaxy for example, hence the term "irregular".

Examples of dIrrs range from systems such as the SMC and NGC 6822, which, with $M_V \approx -16$, are the brightest dIrrs in the Local Group, to galaxies like Phoenix, LGS 3 and GR8, which, with $M_V \approx -10$, are among the faintest galaxies in the Local Group. Interestingly, unlike the dEs and dSphs which, with the exception of Tucana, are all companions to either the Galaxy or M31, the Local Group dIrrs, with the obvious exception of the SMC, are all relatively isolated, lying away from the larger galaxies. This occurrence would seem to have clear implications for the formation and evolution of dwarf galaxies, particularly as regards the ability to retain gas and form stars (e.g. van den Bergh 1994b). However, the existence of the isolated but apparently "standard issue" dSph Tucana provides an intriguing counter-example to any general theory based on this circumstance.

Dwarf irregular galaxies are usually rich in gas with $M_{HI}/L_B > 1.0$ for most systems. Further, the HI gas, although centered on the optical image, is generally much more extended (e.g. Lo et al. 1993, Skillman 1996). Like dSph and dE galaxies, the underlying optical light profile for most dIrrs is approximately exponential (e.g. Vader & Chaboyer 1994). The dIrrs also follow a similar absolute magnitude – surface brightness relation to that of the dE and dSph galaxies (e.g. Binggeli 1986, Vader & Chaboyer 1994) in which the less luminous galaxies have lower surface brightnesses. Again as for the dE and dSph galaxies, there is also a trend for the more luminous dIrrs to have higher abundances (Skillman et al. 1989, Richer & McCall 1995 – see also Skillman, these lectures) though it is important to keep in mind that for dIrrs, "abundance" usually means an oxygen abundance determined from HII region spectroscopy. Indeed, the exact relation between [O/H] for dIrrs and the various metal abundance indicators used for dE galaxies remains problematical (e.g. Richer & McCall 1995; Skillman, these lectures).

The Milky Way dSph companions show evidence for the presence of extensive amounts of dark matter (see Mateo 1997 and the references therein for a review) and this is also the case for dIrr galaxies — there is compelling evidence from HI rotation curves that many dIrrs have extended dark matter halos. For example, Carignan & Freeman (1988, see also Carignan & Beaulieu 1989) have shown that the dIrr DDO 154, which lies somewhat beyond the Local Group, is virtually a "dark galaxy" — at the last point on the rotation curve (at 15 optical scale lengths!) more than 90% of the mass is contributed by the dark component. The global M/L_B value for this dwarf exceeds 75 (Carignan & Beaulieu 1989). Similarly, Lo et al. (1993) have determined M/L_B for the Local Group dIrr LGS 3 as 26 ± 13 in solar units, a value much higher than that expected from the stellar population alone ($M/L_B \approx 1$), implying the presence of dark matter. Preliminary stellar velocity observations reported in Mateo (1997) also support this value. Further, the M/L_B for the dIrr LGS 3 is not that different from the M/L ratios for galactic dSphs of similar luminosity (e.g. Leo II has $M/L_V \approx 12$, Mateo 1997).

Finally, virtually by definition, all dIrr galaxies are either currently forming stars or have done so in the recent past, though the "present-day star formation rates" can vary significantly from dwarf to dwarf. For example, Skillman (these lectures) has pointed out the sizeable difference in recent star formation activity between the dIrrs Sextans and Pegasus. In what follows I will leave the discussion of the young stellar populations in Local Group dIrrs to the other lecturers and concentrate instead on a different question — what can we learn about the star formation histories of Local Group dIrrs on long (>1 Gyr) timescales? In attempting to answer this question we will again rely on the information that can be gained from the resolved stellar populations of these dwarfs, rather than on their integrated properties. Hunter & Gallagher (1985), for example, have shown that the integrated colours, Hα fluxes, and integrated spectra of dIrrs are generally consistent with the assumption of near constant star formation rates over a Hubble time.

5.2. *Local Group dIrr star formation histories on Gyr timescales*

In most dIrrs there is evidence for an underlying background population of red stars whose brightest members have luminosities that correspond to that of the red giant branch tip. It is often tempting then, to conclude that these "background sheet" stars come from a globular-cluster-like population; i.e. one with an age exceeding ~10 Gyr. Yet as we know from Section 2, the observation of a red giant branch is not in itself sufficient evidence from which to conclude anything regarding the age of the stars, other than that it is probably older than 1 or 2 Gyr. Further information must be obtained before the presence of a population with an old age can be correctly inferred. Unfortunately, this requirement for more direct information is frequently overlooked, and the presence of a background sheet of red giant branch stars in a dIrr taken explicitly, but incorrectly, to indicate the occurrence of an old population. We only have to recall the results for the LMC to see the folly of such an unjustified assumption. The LMC possesses a background sheet of red giant branch stars but we know from observations reaching fainter magnitudes that the majority of these red giants are of intermediate-age, and that the true old population is a relatively minor component. We now consider some further examples for Local Group dIrrs.

5.2.1. *IC 1613*

Saha et al. (1992) have determined light curves and periods for a number of RR Lyrae variables in the dIrr galaxy IC1613. The existance of these variable stars is *prima facie* evidence for the occurrence of at least some old stars in this galaxy. But does it mean that the majority of the stars in their field, which was chosen to lie well away from any obvious star forming region and to sample the extended background sheet, are old? Saha et al. (1992) attempted to answer this question quantitatively. To do this they first noted that in the study of Freedman (1988), the c-m diagram for the background sheet region is, as expected, dominated by red giant stars. The tip of red giant branch is well defined and there is only a relatively minor population of stars above the red giant branch tip, together with a sprinkling of blue young main sequence stars (Freedman 1988, Fig. 5). The mean red giant branch color implies <[Fe/H]> ≈ −1.2 (the HII region abundance for IC 1613 is [O/H] ≈ −1.1, Skillman et al. 1989).

The field studied by Saha et al. (1992), who have observations only in one colour, lies further from the central parts of IC 1613 than that of Freedman (1988), and so the assumption of Saha et al. (1992), that their observed luminosity function is dominated by red giants, seems justified. They then define the index μ as the ratio of the number of stars between one and two magnitudes brighter than the nominal mean magnitude of the RR Lyrae stars, to the number of RR Lyraes found. They find a value $\mu \approx 40$ for their IC 1613 field. Since red giant stars can evolve to form either red or blue horizontal branch stars in addition to RR Lyrae variables depending on age and abundance, this index is not constant even for pure old populations. For example, μ is approximately six for RR Lyrae rich globular clusters like M3, but clearly μ can be very large for globular clusters that have only small numbers of RR Lyrae variables, be they metal-rich like 47 Tuc or very metal-poor like M92, for example. Clearly, for the appropriate horizontal branch morphology, the observed value of μ could imply that essentially all the stars in this IC 1613 field are old. However, it is more meaningful to derive a limit by assuming that the old red giants in this IC 1613 field are as efficient at generating RR Lyrae variables as they are in, for example, M3. In this case the observed μ value requires that at least 15% (=6/40) of the stars in this IC 1613 background sheet field are old. While this is a fairly weak constraint on the true fraction of the population that is old, and while it tells us nothing at all about the relative contributions of populations with ages between ∼2 Gyr and ∼10 Gyr (since these stars do not generate RR Lyraes), it is nevertheless a quantitative limit than can be compared with other galaxies.

5.2.2. WLM

In a recent paper Minniti & Zijlstra (1996) have presented CCD observations of WLM, a Local Group dIrr for which $M_V \approx -14$. The galaxy is well resolved into stars in their ESO/NTT data and the majority are confined to the vicinity of the galaxy's major axis. Away from the major axis though, there is a spatially extended population of stars reminiscent of a halo. The c-m diagram for this outer field shows a well defined red giant branch and there are relatively few stars lying above the red giant branch tip. The metal abundance inferred from the mean giant branch colour is [Fe/H] = −1.4 ± 0.2 though again it is unclear how this abundance should be related to the HII region abundance, [O/H] = −1.2 ± 0.15 (Skillman et al. 1989), in the disk of the galaxy. Can we conclude that the red giants observed by Minniti & Zijlstra (1996) represent a dominant old population? If yes, then it would seem to indicate that the early star formation history of WLM must have been rather different than that for the LMC, for example. Minniti & Zijlstra (1996) point out that more data, such as a determination of the horizontal branch morphology in their "halo" field, are required before we can answer this question conclusively. But they do note that, after correction for field star contamination, the giant branch luminosity function drops by more than a factor of 4 at the red giant branch tip. The significance of this number is the following: in globular clusters with [Fe/H] < −1.0, at least observationally, no AGB stars are found above the red giant branch tip so that there is a large drop in the giant branch luminosity function at the RGB tip. On the other hand, at somewhat younger ages, there is approximately one AGB star for each three RGB stars in the vicinity of the RGB tip, so that the total giant branch luminosity function should drop by a factor of approximately 4 at the RGB tip in this case (e.g. Renzini 1992). The fact that the "halo" giant branch luminosity function for WLM does drop by a factor in excess of four then indicates that there is at least a component whose age is comparable to the galactic globular clusters. This is further supported by the fact that the stars above the RGB tip, if interpreted with the Bertelli et al. (1994) isochrones, have ages of approximately 10 Gyr (Minniti & Ziijlstra 1996). With HST it should be possible to reach fainter than the luminosity of the horizontal branch in this dIrr, so it will be very interesting to see if the horizontal branch morphology in the "halo" field studied by Minniti & Zijlstra (1996) conforms to that expected for a dominant old population.

5.2.3. NGC 6822

In the above results for two Local Group dIrrs, the authors have endeavoured to provide some quantitative information on the early star formation histories of these galaxies. But, in general, much more information can be gleaned from an observed c-m diagram if we are prepared to do more than simply overlay isochrones or fit standard globular cluster giant branches. An example of this full quantitative analysis approach is the recent work of Gallart et al. (1996a,b,c) on the Local Group dIrr NGC 6822. In these papers an attempt has been made to reproduce the observations by generating model c-m diagrams for various assumed star formation histories and chemical evolution scenarios. Fundamental to this approach is a complete and thorough knowledge of the photometric errors and completeness corrections that affect the observations. Determining these quantities to the requisite precision necessarily involves extensive tests with artificial stars and simulations of the data, for without such details this type of approach cannot be carried out. However, once these quantities are known, then model c-m diagrams can be appropriately convolved to simulate the observations. Comparisons can then be made with the real data and quantitative goodness-of-fit criteria used to assess the extent to which the assumptions underlying the model data are appropriate. The other requirement for this type of approach is a consistent set of isochrone and luminosity function calculations that cover all stages of stellar evolution — from the main sequence through the red giant branch to core-helium burning and beyond to AGB and post-AGB evolution. To be of use is this type of approach, the isochrone set must also cover a wide range of abundances. Fortunately, the Bertelli et al. (1994) isochrones fulfill all these requirements.

The basic procedure followed by Gallart et al. (1996b, hereafter GABC) is to assume: (a) an initial star formation epoch τ_i, (b) an initial abundance Z_i, (c) a star formation rate as a function of time, and (d) an enrichment law giving the abundance at any epoch. In GABC the enrichment law is taken to be a linear function: abundance (Z) increases linearly with time from Z_i at τ_i to the HII region abundance at the present-day. With these assumptions the isochrone database is then used to generate a model c-m diagram which is then convolved with the completeness corrections and error functions. For the particular case of NGC 6822, the effects of differential reddening across the field must also be included. This process is illustrated in Fig. 5 of GABC.

Not unexpectedly, it is the red part of the c-m diagram that contains information on the star formation history over Gyr timescales, while the blue side provides information on the current and recent star formation history. In this situation the "time resolution" is relatively good for the young populations (see Gallart et al. 1996c) but we expect the time resolution for the older populations to be quite coarse. It should be possible to tell, for example, the difference between $\tau_i = 5$ Gyr and $\tau_i = 10$ Gyr, but not between say $\tau_i = 12$ Gyr and $\tau_i = 15$ Gyr. For this reason the star formation histories investigated in GABC are relatively simple; the five shapes investigated are depicted in their Fig. 4. They basically cover the range from "more star formation early" through "constant star formation rate" to "more star formation later".

GABC isolate two regions in the red side of the c-m diagram for study. The first region, which they refer to as the *red-tangle*, is the region occupied by the red giant branch stars that are fainter than the RGB tip. They define three parameters: the median colour of the red giant branch ∼0.5 mag below the RGB tip, the "red-edge", the colour at which 95% of the red giants are bluer, and the ratio of the number of stars on the blue side to the number on the red side of the giant branch. All these parameters are sensitive to age and abundance — for example, at fixed abundance the median colour becomes bluer with decreasing age. The second region analyzed in the so-called *red-tail*, which is the region above the first giant branch tip where intermediate-age stars predominate. Again a number of parameters are defined which are sensitive to age and abundance. The ratio of the number of stars in the red-tail region to the number in the red-tangle region is also a significant parameter since it will increase, for example, as τ_i decreases. All these quantities can be measured on the synthetic c-m diagrams and on the real data and then compared, allowing a quantitative assessment of the degree of success of any particular model in reproducing the observations.

The basic results given in GABC can be summarized as follows: (a) star formation beginning in NGC 6822 as late as ∼6 Gyr ago can be ruled out. This is basically because the red-tangle becomes too blue for τ_i equal to or less than this value. The best agreement is in fact achieved for a star formation history that begins at early epochs (12 – 15 Gyr ago) from low metallicity gas. (b) From the initial epoch the star formation rate has been approximately uniform, but with a substantial decline in the last few Gyr. A recent "burst" however, is needed to generate the correct numbers of stars in the blue side of the c-m diagram (Gallart et al. 1996c). The results are best illustrated by the "population box" (cf. Hodge 1989) shown in Fig. 18, which is a reproduction of Fig. 21 of GABC. This star formation history is rather different from that for the LMC which had a increase in the star formation rate some 2 to 4 Gyr ago (in contrast to the decrease seen in NGC 6822 at approximately the same time!).

Are we to believe these NGC 6822 results? Fundamentally they are sound but there are two caveats that should be kept in mind. First, recall that they are completely dependent on the isochrone input: any changes in the underlying stellar evolution tracks, or in their conversion from the theoretical quantities M_{bol} and log T_{eff} to observed magnitudes and colours, will modify the results. Inferences made from the red-tail are especially susceptible in this respect because of the sensitivity of this part of the c-m diagram to the details of upper-AGB evolution, particularly as regards the mass-loss law adopted (see Appendix A of GABC for a discussion of this point). Second, in their analysis GABC used a linear enrichment law, whereas a more consistent calculation would couple the star formation and chemical enrichment histories. Their assumption, however, is unlikely to have effected the results in a major way though again it is important to note that the red-tail region has significant abundance sensitivity in addition to that for age.

The major bonus of this type of analysis, however, is that the best fit model can be used to generate explicit predictions for the form of the c-m diagram at fainter magnitudes, especially in the region of the core-helium-burning red clump or horizontal branch stars (see, for example, Aparicio et al. 1996). Hence observations that reach fainter magnitudes, such as with HST/WFPC2, can be used to further constrain the models. Unfortunately, NGC 6822 is not a very good object with which to carry out such a test — the substantial amount of galactic foreground star contamination would prevent the derivation of meaningful results from observations reaching fainter magnitudes.

5.3. *Phoenix and LGS 3*

Two other Local Group dIrr galaxies deserve some comment. These are the low luminosity systems Phoenix ($M_V \approx -10.0$) and LGS 3 ($M_V \approx -10.4$). Phoenix lies approximately 400 kpc from the Galaxy (van de Rydt et al. 1991) while LGS 3 is about 800 kpc away

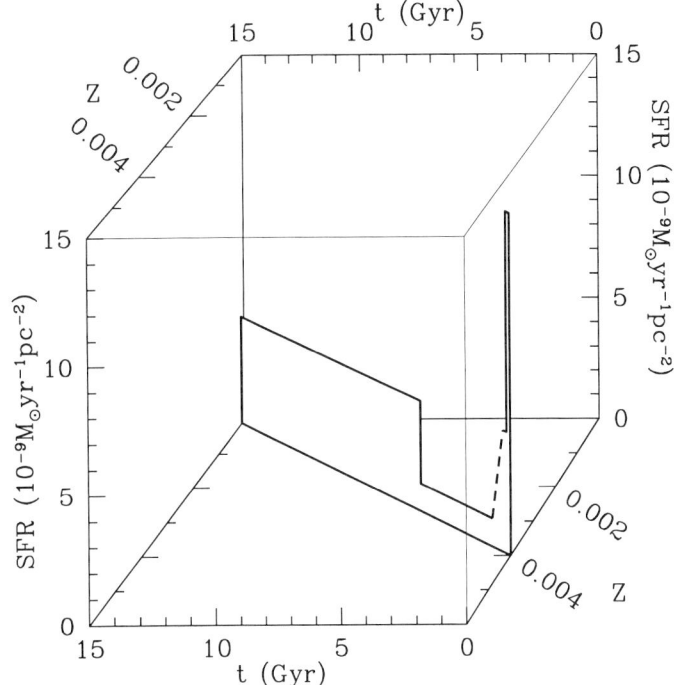

FIGURE 18. A "population box" describing the star formation history of the Local Group dIrr galaxy NGC 6822. The dashed line from 1 to 0.4 Gyr represents the only region where quantitative information has not been derived from the c-m diagram. From Gallart et al. (1996b).

(Lee 1995, Mould 1997). LGS 3 may in fact be a distant companion to either M31 or M33. In appearance both these galaxies lack any obvious signs of current star formation, such as HII regions, so that morphologically they would be classified as dSph systems. Yet both systems contain detectable amounts of HI gas (Phoenix: Carignan et al. 1991, but see also footnote 1 of Oosterloo et al. 1996; LGS 3: Lo et al. 1993). For LGS 3, using the HI mass from Lo et al. (1993), the M_V value from Lee (1995) and assuming B–V $\approx$ 0.7, one derives $M_{HI}/L_B \approx 0.2$ in solar units. Similarly for Phoenix, $M_{HI}/L_B \approx$ 0.1 using the Carignan et al. (1991) HI mass, which may be an overestimate (Oosterloo 1997, *priv. comm.*). Both these M_{HI}/L_B values are significantly smaller than M_{HI}/L_B values for other dIrrs, which typically have $M_{HI}/L_B \approx 1$ or larger. On the other hand, they are also significantly larger than the values (or upper limits) for Local Group dE and dSph systems, which are less than 10^{-2} to 10^{-3} in solar units. So in this sense both Phoenix and LGS 3 are "transition" objects between the gas-rich dIrrs and the gas-poor dSphs and dEs.

The c-m diagrams for both these systems also reveal similarities to both dIrr and dSph systems. For Phoenix, the observations of van de Rydt et al. (1991, see also Ortolani & Gratton 1988) reveal a c-m diagram that is dominated by apparently old, metal-poor red giants. There appears also to be a population of intermediate-age, upper-AGB stars, though the lack of a similar c-m diagram for a nearby field comparison region makes assessing the actual numbers of these stars difficult. Low resolution spectra have been obtained of some of these upper-AGB candidates (see Da Costa 1994) and two have been identified as carbon stars and thus definite Phoenix members. The spectra for the remaining three objects are not easily classified with this combination of low S/N and low resolution, though they are definitely not carbon stars. Their membership (or non-membership) of Phoenix remains uncertain. Based on the V and I band photometry of van de Rydt et al. (1991) and the precepts of Da Costa & Armandroff (1990), the two Phoenix carbon stars have $M_{bol} \approx -3.7$ for a Phoenix distance modulus of 23.1 (van de Rydt et al. 1991). These stars are thus considerably less luminous than the carbon stars in the 1 – 3 Gyr old star clusters in the LMC, in which the brightest carbon stars reach luminosities of at least $M_{bol} \approx -5$ (e.g. Frogel et al. 1990). Even in the SMC cluster Kron 3, which has an age of ~8 Gyr (e.g. Rich et al. 1984), the most luminous carbon star has $M_{bol} \approx -4.2$ (Frogel et al. 1990). These results then suggest that the carbon stars in Phoenix, and this dIrr's intermediate-age population in general, are not likely to have ages much less than 8 – 10 Gyr. There are of course caveats: the number of identified carbon stars is small and as a result it may be that the actual AGB-tip luminosity is brighter, implying a younger age. There may also be bolometrically brighter upper-AGB stars in this dIrr that have not yet been identified. Nevertheless, it seems probable that Phoenix did not make very many stars after the end, perhaps 8 – 10 Gyr ago, of its initial major star formation episode.

The LGS 3 c-m diagram (Lee 1995, Mould 1997) is similar — it also reveals a predominance of apparently old metal-poor red giants. As is the case for Phoenix, the mean abundance for LGS 3 determined from the mean colour of the red giant branch is such that LGS 3 would fall on the (mean abundance, luminosity) relation defined by the Local Group dSph and dE galaxies (e.g. Lee 1995). Further, according to Cook & Olszewski (1989), LGS 3 also contains a population of presumably intermediate-age carbon stars, though no information on the luminosities of these stars is given. Thus is these aspects LGS 3 and Phoenix are again similar to the Local Group dSphs and dEs.

But perhaps the most remarkable thing about Phoenix and LGS 3 is that they both contain modest populations of young blue stars that formed perhaps 150 Myr ago. This is illustrated in Fig. 19 (see Fig. 2 of Mould 1997 for the equivalent plot for LGS 3) where we see, in addition to the dominant red giants, a small population of blue main sequence stars. The ages of these stars is approximately 150 Myr for a Phoenix distance modulus of 23.1 and Mould (1997) suggests a similar age for the blue stars observed in LGS 3. In both dIrrs this young population is a minor component: van de Rydt et al. (1991) suggest that the young stars contribute ~10% of the total integrated light while Mould (1997) gives a similar luminosity fraction for LGS 3.

Thus we have two Local Group dIrr galaxies in which the dominant stellar population is apparently old, yet there are indications of possible intermediate-age populations together with a definite recently formed one. But what we would really like to know is the actual star formation histories in these dwarfs. For example, has the star formation (after some initial event that presumably formed the bulk of the stars) been episodic, like the Carina dSph, as a consequence of, for example, the interplay between heating and cooling of the gas? Or, has the on-going star formation been relatively uniform as it appears to have been in NGC 6822, for example. Or instead, do the younger stars in Phoenix and LGS 3 represent the first epoch of star formation since the end of the initial episode almost a Hubble time in the past? And, if so, what triggered the star formation after such a long dormant period? Ground-based data can provide little further information to answer these questions though it is worth noting that in Phoenix at least, not only is the centroid of the young population offset from the centroid of the old population, but also it occupies a smaller area (Ortolani & Gratton 1988). This is reminiscent of the

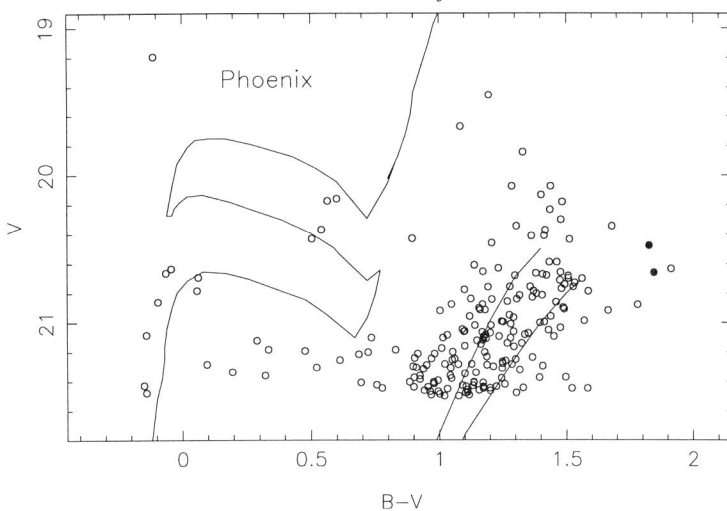

FIGURE 19. A colour-magnitude diagram for the Local Group dIrr Phoenix from the photometry listed in Table 2 of van de Rydt et al. (1991). The two red giants spectroscopically confirmed as carbon stars are shown as filled symbols. The giant branches are those of the globular clusters M92 and M3 plotted for $(m-M)_0 = 23.1$ and $E(B-V) = 0.02$ mag. The isochrone is taken from Bertelli et al. (1994) and is for an age of 158 Myr and an abundance Z=0.0004. It was plotted at the same modulus and reddening as the globular cluster giant branches and is shown to illustrate the presence of young stars in Phoenix.

situation for current star formation in many dIrrs — for example, Sextans A (Skillman, these lectures).

With HST/WFPC2, however, it should be possible to provide considerable new insight into this puzzle. In LGS 3 it should be possible to reach fainter than the horizontal branch whose morphology will provide some coarse age information (cf. Section 2). A limiting magnitude of this order would also correspond to the main sequence turnoff for a 2 – 3 Gyr old population, so the star formation history for ages less than this limit can be directly constrained. In Phoenix, which is much closer than LGS 3, HST/WFPC2 observations should be capable of reaching main sequence stars as faint as the turnoff in a ∼7 Gyr old population. Thus it is in principle possible to constrain the star formation history of this dIrr over much of its lifetime. HST observations of both these galaxies are scheduled for execution during 1997 and it will be very interesting to see what they reveal.

It would be remiss of me, however, if I did not end this section without emphasizing the striking contrast between LGS 3 and Phoenix on the one hand, and the isolated Local Group dE Tucana on the other. All three dwarfs have approximately the same luminosity, all are relatively isolated within the Local Group, and all have apparently old dominant stellar populations. Yet LGS 3 and Phoenix have retained modest amounts of HI over a Hubble time, despite evidence for significant star formation at early times, and have recently formed stars. Tucana, however, is gas free and apparently has not formed any stars since its initial episode. It is relatively easy to conceive of ways whereby gas can be lost from dwarf systems at early times (e.g. Dekel & Silk 1986). The more difficult question is how systems like Phoenix and LGS 3 retain small amounts of gas — there are no obvious mechanisms to prevent the gas from cooling and collapsing to form stars on timescales of a few $\times\ 10^8$ years (e.g. Lo et al. 1993).

One should also not overlook the apparent differences in Gyr-timescale star formation histories among the Local Group dIrr galaxies — we need only consider the difference between the LMC (major *increase* in the star formation rate ~2 – 4 Gyr ago) and NGC 6822 (similarly sized *decrease* in the star formation rate at approximately the same epoch) to realize that the situation for dIrrs is apparently at least as complicated as it is for the dE and dSph galaxies. Clearly we haven't yet solved all the problems revealed by observations of Local Group dwarf galaxies!

6. Dwarf galaxies beyond the Local Group

Dwarf galaxies, of course, are not restricted to the Local Group — they are the most common type of galaxy in the Universe! But this is a Winter School devoted to "Stellar Astrophysics" and so although undoubtedly you can learn much about the stellar population of a dwarf galaxy from observing its integrated colours or spectrum, I will again concentrate on those cases where you can resolve and study the individual stars. In such cases it is generally possible to gain substantially more, and often less ambiguous†, information. However, the restriction to studying the resolved populations in dwarf galaxies necessarily means that only relatively nearby systems can be investigated. Indeed, if we are particularly interested in the old populations within a dwarf galaxy, then the distance restriction is strong, even for observations with HST. Consequently, it is only the dwarfs in the nearby (D $\leq$ ~5 Mpc) groups, such as the Sculptor Group or the M81 Group, that can be studied in this detail. Nevertheless, such studies are important because they allow us to investigate the effect of environments different from that of the Local Group. For example, the M81 Group has clearly had a very different history than the Local Group — there is evidence for a number of substantial interactions in this group in the recent past (e.g. Solinger et al. 1977, Kennicutt et al. 1987). The M81 Group then has provided a very different environment for its dwarf galaxies compared to the relatively benign environment of the Local Group. In the next section I will concentrate on a particular example involving the M81 Group to illustrate the type of studies that can be performed with the resolved stellar populations of dwarf galaxies that lie beyond the Local Group. But many other examples, such as Heisler et al. (1997), Hunter et al. (1996), Tolstoy (1996), Hunter & Thronson (1995), and Greggio et al. (1993), to name but a few targeted at dIrr galaxies, also exist.

6.1. *Low luminosity dE galaxies in the M81 Group*

The M81 Group has been known for some time to contain a large number of dwarf galaxies. For example, Karachentseva et al. (1985) present large scale photographs of more than 30 dwarf galaxy candidate group members of a variety of morphological types. However, such surveys are limited by their use of photographic materials: despite the relative proximity of the M81 Group (distance modulus $\approx$ 27.8, Freedman et al. 1994), surveys of this type are generally unable to detect dE galaxies as faint as, or with surface brightnesses as low as, most Local Group dSph galaxies.

To overcome this situation, my colleagues and I have conducted a new survey of the M81 group using a CCD camera mounted on the Burrell Schmidt telescope at Kitt Peak (Armandroff et al. 1996; see also Caldwell 1997). Each CCD frame covers approximately one square degree with a scale of 2″ per pixel. We have imaged approximately 40 square degrees around the trio of luminous galaxies made up of M81, M82 and NGC 3077. The equivalent radial distance from this trio is ~180 kpc, a volume which, if centered on the Galaxy, would contain all the galactic dSph companions except for the distant Leo

---

† In interpreting integrated spectra, for example, it is often impossible to separate the effects of age from those of metallicity (e.g. Worthey 1994).

systems that lie beyond 200 kpc from the Galaxy. Further, the sensitivity of the CCD survey is such that, were we observing the Galaxy from M81, we would detect all the galactic dSphs in the survey volume with the possible exception of the very extended (and close to the Galaxy) Sagittarius dSph. Not surprisingly, in the CCD data we recover all the previously catalogued M81 group dwarfs in the survey area. In addition, we have found at least 4 previously unknown extremely low surface brightness candidate dE galaxies, systems which are too faint to be detected on photographic plates. The actual number of new candidate dEs is somewhat uncertain since it is necessary to image each object with a larger telescope, seeking (incipient) resolution into stars, in order to be certain that it is a dE galaxy rather than a clump in the "galactic cirrus" that pervades this area of sky (e.g. Sandage 1976).

These faint, low surface brightness objects are difficult to study from the ground but with HST we can measure the magnitudes and colours of stars as faint as two or three magnitudes below the RGB tip, which occurs at I $\approx$ 24.0 at the distance of the M81 Group. At present, we have HST observations for two M81 Group dSph galaxies (Armandroff et al. 1996; see also Caldwell 1997). The first is BK5N, which with $M_V \approx$ –11 (determined from the Schmidt CCD images) is one of the lowest luminosity dE galaxies in the Karachentseva et al. (1985) list. BK5N appears to be analogous to the Local Group dSphs. For example, it falls on the central surface brightness – absolute magnitude relation defined by the Local Group dSph and dE galaxies. The second M81 Group galaxy studied is one of the newly discovered systems, known as F8D1. Although its central surface brightness is in fact somewhat lower than that of BK5N, F8D1 is much larger with an exponential scale length almost 5 times that of BK5N. Consequently, F8D1 is considerably more luminous than BK5N. Indeed with $M_V \approx$ –14.5, F8D1 is brighter than Fornax, the most luminous dSph in the Local Group (with the possible exception of Sagittarius), and it is among the most luminous dwarfs in the M81 Group. It appears that F8D1 is a similar object to the large low surface brightness dwarfs that have been found in the Virgo cluster by Impey et al. (1988), and in the Fornax cluster by Bothun et al. (1991). Given the large numbers of dEs in both Virgo and Fornax, these large low surface brightness dwarfs are in practice relatively rare objects in both galaxy clusters. Whether this is also true for F8D1 and the M81 group remains to be determined. These results are illustrated in Fig. 20 in which the central surface brightness – absolute magnitude relation for dSph and dE galaxies, now including objects beyond the Local Group (cf. Fig. 8), is shown.

With HST we can resolve and measure stars on the red giant branch that lie within two or three magnitudes of the RGB tip in these dwarfs. Aside from intrinsic interest, is there a particular reason why this should be an interesting observation to carry-out? The answer is most definitely "yes". We saw in Fig. 7 (Section 4.1) that the mean abundance of the stars in a dSph or dE galaxy and its total luminosity are correlated, at least for the Local Group systems. In addition, surface brightness measures such as the central value or the mean value inside the radius containing half the total light are also correlated with luminosity for Local Group systems (cf. Fig. 8). Hence, there is also a correlation between mean abundance and surface brightness for the Local Group dSph and dE galaxies. As a result, we then have no way of telling whether total luminosity, a *global* parameter related to the total (baryonic) mass in the dwarf, or surface brightness, a *local* parameter related to the local (baryonic) density, is the fundamental parameter determining the chemical history and thus the mean abundance. Arguments to support either of these quantities can be found in the literature — e.g. Franx & Illingworth (1990) and Phillipps et al. (1990) for local density or Dekel & Silk (1986) and Bender et al. (1993)

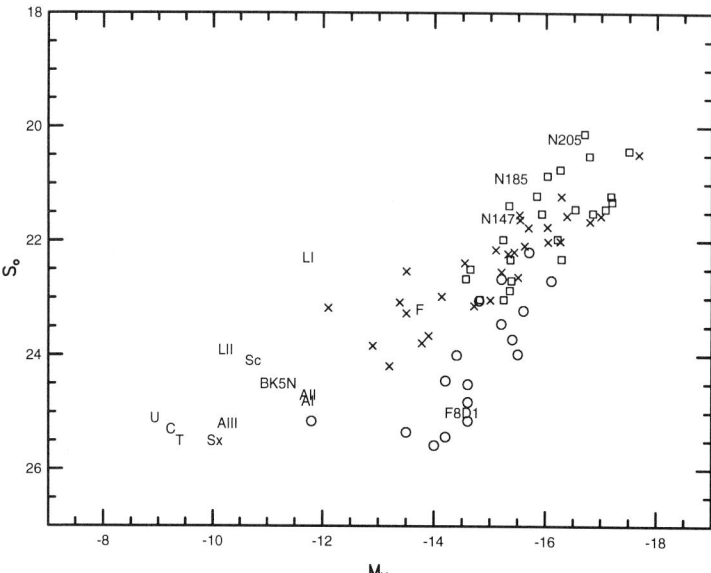

FIGURE 20. The central surface brightness – absolute magnitude relation for dE and dSph galaxies from Caldwell (1997). Local Group galaxies are shown with abbreviated names while crosses and open squares represent dEs in Virgo and Fornax. The open circles are the large low surface brightness galaxies found in Virgo by Impey et al. (1988). The location of the M81 Group dEs BK5N and F8D1 are shown by their names. Note the similarity of BK5N to the Local Group dSphs and that of F8D1 to the Virgo large low surface brightness galaxies.

for total mass. But these two M81 dwarfs, with their similar surface brightnesses but very different luminosities, offer a way to resolve this issue.

Figure 21 shows the c-m diagram for BK5N derived from HST/WFPC2 observations obtained through the F555W and F814W filters. Superposed on the data are giant branches for the standard globular clusters M15 ([Fe/H] = −2.17), M2 ([Fe/H] = −1.58) and NGC 1851 ([Fe/H] = −1.29) from Da Costa & Armandroff (1990). These giant branches have been fitted for a distance modulus of $(m-M)_0 = 27.9$ (derived from the RGB tip luminosity), a value consistent with BK5N's membership of the M81 group. The inferred mean metal abundance for this dwarf is <[Fe/H]> = −1.7 ± 0.3 dex. It is also apparent from Fig. 21 that BK5N contains a small population of stars with luminosities exceeding that of the RGB tip; i.e. like many of the Local Group dEs and dSphs, BK5N contains an intermediate-age population. Comparison with the isochrones of Bertelli et al. (1994) suggests these stars may be as young as perhaps 8 to 10 Gyr though the limited sample size makes assigning an age an uncertain process. There are, however, no indications in Fig. 21 of any bright blue stars similar to those seen in the Local Group systems Phoenix (cf. Fig. 19) and LGS 3; i.e. no stars have formed in BK5N within the last few hundred Myr or so.

In Fig. 22 we see the equivalent c-m diagram for F8D1. Because this galaxy is much larger than BK5N, there are many more stars on the WFPC2 images (BK5N is contained almost completely on the PC frame only). The same giant branches are overplotted with the addition of that for 47 Tuc ([Fe/H] = −0.71). The distance modulus derived from the RGB tip luminosity is $(m-M)_0 = 28.0$, again consistent with M81 Group membership. Further, the derived mean abundance is <[Fe/H]> = −1.0 ± 0.2, or approximately 0.7 dex higher than that of BK5N. As for BK5N there are strong indications of an intermediate-age population in F8D1 with stars perhaps as young as 4 to 5 Gyr present. Further,

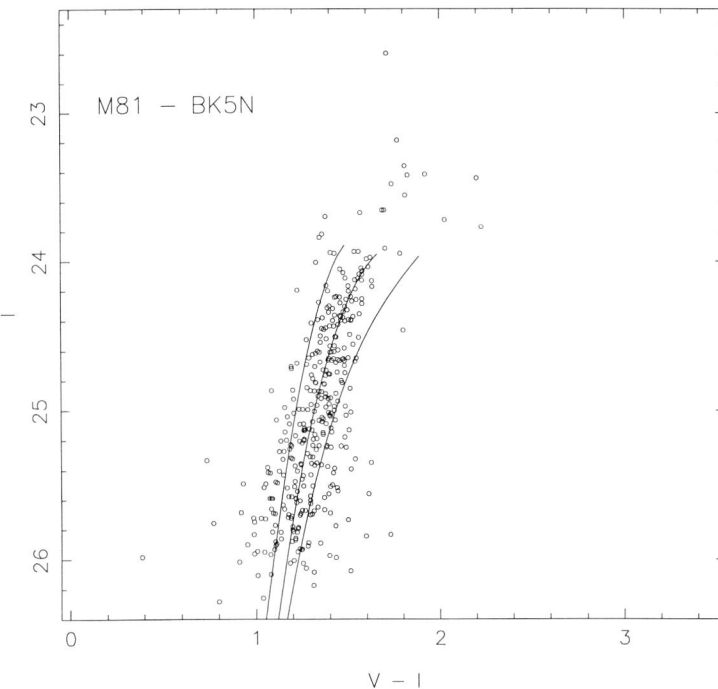

FIGURE 21. The colour-magnitude diagram for the M81 Group dE BK5N from the HST/WFPC2 observations of Armandroff et al. (1996). The standard globular cluster giant branches for M15 ([Fe/H] = -2.17), M2 ([Fe/H] = -1.58) and NGC 1851 ([Fe/H] = -1.29) have been fitted using $(m-M)_0 = 27.9$ and $E(V-I) = 0.065$ mag.

again like BK5N, there are no indications of any young population (age less than a few hundred Myr). At the present time nothing is known about the possible HI content of F8D1 though one can infer from Huchtmeier & Skillman (1994) that HI in BK5N was not detected at the level of $\sim 1 \times 10^6$ solar masses. This implies $M_{HI}/L_B < 0.5$ for BK5N which is not a very strong limit for a dE system.

With these mean metallicities we can now add both galaxies to the mean abundance – luminosity and mean abundance – surface brightness diagrams. This is shown in Fig 23. Not surprisingly, BK5N falls among the Local Group dSph and dE galaxies as it did in Fig. 20. Clearly this M81 group dE closely resembles the Local Group systems. F8D1, however, falls close to the Local Group relation in the (M_V, <[Fe/H]>) plane but lies well off the (surface brightness, <[Fe/H]>) relation defined by the Local Group systems. From these results it is evident that total luminosity is apparently a much better predictor of mean abundance than is surface brightness. Further, if F8D1 is typical of the large low surface brightness dwarf galaxy class, then there is little likelihood of any significant surface brightness – mean abundance relation. In physical terms then, it seems that the larger, more luminous and therefore more massive (at least as regards baryonic mass) dwarf was better able to enrich itself than the smaller less massive (baryonically) system, despite their similar (baryonic) mass density. Presumably F8D1 has a higher velocity dispersion than BK5N in order to support its larger size. These results are clearly of considerable importance for theories of dwarf galaxy formation and evolution. It is also interesting to note that the now known to be fundamental relation shown in the left panel of Fig. 23, is apparently not very dependent on star formation history. As discussed in section 4, the Local Group dSph and dE galaxies have had a wide variety

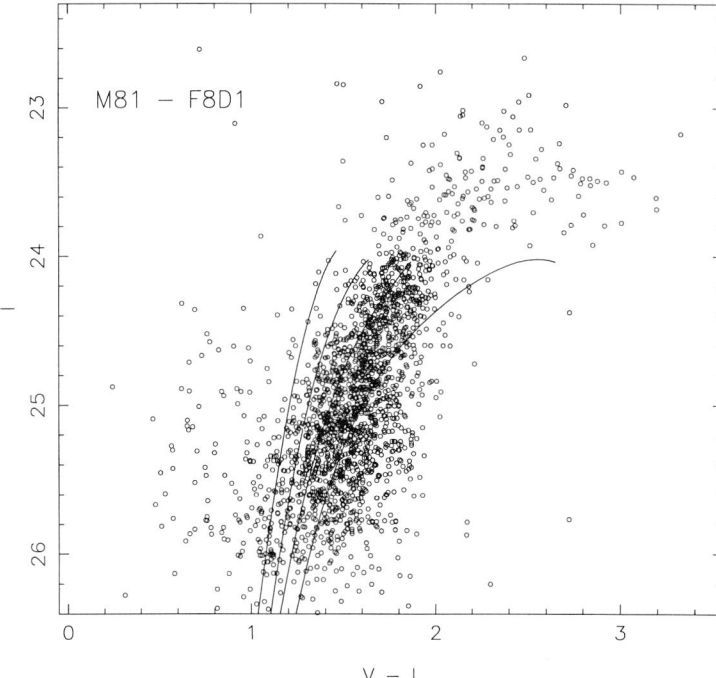

FIGURE 22. The colour-magnitude diagram for the M81 Group dE F8D1 from the HST/WFPC2 observations of Armandroff et al. (1996). The standard globular cluster giant branches for M15 ([Fe/H] = −2.17), M2 ([Fe/H] = −1.58), NGC 1851 ([Fe/H] = −1.29) and 47 Tuc ([Fe/H] = −0.71) have been fitted using $(m–M)_0 = 28.0$ and $E(V–I) = 0.043$ mag.

of star formation histories, yet the accompanying chemical evolution must have been such that the (baryonic) mass – metallicity relation implied by Fig. 23 is still followed. Similarly, the relation does not seem to be strongly influenced by environment, since there is no distinction between the Local Group dwarfs (companions to M31, companions to the Galaxy, and Tucana) and the M81 Group dwarfs. Again these are constraints that detailed formation and evolution models must obey.

One further point concerning these M81 Group dwarf galaxies deserves mention. On the WFPC2 images of F8D1, there is at least one apparently *bona fide* globular cluster that belongs to this dwarf. There may well be more such clusters since the WFPC2 images cover only a fraction of the total extent of F8D1. On the other hand, essentially all of BK5N is contained on the WFPC2 images and there are no globular clusters present. As discussed in Section 4, within the Local Group the frequency of occurrence of globular cluster systems in dSph and dE galaxies increases with luminosity, *and* with central surface brightness since these quantities are correlated for the Local Group systems. For example, the Fornax and Sagittarius dSphs both possess a few globular clusters, but the less luminous, lower surface brightness dwarfs do not. Then, as for the $(M_V, <[Fe/H]>)$ and (surface brightness, $<[Fe/H]>$) relations, the presence of a globular cluster system in the luminous M81 Group dwarf F8D1, but not in the less luminous dwarf BK5N which has a similar central surface brightness to F8D1, suggests strongly that again it is total (baryonic) mass rather than density that determines the occurrence of globular cluster systems. This is an intriguing, if somewhat unexpected, result given the density contrast between the globular clusters and their parent systems. This contrast is particularly marked for F8D1 which becomes the lowest (baryonic) density galaxy in which a globular

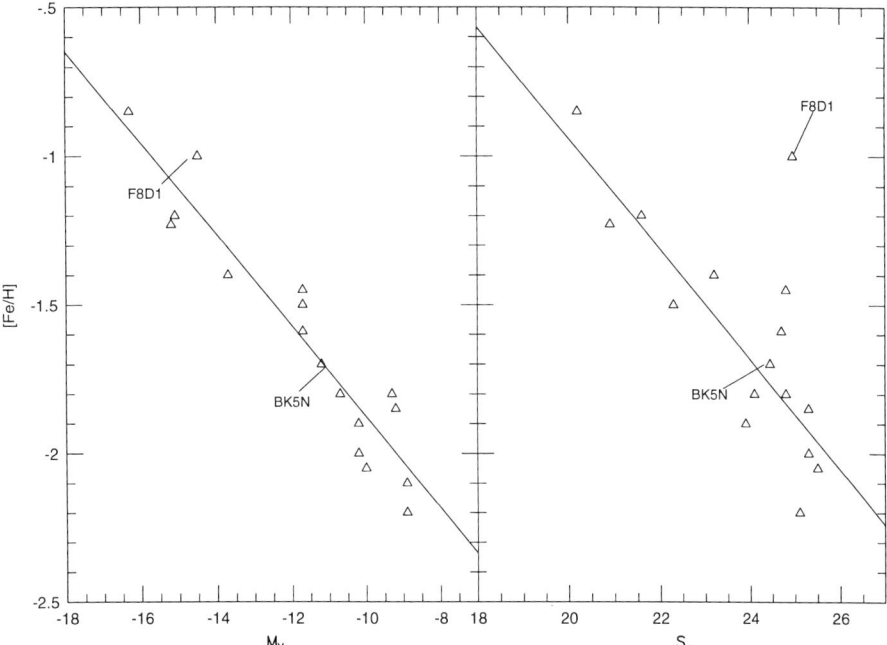

FIGURE 23. Relations between absolute magnitude and mean abundance (*left panel*) and between central surface brightness and mean abundance (*right panel*) for Local Group dSph and dE galaxies together with the new results for the M81 Group dwarfs BK5N and F8D1. From Caldwell (1997).

cluster has formed (the Sagittarius dSph has presumably been distended by its interaction with the Milky Way).

7. Summary

If I have carried out the task set for me by the Winter School organizers then you will have gained from these lectures an idea of the current state of research on the stellar populations of dwarf galaxies. It is an active field and one where we are clearly still a long way from a complete understanding of these supposedly simple systems. Thus there are a number of opportunities, both observationally and theoretically, for newcomers to make important contributions to the field. I urge you also to keep in mind the underlying theme of the Winter School — nearby dwarfs, be they in the Local or neighbouring Groups, are important as stepping stones to the wider Universe. The variety of star formation histories that are seen locally, for example, will have directly visible counterparts when we look back in time at distant systems.

I would like to place on record my thanks to the organizers of the Winter School, Antonio Aparicio and Artemio Herrero, for inviting me to participate in what was an enjoyable experience. I would also like to thank the other lecturers and particularly the students for their interest and attention. Thanks also to Campbell Warden for helping to secure flights from and to Australia that were (relatively) convenient, and to my colleagues and collaborators for allowing the use of material that is either unpublished or in preprint form only.

REFERENCES

Aaronson, M., Gordon, G., Mould, J., Olszewski, E., & Suntzeff, N. 1985, *ApJ*, **296**, L7.

Ables, H.D., & Ables, P.G. 1977, *ApJS*, **34**, 245.

Alard, C. 1996, *ApJ*, **458**, L17.

Alcock, C., Allsman, R.A., Alves, D.R., Axelrod, T.S., Becker, A.C., Bennett, D.P., Cook, K.H., Freeman, K.C., Griest, K., Guern, J.A., Lehner, M.J., Marshall, S.L., Minniti, D., Peterson, B.A., Pratt, M.R., Quinn, P.J., Rodgers, A.W., Stubbs, C.W., Sutherland, S., & Welch, D.L. 1997, *ApJ*, **474**, 217.

Alcock, C., Allsman, R.A., Axelrod, T.S., Bennett, D.P., Cook, K.H., Freeman, K.C., Griest, K., Marshall, S.L., Peterson, B.A., Pratt, M.R., Quinn, P.J., Rodgers, A.W., Stubbs, C.W., Sutherland, S., & Welch, D.L. 1996, *AJ*, **111**, 1146.

Aparicio, A., Gallart, C., Chiosi, C., & Bertelli, G. 1996, *ApJ*, **469**, L97.

Armandroff, T.E. 1994, In *ESO/OHP Workshop on Dwarf Galaxies* (ed. G. Meylan & P. Prugniel). p. 211. ESO.

Armandroff, T.E., Caldwell, N., Da Costa, G.S., & Seitzer, P. 1996, *BAAS*, **28**, 1319.

Armandroff, T.E., Da Costa, G.S., Caldwell, N., & Seitzer, P. 1993, *AJ*, **106**, 986.

Ashman, K.M., & Bird, C.M. 1993, *AJ*, **106**, 2281.

Azzopardi, M. 1994, In *The Local Group: Comparative and Global Properties* (ed. A. Layden, R.C. Smith & J. Storm). p. 129. ESO.

Baade, W. 1944a, *ApJ*, **100**, 137.

Baade, W. 1944a, *ApJ*, **100**, 147.

Baade, W. 1951, *Publ. Univ. Mich. Obs.*, **10**, 7.

Beauchamp, D., Hardy, E., Suntzeff, N.B., & Zinn, R. 1995, *AJ*, **109**, 1628.

Bender, R., Burstein, D., & Faber, S.M. 1993, *ApJ*, **411**, 153.

Bender, R., Paquet, A., & Nieto, J.-L. 1991, *A&A*, **246**, 349.

Bertelli, G., Bressan, A., Chiosi, C., Fagotto, F., & Nasi, E. 1994, *A&AS*, **106**, 275.

Bertelli, G., Mateo, M., Chiosi, C., & Bressan, A. 1992, *ApJ*, **388**, 400.

Bica, E., Alloin, D., & Schmidt, A.A. 1990, *A&A*, **228**, 23.

Binggeli, B. 1986, In *Star-Forming Dwarf Galaxies and Related Objects* (ed. D. Kunth, T. X. Thuan & J. T. T. Van). p. 52. Editions Frontieres.

Binggeli, B. 1994, In *ESO/OHP Workshop on Dwarf Galaxies* (ed. G. Meylan & P. Prugniel). p. 13. ESO.

Bothun, G., Impey, C., & Malin, D. 1991, *ApJ*, **376**, 404.

Buonanno, R., Corsi, C.E., Fusi Pecci, F. 1981, *MNRAS*, **196**, 435.

Buonanno, R., Corsi, C.E., Fusi Pecci, F., Richer, H.B., & Fahlman, G.G. 1995a, *AJ*, **109**, 650.

Buonanno, R., Corsi, C.E., Pulone, L., Fusi Pecci, F., Richer, H.B., & Fahlman, G.G. 1995b, *AJ*, **109**, 663.

Caldwell, N. 1997, In *The Second Stromlo Symposium: The Nature of Elliptical Galaxies* (ed. M. Arnaboldi, G. S. Da Costa & P. Saha). ASP Conf. Series, vol. 116, p. 249. ASP.

Caldwell, N., Armandroff, T.E., Seitzer, P., & Da Costa, G.S. 1992, *AJ*, **103**, 840.

Carignan, C., & Beaulieu, S. 1989, *ApJ*, **347**, 760.

Carignan, C., Demers, S., & Cote, S. 1991, *ApJ*, **381**, L13.

Carignan, C., & Freeman, K.C. 1988, *ApJ*, **332**, L33.

Chiosi, C., Bertelli, G., & Bressan, A. 1992, *ARA&A*, **30**, 235.

Cook, K.H., & Olszewski, E. 1989, *BAAS*, **21**, 775.

Corwin, H.G., de Vaucouleurs, G., & de Vaucouleurs, A. 1985, *Southern Galaxy Catalog*. Univ. of Texas.

CÔTÉ, S., FREEMAN, K., & CARIGNAN, C. 1994, In *ESO/OHP Workshop on Dwarf Galaxies* (ed. G. Meylan & P. Prugniel). p. 101. ESO.

CUDWORTH, K.M., OLSZEWSKI, E.W., & SCHOMMER, R.A. 1986, *AJ*, **92**, 766.

DA COSTA, G.S. 1993, In *The Globular Cluster – Galaxy Connection* (ed. G.H. Smith & J.P. Brodie). ASP Conf. Series, vol. 48, p. 363. ASP.

DA COSTA, G.S. 1994, In *ESO/OHP Workshop on Dwarf Galaxies* (ed. G. Meylan & P. Prugniel). p. 221. ESO.

DA COSTA, G.S. 1997, In *The Second Stromlo Symposium: The Nature of Elliptical Galaxies* (ed. M. Arnaboldi, G. S. Da Costa & P. Saha). ASP Conf. Series, vol. 116, p. 270. ASP.

DA COSTA, G.S., & ARMANDROFF, T.E. 1990, *AJ*, **100**, 162.

DA COSTA, G.S., & ARMANDROFF, T.E. 1995, *AJ*, **109**, 2533.

DA COSTA, G.S., ARMANDROFF, T.E., CALDWELL, N., & SEITZER, P. 1996, *AJ*, **112**, 2576.

DA COSTA, G.S., & MOULD, J.R. 1988, *ApJ*, **334**, 159.

DAVIDGE, T.J. 1992, *ApJ*, **397**, 457.

DAVIDGE, T.J. 1994, *AJ*, **108**, 2123.

DAVIDGE, T.J., & JONES, J.H. 1992, *AJ*, **104**, 1365.

DAVIDGE, T.J., & NIETO, J.-N. 1992, *ApJ*, **391**, L13.

DEKEL, A., & SILK, J. 1986, *ApJ*, **303**, 39.

DEMERS, S., IRWIN, M.J., & GAMBU, I. 1994, *MNRAS*, **266**, 7.

ELSTON, R., & SILVA, D.R. 1992, *AJ*, **104**, 1360.

EGGLETON, P.P. 1968, *MNRAS*, **140**, 387.

FAHLMAN, G.G., MANDUSHEV, G., RICHER, H.B., THOMPSON, I., & SIVARAMAKRISHNAN, A. 1996, *ApJ*, **459**, L65.

FERGUSON, H.C. 1989, *AJ*, **98**, 367.

FRANX, M., & ILLINGWORTH, G. 1990, *ApJ*, **359**, L41.

FREEDMAN, W.L. 1988, *AJ*, **96**, 1248.

FREEDMAN, W.L. 1989, *AJ*, **98**, 1285.

FREEDMAN, W.L. 1992, *AJ*, **104**, 1349.

FREEDMAN, W.L. 1994, In *The Local Group: Comparative and Global Properties* (ed. A. Layden, R.C. Smith & J. Storm). p. 227. ESO.

FREEDMAN, W.L. 1995, In *Stellar Populations*, IAU Symp. 164 (ed. P.C. van der Kruit & G. Gilmore). p. 165. Kluwer.

FREEDMAN, W.L., HUGHES, S.M., MADORE, B.F., MOULD, J.R., LEE, M.G., STETSON, P., KENNICUTT, R.C., TURNER, A., FERRARESE, L., FORD, H., GRAHAM, J.A., HILL, R., HOESSEL, J.G., HUCHRA, J., & ILLINGWORTH, G.D. 1994, *ApJ*, **427**, 628.

FROGEL, J.A., & ELIAS, J.H. 1988, *ApJ*, **324**, 823.

FROGEL, J.A., MOULD, J.R., & BLANCO, V.M. 1990, *ApJ*, **352**, 96.

FROGEL, J.A., & WHITFORD, A.E. 1987, *ApJ*, **320**, 199.

GALLAGHER, J.S., & HUNTER, D.A. 1981, *AJ*, **86**, 1312.

GALLAGHER, J.S., HUNTER, D.A., & MOULD, J. 1984, *ApJ*, **281**, L63.

GALLART, C., APARICIO, A., BERTELLI, G., & CHIOSI, C. 1996b, *AJ*, **112**, 1950. [GABC]

GALLART, C., APARICIO, A., BERTELLI, G., & CHIOSI, C. 1996c, *AJ*, **112**, 2596.

GALLART, C., APARICIO, A., & VILCHEZ, J.M. 1996a, *AJ*, **112**, 1928.

GARDINER, L.T., & HATZIDIMITRIOU, D. 1992, *MNRAS*, **257**, 195.

GRAHAM, J.A. 1975, *PASP*, **87**, 641.

GREEN, E.M., & NORRIS, J.E. 1990, *ApJ*, **353**, L17.

GREGGIO, L., MARCONI, G., TOSI, M., & FOCARDI, P. 1993, *AJ*, **105**, 894.

GRILLMAIR, C.J., LAUER, T.R., WORTHEY, G., FABER, S.M., FREEDMAN, W.L., MADORE,

B.F., AJHAR, E.A., BAUM, W.A., HOLTZMAN, J.A., LYNDS, C.R., O'NEIL, E.J.,JR., & STETSON, P.B. 1996, *AJ*, **112**, 1975.

GRILLMAIR, C.J., MOULD, J.R., WORTHEY, G., HOLTZMAN, J.A., TRAUGER, J., BALLESTER, G.E., BURROWS, C., CLARKE, J., CRISP, D., GALLAGHER, J.S., GRIFFITHS, R., HESTER, J., HOESSEL, J.G., KRIST, J., SCOWEN, P., STAPELFELDT, K., WATSON, A., & WESTPHAL, J. 1997, In *The Second Stromlo Symposium: The Nature of Elliptical Galaxies* (ed. M. Arnaboldi, G. S. Da Costa & P. Saha). ASP Conf. Series, vol. 116, p. 306. ASP.

GUARNIERI, M.D., RENZINI, A., & ORTOLANI, S. 1997, *ApJ*, **477**, L21.

HAN, M., HOESSEL, J.G., GALLAGHER, J.S., HOLTZMAN, J., STETSON, P.B., TRAUGER, J., BALLESTER, G.E., BURROWS, C., CLARKE, J., CRISP, D., GRIFFITHS, R., GRILLMAIR, C., HESTER, J., KRIST, J., MOULD, J.R., SCOWEN, P., STAPELFELDT, K., WATSON, A., & WESTPHAL, J. 1997, *AJ*, **113**, 1001.

HARDY, E., COUTURE, J., COUTURE, C., & JONCAS, G. 1994, *AJ*, **107**, 195.

HARRIS, W.E. 1991, *ARA&A*, **29**, 543.

HEISLER, C.A., HILL, T.A., MCCALL, M.L., & HUNSTEAD, R.W. 1997, *MNRAS*, **285**, 374.

HODGE, P.W. 1963, *AJ*, **68**, 691.

HODGE, P.W. 1973, *ApJ*, **182**, 671.

HODGE, P.W. 1989, *ARA&A*, **27**, 139.

HUCHTMEIER, W.K., & SKILLMAN, E.D. 1994, In *ESO/OHP Workshop on Dwarf Galaxies* (ed. G. Meylan & P. Prugniel). p. 299. ESO.

HUNTER, D.A., & GALLAGHER, J.S. 1985, *ApJS*, **58**, 533.

HUNTER, D.A., LYNDS, R., TOLSTOY, E., & O'NEIL, E. 1996, *BAAS*, **28**, 1319.

HUNTER, D.A., & THRONSON, H.A., JR. 1995, *ApJ*, **452**, 238.

IBATA, R.A., GILMORE, G., & IRWIN, M.J. 1994, *Nature*, **370**, 194.

IBATA, R.A., GILMORE, G., & IRWIN, M.J. 1995, *MNRAS*, **277**, 781.

IMPEY, C., BOTHUN, G., & MALIN, D. 1988, *ApJ*, **330**, 634.

IRWIN, M.J., BUNCLARK, P.S., BRIDGELAND, M.T., & MCMAHON, R.G. 1990, *MNRAS*, **244**, 16P.

JONES, J.H. 1993, *AJ*, **105**, 933.

KARACHENTSEVA, V.E., KARACHENTSEV, I.D., & BORNGEN, F. 1985, *A&AS*, **60**, 213.

KENNICUTT, R., KEEL, W., VAN DER HULST, T., HUMMEL, E., & ROETTIGER, K. 1987, *AJ*, **93**, 104.

KING, I.R. 1966, *AJ*, **71**, 64.

KNAPP, G., KERR, F., & BOWERS, P. 1978, *AJ*, **83**, 360.

KÖNIG, C.H.B., NEMEC, J.M., MOULD, J.R., & FAHLMAN, G.G. 1993, *AJ*, **106**, 1819.

KORMENDY, J. 1985, *ApJ*, **295**, 73.

KORMENDY, J., & BENDER, R. 1994, In *ESO/OHP Workshop on Dwarf Galaxies* (ed. G. Meylan & P. Prugniel). p. 161. ESO.

LAVERY, R.J., & MIGHELL, K.J. 1992, *AJ*, **103**, 81.

LEE, M.G. 1995, *AJ*, **110**, 1129.

LEE, M.G. 1996, *AJ*, **112**, 1438.

LEE, M.G., FREEDMAN, W.L., & MADORE, B.F. 1993a, *ApJ*, **417**, 553.

LEE, M.G., FREEDMAN, W.L., & MADORE, B.F. 1993b, *AJ*, **106**, 964.

LEE, M.G., FREEDMAN, W.L., MATEO, M., THOMPSON, I., ROTH, M., & RUIZ, M.-T. 1993c, *AJ*, **106**, 1420.

LEE, Y.-W., DEMARQUE, P., & ZINN, R. 1990, *ApJ*, **350**, 155.

LEE, Y.-W., DEMARQUE, P., & ZINN, R. 1994, *ApJ*, **423**, 248.

LO, K.Y., SARGENT, W.L.W., & YOUNG, K. 1993, *AJ*, **106**, 507.

LYNDEN-BELL, D., & LYNDEN-BELL, R.M. 1995, *MNRAS*, **275**, 429.

MARIGO, P., BRESSAN, A., & CHOISI, C. 1996, A&A, **313**, 545.

MARTINEZ-DELGADO, D., & APARICIO, A. 1997, ApJ, **480**, L107.

MARTINEZ-DELGADO, D., APARICIO, A., & GALLART, C. 1997, In *The Second Stromlo Symposium: The Nature of Elliptical Galaxies* (ed. M. Arnaboldi, G. S. Da Costa & P. Saha). ASP Conf. Series, vol. 116, p. 302. ASP.

MATEO, M. 1997, In *The Second Stromlo Symposium: The Nature of Elliptical Galaxies* (ed. M. Arnaboldi, G. S. Da Costa & P. Saha). ASP Conf. Series, vol. 116, p. 259. ASP.

MATEO, M., KUBIAK, M., SZYMAŃSKI, M., KALUZNY, J., KRZEMIŃSKI, W., & UDALSKI, A. 1995b, AJ, **110**, 1141.

MATEO, M., OLSZEWSKI, E.W., LEE, M.G., SAHA, A., HODGE, P., KEANE, M., SUNTZEFF, N., FREEDMAN, W.L., & THOMPSON, I. 1994, BAAS, **26**, 1395.

MATEO, M., UDALSKI, A., SZYMAŃSKI, M., KALUZNY, J., KUBIAK, M. & KRZEMIŃSKI, W. 1995a, AJ, **109**, 588.

MIGHELL, K.J. 1990, A&AS, **82**, 1.

MIGHELL, K.J., & RICH, R.M. 1996, AJ, **111**, 777.

MINNITI, D., & ZIJLSTRA, A.A. 1996, ApJ, **467**, L13.

MOULD, J. 1997, PASP, **109**, 125.

MOULD, J., & AARONSON, M. 1983, ApJ, **273**, 530.

MOULD, J., BOTHUN, G.D., HALL, P.J., STAVELY-SMITH, L., & WRIGHT, A.E. 1990, ApJ, **362**, L55.

MOULD, J., & KRISTIAN, J. 1990, ApJ, **354**, 438.

MOULD, J.R., KRISTIAN, J., & DA COSTA, G.S. 1983, ApJ, **270**, 471.

MOULD, J., KRISTIAN, J., & DA COSTA, G.S. 1984, ApJ, **278**, 575.

NEMEC, J.M., & COHEN, J.G. 1989, ApJ, **336**, 780.

O'CONNELL, R.W. 1980, ApJ, **236**, 430.

OLSZEWSKI, E.W., & AARONSON, M. 1985, AJ, **90**, 2221.

OOSTERLOO, T., DA COSTA, G.S., & STAVELEY-SMITH, L. 1996, AJ, **112**, 1969.

ORTOLANI, S., & GRATTON, R.G. 1988, PASP, **100**, 1405.

ORTOLANI, S., & GRATTON, R. 1990, A&AS, **82**, 71.

PHILLIPPS, S., EDMUNDS, M.G., & DAVIES, J.I. 1990, MNRAS, **244**, 168.

PRUGNIEL, P. 1994, In *ESO/OHP Workshop on Dwarf Galaxies* (ed. G. Meylan & P. Prugniel). p. 171. ESO.

RENZINI, A. 1992, In *The Stellar Populations of Galaxies*, IAU Symp. 149 (ed. B. Barbuy & A. Renzini). p. 325. Kluwer.

RICH, R.M., DA COSTA, G.S., & MOULD, J.R. 1984, ApJ, **286**, 517.

RICHER, H.B., CRABTREE, D.R., & PRITCHET, C.J. 1984, ApJ, **287**, 138.

RICHER, H.B., HARRIS, W.E., FAHLMANN, G.G., BELL, R.A., BOND, H.E., HESSER, J.E., HOLLAND, S., PRYOR, C., STETSON, P.B., VANDENBERG, D.A., & VAN DEN BERGH, S. 1996, ApJ, **463**, 602.

RICHER, M.G., & MCCALL, M.L. 1995, ApJ, **445**, 642.

ROSE, J.A. 1994, AJ, **107**, 206.

SAHA, A., FREEDMAN, W.L., HOESSEL, J.G., & MOSSMAN, A.E. 1992, AJ, **104**, 1072.

SAHA, A., & HOESSEL, J.G. 1990, AJ, **99**, 97.

SAHA, A., HOESSEL, J.G., & KRIST, J. 1992, AJ, **103**, 84.

SAHA, A., HOESSEL, J.G., & MOSSMAN, A.E. 1990, AJ, **100**, 108.

SAHA, A., MONET, D.G., & SEITZER, P. 1986, AJ, **92**, 302.

SANDAGE, A. 1976, AJ, **81**, 954.

SANDAGE, A., & BINGGELI, B. 1984, AJ, **89**, 919.

SARAJEDINI, A., & LAYDEN, A.C. 1995, *AJ*, **109**, 1086.

SARAJEDINI, A., LEE, Y.-W., & LEE, D.-H. 1995, *ApJ*, **450**, 712.

SAVIANE, I., HELD, E.V., & PIOTTO, G. 1996, *A&A*, **315**, 40.

SEARLE, L., & ZINN, R. 1978, *ApJ*, **225**, 357.

SEITZER, P., LAVERY, R.J., SUNTZEFF, N.B., WALKER, A.R., & DA COSTA, G.S. 1997, in preparation.

SKILLMAN, E.D. 1996, In *The Minnesota Lectures on Extragalactic Neutral Hydrogen* (ed. E. D. Skillman). ASP Conf. Series, vol. 106, p. 208. ASP.

SKILLMAN, E.D., KENNICUTT, R.C., & HODGE, P.W. 1989, *ApJ*, **347**, 875.

SKILLMAN, E.D., TERLEVICH, R., & MELNICK, J. 1989, *MNRAS*, **240**, 563.

SMECKER-HANE, T.A., STETSON, P.B., HESSER, J.E., & LEHNERT, M.D. 1994, *AJ*, **108**, 507.

SMECKER-HANE, T.A., STETSON, P.B., HESSER, J.E., & VANDENBERG, D.A. 1996, In *From Stars to Galaxies: The Impact of Stellar Physics on Galaxy Evolution* (ed. C. Leitherer, U. Fritze-van Alvensleben & J. Huchra). ASP Conf. Series, vol. 98, p. 328. ASP.

SMITH, E.O., NEILL, J.D., MIGHELL, K.J., & RICH, R.M. 1996, *AJ*, **111**, 1596.

SMITH, H.A., SILBERMANN, N.A., BAIRD, S.R., & GRAHAM, J.A. 1992, *AJ*, **104**, 1430.

SOLINGER, A., MORRISON, P., & MARKERT, T. 1977, *ApJ*, **211**, 707.

STETSON, P.B. 1997, *Baltic Astronomy*, **6**, in press.

STRYKER, L.L. 1984, *ApJS*, **55**, 127.

STRYKER, L.L., DA COSTA, G.S., & MOULD, J.R. 1985, *ApJ*, **298**, 544.

SUNTZEFF, N.B., FRIEL, E., KLEMOLA, A., KRAFT, R.P., & GRAHAM, J.A. 1986, *AJ*, **91**, 275.

SUNTZEFF, N.B., SCHOMMER, R.A., OLSZEWSKI, E.W., & WALKER, A.R. 1992, *AJ*, **104**, 1743.

TIFFT, W.G. 1963, *MNRAS*, **125**, 199.

TOLSTOY, E. 1996, *ApJ*, **462**, 684.

VADER, J.P. 1986, *ApJ*, **305**, 669.

VADER, J.P., & CHABOYER, B. 1994, *AJ*, **108**, 1209.

VALLENARI, A., CHIOSI, C., BERTELLI, G., APARICIO, A., & ORTOLANI, S. 1996b, *A&A*, **309**, 367.

VALLENARI, A., CHIOSI, C., BERTELLI, G., & ORTOLANI, S. 1996a, *A&A*, **309**, 358.

VAN DEN BERGH, S. 1972a, *ApJ*, **171**, L31.

VAN DEN BERGH, S. 1972b, *ApJ*, **178**, L99.

VAN DEN BERGH, S. 1974, *ApJ*, **191**, 271.

VAN DEN BERGH, S. 1994a, In *The Local Group: Comparative and Global Properties* (ed. A. Layden, R.C. Smith & J. Storm). p. 3. ESO.

VAN DEN BERGH, S. 1994b, *ApJ*, **428**, 617.

VAN DE RYDT, F., DEMERS, S., & KUNKEL, W.E. 1991, *AJ*, **102**, 130.

VASSILIADIS, E., & WOOD, P.R. 1993, *ApJ*, **413**, 641.

WALKER, A.R. 1992, *AJ*, **103**, 1166.

WALKER, A.R., & MACK, P. 1988, *AJ*, **96**, 872.

WELCH, G.A., MITCHELL, G.F., & YI, S. 1996, *ApJ*, **470**, 781.

WIRTH, A., & GALLAGHER, J.S. 1984, *ApJ*, **282**, 85.

WORTHEY, G. 1994, *ApJS*, **95**, 107.

YOUNG, L.M., & LO, K.Y. 1997, *ApJ*, **476**, 127.

ZINN, R. 1993, In *The Globular Cluster – Galaxy Connection* (ed. G.H. Smith & J.P. Brodie). ASP Conf. Series, vol. 48, p. 302. ASP.

Resolved Stellar Populations of the Luminous Galaxies in the Local Group

By MARIO MATEO

Department of Astronomy, 821 Dennison Bldg., University of Michigan, Ann Arbor, MI, 48109, USA [mateo@astro.lsa.umich.edu]

The past few years has seen an explosion of new information on the detailed stellar populations of Local Group (LG) galaxies. In this chapter I will focus on the properties of the stellar populations of the more luminous galaxies in the Local Group and (slightly) beyond. My principal aims are to illustrate the rich diversity of populations within individual LG galaxies, how these populations can be isolated and studied in detail, and how the population mixtures differ among the luminous systems of the LG. The story has both reassuring and disturbing implications. Reassuring is the fact that we are now even capable of analyzing stellar populations in some detail out to 1 Mpc and beyond. More disturbing is the growing realization that perhaps fewer generalities can be made about stellar populations in different galaxies than we might have originally hoped – they all have quite unique features that make it seem more difficult than ever to generalize the results from one galaxy to other galaxies. For example, the five most luminous LG galaxies (M31, the Milky Way, M33, and the Magellanic Clouds) appear to have distinctly different cluster and field-star formation rates that do not obviously scale with luminosity, mass or any other global galaxian parameter. But the variety does not end in the LG, and I close the chapter by discussing some features of luminous galaxies in nearby groups. These galaxies not only illustrate some examples of phenomena similar to what we see in the LG, but also some features and galaxy types that are not represented at all locally. It is clear that if we want to use nearby galaxies to understand the rich diversity of galaxy types and parameters in the Universe, we must push out to these nearby groups, exploiting our knowledge of methods and interpretation gained in the Local Group.

In each section of this chapter – which correspond to different lectures in the Winter School – I cite specific examples that attempt to illustrate how the detailed studies of stellar populations in resolved systems can help us begin the complicated job of untangling the formation and evolution of these galaxies in ever greater detail. Finally I wish to convey the excitement that the future holds. Much of what we do now and accept as being routine involves observations, reductions and modelling that was far beyond the wildest dreams of many of the pioneers of this subject only 20-30 years ago. By recounting some of the background that has led us to our impressive present understanding of the stellar populations in galaxies, one may see ways to extrapolate into the future and perhaps even glimpse at what lies ahead.

---

1. Introduction

The overall plan of my Winter-School lectures was to introduce individually the following issues as they relate to the most luminous galaxies of the Local Group:
- Old stellar populations, particularly the presence and form of halos in these galaxies;
- Intermediate-age stellar populations;
- 'Youngish populations', leaving the discussion of the *really* young stars to the experts in this field at this winter school;
- Variable-star populations;
- Star-cluster populations, again excluding the very youngest clusters and associations;
- The derived star-formation histories of these galaxies;
- Luminous galaxies beyond the Local Group, and in particular how these complement and extend what we can learn locally.

Apart from the next section of this chapter, my general plan is to start each section with a basic overview of each topic. I then move on to some specific examples illustrating some of the issues raised in these introductory descriptions. My hope is that the information may prove useful to people starting out in research studies of Local Group (LG) galaxies, yet also provide some helpful research results for more established students or other researchers entering the field. Among the specific topics I address, you will see one theme carried out through all the lectures: tidal interactions. This reflects my interests, and the growing evidence that galaxy encounters have played and continue to play a key role in the evolution of LG galaxies. I will also try to keep the lecture topics focussed on the more luminous nearby galaxies – in particular, the Milky Way, the Magellanic Clouds, M 31 and M33 – but I do occasionally address issues pertaining to the less-luminous dwarfs in the Local Group. Part of reasoning for this goes back to this theme of interactions: because of the intimate relation between dwarfs and the formation of the larger LG galaxies, one cannot really discuss any subset of galaxies in isolation. I hope to convince you of this many times throughout these lectures.

2. Photometric techniques

To do any work on 'resolved' populations in nearby galaxies requires good tools. In particular, these sorts of studies demand that one be able to measure individual stars at faint magnitudes (even the nearest galaxies located beyond the Magellanic Clouds are annoyingly distant); reliably measure these stars in fields that are often very crowded and contaminated by complex features such as emission nebulae; or spiral arms; deal with a wide range of telescopes/detectors from 1m ground-based instruments to HST; obtain good calibration of the photometry, both relative and absolute; perform reliable time-series photometry for variable-star studies; and be able to simulate galaxy evolution on a star-by-star basis using modern, complete stellar-evolutionary models.

What do I mean by 'resolved' photometry? For me this implies deriving the photometric parameters of *all* stars down to the photometric and crowding limits of the data. I emphasize 'all' because it is never safe to simply measure, say, the brightest stars in a field even if they are the principal target scientific target. Unmeasured faint stars do contribute to crowding and sky contamination, and if they are ignored they simply add to the noise in such cases. If *all* the stars are measured, you reduce this source of uncertainty significantly. Remember, a 25th mag G star contributes 1% of the flux of a 20th mag B star in M31, and as Phil Massey emphasizes, even a 1% error in colors for the hottest stars can make a big difference in interpreting their properties.

Let me give you some examples: For the LCO 1m telescope with an excellent, thinned 2048 CCD you get a useful area of 19.6' x 19.6' or about 390 arcmin^2. The V-band 1σ magnitude limit is about 24.0 in one hour with a typical image size of 1.1 arcsec, or about 2 pixels on that detector. The confusion limit sets in at about 350,000 stars per field, or about 900 stars/arcmin^2. For the Planetary Camera of WFPC2 on HST, the area is only 36" x 36" or about 0.36 arcmin^2. But the limiting magnitude is 27.5 (also in one hour; remember, HST is only a 2.4m telescope and the CCDs are not especially sensitive) and the image size is 0.08 arcsec (1.8 pixels). Thus, the confusion limit is more like 180,000 stars on the chip, or a phenomenal 480,000 stars/arcmin^2 !

These are rough estimates. Effects such as undersampling, contamination by faint stars from the underlying luminosity function, bad pixels, saturated objects, complex backgrounds and many other factors further limit the real confusion limit of any CCD image. Nonetheless, one can realistically consider frames with up to 100,000 stars to be 'photometric' and as people push towards more distant and crowded fields, I suspect that such crowding will not even seem very unusual or extreme in the future.

All of this clearly demands high-quality point-spread-function (PSF) fitting photometry. Many excellent photometry programs are available today: DAOPHOT/ALLFRAME (the 'Coca-Cola' of the industry), DoPHOT (with many hybrids), ROMAPHOT, INVENTORY, PAWS, and CAPELLA, to name just a few of the off-brands. A somewhat dated summary of these programs is in the ESO Data-Analysis Workshop (Grosbøl et al. 1989), while a number of recent papers analyzing HST WFPC data (e.g. Ajhar, et al. 1996; Ferrarese et al. 1996; Graham et al. 1997) describe some specific comparisons between DAOPHOT and DoPHOT. A detailed comparison of DAOPHOT and DoPHOT was carried out by Schechter, Mateo and Saha (1993) and I describe this in more detail below.

Poor data cannot produce good photometry, and all of these programs insist on certain minimum input data characteristics, including
- Linear data;
- Good pre-reduction processing (flat fields, bias, etc.);
- Identification and/or removal of cosmic rays (while some of the programs try do deal with cosmic rays themselves, all benefit greatly if cosmic rays can be removed or their effect minimized prior to the photometry reduction);
- Fairly uniform detector/image properties; thus, while variations of 20-40% in the PSF FWHM may be ok, variations by factors of two or more are likely to trip up any reduction program and you would be better off breaking the image into more homogeneous chunks;
- Knowing the detector parameters with some precision, such as the gain, read noise, bad pixel locations, etc.

In this section I will describe some issues regarding PSF photometry using DoPHOT – not for any advertisement purposes, but because I have by far more experience with this code than any other. DoPHOT was first written by Paul Schechter (then at Carnegie Observatories; now at MIT) to reduce tricky CCD scanning data from the LCO 1m telescope (Mateo and Schechter 1989; Schechter, Mateo and Saha 1993). He needed a program that could reduce the photometry quickly, with excellent reproducibility as a function of input parameters, and low operator overhead – preferably completely automatically. Since 1987, Abi Saha, myself, and more recently many others have tinkered extensively with this original code. Some added features include the ability to handle a spatially variable PSF; an option to reduced data in 'fixed-position' mode where the coordinates of stars in the field are known ahead of time (this produces better photometry and saves time once the template positions are determined); the ability to reduce real-array images (initially DoPHOT only would accept short (16-bit) integer images); incorporation of the HST PSF to reduce WFPC2 data more reliably and precisely; use of an empirical, rather than the original analytic, PSF; and hundreds of other minor changes and bug fixes. DoPHOT is written in such a way that additions and changes are strongly encouraged and relatively easy to implement. I find that the people most comfortable with this program have, in fact, tinkered with it extensively, altering specific parts of the code for their needs and in the process learning how DoPHOT works in detail.

The basic operation of DoPHOT differs only in detail from other PSF-fitting codes. For example, DAOPHOT attempts to identify all stars in a first pass, then fit these stars with the known PSF (many users, including myself back in the early days of CCDs would force DAOPHOT to do a second pass to find and fit any stars missed in this operation, but such a second pass is not intrinsic to the code). In DoPHOT's case, the approach is iterative from the start. As input, the program expects a processed FITS image (though some versions can read other formats), the detector parameters (read noise, gain, image size), and a very approximate estimate of the stellar FWHM and sky values. If the PSF is grossly distorted, it is wise to tell DoPHOT also at the start about this distortion. The program takes this information to define a number of 'thresholds' above the sky level. At each threshold, the program

- Finds objects using the current PSF as a filter;
- Fits the PSF (based on χ^2 minimization) with careful weighting by the local noise;
- Digitally subtracts the stars individually from the image;
- Derives the average PSF parameters from the 'best' stars;
- Re-adds each star to the image one at a time, then re-fits the profile using the averaged, updated PSF parameters;
- Either go to the next threshold, or, if this is the last threshold, calculate the aperture corrections for bright, isolated stars.

Figure 1 is a flow chart summarizing these steps and including a bit more of the details.

The threshold concept confuses a lot of people. The global sky estimate is used *only* to find stars, *not* to generate the estimate of the sky for an individual star used to perform the photometry – the local sky is fit for explicitly on a star-by-star basis; you can infer this from the flow chart (Figure 1). DoPHOT has many options on how to deal with the determination of the global sky level, but a popular one calculates a median sky level over the entire frame, then uses the shape and level of the median sky to determine the thresholds. In this way even rather complex fields can be measured, including cores of globular clusters, messy fields near star-forming regions or spiral arms in nearby galaxies, or regions near extensive emission nebulosity.

In general, DoPHOT uses an analytic approximation to a Gaussian as its PSF:

$$e^{-z} \sim (1 + z + \beta_4 z^2 + \beta_6 z^4 + \beta_8 z^6 + \ldots)^{-1}$$

where

$$z = \frac{1}{2}((x/\sigma_x)^2 + 2\sigma_{xy}xy + (y/\sigma_y)^2)$$

and x, y are the CCD coordinates, and σ_j are the quasi-Gaussian sigma's in for $j = x, y, xy$. Usually, only terms up to β_6 are used, but Abi Saha introduced β_8 as a way to better approximate the WFPC2 PSF far from the image center (Saha et al. 1994). There are seven variable parameters in this function: x, y, the underlying sky level, the central intensity, and the three PSF shape parameters (σ_x, σ_y, and σ_{xy}). The β parameters are generally fixed, but they too, with care, can be allowed to vary as part of the fitting procedure. Note that any other seven-parameter function could be used (e.g. a Moffat function if one fixes the power-law slope) without altering the code too drastically. Another feature of DoPHOT is that at most *two* stars can be simultaneously fit – not so severe a limitation as it might seem since the iterative approach identifies and measures neighboring stars very efficiently. Table 1 shows the variation of the parameters from a recent DoPHOT run. You can see that as the thresholds get lower, the program finds more stars and the PSF becomes better defined. However, note that the PSF shape parameters settle towards its final values quite early in the process.

After more than 10 years, DoPHOT retains its original flavor. It remains fast: for images where the stellar FWHM is 2-2.5 pixels, the data reduction rates can exceed 500,000 stars per hour (sph) using a Sparc Ultra 1. Also, the overhead for DoPHOT is very low – in fact, one can readily set up the program to automatically reduce data as they are obtained and processed. This has made DoPHOT the choice for large-scale

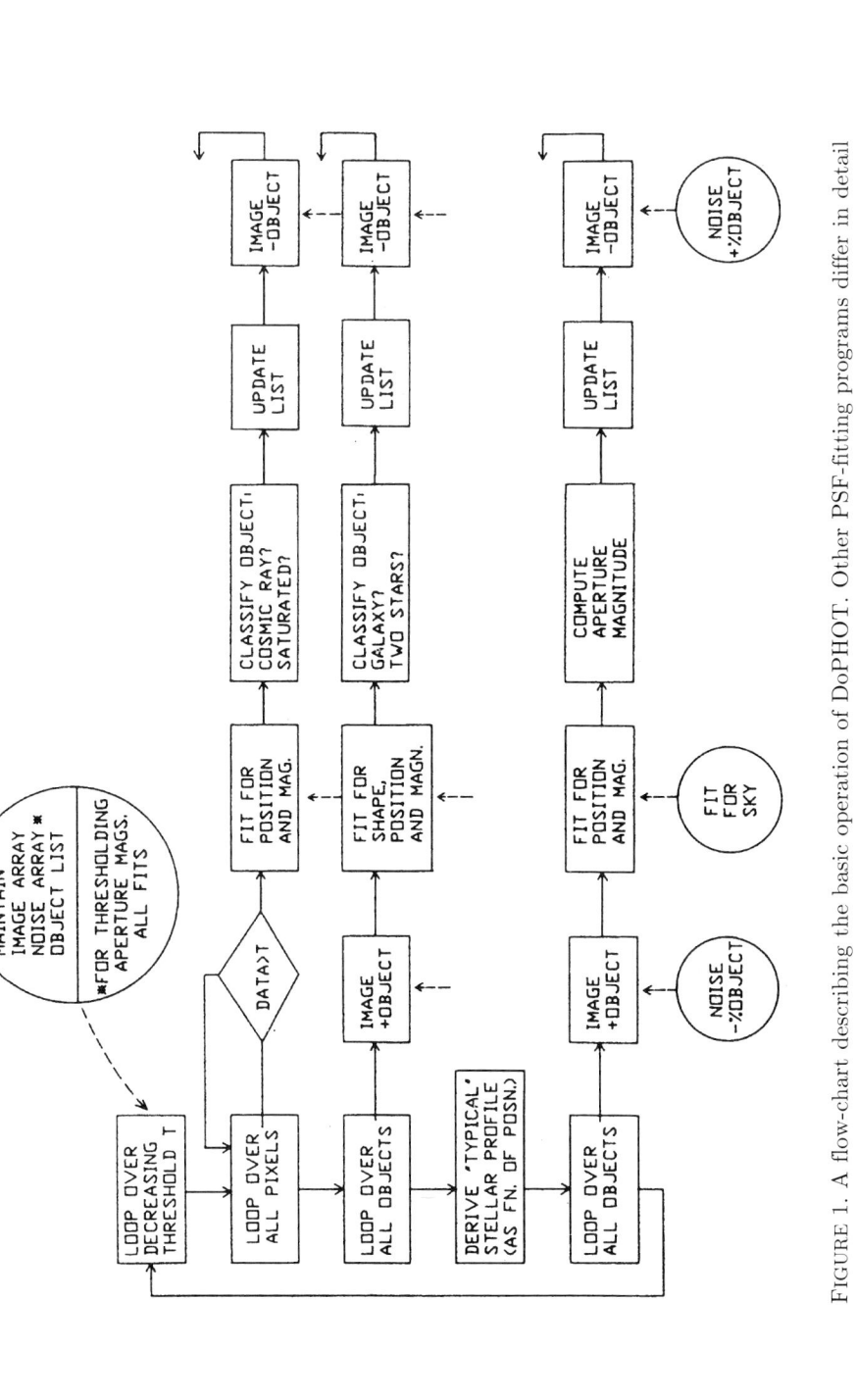

FIGURE 1. A flow-chart describing the basic operation of DoPHOT. Other PSF-fitting programs differ in detail from this, but most do approximately similar things as they find, fit and photometer individual stars or groups of stars in images. The original version of this figure was produced by John Caldwell for Paul Schechter.

TABLE 1. DoPHOT Operation.

Threshold	N_*	N_{PSF}	σ_x pix	σ_{xy} pix^{-1}	σ_y pix
12561.3	278	64	0.336	0.237	0.312
5467.6	476	158	0.341	0.238	0.316
2379.9	757	333	0.341	0.239	0.317
1025.9	1174	535	0.345	0.238	0.321
450.9	2062	809	0.348	0.242	0.322
196.3	3523	1245	0.350	0.241	0.323
85.4	5225	1579	0.351	0.238	0.324
37.2	9493	1569	0.351	0.238	0.324
16.2	15382	1569	0.351	0.238	0.324

photometric surveys such as OGLE and MACHO, as well as other large photometry programs of my own and of others.

How well does DoPHOT work? A few years ago John Tonry generated a globular cluster for us – Tonry 1 – to help us answer this question. The virtual cluster contained about 100,000 stars, about 10,000 of which were readily visible above the sky level. We reduced these data using DoPHOT and DAOPHOT, and found the photometric precision to be very similar (though note that none of us are DAOPHOT experts and it is clear that the photometric quality one gets with that program is more operator-dependent than DoPHOT). We also found that much of the difference in the final photometry came about due to differences in the sky level. As I mentioned above, there have subsequently been many other DoPHOT/DAOPHOT comparisons, especially involving HST observations. The key point to emphasize here is that, yes, different codes produce different results but the differences tend to be systematic at the 3-5% level at worst, and usually only get this bad in very challenging applications, for example near the core of a dense star cluster. Another lesson from Tonry 1 is that while the precise PSF that is used is obviously important at the 1% level, for studies of faint stars in crowded fields – which is what will be the focus of this chapter – the detailed PSF is surprisingly unimportant so long as it is not a gross mismatch to the data. The way in which the local sky is determined for each star, for example, is usually *much* more important!

Let me now turn to some applications of modern photometry programs that are particularly important to contemporary datasets. One has to do with dealing with a spatially-variable PSF in a CCD image. Sadly, not all telescopes/CCDs produce perfectly uniform images. This has become an increasingly common problem as larger CCDs come into use covering larger solid angles in the sky while retaining good image sampling. Any useful reduction program must be able to deal with systematic variations in the the fit quality as a function of position. There are two practical ways to deal with this problem: sub-frame reductions, and incorporating a variable PSF into the reductions.

For sub-frame reductions, one splits the image into an $n \times n$ array. This approach was pioneered for CCD data by Andrezj Udalski and Michal Szymanski for the OGLE project (Udalski et al., 1992). The idea is to choose n so that the PSF is essentially constant within the sub-field, run DoPHOT on each sub-field then stitch the resulting photometry back together to get a list for the entire CCD image by intentionally having some overlap between subfields. This drastically improves the photometry when the PSF is highly (but not too rapidly) variable, it's an easy concept to apply and understand, and it can be readily used by any PSF program by simply specifying the sub-field limits. The problem is that it is a difficult method to adjust once you have set it up for a given value of n, the bookkeeping is horrendous, and you must determine the aperture corrections reliably for at least one sub-field. In crowded fields, this last requirement is difficult to achieve.

Another approach is to allow for a variable PSF – there are many DoPHOT hybrids that do this, and I know this feature is available in DAOPHOT; undoubtedly it is a feature of most other photometry programs too. To implement the variable PSF, DoPHOT assumes the function describing the PSF to be a function of (x,y); in my version of DoPHOT this function is simply a polynomial in both axes, including cross terms. The form of this function is determined after DoPHOT finds and does a first-pass of fitting individual stars at each threshold. The nice features of this approach are that the parameters of the variable PSF are readily and easily adjustable, the bookkeeping is *much* easier, and, in DoPHOT at least, was easy to implement. The drawbacks are that this method cannot handle really gross variations in the PSF (say, more than a 50% variation in FWHM over a frame – though data this poor should be considered suspect in any case), and the resulting aperture corrections are guaranteed to vary with position by up to 30-50% when a variable PSF is used.

I've mentioned aperture corrections a number of times. These corrections allow one to convert the fitted magnitudes from any PSF program (which typically uses on the central region of a stellar image to carry out the fit) to the 'total' magnitudes one would measure for that star if it were isolated and observed through a large aperture. These corrections are typically defined as $m_{tot} - m_{fit}$, and you cannot calibrate photometry without them. In crowded fields, these corrections are *difficult* to determine; when the PSF varies, so do these corrections. Sounds bad. doesn't it? Well, it is bad. It is often the case that, even after the utmost care in performing the PSF photometry, the largest single source of error in CCD photometry turns out to be the aperture corrections. A lot of effort has gone into determining precise aperture corrections with DoPHOT, but the story is long, technical, but not exactly gripping. I discuss this in some detail in the Appendix.

Another increasingly important application of PSF photometry is to carry out real-time measurements. I have mentioned some projects that rely on this capability – OGLE and MACHO routinely monitor millions of stars nightly, yet are able to spot variables on timescales of hours because of their automated reduction facilities. These are obvious applications, but I have recently been involved in another project that benefits enormously from the ability to do real-time photometric measurements. This also gives me a chance to mention some astronomical issues in this otherwise technical section of my chapter.

In 1994, Ibata et al. discovered what is now the closest known external galaxy, the Sgr dwarf. Many studies have addressed the basic properties of this galaxy (Koribalski et al. 1994; Mateo et al. 1995a; Ibata et al., 1995; Sarajedini and Layden 1995; Da Costa and Armandroff 1995; Mateo et al. 1995b; Ibata et al. 1997). One obvious feature of Sgr is that it its extent in the sky is enormous (Mateo et al. 1996; Alard 1996; Fahlman et al. 1996). Just how large it is and mapping its detailed shape, is of interest for models describing how a puny satellite is destroyed as it is pulled into our Milky Way. Sounds easy, until you realize that Sgr has an effective surface brightness of 27 mag/arcsec^2 in its center; it is exceedingly faint in its outer regions.

One way to trace it is to find individual stars. Figure 2 shows an example of the MS of Sgr in a field located very close to the foreground globular cluster M 55. This suggested to us a way to trace the extent of Sgr in the sky. Our idea is to obtain deep CMDs along the projected major axis of Sgr with a 4m class telescope. These data allow us to

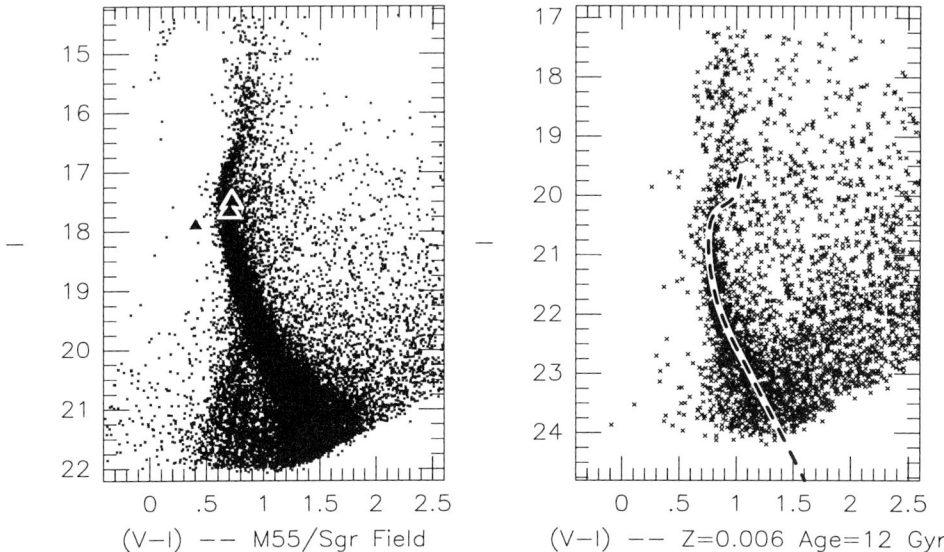

FIGURE 2. Color-magnitude diagram of a field near the center of the globular cluster M 55 (left) and of a field just south of the cluster, but beyond its tidal radius. The excess of faint stars just to the lower left of the M55 main sequence corresponds to the turnoff of the Sgr dwarf galaxy whose center is located over 10 degrees from this field. The sequence evident in the figure on the right is the Sgr main sequence, now uncontaminated by foreground M 55 stars. This confirms the interpretation of the faint stars in the left-hand diagram. Note the change in the scale for the I-band magnitudes in the two plots.

reach the Sgr MS turnoff, at which point Sgr actually may dominate the local star counts (foreground and even halo stars are starting to become rare, and galaxies have not yet begun to dominate the counts as they do at fainter magnitudes; Tyson 1988; Ratnatunga and Bahcall 1985). The practical problem is knowing when – or if – you have reached the end of the galaxy so as not to waste telescope time looking an (unintended) blank fields. To deal with this dilemma I developed a pipeline that can process and carry out DoPHOT photometry of a V-I frame pair *before* the next pair is obtained (at about 10 minutes per exposure using the four-CCD BTC camera at CTIO). If we run out of galaxy, we know not to continue observing in that region, allowing us to more efficiently use the 4m for this exploratory probing of the extremities of Sgr. The real point here is that the capabilities of PSF programs to carry our near-real-time or true real-time reductions can benefit a large range of projects, not merely those interested in variability studies.

Another feature that has become more common in PSF reduction routines is so-called fixed-position mode. The idea here is to obtain measurements of the locations of stars in a field. These can come from, say, coadding and reducing the data from many individual images prior to reducing the images one at a time, or from HST or any other higher-resolution instrument. The program then matches the coordinates of the frame being reduced with the coordinates from the template file, and aligns them (see Groth 1986 and Valdes et al. 1995 for details). Now, using the template coordinates the program reduces the data from the program frame; in the case of DoPHOT, it skips to the last or penultimate threshold because if the template is well made, it contains virtually *all* the stars that can be found in the frame. Though complicated to set up, this approach allows one to reduce data much faster than in non-fixed mode, and with greater precision. If the template list comes from very good-seeing data, this information allows the code to fit blended images as two stars even if DoPHOT would normally want to fit such images as single objects in poor-seeing frames. This approach is particularly useful in variability studies that concentrate on a single field – and hence possible to build a deep template – and makes it easier to find serendipitous objects that show up in an image but are not in the template.

There are many technical issues in the use of any PSF routine, and the only way to really understand them is to use the programs. To help out with this practical requirement, I have included in the Appendix the text of a handout from the winter school that describes the use of a tutorial version of DoPHOT. The information has been updated to allow the reader to obtain and use a version of DoPHOT available as of mid-1997 from the University of Michigan. The tutorial allows you to run DoPHOT in some useful modes described in this section, and it gives you a live version of the code to use in any way you would like. Some interesting updates not found in the version of DoPHOT described here include the HST PSF (Abi Saha, KPNO), ability to reduce real arrays (Knut Olsen and Paul Hodge, University of Washington), and an empirical PSF (Paul Schechter, MIT). Other users have undoubtedly made comparably useful updates/modifications that may be of interest.

3. Star clusters in the Local Group

One thing is clear from studies of the Local Group galaxies: Star clusters are very common. Table 2 provides a census of the total numbers of massive clusters in LG galaxies, categorized into two classes: classic globular clusters that appear similar to the ancient, massive clusters of the Milky Way, and 'other' massive clusters that appear to be younger than globulars (more on these distinctions below).

Table 2 is meant to be comprehensive in that it contains data for all LG galaxies with known cluster populations for all galaxies more luminous than the faintest systems listed (Fornax and Sagittarius). There appears to be be a minimum galaxy luminosity – about $10^7 L_\odot$ – for a galaxy to have massive clusters, and only two galaxies, IC 1613, and the difficult-to-observe IC 10 seems to violate this trend. Also, the specific frequency of cluster (defined as the number of massive clusters per $10^7 L_\odot$) – appears to be around 7-8 (in V-band units) for the old, globular cluster-like objects, but is highly variable for other, massive, but apparently young clusters. For example, the LMC has many young, blue globulars compared to the redder, old systems, in striking contrast to more luminous galaxies such as M31 and the Milky Way *and* a low luminosity galaxy such as Fornax which contains no young, massive clusters at all.

Star clusters in the LG also share some other global properties. First, the luminosity distribution of the red, presumably old star clusters are consistent with a log-normal distribution with $\langle M_V \rangle \sim -7.1$ with $\sigma \sim 1.3$ (van den Bergh 1985; but see Kavelaars and Hanes 1997). I would argue that even the young massive clusters are consistent with this mean luminosity if the fading effects of evolution are taken into account (see Mateo 1993). The relative sizes of the clusters vary with location within individual systems such that core clusters tend to be more compact than clusters found in the

TABLE 2. Star clusters in LG galaxies.

Galaxy Name	M_V	N_{GC}	N_{other}	L_{cl}/L_G	S_{GC}	S_{other}
M 31	−21.1	350	< 5	1.8×10^{-3}	0.15	<0.002
Milky Way	−20.6	160	< 5	1.4×10^{-3}	0.11	<0.003
M 33	−18.9	$\lesssim 20$	$\gtrsim 10$	1.0×10^{-3}	<0.07	$\gtrsim 0.03$
LMC	−18.1	13	$\gtrsim 200$	1.7×10^{-2}	0.09	1.4
IC 10	−17.6	0	0	0	0	0
M 32	−16.4	0-3	0	$< 1.0 \times 10^{-3}$	0-0.10	0
NGC 205	−16.3	9	1	3.6×10^{-3}	0.30	0.03
NGC 6822	−16.3	4(?)	0	1.5×10^{-3}	$\lesssim 0.15$	0
SMC	−16.2	2	> 20	1.2×10^{-2}	0.08	>0.8
NGC 185	−15.2	8	0	8.0×10^{-3}	0.8	0
NGC 147	−15.1	4	0	4.4×10^{-3}	0.4	0
IC 1613	−14.9	0	0	0	0	0
WLM	−14.1	1	0	2.8×10^{-3}	0.28	0
Fornax	−13.7	5	0	2.0×10^{-3}	2.0	0
Sagittarius	−13.7	4	0	1.6×10^{-3}	1.6	0

N_{GC} refers to the number of clusters that are likely similar in morphology and stellar content to the old globular clusters of the Milky Way, while N_{other} refers to the number of massive star clusters that, although morphologically similar to globulars, appear to be composed of significantly younger stars. IC 10 suffers from extensive foreground extinction and the census of its clusters cannot yet be considered very complete.

outer parts of the parent galaxy (van den Bergh 1994a,b; Mateo 1987). This trend is apparent in the star clusters within galaxies ranging from Fornax to the Milky Way (a range of nearly 7 magnitudes in galaxian luminosity). The V-band mass-to-light ratios ((M/L_V)) for old LG clusters range from about 0.5-5, and helps define the standard M/L for a 'pure' old population (Piatek et al. 1994; Pryor et al. 1991, 1989; Djorgovski et al. 1997). Dynamical studies have shown the expected decrease in (M/L) for younger clusters (Fischer, et al. 1992a,b, 1993; Mateo et al. 1991). Finally, the old clusters are nearly interchangeable in that one could put a Fornax cluster in the halo of the M 31 – and vice versa – without easily being able to tell that this had occurred †. This is not to imply that the old cluster populations are completely indistinguishable, but the global properties of old clusters in LG galaxies do seem to greatly overlap. There is even circumstantial evidence that such cluster swapping may have occurred in the LG (Mateo et al. 1986; Lin and Richer 1992, Muzzio 1988), and direct evidence that it *does* occur within our halo as the Sgr dwarf loses its clusters (Da Costa and Armandroff 1995).

Hints that there exists a diversity of massive cluster types has been around for a long time. In his classic book *Star Clusters* (1930), Shapley noted that some LMC clusters

† Van den Bergh (1994a) would disagree, arguing that the Fornax clusters are much too small for their location in the Galactic halo. But these clusters are *not* in the halo, they are in Fornax, and so they are subject to the tidal field of that galaxy. The fact that the Fornax clusters exhibit a clear trend in their size as a function of distance from the center of Fornax itself is clear proof that the sizes of these clusters are controlled by Fornax, not the Galactic halo. It would be of great interest to calculate how rapidly a cluster would expand if its local tidal field (caused by a massive dwarf) were suddenly removed and replaced by that of the Galaxy at large Galactocentric distance.

(which, morphologically were 'globulars' to him) had early A or even B-type spectra, quite distinct from the F to G spectra of galactic globular clusters. In 1952, Gascoigne and Kron compiled the colors of LMC/SMC clusters then studied and found for the LMC that: 'This [color distribution] appears so pronounced as to suggest that we have observed clusters of two types.' In the late 1950's and early 1960's, extensive work by Thackeray, Shapley, Arp, and Hodge proved that this was true: the LMC contains two distinct populations of clusters characterized by different colors and different ages. The term 'Young Populous Cluster' was coined by Hodge (1961) to distinguish these objects – which were morphologically similar to globulars – from their ancient counterparts.

The variety of cluster populations in the LG is even greater than what is seen in the Magellanic Clouds. While M 31 and the Milky Way have cluster systems with similar global properties, M 33 has a cluster system that exhibits as broad a color distribution as seen in the LMC, but apparently uni-modal. NGC 205 contains a number of old clusters, but in its core, one very bright, very blue cluster (Hodge 1973; Lee 1996). The central location and 'cold' kinematic properties (Carter and Sadler 1990) of this cluster suggests a rather unique origin for this object (Jones et al. 1996). WLM has only one very luminous cluster located well away from the galaxy's center. IC 10, and IC 1613, though both more luminous than many LG galaxies that do contain clusters, appears to have none. The LMC alone seems to have a really prominent population of blue populous clusters, although very similar systems are also known to exist in the SMC. Thus, we can conclude that massive clusters are common throughout the LG, but the age distributions of these clusters within individual galaxies vary enormously.

In many ways these LG clusters are very important. It is almost a cliché to say so, but if we cannot understand the detailed photometric and stellar-population properties of these single-aged, mono-metallicity clusters, what real hope do we have for interpreting the global integrated properties of galaxies at any redshift? These clusters offer many important ways to study various stages of stellar evolution – even rather short-lived stages – in detail. For example, calibrating the L_{AGB}-Age relation, determining the minimum age of RR Lyr stars, measuring the masses of Cepheids independently of the pulsational properties of the Cepheids themselves. Clusters are also often the first indication of what the global star-formation history is like in individual galaxies – some examples of the variety we see in the LG were described above.

Aiding us tremendously are the many recent gains in technology that allow us to study these clusters in increasing detail. Prior to the end of the 1950's, about the only information available for LG clusters were their integrated photometric – usually photographically determined – properties, with some integrated spectra available for LMC and SMC clusters. The first usable CMDs showing individual stars in extragalactic clusters became available around that time (e.g. Arp 1958, 1959). The following two decades saw a steady improvement in stellar photometry in extragalactic clusters as photographic techniques were pushed to their limits. The really old clusters in the LMC, for example, revealed giant branches similar to those seen in Galactic globulars. But the 1980's was when the revolution really began: the advent of CCDs suddenly made it possible to reach limiting V-band magnitudes of 24-25, compared to 21 or so with earlier photographic plates. Just one example of the magnitude of the improvement is illustrated in Figure 3 showing state-of-the-art photometry of the LMC cluster Hodge 4 from the early 1960's (Hodge 1960) and early in the CCD era (Mateo and Hodge 1986). More than 100 clusters, mostly in the Magellanic Clouds, had main-sequence photometry available by the end of the 1980's (bibliographic lists for the LMC cluster photometry were compiled by Sagar and Pandey 1989, and Seggewiss and Richtler 1989). Some astronomers even had the nerve to start trying to get stellar photometry in clusters as

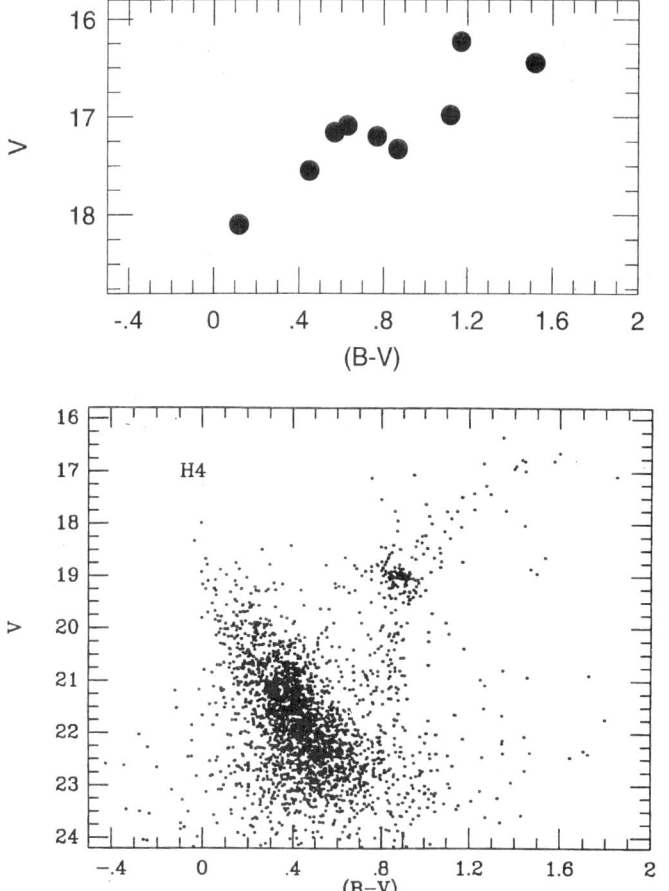

FIGURE 3. Color-magnitude diagrams of the LMC cluster Hodge 4 from Hodge (1960) and Mateo and Hodge (1986). The differences reflect detector technology and telescope improvements over that 26-year time span. It will be very exciting if there will be a similar qualitative improvement in the next 20 years; HST can already deliver a large improvement over the 1986 CCD data.

far away as M 31 (Heasley et al. 1988)! Of course, with HST's excellent resolution it becomes possible to obtain reliable stellar photometry to at least the level of the horizontal branch throughout the LG. The future looks exciting: STIS and the HST Advanced Camera promise to make it possible to reach limiting magnitudes of nearly 30 in V, allowing us to obtain main-sequence photometry of all star clusters (and galaxy fields away from clusters) throughout the LG! If future-generation 8m telescopes perform as expected, considerable photometric and spectroscopic follow-up will be practical from the ground.

I have emphasized the observational side of the picture here, but of course the models have also greatly advanced. These details are described by Cesare Chiosi elsewhere in this volume.

Let me now turn to some specific examples of how star clusters have advanced our understanding of LG clusters – I will restrict myself to the more luminous LG galaxies.

3.1. The minimum age of RR Lyr stars

It was observed long ago that some of the red SMC clusters contain RR Lyr stars, while others that appear very similar in their integrated colors do not. Because RR Lyr stars are typically used as primary tracers of old stellar populations, it would be good to try to understand exactly how old they really are, and whether this age depends on metallicity.

The LMC shows strong evidence of a bimodal distribution of cluster colors, suggesting a possibly bimodal underlying age distribution. In fact, there appears to be a huge range in time when LMC clusters did not form (more on this below). In M 31 and M 33, although the photometry is improving tremendously, we have still only very poor age estimates for old clusters, and very little information about the content of RR Lyr stars in the clusters. On the other hand, the SMC's clusters have colors that show a more continuous spread, and yet are close enough to be studied in detail.

Two clusters in particular help address the age of RR Lyr stars. L 1 was studied by Olszewski et al. (1987) who found it to have an age of 10-12 Gyr, a metallicity of about [Fe/H] = -1.3, yet does not contain any RR Lyr stars. The age and metallicity of the cluster was subsequently revised to 7.3 Gyr and -1.2, respectively (Sarajedini, Lee and Lee 1995). On the same scale as the first paper, NGC 121 has an age of 12 Gyr, a similar metallicity, and contains four RR Lyr stars (Stryker et al. 1985; this cluster was not studied by Sarajedini, Lee and Lee 1995). This pair of clusters implies that the minimum age for the existence of RR Lyr stars in present-day clusters is at least 10 Gyr (or 7.3 Gyr on the newer age scale). Of course, the real interpretation is not so easy: because of their very different luminosities, you would only expect L 1 to contain about 0.6 RR Lyr stars if scaled from the four seen in NGC 121. The lack of these variables in L 1 may only be a stochastic effect; Nature is teasing us here because the SMC does not have any other clusters in this limited age range†. Given the unimodal distribution of cluster colors in M33, that galaxy looks like an excellent candidate to address this issue. We clearly have a lot of work to obtain reliable ages of old clusters at that distance, and a good census of RR Lyr stars in those clusters.

3.2. Clusters as distance indicators

Clusters have long played a central role in getting distances to nearby galaxies. Some of the traditional techniques include using the level of the HB and assuming constancy, or at worst a metallicity dependence. Although applicable throughout the LG, recent work on Carina (Smecker-Hane et al. 1994) would suggest that the L_{HB} may not be so easily characterized as we once believed (see also Hatzidimitriou 1991 to see how age effects the HB luminosity). Another approach involves main-sequence fitting, a method that can be applied in the Magellanic Clouds with improved efficiency using HST. Another, less common method involves equating the pulsational and evolutionary masses of Cepheids found in clusters; it is this technique I will describe in some detail here.

As I mentioned before, the blue populous clusters are fantastic places to observe rare phases of evolution, such as the formation of the extended AGB and RGB as a function of age, or, as I will exploit here, the formation of Cepheids. In fact, over 10 LMC clusters contain from 1 to 20 Cepheids ranging in period from 3-11 days. Compare this to NGC7790 in the Milky Way which contains 3 Cepheids while no other MW cluster is known to contain more than one.

Evolutionary models for these stars suggest that they are located on an extended

† One cluster of comparable age exists in the LMC: ESO-121SC03. Mateo et al. (1986) conclude its age is 10 Gyr (6.4 Gyr on the Sarajedini, Lee and Lee 1995 scale) and [Fe/H] = -1.1. It is not known if it contains RR Lyr stars, but its very red HB would suggest it does not.

blue loop following core He ignition (a good introduction remains Beckers and Mathews 1983). If this loop extends far enough to the blue to intersect the Cepheid instability strip, one may see Cepheids *if* an individual star is observed at just the right evolutionary phase. Since the rate of evolution on the blue loops is fast, one must either have a very populous cluster in order to see a few Cepheids, or, as in the case of the LMC clusters NGC 1866 (20 Cephs) and NGC 2031 (17 Cephs), find examples where the slowest phase of this evolution stage – the extreme blue end of the loop – is located wholly within the instability strip.

The way in which you can exploit the abundance of Cepheids in some LMC clusters is to equate the evolutionary mass of the cluster (M_{evol}) with the pulsational masses of the Cepheids (M_{puls}) (Bertelli et al. 1993). Note that this is not the usual evolutionary mass referred to in the literature of pulsational variables and based essentially only on the location of the variable in the HR diagram, but the mass that explains *all* of the features of the cluster's color-magnitude diagram. In practice, one can trade off distance, metallicity and reddening to derive a rather broad range of evolutionary masses, even at the relatively low level of uncertainty of these quantities for the LMC. The pulsational mass, on the other hand, is derived from appropriate models and depends on the Cepheid period, color, and luminosity. The latter dependence makes M_{puls} quite sensitive to distance, but in a different way than M_{evol}. By requiring that all of the Cepheids have pulsational masses as closely equal to the cluster's evolutionary mass, a unique distance can be obtained. For NGC 1866, we derived $(m - M)_0 = 18.51 \pm 0.21$, while for NGC 2031, we found a true distance modulus of 18.32 ± 0.20. The errors are almost exclusively from uncertainties in the cluster metallicities. This approach is applicable to other LMC clusters with numerous Cepheids (Sebo and Wood 1994, 1995; Mateo et al. 1990).

3.3. *Tracing the halos of LG galaxies*

The oldest clusters are generally easy to identify, and because ancient globulars are rather luminous, they are relatively easy to study spectroscopically to obtain rough estimates of their chemical abundances and good measurements of their radial velocities and internal kinematics. In contrast, field tracers of the halos of *any* LG galaxy – including the MW – are notoriously difficult to study in detail.

In the larger LG galaxies – the MW and M 31 – we see a typically complex, composite structure as traced by the old clusters. In both cases, the outer regions appear to be dominated by metal-poor but possibly youngish clusters that exhibit little systemic rotation and a very large, approximately isotropic velocity dispersion. The inner clusters include metal poor and metal richer systems, and their kinematics are more disk-like, with net rotation and a more anisotropic dispersion tensor. This tendency appears somewhat stronger in M 31 than in our galaxy.

The smaller LG galaxies are intriguing in the ways in which they no longer follow this pattern. In M 33 the reddest clusters do appear to define a kinematically distinct halo with no net rotation and a line-of-sight velocity dispersion of 70 km/s, about what you would expect for an isotropic halo in M 33 given its disk rotation velocity (Schommer and Christian 1991). The LMC – which is of comparable luminosity as M 33 though morphologically much more irregular – possesses a population of old clusters that have more disk-like kinematics: $v_{rot} \sim 70$ km/s, and $\sigma_{los} \sim 20$ km/s (Schommer et al. 1992). There is, surprisingly, no evidence of a classical kinematic halo. Could it be that near $M_V \sim -18$ one reaches a galaxian luminosity that can no longer form and/or retain a classical halo population? What about recent claims (Minniti and Zijlstra 1996) of a possible halo in the low-luminosity LG galaxy WLM? We clearly need more halo tracers to address questions such as these, suggesting that clusters – because of their intrinsically

3.4. Clusters and the age-metallicity relations of the Magellanic Clouds

One thing that clusters remain supremely useful for is the ability to age-date their stars. Though main-sequence photometry is by far the most reliable approach, even integrated colors and spectra can be used to obtain useful age estimates for clusters in the LG. In the Magellanic Clouds, one of the first such age calibrations – the SWB scale – was in fact based on the analysis of spectra and intermediate-band photometry (Searle et al. 1980). Elson and Fall (1985, 1988) noted that even classical UBV broad-band colors can be used to obtain SWB-like age estimates, and introduced the so-called s-parameter which has proven very useful as a moderately precise age estimator of young, and intermediate-age clusters. As expected, this sort of age indicator loses precision for older clusters (Bica et al. 1996; Geisler et al. 1997).

Metallicities of clusters can be obtained from integrated spectra and from some color indices in carefully crafted photometric systems (remember, this must be done in such a way to avoid the more extreme photometric signatures associated with variations in cluster ages). For the Magellanic Clouds, we are even able to obtain colors and spectra of individual member stars, taken out some of the uncertainties involved in interpreting the composite integrated spectra or photometry of the clusters (Girardi et al. 1995). A very small number of MC cluster stars now also have high-dispersion abundances, but this has mostly been limited to young clusters, and consequently high-mass, often evolved, stars. The analysis and interpretation of these stars is non-trivial to say the least, and there has been considerable controversy surrounding these results (e.g. see Bessell 1993; Jasniewicz and Thévenin 1994).

Nonetheless, we have succeeded to a large extent to derive what appears to be a reliable estimate of the age-metallicity relations of the Magellanic Clouds (see Figure 4). It is quite clear that the two Clouds have had distinct chemical histories with the LMC being more evolved. This is consistent with simple chemical evolution models and the fact that the LMC is more massive, and currently has a smaller gas content than the SMC (3% by mass compared to 35% by mass for the SMC). But there are also puzzles in these results. Note that while the SMC has steadily formed some clusters – and perhaps by implication field stars – more-or-less continuously since the oldest clusters were formed, the LMC has clearly had a long hiatus in cluster formation. Interestingly, during this period of no cluster (and field star?) formation, there was an increase in the mean metallicity of the LMC by over a factor of 30. The standard explanation for this sort of behavior is that Type I SNe enrich the gas long after the first halo stars form (Gilmore and Wyse 1991; Köppen 1993), regardless of whether there is star formation going on at intermediate times. Quantitatively, however, the LMC presents a puzzle: if 2% of the baryonic mass of the LMC is in its halo (Suntzeff et al. 1992), the type I supernovae from this population could at most have raised the mean metallicity of the remaining gas up to about [Fe/H] = -1, but not as high as -0.3 as observed for 3-4 Gyr old clusters. Either the rate of SNe in the halo stars was much higher than assumed in typical models, or else the yields of these low-metallicity SNe were much higher than we suppose. Or star formation did proceed during this period, but not in a mode that produced any clusters.

These results on the Magellanic Cloud age-metallicity relations also raise other questions: why are the two Clouds so different? What mechanism induced the resurgence of cluster formation in the LMC 4 Gyr ago? Perhaps the key question is whether the

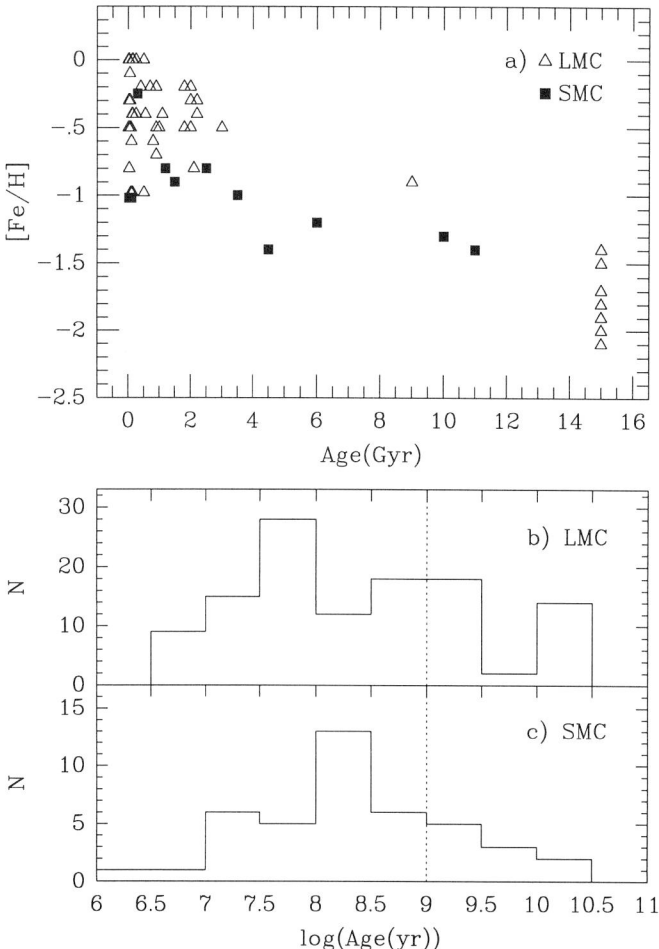

FIGURE 4. The Age-metallicity relations for the Magellanic Clouds (panel a), and the age distributions for clusters with good age estimates from recent ground-based CCD observations for both the Large and Small Clouds (panels b and c, respectively).

cluster formation history and age-metallicity relation is really indicative of the overall star-formation history in any galaxy. We will interrogate the field stars about this below.

3.5. Clusters as tracers of past encounters

Did anyone notice? There is one and only one LMC cluster with an age in the wide chasm betwen about 4-12 Gyr: ESO121-SC03 (Mateo et al 1989; Da Costa 1991 reported a search to find other such clusters, but with no success, and Geisler et al. 1997 repeated the search – again without finding any other old clusters – just to be sure). Its location in Figure 4 is more like that of some SMC clusters than any other LMC cluster. Kunkel (1979) early on suggested a detailed model in which the Magellanic clouds and other LG dwarfs may have had a common ancestry – this idea has most recently been studied in some detail by Lynden-Bell and Lynden-Bell (1995). Or could it be that at some recent past interaction the Magellanic Clouds traded clusters? Gardiner et al. (1994) show the past separation of the Clouds and the Galaxy over the past 15 Gyr according to their model aimed at trying to account for as many of the features of the Magellanic System as possible. There are two very interesting close encounters – 4 Gyr and about 200 Myr ago – that have independently been suggested to be epochs of active cluster formation in the LMC (van den Bergh 1991; Olszewski et al. 1996). The separations of the Clouds are also of interest, reaching as little as 10 kpc on both occasions. The outer regions of the two galaxies would overlap considerably if separated by such a small distance today. Their current separation is approximately 25 kpc. Cluster – and even field star – swapping may have occurred at such times; I think modelers would say 'must have occurred'. This possibility greatly complicates the interpretation of cluster systems and it will be of great interest to determine if such swapping has happened elsewhere in the Local Group. Some prime candidates include swaps between M 31 and M 33, or between M 31 and its smaller companions such as M 32, NGC 205, NGC 147, and NGC 185, all of which appear to have their own globular cluster populations.

4. The old and intermediate-age populations in luminous LG galaxies

The ultimate problem with using clusters to determine fundamental properties of their parent galaxies is that one cannot be sure how representative the cluster systems are of the whole galaxy. One can see from Table 2 that asking whether clusters reliably track the full star-formation and chemical enrichment history of a galaxy may be like asking if 0.1-1% of the biggest, or loudest, or sleepiest, etc. people represent the full properties of humanity. This is not to say the cluster *are* unreliable, only that we don't know for sure one way or the other.

The problem, of course, is that field-star studies are *hard*. Whereas clusters are coeval and mono-metallicity systems (usually), fields are composed of stars with a range of ages and metallicities. Clusters are well-defined kinematically, and although kinematics are often used to help disentangle field-star populations, these usually overlap; e.g. given their kinematic distributions, some individual stars from the Milky Way's thick disk and halo are expected to have nearly identical space motions. Clusters are also easy to find and samples of their stars easy to study – they are all located in one place! But field star studies are deceptively hard, depending on the field population of interest, a given area of the sky may contain relatively few stars of that component. Finally, whereas distance effects are rarely a problem with clusters (at present, this problem only shows up in the Hyades!), distance determinations are difficult, and often crucial in interpreting field-star populations. This is somewhat mitigated in external galaxies where all the stars, even in the field, are essentially at a common distance, but they are still spread over the entire galaxy rather than concentrated in one location. It is little wonder that field-star studies are much less popular than those of clusters.

Despite all these hurdles, it is interesting that meaningful insights into the populations of external galaxies actually first came from field-star studies coupled with what was known about stars in relatively nearby clusters. For example, Baade used observations of M 31 field giants to first identify the distinction between Populations I and II by merging what was then known of field and cluster stars in nearby environments of the Milky Way. In 1955, Salpeter derived his now-famous mass function slope from the observation of a break in the *field star* luminosity function and using knowledge of the location of the MSTO in globular clusters. In the case of Salpeter's study, his reasoning was based on the (then) recent realization that as populations age, the luminosity of the associated main sequence turnoff fades. In a composite system, this has a corollary: as new stars form, they permanently add to the total number of main-sequence stars only for luminosities below the turnoff luminosity of the oldest population in the galaxy. Younger,

TABLE 3. Field-star Tracers of Intermediate-age and Old Populations.

Stellar Type	M_V Range	Age Range
Main Sequence	-8 - $+15$	Any
Subgiant Branch (SGB)	$+4$ - $+1$	3-15 Gyr
Red Giant Branch (RGB)	$+1$ - -2.5	1-15 Gyr
Red Clump (RC)	$+1$	1-15 Gyr
Asymptotic Giant Branch (AGB)	0 - -6	0.1-7 Gyr
Horizontal Branch (HB, RHB, BHB)	$+1$	>10 Gyr

and hence brighter main-sequence stars, may have already evolved off the main sequence depending exactly on when they formed. Thus, though every population contributes to the main sequence for $L < L_{old-TO}$ – where L_{old-TO} is the turnoff luminosity in the oldest population in the galaxy – stellar evolution drains stars more luminous than this limit causing a change in the LF slope at that point.

These insights set the stage for the general view of galaxies that reigned through the 1970's: 'blue' galaxies have a continuous star-formation histories, while 'red' galaxies are predominantly old systems. Searle et al. (1973) in a classic paper noted that some very blue, metal-poor systems were inconsistent with an extended period of star formation, but in fact seemed to be forming stars in significant numbers for the first time at the present epoch. In 1977, Butcher (1977) repeated Salpeter's analysis for field stars in the LMC. He obtained deep (for that time) MS photometry using one of the first plates from the newly-commissioned CTIO 4m telescope. After performing careful photometry of his plates, he then derived a LF for the LMC field. Like Salpeter, Butcher saw a break in the LF except that instead of occurring at $M_V \sim +4.5$, the break occurred fully one magnitude brighter. Following the logic described above, he concluded that the majority of field stars in the LMC are in fact younger – much younger – than in the Milky Way. A few years later, Butcher's student Linda Stryker more-or-less repeated his study, but on a larger scale, and came to the same conclusion (1983). It was clear by 1980 – and emphasized by results from the dwarf galaxies of the LG (see Gary Da Costa's chapter in this volume) – that there is a wide variety of star-formation histories experienced by LG galaxies.

Before I go further in this section, I wish to define a bit more precisely what I mean by old and intermediate-age, and also introduce the stellar players I will employ. By 'old' I mean stars comparable in age to the oldest globular clusters. These are likely to be the halo, or halo-like components of a galaxy. Intermediate-age for me refers to stars ranging from about 1 Gyr to the extreme ages seen in globulars. The 1 Gyr limit corresponds to the onset of the first-ascent giant branch in clusters (Ferraro et al. 1995), representing a fundamental change in the structure of stars that occur at that age – in particular, the boundary between gradual and flash-like ignition of core He that occurs at about 2-3 $M_\odot$. There are a number of specific stellar species associated with both the old and intermediate-age populations as listed in table 3.

Note that these tracers are direct in the sense that simply discovering, for example, an RR Lyr star is very strong circumstantial evidence for the existence of an old population†.

† I ignore here non-age related evolutionary effects such as binary star mass transfer, accretion of large amounts of gas, or dynamical effects in dense stellar regions as the source of any of these

Unfortunately, these tracers all are associated with main sequence progenitors that span a very small range in luminosity from about $M_V \sim +2.0$ to $+4.5$. Note too that these tracers are all rather different in what they actually trace, and their properties are sensitive to different underlying properties apart from age (e.g. mass loss, metallicity, rotation, etc.). Thus, the luminosities of MS stars are directly sensitive to age, and to some extent metallicity. In contrast, subgiants are sensitive to age from below; i.e., you see more low-luminosity SGB stars as a population ages, just the opposite of what happens on the MS. The Red giant branch has notoriously poor age sensitivity – for almost the entire range I define as old and intermediate-age, the luminosities of RGB stars differ very little. However, the RGB is sensitive to metallicity, much more so than the main sequence. The AGB, a long-sought age indicator for intermediate-age stars, is indeed apparently very sensitive to age, but also to mass loss (and hence metallicity) in ways that remain uncertain and poorly modeled from first principles. Observationally, AGB stars are also hard to calibrate as a function of age because the phase is so short-lived that one needs a large progenitor population to see any, even the population would normally contain such stars. Finally, the HB is clearly highly-sensitive to mass-loss, metallicity, and perhaps other stellar parameters (rotation? CNO abundances?). Nonetheless, just about everyone agrees that, like RR Lyr stars, the existence of BHB stars is a strong indicator of very old stars, as old as any globular cluster. But it is important to remember that the lack of such stars does not rule out the existence of an old population, even if it is metal poor.

With these field-star tracers, I will now turn to three more specific issues: The presence and nature of halos in luminous LG galaxies; the Star-Formation History of the LMC, this time without relying on clusters; and tracing tidal encounters in the LG. These intentionally overlap with the topics from Lecture 2 where I used cluster systems to address similar issues. A comparison may help contrast not only the methods used, but also will help show how clusters and field stars differ in their findings for similar scientific questions.

4.1. Halo field populations

Even in our MW, it is very difficult to study the halo using field stars. Because the stellar density of the halo is low, the local number of halo stars is a tiny fraction of the disk stars. In addition, getting distances to individual halo stars is crucial to use them as tracers, but it is also very difficult to derive generally. Contrast that to clusters which are very easy to detect and for which we have many methods of getting distances. Finally, almost by definition field star samples are always contaminated by foreground, background objects, and sometimes by the intrinsic depth of the halo itself for nearer galaxies. Again, this is a much more serious problem than for clusters where local control fields can be observed. The 'controls' for halo fields are often very difficult to define, let alone observe. Many studies have to rely on independent estimates of foreground contamination by Galactic stars, and if deep photometry is required, on studies of galaxy counts that are needed to remove background contamination.

If RR Lyr stars are used as tracers, a complicated bias on metallicity must be understood and removed. Basically, since RR Lyr stars must lie in the instability strip, metal-rich halo stars will have very few as no stars evolve to high enough temperatures to enter the strip, while metal-poor stars live mostly blueward of the strip, only contributing RR Lyr stars during their brief evolutionary excursions through it. For this

tracers in a population; such effects clearly are important in some environments, such as the excess of bluer stars in GC cores (Djorgovski et al. 1991).

reason, BHB stars, again undoubtedly old, are also biased. Thus – and I know I am repeating myself but this is a crucial point – the lack of both BHB and RR Lyr stars does *not* rule out the presence of an ancient population! RGB stars are rare, hard to find, subject to horrible contamination by nearby K and M-type dwarfs, and, apart from some rare cases where luminosities can be inferred, hard to get distances to.

For our Galaxy, the picture that has emerged (largely from studies of clusters) is that the halo can be subdivided into at least two regions, the outer halo characterized by metal-poor clusters that exhibit an apparently rather large age spread, and show no systemic rotation, and the inner halo which is metal-richer, contains (exclusively?) older clusters, and is mildly rotating (Zinn 1996). The situation for the Milky Way halo is very difficult to interpret, principally because the tracers are typically halo stars that just happen to be close to the sun, and thus are subject to kinematic – and possibly other – biases. For example, stars with large R_G never reach the local solar neighborhood, and if there is a metallicity gradient in the halo, this would introduce a bias on metallicity too. Because they are so rare, they are also subject to many observational biases, the details of which are complex, and contentious. You are invited to try to sort this out for yourself in Volume 92 of the ASP Conference Series and edited by Morrison and Sarajedini (1996). Good luck.

In M31, field star studies since the mid 1980's reveal that the inner halo is relatively metal rich, with [Fe/H] ~ -0.6. There is also an abundance spread of about a factor of 2-3, while some studies have identified a tail to very low metallicities in this same inner-halo component (Holland et al. 1996). Because of the moderate distance of M 31, and because metal-rich giants are fainter in the V band than metal-poor giants, there is typically a strong bias against the detection of metal rich stars in M 31 and its satellites (Freedman 1992; Elston and Silva 1992). HST has helped by allowing deeper photometry and by providing the resolution needed to work in really inner halo fields in M 31. What these more recent studies find is that the metallicity distribution is broad, probably bimodal with peaks at [Fe/H] $= -0.6 \pm 0.4$ (75% of the halo stars) and [Fe/H] $= -1.6 \pm 0.6$. This distribution is more metal rich than observed among M 31 halo clusters.

We can also study the field stars in the halo of M 33 – that is, if one believes there is a field halo population. Some surface-photometry studies reveal little or possibly no halo component in this galaxy (Kent 1987; Bothun 1992). While some recent IR photometry reveals populations dominated by mostly young stars (with ages < 1 Gyr) in the inner regions (Minniti et al. 1993), other studies find some evidence for an old, extended component (Regan and Vogel 1994). Suffice to say that as far as field stars are concerned, the M 33 halo is subtle, and somewhat anomalously populous in clusters. Perhaps not surprisingly, the halo of the LMC has also been historically hard to identify. This has to do with the unfavorable face-on orientation of the galaxy, and the fact that many of the old tracers are hard to isolate. For example, because of the large young and intermediate-age field component BHB stars tend to merge photometrically into regions of the CM diagram containing much younger main-sequence stars. Nevertheless, the LMC's halo does reveal itself: metal-poor main sequence stars have been identified in inner inner fields of the galaxy (Elson et al. 1994, 1997) and of course the old clusters imply the presence of an associated halo population (Suntzeff et al. 1992). Spectacular is the recent identification of monumentally large numbers of RR Lyr stars throughout the LMC by the MACHO project – I will say more about these when I discuss variable-star tracers below.

Although halos do seem to exist in most of the luminous MW galaxies, are they all purely old systems; i.e., with ages in excess of 10 Gyr? In the MW it is becoming apparent that even our halo is not exclusively old. Preston and collaborators (Preston et al. 1991,

1994) have identified blue, metal-poor (BMP) stars which they persuasively interpret as stars with ages ranging from 5-10 Gyr, and which appear to comprise as much as 10% of the entire halo population! High velocity A and B stars likewise reveal the presence of apparently young stars either far out in the halo, or with halo-like kinematics. The satellites in our halo, from the smallest dSph galaxies to the LMC mostly experienced complex star formation histories, proving that at least in these regions star formation can proceed within the halo. Likewise, there is evidence of 'young' halo stars in M 31. Morris et al. (1994) find AGB stars there, though contamination by the M 31 disk may partly be to blame. There is preliminary evidence of B and A stars in the M 31 halo, but these stars require confirmation as members of that galaxy (Hambly et al. 1995).

Most of these studies of extragalactic halos have focussed on the inner regions. This should not be surprising: halos and the stars that inhabit them are faint! At a galactocentric distance of 30 kpc and towards the anticenter, even the bright halo tracers are relatively faint: RGB stars would have $m_V \gtrsim 16.5$ and suffer severe contamination by foreground dwarfs, while halo MS stars would have $m_V \sim 22$. In M 31, the surface brightness of its (rather prominent) halo is fainter than 27 mag/arcsec^2 at a projected distance of only 20 kpc from the center. At this position, one would expect to see only 700 RGB and HB stars in a single WFPC2 field, and you would need to reach $V \sim 26$ to do so.

4.2. The star formation history of the LMC

Recall that Butcher and Stryker had identified a clear break in the MS LF of the LMC indicative of a population dominated by stars younger than 4-5 Gyr. Recall too that the LMC clusters show a clear break in their age distribution from about ≥ 12 Gyr to 4 Gyr. Other studies of LMC field stars have also revealed a complex star-formation history in the LMC. Frogel and Blanco (1983) found an extended, bimodal AGB that they interpreted as evidence for a complex SFH over the past Gyr, and which indicated the presence of a strong 'intermediate-age' component. Hardy et al. (1984) used a monumental field-star CMD to identify a 'strong' intermediate-age component near the center of the LMC. But these studies all were working near the very limit of the IR and photographic detectors then available. Because stellar evolution is packed into an increasingly compact region of the HRD for stars older than 1 Gyr, it becomes increasingly difficult to distinguish age components. This also suggests that precision becomes more valuable than simply obtaining huge samples of stars.

Bertelli et al. (1992) tried to improve on past studies of the LMC field star age distribution. They obtained data for three fields which had happened to be used as control regions for nearby LMC clusters. These fields range from 4-5 kpc (projected) from the LMC center, a region that is well within the easily-traced structure of the LMC, but is located mostly beyond the 'young' stellar complexes that dominate blue and emission-line images of the galaxy. As shown in Figure 5, an age mixture is immediately apparent in these data: one simultaneously sees a young main sequence turnoff, and faint subgiant branch stars.

In that study, our goal was to try to characterize the SFH with as few parameters as possible. These turned out to be the age and strength of a star-formation burst, along with the usual mass function slope, metallicity-age relation, etc. We then isolated specific regions in the CMD which exhibit well-understood, pronounced variations as a function of some subset of these parameters. For example, R_1 is an index defined as the ration of MS stars more luminous than $M_V = +3$ to the number of red giant and red clump stars more luminous (in V) than the same limit. With a little thought (and tinkering with the models), one sees that this ratio has weak dependence on the slope of the IMF

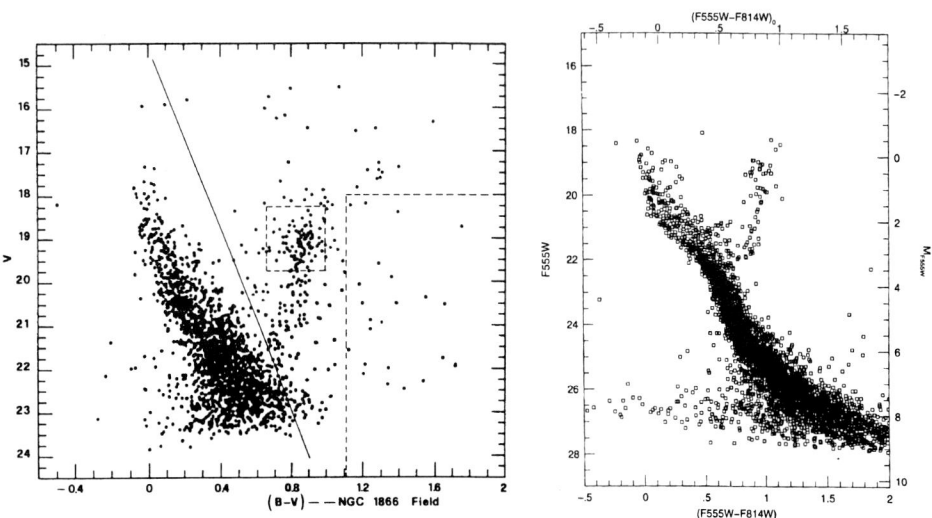

FIGURE 5. The color-magnitude diagram of an LMC field near NGC 1866 from Bertelli et al. (1992; left) and an LMC field observed with HST from Holtzman et al. (1997; right). Although the basic features are not too dissimilar, the considerably greater depth and higher photometric precision of the HST data is readily evident.

(the range of masses on the RGB is very small at a given age), but is very sensitive to when a burst of star-formation occurred and moderately sensitive to the strength of that burst. Other indices were likewise sensitive or insensitive to these and other parameters such as the age of the youngest stars in the field. In practice, the procedures involved measuring the diagnostic indices from the data (after applying whatever completeness and error corrections were required), and comparing the indices with model predictions. For example, one can calculate model indices with fixed burst strength and duration to see how the indices vary as a function of, say, burst age and the IMF slope. In the end, only solutions that could simultaneously account for the observed indices for a common set of parameters were considered to be meaningful. For all three fields, we found that the field-star star formation rate increased by at least a factor of 10 or so, about 4 Gyr ago.

Some recent papers (e.g. Gallart et al. 1996a,b; Aparicio et al. 1996) try a more 'holistic' approach to determine the star formation histories of galaxy field populations (in these cases, for LG dwarfs). This approach is complementary to that of Bertelli et al. (1992), but it does demand more from the models and requires careful weighting of different CMD features. What I like about the index approach is that one can more readily isolate the important features from a problem in which the variables are highly correlated and one can zero-in on regions where you believe the models or data to be particularly good, ignoring regions where the data are poor or the models uncertain. Nonetheless, to the extent one can rely on the models, the global approach is very powerful, and certainly computing and data-reduction techniques are now fully up to dealing with this method's significant numerical demands.

The Bertelli et al. (1992) results seem, on face value, to be consistent with the cluster age distribution, but more recent studies have begun to possibly reveal a more complex situation. For example, Westerlund (1995) claims to see significant 7-10 Gyr stars in some LMC fields, though the data are marginally useful to address this issue. Vallenari et al. (1996a,b) broadly confirm the Bertelli et al. (1992) burst parameters, but they claim that at least one field has a 7 Gyr population. Recent HST observations (Gallagher et al. 1996) find 'evidence for a burst in the SFR approximately 2-3 Gyr ago, similar to other recent studies', while more recent HST results (Holtzman et al. 1997) appear to reveal a significant population of older stars in some LMC fields. So it remains somewhat uncertain if the field stars do indeed mirror the age distribution of the clusters in the LMC. My impression is that they do, but one also sees some evidence that there may be significant field-to-field differences that will make it hard to define a 'global' field SFH. Fortunately, as HST obtains more and more deep images of field regions in the LMC and SMC (which has largely been ignored in recent field-star studies) we should be able to piece together a more detailed picture of the global star-formation in these galaxies.

4.3. *Tidal effects and old and intermediate-age field stars*

It has become increasingly obvious that the structure of the SMC is strongly affected by tidal distortions induced by the LMC and the Milky Way. We see complex structures such as a distorted halo in the SMC, the well-known SMC wing extended like a tidal feature towards the LMC, and as Irwin, Demers and collaborators have found (Irwin et al. 1985; Demers and Irwin 1991), a strong, blue stellar component associated with the so-called Magellanic Bridge connecting the LMC and SMC. Evidence of great depth is also revealed by red clump stars in the SMC. Towards the E/NE, this depth appears to reach 10-15 kpc, more than the projected width of the SMC in the sky (Hatzidimitriou and Hawkins 1989)! In contrast, the western side shows no significant spread, again consistent with a tidal feature, more pronounced on the side facing the LMC. Other examples of tidal effects are seen in NGC 205 which is clearly being disturbed by M 31. Here, the outer isophotes (which are dominated by the light of intermediate-age and old stars) is clearly distorted. Another example is the recently-discovered Sgr dwarf galaxy (Ibata et al. 1994). Here we have a galaxy observed in the throes of dissolving into the halo and for which the intermediate-age and old MS stars are an ideal tracer. Although these last two examples appear to take me away from the more luminous LG galaxies, keep in mind that they are injecting stars into the halos of M 31 and the Milky Way, respectively, and so their tidal evolution is directly relevant to the formation of the halos of their more luminous parents.

5. 'Young' field star populations in luminous LG galaxies

In the last section, I arbitrarily chose the lower age limit for 'intermediate-age' stars to be at 1 Gyr because this is the approximate age at which the first-ascent giant branch appears. Analogously, we can take the onset of the asymptotic giant branch (AGB) as the start of the 'young' era I will talk about here. Recent work suggests that this occurs at an age of about 80-100 Gyr (Testa et al. 1997). With this definition I will then define the 'young' era from approximately 0.1-1 Gyr in the past. What about younger stars? For me these represent ultra-young objects, while for the real experts (Phil Massey and Claus Leitherer) any stars older than 10-20 Myr are likely to already be old, or at least middle-aged! This era of young stars is characterized by large mass loss rates on the AGB and rapid color evolution of composite populations. The latter is quite distinct from the very slow color evolution seen in older populations (see Girardi et al. 1995). This era also

TABLE 4. 'Young' Field Star Population Tracers.

Stellar Type	M_V Range	Age Range
Main Sequence	-8 - $+15$	Any
Asymptotic Giant Branch (AGB)	0 - -6	0.1-7 Gyr
Blue Loop Stars	$+1$ - < -5	25 Myr - 1 Gyr
Red Supergiants	$\lesssim -6$	$\lesssim$ 100 Myr

represents the transition from the star-formation epoch and the steady-state population of a galaxy; thus, there is little gas and few stellar associations found close to the stars I call 'young'.

The young populations have played an important role in understanding the nature of nearby galaxies. For example, Local Group galaxies have been resolved for some time: the LMC/SMC by John Herschel early in the 19th century and probably by others before that. Lord Rosse first resolved individual stars in an even more distant galaxy (M 31) in the 19th century, long before Baade's identification of much fainter Pop II stars. The stars being resolved belonged to the young stellar populations in these galaxies, not the intermediate-age or old populations (much too faint), nor the most luminous stars since, as Phil Massey emphasizes in his chapter, these are inconspicuous (relative to their bolometric luminosities) at optical wavelengths.

The young stars continue to dominate our perception of galaxy structure, and for the most part the luminous component is dominated by these stars in most luminous LG galaxies, while the mass – even if restricted to the stellar mass – is mostly found in the older populations. For example, in M 33, the arms, which are actually quite weak and indistinct, are best traced out by the youngest stars and OB associations. In the SMC, the youngest stars are restricted to a rather small region of the galaxy and to the SMC wing. In the LMC, some of the most prominent features – dubbed 'constellations' by Shapley and covering large regions (many kpc^2 in size) in the galaxy – are the visually most prominent features. These constellations are characterized by their large proportion of young stars, but they do not really trace out the overall structure of the LMC (see ven den Bergh 1981 for an image of the LMC from the moon that graphically illustrates this).

Table 4 lists the specific stellar tracers of the young population. I will defer the discussion of the variable-star tracers to the next section. Note that many of these tracers have analogs among the old and intermediate-age tracers (Table 3), but the luminosities are considerably higher, and the detailed evolutionary stages are also different. Thus, young blue-loop stars are related to old BHB stars in their general behavior in the HRD, but the detailed internal structure is quite different in the two types of stars (as is the reason they are actually visible in the blue part of the CMD). Blue loop stars are also 10 to 100 times more luminous than BHB stars. Also note that the range in luminosities and colors is far larger in this 1 Gyr range that I define as 'young' than for stars spanning the 14 Gyr range of the intermediate-age and old populations. Much finer age resolution – both relative and absolute – is possible when dealing with young stars.

In this section I will deal with four specific topics: the age distribution of stars in the LMC over the past 1 Gyr; the use of young stars as tracers of star formation; the appearance and interpretation of young stars found in unusual places; and finally – you guessed it! – the role of young stars as tracers of the interactions of the Magellanic Clouds.

5.1. *The recent star formation history of the LMC*

As I mentioned before, the color-age relation for integrated systems – such as clusters – provides good age discrimination for objects younger than about 1 Gyr. Girardi et al. (1995) exploit this to determine the age distribution of LMC clusters. If we just focus on the youngest clusters, these results clearly show a burst in the cluster formation rate about 100 Myr ago with a full-width of about ± 50% (a much younger burst is also apparent, but that takes us to ages too low to consider here). We have already seen that Frogel and Blanco (1983) saw evidence for a similar burst in field AGB stars in the LMC. As seems to be the case for the older populations, it appears that the young clusters and field stars do trace the same global star-formation history.

5.2. *The propagation of star formation in specific regions*

Because we do have good age resolution for young stars and clusters, these tracers have been used to try to measure how star formation propagates in regions that have recently or continue to give birth to stars. To the extent that there is a theory for this propagation, there are two basic models in general vogue: self-propagating star formation (sometimes preceded by 'stochastic' to emphasize the apparently random nature of this process and its likely applicability to irregular galaxies), and star formation triggered by spiral density waves. Tidal triggers are also very likely to play an important role in initiating and sustaining star formation in many galaxies (Holtzman et al. 1992; Whitmore and Schweizer 1995; Schweizer et al. 1996); I will discuss this later as it pertains to the LG. The luminous resolved galaxies of the LG offer a few ways in which we can try to see if and how both of these star-formation propagation models actually operate in real systems. I will focus on the LMC and M 33. I note that this exercise will draw me to stars younger than 0.1 Gyr, but the basic idea remains applicable to stars as old as 1 Gyr whose photometric properties make them relatively easy to derive ages for (see also Evan Skillman's chapter for an example, and recent papers of Dohm-Palmer et al. (1997) for details and some wonderful applications in lower-luminosity galaxies).

Isserstedt (1984) developed an approach to study the recent star formation history of the LMC by using the luminosities of LMC red supergiants as an age indicator. His idea was that red supergiants do not vary much in luminosity at a given age – there is no extended giant branch yet until the AGB, then the RGB turns on at 100 Myr and 1 Gyr, respectively – so any variations in L represent differences in age. By merely selecting a magnitude interval and isolating red supergiants, he was able to isolate an age range. Simple! Cepheids (see below), and blue loop stars (see Dohm-Palmer et al. 1997) can also be used in this manner. Not much has been done since then to address the *global* recent star-formation propagation in the LMC, though I think that will soon change.

What about on smaller scales? Dopita et al. (1985) studied star formation and gas dynamics in the well-known super star-forming complex Constellation III (C III) in the northern half of the LMC (see van den Bergh for a nice picture of C III in relation to the rest of the LMC). This region has a roughly circular symmetry, and has long been suspected of displaying outwardly progressing star formation. In this paper, the authors claimed that the HI associated with C III was in fact expanding outward. They also claimed that the ages of clusters in C III became progressively younger as one moved outward from the center (but with considerable error bars), as if the expanding gas somehow triggered star formation in a large, expanding shell around its point of origin. Reid et al. (1987) used large-scale plates of the region to reproduce Isserstedt's approach in C III and concluded that there is no radial age gradient in this region. Frankly, the conclusions of both studies are of marginal significance, since the age effects were demanding the utmost precision from photographic photometric data – but the motivation and the approaches were sound. This would be a fascinating region to re-address this question and determine on what spatial and temporal scales star formation regions can propagate.

M 33 offers a chance to study how star formation propagates as a result of spiral arms. Regan and Wilson (1993) tried to do this recently by objectively identifying associations near the reasonably well-defined Southern spiral arm in that galaxy. From their CCD photometry and using the observed main sequence turnoff point, they derived ages for each association. They found no strong evidence that the star formation associated with this arm propagated *across* the arm, thus concluding that the density wave is not a star formation trigger. Nor did they find any systematic variation in age along the arm, apparently ruling out SPSF *along* the length of the arm. Of course, young stellar complexes do trace the arms in galaxies such as M 33, so there must be *some* trigger involved that can coordinate this process. There are also many observational uncertainties. As Phil Massey has emphasized repeatedly at this school, the ages of any young stellar complexes are suspect if based only on MS photometry in the B and V bands. Reddening corrections are also very difficult to determine reliably. However problematic these issues may be in this particular study, the approach is still of great interest, and I am here trying to emphasize the method more than the (controversial) result. Perhaps in some future winter schools some very similar techniques will provide more robust results on how star formation propagates in nearby galaxies.

5.3. *Young stars in odd places*

Kudritzski has noted elsewhere in this volume that the central cluster of the Milky Way, arguably one of the first locations to form in the evolution of our Galaxy, contains many young stars! No doubt this has to do with the massive object – probably a $> 2 \times 10^6$ $M_\odot$ black hole (Kormendy and Richstone 1995) – located at the Galactic Center. M 33 likewise shows evidence of young stars in its nucleus. The luminous AGB stars in that region appear to have ages under about 1 Gyr. Strong UV emission from the nucleus remains unexplained, but may be associated with hot, young stars. The optical colors and spectra of the nucleus reveal strong evidence of a very young stellar component in the nucleus of M 33, and more recently HST observations directly reveal these stars. Just like our Galaxy, right? The problem is that M 33 does *not* appear to have a nuclear black hole, at least not one more massive than about 5×10^4 $M_\odot$ (Kormendy and Richstone 1995). So, depending on how you look at it, nuclear star-formation either requires a super-massive black hole (the Milky Way) or doesn't care at all if a super-massive black hole is present (M 33). You choose.

Another place where young stars occur, but shouldn't, is at high latitudes far from the plane of the Milky Way. These stars range in spectral type from F to B, and even O! If they are 'normal' stars for their spectral types (and there is never any reason to believe otherwise), they are far too young to have gotten to their location from any reasonable ejection process from the Galactic disk. As an example, consider BD–15°-115, a B-type star with an evolutionary age less than 20 Myr, yet located (and apparently formed) well above the Galactic plane. Specifically, it would have taken 50 Myr for this star to reach its current position if it had been formed in the disk (Conlon et al. 1992). As more examples of such stars were found, it became apparent that they have thick-disk and/or halo-like kinematics. The only reasonable conclusions seem to be that a) these stars formed in the halo or thick disk, well away from the Galactic disk and in the very recent past, or b) they formed in some recently disrupted dwarf galaxy that had either just formed some stars or formed stars during the disruption process. As I mentioned earlier, Preston et al. (1994) claim that as much as 10% of the halo is 'young'. Hambly et al. (1995) found some possible analogs in the M 31 halo, but these are somewhat marginal cases still in need of spectroscopic confirmation. I cannot help but think that these stars – so radically inconsistent with basic tenets of modern Galactic structure – must be fundamentally important. But who will unlock the secret?

5.4. *Young stars and the interaction(s) of the Magellanic Clouds*

There is considerable evidence of interactions between the Clouds. For example, both galaxies are embedded in a common, very complex H I cloud, while the 100 degree-long Magellanic Stream extends away from both galaxies towards the Northern Hemisphere (a good image of the entire system is given by Mathewson and Ford 1984). Models of the interactions of the Magellanic Clouds and the Milky Way inevitably predict strong and long-lasting tidal interactions between the galaxies (Gardiner et al. 1994; Lin et al. 1995). Some of the more prominent episodes of enhanced star formation in the LMC (100 Myr and 4 Gyr ago) seem to correspond to past close mutual encounters between the LMC and SMC. Finally, the SMC is clearly highly distorted, both along the line of sight and projected onto the plane of the sky.

The evidence for interaction is convincing, but can such encounters actually trigger star formation? In the case of the Magellanic system, the answer seems to be 'yes'. Irwin et al. (1985) found a bridge of stars between the LMC and SMC located roughly along the H I bridge that had been known for years. Subsequent work (Irwin et al. 1990; Demers and Irwin 1991; Demers et al. 1991; Grondin et al. 1992) found and studied field star in this region with ages less than 300 Myr with *no* evidence for any older stars. This stellar bridge also contains a number of well-defined clusters/associations with ages less than 20 Myr. The depth of this stellar component is about 10-15 kpc, not too different from the depth observed in the eastern edge of the SMC. Since the streaming time across the bridge is considerably longer than this, these stars must haver formed *in* the bridge region. It would be fascinating to derive reliable abundances for these stars, as this observation might indicate whether the bulk of the material from which they formed came from the LMC (with a mean present-day metallicity of about [Fe/H] ~ -0.3) or the SMC ([Fe/H]$_0 \sim -0.7$).

6. Variable star populations in LG galaxies

The galaxies of the LG have played an important, early role in variable-star research. S And, the first well-observed supernova of the telescopic era, was first seen in 1885 (see the centennial review of this event by de Vaucouleurs and Corwin 1985). As the first extragalactic supernova, it immediately caused problems by proving that M 31 is not a purely nebular system, and by revealing a particularly unusual type of star. Early in the 20th century, Henrietta Leavitt identified and studied the properties of Cepheids in the SMC. Although Cepheids had been well studied for many years before her work, she was the first to notice that they obey a period-luminosity relation (which has proven to be of some use!). She did this by taking advantage of the common distance of the SMC Cepheids, a feature not available for Galactic Cepheids and a very good advertisement of one of the reasons why our LG neighbors are so important. The Milky Way globulars revealed the presence of many short-period pulsating variables – the RR Lyr stars – which have proven to be a useful distance indicator for our Galaxy, and more recently nearby galaxies. Of course, Hubble took many of these results to help establish first that the

TABLE 5. Variable-Star Population Tracers.

Variable Type	M_V Range	Age Range
Classical Cepheids	-2 - -8	20-200 Myr
W Vir Stars	0 - -5	> 10 Gyr
RR Lyr Stars	$+1$ - $+0.5$	> 10 Gyr
Long-Period Variables		
$P < 225$d	-3 - -5	> 10 Gyr
225d $< P < 425$d	-4 - -6	1-10 Gyr
425d $< P$	< -5	< 1 Gyr
Anomalous Cepheids	$+0.5$ - -2	$\gtrsim 5$ Gyr
Novae	< -7	$\gtrsim 5$ Gyr
Supernovae	< -17	< 10 Myr (SN II)
		> 10 Gyr (SN Ia)
Luminous Blue Variables	-8 - -9	< 15 Myr

'nebulae' are often in fact extragalactic stellar systems, and to derive their distance scale. Novae and Cepheids in M 31 and other LG galaxies were crucial in this development. Even one famous failure to spot LG variable stars proved extremely important. When Baade (in the late 1940's) was unable to identify RR Lyr stars in M 31, he knew the distance scale was significantly in error: in fact, Type II Cepheids from globulars had been used to calibrate Type I Cepheids in young populations. The scale of the Universe suddenly got larger by a factor of two and all because of some RR Lyr stars that Baade did *not* see! In the mid-1950's Thackeray finally did find RR Lyr stars in the LMC – the first extragalactic RR Lyrs discovered – providing a beautiful confirmation of Baade's revised distance scale. With the advent of CCDs in the 1980's, there has been an explosion of our knowledge of extragalactic variables.

There are many types of variable stars that play a role in our knowledge of the stellar populations – not to mention the distances – of galaxies in the LG. These are listed in Table 5. In this chapter I will exclusively focus on the relations of these stars to their parent populations, but be aware that they are quite useful for distance determinations, and to help study the basic properties of stars in different evolutionary phases.

In general, there are many features about variable stars that make them particularly useful for population studies. Identification is usually unambiguous – distinctive light curves, periods and and colors often define types uniquely, and none of these depend on distance (apart, of course, from limiting the attainable S/N as more distant galaxies are observed). Most variable types have local, well-known prototypes with reasonably well understood physical properties. Most are stable and thus amenable to long-term study. For example, many pulsating variables in Table 5 have well-defined periods, evolve slowly, and obey well-defined period-luminosity (or possibly period-luminosity-color) relations. These stars are generally also very closely tied to well-defined age groups; thus, RR Lyr stars are always old while Type I Cepheids obey a luminosity-period-age relation and are found only in populations younger than about 200 Myr. Finally, virtually all populations contain some species of variables, making them nearly universal tracers.

There are problems with variables too. Often they present a biased insight into the underlying age and metallicity distribution. The existence of Cepheids says nothing about older stars, while the lack of Cepheid, because of metallicity effects, does not rule out the presence of young stars. The detailed properties of many variables depends sensitively on mass loss – this is especially true of RR Lyr stars and LPVs. There are also cases where unambiguous identifications are *not* easy: RR Lyr stars and anomalous Cepheids can have very similar photometric properties despite having very different luminosities (Nemec et al. 1994). In practice, the underlying period-luminosity relations generally depend on color, the band used to make the observations, and metallicity. Finally, variable-star observations are difficult in that they demand multi-epoch observations, often in difficult fields when extragalactic variables are being observed. This is a method that demands lots of telescope and reduction time.

¿From the ground it is practical to carry out long-term monitoring programs for virtually any of the species listed in Table 5 in many LG galaxies with modest-sized telescopes. The obvious exception of RR Lyr stars which demand 4m class telescopes, or HST beyond the local dwarfs (e.g. Pritchet and van den Bergh 1986a; Saha et al. 1992). Large-scale surveys of the Galactic halo or the halos of other galaxies are also well suited for variable-star detection if multiple epochs are obtained. For example, the Sloan survey intends to repeat measurements of about 100 deg^2 to identify optically variable quasars, and, incidentally, high-latitude variable stars. Various microlensing surveys are monitoring millions of stars nightly in the Magellanic Clouds, M 31 and the Milky Way. Stanek et al. (1996) have carried out extensive surveys of variables in M 31 and M 33. With HST it is possible to identify virtually any type of variable in table 5 throughout the entire LG! Cepheids can be monitored out to the Virgo cluster, and LPVs even further. *Every* LG galaxy is known to contain variable stars; they represent a fundamental population tracer for virtually every known environment.

Let me know turn to some specific applications where variables have played a role in understanding the populations of luminous LG galaxies (and in some other galaxies too). These include variable stars in dwarf spheroidal galaxies, novae in M 31, gravitational microlensing surveys; and – surprise! – variable stars as tracers of tidal interactions.

6.1. *Variable stars in the closest dwarf galaxies*

I will digress a bit from the more luminous LG galaxies, and discuss here the variable star populations of nearby dwarf spheroidal (dSph) galaxies. As described above, these galaxies may play an important role in populating the halo of galaxies such as ours, and so the relation between their stellar populations and those of larger, more luminous galaxies may be more direct than generally believed.

As Gary Da Costa emphasizes in his chapter, the dSph galaxies have experienced surprisingly complex star-formation histories. The first hint of this complexity actually came from studies of the variable stars in these systems. In the late 1970's, Zinn (1980) noted that the mean periods of dSph RR Lyr stars were 'strange'. In particular, while the average periods of the ab-type RR Lyr stars (the ones with asymmetric light curves and periods typically longer than about 0.45 days) in globular clusters tend to clusters about two values: 0.55 days for the so-called Oosterhoff I clusters, and 0.65 days for the Oosterhoff II systems. The Oosterhoff classes are closely related to other cluster properties: OoI clusters tend to have red horizontal branches, and are thus on average more metal-rich than OoII clusters which have mostly blue horizontal branches. Table 6 shows that dSph galaxies cannot be unambiguously classified as either OoI or OoII, but instead typically lie between 0.55 and 0.65 days.

The surprise in all this is that the dSph systems were known to be very metal poor (a conclusion since confirmed and expanded to all the known dSph galaxies of the MW). Thus, the existence of Oosterhoff-intermediate cases implies that some 'second-parameter' must be affecting their HB morphology. Zinn long ago suggested that this

TABLE 6. Mean Periods of Type-ab RR Lyr Stars in LG dSph Galaxies and the LMC.

Galaxy	$N_{RR,ab}$	$\langle P_{ab} \rangle$ days
Draco	133	0.61
Ursa Minor	82	0.64
Carina	57	0.62
Sextans	36	0.61
Leo II	196	0.59
Sculptor	225	0.59
Sagittarius	300+	0.58
LMC	8000+	0.59

parameter may be age, such that the dSph systems are on average younger – and therefore have redder HBs than expected for their metallicity – than globular clusters. Main sequence photometry of dSph galaxies has confirmed this (see Gary Da Costa's chapter in this volume for details and references).

Another odd feature of the dSph variable stars points in the same direction. Anomalous Cepheids are rather rare in all environments *except* dSph. In fact, every dSph system that has been adequately studied has been shown to contain these variables. Although the detailed evolution of these stars remains poorly understood, it seems clear that to produce ACs one requires metal-poor stars with main-sequence masses in the range 1-1.5 $M_\odot$. Contrast this with the usual turnoff mass of 0.7-0.8$M_\odot$ for ancient globular-cluster populations. There are thus two obvious options: either ACs result from the merger of stars either in binaries or due to physical collisions, or else they are *younger* and thus more massive than the older stars in the galaxies. While a merger may have created the one AC known in the entire GC system (V19 in NGC 5466; Mateo et al. 1990), there are so many ACs in the low-density dSph galaxies that the second option – relative youth – seems a more plausible explanation of their existence in those galaxies.

6.2. *Novae in M 31*

Although pulsating stars get most of the press when it comes to determining extragalactic distances, other sorts of variables are useful too. M 31 illustrates how novae can be used to determine distances because of the existence of a reasonably well-defined Luminosity-decline time relation. Specifically, novae that decline most rapidly are more luminous than the slower decliners. Moreover, there is a sharp steepening of the relation that helps make this more suitable for distance determination. Van den Bergh and Pritchet (1987b) illustrate this technique and define the L-Δt relation for M 31 novae as they try to apply the technique to Virgo cluster novae. Because I wish to concentrate on the LG, I only want to emphasize how nicely-defined the relation is for M 31. This suggests that novae may prove useful to get distances to galaxies throughout our group and for nearby groups, especially in cases where these can be monitored with moderate-sized telescopes. The good features of nova is that they are relatively luminous (ranging from $M_V \sim -6$ to -8 at maximum light), they occur frequently (in M 31 the rate is about 15 per year), and they tend to suffer low reddening because they belong to the bulge/thick disk/halo population. Of course, unlike pulsating variables, a certain amount of luck is needed to identify novae (or a dedicated search), and follow-up observations are crucial to determine the decline time.

6.3. *Gravitational microlensing in Local Group galaxies*

This idea, discussed in detail by Paczynski (1986, 1991), involves using intervening lenses (stars, dark-matter objects, planets, etc) to magnify, and brighten resolved background sources towards regions in nearby galaxies with high stellar density. The resulting light curve depends on the relative distances of the source and lens, is usually achromatic (if the source is a close binary or superimposed pair of stars, there actually can be color variations as one star is lensed and the other is not), should only happen once, and has a distinctive light curve shape. The rates are very low: for example, the optical depth to microlensing towards the Galactic center is about 10^{-7}, meaning that at any given time only one star in 10 million will be strongly lensed. Obviously, you have to monitor a lot of stars, and do so frequently and for a long time! Not for the faint-hearted.

There are a number of surveys underway, including MACHO and OGLE (both are now monitoring the Milky Way bulge and the Magellanic Clouds), and other surveys that include observations of M 31. As required by the observational demands of the project, these are large-scale surveys and both OGLE and MACHO have dedicated 50-inch telescopes to carry them out. Over 250 lensing events – mostly towards the Galactic center – have by now been identified out of the 3 billion individual photometric observations that have been carried out.

Of course, one also finds *many* other variables; the numbers from MACHO for the LMC are particularly impressive. In only a central 11 deg^2 region, over 1500 Cepheids (including 45 of the rare multi-mode Cepheids; see Alcock et al. 1995), about 8000 RR Lyr stars, 25000 LPVs and 1200 eclipsing binaries have been identified (Alcock et al. 1997b,c). Some interesting science can be done with these that sheds insight on the parent galaxy populations. For example, in the LMC the ab-type RR Lyr stars are observed to have a mean period of 0.583 days. This is similar to what is seen in the oldest LMC clusters, and, like the dSph galaxies implies an Oosterhoff-intermediate population, though closest to OoI. This is difficult to reconcile with the mean metallicity of the older clusters (and by implication the RR Lyr stars) which is about [Fe/H] = –1.7, closer to the typical metallicity of OoII clusters. It is also well known that the *old* LMC is dominated by a population with a red horizontal branch, though any BHB stars would be difficult to distinguish from the prominent young MS stars found throughout the galaxy. Nonetheless, using the same arguments as for the dSph galaxies, these unusual features of the RR Lyr stars in the LMC imply a relatively young halo. Welch et al. (describe this in detail, and conclude that the LMC RR Lyr stars must be on average about 3-4 Gyr younger than the oldest globular clusters of our halo.

6.4. *Variable stars as tracers of interactions*

As part of an ongoing study to identify and understand the properties of eclipsing binaries in globular clusters, I have been studying the nearby globular cluster M 55 to identify photometric variables along the main sequence. We have indeed found binary stars in that cluster, but we also found a number of other variables that could not have been M 55 members. It turned out that these stars are RR Lyr stars located in the Sgr dwarf galaxy *behind* M 55 (Mateo et al. 1996). At about the same time, Alard (1996) found a large number or RR Lyr stars in the western side of the Sgr dwarf. Again, he found the stars during a search for variables towards the Galactic center; although the bulge RR Lyr stars were visible in great numbers, just as in the case of M 55, the Sgr variables showed up clearly as an enhancement in the numbers of variables at a common distance well behind the bulge. These observations of Sgr RR Lyr stars were an early indication that the galaxy is very much more elongated than originally believed, almost certainly as a result of severe tidal effects by the Milky Way.

An earlier use of variables to study the structure of the SMC likewise has uncovered evidence that that galaxy is severely distorted along the line of sight. In particular, Caldwell and Coulson (1986) and Mathewson et al. (1986, 1988) claimed a significant depth (nearly 20 kpc) in the SMC due to the spread of the galaxy's Cepheids about their best estimate of the Cepheid PLC relation. In the case of the SMC, 20 kpc is extreme, and this result was tantamount to saying that the SMC is being tidally torn apart. Welch et al. (1987) later used IR photometry to claim the depth is 'only' 10 kpc, but the conclusion is the same: the SMC's depth as obtained from analysis of its Cepheids is much larger than expected for a disk-like or even spherical system. I am currently trying to do something similar in Sgr by using the RR Lyr stars in different fields to determine the 3-D structure of that distorted system. Alcock et al. (1997a) have argued that there is a considerable tilt of Sgr relative to the plane of the sky, such that the SE side of the galaxy is considerably more distant than the the NW side that is closest to the Galactic plane.

Variable stars are potentially very powerful probes of the 3-D shape of other nearby galaxies (in the same study, Caldwell and Coulson also found the tilt of the LMC relative to the plane of the sky using its Cepheids, but concluded that there was no evidence of comparable tidal streaming effects as seen in the SMC). High-latitude surveys of 'blank' fields may even be very sensitive to remnant or recently-destroyed dwarfs in a more advanced stage of disruption than Sgr. The Sloan survey's 100 sq. degree field where they intend to obtain repeat observations may help reveal such systems. With HST it may be possible to carry out similar depth studies in more distant systems, or at least to better constrain the relative distances of interacting galaxies such as M 31, NGC 205 and M 32. Because of the short periods of RR Lyr stars in particular, it may not be difficult to tailor HST observations – which often require multiple observations in any case – to search for these variables in nearby systems.

7. Beyond the Local Group

In this last section I want to briefly explore some aspects of nearby groups – and in particular the more luminous galaxies in these groups – that may complement what I have described about the luminous LG galaxies in the earlier sections. To do this, let me start with a summary of what the LG actually is. Many authors define the LG as any galaxies within a certain distance of us or the group's barycenter (the latter is rather insensitive to group membership since the Milky Way and M 31 dominate the local mass distribution; thus, the barycenter is about 2/3rds of the way betwen us and M 31 along the line connecting the two galaxies). It is quite typical in the literature to define all galaxies within 1-1.5 Mpc of the chosen reference point as a LG member. Another way – and perhaps more physically justified – is to only consider galaxies bound to the group as bona fide LG members. This has some disadvantages in that galaxies that have escaped the LG would clearly not be considered members, we still have very incomplete kinematic information on most LG galaxies (in particular, very few have proper motions), and there is usually no very rigorous definition of what is or is not a member kinematically (see Yahil et al. 1977; Zaritsky et al. 1989). One can try a more sophisticated approach based on pattern matching. Karachentsev (1996) uses a specific clustering algorithm to identify groups – including the LG – in the local universe. This is still somewhat arbitrary as it depends on the specific parameters you select for the algorithm and the completeness of your catalogue, but at least it is somewhat deterministic and repeatable. Nevertheless, membership in the LG is still a rather fuzzy concept, particularly near the edge.

If we take a very liberal view and consider the 'maximal' content of the LG, one counts 38 members (compare this to the 29 members listed by van den Bergh 1994, keeping in mind that two galaxies, Sagittarius and Antlia were discovered subsequent to that paper). These galaxies cluster into two dominant subgroups centered around M 31/M 33 (9 galaxies) and the Milky Way/LMC (11 galaxies), respectively. However, 17 of the galaxies appear to be 'free' objects, not clearly associated with either subgroup. As seen in table 7, there are 3 spirals in the LG, 1 dwarf E (M 32), 16 dSph, two galaxies with mixed dSph and Irr characteristics, 4 Magellanic irregulars and 11 generic irregulars. The global LF of the LG is typical to that of 'poor groups' (Ferguson and Sandage 1991) down to $M_B \sim -14$. and the global M/L ratio (in the B-band) is about 50-100 with a velocity dispersion of around 70 km/s, all very typical of other similar galaxy groups. I have provided Table 7 as a summary of the basic properties of all the individual members of the Local Group; see van den Bergh (1994) for a summary of arguments for and against membership of some of these and other galaxies suspected to be LG members. I have added some notes in the table for Sagittarius and Antlia, both discovered after that paper was published. One of the best ways to visualize the distribution of galaxies in the Local Group is to view it in three dimensions; Figure 6 provides three stereo images of the Local Group as it would appear to a large(!) observer located just outside (about 1.5 Mpc from us) and looking in along a set of orthogonal axes.

What about other groups? The nearest ones include the M 81 group which is dominated by (naturally) M 81 and its companion M 82, but also contains the less well-known NGC 2403, NGC 3077, and many smaller irregulars and dwarf spheroidal galaxies. The mean distance of the group is about 3.5 Mpc. The Sculptor group contains a number of intermediate-luminosity galaxies including NGC 7793, NGC 55, NGC 253, NGC 247 and NGC 300, all late-type systems (eg Sb-dIm). Coté (1996) identified many dwarfs that are probably associated with this group, adding to the few previously known. The Sculptor group has a very large depth, ranging from about 1.5-3 Mpc from us (Puche and Carignan 1991). The Maffei group may be the closest one to the LG (Buta and McCall 1983; Karachentsev et al. 1997), but it is located at very low Galactic latitudes and is therefore difficult to observe through the intervening absorption and the very crowded fields of foreground stars. This group's distance is still debated, but is probably about 2-4 Mpc. The Maffei group is particularly important because Maffei I is undoubtedly the closest large elliptical galaxy to us. It can and has been studied in the IR to try to mitigate the absorption and crowding problems (Luppino and Tonry 1993). Maffei II is a large spiral in the group, and IC 342 a famous, face-on late-type spiral. Not surprisingly, there are few dwarfs known in this group (McCall and Buta 1995), though they almost certainly do exist in abundance, but they are difficult to find in this very low latitude line of sight. The Centaurus group is famous for NGC 5128, otherwise known as Cen A, and also NGC 5236. This group is located about 3.7 Mpc away, and in Cen A has the closest giant E and luminous radio source. It is clear even from these pocket descriptions that the luminous galaxies in these groups provide a significant extension of galaxy types, luminosities, and other physical parameters than we see in the LG. Although still challenging, we *must* extend our studies to these groups to learn more about a wider variety of luminous galaxies.

Two basic factors limit this effort. First, of course, is that the greater distances limit the luminosity to which we can identify individual stars (proportional to D^{-2}. The second is crowding, which is proportional to D. In general, the number of stars we can

TABLE 7. Local Group Galaxies

Galaxy	Other Name	α_{2000}	δ_{2000}	Type	v_0 km/s	D Mpc	$M_{B,tot}$
M 31	NGC 224	00 42.7	+41 15	Sb	−59	0.7	−20.6
Galaxy	Milky Way	...	...	Sbc	...	...	−20.3
M 33	NGC 598	01 33.9	+30 39	Sc	3	0.7	−18.3
LMC		05 23.6	−69 45	Im	12	0.05	−18.2
SMC	NGC 292	00 52.7	−72 50	Im	−30	0.06	−16.5
IC 10		00 20.4	+59 18	Im	−83	0.7	−16.2
NGC 3109	DDO 236	10 03.1	−26 10	Irr	130	1.3	−15.9
NGC 205		00 40.3	+41 41	dSph	−1	0.7	−15.6
NGC 6822	DDO 209	19 44.9	−14 49	Im	66	0.7	−15.5
M 32	NGC 221	00 42.7	+40 52	dE	35	0.7	−15.3
NGC 185		00 39.0	+48 20	dSph	39	0.7	−14.8
NGC 147	DDO 3	00 33.2	+48 31	dSph	89	0.7	−14.5
IC 1613	DDO 8	01 04.8	+02 07	Irr	−125	0.7	−14.3
WLM	DDO 221	00 02.0	−15 28	Irr	−42	1.0	−14.2
Sextans A	DDO 75	10 11.1	−04 43	Irr	114	1.3	−13.9
Sextans B	DDO 70	10 00.0	+05 20	Irr	129	1.3	−13.9
IC 5152		22 02.9	−51 17	Irr	74	1.0	−13.6
DDO 210	Aquarius	20 46.9	−12 51	Irr	11	0.7	−13.0
Pegasus	DDO 216	23 28.6	+14 45	Irr	39	1.0	−12.6
Leo A	DDO 69	09 59.4	+30 45	Irr	−25	1.0	−12.5
Sagittarius		18 55.1	−30 29	dSph	173	0.02	−12.5
Fornax		02 39.9	−34 31	dSph	−51	0.15	−11.9
And II		01 16.3	+33 25	dSph	...	0.7	−10.9
And I		00 45.7	+38 00	dSph	...	0.7	−10.8
Leo I	DDO 74	10 08.5	+12 18	dSph	144	0.22	−10.7
Sculptor		01 00.0	−33 42	dSph	73	0.08	−10.5
SclDIG	UKS2323-32	23 26.4	−32 23	dE?	82	1.3	−10.4
Phoenix		01 51.1	−44 27	Irr	...	0.5	−9.9
Antlia		10 03.6	−27 20	dSph/Irr	...	1.3	−9.8
And III		00 35.3	+36 29	dSph	...	0.7	−9.4
SagDIG		19 30.8	−17 41	Irr	30	1.1	−9.4
Leo II	DDO 93	11 13.5	+22 10	dSph	−5	0.22	−9.3
LGS 3	Pisces	01 03.9	+21 53	Irr/dSph	−98	0.7	−9.0
Tucana		22 41.9	−64 25	dSph	...	0.9	−9.0
Sextans		10 13.1	−01 39	dSph	37	0.09	−8.9
Carina		06 41.6	−50 58	dSph	−44	0.09	−8.8
Draco	DDO 228	17 20.0	+57 55	dSph	−43	0.07	−8.0
Ursa Minor	DDO 199	15 08.6	+67 11	dSph	−43	0.07	−7.9

Notes: Sagittarius was discovered in 1994 (Ibata et al.; other references can be found in Section 2). Antlia was discovered in 1996 (Whiting et al. 1997), and a deep photometric survey was carried out by Aparicio et al. (1997). Details of the discovery of the other dwarf galaxies in the table can be found in van den Bergh (1994). I have taken the photometric data from that paper. The kinematic data listed here for the dwarfs comes principally from Zaritsky et al. (1989), updated with more recent data for Sagittarius (Ibata et al. 1997) and Antlia (Whiting et al. 1997). The morphological classifications as either dSph and dE are guided here by the fundamental-plane segregation that appears to occur between these two galaxy types (e.g. see Kormendy 1985, though at that time the term 'fundamental plane' was not yet in vogue).

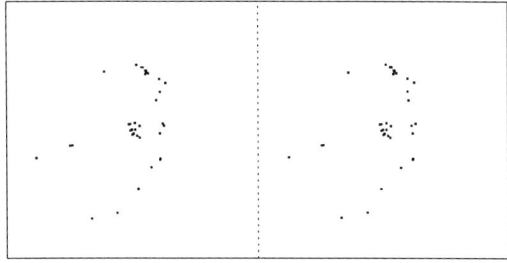

Vernal Equinox Line of Sight

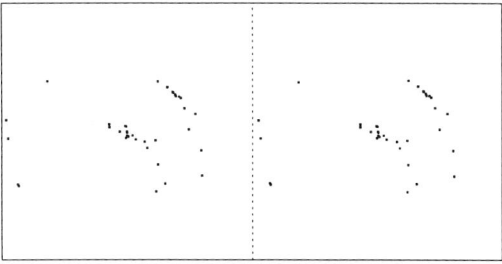

α = 6h Line of Sight

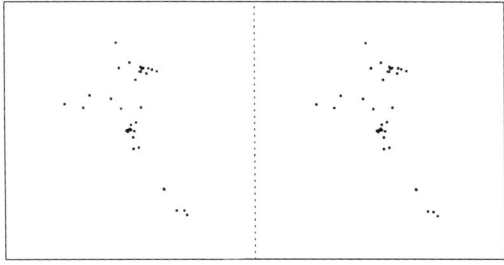

North Pole Line of Sight

FIGURE 6. Stereo images of the Local Group as viewed from a location 1.5 Mpc from the sun along the cardinal directions defined by epoch 2000 equatorial coordinates. Two clumps are clearly visible: near the center is the clumping of galaxies around the Milky Way, while the other elongated clump is associated with M 31. The five galaxies located well away from these clumps (note carefully that two of these are very close to each other, NGC 3109 and Antlia) comprise the outermost (and sometimes debated) members listed in Table 7, NGC 3109, Antlia, Sextans A and B and Leo A. The other 'free floaters' are located much closer to the Milky Way/M 31 complex and include galaxies such as NGC 6822, IC 1613 and WLM. The pronounced elongation of the M 31 subgroup is probably partly an artifact of the imprecise relative distances to those galaxies. To view these stereograms, stare at the figure pairs from about 6 inches away, but try to let your eyes focus on an object very far away. As you do this, the points will begin to merge; once they do, you can then easily refocus without losing the merged configuration. The points will suddenly 'jump out' of the page. Some people find that placing a piece of cardboard between the eyes to help. If you are at all unsure if you are seeing the true stereo effect, you haven't. When you 'get it', the effect is quite striking.

study individually will scale as

$$n = \frac{L_\odot}{L} 10^{\frac{\Sigma - 5\log D + 5}{-2.5}}$$

where n is the number of stars down to a luminosity L per arcsec^2 for a surface brightness Σ. From the ground, n_{max} is about 4 in very good seeing conditions, and for horizontal branch stars $L_{HB} \sim 200 L_\odot$. Thus for $\Sigma < 19.5$ mag/arcsec^2 and at a distance of 2 Mpc, we must reach $V \sim 27$ to identify such stars. Not easy. From HST, n_{max} is about 80, and for $\Sigma < 17.2$ mag/arcsec^2 and a distance of 3.5 Mpc, we must reach V = 28.5. Difficult but doable. In the future, improved sampling and sensitivity should allow one to identify such stars to 4 Mpc. These values are approximate in that I have ignored the crowding by other stars that would be associated with these HB stars: you should not expect to identify easily HB stars in field dominated by OB main sequence stars! But for the detection of old or intermediate-age populations in regions where these are the dominant stellar populations, these numbers should be applicable. As some examples of what we can expect to do, let me consider a few specific cases – both what we can do now, and what we can expect to do in the near future.

7.1. Ages of clusters in Cen A

The complex structure and central dust lane of Cen A has long been interpreted as a region of relatively recent star formation in this large elliptical galaxy, probably associated with some past interaction with a rather massive galaxy. Cen A also contains a large population of globular clusters. As noted earlier, it is increasingly clear that interactions of large, gas-rich galaxies can trigger the formation of mass star clusters; do the Cen A clusters reflect the overall star-formation history of the galaxy, and perhaps the event that led to its distinctive morphology? Minniti et al. (1996) measured the IR colors of five clusters near the core, concluding that they have ages of about 4 Gyr. Though probably too old to be due to or related to the event that formed the dust lane in the galaxy, these clusters are clearly not as old as the oldest population in Cen A. Did they form in a collision, or were they swapped into Cen A from another galaxy? These authors also identified 'young' clusters near the dust lane which are dominated by stars of spectral types earlier than A. These may have indeed formed as part of the merger. These questions and results all suggest that we are seeing something fundamentally different in the Centaurus group compared to anything we see in the LG, and certainly more extreme effects from a tidal encounter than anything going on today locally.

7.2. The old and intermediate-age populations in the Sculptor group

Recall that the tracers for old and intermediate-age populations are faint and suffer from relatively poor age resolution. But also recall that Baade was able to state some far-reaching conclusions merely by identifying the bright red giants in M 31. Graham (1982) did something very similar in the Sculptor group – in particular for NGC 55 and NGC 300 – by resolving their first-ascent giants to confirm that these galaxies are relatively close, and that they contain significant populations of 'old' stars with ages > 1 Gyr. This was certainly not a foregone conclusion, particular for the dIm galaxy NGC 55. Richer et al. (1985) and Pritchet et al. (1987) surveyed the Sculptor group galaxies for C stars, and finding some, were able to refine the ages of the populations to include significant numbers of 'intermediate-age' stars. Here we see examples of how we can start to use skills, techniques and knowledge derived from careful observations in LG galaxies to address basic issues in the luminous galaxies of nearby groups, particularly at the very edge of resolvability for the brightest sorts of stellar tracers. A lot of this work

7.3. 'Young' populations in the M 81 and Sculptor group galaxies

A big advance in our knowledge of these populations has come from the work of Bryan Miller who has studied in detail the current SF rates in many M 81/Scl dwarfs (Miller and Hodge 1995; Miller 1996). The basic idea in the latter paper – based on the basic precepts described by Gallagher et al. 1984 – uses the Hα fluxes of the galaxies to measure the current star-formation rate and to estimate the gas content (from H I measurements and optical emission line mapping) to derive some sort of evolutionary timescale over which star-formation can continue if it proceeds at the current rate. In the M 81 group he found many active irregular (and some dwarf E's too!) galaxies very similar to cases seen in the LG, especially the LMC and SMC. As in these two nearby galaxies, the star formation sites in the nearby groups tend to be concentrated into small regions in the galaxies. The Sculptor group, on the other hand, is far more quiescent. There are very few cases of extremely high star formation rates in the smaller galaxies. What differs in these two groups that can cause such a different mixture of galaxian star formation rates at the present epoch? Here we have an excellent area for future study: to what extent can HST observations help confirm the star-formation rates implied by these results, and can the past star-formation histories be determined more precisely. We need to recall and learn from our experiences in the Local Group. Already it is clear that one can see the most luminous stellar age tracers in these galaxies, but as in the LG, we can also try to enlist the clusters, variable stars, and ultimately fainter stellar tracers to map the star formation more precisely.

7.4. Cepheid populations in galaxies

Recall that Cepheids are young stars, and their distribution can help us map the locations and ages of recent populations in these galaxies over the past 200 Myr. Because Cepheids are relatively rare and occur over a limited age and metallicity range, this approach is restricted to more luminous galaxies with recent star formation. In general, the photometry for these Cepheids is good, and they are typically not too crowded in the color-magnitude diagram or, for the ones with the best photometry, in the sky. One potential problem is that it remains difficult how to take into account metallicity effects on the age estimates for these stars, or even if such effects matter. One beautiful recent application of this has been carried out by (Magnier et al. 1997) who use observations of Cepheids in M 31 to trace the recent star-formation history in the disk of that galaxy. This same approach should be applicable to the Cepheid populations of in more distant galaxies observed with HST.

7.5. Tidal tracers of interactions beyond the LG

As mentioned, Cen A has long been considered a merger. Thomson (1992) pointed out that two nearby galaxies – the Fourcade-Figueroa 'shred' (FFS) and NGC 5237 – may also reveal some interesting clues about the nature of this merger event. He modeled this entire system as a spiral galaxy of moderate luminosity falling into Cen A. He was able to reproduce many of the features of this merger: a small fragment is ejected to become the FFS, while NGC 5237 is the remnant bulge of the original spiral! Does the detailed star-formation histories of these galaxies support this model? Certainly portions of the stellar regions near the Cen A dust lane contain young stars, and the FFS is clearly dominated by recent and on-going star formation, but detailed estimates of the ages of these components remain to be determined. Some of the techniques honed in the LG –

including Cepheids, LPVs, blue MS stars – may be applicable to determine the SFHs of these galaxies. In the M 81 group, there is evidence of a direct stellar bridge between M 81 itself, and M 82, one of the most actively star-forming galaxies in the local Universe. Could it be that this interaction has triggered the incredibly active star formation rate seen now in M 82? Is this a snapshot of what happened between the LMC and the Milky Way 4 Gyr ago?

8. Epilogue

As we try to extend our knowledge of stellar populations and galaxy star-formation histories beyond the LG, the lessons we have learned locally will help guide our efforts. We know now, for example, how much effort is required to obtain reliable age information for old and intermediate-age populations. Detecting the old red giants is a start, but more quantitative results demand far more. Variable stars offer a complementary window on these populations, and one that in many ways is less ambiguous than what we learn from the non-variables. Even a few RR Lyr stars can signal an old population's presence when younger stars dominate a galaxy's more luminous stellar content (eg, see the case of Carina in Saha et al. 1986). Star clusters do appear to reflect the underlying star-formation history, but it would help greatly to try to confirm this in more nearby luminous local group galaxies where both the cluster and field populations can be studied in detail. Looking at nearby groups, however, suggests that mergers could help explain this coupling – triggering high star-formation rates that lead not only to the formation of massive star clusters, but probably numerous field stars as well. But is the opposite true? Certainly in low-luminosity galaxies, field-star formation can proceed at a moderate rate without forming *any* globular clusters, so the detailed relation between the global and cluster formation rate must be complex, and probably dependent on the global star formation rate, and perhaps on the specific mechanism that triggered a give star-formation episode (galaxy collisions, for example, seem to be a good way to form clusters). Finally, and this was a theme I tried to percolate through all the lectures, interactions are everywhere! So much so that I believe it becomes reasonable to ask whether interactions can explain some, most, or even all of the galaxy evolution we see in LG and other nearby galaxies? It should be exciting to find out, and we certainly have a lot of work still to do.

I would like to thank the organizers of the Canary Islands Winter School for hosting such a excellent meeting; Artemio Herrero and Antonio Aparicio were the more visible of these organizers, but I want to also thank all of the many other people who helped make this such an enjoyable and productive experience. As in any school, the quality is only as high as the students, and so I thank the Winter School students for their high level of participation in the school and for sharing with everyone their exciting research in the really first-rate posters we were able to admire during the course of the school. I know the 'pundits' learned as much from the students (if not more!) than the other way around. The intense schedule and the flu epidemic (and maybe too the all-night dancing that some of the participants engaged in) sometimes made the experience grueling for the students, lecturers and organizers, but it was also *extremely* rewarding. Plus, the students were a fun bunch to be with. I also thank the other lecturers for their excellent presentations and for many very enlightening discussions throughout the two weeks in Tenerife. I specifically want to thank Paul Hodge, Ed Olszewski and Nick Suntzeff for discussions relevant to this review, as well as my numerous other collaborators on the various projects I mentioned in this chapter. Finally, I want to thank the IAC for acting as such gracious hosts of this meeting, and, more generally, for supporting the winter schools in general. I cannot fail to mention my gratitude to the vineyards of Tenerife who offered some of the most pleasant wine and cheese (and many other goodies) tasting I've every experienced, along with some of the most hair-raising bus rides getting there!

1. Appendix: stellar photometry examples using DoPHOT

I describe three examples that illustrate different modes of operation of the DoPHOT photometry reduction program. These tests are designed to demonstrate *general* principles of stellar photometry reductions, and also provide information specific to the use of DoPHOT. The tests include
- Reduction of a BV pair of images of a single field;
- Reduction of a BV pair of images with a variable PSF;
- Reduction of a series of V-band images to illustrate the use of 'fixed-position' mode in which a template is used to set the relative positions of stars in the frame.

A full description of how to run the examples and how to access them is provided along with the standard DoPHOT manual for more detailed information about DoPHOT. A tar file containing all the routines and data needed to carry out the examples is also provided.

1.1. Introduction

These exercises provide an introduction on how to use DoPHOT for CCD stellar photometry in a variety of modes. The examples all use data of the LMC cluster NGC 2031 which contains a large number of Cepheid variables. There are three distinct techniques illustrated in these examples:

- 'Simple' two-color stellar photometry in a slightly crowded field.
- Reductions using a variable PSF.
- Time-series photometry using the 'fixed-position' option in DoPHOT.

These basic techniques were described in my first Lecture (Section 2) during the at the Winter School. In all cases, the input data to DoPHOT are processed FITS-format files. These files MUST be in short integer (16 bit) format with the version of DoPHOT used here, but there are versions of the code that can handle real images. Signed or unsigned integers are equally suitable. DoPHOT does NOT check if the format is ok, even if it is written in the FITS header. You may also have to swap bytes to get DoPHOT to run on non-Sparcs; this is controlled by a parameter SWAB in the `diskio.c` routine in the DoPHOT subdirectory (see below). The newest version of DoPHOT available from Michigan has completely replaced this fits reading subroutine with the library of routines available from NASA/Goddard:

`http://heasarc.gov.nasa.gov/docs/software/fitsio/fitsio.html`

I have not included the fitsio library because these may not work with your compiler/computer. If you do want more information about this, contact me via e-mail at the address on the first page of this chapter.

The data and the reduction programs are all located in subdirectories within the tar distribution file. That file is located at

`ftp://ra.astro.lsa.umich.edu/pub/get/mateo/dophot_ws.tar.gz`

If the file is not present, contact me at via e-mail as it may have been moved to save disk space.

The examples described here are located in subdirectories /a1, /a2, and /a3 that will be generated when you gunzip and tar the file. To get started, move to one of these directories. You will find that they contain (a) the data files to be reduced, (b) a file called 'paramdefault' which contains default parameters for DoPHOT, (c) a number of so-called tuneup files (typically in the format tuneup.nnn where nnn is the number of the frame to be reduced), and (d) assorted other files that are described as needed below. You will also see another subdirectory called /results. This contains my results of the reduction of the data frames in the parent directory. These results are provided as a check that you get the same answers on your machine at your institution. If you cannot reproduce the files in the /results subdirectory, you are not running DoPHOT correctly, or the executable file is corrupted or improperly compiled. There are many additional files in the various /results subdirectories that I describe in detail below.

The files in the /aN subdirectories (where N is 1, 2 or 3) include:

• Fits images; these files are fully processed (flat-fielded, bias-subtracted, trimmed, etc.) and are ready to be reduced.

• The file paramdefault which contains *all* of the parameters used by DoPHOT. Look at this file. There are *many* parameters! Most of these are described in the DoPHOT manual. A copy of the manual and all the TeX files associated with it is in the /manual subdirectory. There are two papers you can look up about DoPHOT: Mateo and Schechter (1989), and Schechter, Mateo and Saha (1993). Most of the parameters that are not in the DoPHOT manual are relatively new and some of these are discussed below. Note that you do *not* have to adjust all of these parameters for every file; on the contrary, only a small subset of the parameters need periodic adjustment. Hence, to tailor DoPHOT to your needs you also have ...

• Tuneup files. These files (named here tuneup.nnn) contain only those parameters that you wish to adjust for a given DoPHOT reduction. Things like the sky value, the FWHM and specific file names are often adjusted for each frame, but the tuneup files also contain other parameters that are typically adjusted much less frequently. If a parameter is *not* in this file, its value is taken from the parameter default file (paramdefault in these examples and also the default name, but this file – like any of the files used by DoPHOT – can have any name).

• Each subdirectory also contains a file called explain that provides a very brief description of the data files in that directory. The information in these files is expanded in this document.

• As noted above, each /aN subdirectory contains another subdirectory called results which has the reference reductions for the corresponding data frames. Use these results to check that DoPHOT (and the user!) are working correctly.

Finally, let me point out that I cannot support DoPHOT users as if I were Microsoft – as much as I might like to do so, I cannot readily address all the problems that one might have when applying DoPHOT to data other than the ones provided here. The philosophy behind DoPHOT has always been to 'learn and modify'. Learn to use the program by using it. Learn to fix the program by fixing it. Learn to employ the program to your needs by modifying it. But please, do not expect me or any of the other DoPHOT authors to act like customer support! We just don't have the time, and unlike Bill Gates, it just is not very lucrative!! Also, be very aware that DoPHOT is a living, experimental program. It can and does have bugs. It can and does have strange, sometimes primitive subroutines that could be improved by clever programmers. It can and does have quirks that the current users have learned to live with. Again, if you don't like a feature, if you find a bug, if you want additional capabilities, you should do three things. First, make the changes yourself; second, test that these changes do not significantly alter the output from DoPHOT as defined by the results provided with these examples; and third, please report the change(s) to me so it can be incorporated into future versions of DoPHOT. If the change is extensive, please send the corrected versions of any relevant routines. Do *not* expect me or any other DoPHOT author to do the fixing/changing!

1.2. *Example 1: vanilla DoPHOT*

These data consist of a pair of BV CCD images of NGC 2031 in the LMC. The V images is called fits.009 and the B images is fits.010. Note that you can use *any* names you want for your files. The corresponding tuneup files contain only the most basic information for the data images: approximate sky value, approximate FWHM of the stellar images, the gain and read noise of the CCD, a few flags, and the names of the various input and output files. The flags include AUTOMEDSCALE that tells DoPHOT to scale the median-smoothing parameter for the sky by the FWHM (note that this is not the sky used to do photometry, but rather the sky level used to establish the thresholds; see the manual for details). Another flag is LOGVERBOSITY which controls the log output from DoPHOT. A value of 1 is the minimal verbosity, a value of 4 means you will need a *lot* of disk space for the log file! This can be useful for debugging or filling up disks. The file names are fairly self-explanatory: the input parameter file, PARAMS_DEFAULT *must* be specified, the PARAMS_OUT will list all the parameters you used and can be employed later to run DoPHOT in *precisely* the same manner as you did in an earlier run, IMAGE_IN is the input image name, OBJECTS_OUT is the output data file, LOGFILE is just that, and APCORRFILE is a file with the aperture corrections for selected stars and in a number of concentric apertures.

To run DoPHOT here, just say ../bin/dophot/dophot tuneup.nnn where nnn is the frame number. You can of course alias dophot or put it in some directory in your path, but the point is that the general form of the command line is dophot tuneupfile. If you forget the tuneup file name, DoPHOT will ask (which is inconvenient in batch mode).

After running DoPHOT on both frames you will now have many more files. The principal ones are the dophsum.nnn files (again, you can name these anything you want; these examples reflect my naming convention which has been sculpted more by history and custom than by logic). These files contain, one per line, information about each and every star or other object found in the frames by DoPHOT. The manual describes the details of the output and other options. In these examples I use the most cryptic output format, the so-called INTERNAL format.

To go from raw results to a color-magnitude diagram, you must do the following. Run a program called dtd2 in the /bin/extras subdirectory. This can be run on a single line (dtd2 dophot_output_file new_output_file) or answer the prompts. This program merely alters the format of the files, and loses a bit of information in the process (but information you will likely not want to keep). The output of these files includes the star id, the star coordinates, the measured magnitude and its error, the sky value, the DoPHOT image type, a statistic related to χ^2, and the internal DoPHOT aperture correction when available. Note that the first two lines provide a summary of the DoPHOT reductions; the third line is blank. These first three lines represent the header of these files.

You can then run a program called offset5 in the /bin/ccd subdirectory that can align the two frames (due to filter tilt, refraction or other effects, the stellar coordinates can vary by many pixels even on consecutive frames). The input is fairly straightforward in this semi-interactive alignment routine. You must specify a radius to search for an offset match, and slowly approach the correct value. Use this program and practice to get a feel for how it works; or use your own matching program. I offer `offset5` only for your benefit in this example; it is not part of DoPHOT.

Finally, you should combine the BV photometry for the two colors. To do this run `postdaop` in the directory `/bin/pd`. To the 'Command:' prompt type 'inst', and then answer the questions. Again, `postdaop` is provided to help with this example. Feel free to use any other merging routine. You exit `postdaop` with the command 'exit'. The output file from `inst` now has star id, x, y, V magnitude, V error, V chi-statistic, B magnitude, B error, and B chi-statistic. There are also three header lines, in this case with very little useful information. You can use this file to produce a color-magnitude diagram of the field, most easily with an advanced graphics program such as SUPERMONGO.

Note that one has other options for these post-DoPHOT reduction and analysis steps. For example, I regularly remove stars with relatively large errors compared to other stars in the same magnitude range. I also often extract subregions from the field – in this case there is a cluster near the field center so it is natural to want to isolate either the cluster of field stars separately. There are many other post-DoPHOT processing steps one can imagine. I leave it to you to think about these and to implement them.

Finally, I should mention that DoPHOT can produce a very useful file known as the `FINISHFILE` in the parameter default file. This file provides a thumbnail summary of the CPU time, elapsed time, and DoPHOT results for each frame. DoPHOT appends information from newly reduced files to any existing `FINISHFILE` if the name of the file is the same.

1.3. Example 2: variable PSF

This example again uses a BV pair of images from NGC 2031, but now to illustrate how DoPHOT deals with a variable PSF. As described in the first lecture, DoPHOT calculates at every threshold the average shape parameters for the PSF, then uses these averages to fit all known stars at all thresholds. The difference now is that DoPHOT fits a function of (x,y) to the shape parameters to determine how these vary with location. These functions are then used to determine the PSF at the location of every star before calculating the final fit to the stars. In practice, the variable PSF used here is a polynomial of order n in both axes, and including all cross terms. The current limit is $n = 10$.

You will find the files you need in the `/a2` subdirectory. This directory contains all the files described for the first example. However, note that there are now four tuneup files (but only two images) and there are two files called `run1.com` and `run2.com` that I describe below.

The data files are `fits.052` (the V image) and `fits.053` (the B image). The tuneup files `tuneup.052` and `tuneup.053` are the direct analogs of the tuneup files in the first example. You can run them as described in that example or you can type `source run1.com`. You will see if you print out `run1.com` that it merely contains commands, in sequence, to run DoPHOT on frames 52 and 53.

What about the other two tuneup files? These are called `tuneup.952` and `tuneup.953` and each one contains four additional parameters that correspond to the variable PSF. These parameters are `VARIABLE_PSF` which is the flag that tells DoPHOT you want to turn on the variable PSF, `LOGVPSF` which is the name of the log file with the variable PSF coefficients, `NPSFORDER` which is the order of the 2-D polynomial surface fit to the shape parameters, and `NPSFMIN` is the minimum number of 'perfect' stars (that is, stars suitable for use in constructing the PSF) to be used in the fit. Run DoPHOT in the same way either by hand or using `run2.com`.

All post-DoPHOT processing is the same as in example 1. These frames do not show a severely variable PSF, but you can still see very slight changes between the standard and variable-PSF reductions. Note that you still have to fit the aperture corrections to a surface to determine their variation across the frame. You can use the apcorr.nnn file for this. This contains the star id, x, y, the fitted magnitude, and then m aperture magnitude differences in the sense $m_{ap} - m_{fit}$. The idea is that you can add these aperture corrections to the DoPHOT magnitudes to get the total magnitude out to the maximum aperture you considered. There are parameters in the default parameter file that control the number of apertures, and the size of the maximum aperture used in determining these corrections. I do not supply any routines to calculate the aperture corrections as a function of position; this is a subtle step that I leave to each user to solve for him or herself.

1.4. Example 3: fixed-position mode

The data for this example are located in the /a3 subdirectory; as before, there is a /results subdirectory with my reduction results that you can use as a comparison. This example illustrates how to reduce a series of images in the so-called 'fixed-position'. This allows one to reduce data using previously-measured positions from a good-seeing, deep image, a process that not only speeds up the reductions (since there are fewer iterations within DoPHOT and fewer parameters to fit), but also provides more precise photometry (because blended images can be fit by multiple stellar profiles as needed).

There are 16 FITS files in the /a3 subdirectory; they are summarized in the file called vtimes.dat. The latter file lists on the first line the RA and Dec (and the epoch of the coordinates) of the field. You can list the name of the field after these numbers if you wish. The next lines list: frame number, date (dd,mm,yyyy), ut mid-time of exposure (hh,mm,ss), the exposure time, the observatory code, the filter name, the sky value, the FWHM, and a short description of the data. This file is used by a number of auxiliary programs that I employ – just consider it a useful summary of the data. Note that there are three different observing runs and two different telescopes represented in this data set. In case you are wondering, the full dataset for NGC 2031 consists of over 30 epochs, using four CCDs on three telescopes. Note too that although all the images are V-band exposures, one does *not* have to be restricted to single band in this mode. I regularly use templates from one band to help fix the positions of frames in other bands. One should check to see if atmospheric refraction across the frame or as a function of color becomes significant.

Now look at one of the tuneup files; tuneup.002 would be a good example. Many of the standard parameters are listed, but now there are four new ones not seen in the previous examples. These include the FIXPOS flag, which is here set to 'YES' to signify to DoPHOT that we want fixed-position mode. Also included in the file is SCALERATIO which provide the ratio between the scales of the image and the template image. DoPHOT *may* be able to figure this parameter out when calculating the coordinate transformations, but it is a number that the user usually knows ahead of time, at least approximately. Thus, this parameter provides the opportunity to impose the ratio at the start of the reductions.

The other two new parameters list files that are necessary for fixed-position mode reductions. Before I describe these, let me briefly summarize just what DoPHOT is doing in fixed-position mode. The reductions begin identically to any other DoPHOT reductions: a top threshold is computed and stars are found and fitted that have peak intensities above this limit. After a certain number of good stars are found by DoPHOT, it attempts to determine the offsets between the stars in the frame and the stars listed in a master, or *template* frame. This transformation is carried out using a version of the algorithm described by Groth (1986); a similar but alternative algorithm was recently described by Valdes et al. (1995), and this method will soon be implemented in DoPHOT.

If the transformation is successfully determined, then the coordinates in the template file are transformed to the coordinate system defined by the stars in the frame being reduced. Now DoPHOT has a master list of coordinates of *all* stars in the field. Consequently, it skips to the mth-from-last threshold, where m is defined by the LEVEL_RESTART found in the paramdefault file. This matching causes the fixed-position mode to be faster (in terms of the rate that stars are measured), more precise (because neighboring stars in crowded or blended images are *a priori* known to exist), and more convenient (because at the end of the reductions all stars are transformed to a common coordinate system defined by the template file, if the parameter APPLY_TRANS is set to 'YES'; see the paramdefault file).

The actual matching of the template to the frame's coordinates is tricky, so in practice it is done in a round-about way. One of these new parameters in the tuneup file is thus called AUTOMATCH_FILE, and is the name of a file containing a subset of the brighter, good-quality stars from the template. This file (called the *match* file) must be on the same coordinate system as the template file. Often, DoPHOT can successfully match this file rather than the entire template file, but once the match is made and the transformation is known, the template coordinates can be shifted to the frame being reduced. This is a somewhat black art, and in some cases some care must be taken to get a good match file. For example, it may be unwise to leave the coordinates of stars in a cluster in the match file. Note that in the example given here, the same match file is used for all the frames; however, this is not necessary so long as all match files are on the same coordinate system as the template file. The AUTOMATCH_SUB parameter specifies the template file; the name of the parameter reflects the fact that one of the first things done with the template is to use it to identify and fit stars that are then digitally subtracted from the image.

Note that one should be very careful to use the fixed-position with the variable-PSF option turned off. The reason is that the distortions that cause the PSF to become variable may also cause the centroids of the stars to shift, making it that much harder to match the star positions of the frame and the positions from the template file. Also, at present only a rather low-order transformation is applied to the data; the distortions caused by a variable PSF may be too large to be successfully fit by the adopted function. It is not necessarily impossible to run the variable PSF and fixed-mode simultaneously, but it is generally unreliable and introduces clear, large systematic errors.

You should perform two distinct tests in this example. First, reduce one of the frames with and without fixed-position turned on. Tuneup files tuneup.802 and tuneup.902 are setup to do just this for frame fits.002. When DoPHOT is done, look at the finish file. The listing for the fixed position mode includes the leading coefficients used to transform the coordinates. Note that the fixed-position mode took somewhat more CPU time, but it provided photometry for many, many more stars; thus, the actual reduction rate is about 35% higher. If you look at the output in detail, you will find that the fixed-position mode found many more stars near the cluster center because it was able to take advantage of knowing the coordinates of stars in that region from the template file.

Now you can run DoPHOT on all of the images, noting in particular that (a) the code determines transformation coefficients between the template and the frame, and (b) the frames are shifted to the template coordinates at the end of each DoPHOT run. Results for both examples are summarized in the results subdirectory. You can most conveniently run these examples using the run1.com and run2.com files (simply type 'runN.com' where N = 1 or 2).

The template file itself can be a listing from deep, good-seeing frames of the target field. Or it can be a listing of coordinates from HST or some other high-resolution telescope. Or it can be a combination. The example here combines coordinates from a sum of the good-seeing frames in the dataset, plus coordinates from HST in the cluster core. There is one technical item regarding the template file. DoPHOT uses a code from 1-9 to signify different classes of objects. Objects classified as 1 and 7 are 'good' stars, 3's are binary (or multiply-fit) stars, 2's are galaxies, and so on; details are in the DoPHOT manual. All stars to be used in the template file *must* have their values incremented by 10. Thus, a 1 becomes 11, 2 becomes 12 and so on. There are some exceptions, however, so I have included a little program in /bin/dophot called `maketemplate` that you can use to convert a DoPHOT output photometry file into a template file. The reason for this is so that you (and DoPHOT) can tell which stars were drawn from the template list, and which objects were found independently on the frame being reduced. If necessary, you can re-compile `maketemplate` by typing 'make maketemplate' in the /bin/dophot subdirectory.

One very nice thing about fixed-position mode – mentioned above – is that one does not have to offset the coordinates afterwards since they are automatically shifted to the common system defined by the template file so long as the parameter `APPLY_TRANS` is set to 'YES'. There are many ways to extract data on variable stars from these data – there are no fewer than 15 Cepheids in these frames. But I leave that task to you to solve on your own, since it goes well beyond DoPHOT's jurisdiction. Some papers that might help you if you are interested in searching for variable stars in these data are Kaluzny et al. (1993), and Yan and Mateo (1994).

1.5. *Some final general comments*

You will find all of the DoPHOT code in /bin/dophot. I strongly recommend that you first look at the `TUNEABLE` file and make sure that ncmax, nrmax (maximum numbers of columns and rows in the frame, respectively), and nsmax (maximum numbers of objects) are as large as you will need, but yet compatible with your CPU memory limitations. You should recompile the code if you export it by first deleting all the old object files – 'rm *.o' – then compiling and linking DoPHOT by simply typing 'make dophot'. You will appreciate just how many routines are present when you do this.

On a Sparc Ultra 1, DoPHOT runs about 10 times faster than the examples in the /aN/results directories which were run on a Sun IPX at the IAC in Tenerife. For long time-series photometry, this is a BIG gain. Also, as you will have realized from doing the examples, DoPHOT demands very little human intervention for routine operation. This not only obviously saves time, but also has one very nice implication: If you simply give another person the same data files, the same version of DoPHOT and the same tuneup files, they should get *precisely* the same results that you obtained. Thus, DoPHOT is a code that can produce easily reproducible results. In addition, DoPHOT can readily be run in a batch mode, particularly useful for reducing long time-series photometry.

However, do not be fooled by all this apparent simplicity: you must still carefully and painstakingly assess the quality of your photometry. DoPHOT may run automatically, but it does *not* produce perfect photometry automatically. *That* requires lots of care and testing to be sure the data and the DoPHOT parameters are well-matched.

I intentionally presented this description and these examples in the form of a cookbook. You can get more background on the theory of photometric reductions from calibration to aperture corrections in Stetson and Harris (1988), Stetson (1993) and from the dated, but still useful book edited by Grosbøl, et al. (1989), as well as from the original DoPHOT papers (Mateo and Schechter 1989; Schechter et al. 1994).

Finally, I want to reiterate that DoPHOT is just one of the many excellent photometry reduction programs available today (e.g. DAOPHOT, INVENTORY, ROMAPHOT, etc.). You should carefully assess your needs and try a number of the programs. These examples are only designed to get you started with stellar photometry, and for photometry novices, to allow a users unfamiliar with photometric reduction techniques to get a feel for how these work. All good photometry programs achieve good results, and each program has certain applications that it is particularly best-suited for. Test each, preferably on the data you are most interested in reducing, before you decide. Good luck, and good photometry!

REFERENCES

AJHAR, E. ET AL. 1996 *AJ* **111** 1110.
ALARD, C. 1996 *ApJ* **458** L17.
ALCOCK, C. ET AL. 1995 *AJ* **109** 2280.
ALCOCK, C. ET AL. 1997a *ApJ* **474** 217.
ALCOCK, C. ET AL. 1997b *ApJ* **482** 89.
ALCOCK, C. ET AL. 1997c *AJ* **114** 326.
APARICIO, A., DALCANTON, J. J., GALLART, C. & MARTINEZ-DELGADO, D. 1997 *AJ* **114** 1447.
APARICIO, A., GALLART, C., CHIOSI, C. & BERTELLI, G. 1996 *ApJ* **469** L97.
ARP, H. 1958 *AJ* **63** 273.
ARP, H. 1959 *AJ* **64** 175.
BECKER, S. A. & MATHEWS, G. J. 1983 *ApJ* **270** 155.
BERTELLI, G., BRESSAN, A., CHIOSI, C., MATEO, M. & WOOD, P. R. 1993 *ApJ* **412** 160.
BERTELLI, G., MATEO, M., CHIOSI, C. & BRESSAN, A. 1992 *ApJ* **388** 400.
BESSELL, M. S. 1993 In *New Aspects of Magellanic Cloud Research*, ed. B. Baschek, G. Klare & J. Lequeux (Heidelberg, Springer-Verlag) p. 321.
BICA, E., CLARIA, J. J., DOTTORI, H., SANTOS, J. F. C., JR. & PIATTI, A. E. 1996 *ApJS* **102** 57.
BOTHUN, G. D. 1992 *AJ* **103** 104.
BUTA, R. J., & MCCALL, M. L. 1983 *MNRAS* **205** 131.
BUTCHER, H. R. 1977 *ApJ* **216** 18.
CALDWELL, J. A. R. & COULSON, I. M. 1986 *MNRAS* **218** 223.
CARTER, D. & SADLER, E. M. *MNRAS* **245** P12.
CONLON, E. S., DUFTON, P. L., KEENAN, F. P., MCCAUSLAND, R. J. H. & HOLMGREN, D. 1992 *ApJ* **400** 273.
COTÉ, S. 1996 *PASA* **13** 278.
DA COSTA, G. S. & ARMANDROFF, T. E. 1995 *AJ* **109** 2533.
DA COSTA, G. S. 1991 In *The Magellanic Clouds*, ed. R. Haynes & D. Milne (Dordrecht, Kluwer), p. 183.
DEMERS, S. & IRWIN, M. J. 1991 *A&AS* **91** 171.
DE VAUCOULEURS, G. & CORWIN, H. G. 1985 *ApJ* **295** 287.
DJORGOVSKI, S., PIOTTO, G., PHINNEY, E. S. & CHERNOFF, D. F. 1991 *ApJ* **372** L41.
DJORGOVSKI, S. G., GAL, R. R., MCCARTHY, J. K., COHEN, J. G., DE CARVALHO, R. R., MEYLAN, G., BENDINELLI, O. & PARMEGGIANI, G. 1997 *ApJ* **474** L19.
DOHM-PALMER, R. ET AL. 1997 *AJ* in press.
DOPITA, M., MATHEWSON, D. & FORD, V. 1985 *ApJ* **297** 599.
ELSON, R. A. W. & FALL, S. M. 1985 *ApJ* **299** 211.
ELSON, R. A. W. & FALL, S. M. 1988 *AJ* **96** 1383.

ELSON, R. A. W., FORBES, D. A. & GILMORE, G. F. 1994 *PASP* **106** 632.
ELSON, R. A. W., GILMORE, G. F. & SANTIAGO, B. X. 1997 *MNRAS* **289** 157.
ELSTON, R. & SILVA, D. R. 1992 *AJ* **104** 1360.
FAHLMAN, G. G., MANDUSHEV, G., RICHER, H. B., THOMPSON, I. B. & SIVARAMAKRISHNAN, A. 1996 *ApJ* **459** L65.
FERGUSON, H. C. & SANDAGE, A. 1991 *AJ* **101** 765.
FERRARESE, L. ET AL. 1996 *ApJ* **464** 568.
FERRARO, F. R., FUSI PECCI, F., TESTA, V., GREGGIO, L., CORSI, C. E., BUONANNO, R., TERNDRUP, D. M. & ZINNECKER, H. 1995 *MNRAS* **272** 391.
FISCHER, P., WELCH, D. L., COTÉ, P., MATEO, M. & MADORE, B. 1992b *AJ* **103** 857.
FISCHER, P., WELCH, D. L. & MATEO, M. 1992a *AJ* **104** 1086.
FISCHER, P., WELCH, D. L. & MATEO, M. 1993 *AJ* **105** 938.
FREEDMAN, W. 1992 *AJ* **104** 1349.
FROGEL, J. A. & BLANCO, V. M. 1983 *ApJ* **274** L57.
GALLAGHER, J. S. ET AL. 1996 **ApJ** *466* 732.
GALLAGHER, J. S., HUNTER, D. A. & TUTUKOV, A. V. 1984 *ApJ* **284** 544.
GALLART, C., APARICIO, A., BERTELLI, G. & CHIOSI, C. 1996a *AJ* **112** 1950.
GALLART, C., APARICIO, A., BERTELLI, G. & CHIOSI, C. 1996b *AJ* **112** 2596.
GARDINER, L. T., SAWA, T. & FUJIMOTO, M. 1994 *MNRAS* **266** 567.
GASCOIGNE, S. C. B. & KRON, G. E. 1952 *PASP* **64** 196.
GEISLER, D., BICA, E., DOTTORI, H., PIATTI, A. E. & CLARIA, J. J. 1997 *BAAS* **190** 46.01.
GILMORE, G. & WYSE, R. F. G. 1991 *ApJL* **367** L55.
GIRARDI, L., CHIOSI, C., BERTELLI, G. & BRESSAN, A. 1995 *298* **305** 849.
GRAHAM, J. A. 1982 *ApJ* **252** 474.
GRAHAM, J. A. ET AL. 1997 *ApJ* **477** 535.
GRONDIN, L., DEMERS, S. & KUNKEL, W. E. 1992 *AJ* **103** 1234.
GROSBØL, P. J., MURTAGH, F. & WARMELS, R. H. 1989 *1st ESO/ST-ECF Data Analysis Workshop* (Garching, ESO).
GROTH, E. J. 1986 *AJ* **91** 1244.
HAMBLY, N. C., FITZSIMMONS, A., KEENAN, F. P., DUFTON, P. L., BROWN, P. J. F., IRWIN, M. J. & ROLLESTON, W. R. J. 1995 *ApJ* **448** 628.
HARDY, E., BUONANNO, R., CORSI, C. E., JANES, K. A. & SCHOMMER, R. A. 1984 *ApJ* **278** 592.
HATZIDIMITRIOU, D. 1991 *MNRAS* **251** 545.
HATZIDIMITRIOU, D. & HAWKINS, M. R. S. 1989 *MNRAS* **241** 667.
HEASLEY, J. N., CHRISTIAN, C. A., FRIEL, E. D. & JANES, K. A. 1988 *AJ* **96** 1312.
HODGE, P. 1960 *ApJ* **132** 351.
HODGE, P. W. 1961 *ApJ* **133** 413.
HODGE, P. W. 1973 *ApJ* **182** 671.
HOLLAND, S., FAHLMAN, G. G. & RICHER, H. B. 1996 *AJ* **112** 1035.
HOLTZMAN, J. ET AL. 1992 *AJ* **103** 691.
HOLTZMAN, J. A. ET AL. 1997 *AJ* **113** 656.
IBATA, R. A., GILMORE, G. & IRWIN, M. J. 1994 *Nature* **370** 194.
IBATA, R. A., GILMORE, G. & IRWIN, M. J. 1995 *MNRAS* **277** 781.
IBATA, R. A., WYSE, R. F. G., GILMORE, G., IRWIN, M. J. & SUNTZEFF, N. B. 1997 *AJ* **113** 634.
IRWIN, M. J., KUNKEL, W. E. & DEMERS, S. 1985 *Nature* **318** 160.

ISSERSTEDT, J. 1984 *A&A* **131** 347.
IWRIN, M. J., DEMERS, S. & KUNKEL, W. E. 1990 *AJ* **99** 191.
JASNIEWICZ, G. & THÉVENIN, F. 1994 *A&A* **282** 717.
JONES, D. H. ET AL. 1996 *ApJ* **466** 742.
KALUZNY, J., MAZUR, B. & KRZEMINSKI, W. 1993 *MNRAS* **262** 49.
KARACHENTSEV, I. 1996 *A&A* **305** 33.
KARACHENTSEV, I., DROZDOVSKY, I., KAJSIN, S., TAKALO, L. O., HEINAMAKI, P. & VALTONEN, M. 1997 *A&AS* **124** 559.
KAVELAARS, J. J. & HANES, D. A. 1997 *MNRAS* **285** 31.
KENT, S. 1987 *AJ* **94** 306.
KÖPPEN, J. In *New Aspects of Magellanic Cloud Research*, ed. B. Baschek, G. Klare & J. Lequeux (Heidelberg, Springer-Verlag) p. 372.
KORIBALSKI, B., JOHNSTON, S. & OTRUPCEK, R. 1994 *MNRAS* **270** 43.
KORMENDY, J. 1985 *ApJ* **295** 73.
KORMENDY, J. & RICHSTONE, D. 1995 *ARAA* **33** 581.
KUNKEL, W. E. 1979 *ApJ* **228** 718.
LEE, M.-G. 1996 *AJ* **112** 1438.
LIN, D. C. N. & RICHER, H. B. 1992 *ApJ* **388** L57.
LIN, D. N. C., JONES, B. F. & KLEMOLA, A. R. 1995 *ApJ* **439** 652.
LUPPINO, G. A. & TONRY, J. L. 1993 *ApJ* **468** 519.
LYNDEN-BELL, D. & LYNDEN-BELL, R. M. 1995 *MNRAS* **275** 429.
MAGNIER, E. A., PRINS, S., AUGUSTEIJN, T., VAN PARADIJS, J. & LEWIN, W. H. G. 1997 *A&A*, **326** 442.
MATEO, M. 1987 *ApJ* **323** L41.
MATEO, M. 1993 In *The Globular Cluster-Galaxy Connection*, ed. G. H. Smith & J. P. Brodie (San Francisco, ASP), p. 387.
MATEO, M., HARRIS, H. C., NEMEC, J. & OLSZEWSKI, E. W. 1990 *AJ* **100** 469.
MATEO, M. & HODGE, P. 1986 *ApJS* **60** 893.
MATEO, M., HODGE, P. & SCHOMMER, R. A. 1986 *ApJ* **311** 113.
MATEO, M., KUBIAK, M., SZYMANSKI, M., KALUZNY, J., KRZEMINSKI, W. & UDALSKI, A. 1995b *AJ* **110** 1141.
MATEO, M., MIRABAL, N., UDALSKI, A., SZYMANSKI, M., KALUZNY, J., KUBIAK, M., KRZEMINSKI, W., & STANEK, K. Z. 1996 *ApJ* **458** L13.
MATEO, M., OLSZEWSKI, E. W. & MADORE, B. M. 1990 In *Confrontation Between Stellar Pulsation and Evolution*, ed. C. Cacciari & G. Clementini (San Francisco, ASP), p. 214.
MATEO, M. & SCHECHTER, P. 1989 In *1st ESO/ST-ECF Data Analysis Workshop*, ed. P. J. Grosbøl, F. Murtagh & R. H. Warmels (Garching, ESO), p. 69.
MATEO, M., UDALSKI, A., SZYMANSKI, M., KALUZNY, J., KUBIAK, M., & KRZEMINSKI, W. 1995a *AJ* **109** 588.
MATEO, M., WELCH, D. & FISCHER, P. 1991 In *The Magellanic Clouds*, ed. R. Haynes & D. Milne (Dordrecht, Kluwer), p. 191.
MATHEWSON, D. S. & FORD, V. L. 1984 In *Structure and Evolution of the Magellanic Clouds*, ed. S. van den Bergh & K. S. de Boer (Dordrecht, Reidel), p. 125.
MATHEWSON, D. S., FORD, V. L. & VISVANATHAN, N. 1986 *ApJ* **301** 664.
MATHEWSON, D. S., FORD, V. L. & VISVANATHAN, N. 1988 *ApJ* **333** 617.
MCCALL, M, L. & BUTA, R. J. 1995 *AJ* **109** 2460.
MILLER, B. W. 1996 *AJ* **112** 991.
MILLER, B. & HODGE, P. 1995 *ApJ* **458** 467.

MINNITI, D., ALONSO, M. V., GOUDFROOU, P., JABLONKA, P. & MEYLAN, G. 1996 *ApJ* **467** 221.
MINNITI, D., OLSZEWSKI, E. W. & RIEKE, M. 1993 *ApJ* **410** L79.
MINNITI, D. & ZIJLSTRA, A. A. 1996 *AJ* **112** 590.
MORRIS, P. W., REID, I. N., GRIFFITHS, W. K. & PENNY, A. J. 1994 *MNRAS* **271** 852.
MORRISON, H. & SARAJEDINI, A. 1996 editors of *Formation of the Galactic Halo ... Inside and Out* (San Francisco, ASP).
MUZZIO, J. C. 1988 In *The Harlow-Shapley Symposium on Globular Cluster Systems in Galaxies*, ed. J. E. Grindlay, & A. G. Davis Philip (Dordrecht, Kluwer), p. 297.
NEMEC, J., NEMEC, A. F. L. & LUTZ, T. E. 1994 *AJ* **108** 222.
OLSZEWSKI, E. W., SCHOMMER, R. A. & AARONSON, M. 1987 *AJ* **93** 565.
OLSZEWSKI, E. W., SUNTZEFF, N. B. & MATEO, M. 1996 *ARAA* **34** 511.
PACZYNSKI, B. 1986 *ApJ* **304** 1.
PACZYNSKI, B. 1991 *ApJ* **371** L63.
PIATEK, S., PRYOR, C., MCCLURE, R. D., FLETCHER, J. M. & HESSER, J. E. 1994 *AJ* **109** 1071.
PRESTON, G. W., BEERS, T. A. & SHECTMAN, S. A. 1994 *AJ* **108** 538.
PRESTON, G. W., SHECTMAN, S. A. & BEERS, T. C. 1991 *ApJ* **375** 121.
PRITCHET, C. J., SCHADE, D., RICHER, H. B., CRABTREE, D. & YEE, H. K. C. 1987 *ApJ* **323** 79.
PRITCHET, C. & VAN DEN BERGH, S. 1987a *ApJ* **316** 517.
PRITCHET, C. & VAN DEN BERGH, S. 1987b *ApJ* **318** 507.
PRYOR, C., MCCLURE, R. D., FLETCHER, J. M. & HESSER, J. E. 1989 *AJ* **98** 596.
PRYOR, C., MCCLURE, R. D., FLETCHER, J. M. & HESSER, J. E. 1991 *AJ* **102** 1026.
PUCHE, D. & CARIGNAN, C. 1991 *ApJ* **378** 487.
RATNATUNGA, K. U. & BAHCALL, J. N. 1985 *ApJS* **59** 63.
REGAN, M. W. & VOGEL, S. N. 1994 *ApJ* **434** 536.
REGAN, M. W. & WILSON, C. D. 1993 *AJ* **105** 499.
REID, N., MOULD, J. & THOMPSON, I. 1987 *ApJ* **323** 433.
RICHER, H. B., PRITCHET, C. J. & CRABTREE, D. R. 1985 *ApJ* **298** 240.
SAGAR, R. & PANDEY, A. K. 1989 *A&AS* **79** 407.
SAHA, A. ET AL. 1994 *ApJ* **425** 14.
SAHA, A., HOESSEL, J. G. & KRIST, J. 1992 *AJ* **103** 84.
SAHA, A., SEITZER, P. & MONET, D. G. 1986 *AJ* **92** 302.
SALPETER, E. E. 1955 *ApJ* **121** 161.
SARAJEDINI, A., LEE, Y.-W. & LEE, D.-H. 1995 *ApJ* **450** 712.
SARAJEDINI, A. & LAYDEN, A. C. 1995 *AJ* **109** 1086.
SCHECHTER, P. L., MATEO, M. & SAHA, A. 1993 *PASP* **105** 1342.
SCHOMMER, R. A., OLSZEWSKI, E. W., SUNTZEFF, N. B. & HARRIS, H. C. 1992 *AJ* **103** 447.
SCHOMMER, R. A. & CHRISTIAN, C. A. *AJ* **101** 873.
SCHWEIZER, F., MILLER, B. W., WHITMORE, B. C. & FALL, S. M. 1996 *AJ* **112** 1839.
SEARLE, L., SARGENT, W. L. W. & BAGNUOLO, W. G. 1973 *ApJ* **179** 427.
SEARLE, L., WILKINSON, A. & BAGNUOLO, W. G. 1980 *ApJ* **239** 803.
SEBO, K. M. & WOOD, P. R. 1994 *AJ* **108** 932.
SEBO, K. M. & WOOD, P. R. 1995 *ApJ* **449** 164.
SEGGEWISS, W. & RICHTLER, T. 1989 In *Recent Developments of Magellanic Cloud Research*, ed. K. S. de Boer, F. Spite & G. Stasinska, (Paris, Obs. de Paris) p. 45.

SHAPLEY, H. 1930 *Star Clusters* (New York, McGraw-Hill).

SMECKER-HANE, T. A., STETSON, P. B., HESSER, J. E. & LEHNERT, M. D. 1994 *AJ* **108** 507.

STANEK, K. Z., KALUZNY, J., MATEO, M. & TONRY, J. 1996 *BAAS* **189** 96.09.

STETSON, P. B. 1993 In *Stellar Photometry – Current Techniques and Future Developments*, ed. C. J. Butler, & I. Elliot, (Cambridge: Cambridge University Press).

STETSON, P. B. & HARRIS, W. E. 1988 *AJ* **96** 909.

STRYKER, L. L. 1983 *ApJ* **266** 82.

STRYKER, L. L., DA COSTA, G. S. & MOULD, J. R. 1985 *ApJ* **298** 544.

SUNTZEFF, N. B., SCHOMMER, R. A., OLSZEWSKI, E. W. & WALKER, A. R. 1992 *AJ* **104** 1743.

TESTA, V., MATEO, M. & COTÉ, P. 1997, in preparation.

THOMSON, R. C. 1992 *MNRAS* **257** 689.

TYSON, J. A. 1988 *AJ* **96** 1.

UDALSKI, A., SZYMANSKI, M., KALUZNY, J., KUBIAK, M. & MATEO, M. 1992 *AcA* **42** 235.

VALDES, F. G., CAMPUSANO, L. E., VALÁSQUEZ, J. D. & STETSON, P. B. 1995 *PASP* **107** 1119.

VALLENARI, A., CHIOSI, C., BERTELLI, G., APARICIO, A. & ORTOLANI, S. 1996b *A& A* **309** 367.

VALLENARI, A., CHIOSI, C., BERTELLI, G. & ORTOLANI, S. 1996a *A& A* **309** 358.

VAN DEN BERGH, S. 1981 *A& AS* **46** 79.

VAN DEN BERGH, S. 1985 *ApJ* **297** 361.

VAN DEN BERGH, S. 1991 *ApJ* **369** 1.

VAN DEN BERGH, S. 1994a *AJ* **108** 2145.

VAN DEN BERGH, S. 1994b *ApJ* **432** L105.

WELCH, D. L., MCLAREN, R. A., MADORE, B. F., & MCALARY, C. W. 1987 *ApJ* **321** 162.

WESTERLUND, B. E., LINDE, P. & LYNGÅ, G. 1995 *A& A* **298** 39.

WHITING, A. B., IRWIN, M. J. & HAU, G. K. T. 1997 *AJ* **114** 996.

WHITMORE, B. C. & SCHWEIZER, F. 1995 *AJ* **109** 960.

YAHIL, A., TAMMANN, G. A. & SANDAGE, A. 1977 *ApJ* **217** 903.

YAN, L., AND MATEO, M. 1994 *AJ* **108** 1810.

ZARITSKY, D., OLSZEWSKI, E. W., SCHOMMER, R. A., PETERSON, R. C. & AARONSON, M. 1989 *ApJ* **345** 759.

ZINN, R. 1980 In *Globular Clusters*, ed. D. Hanes & B. Madore (Cambridge, Cambridge University Press), p. 191.

ZINN, R. 1996 In *Formation of the Galactic Halo ... Inside and Out*, ed. H. Morrison & A. Sarajedini (San Francisco, ASP) p. 211.

Chemical Evolution of the ISM in Nearby Galaxies

By EVAN D. SKILLMAN

Astronomy Department, University of Minnesota,
116 Church St. SE, Minneapolis, MN, 55123, USA

Advances in astrophysics are often dependent on inter-connections between different subfields. The goal of this chapter is to serve as an introduction, for a perceived audience of primarily stellar astronomers, into interstellar medium research which is directly related to stellar issues. While there are many such connections, I will focus primarily on topics related to chemical abundances. I start with a brief overview of the methodology of obtaining nebular abundances. This is followed by an overview of simple chemical evolution models. I then review what is known about abundance patterns in dwarf and spiral galaxies, evidence of the possible role of environment, and how these observations fit into our simple ideas of chemical evolution. Finally, I outline what I consider a very long term goal; that of deriving self-consistent star formation histories for galaxies. By this, I mean drawing parallels between the record of star formation as written in the stellar populations and the record as written in the absolute and relative chemical abundances.

1. Introduction and purpose

I would like to start by asking the question: "Why is this chapter in this book?" The emphasis of these lectures is clearly on stellar astrophysics, so why is it necessary to have a chapter dedicated to the interstellar medium?

The answer to that question lies in Antonio's invitation to me to participate in this winter school: "The school will be devoted to what things can be learned about galaxies from their stars, but properties of the gas are an important nexus between stars and galaxy evolution."

I couldn't agree more. An important long term goal of observational extragalactic astronomy is to construct detailed evolutionary histories of galaxies. This goal is best visualized using the "population box" concept introduced by Hodge (1989). Hodge produced schematic representations of star formation rates and chemical abundance evolution for many nearby galaxies (cf. Figure 18 in Gary DaCosta's chapter in this book). Converting these schematic representations into hard data (with error bars) represents a great challenge that all conference participants can fully appreciate. In the ensuing years since Hodge first published his review, there has been tremendous progress in the task, especially with regard to the Local Group dwarf ellipticals (much of this progress due to the capabilities of the Hubble Space Telescope). An excellent example of progress in adding detail to Hodge's diagrams can be found in the study of the dwarf irregular galaxy NGC 6822 by Gallart et al. (1996a,b,c).

I can add another reason for including a chapter on the interstellar medium (ISM). It is important to remind stellar astronomers not to overlook the ISM. The ISM can often add clues which will help to solve the puzzles presented by stellar studies. I give as an example HI observations of IZw18. This is a dwarf galaxy thought by some to be experiencing its first burst of star formation. HI observations reveal a plume of neutral gas connecting IZw18 to a companion (Dufour et al. 1996, Skillman et al. 1996a). This suggests that the present burst of star formation has been triggered by a dynamical

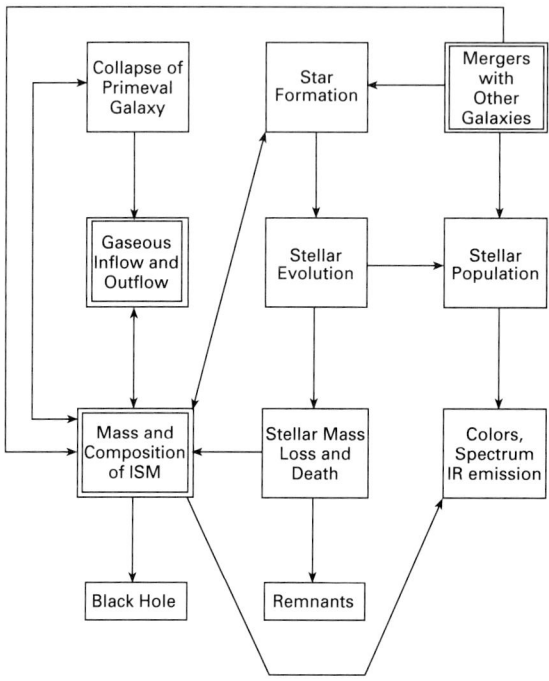

FIGURE 1. A diagram of the photometric and chemical evolution connections between different astronomical entities from Tinsley (1980). This diagram was taken from Knapp (1990) where double-lined boxes emphasize those areas directly relevant to the ISM.

interaction, and reminds us that sorting out the star formation history of this galaxy must include an understanding of its environment.

There are many connections between the studies of stars in a galaxy and the studies of its ISM. A quick perusal of Tinsley's famous flow diagram (Figure 1) elucidates some of these for us. When introductory astronomy textbooks discuss the ISM, they usually introduce the topic in relationship to star formation processes. As the ISM is the raw material for star formation, the link is obvious. When I was originally outlining these lectures, I thought that I would dedicate a good fraction of my time to this connection. However, as my thinking evolved, it became clear that this subject is too large to usefully treat as part of a chapter. Indeed, there are now many, many volumes dedicated to this topic. For example, one of the best is the proceedings of the Third Canary Islands Winter School (Tenorio-Tagle, Prieto, & Sanchez 1992), and probably the most recent is the 1996 Maryland Conference proceedings (Holt & Mundy 1997). It also became clear that the emphasis of this school would be on the understanding of stars that already exist in galaxies, and not on their creation.

A second connection is that of chemical evolution. While this is a problem that rivals star formation in its complexity, the connections to existing stars are more clear. The record of the chemical evolution of galaxies is locked into its low mass stars. The present status of chemical evolution is represented in both the youngest stars and the interstellar medium. As a starting point, we might demand that the abundances measured in those two components agree. We will also see the close connections between stellar evolution theory and the interstellar medium as one compares the theoretical nucleosynthetic yields with the "effective" yields one obtains through measurements of gas mass fractions. It is also possible to constrain nuclear cross sections through measurements of relative abundances, again an area where ISM observations can play a role.

Finally, the ISM can act as a recorder, helping us to better understand stellar physics by demonstrating a star's effects on its environment. I can think of two immediate examples concerning massive stars. The first is a subject which Claus covers in detail in his chapter. The energy deposition of massive stars is best measured by the energy content of the surrounding ISM. The second is a topic which will be covered by Rolf. A long standing problem in nebular photoionization modeling (the Ne III problem) has recently been resolved (Sellmaier et al. 1996). The solution to the problem lay in a better understanding of the effects of winds on the radiative transfer in massive stars. Here we have an example of observations of the ISM helping to bring about an advance in stellar atmospheres.

Thus, I have chosen the following goal for this chapter. I hope to provide an entrance point for stellar astronomers into understanding observations of the ISM and how they might be relevant to helping to understand the studies of resolved stars in other galaxies. I will concentrate on the theme of chemical evolution, as I think that this provides the greatest number of links between stellar and ISM studies.

I note here that the problem of understanding the chemical evolution of galaxies is a messy one. There are many important parameters that are coupled in non-linear ways. Thus, the key to success in this field is the *isolation* of variables. Luckily, we have a wealth of galaxies to study. The *variety* of galaxies in the Local Group and other relatively nearby groups allows us to test for dependences on various galaxian structural properties. It is important to think of the variety in the nearby galaxies as providing baselines in various parameters against which to test the dependences of observables, and later, theories. I have a clear prejudice for the nearby galaxies; I believe that the resolved stellar populations and interstellar medium in these galaxies will eventually allow us our best insights into the processes that determine galaxy evolution.

2. Abundances from HII regions

The purpose of this section is to provide a very brief introduction to abundance determinations from HII regions. My target audience consists of stellar astronomers that might like to connect their abundance studies to those of nebular astronomers, but might be somehow put-off by the trappings of a culture that is somewhat foreign.

Let me insert here a few philosophical statements about nebular abundances and the community of researchers that work on them. It is my impression that the confidence in abundances derived from nebular spectra is vastly different between people that work in the field and those that do not, in the sense that people that work in the field hold nebular abundance measurements to be quite reliable while outsiders consider them suspect. I think that there is a simple misinterpretation which is responsible. For those of us in the field, there is nothing more interesting than a small bit that we don't understand. I put forward as an example temperature fluctuations in nebulae. Temperatures in nebulae cannot be perfectly uniform, there must be fluctuations at some level. Due to the asymmetric temperature dependence of collisional excitation rates, if the fluctuations are large enough, then this will give rise to abundance under-estimates (Peimbert 1967). The problem is to determine the size and importance of this effect. Workers within the field are fascinated with this problem and have dedicated a good deal of work to it. For example, in a wonderful review by Peimbert (1995) several pages are spent describing this problem, while very little is said about the security and utility of the standard abundance measurements. Researchers outside of the field see all of this activity and may believe that all nebular abundance determinations are suspect since the problem of temperature fluctuations is not completely solved.

It is my feeling that, in many cases, this is throwing away the baby with the bath water. Nebular abundances are quite frequently measured with an accuracy more than necessary to provide meaningful constraints on a wide variety of astrophysical problems. I am campaigning to encourage the nebular community to remember to spread this message.

2.1. *It's all in Osterbrock*

The logical starting point for most of us is the wonderful text by Osterbrock (1989; AGN^2). It really is true that "it's all in Osterbrock," but that does not necessarily mean that this text provides a simple, straightforward introduction to the subject of nebular astrophysics. In fact, it can serve in that role, but I think a few words of guidance will help a great deal.

From my experience in teaching from AGN^2, the greatest difficulty arising from first encounters with this text lies in one of the book's great strengths. AGN^2 is organized around physical principles, and is extremely thorough in its treatment. While this allows one to find material easily and to be confident that one really only needs one book, this makes it difficult for the first-time reader. For example, after the discussion of the calculation of spectra (chapter 4) and before the discussion of the comparison of theory with observations (chapter 5), are several sections with relatively detailed descriptions of specific problems (e.g., the Bowen resonance-florescence mechanism for O III).

The very simple solution to this problem is to be aware of the structure of the text, and to be confident that is it possible to skip over many of the chapter sections without fear that the following material will be incomprehensible. Perhaps this is obvious to most, but I have found this simple advice to be invaluable for many AGN^2 neophytes.

A second simple suggestion that I have found very effective is to remind students that an HII region is an extreme example of a non-LTE problem. Since many HII region observables are the result of processes associated with the electron gas which has a thermalized velocity distribution, students are tempted to apply their LTE instincts when approaching HII region problems. Usually this has disastrous results (e.g., naive models of gas temperatures dropping as $distance^2$ from the exciting star). It is important to remember that the energy input source is the absorption of radiation from the exciting star, and to separate highly non-LTE phenomena (e.g., the recombination emission spectrum of H and He) from LTE phenomena (the collisional excitation of the lowest levels of the heavier ions).

2.2. *Emission line abundances for beginners*

In this short overview, I would like to follow along the lines of the review article by Dinerstein (1990). She has divided the abundance analysis of HII region spectra into three different approaches: (1) the "direct" method; (2) the bright-line method; and (3) the photoionization modeling method. I will attempt to emphasize brevity here, as the goal of this section is to act as an entry point, not to be comprehensive overview.

Before beginning a description of the different abundance analysis methods, it is useful to make two introductory remarks. First, I remind the reader that the electron temperature in an HII region reflects the balance between heating and cooling processes. The heating is due to absorption of the ionizing radiation from the exciting star(s), while the cooling is primarily due to collisional excitation of bound electrons (and subsequent radiation) in the ions of heavy elements. Clearly, the electron temperature will have

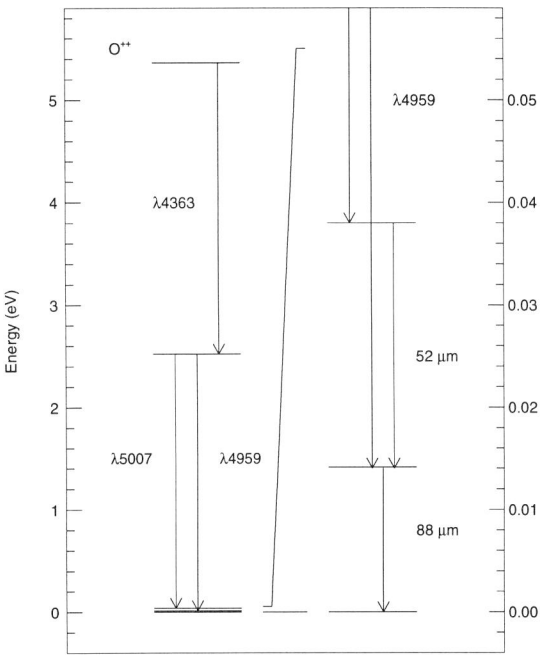

FIGURE 2. A Grotrian diagram of the energy levels in O^{++}. Note that the scale on the left has been expanded by a factor of 100 on the right to show the fine structure lines. For a detailed discussion, see Dinerstein, Lester, & Warner (1985).

a metallicity dependence: At higher metallicity, more coolants are available, and the electron temperature will will be lower.

Second, lets look at a Grotrian diagram and get a feel for the transition energies involved. Figure 2 shows the Grotrian diagram for O^{++}. The transition energies are plotted in eV, which is convenient when remembering that kT is almost 1 eV (0.86) at 10^4K which is typical for HII regions. The left hand side of the diagram shows the transitions that are found at optical wavelengths. The strengths of these emission lines are strongly temperature dependent, and thus strongest in the high temperature (low metallicity) HII regions. The right hand side of the diagram shows the fine structure lines. Note that the scale has changed by a factor of 100. The strengths of these fine structure lines will be temperature independent, but linearly proportional to the abundance of O^{++}.

2.2.1. *The direct method*

Following Dinerstein (1990), the direct method can be broken down into 5 (easy) steps:

I. Measure the observed emission line intensities

This step may seem trivial and self-explanatory. It assumes that you have taken proper care to reduce and calibrate your data. At the telescope, it is important to observe several standard stars (preferably from the HST spectrophotometric standards of Oke 1990). From my experience you can do a fairly good job just by following the standard procedures (with the addition of testing each step before moving on). I would recommend to the reader two guides available from NOAO (ftp: ftp.noao.edu; cd iraf/iraf/docs) "A User's Guide to CCD Reductions with IRAF" (Massey 1997) and "A User's Guide to Reducing Slit Spectra with IRAF" (Massey, Valdes, & Barnes 1992). Typically, one-dimensional spectra are extracted from long-slit (2-D) observations. Special care needs to be taken setting the extraction aperture width. The aperture should be sufficiently wide that small alignment errors do not give rise to systematic errors (this comes at a cost in signal/noise, but its worth the price). Additionally, one should avoid attempting "optimal" extraction. This works for stellar spectra that are inherently point sources and will be normalized, but not for extended nebulosities! And, finally, one should preferably integrate under the emission line profile (as opposed to fitting the line with a Gaussian profile). Fitting procedures can introduce systematic differences between high signal/noise and low signal/noise lines (and you will generally need to measure both accurately).

I'd like to insert yet another editorial comment here. I have had some experience trying to produce accurate emission line ratios from multi-order (echelle) spectrographs. Since the efficiencies change rapidly across the orders and since calibrating the relative efficiencies between orders is non-trivial, my experience tells me that emission line ratios from echelle spectrographs will never be as accurate as those derived from simple, long-slit spectrographs. I think that this is important, especially in the context of the design of spectrographs for the next generation of large telescopes – most of the planned spectrographs are echelle types (because of the ability to achieve high resolution and large wavelength coverage simultaneously). Thus, it is important to plan for long-slit spectrographs at (at least) a few of the large telescopes.

II. Correct the line intensities for reddening

Because we know the theoretical emissivities of the recombination lines of H, and because there are a number of H recombination lines spread through the optical spectrum, it is possible to use the observed line ratios to solve for the reddening of the spectrum. The most recent H emissivity calculations are those provided by Hummer & Storey (1987), and, for the Balmer lines, these are very similar to the values calculated by Brocklehurst (1971). If one assumes a reddening law (f(λ), e.g., Seaton 1979), in principle, it is possible to solve for the extinction as a function of wavelength by measuring a single pair of H recombination lines. Values of C(Hβ), the logarithmic reddening correction at Hβ, can be derived from:

$$log\left[\frac{I(\lambda)}{I(H\beta)}\right] = log\left[\frac{F(\lambda)}{F(H\beta)}\right] + C(H\beta)f(\lambda), \tag{2.1}$$

where $I(\lambda)$ is the intrinsic line intensity and $F(\lambda)$ is the observed line flux corrected for atmospheric extinction. Assuming a reddening law introduces a degree of uncertainty. Studies in our Galaxy show that the reddening law shows large variations between different lines of sight, but these variations are most important in the ultraviolet (Cardelli, Clayton, & Mathis 1989). Note also that the total measured extinction can have both Galactic and extragalactic components.

In practice it is best to measure several H recombination lines as this will provide both an estimate of the effect of stellar absorption underlying the emission lines and an estimate of the error in the measurement of the reddening. Due to the lack of H recombination lines in the near ultraviolet, it is possible to use He recombination lines in the same way. Here one can use the emissivities calculated by Smits (1996), which, in general, agree well with those calculated by Brocklehurst (1972), but *do* differ significantly in some cases (and thus, the Smits values are preferable).

There is a subtle problem with the reddening correction related to the distribution of the dust. The above method assumes a uniform "screen" of dust in front of the HII region. This is probably reasonable for reddening by foreground Galactic dust, but will not work so well if there is dust mixed within the HII region. Israel & Kennicutt (1980) pointed out that the extinction derived from Balmer line ratios tended to give lower values when compared to values derived from comparisons between a Balmer emission line and the thermal radio continuum emission. This implies clumpy extinction, internal extinction, or both. This was supported by detailed observations of Large Magellanic Cloud HII regions by Caplan & Deharveng (1986) and Balmer line — Brackett line — radio continuum emission comparisons of M101 HII regions by Skillman & Israel (1988). Mathis (1983) suggested that albedo variations may also be partially responsible for the optical — radio differences. While this effect is intrinsically interesting, it appears to have only a minor effect on abundance determinations; this is probably because the regions with the lowest amount of reddening will dominate the integral spectrum, so the lower values of reddening are probably appropriate. For a recent discussion of this problem, see Calzetti (1997).

Corrected emission line ratios should carry realistic errors. With CCD detectors, which can be configured to be linear to within measurable limits, it is possible to do a very accurate job of accounting for the different sources of error. Standard errors (δL) for the line strengths (L) can be calculated from the following formula (where all numbers are in accumulated electrons):

$$\frac{\delta L}{L} = \frac{\{C_1 + C_2 + (n_o/\sqrt{n_s})S + nAN^2 + [2.3 f(\lambda)\, \delta C(H\beta)\, L]^2 + [0.01 L]^2 + [\delta F\, L]^2\}^{\frac{1}{2}}}{L},$$
(2.2)

where C_1 is the total number of counts (line plus continuum and sky), C_2 is the continuum, S is the sky background in a single row and n_o and n_s are the number of rows summed over for the object and the number of rows averaged for the sky, nAN^2 accounts for the readout noise, where n is the number of integrations, A is the area summed over in pixels, and N is the r.m.s. readout noise. The fifth term represents the reddening error, where $\delta C(H\beta)$ is the error in $C(H\beta)$, and the sixth term accounts for errors in the flat fielding (estimated at 1%). The seventh term represents errors in the flux calibration, where δF is the r.m.s. relative error for the fit to the flux calibration points for the standard stars (typically 2%). Leaving out the terms accounting for the errors in reddening or flux calibration will result in unrealistically small errors.

III. Derive the local physical conditions

By "local physical conditions" one means the temperature and density of the gas. In fact, since the temperature is governed by the balance between the heating and cooling processes, and since the cooling is governed by different ionic species in different radial zones, one expects different ions to have different mean temperatures (cf. Stasińska 1990; Garnett 1992). While this is best treated with a complete photoionization model, a reasonable compromise is to treat the spectrum as if it arose in two different temperature zones, roughly corresponding to the O^+ and O^{++} zones (since the oxygen ions play a dominant role in the cooling, this is a reasonable thing to do).

The average density can be derived by measuring the relative intensities of two lines which arise from a split upper level. In the "low density regime" collisional de-excitation is unimportant and all excitations are followed by emission of a photon. The ratio of the fluxes then simply reflects the ratio of the statistical weights of the two levels. In the "high density regime", where the level populations are held at the ratio of their statistical

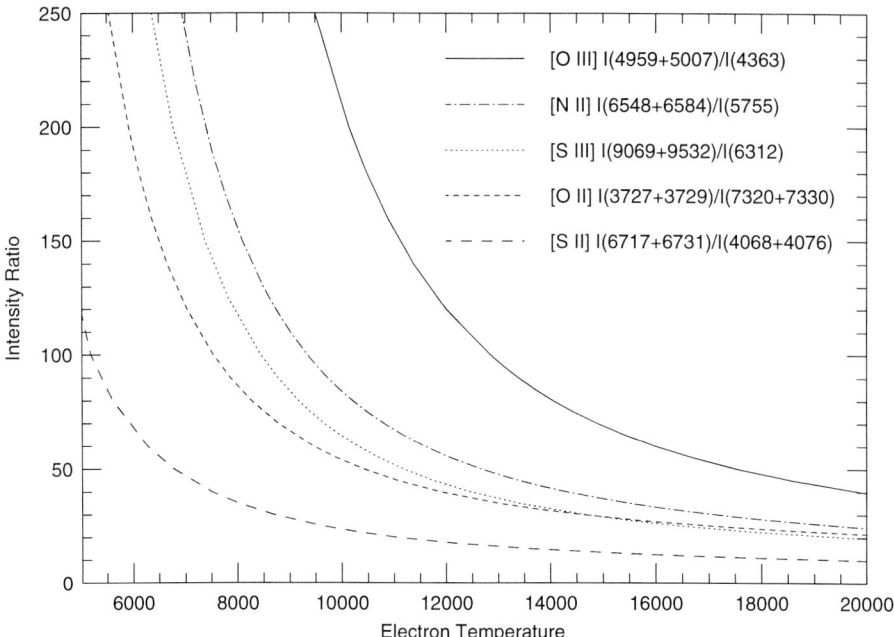

FIGURE 3. A diagram of the temperature sensitive emission line ratios for five ionic species observable in the optical/near-IR. These line ratios were calculated with the five-level atom program of Shaw & Dufour (1995) at an assumed density of 100 cm^{-3}.

weights, the emission ratio becomes the ratio of the product of the statistical weights and the radiation transition probabilities. In the intermediate regime, near the "critical density" the line ratios are excellent density diagnostics. The best known is that of [S II] $\lambda 6717/\lambda 6731$ which is sensitive in the range from 10^2 to 10^4 cm^{-3} and can be observed at moderate spectral resolution. At higher spectral resolution, one can use several other line pairs (e.g., [O II] $\lambda 3726/\lambda 3729$).

In order to convert these line ratios into densities, one needs to know the energy level separations, the statistical weights of the levels, and the radiative and collisional excitation and de-excitation rates. Fortunately, one can use the five-level atom program originally written by De Robertis, Dufour, & Hunt (1987) which has been made generally available within IRAF by Shaw & Dufour (1995; SD95). This program has the additional great advantage that the authors have promised to keep the input atomic data updated. I recommend the SD95 article in *PASP* to any reader interested in abundances from emission lines as it covers much more detail than is allowed here. It is a well written and instructive article.

Deriving temperatures in HII regions is pleasantly simple. A glance at Figure 2 shows that the ratio of the emission from an upper level (an "auroral line", e.g., $\lambda 4363$) relative to the emission from a lower level (a "nebular line", e.g., $\lambda\lambda 4959,5007$) will be highly temperature sensitive. In Figure 3, I have used the program of SD95 to produce a comparison of five different temperature sensitive emission line ratios. Note that I have used observer axes (linear) as opposed to theoretician axes (logarithmic). This rather clumsy representation emphasizes the ranges of utility for the different line ratios (it can be difficult to accurately measure relative emission line ratios with differences exceeding 200!). Clearly the [O III] ratio is most useful at high temperatures, although the $\lambda 4363$ line will always be relatively weak in any HII region. On the other extreme, the [S II] ratio enjoys a wider temperature range of applicability, but actually loses much of its sensitivity to temperature above 15,000K (although there is still a 30% change from 15,000K to 20,000K – maybe the choice of linear axes isn't that great!). Of course, the ionization fraction will affect the absolute line strengths, so [S II] temperatures are rarely available for high temperature HII regions.

All of these ratios share the common problem that they are increasingly difficult to measure at low temperatures (high metallicities). For a sort of record in this game I direct the reader to Kinkel & Rosa (1994). They obtained a measurement of the [N II] $\lambda 6584/\lambda 5755$ ratio of 200, implying a temperature of 6820K and an almost solar oxygen abundance in the HII region Searle 5 in M101.

IV. Derive the ionic abundance ratios

Ionic abundances can be derived by comparing the strength of an ionic emission line to the strength of an H recombination line and then comparing that ratio to the ratio of theoretical emissivities. Stated as an equation:

$$\frac{N(X^i)}{N(H^+)} = \frac{I_\lambda}{I_{H\beta}} \frac{\epsilon_{H\beta}}{\epsilon_\lambda} \qquad (2.3)$$

where ϵ is the emissivity of the respective lines.

The theoretical emissivities can be calculated from the program of SD95. The emissivity for collisionally excited ions in the low–density limit is:

$$\epsilon = N_i N_e h\nu \frac{8.629 \times 10^{-6}}{T^{1/2}} \frac{\Omega_{12}(T)}{g} e^{-\chi/kT} \quad , \qquad (2.4)$$

where N_i is the number density of the ion of interest, N_e is the electron density, h is Planck's constant, ν the frequency of the transition, Ω the average electron collision strength from the lower level, g the statistical weight of the lower level, k Boltzman's constant, T the electron temperature, and χ the energy difference between the two electronic levels. For recombination, the emissivities are significantly more complex, but, in the end, their temperature and density dependences can be fitted with suitable power laws.

Normally, for observations of extragalactic HII regions, the observed recombination emission is restricted to H and He and the observed emission lines from the heavier elements are all collisionally excited. Because of this, absolute abundances (e.g., O^{++}/H^+) will have a strong temperature dependence (and therefore the uncertainty can be dominated by the uncertainty in the electron temperature). Absolute abundances derived from recombination lines (e.g., He^+/H^+) will have very small temperature dependences. Similarly, relative abundances derived from two collisionally excited lines (e.g., N^+/O^+) will have relatively small temperature dependences. Peimbert, Storey, & Torres-Peimbert (1993) have discussed the utility of measuring absolute oxygen abundances using OII recombination lines; unfortunately, given the relative weakness of these lines, this technique will be limited to only the brightness Galactic nebulae.

V. Derive the Elemental Abundance Ratios

When there are emission lines from all of the relevant ionic states, the total elemental abundance is just the sum of the ionic abundances. Unfortunately, this is rarely the case. This problem has given rise to a number of empirical formulae for converting ionic abundances into elemental abundances. Historically, some of these corrections were based on the near coincidences of ionization potentials. For example, consider oxygen:

$$\frac{O}{H} = \left(\frac{O^+}{H} + \frac{O^{++}}{H}\right) \times \left(1 + \frac{He^{++}}{He^+ + He^{++}}\right). \quad (2.5)$$

In this case, O^+ and O^{++} are both observable in the optical ($\lambda 3727$ and $\lambda\lambda 4959, 5007$). The unobserved O^{+3} state is corrected for by assuming that the O^{+3} zone is coincident with the He^{++} zone (based on the coincidence of the second ionization potential of He of 54.4 eV with the third ionization potential of O of 54.9 eV) which is observable via the $\lambda 4686$ line. This is probably relatively safe (but this can be checked with far-IR measurements of [O IV] emission at 26 μm). On the other hand, similar ionization correction prescriptions have failed. This is due, in part, to the fact that photoionization cross-sections are complex functions and therefore ionization potentials can only serve as approximate guides. Thus, modern calculations of ionization correction factors rely on detailed photoionization models (e.g., Mathis 1985; Mathis & Rosa 1991). For very detailed discussions I direct the reader to the cases of sulfur (Garnett 1989), nitrogen (Garnett 1990), carbon (Garnett et al. 1995a) and silicon (Garnett et al. 1995b).

I close this discussion of the "direct method" by attempting to answer one of the questions which came from the class, "Just how accurately can one measure HII region abundances?" My answer: if one has a sufficiently high quality spectrum, it is possible to measure oxygen abundances in HII regions *with accuracies of roughly* 10 – 20%! (even smaller uncertainties are routinely encountered in the literature). This is certainly adequate for most astrophysical applications. Such a tremendously valuable tool can and should be used in a wide range of applications.

2.2.2. *The bright line (or "empirical") method*

What if it is not possible to measure the faint emission lines from the upper levels that are necessary for measuring temperatures in the direct method? It turns out that one can still do pretty well at determining an oxygen abundance with just the brightest lines. Pagel et al. (1979) developed a method to do just this. The method is based on the simple observation that at high metallicities (low HII region temperatures) the bulk of the cooling is done by the IR fine structure lines, while at low metallicities (high HII region temperatures), the bulk of the cooling is done by the high excitation optical lines. Thus, the combined strengths of the optical oxygen lines ($\lambda 3727 + \lambda 4959 + \lambda 5007$) relative to H$\beta$ (the "R_{23}" parameter) increases with decreasing abundance. This relationship was recalibrated by Edmunds & Pagel (1984) and others, but has changed little since then. While relatively uncertain at trans-solar abundances (where there is little change in R_{23} with abundance), it was thought that this method could provide abundances with uncertainties of roughly 0.2 dex.

Of course, the oxygen line strengths can't increase forever with decreasing oxygen abundance, and this relationship turns around at roughly 10% of the solar oxygen abundance. Thus, there is a region of increased uncertainty and the overall relationship is bi-valued. The ambiguity of high versus low-metallicity regimes can be overcome through the observation of additional emission lines (e.g., the [S II] and [N II] lines are relatively strong at high abundances and drop dramatically in strength at low abundances). It turns out that at low abundances (less than 10% of the solar abundance) R_{23} is a fairly accurate measure of the oxygen abundance (Skillman 1989).

McGaugh (1991) constructed a grid of photoionization models, in part, to recalibrate the R_{23} vs. O/H relationship (his Figure 11 gives a comparison of a number of previous calibrations). This work was very important in that it showed that a metallicity dependent initial mass function was not necessary to explain the observed trend of softer ionizing radiation fields with increased metallicity. More work has been done on this since, but McGaugh's conclusion remains true. Despite the importance of this paper, I have reservations about a second conclusion from the work. McGaugh claimed that R_{23} could be calibrated so that it would be comparable in accuracy to the results obtainable by the direct methods not only in the low abundance regime, but also at higher abundances.

I disagree with this second conclusion for three reasons. The first is that the proposed improvement in the calibration, the addition of the [O III]/[O II] ratio as a second parameter in the calibration, does not add sensitivity at higher metallicities (it is no longer sensitive to the ionization parameter). The second reason is that at higher metallicities, R_{23} has a stronger density dependence (since the IR lines are carrying the bulk of the cooling and since increased density leads to more collisional de-excitations, and thus a lower cooling efficiency) which is not true at lower metallicities (Stasińska 1990; Oey & Kennicutt (1993). The third reason is that the role of dust (both through depletion of refractory elements onto grains and through its effects on the heating/cooling balance) is also more important at higher abundances (see Shields & Kennicutt 1995 and the next section). Thus, when using the McGaugh calibration, I would recommend a minimum uncertainty of at least 0.2 dex (and probably larger for trans-solar metallicity objects).

2.2.3. *Photoionization modeling*

Photoionization modeling is a necessary component of any abundance analysis. We have seen that the bright line method relies on calibrations from photoionization models as do ionization correction factors in the direct method. Thus, it might be misleading to give modeling a separate category. I do so to point out that with a high signal/noise spectrum, rich in emission lines, it is possible to derive more accurate abundances and to deduce properties of the exciting stars and the nebular geometry. Often, abundance studies deal with surveys of many objects intended to uncover underlying patterns, but in some cases, the accurate abundances of an individual object are critical for a particular problem. In the latter case, a direct abundance analysis can be put on firmer ground by developing a detailed photoionization model of the nebula.

Really wonderful things can be done when spatially resolved observations are combined with detailed photoionization models. I think a great example of this is the work of Baldwin et al. (1991) on the Orion nebula; this is a real treat with regard to getting the most out of one's observations. In this study it is possible to identify geometrical effects (the relative positions of the exciting stars and the bounding molecular clouds), but the most impressive aspect of this work is the detailed investigation of the possible effects of dust within the nebula. Their appendix C is a valuable monograph on the potential effects of dust on HII regions.

Photoionization modeling also allows us to explore different physical processes which may affect abundance derivations. As mentioned above, Shields & Kennicutt (1995) looked at the effects of dust in high metallicity HII regions. They discovered that the dust acts as an additional heating agent, strengthening the collisionally excited optical forbidden lines, which is why one never observes pure Balmer line optical spectra from high metallicity HII regions (even though they are easy to produce in photoionization models without dust). They also confirmed the suggestion by Henry (1993) that the depletion of refractory elements onto grains (specifically Si and Fe) could lead to hotter nebula as the efficiency of particular fine structure lines as coolants was diminished. From the conclusions of Shields & Kennicutt one can take away the following rule of thumb: HII region abundances are generally more secure for low metallicity regions with high temperature exciting stars and generally less secure for high metallicity regions with low temperature exciting stars.

2.3. Abundances: uniformities and uncertainties

Generally one would like to use the abundances measured in an HII region as a probe of the global ISM abundances. Often, one has the opportunity to measure several HII regions in a galaxy, and, thus, test this assumption. The starting assumption is that the HII region is the result of the photoionization of the surrounding ISM by one or more newly formed massive stars. A second implicit assumption is that the ionized gas of the HII region has not been enriched by stellar winds or ejecta from the exciting star.

There is substantial theoretical and observational evidence that the abundances derived from the HII region surrounding a single star will not be significantly altered by the presence of a stellar wind. Theoretically, it is difficult to get the 10^6 K stellar wind created "bubble" to mix with the surrounding 10^4 K gas (Weaver et al. 1977; Dyson & Smith 1985; Tenorio-Tagle 1996). Also, the observation that many Galactic WR nebulae (Esteban et al. 1992) and WR nebulae in M33 (Esteban et al. 1994) show normal ISM abundances, implies that even in the presence of strong winds, the abundances of the surrounding H II region are not altered. Observations of individual giant HII regions have generally found the abundances to be uniform throughout (e.g., Skillman 1985; Díaz et al. 1987; Rosa & Mathis 1987; Gonzalez-Delgado et al. 1994).

Exceptions have been observed in HII regions excited by single stars in which ejected stellar material is ionized by the exciting star. The signature of this is enhanced N and He (typically factors of 5 – 10 and 20 percent respectively) with an associated O under-abundance (typically a factor of 2 – 5). These have been observed in our Galaxy (see Esteban et al. 1992; and references therein) and in the LMC (Heydari-Malayeri, Melnick, & van Drom 1990; Garnett & Chu 1994). Thus, while there is some observational evidence for local "pollution" associated with a individual WR star or a Luminous Blue Variable star (e.g, η Carina, Davidson et al. 1986), it does not appear to be a large effect when observing nebulae excited by a large cluster of stars.

We will return to this theme in §4.1, where I review the Ph.D. thesis work of Chip Kobulnicky (1997) – an observational exploration into this problem of pollution and mixing timescales within the realm of irregular galaxies.

3. Simple chemical evolution

3.1. Introduction

The topic of chemical evolution of galaxies tends to frighten off even the bravest of astronomers. Some of that can be attributed to uncertainties in abundances determinations (which I hoped that I have addressed above), but I think there is another factor. Most of us, once we have learned a little bit about the subject, acknowledge that there are a large number of factors which affect the chemical evolution of galaxies. This often gives rise to a reaction of surrender – after all, if there are so many free parameters, how can one possibly make progress? I would respond, if ones goals are modest, I think that there is a great deal of potential for progress. The purpose of this next section is to lay out some of the simplest ideas concerning chemical evolution.

3.2. Absolute abundances: some simple formulae

The formulae for chemical evolution have developed over the years. If I wanted a student to fully appreciate the development of these formulae, I would direct them to read Schmidt (1963), Searle & Sargent (1972), Larson (1972), Audouze & Tinsley (1976),

Pagel & Edmunds (1981), Peimbert & Serrano (1982), Chiosi & Matteucci (1982), Matteucci & Chiosi (1983), and Edmunds (1990). If they ever returned to my office, I would further direct them to concentrate on our galaxy and read Pagel & Patchett (1975), Tosi (1988), Pagel (1989), and Matteucci & Francois (1989).

However, I do not expect the reader of this chapter to perform this exercise. Instead, I would direct them to the concise summary given in Peimbert, Sarmiento, & Colin (1994) (also found in Peimbert, Colin, & Sarmiento 1994). I will run through their equations, adding my own personal notes. I am suggesting that this set of equations should be considered as the "standard set". That is, I regard this set of equations as a good background from which to consider many of the possible routes of chemical evolution.

Starting with the definition of the yield (often referred to as the net, real, or true yield):

$$y(Z) = M(Z)/M_\star, \tag{3.6}$$

where the numerator is the mass of heavy elements returned to the ISM (not just created) and the denominator is the entire mass converted into stars which is not returned to the ISM (low mass stars, remnants, planets); integrated over the entire IMF.

An element specific yield can be written as (for example, here we use oxygen):

$$y(O) = M(O)/M_\star \tag{3.7}$$

For a specific volume (e.g., defining a volume surrounding a galaxy or a zone of a galaxy), the mass in baryons will change if there is inflow or outflow.

$$dM_b/dt = f_I - f_O \tag{3.8}$$

The total mass in gas is reduced by the star formation rate (ψ) and increased by the return from the various phases of stellar evolution:

$$dM_g/dt = -(1-R)\psi + f_I - f_O \tag{3.9}$$

where R is the mass fraction returned to the ISM. Note the difference between R, which is *all* of the mass which has been incorporated into stars and then returned to the ISM and y which is the mass of an element that is *produced* in the stars and then returned to the ISM.

Thus, the change in the mass of heavy elements can then be represented as:

$$dM_Z/dt = -Z(1-R)\psi + y(Z)(1-R)\psi + Z_I f_I - Z f_O \tag{3.10}$$

where the subscript on Z_I indicates that it is the metallicity of the inflowing gas (in some cases assumed to be zero).

Meanwhile, the heavy element mass fraction Z will change as:

$$M_g dZ/dt = y(Z)(1-R)\psi + (Z_I - Z)f_I. \tag{3.11}$$

Then from the equation above and introducing $\mu = M_g/M_b$:

$$M_b d\mu/dt = -(1-R)\psi + (f_I - f_O)(1-\mu). \tag{3.12}$$

Dividing the last two equations yields the following relationship:

$$dZ/dln\mu = \frac{y(Z)(1-R)\psi + (Z_I - Z)f_I}{-(1-R)\psi + (f_I - f_O)(1-\mu)}. \tag{3.13}$$

Peimbert et al. emphasize that the above equations hold for the "instantaneous recycling approximation" (IRA). The IRA is deemed suitable for elements that are created primarily in high mass stars so that the newly created elements are returned to the ISM in a period of time which is short compared to the lifetime of an "average" star. Presently, there is some debate concerning the applicability of the IRA even for elements (like oxygen) which fit this bill. The problem is the uncertainty of detecting these elements. If one wants to apply an observational check of the chemical evolution under this prescription, it is important not only that the elements are returned to the ISM in a short time, but also that they show up in an observable phase within that short time.

A very special case of the above equation results in the absence of flows. The equation then reduces to the case of the "simple closed box" model (Schmidt 1963; Searle & Sargent 1972):

$$y(Z) = \frac{Z}{ln\mu^{-1}} \qquad (3.14)$$

If galaxies were well described as simple closed boxes, then the diagnostic power of this equation would be tremendous. Measuring the gas mass fraction and the ISM abundance would measure the yields of each of the observable elements! There are, unfortunately, a number of pitfalls. The first, and obvious one, is that galaxies may not behave as simple closed boxes. There is evidence that infall is important for large galaxies (Pagel & Patchett 1975) and it is very popular to consider outflow as likely for dwarf galaxies (e.g., Dekel & Silk 1986; Marlowe et al. 1995).

Even if galaxies behaved as simple closed boxes to first order, there are other complications. Measuring the appropriate gas mass fraction is problematic. This is the ratio of the baryonic mass in gas to the total baryonic mass (stars + gas). The ubiquitous presence of dark matter makes it difficult to identify the precise mass in stars. There is also the question of the appropriate measurement of gas mass. Not only is there uncertainty regarding the molecular gas masses in galaxies (cf. Israel 1988a,b; Maloney & Black 1988), but galaxies show a large range in gas distribution size relative to stellar distribution size. Should the gas at large radii (which may not be available for star formation and may not be participating in the enrichment) be included in the calculation of gas mass? Finally, we must consider the possibility that newly synthesized elements are retained in a hot phase and not well mixed into the more easily observable cool phases. The above is not meant as a discouragement, only a warning.

Note that an optimistic observer can still salvage something from this. By measuring Z and μ and calculating y from the above formula, it is possible to obtain an "effective yield." While this may not be directly related to the yield that one would calculate from a detailed understanding of stellar nucleosynthesis, stellar evolution, and the IMF, nonetheless, effective yields may prove to be a very useful tool in comparing galaxies and their chemical evolutionary states.

Undaunted, we then follow Peimbert et al. as they consider the complication of adding inflow. When this is set equal to the star formation rate, one gets:

$$y(Z) = \frac{Z}{1 - exp(1 - \mu^{-1})}. \qquad (3.15)$$

Setting the star formation rate equal to the inflow rate is not only a neat trick that results in a relatively simple formula, but it may have a good physical motivation. If galaxies obey some sort of star formation law which requires a threshold gas density (e.g., Kennicutt 1989) then it makes good sense to equate the inflow rate with the star formation rate.

Dropping the assumption that the inflow equals the star formation rate results in a significantly more complex relationship. If the inflow rate is parameterized as $f_I = \alpha(1-R)\psi$, then the result is:

$$y(Z) = \frac{Z\alpha}{1 - [\alpha - (\alpha-1)\mu^{-1}]^{\alpha/(1-\alpha)}} \qquad (3.16)$$

(Things are getting considerably complex at this point).

If, on the other hand, the system is losing gas (parameterized as $f_O = \lambda(1-R)\psi$), then one has:

$$y(Z) = \frac{Z(\lambda+1)}{ln[(\lambda+1)\mu^{-1} - \lambda]}. \tag{3.17}$$

And, finally, combining inflow, outflow and allowing for a linear metallicity dependent yield (a), one has:

$$y(Z) = Z\left\{a + \frac{\alpha - a}{1 - [(\alpha - \lambda) + \mu^{-1}(1 - (\alpha - \lambda))]^{\frac{-\alpha+a}{1-(\alpha-\lambda)}}}\right\} - \alpha Z_I \tag{3.18}$$

OK, at this point, I believe that the equation has become so complex that it is not possible to tell by inspection how the yield and the gas mass fraction are related over the large range of possible parameters. Is this equation unnecessarily complex? I don't think so. Inflow and outflow are well motivated ideas, and metallicity dependent yields are also (Maeder 1992; see next section). Fortunately, this equation is still simple enough that it can be coded relatively easily and played with by running various numerical models. This is where I will leave off, encouraging the motivated reader to do so. At this point, I will reiterate and reinforce my earlier point - the key to observational progress in this field lies in cleverly isolating variables (i.e., using the vast diversity of galaxies or focussing on elements of different origins as described in the next section).

3.3. Relative abundances

It is my impression that the study of relative abundances holds the promise for solving many of the problems of the chemical evolution of galaxies. Since different elements are formed in different processes corresponding to different stellar mass ranges, focussing on different elemental abundances can lead to the isolation of variables discussed above. Historically, there have been some very simple ideas and experiments along these lines, but the future promises a great degree of sophistication. For example, I expect that nucleocosmochronology will see a vast increase in diagnostic power in the next decade and comparison of r- and s- process elements will enjoy a greater range of applicability (e.g., Pagel & Tautvaišienė 1995, 1997).

Here I will try to stick to the simple basics. There are three important concepts to discuss when considering relative abundances. These are differences in nucleosynthetic origin (I take as an example primary versus secondary processes), time delays (or the breakdown of the IRA), and changing yields (e.g., metallicity dependent yields).

3.3.1. Nucleogenesis differences

Relative abundances are usually not compared in the absolute, but generally compared along some baseline. For example, C/N or N/O may be studied as a function of metallicity. Metallicity is generally held to be O/H for nebular astronomers and Fe/H for stellar astronomers. In both cases (O and Fe) the elements are thought to be predominantly primary, i.e., built up chiefly from the hydrogen that was contained in the protostar during its initial collapse.

A secondary element is one that is built up from some other element besides H. For example, the C-N-O bi-cycle process that converts H into He in intermediate mass stars converts most of the C and O into N (since the $^{14}N(p,\gamma)^{15}O$ reaction is the rate determining step). When this reaction chain has reached equilibrium, the amount of N produced will be proportional to the metallicity of the star. Thus, when plotted versus metallicity, ratios of secondary elements to primary elements (e.g., N/O or N/C) would be expected to increase linearly proportional to metallicity. This is in contrast to ratios of primary elements (e.g., Fe/O or C/O) which would be expected to be independent of metallicity.

3.3.2. *Time delays*

Given the above, one might naively expect Fe/O to be constant everywhere. In fact, it is not. Older stars in our galaxy show lower values of Fe/O (see Wheeler, Sneden, & Truran 1989). Why is this? Because stars have non-negligible lifetimes. As a result, a temporal variation in the ratio of two elements will be observed if the two elements are produced in stars with different lifetimes (Tinsley 1980). Thus, oxygen, which is produced primarily in high mass stars, and returned to the ISM in type II supernovae, will show an enhancement at early times, and Fe, which is produced primarily in the type I supernovae of intermediate and low mass stars, will show an enhancement at later times. This is just one example of a breakdown of the IRA.

3.3.3. *Changing yields*

If stellar nucleosynthesis yields change as a function of metallicity, then this will have a resultant effect on relative abundances. For example, recent stellar evolution models including mass loss via winds predict much larger yields of C relative to O (Maeder 1992, Woosley et al. 1993). If stellar winds are primarily radiatively driven, and dependent on opacity, then the stellar mass loss rates should depend on metallicity. Maeder (1992) presents an analysis of the expected yields from stellar evolution models which incorporate metallicity-dependent stellar mass loss. These models result in an increasing yield of carbon, and a decreasing yield of oxygen, with increasing metallicity.

Another possibility for changing yields would result from a variable IMF. In the present climate, it is not fashionable to consider IMF variations. First off, the IMF police will come and take you away, and no one wants that! But, seriously, there are many good observational arguments supporting an environment and metallicity independent IMF (Massey et al. 1995a,b, and Phil's chapter in this volume). However, a variable stellar initial mass function has been invoked to explain some properties of giant H II regions and starburst galaxies (e.g., Melnick, Terlevich, & Eggleton 1985). If one wanted to consider the effect of a metallicity dependent IMF (this is all hypothetical, of course), then it is easy to see that an increase in the proportion of high mass stars or an increase in the upper mass cut-off would lead to more heavy element production from a given population of stars. If the IMF is weighted toward more massive stars at low metallicities, then an increasing C/O ratio with metallicity would result from a change in the relative contributions of massive and intermediate mass stars. Now, I'm not suggesting that you consider this, but I thought I should mention it for completeness.

4. Abundance patterns in dwarf galaxies

4.1. *Uniform abundances?*

One outstanding question concerning abundances in dwarf galaxies is the variation of abundance within a galaxy. Since star formation in dwarf irregulars is thought to be episodic, and the solid-body nature of the rotation inhibits efficient mixing, it seems plausible that there might be large variations in abundance from region to region. However, the best studied irregular galaxies (SMC & LMC: Peimbert & Torres-Peimbert 1974, 1976; Dufour 1975; Dufour & Harlow 1977; Pagel et al. 1978; NGC 6822: Pagel, Edmunds, & Smith 1980; Sextans A: Skillman, Kennicutt, & Hodge 1989) indicate that the dispersion in oxygen abundance is generally quite small ($\approx 25\%$) and consistent

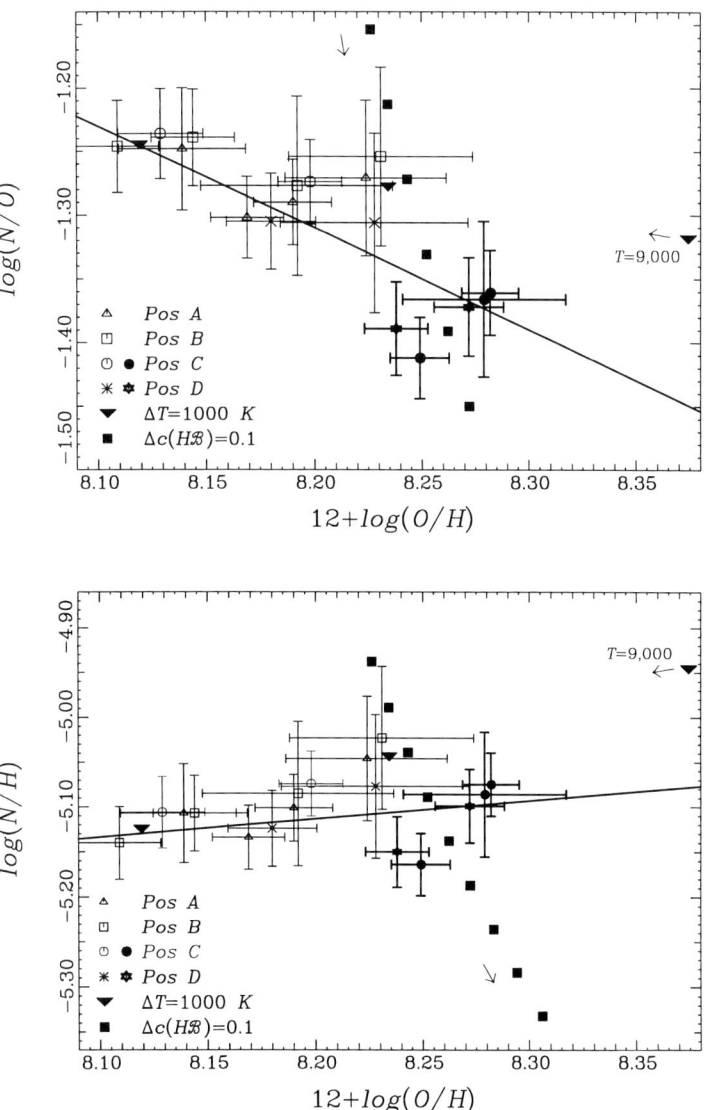

FIGURE 4. (upper) O/H versus N/O abundances at 18 positions in NGC 4214 using wide (12.5″) apertures. Solid circles and stars denote regions near Knot 5 which show higher O/H abundances and lower N/O ratios (0.1 dex) than the rest of the surveyed region. Filled triangles/squares illustrate the effect on the derived abundances when the adopted $T_e/c(H\beta)$ is artificially displaced by 1000 K/0.1. The solid line is a least squares fit treating O/H as the independent variable. (lower) Log(N/H) versus 12+log(O/H). No evidence of N/H variations is seen. (From Kobulnicky & Skillman 1996)

within observational errors. These observations support the picture discussed in §2.3 – that the newly synthesized elements are returned globally and not locally.

However, Kunth & Sargent (1986), in an attempt to explain why no galaxies more metal–poor than IZw18 (2% of the solar value; Skillman & Kennicutt 1993 and references therein) are seen, suggested that H II regions may "pollute" themselves on short (10^6 yr) timescales with nucleosynthetic products from the current burst of star formation. Although it seems certain that the heavy elements produced inside the short-lived, massive stars must contribute to the enrichment of the interstellar medium, the spatial and temporal scales on which this happens are poorly known.

Understanding element enrichment processes is especially important for interpreting the chemical evolution of low–mass, low–metallicity galaxies which have been used to derive primordial helium abundances and study the effects of supernova-driven galactic winds. Pagel et al. (1986) noticed that galaxies with spectral signatures of Wolf–Rayet (W-R) stars often had higher He and N abundances than other galaxies at similar metallicity, as measured by the O/H abundance. They suggested that N– and He–rich Wolf–Rayet star winds enrich the nebulae on timescales short compared to the lifetime of the H II region ($< 10^7$ yr).

Kobulnicky (1997) has made a concentrated observational effort to test whether HII regions are polluted by their exciting clusters. A recent optical spectroscopic study of the Magellanic irregular NGC 4214, which contains multiple starburst knots of different ages and varying W-R star content, revealed no N or He enrichments and no abundance fluctuations in proximity to young stellar clusters (Kobulnicky & Skillman 1996, KS96). Figure 4 is taken from that reference and shows the high degree of homogeneity in the abundances in NGC 4214. The oxygen abundances all fall within a range of about 0.15 dex (including observational errors) and the nitrogen abundances show a very similar range. Additionally, KS96 reanalyzed abundance data from the literature and found no systematic differences between galaxies with W-R star features and those without such features.

Kobulnicky & Skillman (1997) conducted similar observations of NGC 1569 and found similar results (supporting the finding of Devost et al. 1997 based on empirical abundance measurements). Despite all of the evidence pointing to NGC 1569 being a post-starburst galaxy (e.g., Israel & DeBruyn 1988), there is no evidence of abundance anomalies in the vicinities of the aging clusters of massive stars. No localized chemical self-enrichment ("pollution") from massive star evolution is found, even though the data are sensitive to the chemical yields from as few as two or three massive stars.

Kobulnicky (1997) argues that under the assumption that the stellar clusters are depositing their freshly created oxygen locally in the ISM, strong chemical signatures in the surrounding interstellar material should be detected unless one or more of the following are true: 1) Different star forming regions throughout the studied galaxies "conspire" to keep star formation rates and global abundances uniform at all times, 2) ejecta from stellar winds and supernovae are transported to all corners of the galaxy on timescales of $< 10^7$ yr, *and* are mixed instantaneously and uniformly, or 3) freshly synthesized elements remain unmixed with the surrounding interstellar medium (possibly residing in a hard-to-observe hot 10^6 K phase or, perhaps less likely, a cold, dusty, molecular phase). This reasoning supports the hypothesis that the newly synthesized elements are predominantly returned to the hot ISM where they can travel significant distances before cooling and becoming visible in the warm and cool phases of the ISM. This implies that the instantaneous recycling approximation often used in galactic chemical evolution modeling may not be generally applicable, even for oxygen!

Nonetheless, it is interesting to study exceptional cases. The best candidate for localized pollution in an extragalactic H II region is the amorphous galaxy NGC 5253 where a region of enhanced N (Welch 1970; Walsh & Roy 1987; 1989, WR89) and possibly He (Campbell, Terlevich, & Melnick 1986, CTM) coincides with the spectral signature of Wolf-Rayet stars. NGC 5253 contains a very young starburst (Rieke, Lebofsky, & Walker 1988; Beck et al. 1996), making it an ideal laboratory for testing theories of nucleosynthesis and dispersal of heavy elements by massive stars in giant H II regions. In order to investigate the one well–established case of N enrichment in NGC 5253 and its re-

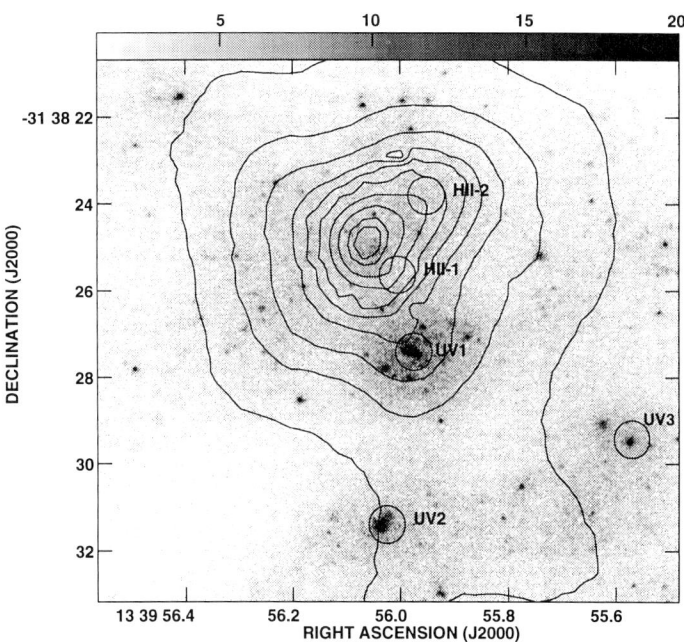

FIGURE 5. Hα (contours, from Martin & Kennicutt 1995) and 2200Å continuum (greyscale, from Meurer et al. 1995) image of the central region of NGC 5253. Five 0.86″ apertures from our FOS program are marked. Contours are 95%, 90%, 80%, 70%, 60%, 50%, 40%, 30%, 20%, 10% of the peak Hα flux. N is overabundant by a factor of 3 at locations HII-1 and HII-2 compared to UV-1 and other galaxies of similar metallicity. The peak of the nebular emission is 40 pc from the UV-1 star cluster, and we show that UV-1 is unlikely to be the dominant source of ionizing photons. (From Kobulnicky et al. 1997)

cent star formation history, IMF, and extinction in more detail, Kobulnicky et al. (1997) conducted ultraviolet and optical spectroscopy at five locations in NGC 5253 using the Hubble Space Telescope. In particular, we wanted to measure the carbon abundance in the region of N enhancement, and determine whether C, too, showed signs of localized enrichment that might be indicative of the nucleosynthetic origin of both elements.

Figure 5 shows the locations of the five observations. Slit locations UV-1, UV-2, and UV-3 are centered on three bright, young stellar clusters. Location HII-1 was intended to sample the gas at the location of the suspected N enrichment south of the emission line peak but close to the cluster of WR stars at UV-1, while position HII-2 was intended to sample a "normal" unenriched region nearby.

The abundance measurements at HII-1, HII-2, and UV-1 reveal that the He, C, O, S, and Si abundances are consistent with those in most low–metallicity systems. Surprisingly, N, appears elevated by a factor of 3 ($\log(N/O)=-0.85$) at *both* locations HII-1 and HII-2 while the N/O at position UV-1 is typical of most metal–poor galaxies ($-1.5 < \log(N/O) < -1.3$). The N/O ratios at locations HII-1 and HII-2 are consistent with one another, and appear even slightly higher than any of those found by Walsh & Roy (1989). At the position of UV-1, *despite the presence of WR stars as evidenced by its spectrum*, no extraordinary enrichment is evident.

In contrast to the elevated N abundances, the He/H abundances appear to be typical

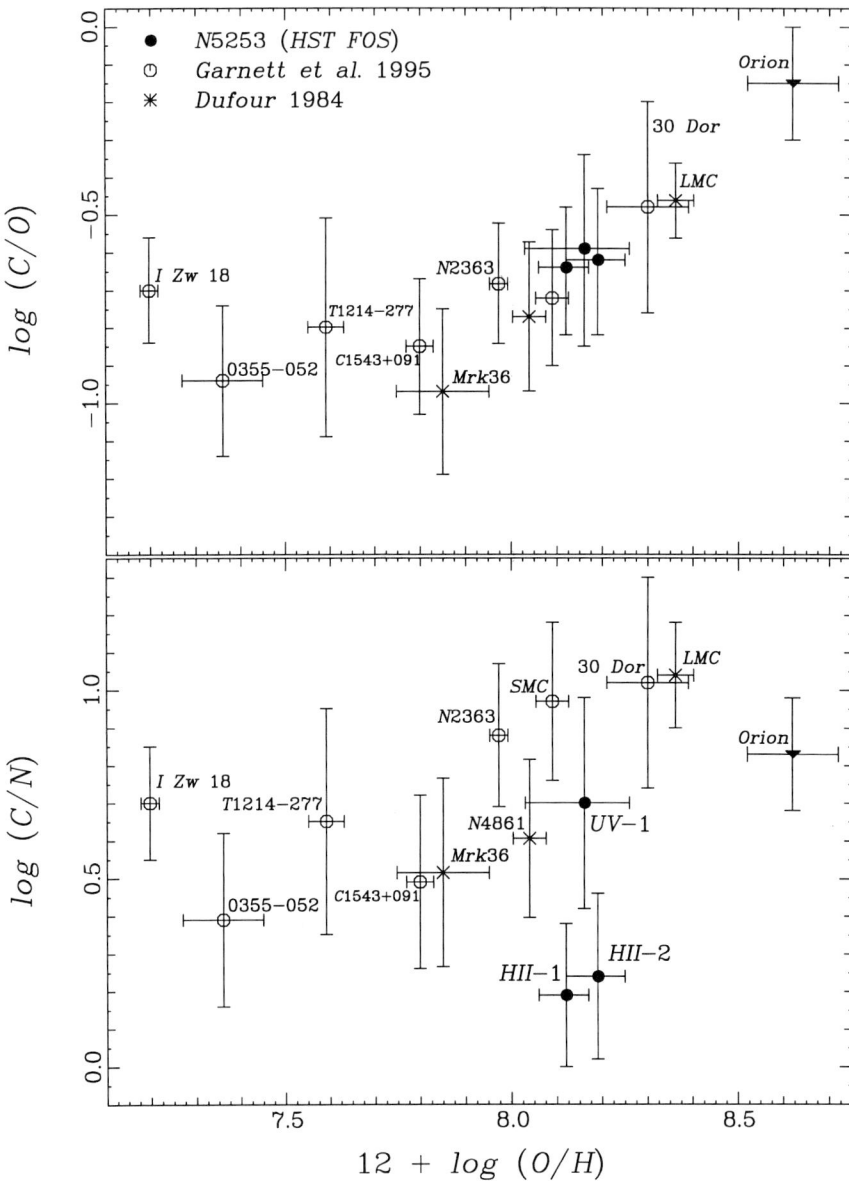

FIGURE 6. (Upper panel): log(C/O) versus oxygen abundance for 9 galaxies with reliable carbon abundance measurements. The three positions in NGC 5253 from this work are plotted with solid symbols, and are consistent with uniform C abundance throughout. (Lower panel): log(C/N) versus oxygen abundance for 9 galaxies with reliable carbon abundance measurements. The three positions in NGC 5253 from this work are plotted with solid symbols, and demonstrate the remarkable N overabundance relative to C at two locations in NGC 5253. (From Kobulnicky et al. 1997)

of those observed in metal–poor galaxies, although at the upper edge of the distribution (Pagel et al. 1992). C/O ratios, likewise, appear to be typical of similarly metal–poor systems. Figure 6 (upper panel) shows the C/O versus O/H for NGC 5253 along with 10 other points taken from Garnett et al. (1995a) and Dufour (1984) (see §4.3.2). N appears to be the only element showing elevated abundances. This results in the deviant positions of HII-1 and HII-2 in Figure 6 (lower panel) which shows C/N versus $12+\log(O/H)$.

Thus, it appears that at the location of the strongest Hα emission, the area of the youngest starburst, there is an N enrichment with otherwise normal abundances. This may be the result of pollution from the young cluster (which is not easily visible in the UV image shown in Figure 5). It is possible that the most massive stars have released a few solar masses of N through stellar winds or ejection events. Future high spatial resolution spectroscopy of this region will help us to determine if there is a pattern of pollution.

4.2. *The metallicity–luminosity relationship*

Since the nebular abundance studies of the Magellanic Clouds in the mid 1970's (e.g., Peimbert & Torres-Peimbert 1974, 1976), it has been suggested that there might be a correlation between galaxian mass and the metallicity of the interstellar medium. Surveys of the abundances in a number of H II regions in irregular galaxies by Lequeux et al. (1979), Talent (1980), and Kinman & Davidson (1981) all produced a clear correlation of oxygen abundance with both galaxy mass and luminosity.

My collaborators and I have been using this relationship to find very low metallicity objects for the study of relative abundances (Skillman et al. 1988a, 1989a,b, 1994, 1997). Figure 7 is a plot of oxygen abundance versus luminosity for a collection of nearby dwarf irregular galaxies. Under the assumption of comparable mass/light (M/L) ratios, the correlation between metallicity and luminosity would also produce a mass – metallicity relationship. Dwarf galaxies from the M81 group (Miller & Hodge 1996) and the Sculptor group (Miller 1996) have also been found to lie on this relationship. Here I have plotted oxygen abundance versus luminosity instead of versus mass because luminosities are generally available, but reliable masses are more difficult to obtain. For those galaxies with reliable mass estimates, a correlation is also found between abundance and mass.

Not all dwarf galaxies comply with the metallicity – luminosity relationship shown in Figure 7. For example, many blue compact dwarf galaxies (BCDGs), which derive a significant fraction of their total luminosity from their high surface brightness, active star forming regions, lie to the left in Figure 7 (e.g., Roennback & Bergvall 1995). If the underlying correlation is between mass and metallicity, then this would be expected since the BCDGs are likely to have lower M/L ratios. While the correlation in Figure 7 is clear, it is also clear that there is significant scatter at any given absolute magnitude. The uncertainty in the oxygen abundance is usually less than 0.1 dex, and distance uncertainties may produce uncertainties in the luminosity as large as a magnitude (or, in the case of IC 5152, larger). Richer & McCall (1995) defined a subset of galaxies with accurate distances and abundances and found essentially the identical relationship. They also found that the scatter was smaller for the higher luminosity galaxies and larger for the lower luminosity galaxies (as would be expected if the scatter is mainly due to luminosity variations from different levels of recent star formation). Given a reasonable scatter in the M/L ratios, a tighter correlation between mass and metallicity may underlie the relationship in Figure 7.

On the other hand, the fundamental relationship may not be between mass and metallicity, but between surface density and metallicity. Mould, Kristian, & DaCosta (1983) first discovered that dwarf elliptical galaxies also show a strong correlation between metal-

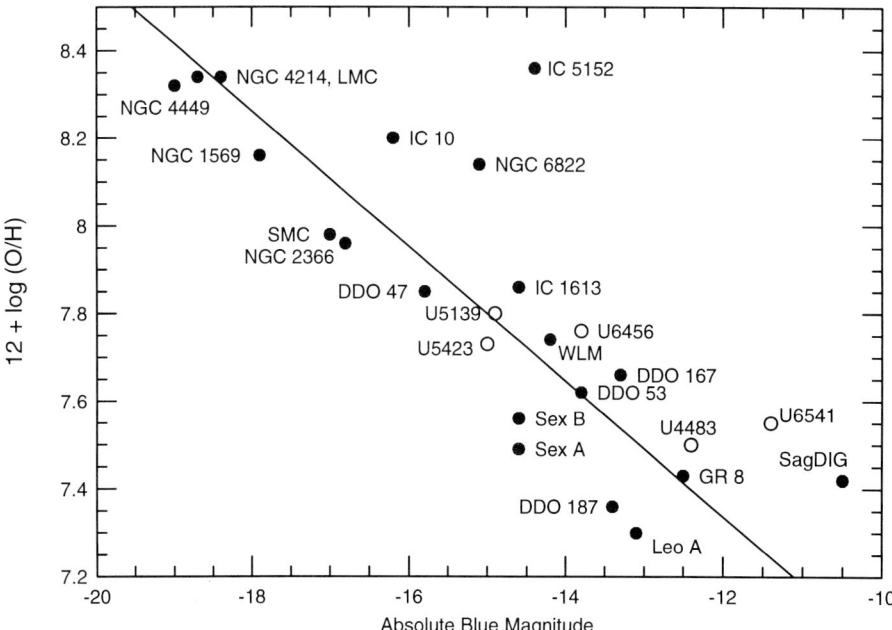

FIGURE 7. Oxygen abundance plotted versus galaxy luminosity for dwarf irregular galaxies (from Skillman, Kennicutt, and Hodge 1989). New observations of dwarf galaxies in the M 81 group have been added as open circles.

licity and luminosity, and Aaronson (1986) showed that the metallicity vs. luminosity relationships for the two classes of galaxies are roughly identical in both slope and zero-point. Since low luminosity, high surface brightness, and relatively metal rich dwarf ellipticals have been observed, Bothun & Mould (1988) have suggested that surface density may be the fundamental parameter determining metallicity (see also Edmunds & Phillips 1989). The same may hold true for the dwarf irregulars, and is worth investigating observationally. To date, Van Zee (1996) has found that her sample of low surface brightness dwarf galaxies obey the same metallicity – luminosity relationship, and note the result described by Gary DaCosta in his lecture on the dwarf galaxy F8D1 in the M81 group.

Additional measurements of dynamical masses and mass surface densities will hopefully clarify the situation. In my opinion, there is a shortage of reliable mass estimates for dwarf irregulars. What we really want for this type of work are synthesis HI observations which provide a gas distribution and a rotation curve, from which it is possible to derive the mass distribution. Unlike the case for most spiral galaxies, a single dish line profile cannot be converted into a reliable total mass estimate because estimates of the appropriate inclination and radius from optical images can be specious. Since reliable mass estimates must be based on the more time consuming interferometric observations, results are lagging behind abundance measurements, but there is progress (e.g., Côté 1995).

4.3. Relative abundance patterns

4.3.1. C/O

The relative abundances of carbon and oxygen are important for a variety of problems, including: (1) C and O are important sources of opacity in stars. Knowledge of the time evolution of C and O abundances is necessary to properly model the structure and evolution of stars at different ages and abundances, (2) The formation of the CO molecule in the ISM should be regulated by the relative abundances of C and O at some level, so an understanding of the variation of C/H and C/O in the ISM is relevant for modeling the formation of CO molecules and the possible effects of abundance variations on the $I(CO)/N(H_2)$ relation, (3) There is evidence that the interstellar dust-to-gas ratio is related to the interstellar carbon abundance (Mathis 1990), so knowledge of C abundances in a variety of environments is relevant to understanding the amount, composition, and evolution of dust in those environments, (4) In theory, oxygen is synthesized almost entirely in massive stars ($M > 10\ M_\odot$), while carbon is produced in both massive and intermediate mass stars. Thus, the ejection of some carbon is delayed in time with respect to oxygen, so the C/O ratio offers a potential "clock" for determining the relative ages of stellar systems.

Garnett et al. (1995a) have used the HST to observe HII regions spanning a range in abundance. The resulting carbon and oxygen abundances for the targets from our sample are displayed in Figure 8 (upper), plotted as log (C/O) vs. log (O/H). The data show an apparently continuous increase in C/O with increasing O/H in these metal-poor systems. By comparison, a simple chemical evolution model with instantaneous recycling predicts constant C/O if both C and O are primary elements, or C/O $\propto$ O/H if O is primary and C secondary. Since only primary sources of C are known to exist, the trend in Figure 8 suggests that either the instantaneous recycling approximation does not hold for both C and O or that the yield of C varies with respect to O (or both). This is discussed in detail below.

Figure 8 (lower) also shows the variation of C/N with O/H, with nitrogen abundances obtained from the sources of optical spectra listed above. If we exclude the Galactic points (Orion, B stars, and solar system), then a trend of increasing C/N with O/H is apparent. A larger sample of galaxies is needed to confirm and delineate the trend more clearly, but it appears that the pattern of chemical evolution seen in the dwarf galaxies does not fit in with that seen in our own galaxy.

Figure 9 presents a comparison of our new results with previous results found in the literature. Here we are interested in two questions, (1) do these new results agree with previous results on C/O measurements of H II regions in dwarf galaxies and (2) is the pattern of C/O versus O/H seen in the dwarf galaxies similar to that seen in our Galaxy?

The top panel of Figure 9 addresses the first question. Here the points from Figure 8 are plotted along with results from IUE observations taken from the literature. Points which represent observations of the same galaxy are connected, showing a good agreement between the new and old observations.

The bottom panel of Figure 9 presents C/O and O/H measurements of Galactic stars taken from the literature. There is considerable scatter in log (C/O) for the disk stars, with an average value of about -0.4. For the halo stars, log (C/O) is fairly constant at -0.9 (with a scatter consistent with observational errors). If the trend of increasing C/O with O/H seen in the H II regions of dwarf galaxies is supported by more observations, then *a clear difference is emerging between the abundance pattern seen our Galaxy and the dwarf galaxies.*

Figure 9 prompts one more philosophical aside. Plots of relative abundance versus

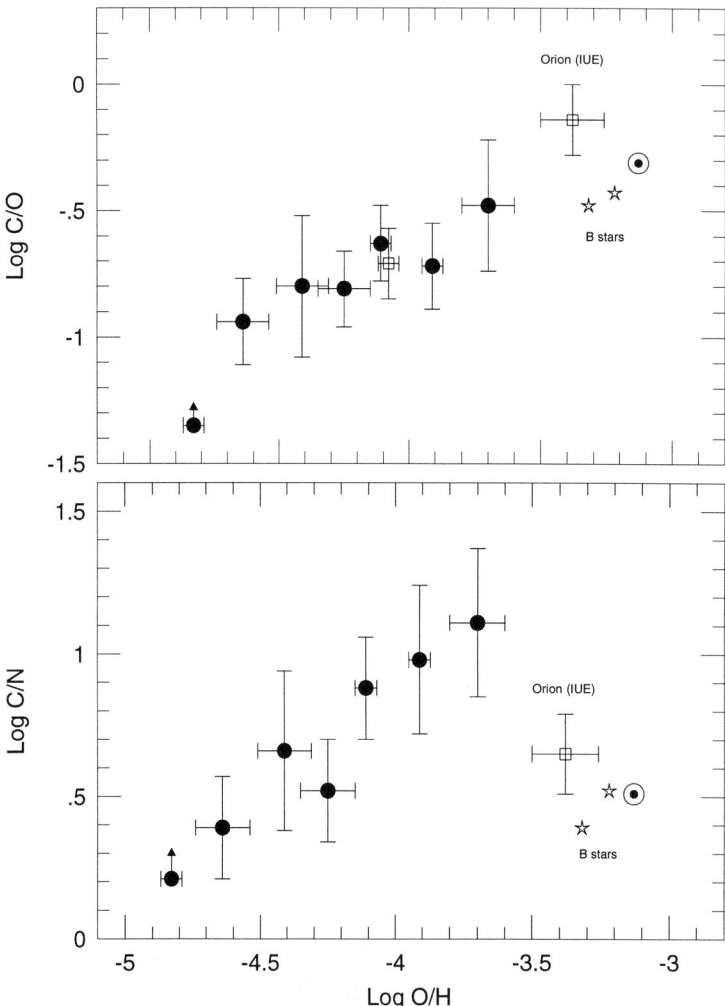

FIGURE 8. (upper) The C/O abundance ratio in irregular galaxies vs. O/H. Filled circles are from FOS spectra; unfilled squares are from IUE observations for Orion (average from the literature) and NGC 2363 (Peimbert et al. 1986). The stars represent the mean abundances in Galactic B stars determined by Gies & Lambert (1992) and Cunha & Lambert (1993). The solar system value is also shown (Grevesse & Noels 1993). (lower) C/N abundance ratio vs. O/H in irregular galaxies. Symbols are the same as above. (From Garnett et al. 1995a)

metallicity in our Galaxy (lower panel) hopefully allow us to trace the relative abundances in a single system with time. Is this true for the upper panel? Obviously not. These are different systems, and they are all sampling the abundances at the present epoch. From the metallicity-luminosity relationship, we know that the abcissa of the upper panel of Figure 9 is really a ranking in the sizes (masses) of the systems (in addition to the chemical maturity), so *it is unlikely that the low metallicity systems will evolve into the high metallicity systems* (i.e., that they will evolve from low mass to high mass). On the other hand, if different galaxies evolved with vastly different star formation histories

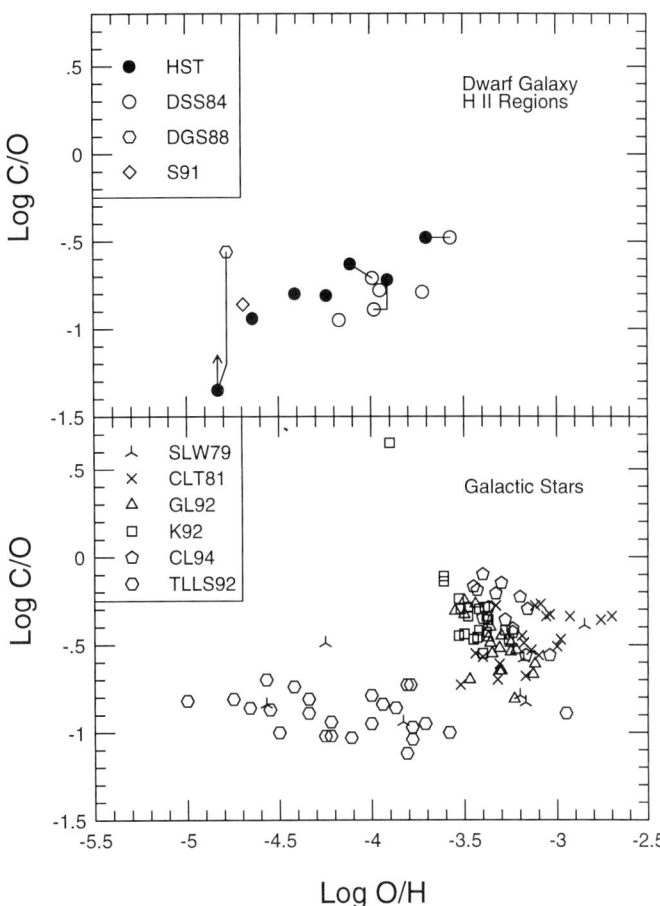

FIGURE 9. The C/O abundance ratio versus O/H for H II regions in irregular galaxies and Galactic stars. (Upper) New HST HII region abundances are compared with previous results from IUE observations. The open circles come from Dufour, Schiffer, & Shields (1984), the open hexagon represents IZw18 as from Dufour, Garnett, & Shields (1988) as adjusted by Dufour & Hester (1989), and the open diamond represents UGC 4483 as reported by Skillman (1991). (Lower) Galactic stars are presented. The data are taken from Sneden, Lambert, & Whitaker (1979), Clegg, Lambert, & Tomkin (1981), Gies & Lambert (1992), Kilian (1992), Tomkin et al. (1992), and Cunha & Lambert (1994). (From Garnett et al. 1995a)

and/or vastly different IMFs, we would not expect to see the smooth relationships we often find when comparing different systems in this way. This is probably telling us that galaxies all evolve obeying some simple (universal?) physical principles.

Before proceeding with an interpretation of these trends, I would like to make a small aside to return to one of the themes I discussed in my introduction – the possibility of using ISM abundances to give insight to better understand stellar physics. ^{16}O is predominantly a product of α-particle captures onto ^{12}C during the He burning phase of stellar evolution. The amount of O ejected by a star depends on its mass. Stellar evolution models generally show that 10 $M_\odot$ is the smallest star that can produce and eject new oxygen. ^{12}C is also produced during He burning through the well-known "triple-α" reaction. Theoretical models indicate that carbon can be ejected not only by

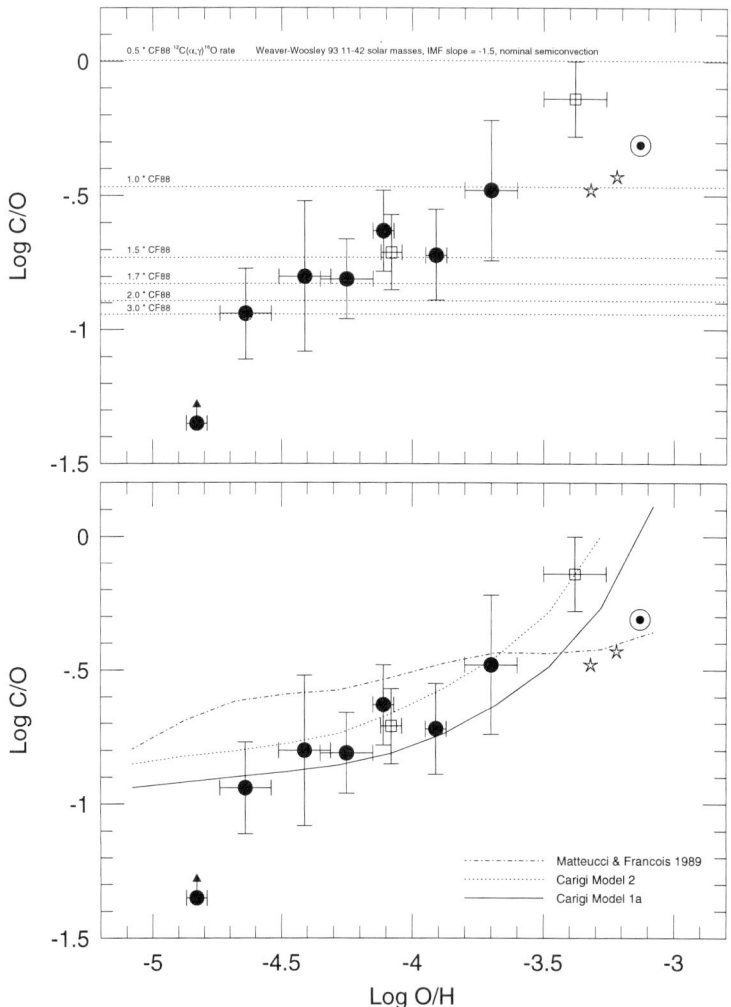

FIGURE 10. (upper) Comparison of our C/O abundance ratios in irregular galaxies with the results of stellar nucleosynthesis calculations (Weaver & Woosley 1993). The horizontal dotted lines indicate the value of the C/O ratio one obtains by integrating the C and O produced by massive stars over a mass range 11 - 42 $M_\odot$ and an IMF with slope -1.5, for various values of the $^{12}C(\alpha,\gamma)^{16}O$ reaction rate. Each line is labeled with the magnitude of the reaction rate, as defined in WW93; the reaction rate increases from 0.5 times the CF88 rate for the top line to 3.0 times CF88 for the bottom line. (lower) Comparison of C/O abundances in irregular galaxies with chemical evolution models from Carigi (1994) and Matteucci & François (1989); two of Carigi's models are represented by the solid and dotted lines; model 1a represents evolution for a Salpeter IMF, while model 2 is for a Scalo IMF. Both models use massive star yields from Maeder (1992). The Matteucci & François model is shown by the dot-dash line; this model uses massive star yields from Woosley (1986). Note that the chemical evolution models were designed to simulate the evolution of the solar neighborhood. (From Garnett et al. 1995a)

massive stars, but also by intermediate mass stars ($\approx$ 2-8 $M_\odot$) through the convective dredge-up of freshly-synthesized carbon during the asymptotic giant branch evolution phase of these stars. Measurements of carbon abundances in planetary nebulae show that some of them have enrichments of carbon, confirming that dredge-up does occur in at least some intermediate mass stars.

The relative amounts of C and O produced during He burning are sensitive to the $^{12}C(\alpha,\gamma)^{16}O$ nuclear reaction rate, as demonstrated by the models of Weaver & Woosley (1993; hereafter WW93). The value of the cross section for this reaction has been a source of considerable uncertainty (cf. Rolfs & Rodney 1988). Some attempts have been made to determine the rate empirically. Arnett (1971), for instance, tried to constrain the reaction rate by comparing the solar system C/O ratio with the predictions of stellar nucleosynthesis. This analysis assumed that both C and O are produced in massive stars only and that the C/O ratio does not evolve with time. Clearly, however, the observations of C/O in Galactic stars and extragalactic H II regions show that the C/O ratio has evolved. WW93 tried a similar approach to the problem, comparing their results for massive star nucleosynthesis with the solar system abundance pattern, but *excluding* carbon in the comparison. They found that a reaction rate of 1.7±0.5 times the Caughlan & Fowler (1988, CF88) rate produced the best match to the solar system abundance distribution for elements between O and Fe. New laboratory measurements of the $^{12}C(\alpha,\gamma)^{16}O$ cross section have been reported by Buchmann et al. (1993) and Zhao et al. (1993). In terms of the commonly used astrophysical S factor, S(E), Buchmann et al. measured a value $S_{E1}(0.3\text{ MeV}) = 57\pm13$ keV-b, while Zhao et al. obtained a value $S_{E1}(0.3\text{ MeV}) = 95\pm32$ keV-b for the same reaction. For comparison, WW93's best estimate corresponds to $S_{E1}(0.3\text{ MeV}) = 102\pm30$ keV-b.

With these new observations, it is possible to estimate the $^{12}C(\alpha,\gamma)^{16}O$ reaction rate by comparing the stellar nucleosynthesis results of WW93 with the C/O ratios measured in the most metal-poor galaxies. If we assume that (1) the most metal-poor galaxies in our sample have abundance ratios dominated by nucleosynthesis from massive stars and (2) the stellar IMF does not change significantly with metallicity, then the asymptote of the C/O abundance ratio can provide a measure of the reaction rate.

Figure 10 shows our observed C/O ratios overlaid with the C/O ratios obtained by integrating the WW93 results over the stellar mass range 11-42 $M_\odot$ with an IMF slope of −1.5 (taken from their Table 7). The horizontal dotted lines show the theoretical integrated C/O ratios obtained for different values of the $^{12}C(\alpha,\gamma)^{16}O$ rate in units of the CF88 rate. The observed C/O ratios in our three most metal-poor galaxies (not including IZw18) are consistent with WW93's estimate for $^{12}C(\alpha,\gamma)^{16}O$. While this comparison should be viewed with caution because of a number of additional uncertainties which can affect the results (e.g., the effects of stellar mass loss and mixing on stellar evolution and nucleosynthesis have only recently been explored extensively, e.g., Maeder 1992; Woosley, Langer & Weaver 1993), it is instructive to see what types of observations can be planned to cross the bridge between ISM and stellar physics.

Returning to the interpretation of the observed abundance ratios, it is clear that the C/O abundance ratio in low-metallicity systems is lower than in the solar neighborhood. The low C/O ratios we observe in our most metal-poor galaxies are consistent with the C/O ratios observed by Reimers et al. (1992) in Lyman-limit and metal-line absorption systems toward the QSO HS1700+6416. These results combined with ours clearly show that the C/O ratio has evolved from low values at early times to the present-day "cosmic" ratio.

Lets consider the trend in Figure 8 within the framework of the simple closed box model of chemical evolution and the instantaneous recycling approximation. Since both

C and O should be primary elements, the simple model predicts that C and O should maintain a constant ratio. This prediction is clearly at odds with the observations.

There are several possible ways to explain the variation in C/O with O/H:

(1) *Breakdown of the instantaneous recycling approximation.*

While the IRA may apply to oxygen, it is less likely to apply to carbon. The low C/O ratios in the dwarfs may indicate that their stellar populations are, on average, much younger than more massive galaxies. This is another way of saying that the star formation histories of the galaxies may be different.

(2) *Variable IMF.*

A comparison of the upper and lower panels of Figure 8 proves interesting. Note the abrupt drop in C/N between the solar neighborhood objects and the more metal-rich dwarf galaxies (NGC 2363 and LMC), which is attributable to much larger nitrogen abundances in the Galactic nebulae. In contrast, the C/O ratio maintains a smooth upward trend including the Galactic objects. Taken together, these features suggest that either the nucleosynthesis of C is largely decoupled from that of N or that the chemical evolution histories of the dwarfs and the Galaxy are very different. For this to be accounted for by changes in the IMF, one possibility is that C and O production is dominated by massive stars, while N comes mainly from intermediate mass stars (IMS). Another possibility is that N and C are both produced mainly in intermediate mass stars, but in different mass ranges. (But remember that there is a lack of evidence for variable IMFs.)

(3) *Variable yields.*

Carigi (1994) has calculated chemical evolution models for the Galaxy which take the Maeder yields for massive stars into account. Her models show a significantly steeper increase in C/O with O/H than the older models of Matteucci & François (1989), which used massive star yields from Woosley (1986). The results for Carigi's models 1a and 2 are plotted with our data in the lower plot of Figure 10. The models reproduce the trend in the data fairly well. Prantzos, Vangioni-Flam, & Chauveau (1994) have computed similar models for the solar neighborhood also, and find similar results.

While the present data do not allow for a definitive choice between the above alternatives, I think that choice (3) is the most probable at this time. Future observations of objects which show a range in C/O at a constant O/H may help in constraining the choices.

4.3.2. *N/O*

Figure 11 shows the N/O abundances versus O/H for a sample of dIs and HII galaxies compiled by Kobulnicky & Skillman (1996). While the production of O in galaxies is dominated by nucleosynthesis in massive stars and relatively prompt return to the ISM via supernovae explosions, the dominant processes leading to the enrichment of N in galaxies is still not well understood. It seems clear that the secondary production of N from C and O in the CNO cycle in intermediate mass stars, with subsequent release into the ISM via red giant winds and planetary nebula, must play an important role. In the more massive spiral galaxies, at higher metallicities, N/O increases roughly linearly with O/H, as would be expected if this secondary production mechanism dominated N production (Vila-Costas & Edmunds 1993).

In contrast, it has been known for some time that N/O is relatively constant over a large range in metallicity in the stars in our galaxy (Laird 1985; Carbon et al. 1987) and that the dwarf irregular galaxies also show relatively constant N/O (although with large scatter; Garnett 1990, Thuan et al. 1995). A constant value of N/O over a large range in metallicity could be taken as evidence of primary production of N (Pagel 1985).

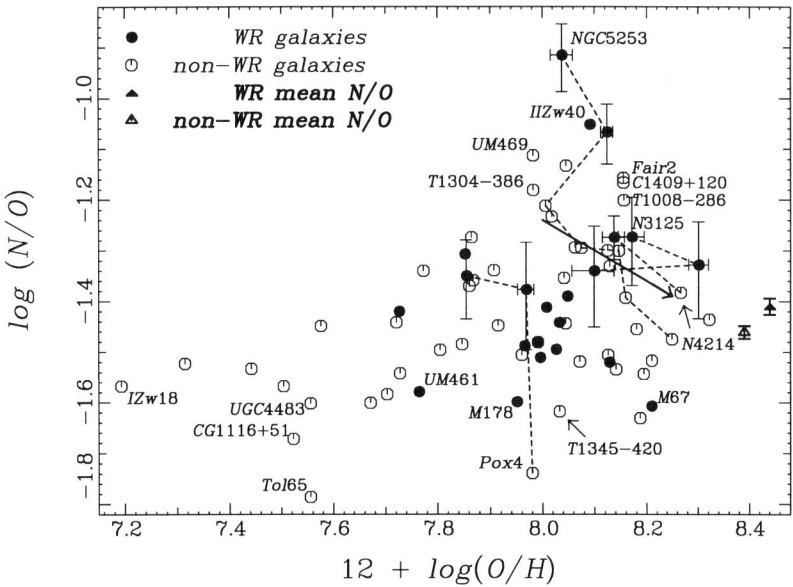

FIGURE 11. Log(N/O) versus 12+log(O/H) for 72 measurements of 60 low–metallicity galaxies from the literature. Abundances have been recomputed in a self–consistent manner from published line strengths. Triangles mark the average values of log(N/O) for galaxies with and without WR features when NGC 5253 is removed from the sample. The difference between the two subsamples becomes negligible if galaxies with 12+log(O/H) < 7.6 are excluded from the sample. A heavy vector shows the theoretical evolution of metal–poor galaxies during a period of oxygen pollution (Garnett 1990, Fig. 7). (From Kobulnicky & Skillman 1996)

Two processes leading to primary N production have been suggested. Renzini & Voli (1981) showed that increasing the convective scale length over the pressure scale length resulted in freshly synthesized ^{12}C being brought up to the convective envelope whereupon it can by converted to N via the CNO cycle. This "hot bottom burning" mechanism (Iben & Renzini 1983) will be most effective in the mass range of 4 to 5 solar masses. Alternatively, Timmes, Woosley, & Weaver (1995) and Woosley & Weaver (1995) have suggested that primary N might be produced in massive stars (heavier than 30 solar masses) of low metallicity. Again, if convection is enlarged beyond the standard models, a convective, helium burning shell penetrates into the hydrogen burning shell, and freshly synthesized C is converted into N.

These two primary N production processes would have substantially different delivery times. The primary N produced in hot bottom burning in intermediate mass stars would be returned to the ISM only after a few hundred million years. The primary N produced in the low metallicity, high mass stars would be returned in less than 10 million years. Note that the secondary N produced in intermediate mass stars would have the same delayed delivery as the primary N produced in intermediate mass stars.

Garnett (1990) discussed the possibility that the spread in the values of N/O at a given O/H as measured in dwarf irregulars could be attributed to a delay between the delivery of O and N to the ISM (as proposed by Edmunds & Pagel 1978). If a dwarf galaxy experiences a dominant global burst of star formation, then the ISM O abundance will increase after roughly 10 million years, with a resulting decrease in N/O (as perhaps in the case of NGC 6822; Skillman, Terlevich, & Melnick 1989). Then, over the period of several hundred million years, the N/O abundance ratio will increase at constant O/H (given the absence of a subsequent burst of star formation in that time interval, as perhaps in the case of the Pegasus dwarf irregular galaxy; Skillman, Bomans, & Kobulnicky 1997, see §6.2). Under these two assumptions (dominant bursts of star formation separated by quiescent periods and delayed N delivery to the ISM) the N/O ratio becomes a clock, measuring the time since the last major burst of star formation. Low values of N/O imply a very recent burst of star formation, while high values of N/O imply a long quiescent period.

4.3.3. S/O

Garnett (1989) discusses the production of sulfur, and concludes that it should come predominantly from stars with masses larger than 10 solar masses. Thus, sulfur and oxygen are both produced in massive stars, and one might expect the abundance of sulfur to follow that of oxygen quite closely (i.e., there is unlikely to be a significant "time delay" problem between the two). Garnett points out that potential variations between S and O would point to variations in the IMF and that observed S/O ratios may be useful in constraining nucleosynthesis calculations.

Woosley & Weaver (1995) have produced an extensive set of nucleosynthesis calculations for massive stars varying both stellar mass and metallicity. The oxygen yields show little dependence on metallicity, but the sulfur yields do change. However, the important factor for sulfur yields is the treatment of the hydrodynamics, specifically the degree of "fall back". With changes in the hydrodynamics it is possible for the sulfur production to change dramatically from all of the newly produced sulfur being ejected to all of it falling back into the remnant.

Overall, Garnett found S/O measurements from dwarf galaxies averaged log (S/O) = -1.55±0.1 which is close to the solar value of log (S/O) = -1.66 (Anders & Grevesse 1989), and no evidence for a systematic variation in S/O with O/H. However, he cautioned that more observations at higher and lower O/H were needed. Observations of the very low abundance dwarfs IZw18 and UGC 4483 also found nearly solar ratios of S/O (Skillman & Kennicutt 1993; Skillman et al. 1994), and Thuan et al. (1995) found a very narrow range in S/O for a sample of blue compact galaxies, consistent with that found by Garnett, and slightly higher than the solar value. From this observational evidence, it would appear that either the IMFs are very similar in dwarf galaxies or that sulfur and oxygen production in massive stars have very similar mass and metallicity dependences (or both).

4.3.4. Si/O and Fe/O

Interstellar dust grains play an important role in the ISM; they absorb ionizing radiation, absorb and scatter starlight, remove heavy elements from the gas phase, and are the reputed sites for the formation of molecules. One of the major uncertainties in our understanding of grains is their composition, and how the composition and abundance of dust vary with environment. Much of our knowledge of the composition of dust comes from the study of features in the interstellar extinction curve (such as the 2175 Å absorption bump), and from measurements of interstellar absorption lines. The absorption line measurements reveal that many heavy elements have ISM abundances that are smaller than their "cosmic" values (defined by the atomic abundances measured in the solar photosphere or meteorites), and thus it is inferred that the unseen portion

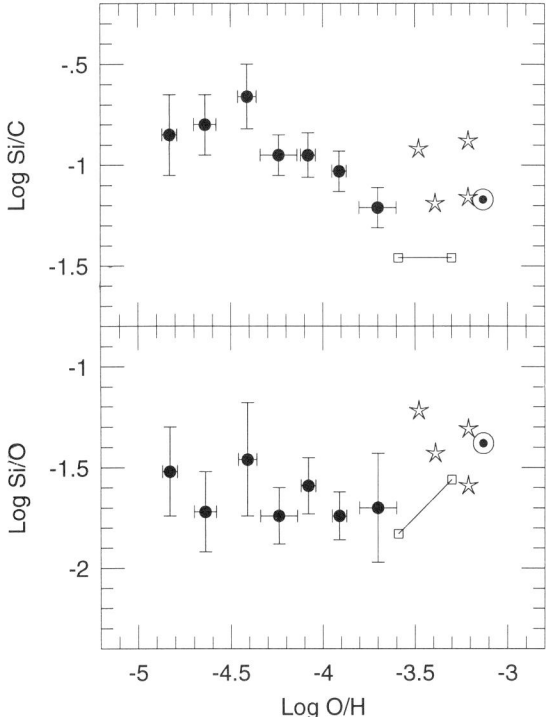

FIGURE 12. Top: The abundance ratio log Si/C plotted against log O/H for our extragalactic HII region sample. Filled circles represent our FOS measurements. Open squares represent abundances in the Orion Nebula from IUE spectra. The open stars represent the average Si/C ratios from different studies of Galactic B stars and A supergiants (Kilian 1992, Kilian et al. 1994, Kilian-Montenbruck et al. 1994, Cunha & Lambert 1994, and Venn 1995) Bottom: Log Si/O vs. log O/H for our HII regions. Symbols are the same as above. (From Garnett et al. 1995b)

is incorporated, or depleted, onto dust grains (see reviews by Jenkins 1987, 1989). In most cases, the depletions show a dependence on the average gas density along the line of sight. It is also possible to observe emission lines from ions of refractory elements in ionized nebulae, from which one can infer the properties of grains in the ionized gas and study how grains may be modified in the vicinity of hot stars.

Figure 12 displays the variation of Si/O and Si/C versus O/H for a sample of HII regions with a range of metallicities (from Garnett et al. 1995b). Also shown are Si/O and Si/C in the Orion Nebula, the Sun (Grevesse & Noels 1993), and in solar neighborhood B stars and A supergiants. Figure 12 shows (1) Si/C in the HII regions declines with increasing O/H by about 0.5 dex over the range $-4.8 < \log$ O/H < -3.1, and (2) Si/O is remarkably constant over the same metallicity range. The weighted mean value for the seven objects is log Si/O = -1.59 ± 0.07. For comparison, log Si/O = -1.37 in the Sun. The decline in Si/C reflects the increasing C/O ratio with O/H discussed in §4.3.2.

Measurements of gas-phase abundances for Si from interstellar absorption lines in the Galactic ISM typically show that Si is mostly depleted onto grains, with the depletion dependent on the mean density of the absorbing clouds (Jenkins 1987; Sofia, Cardelli, &

Savage 1994). Si should also be depleted in our HII regions, which are relatively dense gas associated with regions of star formation, and we attempt to use our Si/O measurements to estimate the depletion of Si in the ionized gas.

From Figure 12 it can be seen that the range of average values for Si/O determinations in Galactic stars implies an uncertainty of approximately 0.3 dex in the reference value. If we compare the average Si/O in our HII region sample with the averages for the different samples of Galactic stars, the implied depletions for Si lie between -0.4 dex and 0.0 dex, that is, 40-100% of the Si in the HII regions is in the gas phase. If we account for about $0.1 - 0.2$ dex depletion of oxygen in the HII regions, the inferred Si depletions lie between -0.6 dex and -0.1 dex (assuming Si/O is constant with metallicity, Timmes, Woosley, & Weaver 1995).

There are few measurements of refractory element depletions in HII regions for comparison. Iron abundances have been measured in three Galactic HII regions: Orion (Osterbrock, Tran, & Veilleux 1992), M17 (Peimbert, Torres-Peimbert, & Dufour 1993), and M8 (Peimbert, Torres-Peimbert, & Ruiz 1992). These studies inferred Fe depletions of -0.5 to -1.2 dex. These depletions are consistent with our measured Si depletions, if the trend of Fe depletion as a function of Si depletion shown by Sofia et al. (1994) holds everywhere. However, only one object (Orion) has had both Si and Fe abundances measured; a larger sample of measurements of both Si and Fe in HII regions is needed. Thuan et al. (1995) observed Fe III emission in a number of low metallicity blue compact galaxies and derived iron abundances which showed essentially no depletion. However, these were based on a single iron line (with a potentially uncertain emission coefficient), and I would urge caution in interpreting these abundances at present.

The fact that we see an average Si depletion of only -0.3 dex in the observed HII regions suggests that the grains in the ionized gas have been modified. Draine & Salpeter (1978) showed that thermal sputtering of grains in a hot gas is very inefficient at temperatures below 10^5 K, so this process is unlikely to be significant in photoionized gas at 10^4 K. Therefore, it may be that shocks are modifying the grains within the HII region. These could arise from stellar winds from O and Wolf-Rayet stars impinging upon the ionized gas, or from supernovae. Direct evidence for shocks in some giant HII complexes comes from detections of supernova remnants associated with them (Skillman 1985, Chu & Kennicutt 1986) and of diffuse X-ray gas within giant HII complexes (Williams & Chu 1995). The low Si depletions we see in the HII regions suggest that grain modification may be a general phenomenon in HII regions. It should be noted, however, that grain destruction appears to be rather incomplete within the HII regions: although relatively low depletion factors are observed for Si and Fe in the HII regions compared to the diffuse ISM, the measurements indicate that most of the Fe (and presumably, other elements such as Ca and Mg) is still in grains.

Our measurements of Si depletions appear to be consistent with the results of Sofia et al. (1994), who found that the Si/Fe and Si/Mg ratios in dust cores were smaller than expected if most of the Fe and Mg in dust cores is in silicates. They inferred that the dust cores must be predominantly metal and metal-oxide grains, while Si resides mainly in grain mantles. The low Si depletions we measure within HII regions appear to support this picture; we suggest that grains within HII regions have their mantles eroded, releasing a large fraction of the Si into the gas phase, while Fe-Mg grain cores may survive largely intact. Additional measurements of the abundances of silicon and other refractory elements in a variety of HII region environments will be needed to substantiate this idea. Remember that grain destruction or erosion within HII regions has important consequences for photoionization modeling (see §2.2).

4.4. Environmental effects?

Dwarf galaxies are known to exhibit a strong morphology–density relationship. Binggeli et al. (1990) have shown that dE galaxies are almost always found in cluster environments or as companions to larger galaxies, while dI galaxies are more commonly found in lower density environments (see Ferguson & Binggeli 1994 for a more comprehensive review). Given this background, it seems logical to ask whether environment can play a role in the chemical evolution of dwarf galaxies.

The single concerted effort to answer this question has been conducted by Vílchez (1995). He has assembled four different samples of dwarf galaxies: Virgo Cluster core dwarfs; Virgo "Clump" dwarfs; Void dwarfs; and Virgo Cluster peripheral dwarfs. The HII regions in these galaxies have been studied spectroscopically, and their properties (excitation, Balmer line equivalent width, oxygen abundance) have been compared. The different groups show differences in excitation, but it is not clear whether this is due to intrinsic differences in the populations or differences in the selection criterion. Thus, interpretation at this point must be a bit guarded. However, when using the metallicity–luminosity relationship as a diagnostic, the dwarfs from the low density environments appear to show a relationship which is offset from that of the dwarfs from high density environments. Unfortunately, the abundances are based on empirical, and not direct, measurements, and therefore the uncertainties are significant (see discussion in §2). Vílchez is careful to emphasize these uncertainties, but the possibility of a real difference between the chemical properties of low density and high density environment dwarfs exists. Follow-up work on this question is definitely to be encouraged!

5. Abundance patterns in spiral galaxies

It has been known for about 20 years that our Galaxy has a radial abundance gradient as evidenced in the composition of stars and the ISM (see Pagel & Edmunds 1981 for discussion and references). The abundance gradient in the ISM is comparable to those measured in similar extragalactic systems and has become very well defined over the last few years (cf. Esteban & Peimbert 1995), but there was some doubt concerning the stellar abundance gradient (e.g., Gehren et al. 1985). The perceived difference between the Galactic abundance gradient as measured in the stars and as measured in the ISM has led to a considerable lack of confidence in our ability to accurately measure abundances. Fortunately, the situation has changed recently, due, in part, to one of our winter school participants. Smartt & Rolleston (1997) have shown that Galactic B stars do, in fact, show the same abundance gradient as measured in the ISM. This very important result should now lend credibility to future abundance work.

Nonetheless, some basic aspects of the systematic behavior of composition gradients in spiral galaxies remain poorly understood (see, for example, Vila-Costas & Edmunds 1992, VCE, and Zaritsky, Kennicutt & Huchra 1994, ZKH). Among these is the question of the differences in gradients among galaxies and the correlation of gradients with the structural properties of galaxies. Abundances in spiral galaxies have been determined to correlate with Hubble type, galaxy luminosity, and maximum circular velocity (e.g., Pagel & Edmunds 1981, McCall 1982, Garnett & Shields 1987, VCE, Oey & Kennicutt 1993, ZKH), although since these parameters correlate with each other, the fundamental correlation is as yet unclear (Garnett & Shields 1987, ZKH). It is still true that most spirals included in these studies typically include only a handful of observed HII regions, and this can lead to biases (Garnett & Shields 1987; ZKH; Kennicutt & Garnett 1996). For example, Kennicutt & Garnett (1995), from a spectroscopic sample of 40 HII regions

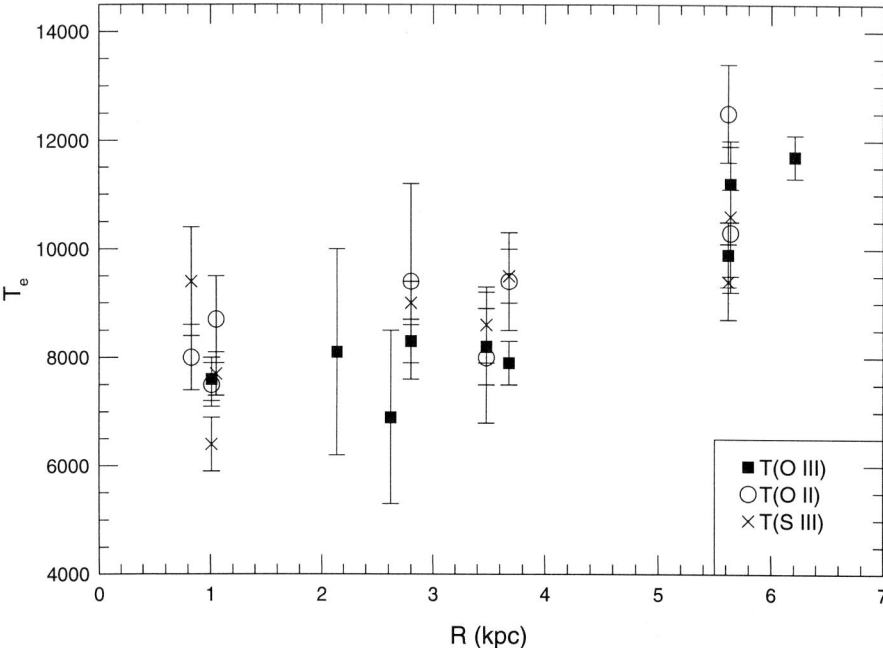

FIGURE 13. Electron temperature measurements vs. galactocentric radius for HII regions in NGC 2403. The filled squares are temperatures measured from the [O III] lines; open circles from [O II] measurements, and crosses from [S III] measurements. (From Garnett et al. 1997b)

in M101, discovered an asymmetry in derived abundances across that galaxy, which had not been noticed in previous studies using smaller HII region samples. Thus, detailed spectroscopic studies of large samples of HII regions in spirals are still needed to improve the galaxy comparisons.

Recently, Garnett et al. (1997b) have produced a detailed study of NGC 2403. I will use this study to highlight some of the important topics in spiral galaxy abundance studies. NGC 2403 is a nearby spiral galaxy of Hubble type SAB(s)cd (de Vaucouleurs et al. 1991; RC3) in the M81 group of galaxies. NGC 2403 is a morphological twin to the well studied Local Group Scd spiral M33, and, therefore, comparisons between the two galaxies are of interest. Both galaxies also have favorable inclinations, allowing good observations of both the contents of the disks and the internal dynamics.

5.1. Radial gradients

Because of their, on average, elevated metallicities, studies of HII region abundances in spiral galaxies are dominated by empirical abundances (cf. §2.2.2). For nine of the twelve HII regions in our NGC 2403 sample, we have detected at least one emission line ratio that is sensitive to electron temperature, (T_e). We were able to measure three of the T_e-diagnostic ratios discussed in §2.2.1:

(1) the commonly used [O III] $\lambda\lambda 4959,5007/\lambda 4363$ ratio. This ratio is easiest to measure at lower abundances where the [O III] lines are strong.

(2) [O II] $\lambda 3727/\lambda 7325$: This line ratio is often relatively easy to measure in low-excitation HII regions, because of the weaker dependence on electron temperature than

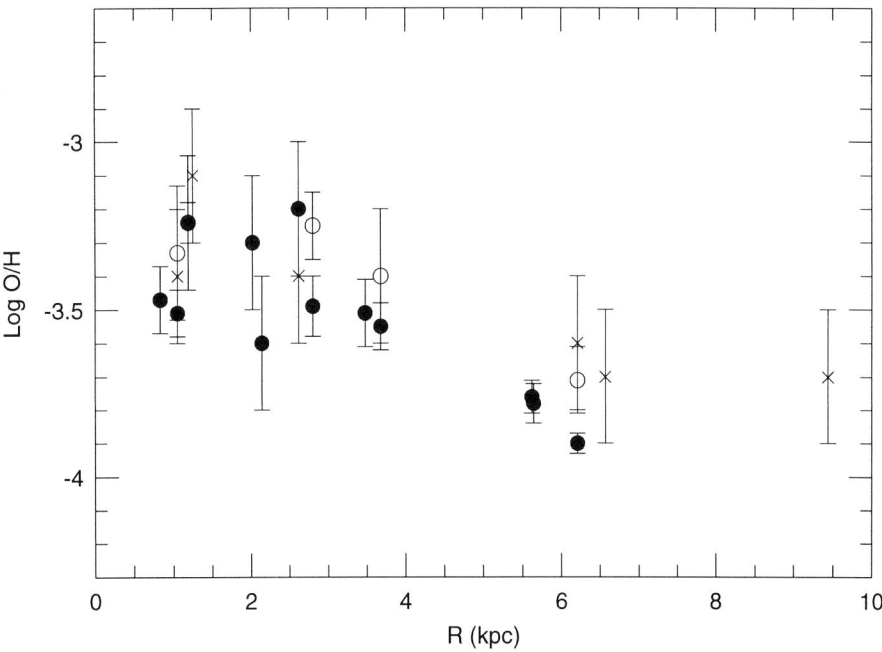

FIGURE 14. Log O/H vs. radius for NGC 2403 HII regions. Filled circles show the regions observed in this study, open circles represent measurements from Fierro et al. (1986), and crosses data from McCall et al. (1985). A weighted least-squares fit yields a gradient of -0.102±0.009 dex/kpc. (From Garnett et al. 1997b)

the [O III] ratio. However, the long wavelength baseline makes the ratio much more sensitive to reddening.

(3) [S III] $\lambda\lambda 9069, 9532/\lambda 6312$: A systematic uncertainty arises because the $\lambda\lambda 9069$, 9532 lines can suffer from telluric absorption. An evaluation of the degree to which these lines are affected can be made from the $\lambda 9532/\lambda 9069$ intensity ratio, which has a theoretical value of 2.44 (based on the ratio of the Einstein A-values). For our purposes, we evaluate the [S III] temperatures by revising either the $\lambda 9069$ or the $\lambda 9532$ upward so that the $\lambda 9532/\lambda 9069$ ratio equals 2.44.

Figure 13 shows a plot of the various measured values of T_e vs. galactocentric radius in NGC 2403. The plot clearly shows that there is a radial gradient in electron temperature across the galaxy, an indication of a corresponding gradient in the composition of the ionized gas. There is evidence that the [O II] temperatures are systematically higher than the [O III] temperatures; in agreement with photoionization model predictions of the temperature of cool, metal-rich HII regions (e.g., Stasińska 1990, Garnett 1992).

Figure 14 shows O/H as a function of radius in NGC 2403 for twelve HII regions. We have confirmed that there is a gradient in O/H across NGC 2403 with a slope of -0.1 dex/kpc, similar to gradients found in spiral galaxies of similar de Vaucouleurs T type (VCE; ZKH). The steep slope is consistent with the lack of a bar in NGC 2403 (cf. Martin & Roy 1994). An important test of the gradient would be an improved measurement of the outermost point, H109, at 9.4 kpc ($R/R_0 = 1.18$) from the nucleus. The forbidden line strengths suggest a relatively high O/H based on the Edmunds &

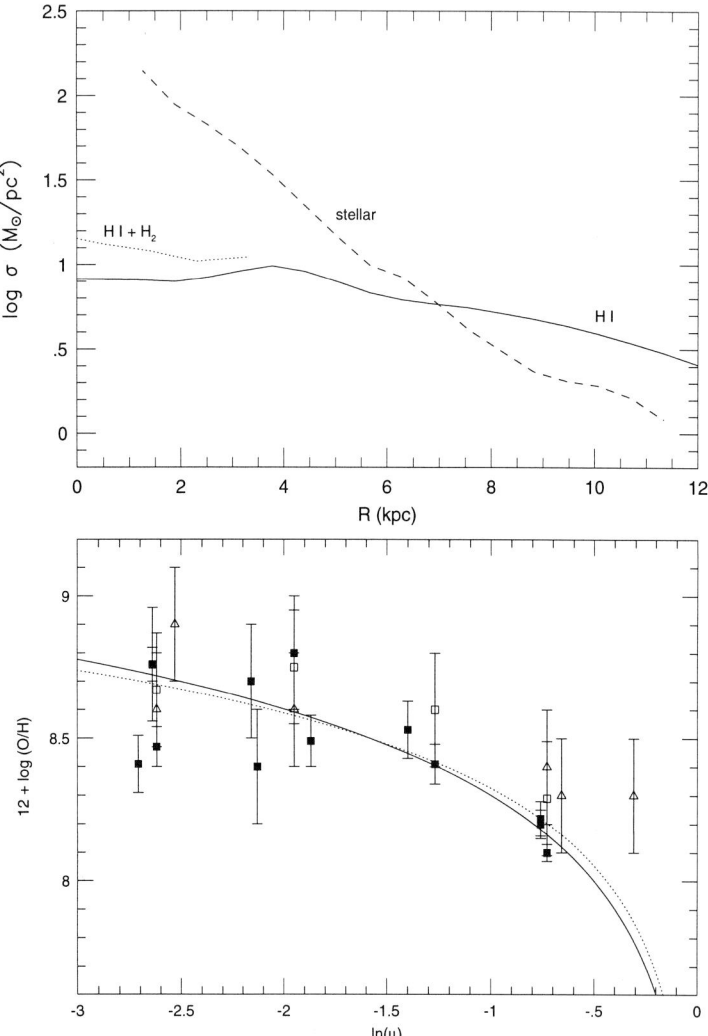

FIGURE 15. Top: Radial variations of surface mass densities of HI, H_2, and stars in NGC 2403. Bottom: O/H vs. ln(gas fraction) in NGC 2403. Solid line is the simple closed box chemical evolution model with oxygen yield by mass $y(O) = 0.0022$; dotted line is Clayton's parameterization of models with infall of metal-free gas with $y(O) = 0.0030$. (From Garnett et al. 1997b)

Pagel (1984) calibration, but they are also in the regime where the calibration is double-valued. A good measurement of O/H in H109 would resolve the ambiguity in the slope of the composition gradient, and provide important information on the conditions in the extreme outer disk of NGC 2403. The value of accurate abundance determinations in extreme outer disk HII regions was demonstrated by Garnett & Shields (1987), Garnett, Odewahn, & Skillman (1992) and Kennicutt & Garnett (1996).

Let us now compare the abundances in NGC 2403 with the simple, closed box model of chemical evolution. We assume that a ring at fixed galactocentric radius of a galaxy can be described by the closed box model, and compute a model for O/H as a function of gas fraction which matches the observations.

The gas fraction as a function of radius was determined from HI 21 cm measurements (Wevers et al. 1986) and H_2 determinations from CO measurements (Thornley & Wilson 1995). There is a well known uncertainty in converting CO observations to H_2 distributions (cf. Israel 1988a,b; Maloney & Black 1988), but the inner part of NGC 2403 has metallicities close to the solar value, so the CO-to-H_2 conversion may be expected to be similar to solar neighborhood values.

Stellar surface mass densities for the disk were determined from the radial disk surface brightness distribution (Wevers et al. 1986). The run of disk mass surface density as a function of radius for the various components is displayed in Figure 15 (top). Models were computed with various values for the oxygen yield, $y(O)$, and the best model, with $y(O) = 0.0024 \pm 0.0001$ (corresponding to $y = 0.0047$ if O comprises 45% of Z by mass), is shown as the solid curve in Figure 15 (bottom).

As noted previously by Garnett & Shields (1987) for M81, the simple model provides a surprisingly good fit to the observations. However, Twarog (1985) and Clayton (1987) have pointed out that the Z-ln μ diagram distinguishes poorly between the simple model and models incorporating infall. This is illustrated in Figure 15 by the dotted curve, which is a model based on the alternate relation

$$Z^*(\mu) = -y/2[-ln\ \mu + ln(-ln\ \mu) + 1] \tag{5.19}$$

from Clayton (1987), who presented it as a useful approximation to a number of models with infall of metal-free gas that decays exponentially with time. The dotted curve plotted in Figure 15 has $y(O) = 0.0030$. It is clear that this model also adequately explains the data, ignoring the outermost point (H109 from MRS). The yield derived for this model is still significantly smaller than that obtained for the solar neighborhood, but this should not be taken to be a reliable estimate for the yield in NGC 2403. From this exercise, we do not see evidence in NGC 2403 for a systematic variation in effective yield with metallicity.

So, why do spiral galaxies have gradients? That most spiral galaxies have chemical gradients has been known for quite some time (e.g., Pagel & Edmunds 1981). Unfortunately, theorists have thought of many different ways to create abundance gradients. These include: radial variations in the star formation rate (the basis for the simple model fit above, e.g., Phillipps & Edmunds 1991); inflow, radial flows, or a combination (e.g., Mayor & Vigroux 1981; Lacey & Fall 1985; Clarke 1989; Sommer-Larson & Yoshii 1989, 1990; Götz & Köppen 1992; Zaritsky 1992), or IMF variations (e.g., Güsten & Mezger 1982).

With so many different theories, and with so few features in the gradients, it may be some time before we have a good understanding of the origins of the gradients in spirals. I take hope in two different facts. First, the study by Martin & Roy (1994) represents a true step forward in our understanding of spirals – they were able to show that the shallower gradients in barred spiral galaxies correlated well with the strength of the bar. Secondly, the sampling in spiral galaxies is reaching the point that questions concerning the homogeneity of the abundances in radial zones (e.g., Kennicutt & Garnett 1996) and the degree to which the gradient is well described as an exponential (e.g., Scowen et al. 1992; Zaritsky 1992) can be answered. Optimistically, as these questions are answered, fundamental constraints on mixing and the role of galaxian dynamics may produce a true understanding of the gradients in spirals.

To support my optimistic outlook, I return to the study of NGC 2403 by Garnett et al. (1997b). In that study, Don assembled a sample of spiral galaxies for which he felt that

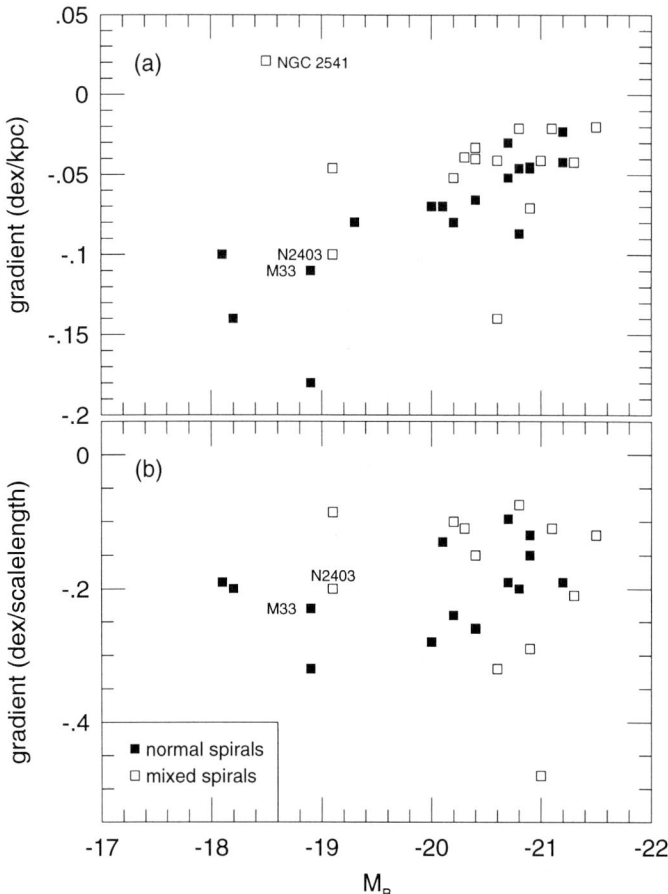

FIGURE 16. (a) Oxygen abundance gradient per unit physical length (dex/kpc) versus galaxy luminosity. A good correlation is seen. (b) The correlation in (a) disappears when the gradient per unit disk scale length is plotted instead. (From Garnett et al. 1997b)

the gradients were well determined and compared their gradients with their luminosities. The results are shown in Figure 16. There is quite a marked correlation between the slope of the abundance gradient G (in dex/kpc) and galaxy luminosity (Figure 16a). This correlation arises because bigger galaxies have longer exponential scale lengths. Figure 16b illustrates this by showing the abundance gradients in dex per disk scale length (G_S) as a function of M_B. G_S shows no correlation with luminosity. Indeed, the G_S values appear to occupy a relatively narrow range, with one outlying point. It would be interesting to determine whether the scatter in G_S is greater than observational scatter (assuming a 25% uncertainty seems reasonable — ZKH obtained uncertainties ranging between 10% and 100% for gradient slopes). Figure 16b suggests that non-barred spirals are homologous in their chemical properties, with the global metallicity scaling as the galaxy mass. It also indicates that the gas abundances are tied to the local mass surface density in the disk.

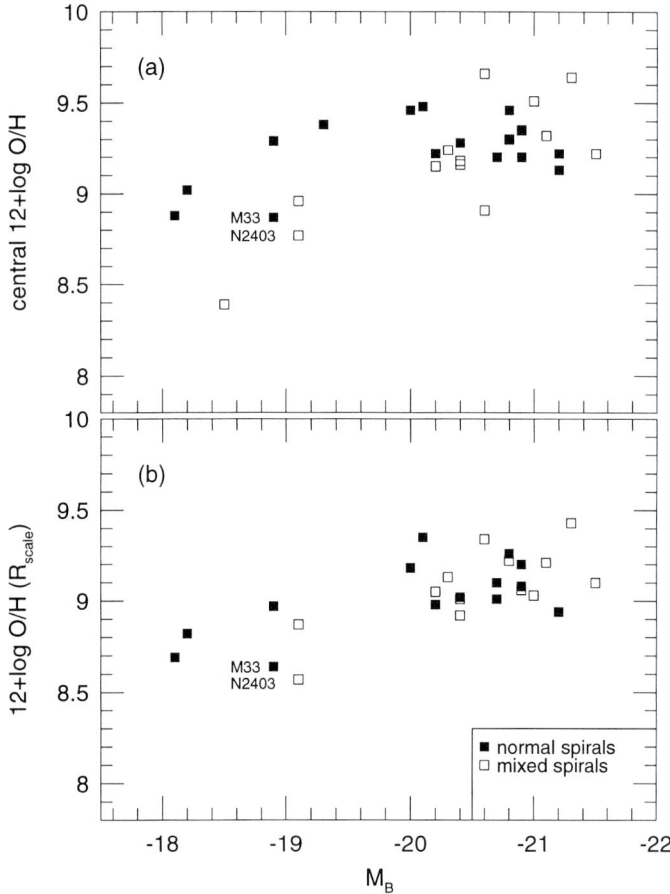

FIGURE 17. Fiducial interstellar O/H versus galaxy luminosity, for selected spiral galaxies. The upper panel shows the central abundance extrapolated from the abundance gradient; lower panel shows the characteristic O/H at one disk scale length from the nucleus. (From Garnett et al. 1997b)

5.2. *The metallicity–luminosity relationship*

As noted before, Garnett & Shields (1987) showed that spiral disk abundances correlated very well with galaxy luminosity, which was later confirmed by VCE and ZKH. Figure 17 shows both the central oxygen abundance extrapolated to zero radius from the abundance gradient, and the characteristic mean abundance at one disk scale length for the sample of well studied spirals assembled by Garnett et al. (1997b). This metallicity-luminosity correlation extrapolates smoothly to include low luminosity irregular galaxies all the way down to $M_B \approx -10$ (Skillman, Kennicutt, & Hodge 1989, ZKH, see §4.2). This remarkable correlation tells us that galaxy gravitational potential plays a key role in determining the subsequent evolution of a system. The characteristic metallicity of a galaxy shows a mild correlation with Hubble type, reflecting the tendency for early T-types to be more massive than late T-types (ZKH).

Is there more to the luminosity–metallicity relationship? McCall (1982) noted a correlation between the stellar mass surface density, (σ_d), and oxygen abundance in spiral

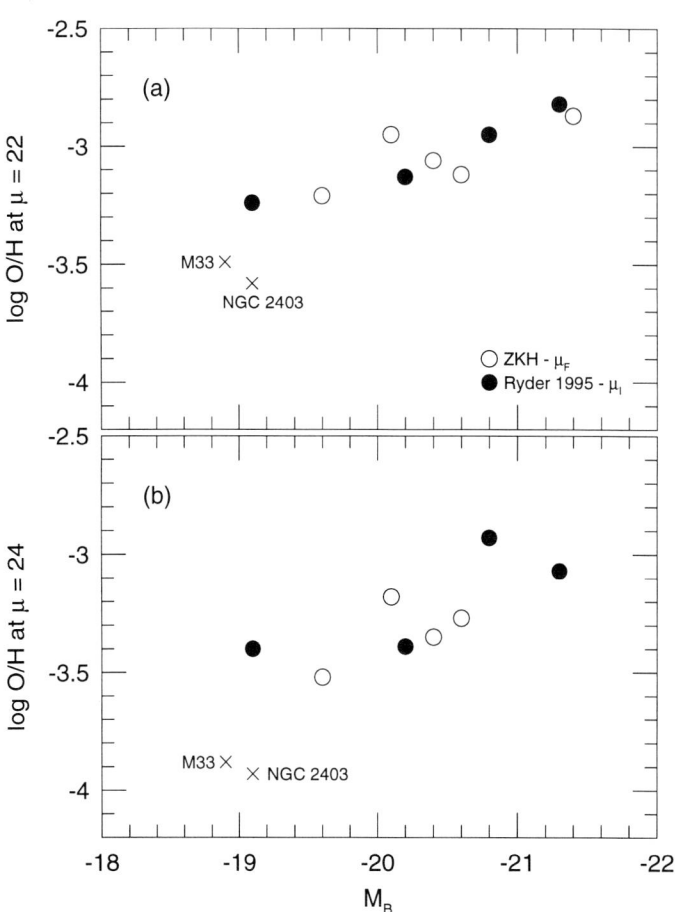

FIGURE 18. Characteristic abundance at a given value of surface brightness versus galaxy luminosity. Filled circles represent galaxies observed by Ryder (1995), with I-band surface brightness measurements. Unfilled circles are galaxies observed by ZKH, with photograph F-band surface brightness data. All data except for NGC 2403 and M33 were taken from Ryder (1995). (From Garnett et al. 1997b)

galaxies. Edmunds & Pagel (1984) reiterated this point and showed an impressive correlation between σ_d and O/H for late type (Scd) spiral galaxies. Garnett et al. (1997b) compared O/H vs. stellar mass surface density for both NGC 2403 and M33 found their gas abundances follow the same metallicity-surface density correlation as seen in other Scd spirals by Edmunds & Pagel. This may reflect fundamental similarities in the evolution of these galaxies. For example, Ryder (1995) argued that the surface brightness-metallicity correlation can be understood in terms of a galaxy evolution model which includes self-regulating star formation (see also Phillips & Edmunds 1991). In these models energy input from newly formed stars as well as older stars in the galaxy is assumed to feed energy back into the interstellar medium so as to inhibit further star formation. Such models appear to reproduce the correlation between current star formation rate and surface brightness (Dopita & Ryder 1994), and between metallicity and

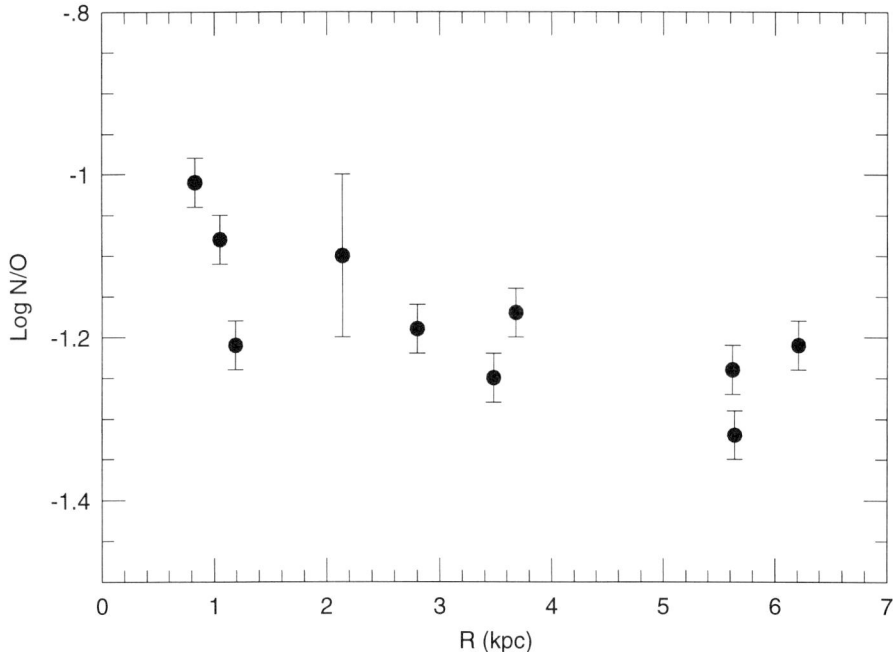

FIGURE 19. N/O gradient in NGC 2403. (From Garnett et al. 1997b)

surface mass density (Phillips & Edmunds 1991; Ryder 1995), better than models based on a simple Schmidt (1959) law for star formation.

Is this relation universal? To test this, Don compared the abundances in different galaxies at similar disk optical surface brightness. The results are shown in Figure 18. NGC 2403 and M33 have been added to galaxies studied by Ryder and ZKH. The result is rather remarkable. There is a clear correlation between the abundance at a given surface brightness and the total luminosity of the galaxy. Figure 18 suggests that massive spirals are more efficient at enriching their ISM *at fixed surface brightness* than low mass spirals. Note however, that the sample is small, and probably contains a limited range in overall galaxy surface brightness. Nonetheless, I find this result very exciting and hope that there will be more follow-up work. This diagram must be telling us something fundamental about galaxy evolution. Perhaps we are closer to defining a fundamental set of parameters which govern spiral galaxy evolution (e.g., Mollá, Ferini, & Díaz 1996, and references therein).

5.3. *Relative abundance patterns*

5.3.1. *N/O*

Figure 19 shows the variation of log N/O with radius in NGC 2403 for our HII region sample. The weighted least-squares fit to the data yields a gradient of -0.032 ± 0.005 dex/kpc (to be compared with the O/H gradient of -0.201 ± 0.009 dex/kpc). This is very similar to the gradient found by Víchez et al. (1988) for N/O in M33.

When discussing relative abundance gradients in spiral galaxies, one has to be careful to distinguish between gradients in terms of galactic radii, and gradients relative to metallicity baselines. I will try to follow my own advice. The rather shallow gradient in

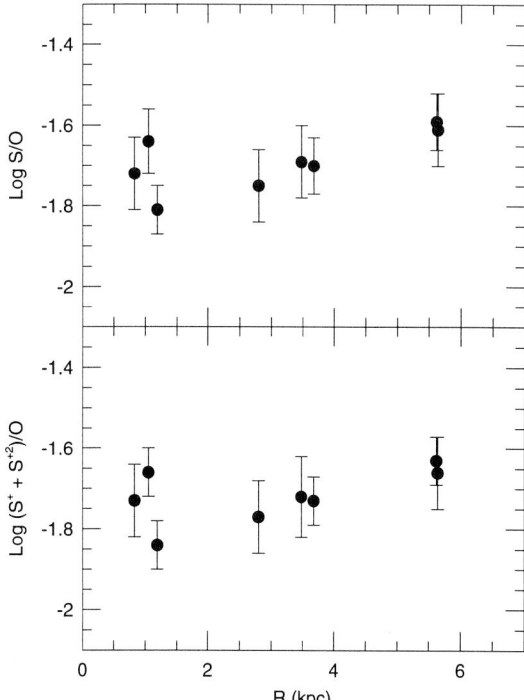

FIGURE 20. Top: S/O vs. radius in NGC 2403. Bottom: $(S^+ + S^{+2})$/O vs. radius in NGC 2403. (From Garnett et al. 1997b)

N/O is at odds with the notion that N is a purely secondary element: in such a situation N/O should vary linearly with O/H and the N/O radial gradient should have the same slope as the O/H radial gradient. A shallower slope is expected if N has a primary nucleosynthesis component (Matteucci & Tosi 1985; Díaz & Tosi 1986), as predicted in some models for the third dredge-up on the asymptotic giant branch of intermediate mass stars (e.g., Renzini & Voli 1981). Such a primary component has been invoked to explain the nearly constant N/O vs. O/H observed in dwarf irregular galaxies (§4.3.2). The N/O gradient can be explained in this picture if there is a primary N component at a level log N/O = −1.4, similar to the mean value observed in the dwarf irregulars.

However, there is a possible systematic uncertainty in the gradient due to the poorly known correction for N^{+2}, which is predicted to be the dominant ionization state of N in many cases (e.g., Garnett 1990). Far-infrared spectroscopic measurements of [N III] and [O III] in Galactic HII regions (e.g., Lester et al. 1987) suggest that higher N/O ratios are obtained from these lines than from the optical [N II] and [O II] lines. Until the optical/IR discrepancies for nitrogen are resolved, the true gradient for N in NGC 2403 (and in spiral galaxies in general) remains uncertain.

5.3.2. *S/O*

Shields & Searle (1978) pointed out that sulfur is an important secondary coolant in metal-rich HII regions, and that measurements of [S III] could help constrain photoionization models for such nebulae. Although it is well established that S/O in HII regions is generally near the solar system value, log S/O ≈ −1.7 (Anders & Grevesse 1989), the question of systematic variations with O/H remains unclear and controversial from several recent studies. Shaver et al. (1983), Vílchez et al. (1988) and Díaz et al. (1991) derived an apparent decline in S/O by approximately 0.5 dex from log O/H = −3.7 to log O/H = −2.6 in the Galaxy, M33, and M51. On the other hand, Garnett (1989) and Torres-Peimbert et al. (1989) discern no systematic variation in S/O from the solar system value over the range −4.8 < log O/H < −3.3 in dwarf irregular galaxies and M101. Meanwhile, measurements of IR [S III] lines (Pipher et al. 1984, Simpson et al. 1995) indicate that the sulfur abundance gradient in the Milky Way is much steeper than the gradient derived by Shaver et al. (1983), and is essentially identical to the oxygen gradient. Note that Garnett (1989) and Vílchez et al. (1988) disagree by a factor of two in the S abundance for the region NGC 588 in M33, a critical low-abundance point in the spiral galaxy sample. A new observation of this HII region would be very valuable.

Our new [S III] measurements for the NGC 2403 sample provide a new homogeneous set of data to address this question. Our derived S/O abundance ratios are displayed in Figure 20; the top panel shows the S/O with full ionization corrections for S (§3.4); the lower panel shows $S^+ + S^{+2}$ alone. The ionization corrections we apply for S are minor, but they do introduce a small systematic effect. Figure 20 shows that there is little variation in S/O across the disk of NGC 2403. Formally, the observed gradient in S/O is barely significant, and the total range in S/O in NGC 2403 is only about 0.2 dex between 1 and 6 kpc, essentially the same as the observational scatter.

This would seem to indicate a universal value for S/O consistent with the solar value. However, much of the evidence for systematic variation in S/O is based on measurements of HII regions more metal-rich than those in the NGC 2403 sample; this remains a critical area for future study.

5.4. Environmental effects?

Spiral galaxies are known to be modified by the environment within rich clusters of galaxies (see reviews by Haynes 1990 and Whitmore 1990). Tidal stripping, ram pressure stripping, and ablation by the intracluster medium can all play a role and may lead to significant effects on galaxy evolution. The cluster environment clearly influences galaxy morphology (Dressler 1980), which leads to the natural question of whether it also affects the star formation and chemical evolution histories of galaxies. Does the cluster environment predetermine galaxy morphology by solely influencing initial conditions or does it continually influence the evolution of a member galaxy? Comparing the chemical abundances of field and cluster members may provide clues to these fundamental questions. Moreover, a systematic abundance differential between cluster and field galaxies would have implications for the stellar populations and could impact a number of currently popular distance indicators.

My collaborators and I have been pursuing this question by studying the Virgo cluster (Shields, Skillman, & Kennicutt 1991; Skillman et al. 1996). The Virgo cluster, due to its proximity, provides the logical starting point for the study of environmental effects on abundances in cluster spirals. The Virgo cluster is irregular, and in some respects this compromises the study of environmental effects. However, the extensive catalog of the positions, morphologies, and radial velocities of 1277 galaxies by Binggeli, Sandage, & Tammann (1985) allows a definition of the cluster's structure. In addition, the structure of the hot gas component, which has been revealed through sensitive mapping of the X-ray emission made possible by ROSAT (Böringer et al. 1994), is very similar to the galaxy distribution.

Substantial evidence points to differences between the spiral galaxies in the core of the Virgo cluster and field spiral galaxies. Davies & Lewis (1973) first pointed to a deficiency in HI gas in Virgo spirals when compared to field spirals. Giovanelli & Haynes (1985) showed that galaxies in the outer parts of the Virgo cluster (and eight other clusters) were less deficient in HI than the galaxies in the core of the cluster. From HI synthesis observations of Virgo cluster spirals, Warmels (1986) concluded the HI deficiencies could be attributed to a depletion of the HI at large galactocentric radii. This was supported by Cayatte et al. (1994, CKBG).

Here I will review the findings from observations of chemical abundances in HII regions in spiral galaxies of the Virgo cluster and a comparison of Virgo galaxies and field spirals (Skillman et al. 1996b). With these new data there now exist nine Virgo spirals with abundance measurements for at least four HII regions. Our sample of Virgo galaxies ranges from HI deficient objects near the core of the cluster to galaxies with normal HI properties, far from the cluster core. We looked at the relationship between the HI disk characteristics and the chemical abundances to determine whether dynamical processes that remove gas from the disk, such as ram pressure stripping by the intracluster medium, also affect the chemical abundances.

The radial HI surface profiles of the Virgo spirals are shown in Figure 21 (upper). The points represent the azimuthally averaged HI surface densities, while the solid lines represent the mean distributions for field spirals of the same morphological type (from CKBG). CKBG demonstrate that the radial HI surface density distribution of a field spiral galaxy is chiefly determined by its morphological type, if it is normalized to the optical size of the galaxy. We have normalized the observed profiles of the Virgo spirals in the same manner, using the D_O diameters from the RC3. We have ordered the panels in Figure 21 in increasing D_H/D_O relative to the mean for the appropriate T-type, with the result that the galaxies divide into 3 deficiency groups: those with strong HI deficiencies, intermediate cases, and those with no HI deficiencies. Note the large variation in HI deficiency within the Virgo sample, especially at the extremes (e.g., compare NGC 4501 and NGC 4689 with NGC 4651).

For comparison to field galaxy abundances, we have chosen the large sample of Zaritsky, Kennicutt, & Huchra (1994, ZKH). To ensure consistency among the abundances measured for the various field galaxies and the ZKH field galaxy sample, we have applied the same empirical R_{23} abundance calibration as used by ZKH for all 70 of the Virgo cluster HII regions. The derived oxygen abundances are plotted as a function of galactocentric radius (normalized to the effective radius, following McCall 1982) in Figure 21 (lower) and the ordering of the galaxies is identical to that used in the upper panel.

Comparing the HI data with the abundance patterns in Figure 21, we note that the peripheral galaxies have strong radial gradients in O/H and reach rather low values, 12 + log (O/H) $\leq$ 8.7, at the largest radii. In contrast, the three most HI deficient galaxies show high abundances with little evidence for radial gradients (but note that the radial range of HII regions is very limited in the HI deficient spirals).

This is further illustrated in Figure 22, which shows the characteristic O/H values and gradients for the Virgo galaxies as a function of D_H/D_O. In the top panel of Figure 22, a trend of decreasing O/H with increasing D_H/D_O is evident for the Virgo galaxies. It also shows that our grouping into three categories is somewhat arbitrary as all nine Virgo spirals display a continuum of decreasing mean O/H with decreasing HI deficiency. The bottom panel of Figure 22 illustrates the relationship between oxygen abundance gradient and HI deficiency. Here we tentatively see weak evidence for a trend of stronger gradient with decreasing HI deficiency (with the caveat that the radial sampling is small).

The main result of the intra-Virgo comparison is that the three most HI deficient Virgo spirals have larger mean abundances (0.3 to 0.5 dex in O/H) than the spirals on the periphery of the cluster. This suggests that dynamical processes associated with

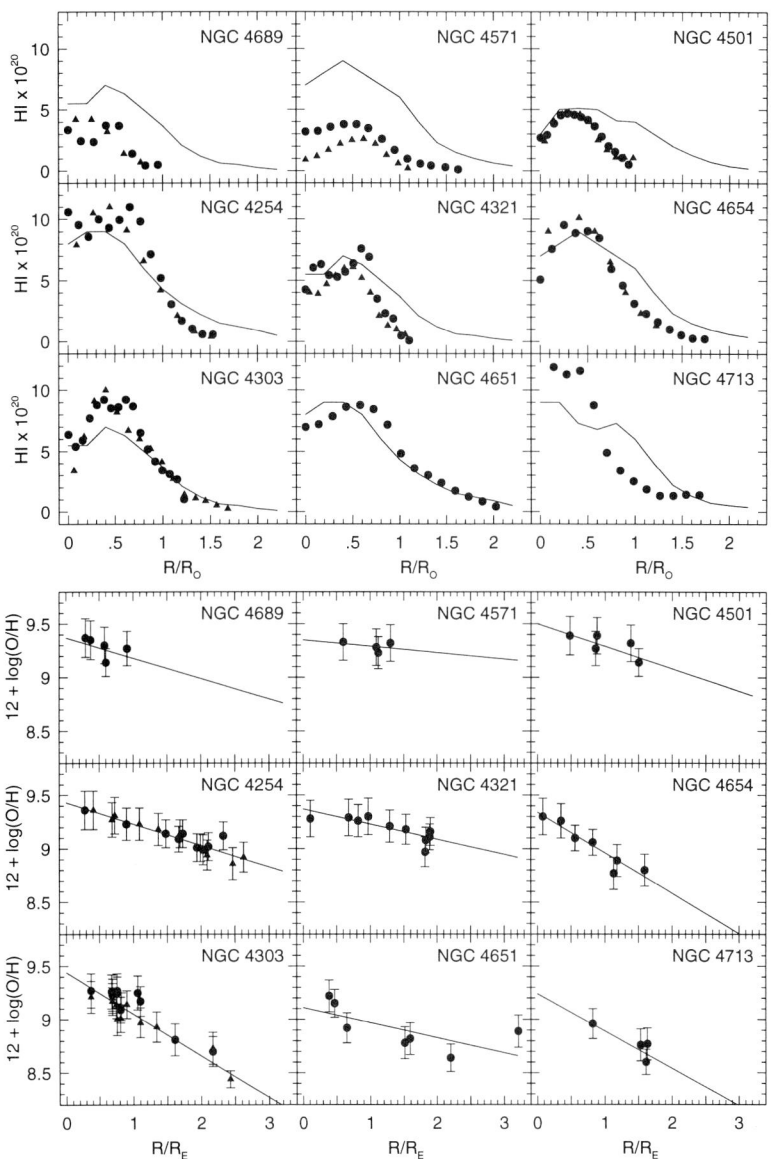

FIGURE 21. (upper) A comparison of the radial HI profiles of the Virgo spiral galaxies in our sample with the average radial HI profile for the field galaxies of corresponding Hubble type as derived by Cayatte et al. (1994). The galaxies are ordered in increasing D_H/D_O; the top row shows the HI-deficient (cluster core) galaxies and the bottom row shows the HI-normal (peripheral) galaxies. The individual points represent the radially averaged HI column densities, while the solid lines represent the average radial HI profile for the corresponding Hubble type. The filled circles represent the WSRT observations of Warmels (1988), while the filled triangles represent the VLA observations of Cayatte et al. (1990), and, in the case of NGC 4571, van der Hulst et al. (1987). (lower) Oxygen abundances from HII regions are plotted versus galactic radius for the nine Virgo spiral galaxies. The oxygen abundances are determined from measurements of [O II] and [O III] in individual HII regions and calibrated empirically as described in the text. The radial positions are normalized to the effective optical radius as defined in the RC3. The filled triangles are taken from Henry et al. (1992, 1994). (From Skillman et al. 1996b)

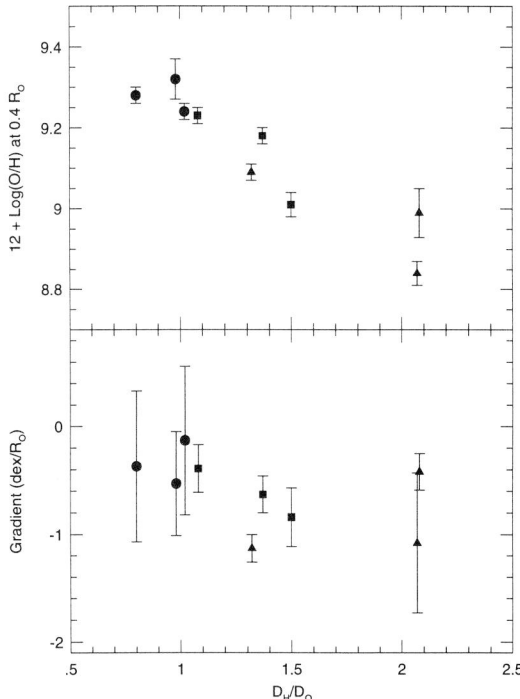

FIGURE 22. (upper) A plot of mean O/H as a function of D_H/D_O for the Virgo spirals. (lower) Oxygen abundance gradient versus D_H/D_O. The filled circles represent the HI deficient galaxies, the filled squares represent the intermediate galaxies, and the filled triangles represent the galaxies with normal HI disks. (From Skillman et al. 1996b)

the cluster environment are more important than cluster membership in determining the current chemical properties of spiral galaxies.

Next we compared the abundance properties of our Virgo sample to the large sample of field spirals studied by Zaritsky, Kennicutt, & Huchra (1994). Since the mean abundances of disks are systematically correlated with galaxy type, luminosity, and circular velocity it is important to check that the patterns in Figure 22 are not due to underlying variations in those properties. For example, many of the Virgo spirals are very luminous galaxies, and thus, they might be metal-rich objects solely on that basis, independent of cluster environment.

Following ZKH, we have fitted a linear relation in log O/H versus radius to the abundances of the individual HII regions to determine the "mean" value of O/H at a fiducial radius. Like ZKH, we use a fraction of the isophotal radius, $0.4\,R_O$ as the fiducial radius. In Figure 23 we added the Virgo spirals to the field spirals from Figure 10 of ZKH. We excluded strongly barred spirals (RC3 "B" type) from the comparison because bars have been shown to affect the gradient slope (Pagel et al. 1979, VCE, ZKH, Martin & Roy 1994, Friedli, Benz, & Kennicutt 1994). In Figure 23(a) we plot the mean O/H as a function of M_B. The HI deficient Virgo galaxies have higher oxygen abundances than the field galaxies. This is seen again in Figure 23(b) where the mean O/H value is plotted against V_C and in Figure 23(c), where the galaxies are plotted as a function of T-type.

The outstanding impression rendered by Figure 23 is that the three HI *deficient Virgo*

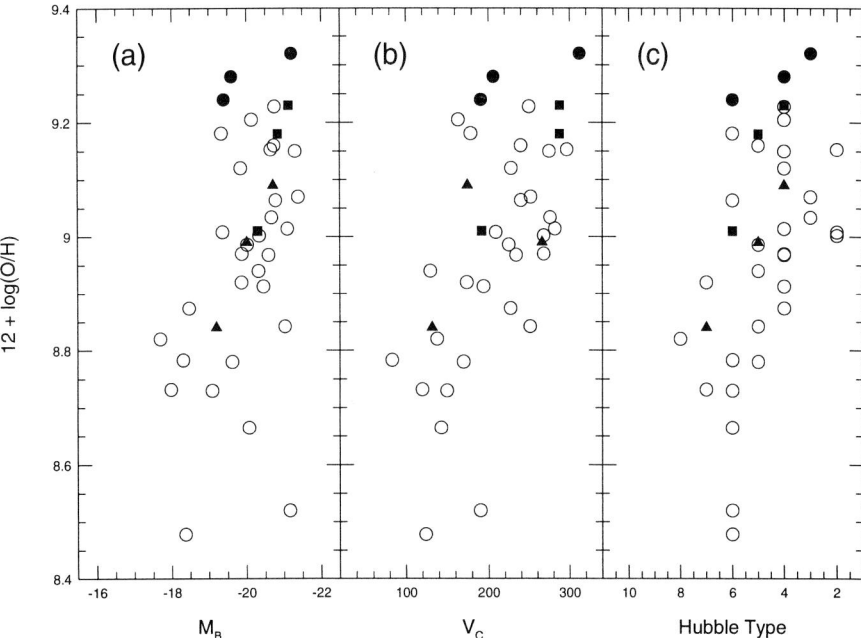

FIGURE 23. Mean O/H as a function of (a) absolute blue magnitude, (b) maximum rotation curve velocity, and (c) Hubble type for the Virgo spirals and the non-barred spirals of the ZKH sample. The open circles correspond to the field sample of ZKH, while the points for the Virgo spirals have been coded as in Figure 22. (From Skillman et al. 1996b)

spirals are all at the top of the abundance distributions of galaxies with similar properties. The dispersion in the properties of field galaxies and the small size of the Virgo sample make it difficult to draw definitive conclusions about any systematic difference between the field and Virgo spirals. Nevertheless, the HI deficient Virgo galaxies have larger mean abundances than field galaxies of comparable luminosity or Hubble type, while the spirals at the periphery of the cluster are indistinguishable from the field galaxies.

Having empirically established the abundance differential and provisionally quantified the magnitude of the effect, it is reasonable to ask if the observed effect is consistent with the physical processes known to be operating in the Virgo cluster environment. Our framework is that spiral galaxies are falling into the Virgo cluster (Tully & Shaya 1984), and that removal of gas from the infalling spirals by the cluster environment is primarily responsible for the HI deficiencies observed (Warmels 1986).

Simple, illustrative chemical evolution models with infall of metal-poor gas were constructed and compared to models in which the infall is terminated. The models were constrained by comparison with observed gas mass fractions, current star formation rates, and gas consumption times (e.g., Kennicutt, Tamblyn, & Congdon 1994). The model results indicated that the curtailment of infall of metal-poor gas onto cluster core spirals may explain part of the enhanced abundance. However, additional work is needed, particularly modeling of the effects of truncating the outer gaseous disk within the context of models with radial gas transport.

I think that this work has implications beyond a better understanding of the chemical evolution of field and cluster spiral galaxies. Foremost are implications for distance determination studies. The elevated abundances of HI deficient cluster spirals may impact both Tully-Fisher and Cepheid variable distance determinations. For example, if the dust content of spiral galaxies scales with abundance (van den Bergh & Pierce 1990), and the opacity corrections for galaxy luminosity are important (see Giovanelli et al. 1994 for a recent discussion), then there will be a systematic offset between cluster and field spiral galaxy Tully-Fisher relationships. Additionally, if the distances derived from Cepheid variable stars have a metallicity dependence (cf. Gould 1994), then there could be a systematic offset between distances derived for core cluster galaxies and field galaxies. It is also possible that chemical abundance studies will provide insight into the processes by which gas is removed from spiral galaxies as they enter the cluster environment. This may, in turn, lead to a more secure interpretation of current studies of the evolution of cluster galaxies (e.g, Dressler et al. 1994), and, thus, a better understanding of the chemical abundance patterns in both field and cluster spirals.

6. Self-consistent star formation histories

Now I will attempt to make the previous five sections relevant to an audience of stellar astronomers. In this last section I will show some directly observable connections between stellar evolution as observed in stars and chemical evolution as observed in the ISM. The goal is to obtain constraints on the star formation histories of galaxies. When the constraints are in agreement, they merit the label "self-consistent". The examples I am going to show come from dwarf galaxies, which are, hopefully, simple enough systems that we can make some progress.

6.1. *Star formation histories of dwarf irregular galaxies*

Establishing the detailed star formation history of a galaxy is a very difficult task. Two-point star formation histories, like those determined from Hα equivalent widths (Kennicutt 1983; Kennicutt, Tamblyn, & Congdon 1994) give us clues to the question of whether the current star formation rate is comparable to the past average rate, but are unable to answer detailed questions (e.g., was there an early dominant burst of star formation). Gallagher, Hunter, & Tutukov (1984) proposed a three-point scheme based on the dynamical mass, the blue luminosity, and the Hα luminosity, and found that these measures were consistent with roughly constant star formation histories for the irregular galaxies in their sample. Unfortunately, the uncertainties in the conversion of these observables into constraints allows consistency with a large range of star formation histories.

Measures of resolved stars offer another avenue. Indeed, there have been many observational programs that have been very successful at modeling the *recent* star formation histories of dIs (e.g., Hodge 1980; Aparicio et al. 1987). However, constraining the *early* star formation histories of galaxies ($\geq$ 1 Gyr) is a very difficult problem. As an example of what can be done from the ground, one can consider the impressive studies of Sextans B (Tosi et al. 1991) and NGC 3109 (Greggio et al. 1993). These galaxies have distance moduli of about 26.6, so with a V-band limit of roughly 23, stars with absolute magnitudes brighter than -3.5 can be reliably recorded. Their method of comparing synthetic color magnitude diagrams (CMDs) to the observations is successful in re-creating the distribution of the stars, but the comparisons are not very sensitive to the star formation histories (i.e., in the upper CMD, the models for constant star formation look very similar to models of exponentially decreasing star formation and models of two distinct bursts). Gallart et al. (1996a,b,c) have pushed this method to its limits (for ground based data) in their study of NGC 6822, and Tolstoy & Saha (1996) have developed a statistical method to determine the goodness of the fit of Monte Carlo simulations of data to observed CMDs; this allows them to quantify the errors in star formation history models (e.g., Tolstoy 1996).

In the next sections I will describe different projects which (in my opinion) represent new steps forward toward reaching the goal of detailed star formation histories. One advance comes from the fantastic imaging abilities of the Hubble Space Telescope. This has allowed great improvements in the studies of resolved stellar photometry. I will argue that relative chemical abundances can also be used to constrain recent star formation histories. When these two techniques are in agreement, I think that we have achieved "self-consistent star formation histories".

6.2. Example 1: the Sextans A and Pegasus dwarf irregular galaxies

Sextans A (DDO 75, A 1008-04) is a gas-rich dI with active star formation as evidenced by an abundance of HII regions (Hodge, Kennicutt, & Strobel 1994; Aparicio & Rodriguez-Ulloa 1992). It is low surface brightness with an apparent magnitude $m_B = 11.86$, and a color index $B - V = 0.26$ (Hunter & Plummer 1996). Its metallicity has been measured to be ∼4% solar from HII region spectroscopy (Skillman et al. 1989b). The HI shows solid body rotation and is concentrated in two clumps which correspond to the major star forming regions (Skillman et al. 1988b). It is located on the periphery of the Local Group with a distance of roughly 1.44 Mpc as determined from Cepheid variable stars and the tip of the red giant branch (Piotto et al. 1994; Sakai et al. 1996; and references therein). Sextans A has a diameter $D_{25} = 5'.9$ (de Vaucoulers et al. 1991), corresponding to 2.5 kpc at this distance.

Sextans A is one of four nearby dwarf irregular galaxies observed in a cycle 5 HST program to conduct resolved stellar photometry. Program collaborators include Chiosi (Padua), Dufour (Rice), Gallagher, Hoessel (Wisconsin), Mateo (Michigan), Saha (STScI), and Tolstoy (ESA). I will report here some of the results from this study (Dohm-Palmer et al. 1997). Each galaxy was observed in three filters: F439W (4000 s), F555W (1800 s) and F814W (1800 s). Note the very modest integration times, which provide absolutely stunning results. The stellar photometry was extracted using a modified version of the PSF fitting program DoPHOT (Schecter, Mateo & Saha 1993; Saha et al. 1996) and the calibration to B, V, and I was done with the equations of Holtzman et al. (1995).

The CMD in V and I for Sextans A is shown in Figure 24. Clearly visible are the red giant branch (RGB) and the red clump stars just at the limit of photometry. In addition to these older stars, there is a prominent population of young blue stars. These stars, which have been labeled the "blue plume" in ground based observations, have been resolved by the HST into two separate populations. There is a well defined main sequence (MS), indicating very recent star formation. Just redward of the main sequence is a clearly separate population that corresponds to massive He-Burning (HeB) stars at the bluest point in their "blue-loop" phase of evolution (e.g., Bertelli et al. 1994; B94). The corresponding red end of this phase can also be seen (RSG).

To interpret the recent SFH of Sextans A, we rely upon our knowledge of stellar evolution. There are two main populations which we can use to glean the recent SFH. Traditionally the MS is used, specifically using the strength and position of MS turnoffs. This works very well for isolated generations of stars, such as in clusters. However, the MS in galaxies with continuous star formation can be difficult to interpret because each successive generation lies on top of the previous. We can, however, examine the luminosity function of the MS to look for gaps, flat regions and other features that may give a clue to the SFH (e.g., Butcher 1977). In addition, we can use statistical methods

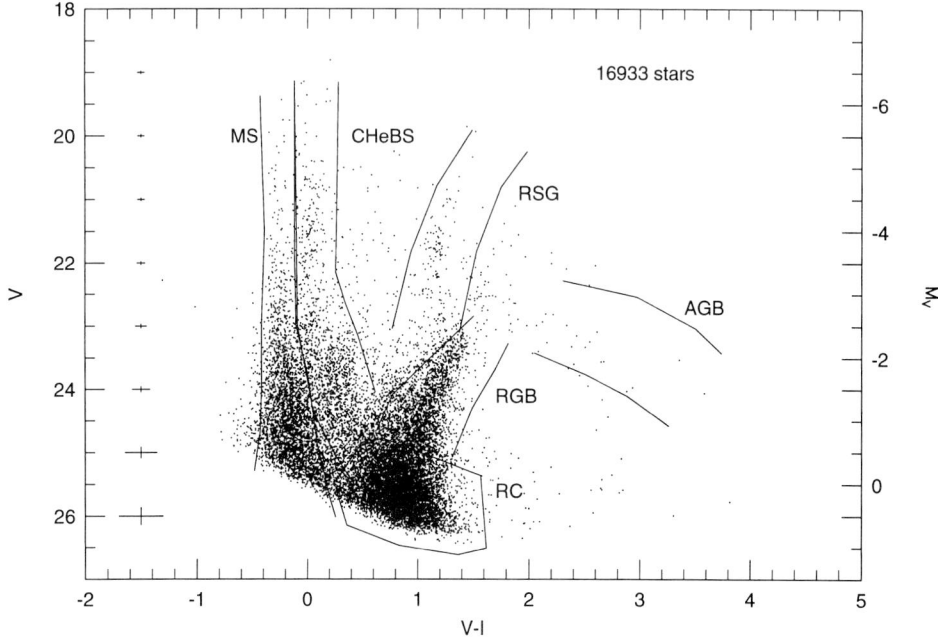

FIGURE 24. The CMD in V and I for Sextans A. The points plotted have errors less than 0.2, and they have been corrected for interstellar reddening. The crosses on the left indicate the average error at the given V magnitude. The right axis shows the absolute V magnitude assuming a distance modulus of 25.5. The major populations in the CMD have been labeled. One is the main sequence (MS) on the blue side of the diagram. Just redward of the main sequence is a separate population of core Helium burning stars. On the red side, the RGB is well defined, and there are hints of the red clump just at the photometric limits. (From Dohm-Palmer et al. 1997)

for extracting SFH's from these chronologically over-lapping populations (e.g., Tolstoy & Saha 1996).

The blue HeB stars provide a parallel track to the MS in which to observe star formation events. From the number of stars at each magnitude, we can calculate the SFR for the age corresponding to the time it takes stars to reach this phase of evolution. There are two advantages to using the blue HeB as an indicator of SFH: (1) The blue HeB stars are about 2 mag. brighter than the MS turnoff stars of the same age (e.g., B94). This allows us to probe the recent SFH further back in time (for the same photometric limits) than we can from MS turnoffs, (2) There is little confusion from overlapping generations. All of the blue HeB stars of a certain magnitude come from the same generation of stars (there is confusion from stars leaving the MS, but only at the few percent level). In practice, the blue HeB stars can probe the SFH back to about 600 Myr. At older ages, the blue HeB stars blend with the red clump and horizontal branch, becoming degenerate in time.

To determine the SFH from the MS stars, we analyze the MS in discrete bins. Within each bin, we assume that the mass and the SFR are constant. We also assume the IMF is constant with time. However, each bin will have stars from all generations younger than the turnoff age (TO) for that mass. Thus, we recursively find the SFR by starting

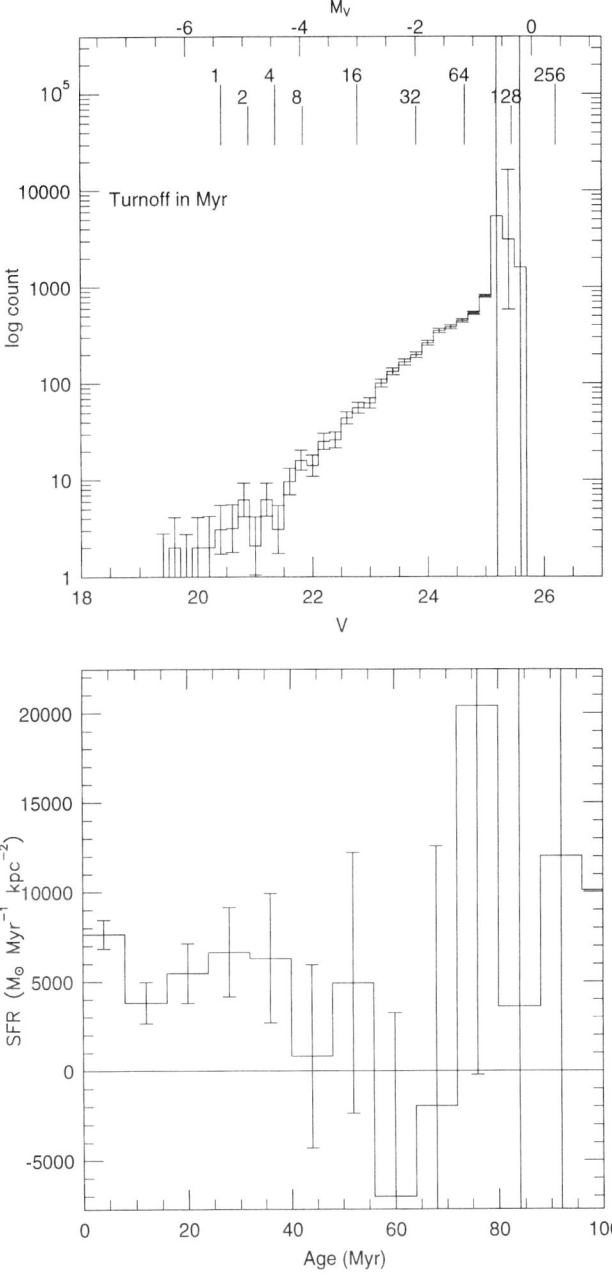

FIGURE 25. (upper) The V luminosity function of the MS stars. The errors reflect both Poisson counting noise and errors due to the incompleteness correction. At the top are the ages for the MS turnoff. Based on the MS turnoff alone, we are only able to probe back 100 Myr of SFH. There are no apparent features over this time period. (lower) The global SFR for the area of Sextans A within the field of view determined from the MS luminosity function. A Salpeter IMF has been assumed. This is a result of a recursive process that starts with the youngest bin. The low counts in the youngest bins contribute large uncertainties to all the bins. At several points, the SFR drops below zero, however, within the noise of the counting statistics these are consistent with zero. (From Dohm-Palmer et al. 1997)

with the youngest bin, and working backward through time. One difficulty with this procedure is that the calculated SFR of the oldest bins then depends on that determined for the youngest bins, and in the youngest bins, the counting statistics are low. The errors associated with this, therefore, propagate through the entire SFH determined from the MS. Finally, we convert the rate into units of $M_\odot$ Myr^{-1} by multiplying by the average initial mass, which is found by determining the average mass weighted by the IMF. For comparison with other galaxies, we convert the SFR into a SFR/area by dividing by the area covered by the observation (0.92 kpc^2).

Fig 25 (upper) shows the MS luminosity function in Sextans A. We have included the turnoff age scale at the top. Using the MS alone, we are only able to look back $\approx$ 100 Myr. The lack of obvious features in the luminosity function indicate there has been relatively constant (globally averaged) star formation over the past 100 Myr within the field of view. The presence of the most massive stars also indicates that there is SF younger than 10 Myr, in agreement with the presence of HII regions.

We have converted the MS luminosity function counts into a SFR as a function of time in Figure 25 (lower). We have re-binned the data into linear time bins over the past 100 Myr. Note that at several points the SFR drops below zero. The negative points result from over-subtracting based on the large SFR determined from the youngest bins. This highlights the great difficulty of using the MS to directly determine the SFR when the SFR is relatively constant with time. For the region within the field of view, we see a nearly constant SFR over the past 50 Myr, at $\approx$ 6000 $M_\odot$ Myr^{-1} kpc^{-2} (assuming a Salpeter IMF). Older than 50 Myr, there are large variations in the SFR, but the errors are so large that the rates are consistent with a constant SFR over the past 100 Myr.

Next, we do the same with the blue HeB stars. Again, we analyze the data in discrete bins, and assume that the mass and SFR are constant over these bins. For the blue HeB stars, there are no overlapping generations, and we can assign a single mass and age to each magnitude. Figure 26 (upper) shows the luminosity function for the blue HeB stars. Also plotted are the ages for each magnitude adopted from B94. Note that we are able to extend the SFH back in time a factor of ten further than we could with the MS.

The HeB luminosity function of Sextans A shows a depressed region at $M_V \sim -4$, between the ages of 80 and 180 Myr. This indicates a quiescent period, where the SFR drops off by a factor of 12. This was not obvious in the MS luminosity function because our data begins to become incomplete at about the magnitude corresponding to 100 Myr MS turnoff. This indicates that the SFR in Sextans A may not have been constant over the last 1 Gyr. An important caveat is that the field of view covers only a quarter of Sextans A. There may have been quite active star formation in a region outside the field of view during this time.

The resultant SFR for a Salpeter IMF is shown in Figure 26 (lower). There are four events which are spatially distinct and whose effects can be detected within the field of view. Note that although we can identify star formation events based on the spatial information, the global SFR appears to be relatively constant over the entire field of view.

Sextans A can now be contrasted with the Pegasus DIG. The Pegasus dwarf irregular is outstanding for two reasons. First, it is one of the least luminous star forming galaxies in the Local Group with $M_B \approx -12.5$, and second, this system is very gas poor ($\log(M_{HI}/L) = -0.53$; Fisher & Tully 1975). VLA HI imaging confirmed that the Pegasus dwarf is indeed very gas poor, with both very low HI column density over the entire face of the galaxy and a small HI/optical size ratio (Lo et al. 1993).

Our CMD for Peg DIG is shown in Figure 27. Sextans A and Peg DIG are at very similar distances, so the CMDs have similar photometric limits and covered areas and

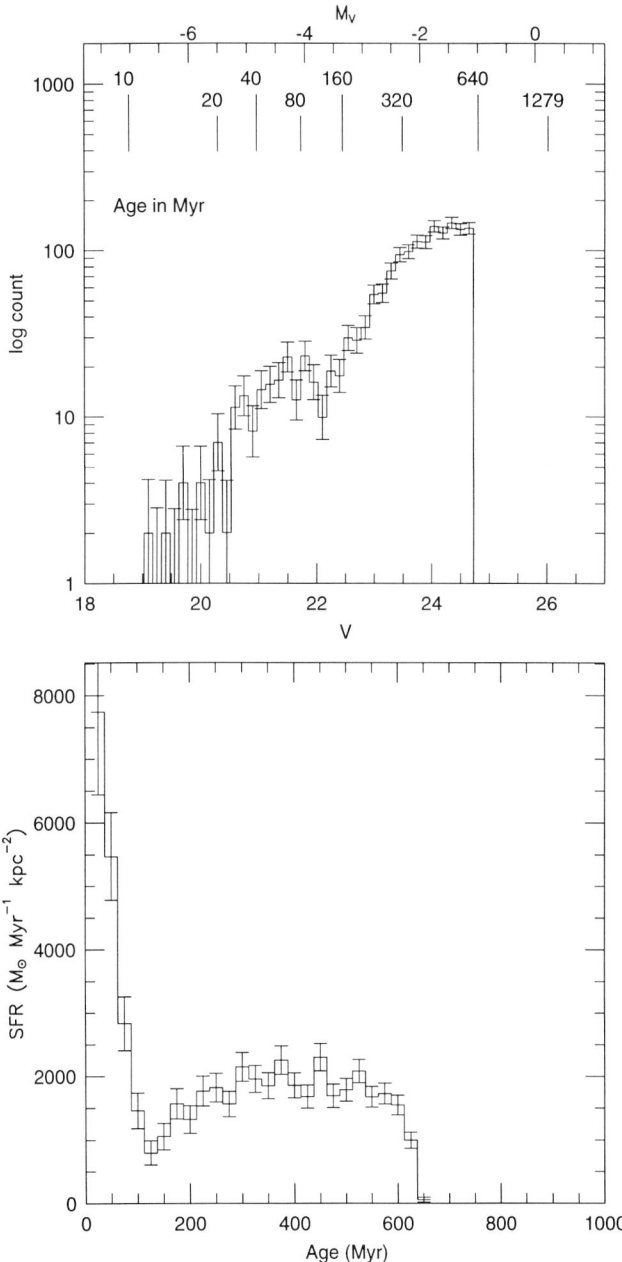

FIGURE 26. (upper) The V luminosity function of the HeB stars. The errors reflect both Poisson counting noise and errors due to the incompleteness correction. The ages corresponding to a given magnitude (assuming a metallicity) are plotted above the luminosity function. Using this dating method, we are able to extend the SFH to almost 1 Gyr. At 80 Myr there is a flattening of the luminosity function, indicating a quiescent period with little star formation. (lower) The SFR for Sextans A over the last 600 Myr, based on the blue HeB stars. The blue HeB luminosity function has been normalized to account for the IMF and the changing lifetime in this phase with mass. For this plot we used the Salpeter IMF slope. There are four events that can be identified due to their spatial coherence. For ages older than ∼600 Myr, there is likely contamination from RGB stars scattering into that region of the CMD. (From Dohm-Palmer et al. 1997)

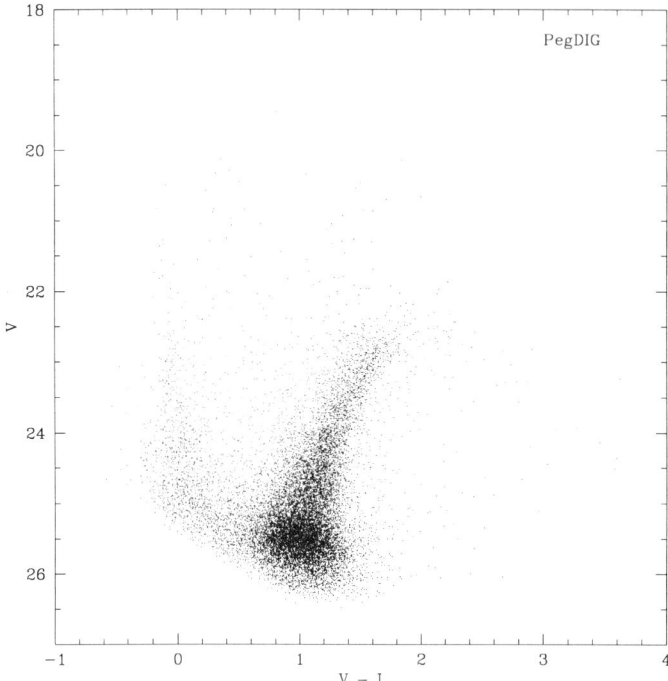

FIGURE 27. The CMD in V and I for the Pegaus DIG. The points have been corrected for interstellar reddening. Note the paucity of the MS and HeB populations compared to Sextans A. (From Gallagher et al. 1997, in prep.)

thus, direct comparisons are possible. The relatively faint MS and the near absence of blue HeB stars shows that while Peg DIG has had star formation over the last 100 Myrs, the recent star formation rate is clearly depressed relative to that in Sextans A.

This difference can be quantified. In Figure 28, we have constructed the star formation history over the last 600 Myr for the Pegasus dwarf from its HeB stars. Comparing Figure 28 with Figure 26, we see that the average star formation rate in the Pegasus dwarf, over the last 400 Myr, has been roughly about one-tenth that observed in Sextans A.

It is probably time to add a very important caveat. In Figures 25, 26, and 28 we have estimated stellar ages from models. From Cesare Chiosi's lectures you can understand better the uncertainties in these models. As better models incorporate more sophisticated treatments of physical processes and are better constrained by observations, the models will change. It is also clear that in assuming an IMF, our numbers are vulnerable to a significant systematic uncertainty. Estimating the effects of these uncertainties is not simple.

I would now like to look at Peg DIG in view of the discussion of relative abundances in §4.3. The discovery of faint, resolved Hα emission through deep narrow-band imaging with the Calar Alto 2.2m prompted spectroscopy in order to obtain an ISM abundance (Skillman, Bomans, & Kobulnicky 1997). The brightest HII region is too faint for a direct abundance measurement, but we were able to derive an oxygen abundance of roughly 10% of the solar value from photoionization modeling.

Figure 29 shows the position of Peg DIG in the abundance – luminosity plane in comparison to the well defined sample of dIs assembled by Richer & McCall (1995;

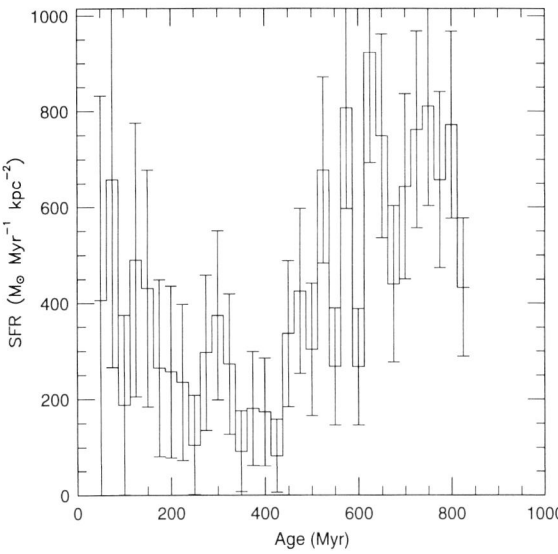

FIGURE 28. The SFR for the Pegasus DIG over the last 600 Myr, based on the blue HeB stars. The blue HeB luminosity function has been normalized to account for the IMF and the changing lifetime in this phase with mass. For this plot we used the Salpeter IMF slope. Note the overall lower level of star formation relative to that found in Sextans A. (From Dohm-Palmer et al. 1997)

RM95). The solid line in Figure 29 is the least squares fit to the more luminous galaxies calculated by RM95. (note that a new distance for Leo A determined from our WFPC2 imaging places it closer to the general trend, Tolstoy et al. in prep.). In Figure 29 it can be seen that Peg DIG fits well in the trend established by other, well studied dwarf irregular galaxies, but lies near the top of the distribution. If the metallicity – luminosity relationship is physically based on a metallicity – mass relationship, then the position of Peg DIG may be due to the lack of recent star formation. This results in a lower blue luminosity per unit mass when compared to more actively star forming galaxies.

Such a scenario is most likely for the lowest mass dwarf galaxies. The higher mass galaxies have, in general, higher metallicities, higher surface brightnesses, and lower gas mass fractions, implying that a current generation of star formation will have a relatively small effect on the present luminosity. However, a burst of star formation can significantly enhance the total luminosity of an extreme dwarf galaxy. Due to the lack of a strong underlying stellar population, as the current burst ages, the fading of an extreme dwarf galaxy will be greatest, thus, leading to a larger scatter in the low luminosity systems (as noted by RM95).

In order to test whether fluctuations in the mass/light ratios of the dIs are the dominant source of the scatter, galaxy colors can be compared versus their positions relative to the mean relationship in the metallicity – luminosity plane. Figure 30 shows such a comparison. Two things are immediately apparent. First, there is a cluster of points with B–V colors between roughly 0.3 and 0.4 with small differences between the predicted and observed blue luminosities (which includes Sextans A). Secondly, Peg DIG lies offset from this group with a redder color and a lower than predicted luminosity. Within the cluster of blue galaxies, there is no evidence for the predicted trend of redder galaxies having fainter luminosities than predicted by the RM95 relationship. However, note that

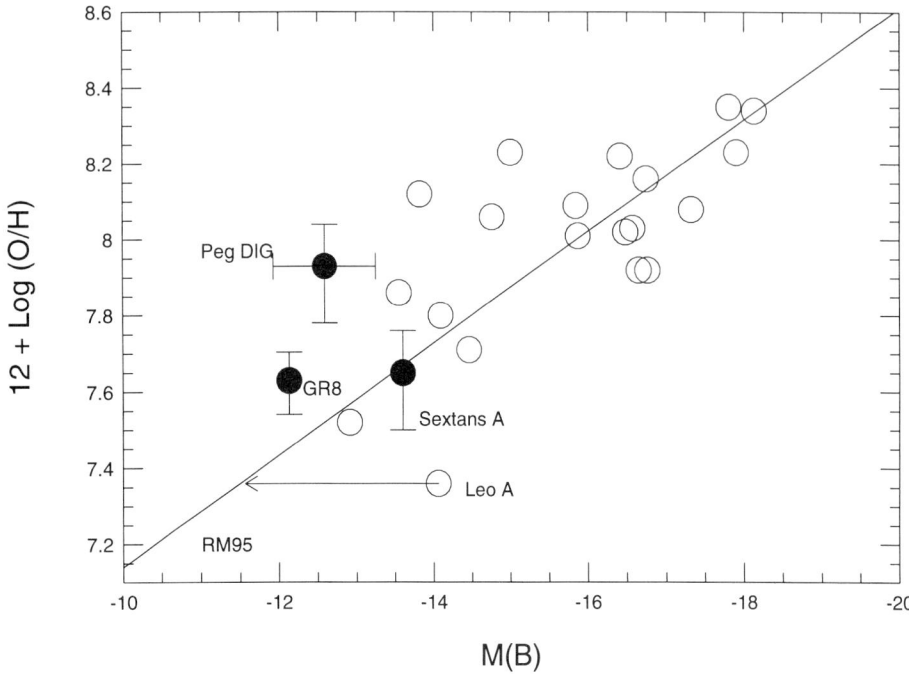

FIGURE 29. Plot of HII region oxygen abundance versus absolute blue magnitude following Richer & McCall (1995) (with alterations described in the text). Note the position of Peg DIG. The error bars in oxygen are those implied by the comparison of the observations with photoionization modeling. The new position of Leo A is based on our new WFPC2 CMD.

most of the dispersion in the points could be due simply to observational errors (a 0.1 error in log (O/H) translates to an error in the predicted luminosity of 0.7 magnitudes and errors of 0.1 in the color should be typical).

One of the most important features in Figure 30 is the paucity of "reddish" dIs (B−V ≥ 0.5). This means that there are very few galaxies with which to test the hypothesis. The lack of red dwarf galaxies in the RM95 sample is well understood as a selection effect; observations of bright HII regions are favored for abundance study work. Clearly more abundance studies of red dIs are needed.

Since the metallicity of Peg DIG appears to make sense, lets check on the relative abundances. Figure 31 shows the N/O abundances versus O/H for the sample of dIs and HII galaxies compiled by Kobulnicky & Skillman (1996). Only the points for those galaxies which do not show evidence of WR features in their spectra are plotted. Here we see that Peg DIG stands at the upper envelope of the values of N/O. Recent Calar Alto observations of Sextans A have allowed us to determine an N/O (Skillman et al. in prep.), and note that Sextans A appears near the lower envelope of values.

In §4.1, I argued that ISM abundances mainly reflect the past chemical enrichment history and not current "pollution." In this interpretation, it may be possible to see depressed values of N/O in galaxies which have had recent episodes of star formation (O production happens quickly in massive stars and is over within a few tens of millions of years while N enrichment is delayed because it forms in more slowly evolving intermediate mass stars with ages in excess of 100 Myr). On the other hand, in galaxies that have

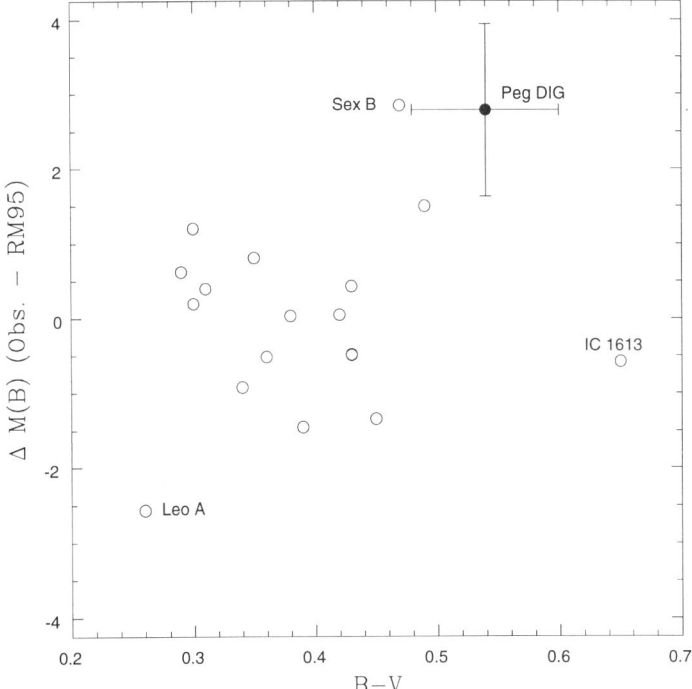

FIGURE 30. A comparison of the difference between the observed absolute blue magnitude and that predicted by the metallicity – luminosity relationship as parameterized by RM95 versus the reddening corrected B–V color of the galaxy for the sample of galaxies shown in Figure 29. Peg DIG is seen to be significantly separated from the cluster of points with blue colors and relatively small luminosity differences. Illustrative error bars have been added to the point representing Peg DIG. While no trend is seen in the cluster of blue galaxies, we suggest that such a diagnostic diagram (preferably with a longer wavelength difference for the color measurement) may provide a strong test for the origin of the metallicity – luminosity relationship. (From Skillman et al. 1997)

been relatively quiescent for a long period of time one would expect relatively high values of N/O (Edmunds & Pagel 1978).

The positions of Peg DIG and Sextans A in Figure 31, and their recent star formation histories as recorded in their CMDs (Figures 24 and 27) appear to support this simple picture. Maintaining some skepticism, I could point out that there is a smaller range in N/O at the very low metallicity end of the scale, and, thus, maybe the position of Sextans A is not telling us that much. However, I return to my point that there are very few abundance studies on quiescent or reddish dIs. If we can study some of these systems at low metallicity, we may turn up some higher values of N/O (but note that these are not predicted in the simple scheme of nitrogen production being chiefly primary at low metallicity).

At this time, it is not possible to compile an extensive list of objects to test this hypothesis, but there are some other important test cases worth mentioning. NGC 6822 has one of the lowest values of N/O (log (N/O) = -1.66; Pagel, Edmunds, & Smith 1980). In NGC 6822, Hodge (1980) found evidence for stellar cluster formation over the last 100 Myr, with a strongly enhanced period of stellar cluster formation in the interval 75 to 100 Myr ago. Marconi et al. (1995) found evidence for episodic star formation in

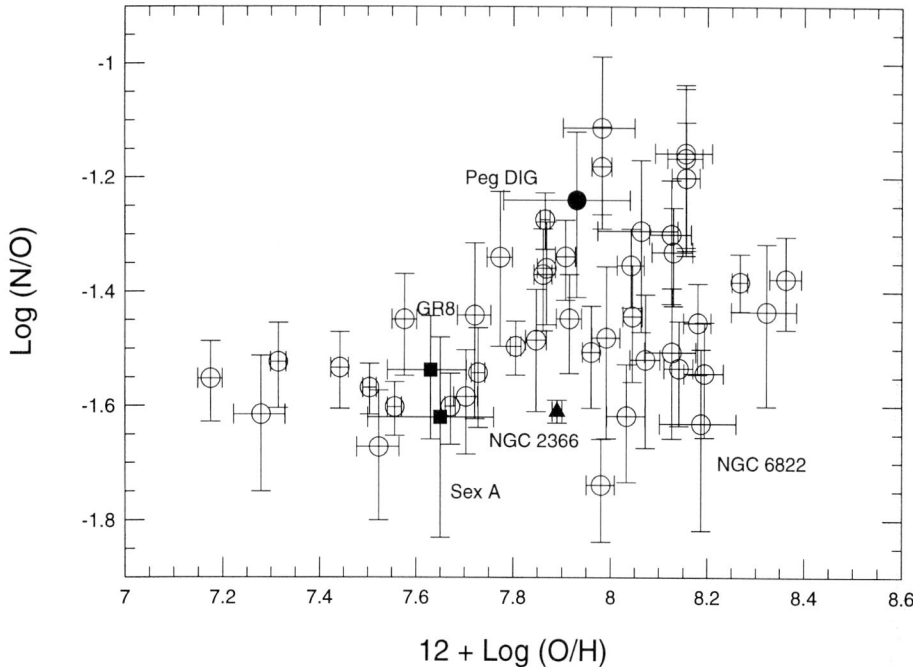

FIGURE 31. A comparison of the N/O and O/H in Peg DIG with the collection of dwarf irregular galaxies and HII galaxies assembled by Kobulnicky & Skillman (1996; see their Table 5 and Figure 15 for identification of individual points). Only galaxies without WR emission features and errors in log (N/O) less than 0.2 have been plotted. Note the positions of Peg DIG and Sextans A.

NGC 6822, but the resolution of their modeling does not allow for a precise star formation history over the last few 100 Myr. Gallart et al. (1996c) found an enhancement in the star formation rate over the last 100-200 Myr. If the stellar cluster formation history reflects the overall star formation history, then it appears that NGC 6822 has experienced a productive period of star formation over the last 50 - 200 Myr, which has not yet resulted in the elevation of the ISM N abundance. This could be consistent with N production delayed by a few 100 Myr.

Finally, we consider the case of NGC 2366. This galaxy currently shows a relatively high rate of massive star formation as evidenced by its $H\alpha/L(B)$ ratio (Hunter et al. 1993). Aparicio et al. (1995) present evidence for a dominant burst of star formation roughly 20 – 50 million years ago. On the other hand, based on deeper photometry, Tolstoy (1995) finds a relatively constant star formation rate over the last 300 million years. Although it is important to sort out this discrepancy, both studies support the view of vigorous star formation during the last few hundred million years in NGC 2366. This may be consistent with the relatively low N/O observed (log (N/O) = -1.61; González–Delgado et al. 1994).

In the future, with deeper stellar photometry (using HST) and statistical treatment of CMDs like those presented by Greggio (1994), Tolstoy (1995), and Aparicio et al. (1996), it should be possible to determine much more accurate star formation histories for the last 0.5 Gyr for a large sample of the nearby dIs. If the preliminary evidence presented here for delayed N production holds up, then it should be possible, in principle, to calibrate the length of that delay. This would be of great value in understanding the chemical evolution of galaxies. For example, Pettini et al. (1995) and Lipman (1995) find very low values for log (N/O) of ~ -2 for several damped Lyman-α system of low metallicity. They favor an interpretation of delayed N production which means that they are observing these systems within a few hundred million years of dominant star formation episodes. Since the nearby dIs have similar metallicities to these damped Lyman-α systems, it is reasonable to expect that the stellar populations are similar. Thus, more studies of the nearby dIs can help us to understand galaxy formation at redshifts of 2 to 3 seen in the damped Lyman-α systems.

6.3. Example 2: IZw18

The dwarf emission-line galaxy IZw18 plays an important role in studies of the properties and evolution of dwarf irregular galaxies because of its extreme properties. Searle & Sargent (1972) determined early on that IZw18 had a very low oxygen abundance; since then the O/H value has been refined to the current best estimate of $1.47\pm0.15 \times 10^{-5}$ in the NW HII region (Skillman & Kennicutt 1993, SK93), the lowest value measured in any emission-line galaxy. The colors of IZw18 are very blue, dominated by young massive stars, and, to date, no evidence has been found for an older red stellar population (see Hunter & Thronson 1995). The oxygen deficiency and blue colors lead to the conclusion that IZw18 and similar galaxies are either very young, experiencing their first episodes of star formation, or have experienced only sporadic episodes of star formation over their history (Searle, Sargent, & Bagnuolo 1973).

If the ionized gas in IZw18 is pristine material which has recently been contaminated by Type II supernova ejecta, the heavy element abundance ratios should show values characteristic of massive star nucleosynthesis. In particular, the gas should be relatively deficient in elements produced mainly in low- and intermediate-mass stars, such as carbon and nitrogen, compared to oxygen, which is produced in massive stars alone. Indeed, Kunth et al. (1994) inferred from high-resolution GHRS spectra an oxygen abundance in the neutral gas that was much smaller than in the HII region in IZw18, and concluded that the ionized gas has been enriched by the present starburst (but see Pettini & Lipman 1995 for a criticism of their analysis).

With this background in mind, Garnett et al. (1997a) obtained new HST observations of IZw18. Observations were obtained in two separate components of IZw18. Figure 32 shows our new results for C/O in IZw18 compared with the data from Garnett et al. (1995a; G95a, discussed in §4.3.1) plus new observations of NGC 5253 by Kobulnicky et al. (1997, discussed in §4.1). Small number statistics still inhibit any interpretation of the measured dispersion in C/O, so we shall rely on comparison with the most metal-poor galaxies to interpret the IZw18 measurements. The three most metal-poor galaxies from the G95a sample have a mean log C/O = -0.85 ± 0.07 (mean error); by comparison, the mean log C/O = -0.60 ± 0.09 in IZw18 is 2.9σ above this average (and even higher than expected from extrapolating the trend of C/O vs O/H observed by G95a). Similarly, the mean log C/N = $+0.98$ in IZw18 is also well above the values determined for the other very metal-poor dwarf galaxies (G95a).

The elevated C/O abundance ratio in IZw18 suggests that the galaxy has experienced enrichment in carbon from an older generation of stars, and that the current burst of star formation is not the first. G95a noted that nucleosynthesis models for 10-42 $M_\odot$ stars by Weaver & Woosley (1993) predict an integrated log C/O = -0.83 for their "best estimate" of the $^{12}C(\alpha,\gamma)^{16}O$ nuclear reaction rate, corresponding quite well with the values measured by G95a in their most oxygen-poor galaxies. If this correctly represents

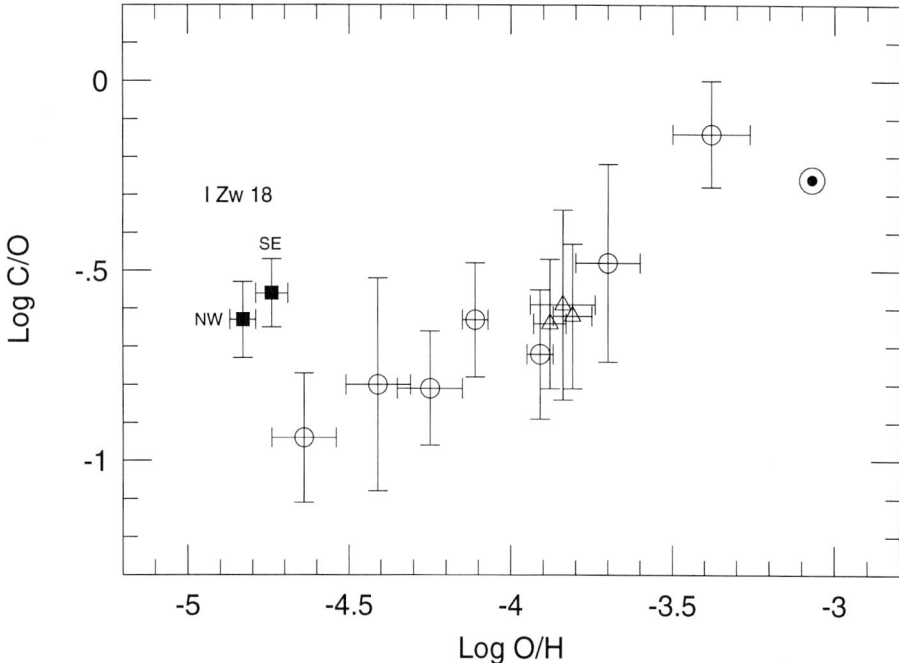

FIGURE 32. Log C/O vs Log O/H from FOS spectroscopy of IZw18 (filled squares), compared with data from Garnett et al (1995a; unfilled circles) and Kobulnicky et al. (1996; unfilled triangles). Solar value is from Grevesse & Noels (1993). (From Garnett et al. 1997a)

the nucleosynthesis contribution from massive stars alone, then the additional carbon observed in two widely separated locations in IZw18 must come from lower mass, long-lived carbon star and planetary nebula progenitors with lifetimes $\gg 10$ Myr.

This is, essentially, the "time delay" described in §3.3. Heavy elements that are produced in massive stars, such as oxygen, are injected into the ISM by a starburst early and can enrich the gas on relatively short timescales; supernova-driven galactic winds may also eject much of the oxygen-rich SN ejecta into the galaxy halo and reduce the effective oxygen yield. After the massive stars have died away, lower-mass stars can enrich the ISM in carbon (and nitrogen) in more gentle mass loss events, leading to higher C/O and N/O than one would expect in a young galaxy which has been enriched by massive stars alone. The relative abundances of C, N, and O then provide an indication of the time elapsed since the last major starburst in a dwarf galaxy (Edmunds & Pagel 1978).

In Figure 33 we compare the observed C/O ratio in IZw18 with the predictions of chemical evolution models for blue compact dwarf galaxies as a function of galaxy age. The filled squares show models computed by Kunth, Matteucci, & Marconi (1995; KMM) specifically to explain the abundances in IZw18. These are all one-burst models of short age, except for their model 6, which is a model consisting of two bursts, one that occurred 1 Gyr ago and one beginning only 10 Myr ago. We show their predicted abundances for IZw18 at the beginning of the second burst, after 990 Myr. We also show the results of general models for metal-poor blue compact dwarf galaxies by Carigi et al. (1995; CCPS). These models were computed assuming continuous star formation rather than starbursts, and different massive star yields. Both sets of models include the effects of

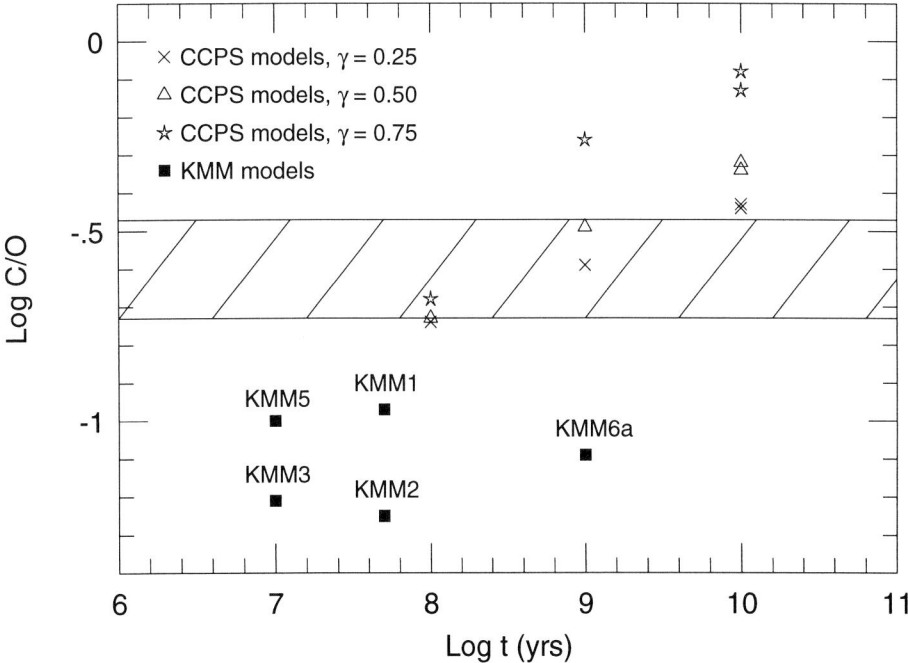

FIGURE 33. Log C/O vs. log age from chemical evolution models for IZw18 (KMM) and blue compact dwarf galaxies (CCPS). The KMM models are labeled according to model number; KMM6a shows abundances from the KMM two-burst model 6 after 990 Myrs, at the beginning of the second burst. For the CCPS models, γ refers to the strength of the SN-driven galactic wind. The hatched area shows the range occupied by the observed values of C/O in IZw18 and corresponding errors. (From Garnett et al. 1997a)

differential (heavy element enriched) winds. The hatched region shows the values of C/O encompassed by the observations of IZw18 and corresponding error bars.

None of the KMM models are able to account for the high C/O in IZw18, suggesting that an additional source of carbon is needed in their models. The CCPS models can explain our observed C/O ratios for galaxy ages of the order 1 Gyr. From these results it could be inferred that IZw18 had an episode of star formation that occurred at least a few hundred million years ago that led to the presently observed levels of carbon and nitrogen. Interestingly, this corresponds roughly to the age inferred from stellar photometry in the companion irregular galaxy NW of the main body of IZw18 (Dufour et al. 1996), which may suggest that an older population may still lay obscured by the light of the luminous present-day burst in the main body.

The agreement in N/O and O/H between the NW and SE components of IZw18 led SK93 to question the validity of the "self-pollution" model for IZw18. The agreement between our two high values for C/O further strengthen this argument. Thus, it seems highly improbable that these two separate star formation events could lead to essentially identical abundances in a self-pollution model. At this point, the simplest explanation appears to be that IZw18 is not a young galaxy, but rather has had one or more previous episodes of star formation. Future deep HST photometry with the HST will be needed to resolve this unambiguously. In the end, the best way to determine the presence of an older generation of stars will be to detect it directly. However, if we are correct, and the ISM abundance is indicative of an, as yet, unseen population, we will be adding an important diagnostic to our tool kit for galaxy analysis.

7. Summary

Throughout my lectures I tried to stress the accuracy and utility of nebular abundances and to show some sense of their wide range of applicability. As stellar astronomers explore the nearby galaxies, I believe nebular abundances will serve as a great aid in these endeavors. I have also tried to promote the notion that while a detailed understanding of the chemical evolution of galaxies remains a long term goal, there is much to be gained from trying to acquire observations which provide simple clues and constraints.

Only the nearest galaxies will allow us the observations necessary to resolve the details which will lead to a deeper understanding of how galaxies evolve.

The work of my Minnesota colleagues Don Garnett, Chip Kobulnicky, and Robbie Dohm-Palmer has been highlighted in this chapter and I wish to thank them for very enjoyable collaborations (and for helpful editing of this chapter). I would also like to thank Kim Venn for giving me many valuable comments on this chapter. I wish to thank collaborators and friends B. Benjamin, R. Clegg, S. Côté, R. Dufour, J. Gallagher, P. Hodge, J. Hoessel, R. Kennicutt, N. Langer, M. Mateo, B. Miller, B. Pagel, M. Peimbert, M. Rosa, J.-R. Roy, A. Saha, G. Shields, J. Shields, E. Terlevich, R. Terlevich, E. Tolstoy, S. Torres-Peimbert, J. Walsh, and D. Zaritsky for helping to make this so much fun. Partial support from a NASA LTSARP grant No. NAGW-3189 and from the graduate school of the University of Minnesota is gratefully acknowledged. I would especially like to thank the organizers for their invitation to this winter school and for all of their efforts in making it such a great experience.

REFERENCES

AARONSON, M. 1986 In *Star Forming Dwarf Galaxies and Related Objects* (eds. D. Kunth, T.X. Thuan, and J.T.T. Van) Editions Frontieres, 125.

ANDERS, E., & GREVESSE, N. 1989 *Geochim. Cosmochim. Acta* **53**, 197.

APARICIO, A., ET AL. 1995 *AJ* **110**, 212.

APARICIO, A., GALLART, C., CHIOSI, C., & BERTELLI, G. 1996 *ApJL* **469**, 97.

APARICIO, A., GARCIA-PELAYO, J.M., MOLES, M., & MELNICK, J. 1987 *A&AS* **71**, 297.

APARICIO, A. & RODRIGUEZ-ULLOA, J. A. 1992 *A&A* **260**, 77.

ARNETT, W. D. 1971 *ApJL* **170**, L43.

AUDOUZE, J., & TINSLEY, B. M. 1976 *ARA&A* **14**, 43.

BALDWIN, J. A., FERLAND, G. J., MARTIN, P. G., CORBIN, M R., COTA, S. A., PETERSON, B. M., & SLETTEBAK, A. 1991 *ApJ* **364**, 580

BECK, S., TURNER, J. L., HO, P. T. P., LACY, J. H., & KELLY, D. M. 1996 *ApJ* **457**, 610.

BERTELLI, G., BRESSAN, A., CHIOSI, C., FAGOTTO, F. & NASI, E. 1994 *A&AS* **106**, 275, (B94).

BINGGELI, B., SANDAGE, A., & TAMMANN G. A. 1985 *AJ*, **90**, 1681.

BINGGELI, B., TARENGHI, M., & SANDAGE, A. 1990 *A&A*, **228**, 42.

BOTHUN, G. D., & MOULD, J. R. 1988 *ApJ*, **324**, 123.

BROCKLEHURST, M. 1971 *MNRAS* **153**, 471.

BROCKLEHURST, M. 1972 *MNRAS* **157**, 211.

BÖHRINGER, H., BRIEL, U. G., SCHWARZ, R. A., VOGES, W., HARTNER, G., & TRÜMPER, J. 1994 *Nature* **368**, 828.

BUCHMANN, L., ET AL. 1993 *Phys. Rev. Letts.* **70**, 726.

BUTCHER, H. 1977 *ApJ*, **216**, 372.

CALZETTI, D. 1997 *AJ* **113**, 162.

CAMPBELL, A., TERLEVICH, R., & MELNICK, J. 1986 *MNRAS* **223**, 811.

CAPLAN, J., & DEHARVENG, L. 1986 *A&A* **155**, 297.

CARBON, D. F., BARBUY, B., KRAFT, R. P., FRIEL, E. D., & SUNTZEFF, N. B. 1987 *PASP* **99**, 335.

CARDELLI, J. A., CLAYTON, G. C., & MATHIS, J. S. 1989 *ApJ* **345**, 245.

CARIGI, L. 1994 *ApJ* **424**, 181.

CARIGI, L., COLIN, P., PEIMBERT, M., & SARMIENTO, A. 1995 *ApJ* **445**, 98.

CAUGHLAN, G. A., & FOWLER, W. A. 1988 *Atom. Data Nucl. Data* **40**, 238.

CAYATTE, V., VAN GORKOM, J. H., BALKOWSKI, C., & KOTANYI, C. 1990 *AJ* **100**, 604.

CAYATTE, V., KOTANYI, C. BALKOWSKI, C., & VAN GORKOM, J. H. 1994 *AJ* **107**, 1003, (CKBG).

CHIOSI, C., & MATTEUCCI, F. 1982 *A&A* **110**, 54.

CHU, Y.-H., & KENNICUTT, R. C. 1986 *ApJ* **311**, 85

CLARKE, C. J. 1989 *MNRAS* **238**, 283.

CLAYTON, D. D. 1987 *ApJ* **315**, 451.

CLAYTON, D. D., & PANTELAKI, I. 1993 *Phys. Rep.* **227**, 293.

CLEGG, R. E. S., LAMBERT, D. L., & TOMKIN, J. 1981 *ApJ* **250**, 262.

CÔTÉ, S. 1995 Ph.D. Thesis, Australia National University.

CUNHA, K., & LAMBERT, D. L. 1994 *ApJ* **426**, 170.

DAVIDSON, K., DUFOUR, R. J., WALBORN, N. R., & GULL T. R. 1986 *ApJ* **305**, 867.

DAVIES, R. D., & LEWIS, B. M. 1973 *MNRAS* **165**, 231.

DEKEL, A., & SILK, J. 1986 *ApJ* **303**, 39.

DE ROBERTIS, M., DUFOUR, R. J., & HUNT, R. 1987 *JRASC* **81**, 195.

DEVOST, D., ROY, J.-R., & DRISSEN, L. 1997 *ApJ* **482**, 765.

DE VAUCOULERS, G., DE VAUCOULERS, A., CORWIN, H. G. JR., BUTA R. J., PATUREL, G., & FOUQUÉ, P. 1991 *Third Reference Catalog of Bright Galaxies,* Springer-Verlag, (RC3).

DÍAZ, A. I., TERLEVICH, E., PAGEL, B. E. J., VÍLCHEZ, J. M., & EDMUNDS, M. G. 1987 *MNRAS* **226**, 19.

DÍAZ, A. I., TERLEVICH, E., VÍL CHEZ, J. M., PAGEL, B. E. J., & EDMUNDS, M. G. 1991 *MNRAS* **253**, 245.

DÍAZ, A. I., & TOSI, M. 1986 *A&A* **158**, 60.

DINERSTEIN, H. L. 1990 In *The Interstellar Medium in Galaxies* (eds. H. A. Thronson, Jr., & J. M. Shull), Kluwer, 257.

DINERSTEIN, H. L., LESTER, D. F., & WARNER, M. W. 1985 *ApJ* **291**, 561.

DOHM-PALMER, R. C., SKILLMAN, E. D., SAHA, A., TOLSTOY, E., MATEO, M., GALLAGHER, J., HOESSEL, J., & DUFOUR, R. J. 1997 *AJ*, submitted.

DOPITA, M. A., & RYDER, S. D. 1994 *ApJ* **430**, 163.

DRAINE, B. T., & SALPETER, E. E. 1978 *ApJ* **231**, 77.

DRESSLER, A. 1980 *ApJ* **236**, 351.

DRESSLER, A., OEMLER, A. JR., SPARKS, W. B., & LUCAS, R. A. 1994 *ApJ* **435**, 23.

DUFOUR, R.J. 1975 *ApJ* **195**, 315.

DUFOUR, R.J. 1984 In *Future of Ultraviolet Astronomy Based on Six Years of IUE Research* (eds. J. Mead, R. Chapman, & Y. Kondo) NASA, 107.

DUFOUR, R.J., GARNETT, D.R., & SHIELDS, G.A. 1988 ApJ **332**, 752.

DUFOUR, R.J., GARNETT, D.R., SKILLMAN, E.D., & SHIELDS, G.A. 1996 In *From Stars to Galaxies: The Impact of Stellar Physics on Galaxy Evolution* (eds. C. Leitherer, U. Fritze-von Alvensleben, and J. Huchra) ASP Conf. Ser., **98**, 358.

DUFOUR, R.J., & HARLOW, W. W. 1977 ApJ **216**, 706.

DUFOUR, R.J., & HESTER, J. J. 1989 ApJ **350**, 149.

DUFOUR, R.J., SCHIFFER, F. H. & SHIELDS, G.A. 1984 In *Future of Ultraviolet Astronomy Based on Six Years of IUE Research* (eds. J.— Mead, R.— Chapman, & Y. Kondo) NASA CP-2349, p. 111.

DYSON, J. E., & SMITH, L. J. 1985, In *Cosmic Gas Dynamics* (ed. F. D. Khan) VNU Science Press, 173.

EDMUNDS, M. G. 1990 MNRAS **246**, 678.

EDMUNDS, M. G., & GREENHOW, R. M. 1995 MNRAS **272**, 241.

EDMUNDS, M. G., & PAGEL, B. E. J. 1978 MNRAS **185**, 78p.

EDMUNDS, M. G., & PAGEL, B. E. J. 1984 MNRAS **211**, 507.

EDMUNDS, M. G., & PHILLIPS, S. 1989 MNRAS **241**, 9p.

ESTEBAN, C., & PEIMBERT, M. 1995 RMA&ASC, **3**, 133.

ESTEBAN, C., VÍLCHEZ, J. M., SMITH, L. J., & CLEGG, R. E. S. 1992 A&A **259**, 629.

ESTEBAN, C., VÍLCHEZ, J. M., & SMITH, L. J. 1994 AJ **107**, 1041.

FERGUSON, H. C., & BINGGELI, B. 1994 A&ARv **6**, 67.

FISHER, J. R., & TULLY, R. B. 1978 A&A **44** 151.

FRIEDLI, D., BENZ, W., & KENNICUTT, R. C. 1994 ApJ **430**, L105.

GALLAGHER, J.S., HUNTER, D.A., & TUTUKOV, A.V. 1984 ApJ **284**, 544.

GALLART, C., APARICIO, A., & VÍCHEZ, J. M. 1996a AJ **112**, 1928.

GALLART, C., APARICIO, A., & VÍCHEZ, J. M. 1996b AJ **112**, 1950.

GALLART, C., APARICIO, A., BERTELLI, G., & CHIOSI, C. 1996c AJ **112**, 2596.

GARNETT, D. R. 1989 ApJ **345**, 282.

GARNETT, D. R. 1990 ApJ **363**, 142.

GARNETT, D. R. 1992 AJ **103**, 1330.

GARNETT, D. R., DUFOUR, R. J., PEIMBERT, M., TORRES-PEIMBERT, S., SHIELDS, G. A., SKILLMAN, E. D., TERLEVICH, R., & TERLEVICH, E. 1995b ApJL **449**, L77.

GARNETT, D. R., & CHU, Y.-H. 1994 PASP **106**, 626.

GARNETT, D. R., ODEWAHN, S. C., & SKILLMAN, E. D. 1992, AJ, **104**, 1714.

GARNETT, D. R., & SHIELDS, G. A. 1987 ApJ **317**, 82.

GARNETT, D. R., SHIELDS, G. A., SKILLMAN, E. D., SAGAN, S. P., & DUFOUR, R. J. 1997b ApJ, in press.

GARNETT, D. R., SKILLMAN, E. D., DUFOUR, R. J., PEIMBERT, M., TORRES-PEIMBERT, TERLEVICH, R., & TERLEVICH, E., & SHIELDS, G. A. 1995a ApJ **443**, 64, (G95a).

GARNETT, D. R., SKILLMAN, E. D., DUFOUR, R. J., & SHIELDS, G. A. 1997a ApJ, **481**, 174.

GEHREN, T., NISSEN, P. E., KUDRITZKI, R. P., & BUTLER, K. 1985, In *Production and Distribution of C, N, & O Elements* (eds. I. J. Danziger, F. Matteucci, & K. Kjär) ESO, p171.

GIES, D. R., & LAMBERT, D. L. 1992 ApJ **387**, 673.

GIOVANELLI, R., & HAYNES, M. 1985 ApJ **292**, 404.

GIOVANELLI, R., HAYNES, M., SALZER, J. J., WEGNER, G., DA COSTA, L. N., & FREUDLING, W. 1994 AJ **107**, 2036.

GONZALEZ-DELGADO, R. M., PÉREZ, E., TENORIO-TAGLE, G., ET AL. 1994 ApJ **437**, 239.

GÖTZ, M., & KÖPPEN, J. 1992 A&A **262**, 455.

GOULD, A. 1994 *ApJ*, **426**, 542.

GREGGIO, L. 1994 In *The Local Group: Comparative and Global Properties* (eds. A. Layden, R. C. Smith, & J. Storm) ESO, p. 72.

GREGGIO, L., MARCONI, G., TOSI, M., & FOCARDI, P. 1993 *AJ* **105**, 894.

GREVESSE, N., & NOELS, A. 1993 In *Origin and Evolution of the Elements* (eds. N. Prantzos, E. Vangioni-Flam, & M. Casse) Cambridge University Press, 15.

GÜSTEN, R., & MEZGER, P. G. 1982 *Vistas Astr.* **26**, 159.

HAYNES, M. P. 1990 In *Clusters of Galaxies* (ed. W. R. Oegerle, M. J. Fitchett, & L. Danly) Cambridge University Press, 177.

HENRY, R. B. C. 1993 *MNRAS* **261**, 306

HENRY, R. B. C., PAGEL, B. E. J., & CHINCARINI, G. L. 1994 *MNRAS* **266**, 421.

HENRY, R. B. C., PAGEL, B. E. J., LASSITER, D. F., & CHINCARINI, G. L. 1992 *MNRAS* **258** 321.

HEYDARI-MALAYERI, M., MELNICK, J., & VAN DROM, E. 1990 *A&A* **236**, L21.

HODGE, P.W. 1980 *ApJ* **241**, 125.

HODGE, P.W. 1989 *ARA&A* **27**, 139.

HODGE, P., KENNICUTT, R. C. & STROBEL, N. 1994 *PASP* **106**, 765.

HOLT, S. S., & MUNDY, L. G. 1997 *Star Formation Near and Far*, AIP Conference Proceedings, **393**.

HOLTZMAN, J. A., BURROWS, C. J., CASERTANO, S., HESTER, J., TRAUGER, J. T., WATSON, A. M., & WORTHEY, G. 1995 *PASP* **107**, 1065.

HUMMER, D. G. & STOREY, P. J. 1987 *MNRAS* **224**, 801.

HUNTER, D. A., HAWLEY, W. N., & GALLAGHER, J. S. 1993 *AJ* **106**, 1797.

HUNTER, D. A., & PLUMMER, J. D. 1996 *ApJ* **462**, 732.

HUNTER, D. A., & THRONSON, H. A. 1995 *ApJ* **452**, 238.

IBEN, I. JR., & RENZINI, A. 1983 *ARA&A* **21**, 271.

ISRAEL, F. P. 1988a In *Millimetre and Submillimetre Astronomy* (ed. R. D. Wolstencroft & W. B. Burton), Kluwer, p. 281.

ISRAEL, F. P. 1988b In *Molecular Clouds in the Milky Way and External Galaxies* (ed. R. L. Dickman, R. L. Snell, & J. S. Young), Springer-Verlag, p. 428.

ISRAEL, F. P., & DE BRUYN, A. G. 1988 *A&A* **198**, 109.

ISRAEL, F. P., & KENNICUTT, R. C., JR. 1980 *Astr. Letts.* **21**, 1.

JENKINS, E. B. 1987 In *Interstellar Processes* (eds. D. J. Hollenbach & H. A. Thronson) Kluwer, 533.

JENKINS, E. B. 1989 In *IAU Symposium 135, Interstellar Dust* (eds. L. J. Allamandola & A. G. G. M. Tielens) Kluwer, 23.

KENNICUTT, R. C., JR. 1983 *AJ* **88**, 483.

KENNICUTT, R. C., JR. 1989 *ApJ* **344**, 171.

KENNICUTT, R. C., JR., & GARNETT, D. R. 1996 *ApJ* **456**, 504.

KENNICUTT, R. C., JR., TAMBLYN, P., & CONGDON, C. W. 1994 *ApJ* **435**, 22.

KILIAN, J. 1992 *A&A* **262**, 171.

KILIAN, J., MONTENBRUCK, O., & NISSEN, P. 1994 *A&A* **284**, 437.

KILIAN-MONTENBRUCK, J., GEHREN, T., & NISSEN, P. 1994 *A&A* **291**, 757.

KINKEL, U., & ROSA, M. R. 1994 *A&A* **282**, L37.

KINMAN, T. D., & DAVIDSON, K. 1981 *ApJ* **243**, 127.

KNAPP, G. R. 1990 In *The Interstellar Medium in Galaxies* (eds. H. A. Thronson & J. M. Shull), Kluwer, p. 3.

KOBULNICKY, H. A. 1997 Ph.D. Thesis, University of Minnesota.

KOBULNICKY, H. A., & SKILLMAN, E. D. 1996 ApJ **471**, 211.

KOBULNICKY, H. A., & SKILLMAN, E. D. 1997 ApJ in press.

KOBULNICKY, H. A., SKILLMAN, E. D., ROY, J.-R., WALSH, J., & ROSA, M. 1996, ApJ **477**, 679.

KUNTH, D., LEQUEUX, J., SARGENT, W. L. W., & VIALLEFOND, F. 1994 A&A **282**, 709.

KUNTH, D., MATTEUCCI, F., & MARCONI, G. 1995 A&A **297**, 634, (KMM).

KUNTH, D., & SARGENT, W. L. W. 1986 ApJ **300**, 496.

LACEY, C. G., & FALL, S. M. 1985 ApJ **290**, 154.

LAIRD, J. B. 1985 ApJ **289**, 556.

LARSON, R. B. 1972 Nature **236**, 21.

LEQUEUX, J., PEIMBERT, M., RAYO, J. M., SERRANO, A., & TORRES-PEIMBERT, S. 1979 A&A **91**, 269.

LESTER, D. F., DINERSTEIN, H. L., WERNER, M. W., WATSON, D. M., GENZEL, R., & STOREY, J. W. V. 1987 ApJ **320**, 573.

LIPMAN, K. 1995 Ph.D. Thesis, Cambridge University.

LO, K. Y., SARGENT, W. L. W., & YOUNG, K. 1993 AJ **106**, 507.

MAEDER, A. 1992 A&A **264**, 105.

MALONEY, P., & BLACK, J. H. 1988 ApJ **325**, 389.

MARCONI, G., TOSI, M., GREGGIO, L., & FOCARDI, P. 1995 AJ **109**, 173.

MARLOWE, A. T., HECKMAN, T. M., WYSE, R. F. G., & SCHOMMER, R. 1995 ApJ **438**, 563.

MARTIN, C. L. 1996 ApJ **465**, 680.

MARTIN, C. L., & KENNICUTT, R. C., JR. 1995 ApJ **447**, 171.

MARTIN, P., & ROY, J. R. 1994 ApJ **424**, 599.

MASSEY, P., JOHNSON, K. E., & DEGIOIA-EASTWOOD, K. 1995a ApJ **454**, 151.

MASSEY, P., LANG, C. C., DEGIOIA-EASTWOOD, K., & GARMANY, K. 1995b ApJ **438**, 188.

MATHIS, J. S. 1983 ApJ **267**, 119.

MATHIS, J. S. 1985 ApJ **291**, 247.

MATHIS, J. S., & ROSA, M. R. 1991 A&A **245**, 625.

MATTEUCCI, F., & CHIOSI, C. 1983 A&A **123**, 121.

MATTEUCCI, F., & FRANCOIS, P. 1989 MNRAS **239**, 885.

MATTEUCCI, F., & TOSI, M. 1985 MNRAS **217**, 391.

MAYOR, M., & VIGROUX, L. 1981 A&A **98**, 1.

MCCALL, M. L. 1982 Ph.D. Thesis, University of Texas.

MCGAUGH, S. S. 1991 ApJ **380**, 140.

MELNICK, J., TERLEVICH, R., & EGGLETON, P. P. 1985 MNRAS **216**, 255.

MEURER, G. R., FREEMAN, K. C., DOPITA, M. A., & CACCIARI, C. 1992 AJ, **103**, 60.

MILLER, B. W. 1996 AJ **112**, 991.

MILLER, B. W., & HODGE, P. 1996 ApJ **458**, 467.

MOLLÁ, M., FERINI, F., & DÍAZ, A. I. 1996 ApJ **466**, 668.

MOULD, J. R., KRISTIAN, J., AND DA COSTA, G. S. 1983 ApJ **270**, 471.

OEY, M. S. & KENNICUTT, R. C. 1993 ApJ **411**, 137.

OKE, J. B. 1990 AJ **99**, 1621.

OSTERBROCK, D. E. 1989 *Astrophysics of Gaseous Nebulae and Active Galactic Nuclei*, University Science Books (AGN^2).

OSTERBROCK, D. E., TRAN, H. D., & VEILLEUX, S. 1992 ApJ **389**, 305.

PAGEL, B. E. J. 1985 In *Production and Distribution of C, N, & O Elements* (eds. I. J. Danziger, F. Matteucci, & K. Kjär) ESO, p155.

PAGEL, B. E. J. 1989 *RMxA&A* **18**, 161.
PAGEL, B. E. J., & EDMUNDS, M. G. 1981 *ARA&A* **19**, 77.
PAGEL, B. E. J., EDMUNDS, M. G., BLACKWELL, D. E., CHUN, M. S., & SMITH, G. 1979 *MNRAS* **189**, 95.
PAGEL, B. E. J., EDMUNDS, M. G., FOSBURY, R. A. E., & WEBSTER, B. L. 1978 *MNRAS* **184**, 569.
PAGEL, B. E. J., EDMUNDS, M. G., & SMITH, G. 1980 *MNRAS* **193**, 219.
PAGEL, B. E. J., & PATCHETT, B. E. 1975 *MNRAS* **172**, 13.
PAGEL, B. E. J., SIMONSON, E. A., TERLEVICH, R. J., & EDMUNDS, M. G. 1992 *MNRAS* **255**, 325.
PAGEL, B. E. J., & TAUTVAIŠIENĖ, G. 1995 *MNRAS* **276**, 505.
PAGEL, B. E. J., & TAUTVAIŠIENĖ, G. 1997 *MNRAS* **288**, 108.
PAGEL, B. E. J., TERLEVICH, R. J., & MELNICK, J. 1986 *PASP* **98**, 1005.
PANTELAKI, I., & CLAYTON, D. D. 1987 In *Starbursts and Galaxy Evolution*, (eds. T.X. Thuan, T. Montmerle, and J. Tran Thanh Van), Editions Frontieres, p. 145.
PEIMBERT, M. 1967 *ApJ* **150**, 825.
PEIMBERT, M. 1995 In *The Analysis of Emission Lines*, (eds. R. E. Williams & M. Livio), Cambridge University Press, p. 165.
PEIMBERT, M., COLIN, P., & SARMIENTO, A. 1994 In *Violent Star Formation* (ed. G. Tenorio-Tagle), Cambridge University Press, p. 79.
PEIMBERT, M., PENA, M., & TORRES-PEIMBERT, S. 1986 *A&A* **158**, 266.
PEIMBERT, M., & SERRANO, A. 1982 *MNRAS* **198**, 563.
PEIMBERT, M., SARMIENTO, A., & COLIN, P. 1994 *RMxAA* **28**, 181.
PEIMBERT, M., STOREY, P. J., & TORRES-PEIMBERT, S. 1993 *ApJ* **414**, 626.
PEIMBERT, M., & TORRES-PEIMBERT, S. 1974 *ApJ* **193**, 327.
PEIMBERT, M., & TORRES-PEIMBERT, S. 1976 *ApJ* **203**, 581.
PEIMBERT, M., TORRES-PEIMBERT, S., & DUFOUR, R. J. 1993 *ApJ* **418**, 760.
PEIMBERT, M., TORRES-PEIMBERT, S., & RUIZ, M. T. 1992 *RMxA&A* **24**, 155.
PETTINI, M., & LIPMAN, K. 1995 *A&A* **297**, L63.
PETTINI, M., LIPMAN, K., & HUNSTEAD, R. W. 1995 *ApJ* **451**, 100.
PHILLIPPS, S., & EDMUNDS, M. G. 1991 *MNRAS* **251**, 84.
PIPHER, J., HELFER, H. L., HERTER, T., BRIOTTA, D. A., HOUCK, J. R., WILLNER, S. P., & JONES, B. 1984 *ApJ* **285**, 174.
PIOTTO, G., CAPACCIOLI, M., & PELLEGRINI, C. 1994 *A&A* **287**, 371.
PRANTZOS, N., VANGIONI-FLAM, E., & CHAUVEAU, S. 1994 *A&A* **285**, 132.
REIMERS, D., VOGEL, S., HAGEN, H.-J., ENGELS, D., GROOTE, D., WAMSTEKER, W., CLAVEL, J., & ROSA, M. J. 1992 *Nature* **360**, 561.
RENZINI, A., & VOLI, M. 1981 *A&A* **94**, 175.
RICHER, M. G., & MCCALL, M. L. 1995 *ApJ* **445**, 642, (RM95).
RIEKE, G. H., LEBOFSKY, M. J., & WALKER, C. E. 1988 *ApJ* **325**, 679.
ROENNBACK, J., & BERGVALL, N. 1995 *A&A* **302**, 353.
ROLFS, C. E., & RODNEY, W. S. 1988 *Cauldrons in the Cosmos*, University of Chicago Press.
ROSA, M., & MATHIS, J. S. 1987 *ApJ* **317**, 163.
RYDER, S. D. 1995 *ApJ* **444**, 610.
SAHA, A., SANDAGE, A., LABHARDT, L. TAMMANN, G. A., MACCHETTO, F. D., & PANAGIA, N. 1996 *ApJ* **466**, 55.
SAKAI, S., MADORE, B. F., & FREEDMAN, W. L. 1996 *ApJ* **461**, 713.
SCHECTER, P. L., MATEO, M. L., & SAHA, A. 1993 *PASP* **105**, 1342.

SCHMIDT, M. 1959 *ApJ* **129**, 243.
SCHMIDT, M. 1963 *ApJ* **137**, 758.
SCOWEN, P. A., DUFOUR, R. J., & HESTER, J. J. 1992 *AJ* **104**, 92.
SEATON, M. J. 1979 *MNRAS* **187**, 73p.
SEARLE, L., & SARGENT, W. L. W. 1972 *ApJ* **173**, 25.
SEARLE, L., SARGENT, W. L. W., & BAGNUOLO, W. G. 1973 *ApJ* **179**, 427.
SELLMAIER, F. H., YAMAMOTO, T., PAULDRACH, A. W. A., & RUBIN, R. H. 1996 *A&A* **305**, L37.
SHAVER, P. A., MCGEE, R. X., NEWTON, L. M., DANKS, A. C., & POTTASCH, S. R. 1983 *MNRAS* **204**, 53.
SHAW, R. A., & DUFOUR, R. J. 1995 *PASP* **107**, 896, (SD95).
SHIELDS, J. C., & KENNICUTT, R. C., JR. 1995 *ApJ* **454**, 807.
SHIELDS, G. A., SKILLMAN, E. D., & KENNICUTT, R. C. JR. 1991 *ApJ* **371**, 82.
SHIELDS, G. A., & SEARLE, L. 1978 *ApJ* **222**, 821.
SIMPSON, J. P., COLGAN, S. W. J., RUBIN, R. H., ERICKSON, E. F., & HAAS, M. R. 1995 *ApJ* **444**, 721.
SKILLMAN, E. D. 1985 *ApJ* **290**, 449.
SKILLMAN, E. D. 1989 *ApJ* **347**, 883.
SKILLMAN, E. D. 1991 *PASP* **666**, 919.
SKILLMAN, E. D., BOMANS, D. J., & KOBULNICKY, H. A. 1997 *ApJ* **474**, 205.
SKILLMAN, E. D., & ISRAEL, F. P. 1988 *A&A* **203**, 226.
SKILLMAN, E. D., & KENNICUTT, R. C., JR. 1993 *ApJ* **411**, 655.
SKILLMAN, E. D., KENNICUTT, R. C., & HODGE, P. W. 1989b *ApJ* **347**, 875.
SKILLMAN, E. D., KENNICUTT, R. C. JR., SHIELDS, G. A., & ZARITSKY, D. 1996b *ApJ* **462**, 147.
SKILLMAN, E. D., MELNICK, J., TERLEVICH, R., & MOLES, M. 1988a *A&A* **196**, 31.
SKILLMAN, E. D., PALMER, R., GARNETT, D. R., & DUFOUR, R. J. 1996a In *From Stars to Galaxies: The Impact of Stellar Physics on Galaxy Evolution* (eds. C. Leitherer, U. Fritze-von Alvensleben, and J. Huchra) *ASP Conf. Ser.*, **98**, 366.
SKILLMAN, E. D., TERLEVICH, R. J., KENNICUTT, R. C., GARNETT, D. R., & TERLEVICH, E. 1994 *ApJ* **431**, 172.
SKILLMAN, E. D., TERLEVICH, R., AND MELNICK, J. 1989a *MNRAS* **240**, 563.
SKILLMAN, E. D., TERLEVICH, R. J., TEUBEN, P. J., & VAN WOERDEN, H. 1988b *A&A*, **198**, 33.
SMARTT, S. J., & ROLLESTON, W. R. J. 1997 *ApJL* **481**, L47.
SMITS, D. 1996 *MNRAS* **251**, 316.
SNEDEN, C., LAMBERT, D. L., & WHITAKER, R. W.) 1979 *ApJ* **234**, 964.
SOFIA, U. J., CARDELLI, J. A., & SAVAGE, B. D. 1994 *ApJ* **430**, 650.
SOMMER-LARSON, J., & YOSHII, Y. 1989 *MNRAS* **238**, 133.
SOMMER-LARSON, J., & YOSHII, Y. 1990 *MNRAS* **243**, 468.
STASIŃSKA, G. 1990 *A&AS* **83**, 501.
TALENT, D. L. 1980 PH.D. THESIS, RICE UNIVERSITY.
TENORIO-TAGLE, G. 1996 *AJ* **111**, 1641.
TENORIO-TAGLE, G., PRIETO, M., & SANCHEZ, F. 1992 *Star Formation in Stellar Systems*, CAMBRIDGE UNIVERSITY PRESS.
THORNLEY, M. D., & WILSON, C. D. 1995 *ApJ* **447**, 616.
THUAN, T. X., IZOTOV, Y. I., & LIPOVETSKY, V. A. 1995 *ApJ* **445**, 108.
TIMMES, F. X., WOOSLEY, S. E., & WEAVER, T. A. 1995 *ApJS* **98**, 617.

TINSLEY, B. M. 1980 *Fund. Cosm. Phys.* **5**, 287.
TOLSTOY, E. 1995 PH.D. THESIS, UNIVERSITY OF GRONINGEN.
TOLSTOY, E. 1996 *ApJ* **462**, 684.
TOLSTOY, E., & SAHA, A. 1996 *ApJ* **462**, 672.
TOMKIN, J., LEMKE, M., LAMBERT, D. L., & SNEDEN, C. 1992 *AJ* **104**, 1568.
TORRES-PEIMBERT, S., PEIMBERT, M., & FIERRO, J. 1989 *ApJ* **345**, 186.
TOSI, M. 1988 *A&A* **197**, 47.
TOSI, M., & DÍAZ, A. I. 1985 *MNRAS* **217**, 571.
TOSI, M., GREGGIO, L., MARCONI, G., & FOCARDI, P. 1991 *AJ* **102**, 951.
TULLY, R. B., & SHAYA, E. J. 1984 *ApJ* **281**, 31.
TWAROG, B. A. 1985 IN *IAU Symposium 106, The Milky Way Galaxy* (EDS. H. VAN WOERDEN, W. B. BURTON, & R. J. ALLEN) REIDEL, P. 587.
VAN DEN BERGH, S., & PIERCE, M. J. 1990 *ApJ* **364**, 444.
VAN DER HULST, J. M., SKILLMAN, E. D., KENNICUTT, R. C., & BOTHUN, G. D. 1987 *A&A* **177**, 63.
VAN ZEE, L. 1996 PH.D. THESIS, CORNELL UNIVERSITY.
VENN, K. A. 1995 *ApJS* **99**, 659.
VÍLCHEZ, J. M. 1995 *AJ* **110**, 1090.
VÍLCHEZ, J. M., PAGEL, B. E. J., DÍAZ, A. I., TERLEVICH, E., & EDMUNDS, M. G. 1988 *MNRAS* **235**, 633.
VILA-COSTAS M. B., & EDMUNDS, M. G. 1992 *MNRAS* **259**, 121 (VCE).
VILA-COSTAS M. B., & EDMUNDS, M. G. 1993 *MNRAS* **265**, 199.
WALSH, J. R., & ROY, J-R. 1987 *ApJ* **318**, L57.
WALSH, J. R., & ROY, J-R. 1989 *MNRAS* **239**, 297.
WARMELS, R. H. 1986 PH.D. THESIS, UNIVERSITY OF GRONINGEN.
WEAVER, R., MCCRAY, R. A., CASTOR, J., SHAPIRO, P., & MOORE, R. 1977 *ApJ* **218**, 377.
WEAVER, T. A., & WOOSLEY, S. E. 1993 *Phys. Rep.* **227**, 65.
WELCH, G. A. 1970 **ApJ 161**, 821.
WEVERS, B. M. H. R., VAN DER KRUIT, P. C., & ALLEN, R. J. 1986 *A&AS* **66**, 505.
WHEELER, J. C., SNEDEN, C., & TRURAN, J. W. 1989 *ARA&A* **27**, 279.
WHITMORE, B. C. 1990 IN *Clusters of Galaxies*, (ED. W. R. OEGERLE, M. J. FITCHETT, & L. DANLY), CAMBRIDGE UNIVERSITY PRESS, 139.
WILLIAMS, R. M., & CHU, Y.-H. 1995 *ApJ* **439**, 132.
WOOSLEY, S. E. 1986 IN *Nucleosynthesis and Chemical Evolution* (EDS. B. HAUCK, A. MAEDER, & G. MEYNET) GENEVA OBSERVATORY, P. 1.
WOOSLEY, S. E., LANGER, N., & WEAVER, T. A. 1993 *ApJ* **411**, 823.
WOOSLEY, S. E., & WEAVER, T. A. 1995 *ApJS* **101**, 181.
ZARITSKY, D. 1992 *ApJL* **390**, L73.
ZARITSKY, D., KENNICUTT, R. C., & HUCHRA, J. P. 1994 *ApJ* **420**, 87, (ZKH).
ZHAO, Z., FRANCE III, R. H., LAI, K. S., RUGARI, S. L., & GAI, M. 1993 *Phys. Rev. Letts.* **70**, 2066.

Populations of Massive Stars and the Interstellar Medium

By CLAUS LEITHERER

Space Telescope Science Institute †, 3700 San Martin Drive, Baltimore, MD 21218, USA

The properties of massive stars and their stellar winds are reviewed. Winds are crucial for stellar evolution and for the energization of the interstellar medium. The lectures cover the properties of individual stars, which serve as our local calibrators. More and more distant populations can be used as cosmic yardsticks to trace massive star populations out to cosmological distances. Techniques to determine the properties of these populations are introduced. Observed wind phenomena range from individual stellar ejecta to a large-scale blow-out of the interstellar medium in galaxies.

1. Introduction

Massive stars are an important — and sometimes the dominant — energy source for the evolution of a galaxy. Their high luminosity, both in terms of light and mechanical energy output, makes them detectable up to cosmological distances.

In these lectures I will attempt to give an overview of the properties of individual massive stars and stellar populations, and their effects on the surrounding interstellar medium (ISM). Stellar winds play a key role: they control stellar evolution, shape the ambient ISM, and affect the dynamical and chemical evolution of galaxies, and ultimately of the universe.

There will be five lectures. The first lecture (Section 2) sets the stage from an observational point of view. I will present a survey of observations of nearby and distant massive stars and populations. The Galaxy will be the starting point, where regions of high-mass star formation can be studied in great detail. Moving farther away to the Magellanic Clouds and other Local Group galaxies, we will see how some details are lost but how the large-scale properties can be better gauged. Once we are outside the Local Group, global aspects of galaxy evolution begin the evolve from "stellar" astrophysical considerations. At the conclusion of this lecture I will briefly present the latest observations of the most distant known massive star populations and compare them to their local counterparts. The goal of this lecture is to put into perspective the range of scales and sizes available to delineate the properties of massive stars and their interaction with the ISM.

The second part (Section 3) discusses the properties of nearby, resolved stellar populations. Observations of individual massive stars, as well as of entire stellar populations are feasible in our Galaxy, the Local Group, and somewhat beyond. I will introduce the concept of evolutionary synthesis analysis and discuss how theoretical Hertzsprung-Russell diagrams and color-magnitude diagrams can be constructed. Comparison with models of stellar atmospheres and evolution taking into account the winds allows one to constrain the initial mass function (IMF) and the star-formation history of the parent galaxy.

Part three (Section 4) performs one step further out: I will show how the integrated properties of stars and gas can be utilized to study the stellar content and the amount of light and particles released into the ISM. Evolutionary synthesis techniques are generalized to integrated properties, both for stellar line spectra and for nebular recombination

† Operated by AURA, Inc., for NASA under contract NAS5-26555

lines. Stellar winds are again crucial: they provide a powerful method to infer stellar masses, they modify the stellar far-ultraviolet energy distribution, and they affect the H II region spectrum by non-thermal heating. The goal of this and the preceding lecture is to present the techniques to constrain the stellar populations and to discuss what has been derived. This then leads to the next lectures: given the observed *radiant* properties of stellar populations — resolved and unresolved, how do the *non-radiant* properties affect the ISM?

In Section 5 I will cover the details of the stellar wind interaction with the ISM. A quantitative description of the mass and energy release by individual massive stars is given and the change of the wind properties during the course of evolution is outlined. By combining the wind properties of single stars with evolutionary models of populations, the total wind output can be calculated. Stellar winds can rival supernovae in their output of mass and momentum. I will discuss how a hot star population transits from an early photon-dominated stage to a subsequent matter-dominated phase. The purpose of this lecture is to present the *output* of non-radiant energy by stars from a stellar perspective. The interstellar perspective follows in Section 6.

The final lecture (Section 6) deals with the *input* of mass and energy into the ISM of galaxies. Single stars and entire populations can form cavities in the surrounding ISM due to the snow-plow effect of the winds. I will demonstrate how the collective effect of stars and their winds affect the evolution of galaxies and the entire universe. Such conditions are met in regions of violent star formation. If star formation is intense and/or the gravitational well of the parent galaxy is small, winds and supernovae can produce a blow-out, leading to a large-scale expulsion of the ISM from the galaxy. The implications of these effects on a cosmic evolutionary scale are briefly mentioned.

2. Regions of high-mass star formation

The mass range of stars drawing their energy supply from nuclear fusion covers about three orders of magnitude. The least massive stars known have masses around 0.1 $M_\odot$ (Liebert & Probst 1987; Bessell & Stringfellow 1993), and the most massive examples are around 100 $M_\odot$ (Maeder & Conti 1994). Stars on, or close to the location in the Hertzsprung-Russell diagram (HRD) where core-hydrogen burning is initiated (the "zero-age-main-sequence" or ZAMS), have a well-defined relation between stellar mass (M), luminosity (L), and effective temperature ($T_{\rm eff}$). Excellent reviews on the subject have been written by Chiosi & Maeder (1986) and Chiosi, Bertelli, & Bressan (1992). A recent theoretical HRD is reproduced in Figure 1. The figure illustrates the steep relation between mass and luminosity: $L \propto M^3$. Stars one hundred times more massive than the Sun are one million times more luminous! Except for stars of transient brightness, like novae and supernovae, hot, massive stars are the most luminous stellar objects in the universe.

Massive stars are, however, extremely rare: The number of stars formed per unit mass interval is roughly $\propto M^{-2.35}$ (Salpeter 1955). Therefore we expect to find only extremely few massive stars in comparison with solar-type stars. This is consistent with observations in our solar neighborhood: the closest ~50 $M_\odot$ star is the O4 supergiant ζ Puppis at a distance of about 500 pc, and large volumes need to be sampled to observe significant numbers of massive stars.

The definition of a massive star is somewhat arbitrary. For the purpose of these lectures I am defining massive stars as having ZAMS masses of more than 5 $M_\odot$. Note that the masses decrease during the stellar lifetime, as I will discuss later. This definition takes into account the capability of massive stars to affect their environment by their output

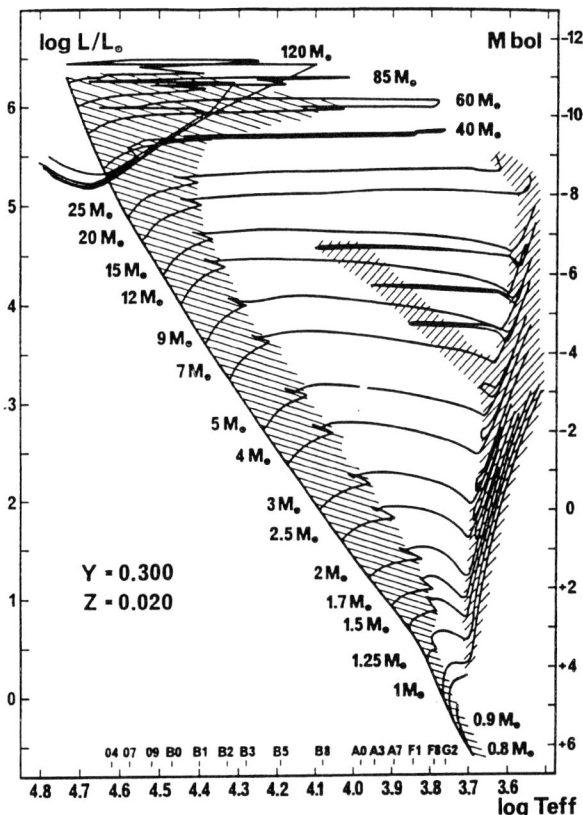

FIGURE 1. Theoretical HRD for solar metallicity stars. The thick lines indicate the evolutionary tracks of stars in the mass range 0.8 M$_\odot \leq M_\circ \leq$ 120 M$_\odot$, where $M_\circ$ is the ZAMS mass. From Schaller et al. (1992).

of radiation and matter. Stars with $M_\circ > 5$ M$_\odot$ (or equivalently, spectral type mid-B, see Figure 1) can *ionize* the surrounding gas and they can deposit *mechanical energy*, both via winds and later as supernovae.

2.1. *High-mass star formation in the Galaxy*

Our Galaxy does not classify as a starburst — a term denoting powerful star formation events in galaxies that will be discussed further below. Even the most luminous star-forming regions in our Galaxy are tiny on a cosmic scale. They are not dominated by the properties of an entire population but by individual stars. Therefore *stochastic effects* prevail. Extinction represents a severe problem when a reliable census of the Galactic high-mass star-formation history is required. Massive stars belong to the extreme population I, with correspondingly small vertical scale heights. Any line of sight between the Sun and high-mass star formation regions suffers heavy obscuration. Moreover, the proximity of Galactic regions — although advantageous for detailed studies of individual stars — makes it often difficult to obtain integrated properties, such as total emission-line fluxes of the ionized gas. Nevertheless, observations in the Galaxy allow us to study individual massive stars in great detail — an important consideration before venturing out to distant populations.

HD(E)/CPD	Sp. Type	E_{B-V}	M_V	$R=3$	$V_0 - M_V$ $R=4$	$R=5$
Tr 14						
93128	O3 V((f))	0.56	(−5.5)	12.59	12.03	11.47
93129A	O3 If*	0.54	−6.4	12.1	11.5	11.0
93129B	O3 V((f))	0.54	(−5.5)	12.8	12.2	11.7
−58°2611	O6 V((f))	0.60	(−5.3)	13.13	12.53	11.93
−58°2620	O6.5 V((f))	0.50	(−5.3)	13.07	12.57	12.07
Average		0.55		(12.90)	(12.33)	(11.79)
				±0.13	±0.13	±0.13
Tr 16						
93027	O9.5 V	0.30	−4.1	11.92	11.62	11.32
93028	O9 V	0.25	−4.3	11.91	11.66	11.41
93130	O6 III(f)	0.54	−5.6	12.04	11.50	10.96
93146	O6.5 V((f))	0.34	−5.3	12.72	12.38	12.04
93160	O6 III(f)	0.49	−5.6	11.94	11.45	10.96
93161AB	O6.5 V((f))	0.54	−6.05	12.27	11.73	11.19
93204	O5 V((f))	0.42	−5.5	12.66	12.24	11.82
93205	O3 V	0.37	−5.5	12.14	11.77	11.40
93222	O7 III((f))	0.37	−5.6	12.60	12.23	11.86
93250	O3 V((f))	0.47	−5.5	11.47	11.00	10.53
93343	O7 V(n)	0.61	−4.8	12.44	11.83	11.22
93403	O5 III(f) var	0.53	−6.4	12.09	11.56	11.03
303308	O3 V((f))	0.45	−5.5	12.32	11.87	11.42
305536	O9 V	0.36	−4.3	12.16	11.80	11.44
−59°2600	O6 V((f))	0.53	−5.3	12.32	11.79	11.26
−59°2603	O7 V((f))	0.46	−4.8	12.19	11.73	11.27
−59°2635	O7 Vnn	0.55	−4.8	12.46	11.91	11.36
−59°2636	O8 V	0.62	−4.4	11.85	11.23	10.61
−59°2641	O5 V	0.64	−5.5	12.87	12.23	11.59
Average		0.47		12.23	11.76	11.30
				±0.08	±0.08	±0.09
Distance (pc)				2800	2250	1800

TABLE 1. Stellar content of Trumpler 14 and 16 (from Walborn 1995).

Galactic massive stars are concentrated in loose associations or gravitationally bound open clusters. Humphreys (1978) and Garmany & Stencel (1992) published catalogs of known optical OB associations and clusters. Smith, Biermann, & Mezger (1978) give a list of radio-selected H II regions. These harbor clusters having too high interstellar extinction to be visually observable.

2.1.1. *The Carina region*

The Carina region (= NGC3372) hosts one of the more spectacular high-mass star formation regions in the Galaxy. The observed properties have been reviewed by Walborn (1995). Its proximity of about 2 kpc permits detailed studies of the stellar content. Table 1 lists spectral types, magnitudes and distances of known members of the two dominant clusters Trumpler 14 and 16. Three distances are listed, depending on the choice of the extinction law. Distances are a major source uncertainty for deriving parameters of Galactic stars. Several O3 stars have been found in Trumpler 14/16. They are the earliest and most massive species in the MK classification scheme (Walborn 1982). HD93128, HD93129A, HD93205, and HDE303308 have been analyzed by Kudritzki et al. (1992), who found masses around 100 $M_\odot$. The enigmatic object η Carinae is located in this region as well. η Car is a luminous blue variable (LBV), a short evolutionary phase when a substantial fraction of stellar mass is shed and ejected into the ISM (see Davidson 1989 for a review of its properties). LBV's will be discussed in more detail in Section 5.1.3.

2.1.2. NGC3603 and W51

NGC3603, the Galaxy's most massive *visible* H II region, has been called the Galactic clone of the LMC mini-starburst 30 Dor (Moffat et al. 1994). Drissen et al. (1995) have obtained high-quality spectra of individual stars. Their spectral classification work is essential for a complete census of the region. 14 massive O stars are found within about 1 pc^3. The output of ionizing photons can be calculated with, e.g., the tables published by Panagia (1973) or Vacca, Garmany, & Shull (1996). See also Table 6 below. These stars are essentially sufficient to provide the $\sim 10^{51}$ photons s^{-1} counted from the recombination luminosity of the gas (Kennicutt 1987). At the same time the stellar winds transfer about 10^{38} erg s^{-1} of mechanical luminosity to the ISM. The observed supersonic velocity structure of the ISM (up to 70 km s^{-1}; Balick et al. 1980; Clayton 1986) may result from energy injection by stellar winds.

W51 is among the most powerful star forming regions in the Galaxy. It is an H II region/molecular cloud complex 7.5 kpc from the Sun that has been extensively studied at radio and mid-infrared wavelengths (e.g., Bieging 1975; Mehringer 1994; Genzel et al. 1982). More than 40,000 $M_\odot$ is forming into stars in W51. However, the foreground visual obscuration is $A_V \approx 20$ magnitudes, and so these stars are visible only in near-infrared observations, such as the work of Goldader & Wynn-Williams (1994).

Another nearby dusty region of massive star formation is M17. Murray-Hanson & Conti (1995) used infrared spectroscopy to unveil the obscured stellar population. Despite the large observational efforts required, dusty Galactic sites of high-mass star formation are important as they may in some aspects resemble distant, dust-enshrouded starbursts.

2.1.3. The Galactic Center

The center of our Galaxy is a unique, nearby laboratory which allows us to study in detail processes presumed to be occurring in the nuclei of distant galaxies. Excellent reviews of the central region of our Galaxy have been given by Genzel & Townes (1987), Genzel, Hollenbach, & Townes (1994), and Morris & Serabyn (1996). A conference volume on the Galactic Center has been edited by Gredel (1996).

The Galactic Center is far too complex to be covered in full detail here. The reader is referred to the above mentioned reviews. Three aspects are of particular relevance to the present subject: the formation of massive stars in a very dense environment, the effects of winds from a dense star cluster on the ISM, and the energetics of the Galactic Center in comparison with AGN's.

The central few parsecs of the Galaxy contain a cluster of hot stars whose ultraviolet radiation and stellar winds energize the ambient ISM, both in terms of photoionization and mechanical energy transfer. Figure 2 is a 2.2 μm speckle image of the central 3" obtained by Eckart et al. (1995). 1" corresponds to 0.041 pc at the distance of the Galactic Center. The image reaches the diffraction limit of $\sim 0.15''$. The dense star cluster in the Galactic Center contains about 25 so-called He emission-line stars, young (~ 5 Myr), massive (50 – 100 $M_\odot$) stars formed during a recent star formation event (Najarro et al. 1994; Krabbe et al. 1995). High dust obscuration makes infrared spectroscopy necessary for analysis (Blum, DePoy, & Sellgren 1995; Figer, McLean, &

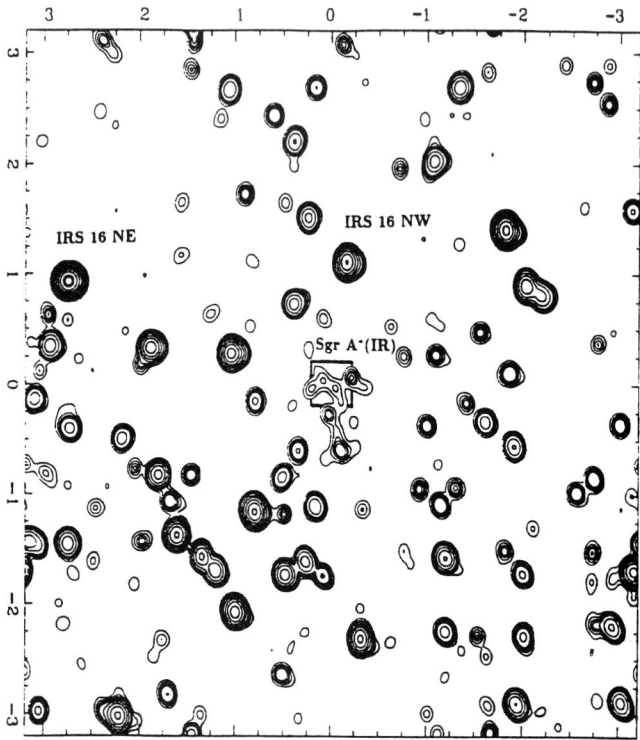

FIGURE 2. 2.2 µm speckle image of the Galactic Center. Field size is $3.2'' \times 3.2''$. From Eckart et al. (1995).

Morris 1995; Tamblyn et al. 1995) — a situation similar to that encountered in the study of infrared-luminous starburst galaxies (e.g., Rieke et al. 1980 for M82).

Observations of stellar proper motions near the Galactic Center by Eckart & Genzel (1996) give evidence for a massive black hole in the center of our Galaxy. Stellar velocity dispersions suggest a central dark mass density of $>6.5 \times 10^9$ $M_\odot$ pc^{-3}, consistent with a core collapsed cluster of stellar black holes or a single massive black hole. Is the Galactic Center a miniature counterpart of the nearest AGN, NGC1068 ($d = 12$ Mpc)?

2.2. 30 Doradus: the Rosetta Stone

High-mass star-formation regions in the Local Group of galaxies are excellent laboratories to study starbursts: their proximity (with respect to galaxies in the Hubble flow) permits detailed studies of *individual* stars, yet their distance (with respect to Milky Way clusters) makes it possible to obtain integrated properties as well. Although more luminous than the Galactic regions of Section 2.1, they are still underluminous in comparison with the starburst prototypes in Sections 2.3. Therefore I will refer to these regions, which are mostly giant H II regions, as "mini-starbursts".

Kennicutt (1984) gives a catalog of the most prominent giant H II regions in nearby galaxies, including their nebular properties. Regions studied in some detail are NGC595 and NGC604 in M33 (Hunter et al. 1996; Malumuth, Waller, & Parker 1996; Terlevich et al. 1996) and 30 Dor in the LMC (see Walborn 1991 for a review). I will take 30 Dor as the Rosetta Stone of a mini-starburst for a more extended discussion.

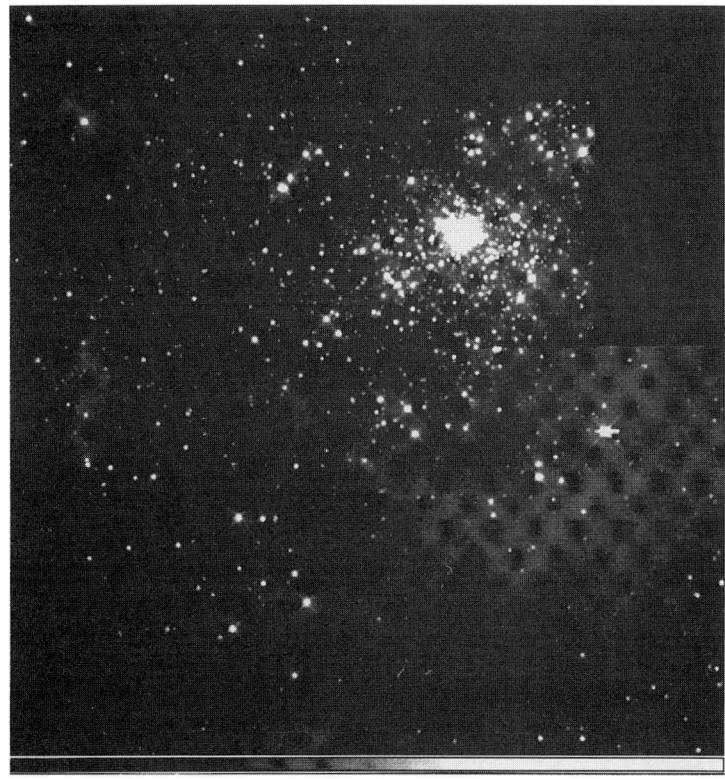

FIGURE 3. HST WFPC2 V image of 30 Dor. Field size is $100''$, corresponding to 27 pc. 30 Dor is the most powerful giant H II region in the Local Group of galaxies.

2.2.1. *The stellar content*

Numerous studies of 30 Dor (with its ionizing cluster NGC2070, whose massive center is R136) exist. Melnick (1985) classified the brightest hot stars. His work has been advanced by Walborn & Blades (1997) in their spectral classification study of 106 OB stars. A Hubble Space Telescope (HST) WFPC2 image of 30 Dor taken in the visual passband is shown in Figure 3. The image illustrates the density of stars and the complexity of the ionized gas. Crowding becomes severe for stellar spectroscopy in the central R136 region. HST allows spatially resolved aperture spectroscopy of individual stars: ultraviolet spectra of three hot, luminous stars obtained by Heap et al. (1994) are in Figure 4. The stars have temperatures above 40,000 K, luminosities around 10^6 $L_\odot$, and masses above 50 $M_\odot$. Their wind properties are particularly outstanding. The ultraviolet spectra in Figure 4 show strong P Cygni profiles of C IV $\lambda 1550$, N V $\lambda 1240$, and He II $\lambda 1640$, indicating high mass loss at rates of 10^{-5} $M_\odot$ yr^{-1} and above. These wind lines are strong enough to be detectable in even the most distant starburst galaxies, as I will discuss below.

Hunter et al. (1995) made use of the refurbished HST to detect stars down to 23.5 mag. Allowing for crowding and reddening, this corresponds to a minimum M_V of $\sim +1$, or a mass of 2.5 $M_\odot$. There is no indication for a turn-over of the luminosity function: intermediate-mass stars are found at the same frequency as high-mass stars, indicating that 30 Dor is forming stars with masses down to the detection limit. The star-formation

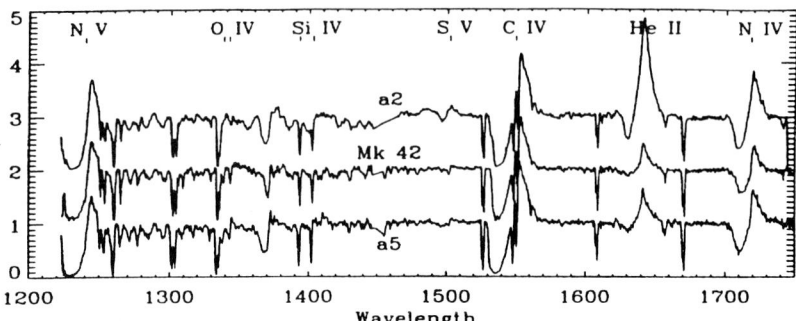

FIGURE 4. HST ultraviolet spectra of the central 30 Dor stars R136a5, Mk42, and R136a2. Note the broad stellar wind features of, e.g., N V $\lambda1240$, C IV $\lambda1550$, and He II $\lambda1640$. From Heap et al. (1994).

history in the greater 30 Dor region is complex. Walborn & Blades (1997) identify five individual stellar groups which they associate with discrete star-formation epochs. The energy output (both radiant and mechanical) is dominated by the central R136 cluster. In addition, younger ($< 10^6$ Myr) and older ($> 10^7$ Myr) populations are present (see also Walborn 1991), some of which may be related due to sequential star formation.

2.2.2. *Properties of the gas*

Kennicutt & Chu (1994) have reviewed the violent interstellar medium in 30 Dor. Supersonic velocity dispersions of about 40 km s^{-1} are measured. This is consistent with observations of other giant H II regions having comparable luminosity. In addition, spatially resolved spectra reveal the presence of multiple shells having scales of up to 100 pc and expanding at velocities of up to 300 km s^{-1} (Chu & Kennicutt 1994). The Einstein Observatory detected several sources or X-ray emitting gas which is probably shocked by stellar winds and supernovae (Wang & Helfand 1991a,b).

The interstellar environment in 30 Dor differs from that of smaller Galactic H II regions. Galactic regions, such as Trumpler 14 and 16 are not massive enough to form significant numbers of massive stars having strong stellar winds at the same evolutionary epoch, whereas in 30 Dor the *collective* effort of winds and supernovae is capable of initiating and maintaining large interstellar bubbles and supershells. The shell masses range from a few hundred to several thousand M$_\odot$. The mechanical energy requirements are around 10^{51} erg. Supernovae, and possibly stellar winds, are the likely sources of energy. The energy source of the observed macroscopic motion has profound implications. If winds and supernovae are responsible, the measured velocities are mostly indicative of the evolutionary state of the stellar population. An alternative suggestion for the observed supersonic velocities is gravitational bulk motion of virialized material (Melnick 1977). In the latter case, the velocity dispersion traces the mass of the stars and gas and could in principle serve as a powerful tool to determine luminosity, and therefore distance (Terlevich & Melnick 1981). It is under debate which of the two mechanisms prevails, but it is most likely a combination of both (Yang et al. 1996).

2.2.3. *30 Dor in perspective*

It is instructive to compare different scales within the greater 30 Dor region and relate them to the LMC as a whole. This has been done by Walborn (1991) and the results are listed in Table 2. The 30 Dor nebula extends over a few hundred parsecs and is

	Linear	Angular 52.5 kpc	1 Mpc	10 Mpc
LMC	5 kpc	5°	15'	1.5'
30 Doradus Region	1 kpc	1°	3'	20''
30 Doradus Nebula	200 pc	15'	50''	5''
30 Doradus Cluster	40 pc	3'	10''	1''
R136	2.5 pc	10''	0.5''	0.05''
R136a	0.25 pc	1''	0.05''	0.005''

TABLE 2. Characteristic dimensions in the greater 30 Doradus region. From Walborn (1991).

Region	t (Myr)	$L_{\rm UV}$ (erg s^{-1} Å^{-1})	$M_{\rm B}^{\rm clu}$	$M_{\rm B}^{\rm gal}$	$M_{\rm B}^{\rm 3Myr}$
R136	3	6×10^{37}	-11.3	-17.9	-11.3
NGC4214#1	5	2×10^{38}	-13.0	-18.8	-13.5
NGC1741B1	5	6×10^{39}	-16.7	-20.3	-17.2
NGC1569A	10	3×10^{38}	-14.1	-16.2	-15.1
NGC1705A	15	6×10^{38}	-14.0	-17.0	-15.5

TABLE 3. Comparison of R136 with other starburst regions. From Leitherer (1997).

ionized by a star cluster of linear size 40 pc. The core of the cluster is the dense stellar group R136a within 0.25 pc. These structures are resolvable from space or with image restoration techniques from the ground. Previous inadequate spatial resolution led to the claim that R136a is a single super-massive star, instead of a dense star cluster (see Walborn 1991 for a discussion). If 30 Dor were observed at 10 Mpc, only HST could perform spatially resolved imaging and spectroscopy of R136, which would subtend an angle of only 0.05''.

How does the brightness of R136 rank in comparison with other starburst regions? The ranking is in Table 3 (from Leitherer 1997). The regions are arranged by increasing age. Columns 3 – 6 list: $L_{\rm UV}$, the cluster luminosity at 1500 Å, $M_{\rm B}^{\rm clu}$, the absolute B magnitude, $M_{\rm B}^{\rm gal}$, the total absolute B magnitude of the host galaxy, and $M_{\rm B}^{\rm 3Myr}$, the B magnitude the cluster would have if observed at an age of 3 Myr. (Clusters fade with age and older clusters would be brighter if observed at an earlier epoch). R136 ranks at the bottom of the list. Note that a typical ionizing O star has an $L_{\rm UV}$ of a few times 10^{35} erg s^{-1} Å^{-1}. The total ultraviolet luminosity of R136 is produced by a few hundred such stars. NGC1741B1, a massive star cluster in NGC1741, is two orders of magnitude more luminous than R136, the mini-starburst. The two clusters NGC1569A and NGC1705A are bright but not outstanding in terms of their luminosity (compare with NGC1741B1 — but note the caveat that NGC1741B may itself be composed of individual clusters). What makes them extraordinary is their brightness relative to the host galaxy (col. 4 and 5): they provide a significant fraction of the blue and ultraviolet light of the galaxy. They become even more impressive if fading due to age effects is taken into account.

A ranking in terms of nebular properties is done in Table 4. Quantities listed are the diameter, the Hα luminosity, the corresponding number of ionizing photons, $N_{\rm L}$ ($L(H\alpha)$ [erg s^{-1}] $= 1.36 \times 10^{-12} N_{\rm L}$), the number of ionizing stars, and the mass of

	Orion	30 Dor	GEHR	NGC7714
Diameter (pc)	10	400	100–1000	600
$\log L(H\alpha)$ (erg s^{-1})	37	40.2	38–41	42
$\log N_L$ (photons s^{-1})	49	52	50–53	54
Ionizing O Stars	6	1000	10–10,000	>10,000
$\log M(HII)$ (M$_\odot$)	2–3	5.9	3–7	7

TABLE 4. Range of physical parameters in 30 Dor and other H II regions. From Kennicutt (1991).

ionized gas. The local comparison H II region is the Orion nebula, a well-studied star-formation region with only a handful of O stars (Kennicutt 1984). 30 Dor exceeds Orion by three orders of magnitude in all parameters listed. Column 4 of Table 4 gives the range of values observed for giant extragalactic H II regions. In terms of distance, most of them are within or close to the Local Group of galaxies. 30 Dor is fairly typical on this scale. The entries in the last column are for the nuclear starburst NGC7714. This galaxy is 1 to 2 orders of magnitude more powerful than 30 Dor.

2.3. Starburst galaxies: stars, clusters, gas

Detailed spectroscopic studies of large numbers of individual stars are no longer feasible beyond a few Mpc: even the visually brightest star in Trumpler 14, HD93129A (see Table 1), would appear at 1 Mpc as a $V = 18.6$ star (assuming no extinction). Obtaining data sufficient for a quantitative spectroscopic analysis of many stars would by far exceed the available resources of large telescopes. Crowded-field photometry is still feasible for distances up to about 10 Mpc, where the field-star population and (for the most nearby cases) and some of the less compact clusters can be resolved.

The closest examples of the so-called starburst galaxy class are found in this distance range. Observationally, starburst galaxies are typically selected on the basis of their emission-line spectrum (e.g., Sargent & Searle 1970), their blue optical colors (e.g., Balzano 1983), or their far-infrared colors (Soifer et al. 1987). Three major categories of starbursts emerge from this selection: (i) Galaxies and stellar systems whose optical spectrum is *H II region*-like, sometimes with an indication of an older, underlying population. Examples are H II galaxies, which may be considered scaled-up versions of 30 Dor (Terlevich et al. 1991), and amorphous irregular and blue compact dwarf galaxies (Thuan 1991). Typical masses are between 10^6 and 10^9 M$_\odot$. (ii) *Nuclear starbursts*, such as the proto-type NGC7714 (Weedman et al. 1981), whose optical morphology is similar to that shown by Seyfert galaxies but the primary energy source being a young stellar population. The main discriminant from (i) is the absence of wide-spread massive-star formation over the whole galaxy. Nuclear starbursts are typically more massive than H II galaxies. Representative values are in the 10^8 to 10^{10} M$_\odot$ range. (iii) *Infrared-luminous galaxies*, whose stellar population is enshrouded in dust. The observed infrared luminosity results from conversion of ultraviolet into infrared photons by dust absorption. These objects have the highest masses (> 10^{10} M$_\odot$), and the most extreme cases can rival QSO's in their luminosities (see Heckman 1991 for a discussion).

The observed phenomena are understandable as an episode of massive-star formation in an evolutionary phase when a significant amount of gas is converted into stars. The timescale τ_o of the exhaustion of the gas reservoir is often used for a rigorous definition of a starburst: starburst galaxies are those galaxies in which τ_o is significantly shorter

FIGURE 5. WFPC2 (PC) V image of NGC1569 obtained by the author, in collaboration with M. Clampin, G. De Marchi, L. Greggio, A. Nota, and M. Tosi. Field size is 35". The bright object in the center is the super star cluster NGC1569A.

than a Hubble time (Weedman 1987). Characteristic values for $\tau_\circ$ between 10^7 and 10^{10} yr have been estimated for blue compact dwarf galaxies (Thuan 1991) and infrared-luminous galaxies (Heckman, Armus, & Miley 1990). Such estimates are rather uncertain for several reasons. Observationally, a determination of the total star-formation rate and the gas mass are required, both of which are far from trivial to obtain. On the theoretical side, the mass return from stars to the interstellar medium (ISM) and gas infall into the starburst region from outside must be taken into account (Kennicutt 1992). Last but not least, a highly uncertain extrapolation must be made from the number of high-mass stars formed, which are accessible to direct observations, to the low-mass stars, which are often not constrained by direct measurements. Yet, it is the low-mass star-formation rate which determines $\tau_\circ$. As a consequence, a definition of a starburst via $\tau_\circ$ is often not very practical — despite its astrophysical appeal.

2.3.1. *NGC1569 — the closest starburst*

NGC1569 is often considered to be the closest local analog of an evolved starburst galaxy (e.g., Waller 1991; Vallenari & Bomans 1996). The galaxy is a Magellanic-type irregular in the M81/IC342/Maffei group (Tully & Fisher 1987). Waller (1991) gives a comprehensive summary of the main properties of this galaxy. The brightest blue stars can be resolved with ground-based telescopes (Ables 1971), leading to a distance estimate of 2.2 ± 0.6 Mpc. An HST V image of NGC1569 is reproduced in Figure 5.

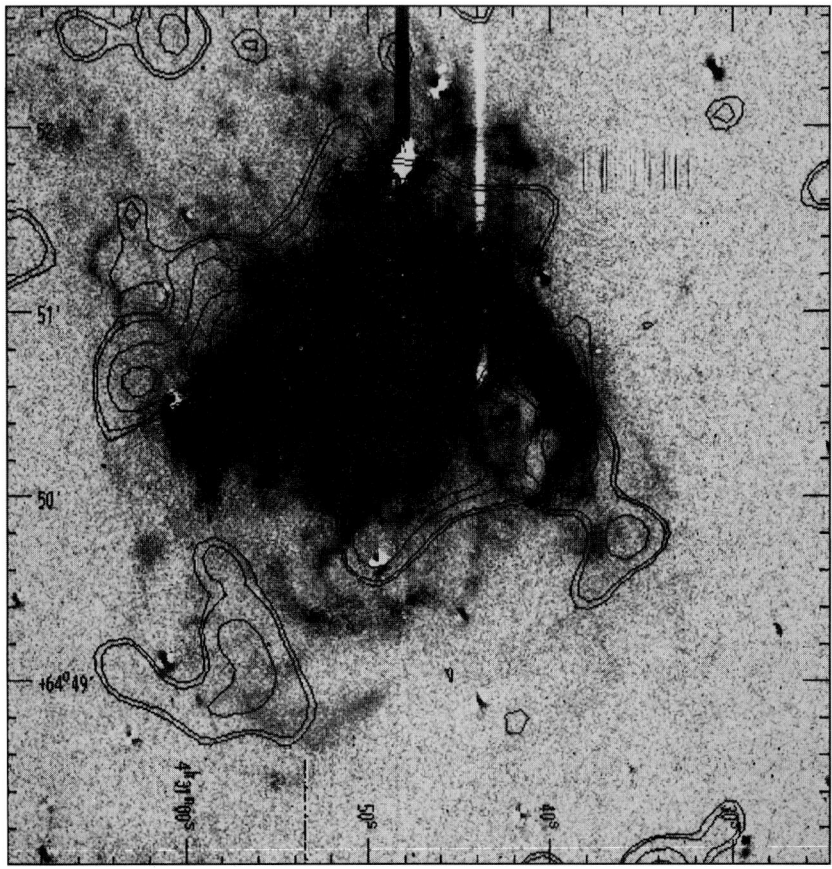

FIGURE 6. Overlay of a ROSAT HRI image (contours) on a continuum-subtracted Hα image of NGC1569 (from Della Ceca et al. 1996). The extended emission is due to cool (Hα) and hot (ROSAT) ionized gas of a galactic superwind.

NGC1569 hosts several high-density clusters, called "super star clusters" (Arp & Sandage 1985; Melnick, Moles, & Terlevich 1985), which have been suggested to be a particular mode of massive star formation (O'Connell, Gallagher, & Hunter 1994). NGC1569A of Table 3 is the brightest super star cluster in NGC1569. Super star clusters may have properties similar to those expected for young globular clusters (Meurer 1995) if they contain a substantial number of less massive (~ 1 M$_\odot$) stars as well. HST imaging and ground-based spectroscopy of the super star clusters are discussed in De Marchi et al. (1997) and González-Delgado et al. (1997).

Deep Hα and X-ray images of NGC1569 (Figure 6) show kiloparsec-scale emission along its minor axis (Heckman et al. 1995; Della Ceca et al. 1996). The starburst in NGC1569 (and the central super star cluster in particular) appears to be driving an outflow of the interstellar medium on a global scale. The dynamical age is $\sim 10^7$ yr. The energetics of the outflow are such that most of the galaxy's ISM could be ejected. The starburst in NGC1569 is truly global, as opposed to the 30 Dor mini-starburst, which does not affect the overall ISM of the LMC.

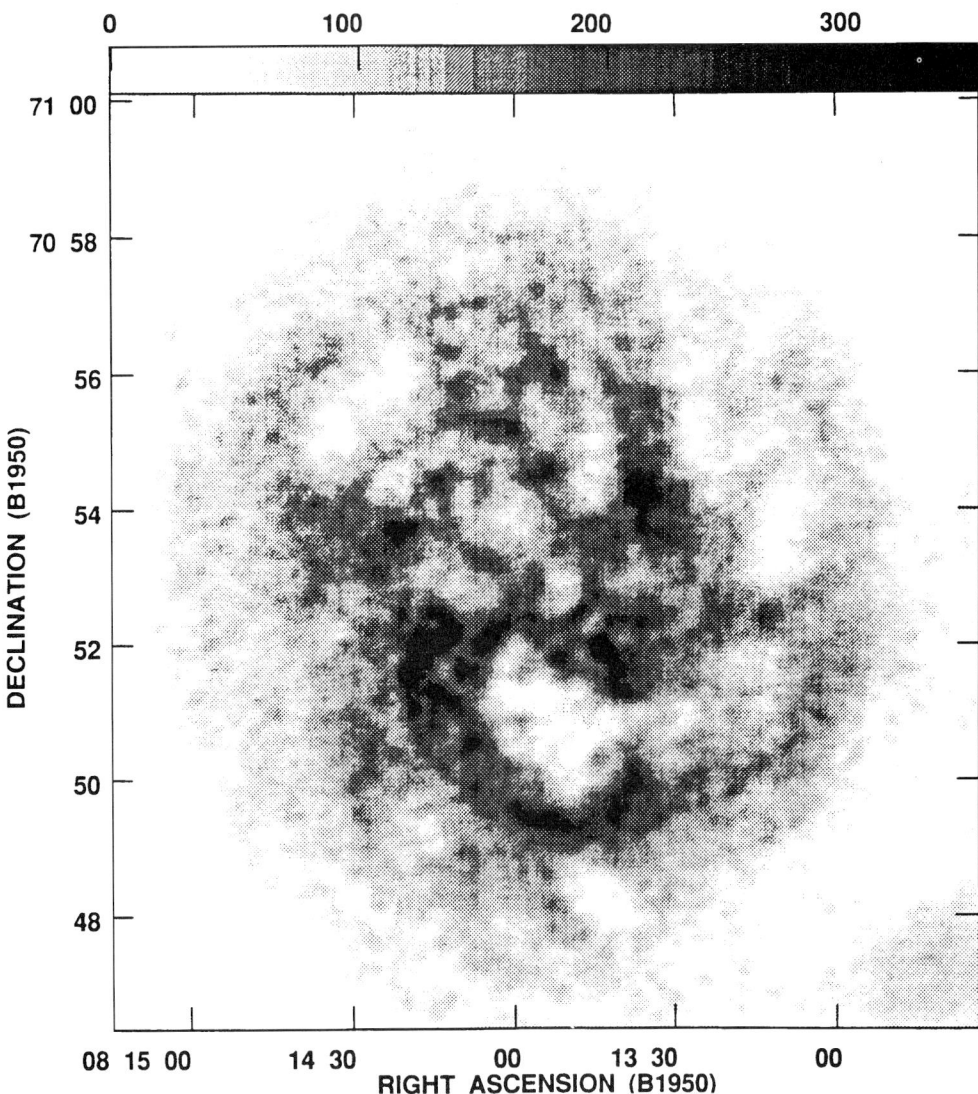

FIGURE 7. H I map of Holmberg II (= UGC4305) obtained with the VLA by Puche et al. (1992). About 50 individual H I holes are detected. The most prominent hole at $\alpha = 8^h 13^m 50^s$ and $\delta = 70°51'$ has a characteristic energy of 2×10^{53} erg.

2.3.2. Stars and ISM in nearby dwarf galaxies

There is increasing observational evidence for the ubiquitous presence of outflows from galaxies driven by massive stars. NGC1705 is another well-studied prototype. A case study has been performed by Meurer et al. (1992). NGC1705 shares two key characteristics with NGC1569: nebular emission around the galaxy suggests a galactic superwind, and the nucleus of NGC1705 hosts a luminous super star cluster (Meurer et al.; O'Connell et al. 1994). Ground-based work reveals a ~1 Gyr old field population which defines the NGC1705 morphology. Nearly half of the optical/ultraviolet light is contributed by the central super star cluster (NGC1705A in Table 3). The cluster is the likely power source for the observed bipolar galactic outflow. Although details depend on the specific star formation history, it is likely that the bipolar outflow is enriched in metals by supernova explosions and stellar winds.

Marlowe et al. (1995) did a detailed multi-waveband search for outflows in nearby dwarf galaxies with centrally concentrated star formation. Kiloparsec-scale superbubbles and filaments oriented along the galaxy minor axis were generally found. In each case the mechanical energy released by stellar winds and supernovae appears to power the outflows.

Neutral hydrogen maps of galaxies provide further evidence for massive stars/ISM interactions. *Edge-on* galaxies are ideally suited to search for H I or H II cloud complexes at large vertical distances. Results for NGC891 (Dettmar 1990; Rand, Kulkarni, & Hester 1990; Rupen 1991) suggest structures very similar to those found optically. Holes and cavities in the H I layer or high-velocity gas are seen in more *face-on* galaxies such as M31 (Brinks & Bajaja 1986), M33 (Deul & den Hartog 1990) or NGC6946 (Kamphuis & Sancisi 1993). Figure 7 is an H I map of the dwarf galaxy Holmberg II. Puche et al. (1992) studied the properties of about 50 H I holes detected in this galaxy. 1' corresponds to 900 pc at the distance of Holmberg II so that the holes have typical diameters of several hundred pc. Most likely, the holes are the result of sequential star-formation events whereby photoionization, winds, and supernovae shape the ISM. Van der Hulst & Kamphuis (1991) summarized the parameters of typical holes in nearby galaxies. Ages of a few times 10^7 yr and energies of $10^{52} - 10^{54}$ erg are consistent with a blow-out due to the effect of massive stars.

2.3.3. *Unresolved populations: luminous starbursts*

The galaxies I discussed so far have typical distances of a few Mpc — a factor of 100 more distant than the local Rosetta Stone 30 Dor. The 30 Dor cluster (apparent size 3', see Table 2) would subtend an angle of several arcseconds if observed at that distance. What would 30 Dor and the LMC look like if they were yet another factor of 10 farther away? Figure 8 gives the answer. NGC1741 is a luminous irregular at a distance of 50 Mpc, highly reminiscent of the LMC when seen on wide-angle photographs. It has two luminous H II regions at the NW end of the bar (NGC1741B1 of Table 3 is one of them). The bar population is largely unresolved. If stars were projected from NGC1569 to NGC1741, they would appear about 6.5 mag fainter; an isolated 100 $M_\odot$ star would be barely detectable. NGC1741 is classified as a Wolf-Rayet (WR) galaxy: it shows broad *stellar* He II lines in its spectrum (Conti 1991; Vacca & Conti 1992). WR stars are a brief (10^5 to 10^6 yr), late evolutionary phase of a massive star. They will be discussed in full detail in Section 5.1.2. NGC1741 contains about 10^3 WR stars, based on an analysis of its spatially integrated light by Vacca & Conti (1992) — more WR stars than known individually in all galaxies of the Local Group combined. The galaxy has a highly disturbed optical morphology with two starburst centers, possibly arising from a galaxy merger, most likely as a result of interaction with other members of Hickson Compact Group 31 (Hickson, Kindl, & Auman 1989; Rubin, Hunter, & Ford 1990).

Individual stars (or even clusters) can no longer be investigated at distances of ~50 Mpc and beyond. Stellar and nebular properties are derived from the integrated light. Such techniques will be discussed in Section 4. Generally the results point at the same phenomena one observes more locally, but often at a grander scale. This is of course a selection effect: Galaxies such as NGC1569 would be rather inconspicuous at 100 Mpc, and more luminous and spectacular species are studied instead.

Lehnert's (1992) census suggests that all galaxies in the local universe with high infrared luminosities ($\log(L_{\rm IR}/L_\odot) > 10.5$), large infrared excesses ($L_{\rm IR}/L_{\rm opt} > 2$), and

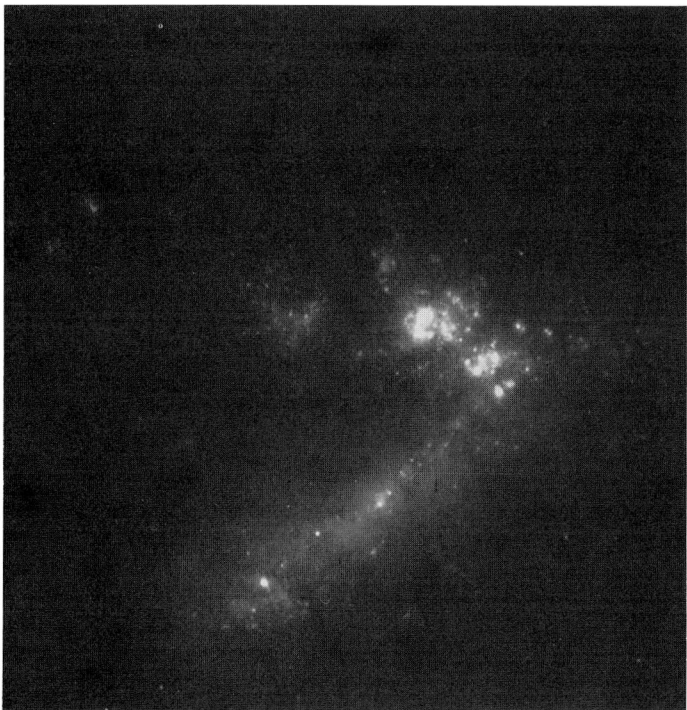

FIGURE 8. Composite B, V, I WFPC2 (PC) image of NGC1741 obtained by P. Conti, in collaboration with the author and W. Vacca. Field size is 35″. North is up and east to the left. NGC1741 is luminous irregular galaxy at a distance of 50 Mpc.

warm far-IR colors ($L_{60}/L_{100} > 0.5$) have outflows, often called galactic superwinds. These are scaled-up versions of the phenomenon observed in NGC1569. A detailed study of the superwind properties and the underlying massive-star population in a sample of infrared-luminous galaxies was done by Heckman, Armus, & Miley (1990). The most luminous object in their sample is the IRAS galaxy IRAS00182-7112 at a distance of $\sim$1.3 Gpc. Using its infrared luminosity, Heckman et al. estimate a star-formation rate of $\sim$2000 $M_\odot$ yr^{-1}. In order to gauge this value, a comparison with NGC1569A is instructive. NGC1569A has a total mass of approximately $10^5 - 10^6$ $M_\odot$ (Ho & Filippenko 1996). Assuming star formation occurred quasi-instantaneously within 1 Myr (a likely assumption given the small size of the cluster and the observed timescale, e.g., in 30 Dor; Malumuth & Heap 1994), a star-formation rate of $\sim$1 $M_\odot$ yr^{-1} is found. IRAS00182-7112 forms stars at a rate three orders of magnitude higher! (Note that this rough estimate of the star-formation rate in NGC1569A is very uncertain!) However, it is not known how concentrated the star formation in IRAS00182-7112 is. Most likely it is spread over hundreds of parsecs whereas in NGC1569A star formation is concentrated within about 1 pc. In terms of its star-formation *density*, NGC1569A may be spectacular after all.

2.4. *Implications for distant galaxies*

I will conclude with a brief look at the most distant star-forming galaxies known. They are the endpoint of this section's cosmic voyage, which began in the Carina star forming region. Although beyond the principal scope of these lectures, application of stellar

FIGURE 9. Composite of the Hubble Deep Field, constructed from individual F450W, F606W, and F814W images. From Williams et al. (1996).

physics to local galaxies, they are the final milestones that help us better define and understand the sizes and scales involved in the subsequent sections.

IRAS00182-7112 is the most distant object I discussed so far. It has a redshift of 0.33 (Armus, Heckman, & Miley 1990). Few starburst galaxies have been studied in detail at larger redshifts. The results of the major redshift surveys, such as, e.g., the Universidad Complutense de Madrid Survey (Zamorano et al. 1994, 1996), the Canada-France Redshift Survey (Lilly et al. 1995), or the Southern Sky Redshift Survey (da Costa et al. 1988, 1989, 1991) are limited to the redshift range $0 < z < 1$. On the other hand, blank-sky surveys for strong Lyα-emitting primeval galaxies at high redshift ($z \approx 3$) have not produced large numbers of star-forming galaxy candidates (Pritchet & Hartwick 1990; Djorgovski & Thompson 1992).

Steidel & Hamilton (1992, 1993) have obtained multicolor broadband observations of the restframe ultraviolet stellar continuum of Lyman break candidates. Deep spectroscopy has led to the discovery of a group of bona fide star-forming galaxies at cosmological distances around $z \approx 3$ (Steidel et al. 1996). The luminosity of the star-forming galaxies at $z \approx 3$ can be compared to that of NGC1741B1: Steidel et al. find $L_{\rm UV} = 10^{41} h_{50}^{-2}$ erg s^{-1} Å^{-1} for $q_0 = 0.5$, and somewhat larger for smaller q_0. The value for NGC1741B1 is $L_{\rm UV} \approx 6 \times 10^{39}$ erg s^{-1} Å^{-1} (Table 3), about 10 times fainter. Note

C. Leitherer: *Massive Stars and the ISM*

that Steidel et al. call these galaxies "normal" star-forming galaxies — they are not a powerful starburst episode when a significant fraction of gas is converted into stars. Other star-forming galaxies at cosmological distances have been detected by Yee et al. (1995) and Ebbels et al. (1996).

The Hubble Deep Field (HDF) is an HST program to image an indistinguished field at high Galactic latitude in four passbands as deeply possible (Williams et al. 1996). A composite of three passbands is reproduced in Figure 9. The image shows a plethora of distant field galaxies, many of them having complex and disturbed morphologies. Madau et al. (1996) applied color selection criteria to construct a sample of star-forming galaxies in the redshift range $2 < z < 4.5$. Spectroscopy of these star-forming candidates is required for analysis, an effort reaching the limits of our current observational capabilities. Possibly these observations will find the first generation of massive stars ever formed in the universe.

3. Massive stars in resolved populations

In this section I will discuss techniques to determine the properties of stars in galaxies which are close enough that their populations can still be resolved into individual stars. I will focus on the radiative energy output of stars, in other words, on their spectral energy distribution. The discussion on the mechanical energy output is deferred to Section 5 for reasons of clarity. The methods described here do of course simultaneously address stellar-wind properties as well.

3.1. *Initial mass function and star-formation history*

A new-born star is completely determined by its chemical composition, mass, and rotation velocity. The chemical composition is separated into hydrogen, helium, and heavy element mass fractions X, Y, and Z, respectively. The Sun has approximately $X = 0.68$, $Y = 0.30$, and $Z = 0.02$ (e.g., Schaller et al. 1992). Z includes all elements heavier than helium. There is a fixed relation between Y and Z due to the condition that the first generation of stars had a primordial helium abundance of $Y_\mathrm{p} \approx 0.24$ (Terlevich, Skillman, & Terlevich 1993) and $Z = 0$. Subsequent stellar generations increased both the helium and heavy-element abundance such that the present-day abundances of Y and Z are produced. The relation between Y_p and the present-day helium abundance can be parameterized as

$$Y = Y_\mathrm{p} + \frac{dY}{dZ} Z. \qquad (3.1)$$

$dY/dZ = 3$ reproduces the chemical composition of the Sun given above. Therefore the chemical abundance can (in principle!) be reduced to the heavy element abundance Z alone. Note that the value of dY/dZ is under debate (e.g., Pagel 1992).

Currently no self-consistent grid of evolutionary models for massive stars are available that take rotation into account. Therefore rotation can not be considered here. Evidence is increasing that some aspects of stellar evolution, such as heavy-element abundances, are rotation-dependent (e.g., Fliegner, Langer, & Venn 1996) but the principal parameters governing stellar evolution are M and Z.

For a given metallicity (as Z is often referred to), the distribution of stellar masses at birth (the IMF) is the fundamental quantity for the evolution of a stellar population. Excellent reviews of the conceptual approach to describe the IMF have been given by Tinsley (1980) and Scalo (1986). I am essentially following these papers in my discussion.

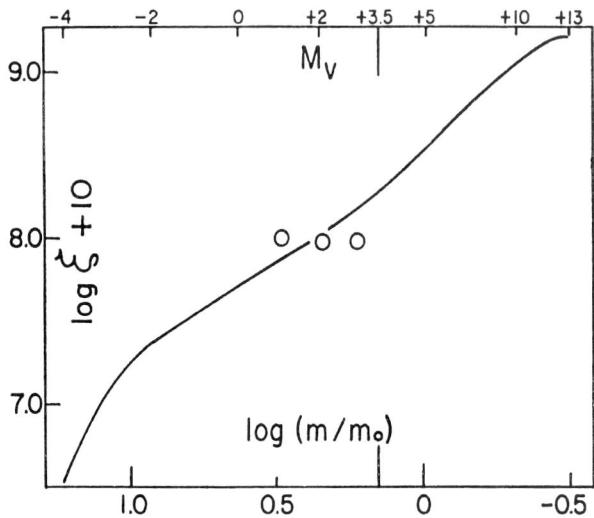

FIGURE 10. Salpeter's (1955) determination of the initial mass function in the solar neighborhood. Plotted is the IMF parameterized as $\xi(m) \propto m^{-x}$. Salpeter found an exponent of $x = 1.35$.

The number of stars formed in the mass interval dm and time interval dt is

$$\phi(m)\psi(t)dm dt. \tag{3.2}$$

$\psi(t)$ is the total mass of stars formed per unit time, also called the star-formation rate. It has units $M_\odot$ yr^{-1}. $\phi(m)$ can be time-dependent and is often normalized such that

$$\int_0^\infty m\phi(m)dm = 1. \tag{3.3}$$

$\phi(m)$ is called the IMF. It is customary to parameterize the $\phi(m)$ as a power law

$$\phi(m) \propto m^{-\alpha}, \tag{3.4}$$

where α is called the slope of the IMF. The stellar numbers are counted on *linear* mass intervals. There are alternative ways to express the IMF. Sometimes the number of stars is counted per *logarithmic* mass interval. For clarity I call this function $\xi(m)$. The relation between $\phi(m)$ and $\xi(m)$ is

$$\phi(m) = \log e\, m^{-1}\xi(m). \tag{3.5}$$

If $\xi(m)$ is parameterized in terms of a power law, the common notation is

$$\xi(m) \propto m^{-x} \tag{3.6}$$

with $x = \alpha - 1$. Note that the constant of proportionality in eqs. (3.4) and (3.6) differs by a factor of $\log e$. The power law exponent x is used in Tinsley's (1980) work. Finally, and to add to the confusion, eq. (3.6) can be redefined via a power law exponent parameter of opposite sign, Γ, with $\Gamma = -x$.

In a seminal paper, Salpeter (1955) derived the IMF in the solar neighborhood. (He called the IMF "original" mass function). His paper — only seven pages long — had an enormous impact on the field. In Figure 10 his IMF is reproduced. Salpeter found an initial mass function with $\alpha = 2.35$ (or $x = 1.35$, or $\Gamma = -1.35$). This value is close to modern results for the IMF in local high-mass star formation regions and also in starbursts (see Moffat 1997; Leitherer 1997).

Massive stars have evolutionary timescales that are short on a galactic scale. Therefore the IMF is usually considered time-invariant when massive stars are studied, and any variability of the number of stars formed is expressed in terms of a variable star-formation rate. The royal goal then is to reconstruct the IMF and the star-formation rate from the observed HRD or color-magnitude diagram (CMD).

3.2. Evolutionary synthesis of resolved populations

The modeling technique used to describe the properties of stellar populations is called evolutionary synthesis. Numerous reviews on the subject are available. As a starting point, the conference proceedings edited by Leitherer, Fritze-v. Alvensleben, & Huchra (1996a) and the database described by Leitherer et al. (1996b) are recommended. The concept of evolutionary synthesis is simple (but details can be complicated). The goal is to generate a theoretical HRD/CMD and related parameters that matches the observations in as many aspects as possible. The models are essentially structured in four components:

• Stars form at a specified rate (Section 3.2.1).
• They are distributed in the HRD/CMD along the ZAMS with a mass-dependent weighting function (Section 3.2.2).
• Evolutionary models describe the time dependence of the stellar mass, luminosity, effective temperature, and chemical composition at the surface (Section 3.2.3).
• Stellar energy distributions and other properties are assigned to each position in the HRD or CMD (Section 3.2.4).

The output product is a synthetic HRD/CMD for comparison with observations. In Section 4, I will introduce one further step in the concept: integration over the HRD/CMD to simulate the properties of unresolved populations.

3.2.1. Star-formation rate

Following the notation I introduced in Section 3.1, I parameterize the time-variable star-formation rate. Stars form out of interstellar gas at a rate

$$\psi(t) = \psi(t_\circ) e^{-t/\tau}, \qquad (3.7)$$

where t and τ are the age and the duration of the star-formation event (also called "burst"), respectively. $t_\circ$ denotes the epoch of the onset of the star-formation event. Two limiting cases are of particular interest. If the burst duration is short compared to the age ($\tau < t$), most stars are formed early at some epoch, with little subsequent star formation. Elliptical galaxies are close to this case. Most of their gas was converted into stars with high efficiency in an initial starburst. Subsequent star formation decreased exponentially to the currently observed low levels. See Kennicutt's (1995) review of star-formation rates in different types of galaxies.

Massive stars have evolutionary timescales on the order of a few million years. If $\tau \ll t$, even the most massive stars can be considered to have formed coevally at $t \approx 0$. This is called an instantaneous burst. For all practical purposes, star formation is negligible at $t > 0$ in this case so that the strength of the burst is more conveniently parameterized by the total mass converted into stars at $t = 0$. This would be the case $\tau \to 0$ in eq. (3.7). A star-formation law of this type is generally found to apply in individual clusters of old, intermediate, or young age (Meynet, Mermilliod, & Maeder 1993).

Another limiting case of the star-formation law assumes that the star-formation rate is constant with time. Such a law is characteristic for large systems, like the disk of our Galaxy (Mezger 1987). Star formation proceeding at a constant rate can always be approximated theoretically as a series of instantaneous events. Observationally, this may resemble a system composed of numerous individual starburst regions with an age range comparable to τ.

Irregular galaxies are often more complicated in terms of their star-formation history than ellipticals ($\psi(t) \propto e^{-t/\tau}$), large spirals galaxies ($\psi(t) \approx$ constant), or starburst clusters ($\psi(t) \propto \delta$-function). They can experience short episodes of elevated star formation after long periods of quiescence (Gallagher 1996).

For modeling purposes, the two cases of an instantaneous burst and a constant star formation will always bracket the real astrophysical case.

3.2.2. Initial mass function

The time-independent IMF of the stellar population is parameterized via eq. (3.4), and the power law is defined between the upper and lower cut-off masses, $M_{\rm up}$ and $M_{\rm low}$, respectively. $M_{\rm up}$ and $M_{\rm low}$ are identified with the most and least massive stars formed in the population, respectively. Typical values are 100 $M_\odot$ for $M_{\rm up}$ and 0.1 $M_\odot$ for $M_{\rm low}$, respectively (see Section 2). The normalization is determined by the total mass converted into stars. If C is the proportionality factor in eq. (3.4), the total number of stars between $M_{\rm up}$ and $M_{\rm low}$ is found from

$$N_{\rm tot} = C \int_{M_{\rm low}}^{M_{\rm up}} m^{-\alpha} dm. \qquad (3.8)$$

Correspondingly, the total mass of stars in the same mass interval is obtained by integrating over the IMF-weighted stellar mass:

$$M_{\rm tot} = C \int_{M_{\rm low}}^{M_{\rm up}} m^{1-\alpha} dm. \qquad (3.9)$$

Note that most of the mass which is converted from gas in stars is locked into low-mass stars due to the functional behavior of the IMF. For an IMF with $\alpha = 2.35$, stars with masses between 1 and 10 $M_\odot$ contribute six times more mass to $M_{\rm tot}$ than all stars above 40 $M_\odot$. A useful tabulation of the number of stars distributed over different mass intervals at high and low masses for a Salpeter IMF has been published by Zinnecker (1995).

Conversely, most of the stellar *light* is provided by the most massive stars. If $L \propto m^3$ (the average slope between 1 and 100 $M_\odot$, but note that the relation flattens at the highest masses), one has $L \propto \int m^{3-\alpha} dm$, with $\alpha = 2.35$. This estimate applies to a ZAMS population with no evolution. If evolved stars are present as well, the luminosity depends in addition on an integral over lifetime. In anticipation of the results of Section 5, I include an estimate for the mechanical energy output on the ZAMS. O-star mass-loss rates ($\dot{M}$) are heavily weighted towards the highest luminosities ($\dot{M} \propto L^{1.6}$; Lamers & Cassinelli 1996, their eq. (8)). Therefore the energy output from a cluster of O stars following a Salpeter IMF is dominated by the most massive stars.

3.2.3. Stellar structure and evolution models

Stellar structure and evolution models are reviewed in this volume by C. Chiosi. Therefore I can be brief. For the purpose of evolutionary synthesis, such models provide a description of the time-dependence of L, M, $T_{\rm eff}$, and surface chemistry for a set of initial stellar masses and metallicities. Starting on the ZAMS, the models follow the stars in the HRD until they form either a black hole or explode as a supernova. Important issues are:

- The models include convective overshooting from the convective cores and stellar mass loss.
- Mass-loss rates are crucial in almost all stellar phases.
- The influence of metallicity enters directly, viz. in the form of modified opacities, and indirectly, viz. by metallicity dependent mass-loss rates.
- WR stars are the low-mass descendants of O stars. Their positions in the HRD must be corrected for the effects of stellar winds. The effective radius of WR stars becomes larger due to optical depth effects, and T_{eff} is smaller.
- In general, three evolutionary scenarios can occur. The most massive stars evolve from the main-sequence (MS) into blue supergiants (BSG) and LBV's, and then become WR stars:

$$MS \to BSG \to LBV \to WR.$$

Less massive stars having smaller mass-loss rates evolve into red supergiants (RSG) after their main-sequence, BSG, and LBV phases before becoming WR stars. This evolutionary channel is quite uncertain:

$$MS \to BSG \to LBV \to RSG \to WR.$$

Supergiants of even lower mass do not lose enough surface material to make the products of nuclear processing visible at the stellar surface such that the spectral appearance would change significantly. They never enter the WR phase:

$$MS \to BSG \to RSG.$$

The critical masses dividing the evolutionary models into these scenarios are Z dependent. Notice that WR stars do not descend from red supergiants if $Z < Z_\odot$ in these models, and that observational evidence for any evolutionary connection between RSG's and WR's is weak. These sequences can be easily identified in Figure 1. A full discussion of these and other models can be found in the conference proceedings edited by Leitherer et al. (1996a).

Some central stars of planetary nebulae resemble "classical" WR stars in their spectroscopic properties (e.g., Tylenda, Acker, & Stenholm 1993). They are less luminous and are supposed to evolve from red giants. They are unrelated to high-mass populations.

3.2.4. *Stellar energy distributions*

Stellar energy distributions are required to compute the "light output" of each stellar position in the HRD. "Light output" can be the line spectrum or simply a color, such as $(B-V)$. For instance, transformation from the theoretical HRD to the observed CMD makes it necessary to convolve the computed stellar energy distribution with the appropriate filter profiles (e.g., Aparicio et al. 1996).

R. Kudritzki covers the state of the art of stellar atmospheres in his contribution to the volume. Olofsson (1996) reviewed stellar atmospheres from the point of view of evolutionary synthesis. Hot star atmospheres have to face two challenges: line-blanketing and non-LTE effects, both complicated by strong stellar winds. In the simplest case, the light output is parameterized in terms of a spectral type, and the model predicts stellar numbers per mass interval and their spectral types during evolution.

The spectra of stars on the main-sequence and in pre-WR phases can often be represented by Kurucz's (1992) plane-parallel blanketed LTE atmospheres. They are the most widely used model set for synthesis work. Kurucz's models are not optimized for the coolest stars with significant molecular opacities. A hybrid grid with Kurucz's (1992) models and tailored cool-star atmospheres was prepared by Lejeune, Buser, & Cuisinier (1997).

The continuum of stars with very strong winds (WR stars) cannot be described by static, plane-parallel model atmospheres. Expanding, spherically extended non-LTE models are required. Examples for such models are those of Hillier (1988, 1989), Gabler et al. (1989), Schmutz, Hamann, & Wessolowski (1989), or Schaerer & de Koter (1997).

As an alternative to theoretical models, observed spectral energy distributions can be used at wavelengths that are accessible for ground-based observations. Examples are libraries for the ultraviolet by Fanelli et al. (1992), for the optical by Jacoby, Hunter, & Christian (1984), and for the infrared by Lançon & Rocca-Volmerange (1992).

3.2.5. *Synthesis*

Evolutionary synthesis makes a prediction for the spectrum (and other observables) of a stellar population, with the star-formation history (IMF, star-formation rate, age, etc.) as the free parameter. If agreement with observations is found, constraints on the star-formation history can immediately be derived. This makes evolutionary synthesis complementary to *population synthesis*, which attempts to maximize the agreement between the synthetic and observed spectrum — irrespective of the astrophysical plausibility of the solution. A comparison of the two methods is given by Fritze-v. Alvensleben (1993).

The predictive power of evolutionary synthesis is largely dependent on the reliability of the stellar models used as ingredients. There is, however, an inverse feedback between evolutionary synthesis and stellar astrophysics. Failure of evolutionary synthesis predictions can sometimes be traced back to a deficiency in the stellar models. An example is the ratio of WR over O stars in Local Group galaxies which is observed to be higher than predicted by evolution models with standard mass-loss rates. This may suggest, e.g., that WR mass-loss rates higher than observed are required (Meynet 1995) or that the effect of binary evolution must be included (Vanbeveren, Van Bever, & De Donder 1997; see also Schaerer & Vacca 1996).

Of course there are caveats as well. Evolutionary synthesis is a hybrid approach, partly theoretical and partly empirical, and it is often difficult to make an objective assessment of the uncertainties. It is also worth recalling that evolutionary and spectral synthesis can never be unique. Agreement between model and observations does *not prove* a particular model is correct. All that can be derived is *consistency* between model and observations. Often, however, a number of independent constraints consistently point at the same solution, lending some credibility to the predictive power of the method.

3.3. *HRD and CMD analysis of single stellar populations*

I define single stellar populations as having no significant age spread during formation. This is the case $\tau \to 0$ in eq. (3.7). The duration of the star-formation event must be less than the evolutionary timescales of the most massive stars, which is about 3 Myr. This condition is met in compact stellar clusters. Clusters in the Milky Way and in nearby galaxies can be studied photometrically and spectroscopically. Spectra allow the determination of reliable effective temperatures and bolometric luminosities, and observational HRD's can be generated. Photometry alone is used for fainter populations to construct color-magnitude diagrams.

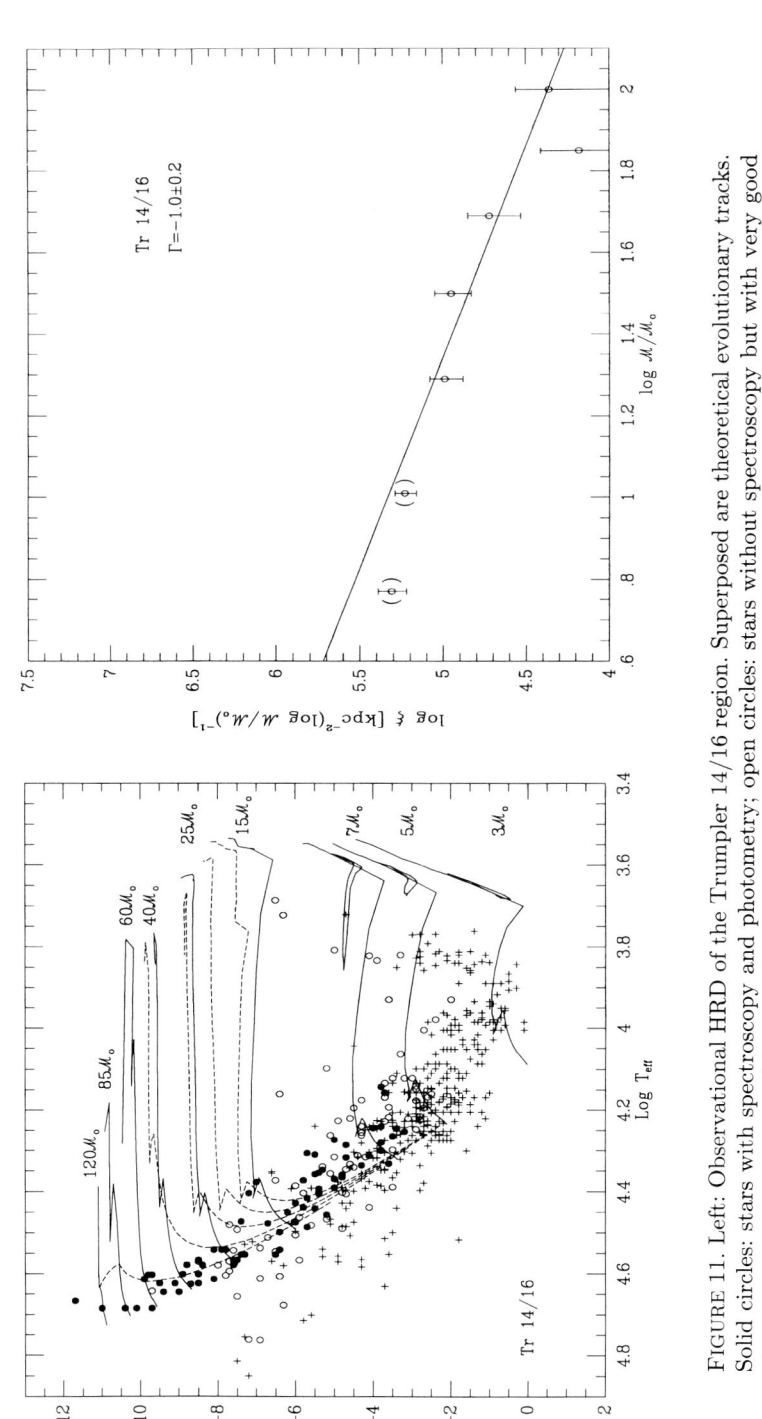

FIGURE 11. Left: Observational HRD of the Trumpler 14/16 region. Superposed are theoretical evolutionary tracks. Solid circles: stars with spectroscopy and photometry; open circles: stars without spectroscopy but with very good photometry; crosses: stars with poor photometry only. Right: Derived IMF for Trumpler 14/16. From Massey *et al.* (1995a).

3.3.1. High-mass star formation in the Galaxy

The most comprehensive survey of high-mass star formation regions in the Galaxy was done by Massey, Johnson, & DeGioia-Eastwood (1995a). They obtained wide-field CCD imaging and spectroscopy to do photometry and spectral classification of massive stars in 13 Galactic clusters. An example is in Figure 11. The figure shows the observed HRD and the derived IMF for Trumpler 14/16 (see also Table 1). Trumpler 14/16 is one of the richest Galactic young clusters, yet the upper part of the HRD is rather sparsely populated. Massey et al. find 82 stars more massive than 10 $M_\odot$ and only a few above 60 $M_\odot$. Massive stars are rare! The most massive stars have masses around 100 $M_\odot$. Essentially all stars with good data are close to the ZAMS — consistent with the young age of the cluster.

The derived IMF has a slope of $\Gamma = -1.0 \pm 0.2$, which is close to Salpeter's value. The derived age is less than 3 Myr, suggesting that even the most massive stars theoretically expected have not yet evolved into supernovae or black holes. Note the large statistical errors of $\xi(m)$ for the highest masses due to small number statistics. Other clusters in the Milky Way give similar results: the IMF is Salpeter-like, and most massive stars are born during a period $\tau < 3$ Myr.

3.3.2. 30 Doradus

Next I turn to the Rosetta Stone 30 Doradus. Parker & Garmany (1993) established an observational HRD (Figure 12). 30 Dor is more than an order of magnitude more massive (in terms of its high-mass star content) than Trumpler 14/16 and has a correspondingly larger number of stars. The upper end of the HRD is well-defined. The richness of 30 Dor allows IMF studies of sub-regions. Parker & Garmany find indications for spatial variations of the IMF slope, but uncertainties are large. The mean IMF slope is $\alpha = 2.5 \pm 0.2$, again similar to the Salpeter value. Ground-based spectroscopy is primarily sensitive to the most massive stars with masses above ~ 30 $M_\odot$. Deep HST imaging and crowded-field photometry is complementary: the lack of spectroscopic information prevents mass determinations of the most massive stars due to color degeneracy (see Section 3.4.1) but photometric analysis of fainter, less massive stars is possible with HST. The CMD study of Hunter et al. (1995) gives evidence for a continuation of the high-mass IMF towards intermediate masses: a Salpeter-like IMF was found down to 2.5 $M_\odot$.

Malumuth & Heap (1994), using HST UBV photometry, compared the properties of 30 Dor within and outside R136. They find a flatter IMF in the central region of R136. ("Flatter" means a shallower slope and therefore an excess of very massive stars). This could suggest mass segregation due to dynamical evolution (cf. Binney & Tremaine 1987). The relaxation timescale (i.e. the timescale for the most massive stars to diffuse to the cluster center) is rather uncertain but can be shorter than the age of the R136 cluster (~ 3 Myr). Therefore mass segregation as a consequence of dynamical evolution is possible. Alternatively, the star-formation process itself may favor the center of the cluster as the location of the most massive stars if the most massive stars form out of the densest molecular clouds (Larson 1991).

3.3.3. Young clusters in the Magellanic Clouds

The Magellanic Clouds contain numerous spectacular young, massive clusters other than 30 Dor itself. The giant H II region N11 has the second-largest Hα luminosity in the LMC after 30 Dor. A detailed study of the stellar and nebular morphology and of the stellar content was performed by Walborn & Parker (1992) and Parker et al. (1993). A reproduction of an Hα photograph is in Figure 13. Two massive clusters shape the

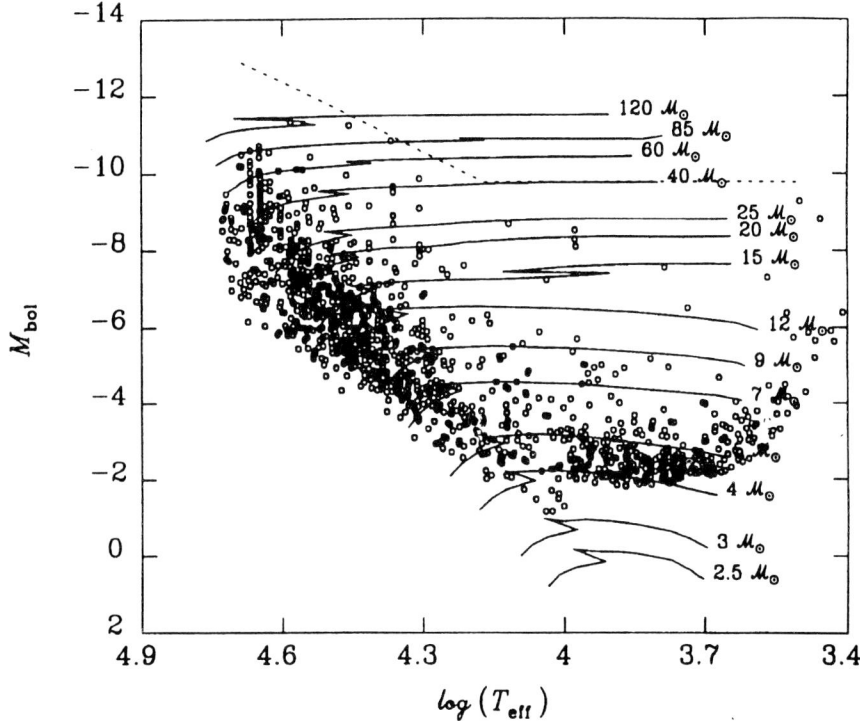

FIGURE 12. Observed HRD of 30 Dor (from Parker & Garmany 1993). 1230 stars are plotted. The diagram contains stars with photometry and spectroscopy, and with photometry only.

morphology in terms of the light (via ionizing photons) and mass (via mechanical wind power) distribution: LH9 is located in a central cavity. Most likely, the cavity is the result of the combined action of stellar winds sweeping out the interstellar medium in N11. In contrast, LH10, the second cluster is completely embedded in gas, suggesting that little of the ambient interstellar gas has been removed by winds.

Parker et al. (1993) analyzed the stellar content of both clusters by comparing them to evolutionary models (Figure 13). LH9 is clearly more evolved than LH10 and has fewer very massive stars. The age difference and the morphology suggest a sequential star-formation event. LH9 formed first and created the cavity due to strong stellar winds when the first post-main-sequence stars appeared. A second generation of stars (LH10) was triggered by LH9 in the periphery of its H II region. LH10 has an age of less than 3 Myr and is still too young to develop powerful post-main-sequence winds: its ambient gas has not yet been swept away. A key point of this study is the determination of an age scale. Analysis of the HRD population provides an evolutionary "clock" which can be used to time sequential star formation.

Hill, Madore, & Freedman (1994) and Massey et al. (1995b) give a summary of the massive stellar content of other prominent young clusters in the Magellanic Clouds. Again, rather similar IMF slopes are found.

3.4. Composite populations

So far I considered *single* populations. They are observed, e.g., in dense clusters, whose star formation event is short in comparison with the lifetime of the stars. More complex star-formation histories are observed in larger stellar systems. *If* star formation proceeds

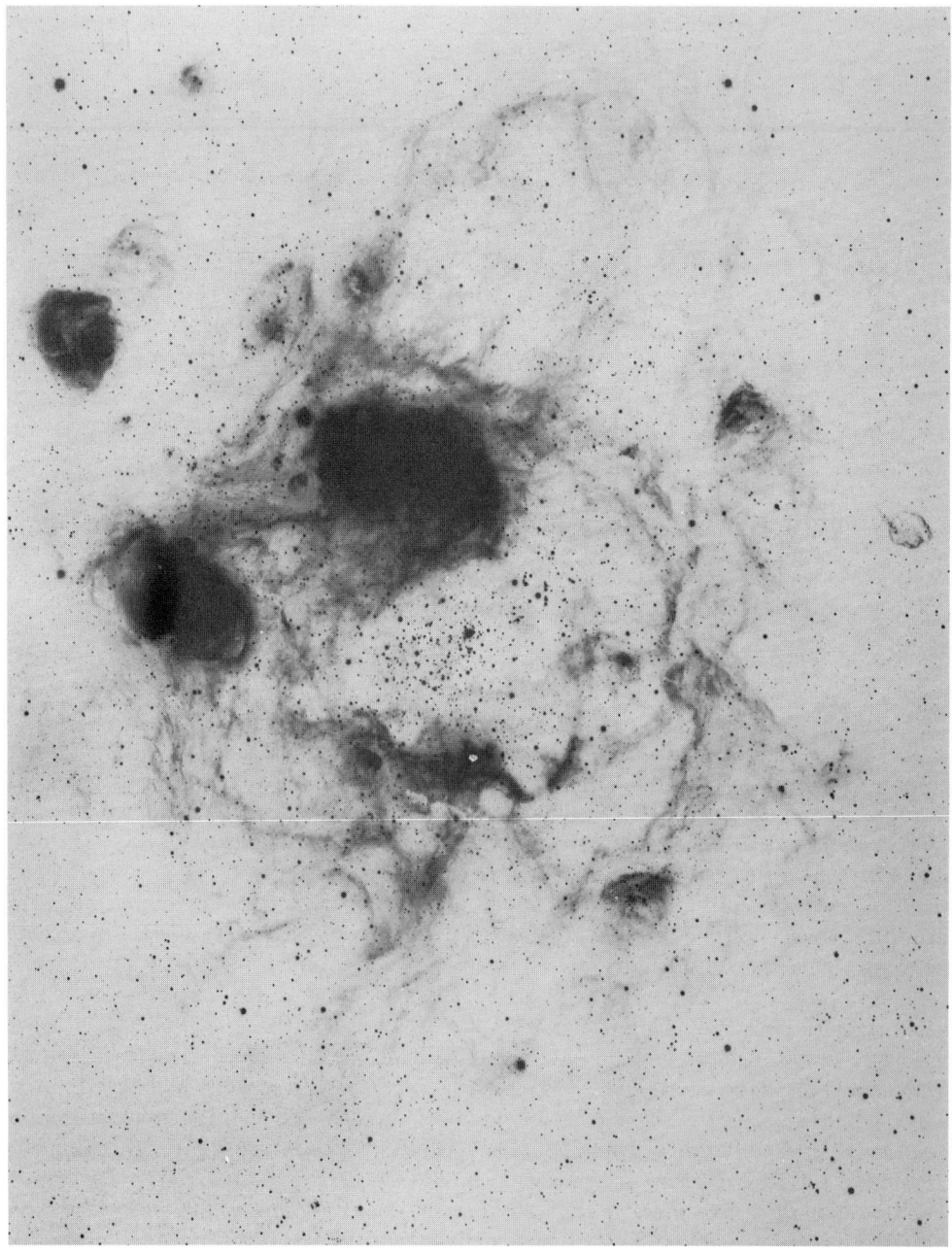

FIGURE 13. Hα photograph of the N11 region in the LMC (from Walborn & Parker 1992). The central cavity (size $5.3' \times 4.0'$ or 80×60 pc) contains the association LH9. The large (overexposed) region north of LH9 contains the association LH10. LH9 is several Myr older than LH10, and may have induced star formation in LH10.

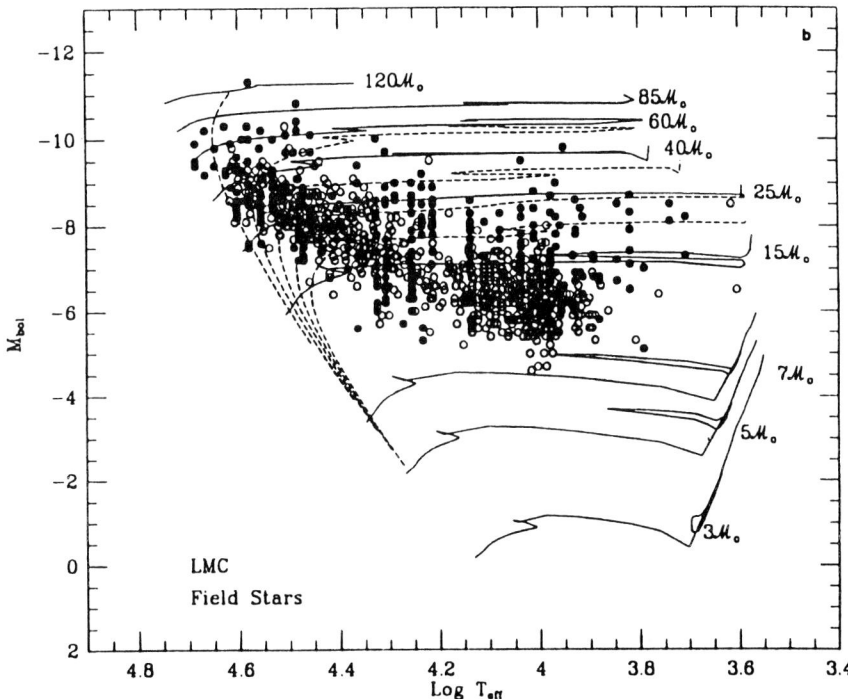

FIGURE 14. HRD of the massive *field*-star population in the LMC (from Massey et al. 1995b). Solid and open symbols denote stars for which spectroscopy and photometry, and photometry only exist.

as a shock or sound wave, velocities on the order of 10^2 km s^{-1} are expected. A timescale is set by the crossing time since information cannot propagate faster than the speed of the shock or sound wave. Satyapal et al. (1995) measured an age gradient along starburst clusters in the galaxy M82 and interpreted this in terms of star formation propagating at a velocity of ∼50 km s^{-1}. Generally, when observing the stellar population of an entire galaxy, *multiple* populations will be present. The goal of the analysis is to decompose the individual populations.

3.4.1. *The color degeneracy of hot stars*

Spectroscopy of large numbers of individual stars for stellar population studies becomes impractical at distances beyond the Magellanic Clouds. Multi-band photometry to determine the spectral energy distribution is used instead. Hot stars emit a substantial fraction of their radiation below 912 Å (Panagia 1973; Vacca et al. 1996) so that only the Rayleigh-Jeans part of their spectrum can be observed directly. As a result, the optical colors of all stars more massive than 20 M$_\odot$ are degenerate and of limited value to determine the stellar population. An impressive demonstration of this particular property of hot stars can be found in the paper by Massey et al. (1995b; their Figure 1). These authors simulated the uncertainty in $T_{\rm eff}$ and L resulting from a photometric error of only ±0.02 mag in $(U - B)$ and $(B - V)$. The uncertainty translates into a mass versus color degeneracy which makes it nearly impossible to distinguish between the most massive stars in the mass range 40 – 100 M$_\odot$ on the basis of colors alone. Voyager observations of the ultraviolet continuum in O stars longward of 912 Å by Longo et al. (1989) indicate that a color degeneracy exists even in that wavelength range: the observed stars have rather similar continua. This caveat should be kept in mind in the following discussion.

3.4.2. Clusters versus field stars

The observed HRD for all LMC field stars for which temperature and luminosity information could be derived is in Figure 14 (Massey et al. 1995b). Clusters are not included in this diagram. This HRD is strikingly different from its counterpart in 30 Dor (Figure 12). The LMC diagram shows numerous stars around 15,000 K and $M_{\rm Bol} \approx -8$, corresponding to late-O/early-B supergiants whereas in 30 Dor these stars are infrequent as compared to less evolved stars. The LMC HRD reflects the integrated star-formation history over the past 20 Myr (corresponding to the lifetime of the faintest stars included on the diagram, which have $M \approx 10~M_\odot$). The population density at the top of the diagram is smaller than in 30 Dor, suggesting there are fewer extremely massive stars in the field than in 30 Dor (and in other LMC clusters).

The initial mass function for the field stars population is much steeper than for stars in clusters: $\alpha \approx 5$ versus $\alpha \approx 2.35$. A similar result is found in the Galaxy: although massive stars can form in the field, they are more likely to form within dense clusters (Massey et al. 1995a).

3.4.3. How to interpret color-magnitude diagrams

The observed HRD or CMD of a galaxy can be fairly complex. The distribution of the stars in the CMD reflects, e.g., the past star-formation history, the possibility of IMF variations, and the chemical evolution of the galaxy over many stellar generations. Interpretation of such a CMD requires much more sophisticated methods than for the interpretation of a diagram of a single stellar population. Simulations to recreate the stochastic nature of the star-formation process usually employ Monte Carlo-type calculations. Tosi et al. (1991) and Greggio et al. (1993) applied this technique to interpret the CMD's of the nearby galaxies Sextans B, DDO210, and NGC3109. Apart from addressing stochastic processes, numerical methods can account for small-number statistics, reddening effects, or binary contributions. Gallart, Aparicio, & Vílchez (1996a), Gallart et al. (1996b), and Gallart et al. (1997) did a detailed study of the past and recent star-formation history of the Local Group irregular NGC6822. Their simulation of crowding and reddening effects is shown in Figure 15. Inclusion of these effects results in a theoretical CMD which closely resembles the observed one.

How would one distinguish between a "good" and a "bad" solution among all models? Obviously an unbiased, quantitative method is desired, which determines the best match to the observations among all synthetic CMD's. Tolstoy & Saha (1996) discuss a statistical method to test the likelihood for a model being a good match for the observations. The method makes use of Bayesian inference which holds that the probability of a hypothesis being correct increases with the amount of information it can be supported. This method allows one to assign likelihoods to different solutions, and the solution having the largest likelihood is considered the "best" model.

Tolstoy (1995) applied this technique to infer the star-formation history of the irregular galaxy NGC2366. NGC2366 contains numerous bright H II regions, the brightest being the giant star forming region NGC2363. Drissen, Roy, & Moffat (1993) suggest a morphology similar to that of N11 (Section 3.3.3): a two-stage starburst separated in age by a few Myr is responsible for an expanding superbubble on the one hand, and quiescent nebular gas on the other. Tolstoy (1995) recreated the star-formation history of NGC2366 by determining the star-formation histories of individual regions and superposing them via Bayesian inference. The observed CMD and the most likely model are

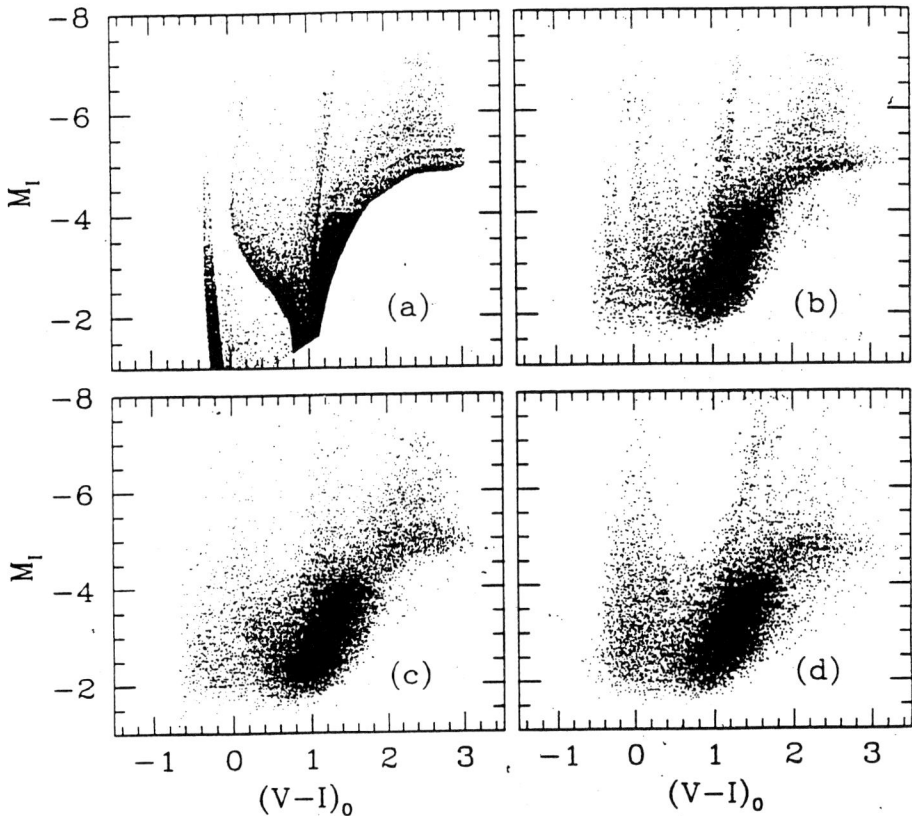

FIGURE 15. Simulation of crowding effects and differential reddening in the CMD of NGC6822. a) Synthetic CMD; b) synthetic CMD with crowding; c) synthetic CMD with crowding and reddening; d) observed CMD. From Gallart et al. (1996b).

in Figure 16. NGC2366 is characterized by a relatively constant star formation over the last 0.3 Gyr, and in addition a present episode of strong star-formation activity (Aparicio et al. 1995; Tolstoy 1995).

3.5. Key results

I am briefly summarizing the key properties of local massive star populations:
- Clusters form nearly instantaneous. Individual populations show little evidence for an age spread but there is evidence for triggering of subsequent star formation.
- Stars with masses as high as 100 $M_\odot$ are observed. Stars more massive — if they exist — are extremely rare and not important for the ionization and non-thermal heating of the ISM.
- Stars with masses below a few solar masses are difficult to detect in high-mass star formation regions. They have been found where searches have been pushed to the limits, suggesting they may generally be present.
- The initial mass function in regions of high-mass star formation is surprisingly uniform and close to Salpeter's (1955) classical result.

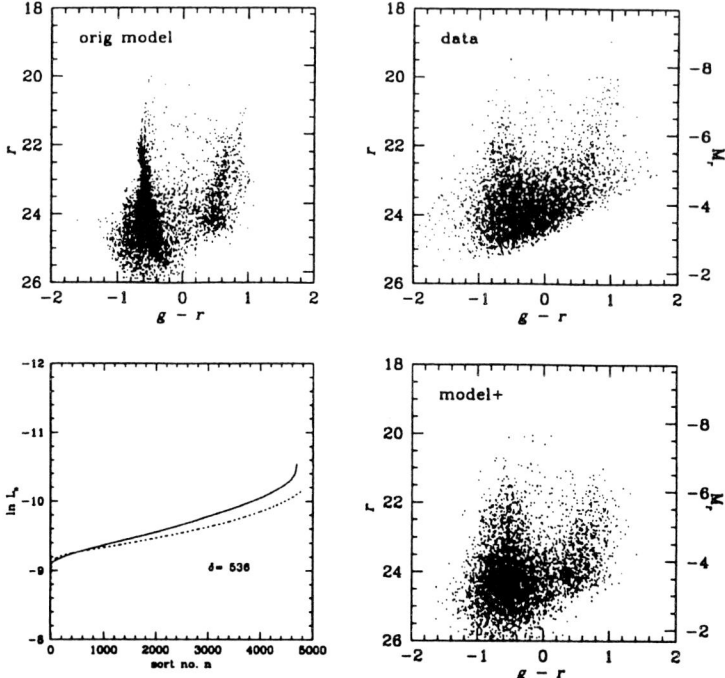

FIGURE 16. Comparison between the observed and theoretical CMD for NGC2366 (from Tolstoy 1995). The four diagrams are (counter-clockwise from upper right): the observations, the most likely model before adding noise, the sorted probability-cumulative likelihood curve, and the most likely model after adding noise.

- The environment, such as metallicity, galaxy type, or location within the galaxy appears to have little influence on the star-formation process.

4. Evolutionary synthesis of unresolved high-mass populations

HRD's and CMD's contain essentially two fundamental pieces of information: temperatures and luminosities of individual stars, derived from the location within the diagrams, and the star-formation history, derived from the population density. The important point is that an observed HRD or CMD allows the separation of these two aspects (at least in principle — issues such as, e.g., reddening complicate the analysis). Galaxies at larger distance are no longer resolved into individual stars, and spatially integrated spectra of their populations must be used to constrain their properties. From a technical point of view, this corresponds to collapsing the information in an HRD/CMD (position and density) into one dimension: an integrated spectrum. The (ambitious!) goal is to reconstruct the star-formation history from the spectrum, assuming all aspects of stellar astrophysics can be predicted by atmosphere and evolution models of individual stars. The difference between the technique described here and in Section 3 is the additional integration over all stellar generations and masses. Individual stellar parameters are no longer relevant, but the integrated properties of the entire population.

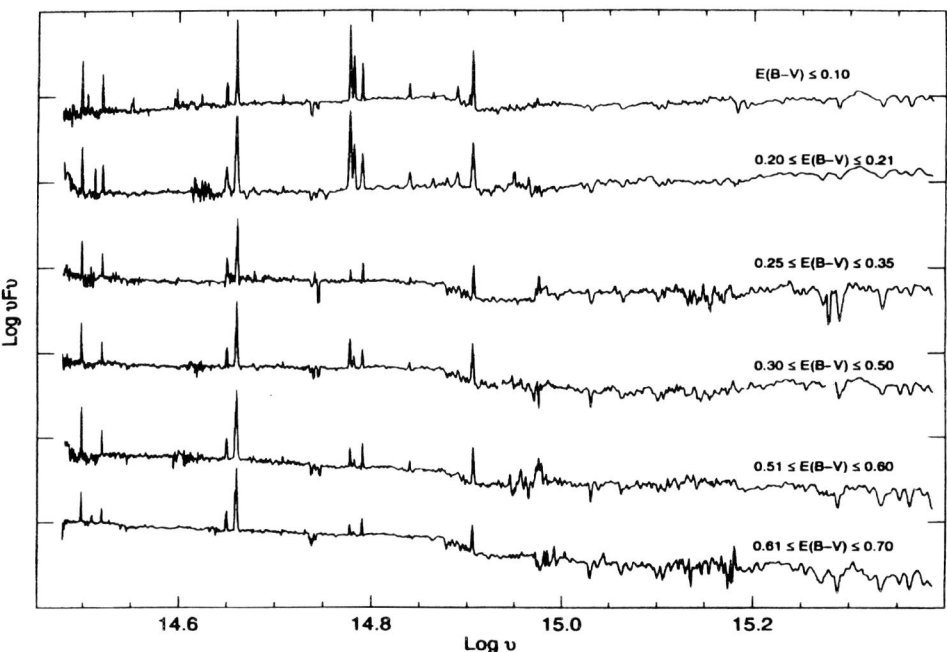

FIGURE 17. Templates of starburst galaxies observed through similar optical and ultraviolet apertures (from Kinney et al. 1996). The six groups have different interstellar reddening. The wavelength range is approximately 10,000 to 1200 Å. Note the *spectral dichotomy*: an almost pure emission-line spectrum is seen longward of the Balmer jump, whereas an absorption-line spectrum is observed at shorter wavelengths.

4.1. *A starburst spectrum between 1200 Å and 10,000 Å*

In Figure 17 a series of average starburst spectra is reproduced. The difference between the optical and ultraviolet is striking. Superposed on a nearly flat continuum due to B and A stars, a rich emission-line spectrum is observed in the optical. In contrast, the ultraviolet shows essentially only stellar and interstellar absorption lines. The behavior of the spectra in Figure 17 can easily be understood from the fact that massive stars have stellar winds and are embedded in left-over gas from the SF process, together with the photon energies resulting from the atomic structure.

To be more explicit, consider this (see also Table 5): Ionizing photons with wavelengths of a few hundred Å are emitted by stars having masses higher than $10 - 20$ $M_\odot$. The photons are absorbed and re-emitted in the stellar wind within about 10^{13} cm (or 10 $R_\star$) from the star. A few percent of the photon luminosity are converted into mechanical luminosity. (Recall that O and WR stars have bolometric luminosities of $10^5 - 10^6$ $L_\odot$, emit about one third of their flux below 912 Å, lose mass at rates of $10^{-6} - 10^{-5}$ $M_\odot$ yr^{-1}, and have wind velocities of a few thousand km s^{-1}; see Section 5.1 for the numbers.) The ionization and excitation state and the emission measure are such that the optical recombination lines of the wind are relatively weak. WR stars are a notable exception. Their winds are dense enough that their He II $\lambda 4686$ (and other lines) can sometimes be observed in the integrated spectra of stellar populations (e.g., Conti 1991). It is more common that only ground-transitions (due to the wind density) of ions in the $10 - 100$ eV range (due to the stellar radiation field) are strong. These lines are in the

Wavelength range	3600 – 10,000 Å	912 – 3000 Å
Spectral lines	mostly emission	mostly absorption
Mass range	> 20 M$_\odot$	> 10 M$_\odot$
λ of photons	a few 10^2 Å	a few 10^2 Å
Distance of line formation	10^{20} cm	10^{13} cm
v of absorbing gas	mildly supersonic	highly supersonic
Technique to measure L^{912^-}	heating of ISM	wind momentum flux

TABLE 5. Comparison of massive-star diagnostics in the optical and ultraviolet regions.

satellite-ultraviolet and below, as determined by atomic physics. At this point we have a rough outline of a hot-star spectrum: weak emission and absorption lines in the optical and strong absorption lines in the ultraviolet.

The stellar wind is optically thin to a significant fraction of the far-ultraviolet photons so that their absorption occurs further away from the star, typically at radii of 10^{20} cm (or tens of parsecs). This leads to the formation of a classical H II region spectrum. The ionization conditions and the emission measure are rather different from those prevailing in stellar winds. Ultraviolet resonance lines in absorption are present in H II regions as well, but they are generally much weaker than lines formed in stellar winds (see Kinney et al. 1993). However, optical recombination lines are very strong due to the large emitting volume. This then leads to the complete, *unified* picture of a massive stellar population with stellar winds and embedded in interstellar gas: a strong emission-line spectrum in the optical and an absorption spectrum in the ultraviolet.

Early-type stars of different masses contribute to the continuous spectrum in Figure 17, with less massive stars becoming more important at longer wavelengths. For a ZAMS population, the stellar masses "seen" *in the continuum* at 1200 Å, 2000 Å, 5000 Å, and 10,000 Å have masses of about 20 M$_\odot$, 15 M$_\odot$, 5 M$_\odot$, and 2 M$_\odot$, respectively. These numbers change drastically when individual stellar *lines* are considered: for instance the C IV λ1550 and Si IV λ1400 lines come from stars in the 50 M$_\odot$ range. In other words, the lines trace a mid-O population superposed on a late-O/early-B population.

My goal is to emphasize the *unified* point of view. Since stars, winds, and the surrounding interstellar gas are intimately connected and their spectral signatures often can not be separated, photoionization models and stellar-line-synthesis models should be considered conceptually the same.

4.2. Nebular diagnostics as stellar tracers

Numerous studies exist to derive the properties of the stellar population by comparison with photoionization models where the ionizing source is a collection of stars. Early attempts assumed single stars as the source of the ionizing spectrum (e.g. Stasińska 1980; Campbell 1988). Important progress was made by McGaugh (1991), who used the stellar evolution tracks of Maeder (1990) to compute the ionizing radiation field corresponding to a zero-age burst assuming a Salpeter IMF. Doing so, he showed that the ionizing radiation field softened for higher abundances of the elements due to the dependence of the location of the ZAMS: higher metal content results in significantly lower effective temperatures of the ionizing stars. However, this study did not yet take into account the time evolution of the stellar population.

Massive, hot stars undergo dramatic temperature changes on timescales of a few Myr. The emitted far-ultraviolet flux reflects these variations (Figure 18). Due to ageing

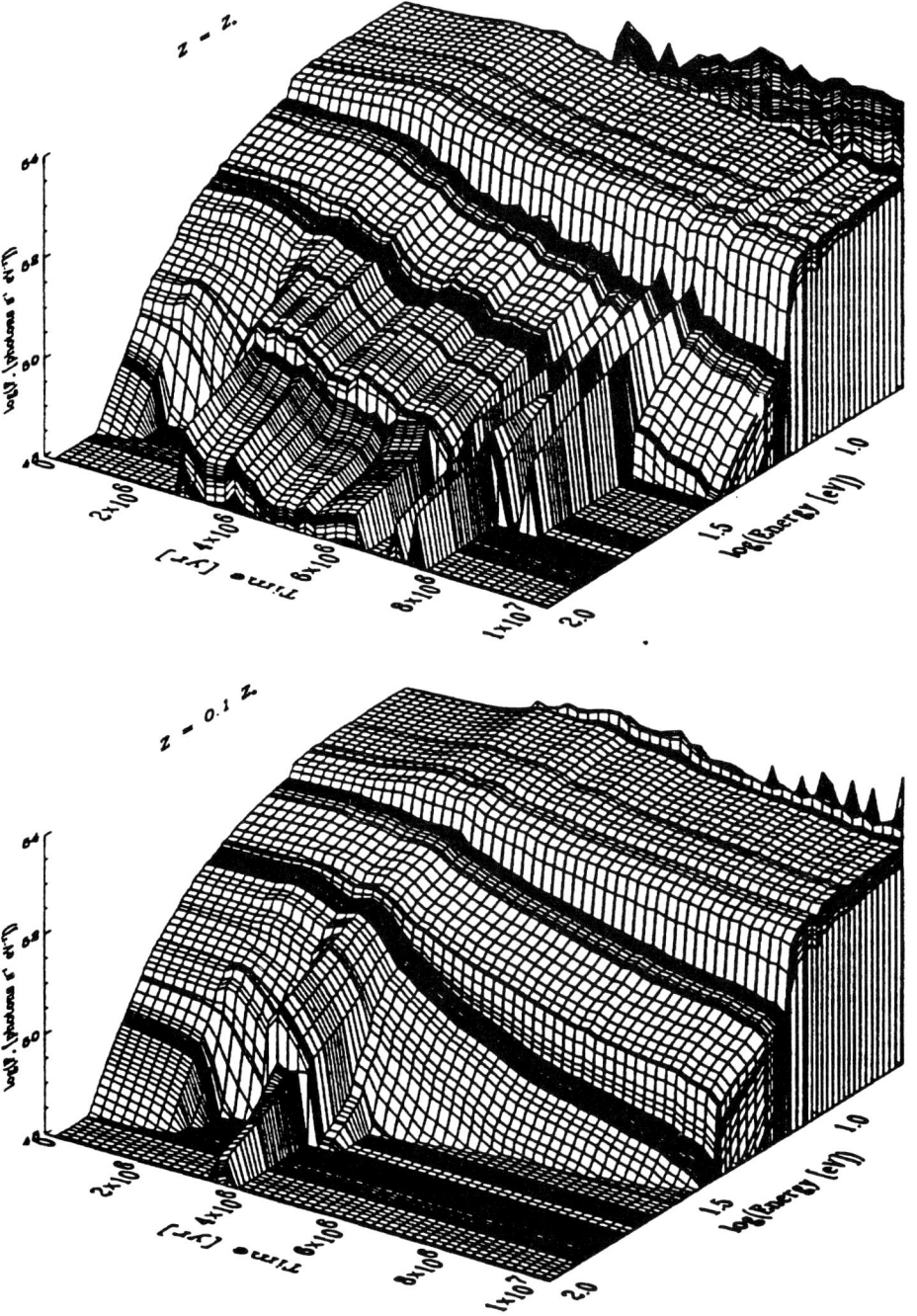

FIGURE 18. Three-dimensional representation of a far-ultraviolet spectrum of a young, evolving single population. Plotted quantity is F_ν in photons s^{-1} eV^{-1} versus time (from 0 to 10 Myr) and photon energy. Note the H I, He I, and He II discontinuities at 13.6 eV, 24.6 eV, and 54.4 eV, respectively. The two figures are for $Z_\odot$ and 0.1 $Z_\odot$ (from Leitherer, Gruenwald, & Schmutz 1992a). High flux levels above 24.6 eV indicate the presence of WR stars. A Salpeter-type IMF with upper cut-off mass of 120 $M_\odot$ was used.

effects, the number of hydrogen ionizing photons close to 912$^-$ Å declines smoothly between 0 and 10 Myr when the hottest and most massive stars evolve from the ZAMS and eventually explode as supernovae or form black holes. The spectrum at even higher energies has a more complex temporal behavior. A flux excess shortward of 24.6 eV appears at about 3 Myr. It is produced by a population of hot stars with strong winds. Observationally, most of them are identified as WR stars. If $Z = Z_\odot$, this phase lasts for about 5 Myr (upper panel in Figure 18). It is shorter at low metallicity when wind effects become less important and fewer stars with strong winds form. The complexity of the far-ultraviolet spectrum of massive stars during their evolution suggests a correspondingly complex spectrum of the interstellar gas that is ionized by the stars.

The important step of addressing this time evolution was made by Terlevich & Melnick (1985), Olofsson (1989), Cid Fernandes et al. (1992), Bernlöhr (1993), Cerviño & Mas-Hesse (1994), García-Vargas & Díaz (1994), García-Vargas, Bressan, & Díaz (1995a,b), and Stasińska & Leitherer (1996). Such models are applicable to giant H II regions containing a mix of stars that have already evolved off the ZAMS. García-Vargas (1996) and Stasińska (1996) give comprehensive reviews of the subject.

Recent attempts to model H II regions with time-dependent evolutionary synthesis are remarkably successful. A time-series of ionizing spectra for different chemical composition calculated by García-Vargas et al. (1995a) is shown in Figure 19. When combined with a photoionization code, these models allow detailed predictions for the nebular emission-line spectrum on the one hand, and for stellar properties on the other. Stellar evolution models by Bressan et al. (1993), Fagotto et al. (1994a,b), and the photoionization code CLOUDY (Ferland 1991) were used. A comparison between the model predictions for metal-rich H II regions and observations suggests that the nebular properties of metal-rich H II regions can be understood as produced by a stellar population having a Salpeter IMF in the mass range 20 – 100 $M_\odot$. At the same time the mass of the stellar cluster responsible for the ionization can be derived from the Hα luminosity. If an assumption for the mass locked in non-ionizing stars is made (which is equivalent to extrapolating the derived IMF to below about 20 $M_\odot$), the total mass of stars formed can be obtained. Such mass determinations are of interest with respect to the relation between very young star clusters containing O and WR stars and old globular clusters. Will the most massive young clusters in H II regions resemble Galactic globular clusters after a Hubble age (Meurer 1995)?

Stasińska & Leitherer (1996) performed a comparison between models and observations for a sample of 100 metal-poor ($Z < 0.5\ Z_\odot$) H II regions. The modeling concept was similar to that of García-Vargas et al. (1995a) but different atmospheres and evolution models were used to address the importance of winds for the resulting energy distribution. The optically thick winds of Maeder's (1990) evolution models were removed and replaced by the expanding wind models of Schmutz, Leitherer, & Gruenwald (1992). This was a first (crude!) attempt of using "unified models" that has been brought closer to perfection by Schaerer et al. (1996a,b) and Schaerer & de Koter (1997).

Stasińska & Leitherer (1996) imposed the condition of a detection of the [O III] λ4363 line on their sample in order to have a reliable electron temperature, and therefore metallicity determination. Three diagnostic diagrams from their work are reproduced in Figure 20. The observed distribution of the data points results from the physical behavior of the lines, after convolution with observational selection effects. Most importantly, the condition of detectable [O III] λ4363 introduces a Malmquist bias which needs to be accounted for in the models. The model predictions (right column) are plotted with large symbols and with small points. Evolutionary phases denoted by small points are charac-

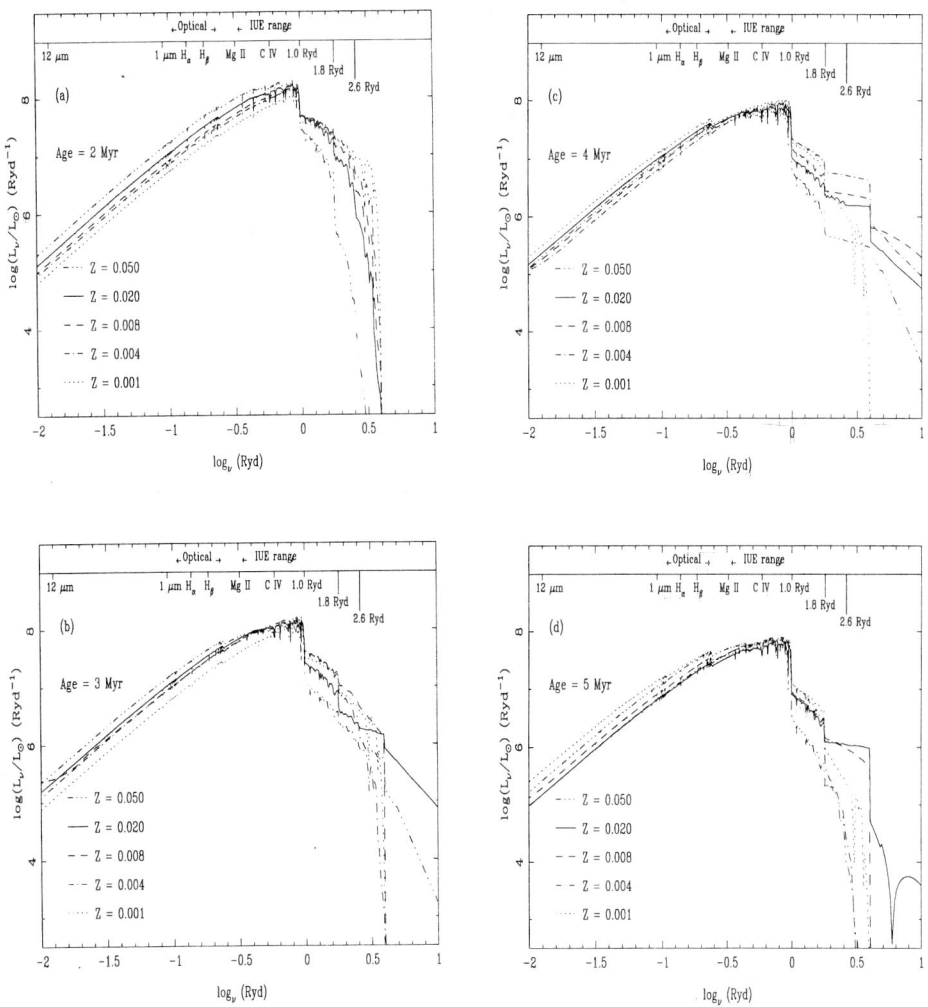

FIGURE 19. Spectral energy distributions of an ionizing stellar cluster of age 2, 3, 4, and 5 Myr and different chemical composition (from García-Vargas et al. 1995a). Note the strong time-dependence of the far-ultraviolet spectrum below 4 Ryd (228 Å).

terized by unobservable [O III] λ4363, and the corresponding section of the observational diagrams should be unoccupied.

[O III] λ5007/Hβ versus the Hβ equivalent width is essentially the relation between the hardness of the spectrum and age. The Hβ equivalent width closely traces the age of an H II region (Copetti, Pastoriza, & Dottori 1986) if there is no serious contamination by light from an older population and if there is no significant internal dust. After 5 Myr, the most massive stars have evolved to lower temperatures and/or have disappeared, and [O III] λ5007 can no longer be excited. For younger ages, models predict that [O III]/Hβ is little dependent on age. The observations agree quite well with these predictions.

The middle diagrams in Figure 20 address a key issue. Plotted in this figure is [O III] λ4363/[O III] λ5007, which is essentially the electron temperature of the H II region. The tight observational relation between electron temperature T_e and O/H is the well-

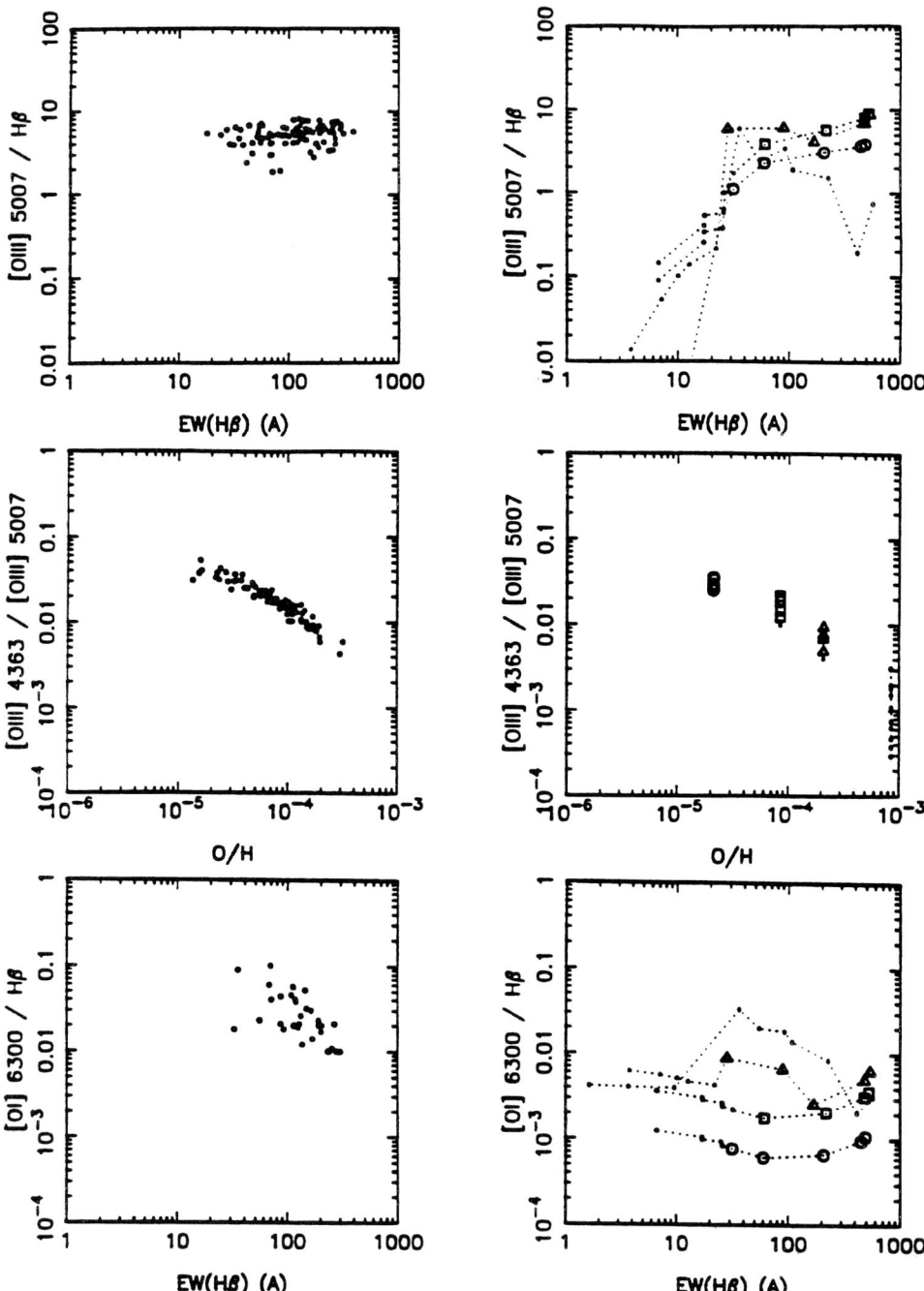

FIGURE 20. Diagnostic diagrams of Stasińska & Leitherer (1996) for observations (left) and models (right). The top diagrams give the ratio [O III] $\lambda 5007$/Hβ vs. Hβ equivalent width, the middle diagrams [O III] $\lambda 4363$/[O III] $\lambda 5007$ vs. oxygen abundance, and the bottom diagrams [O I] $\lambda 6300$/Hβ vs. Hβ equivalent width, respectively. Individual points in the theoretical diagrams correspond to age-steps from 1 to 9 Myr, at intervals of 1 Myr. Triangles, squares, and circles are for Z$_\odot$, 0.25 Z$_\odot$, and 0.1 Z$_\odot$, respectively. A standard Salpeter IMF was used for all models.

known decrease of the electron temperature with increasing metallicity (Stasińska 1980). A decrease of T_e with increasing O/H is the natural consequence of higher cooling rates in a metal-rich environment since metal lines are the main coolants of the ISM (e.g., Osterbrock 1989). Earlier single-star models failed to reproduce the observed relation unless the temperature of the ionizing stars had an *additional* metallicity dependence: a high-metallicity environment appeared to require a softer ionizing radiation field, or alternatively, a bias against very massive stars. Taking into account metallicity-dependent stellar atmospheres, winds, and evolution obliviates the need to postulate a metallicity dependence of the IMF — in agreement with results for nearby resolved star-forming regions.

[O I] $\lambda 6300/H\beta$ is plotted in the bottom diagrams of Figure 20. Even the most extreme models do not produce cases with [O I] $\lambda 6300/H\beta > 0.02$, which are commonly observed. However, shocks, which are naturally expected to be present in starbursts as a result of stellar winds and supernovae, can account for the observed intensities of the [O I] line. I will return to this point in Section 4.5.

Despite the success of the evolutionary synthesis models described in this section, many problems remain. Issues such as, e.g., the geometry of the H II region or the presence of dust are only beginning to be addressed (Calzetti et al. 1994, 1996; Shields & Kennicutt 1995).

4.3. *Ultraviolet line synthesis*

Analysis of the massive stellar content of distant populations from photoionization models is inexpensive from an observational point of view. Optical spectra of H II regions are easy to obtain and a significant database can be found in the literature. However, as pointed out at the end of the previous section, their interpretation is often complicated by higher-order effects that are not easy to account for in spectral synthesis models. The fundamental problem is the emission-line formation some 10^{20} cm from the location of the ionizing star (see Table 5). Even worse, a significant fraction of the ultraviolet radiation may leak out of H II regions and ionize the diffuse interstellar gas (Ferguson et al. 1996). Preferably one would like to utilize lines originating in the immediate stellar vicinity, such as stellar-wind lines observed in the ultraviolet. These lines are the result of strong stellar winds due to radiation pressure (Morton 1967; Lucy & Solomon 1970). Typical wind velocities in O stars are about 2000 km s^{-1} to 3000 km s^{-1} (Groenewegen, Lamers, & Pauldrach 1989; Prinja, Barlow, & Howarth 1990). The wind properties of O stars will be discussed in Section 5 and are summarized in Table 6. All strong ultraviolet lines in the spectra of O stars originate predominantly in the outflow and have blueshifted absorptions. These are the *only* strong stellar lines from hot stars visible in the spectrum of a stellar population (Figure 17).

It is clear that stellar-wind lines contain information on the stellar mass. Stellar winds are driven by radiation pressure (see R. Kudritzki's contribution). Radiative momentum is transformed into kinetic momentum with an efficiency η:

$$\eta = \frac{\dot{M} v_\infty}{L c^{-1}}, \qquad (4.10)$$

where $\dot{M}$, v_∞, L, and c are mass-loss rate, wind velocity, bolometric (radiant) luminosity, and speed of light, respectively. η is remarkably constant for O stars: $\eta \approx 0.3$ for stars with solar chemical composition (Lamers & Leitherer 1993). Equally important, this result can be understood and (to within a factor of 2) reproduced by stellar-wind models (Puls et al. 1996). This forms the basis for the concept to use stellar-wind lines in the ultraviolet as a tool to constrain massive stellar populations: the profiles contain

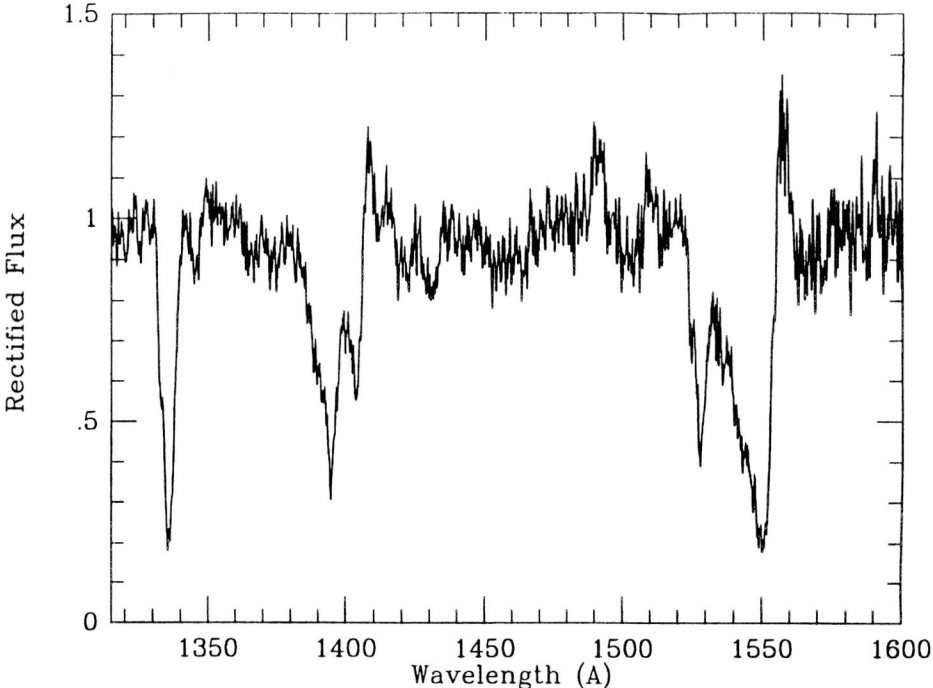

FIGURE 21. HST ultraviolet spectrum of the starburst galaxy He2-10, taken by P. Conti, in collaboration with the author and W. Vacca.

information on $\dot{M}v_\infty$, and therefore on L via eq. (4.10). η is known empirically, or from wind models. Evolution models predict a stellar mass-luminosity relation so that the line profile is eventually tied to the stellar mass. Superposition of the individual profiles of stars in a hot-star population leads to a prediction for the ultraviolet spectrum of a population of massive stars.

The credit for recognizing the potential of this method to constrain IMF parameters goes to Sekiguchi & Anderson (1987). An *IUE* low-dispersion library consisting of a non-evolving main-sequence population of OB stars was used to predict C IV $\lambda 1550$ and Si IV $\lambda 1400$ equivalent widths in starburst galaxies. The limitation to a main-sequence population was relaxed by Mas-Hesse & Kunth (1991) who computed detailed evolutionary synthesis models of C IV $\lambda 1550$ and Si IV $\lambda 1400$ equivalent widths for evolving starbursts. This method can be adequate for the interpretation of low-dispersion *IUE* data, whose resolution is often insufficient to provide structure in the lines of starburst spectra.

HST is capable of obtaining ultraviolet spectra of starbursts at a spectral resolution permitting not only measurements of equivalent widths but also *line-profile studies*. An example is in Figure 21 where a recent HST spectrum of the blue compact dwarf galaxy He2-10 is reproduced. From ultraviolet imaging, Conti & Vacca (1994) could resolve the star-formation activity into numerous small sub-units, whose characteristic sizes are 10 pc or less. The sizes and masses are typical for globular clusters. He2-10 may be forming extremely young globular clusters as the result of violent star formation. The ultraviolet spectrum is consistent with this suggestion. It shows pronounced P Cygni profiles of C IV $\lambda 1550$ and Si IV $\lambda 1400$ due to massive stars (compare with Figure 4). He2-10 is

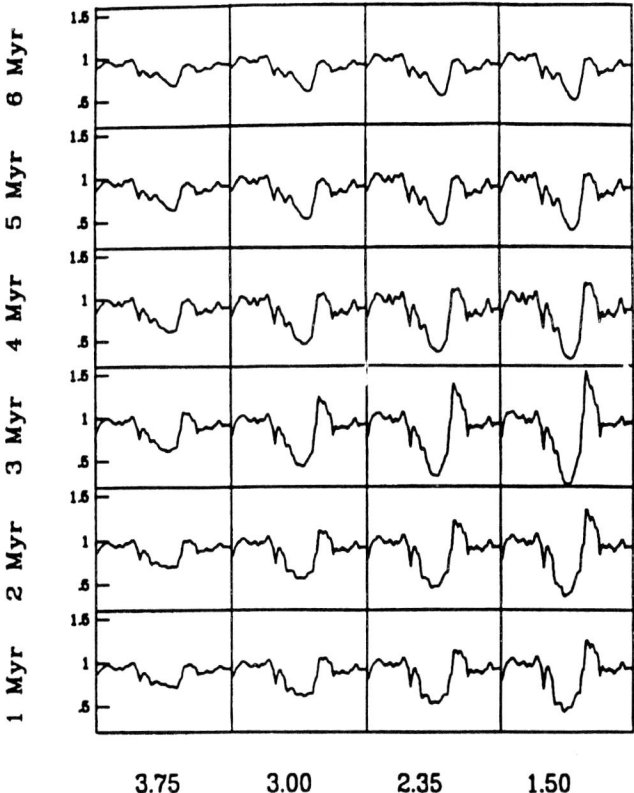

FIGURE 22. Synthetic C IV λ1550 profiles for a hot-star population evolving as a single burst from 1 to 6 Myr (ordinate) and IMF slopes α between 3.75 and 1.5 (abscissa). α = 2.35 corresponds to Salpeter. From Leitherer et al. (1995).

classified as a WR galaxy, based on the λλ4650/86 feature in the optical spectral region (Conti 1991). The corresponding feature in the ultraviolet is N IV] λ1486, which is strong only in the winds of WR stars, but not in O stars (Walborn, Nichols-Bohlin, & Panek 1985).

Robert, Leitherer, & Heckman (1993) and Leitherer, Robert, & Heckman (1995) computed a grid of evolutionary synthesis models to predict the behavior of several strategic ultraviolet lines from a population of massive stars, including C IV λ1550 and Si IV λ1400. The case of C IV λ1550 is in Figure 22. The general trend in this figure is easy to understand: the larger the fraction of early-O stars (responsible for the stellar-wind lines) relative to late-O/early-B stars (responsible for the continuum), the more pronounced the observed P Cygni profile. A larger fraction of early-O stars can be generated either by a flatter IMF (abscissa) or by a younger age (ordinate).

Si IV λ1400 differs from C IV λ1550: In addition to being sensitive to stellar mass, it is also dependent on the evolutionary phase of star since it is much enhanced in supergiants in comparison with main-sequence stars. If the ratio of supergiants over main-sequence stars is high, such as in an instantaneous burst observed at $t \approx 3$ Myr, the spectrum of the population will show a Si IV λ1400 P Cygni profile. Si^{3+} is a trace ion in most

O-star winds (much more than C^{3+}), whose physical conditions favor higher ionization states than Si^{3+}. Therefore stellar winds in O-main-sequence stars are not dense enough to generate enough opacity to produce a strong Si IV λ1400 P Cyg profile. As soon as O main-sequence stars evolve into supergiants, their mass-loss rates and wind densities increase. Higher density favors lower ionization states via recombination and the number of Si^{3+} ions becomes large enough to produce an observable Si IV λ1400 line (Drew 1990; Pauldrach et al. 1990).

The concept of utilizing stellar-wind lines to probe starburst regions is a generalization of the wind-momentum/luminosity relation for individual stars (Kudritzki et al. 1996) to an unresolved population. The basic concept is the tight relation between stellar-wind properties (i.e. the line shape and velocity) and stellar luminosity, which is an immediate consequence of a radiatively driven wind. Since there exists a mass-luminosity relation as well, the profiles are sensitive to the mass function. Additionally, an age dependence is found since the stellar mass-loss rates are strongly dependent on luminosity and temperature, both of which vary drastically over a few Myr. This method has been successfully applied, e.g., to the nearby starburst galaxy NGC4214 (Leitherer et al. 1996c) and to the distant protogalaxy candidate MS 1512-cB58 (Yee et al. 1996).

4.4. *The infrared: gas and evolved stars*

Ultraviolet observations give *direct* sensitivity to the photospheres of the hottest, most massive O stars. Ultraviolet spectra taken with HST exhibit strong emission and absorption features from the gas near the stellar surfaces, while ultraviolet imaging observations reveal luminous young star clusters in starbursts (Conti & Vacca 1994; Meurer et al. 1995; Leitherer et al. 1996c). In contrast, optical observations are only *indirectly* sensitive to massive stars, via strong nebular emission lines; the optical continuum is dominated by less massive B & A stars. Almost paradoxically, near-infrared observations, which are best known for their sensitivity to cool stars, once again begin to give direct sensitivity to the most massive stars in starbursts. A complete review of massive-star populations observed in the infrared has been given by Moorwood (1996).

Red supergiants dominate the near-infrared spectrum for starbursts more than a few Myr old, and remain the dominant contributor as long as any are present (e.g., Rieke 1991). The strongest features in the 2 μm spectra of starburst galaxies are the strong CO absorption bands due to RSG's. This feature is frequently observed in starbursts that are old enough to produce red supergiants (Oliva et al. 1995; Goldader 1996). Since red supergiants are the cool descendants of hot O stars, the CO band can be used to constrain the hot-star population. CO absorption is also associated with less luminous red giants, and an underlying older population may contaminate the information. Furthermore, reliable stellar evolution models are required to provide the link between hot and cool stars. While such models exist for metal-rich ($Z \approx Z_\odot$) stars, all current stellar evolution models fail to predict the correct temperature of metal-poor red supergiants (Goldader, Leitherer, & Schaerer 1997). The reason is not fully understood. As a result, infrared diagnostics which rely on metal-poor red supergiants are not suitable to constrain the properties of the stellar population.

The most prominent emission lines of hydrogen and helium are Brγ and He I $\lambda 2.1 \mu$m. These lines reveal the presence of the hottest stars. Doyon, Puxley, & Joseph (1992) and Doherty et al. (1995) used the ratio of the two lines to measure the relative sizes of the hydrogen and helium Strömgren spheres. However, the behavior of this line ratio is complex and predominantly a function of local conditions, such as geometry and dust (Shields 1993).

The ultraviolet, optical, and infrared wavelength range by themselves offer powerful

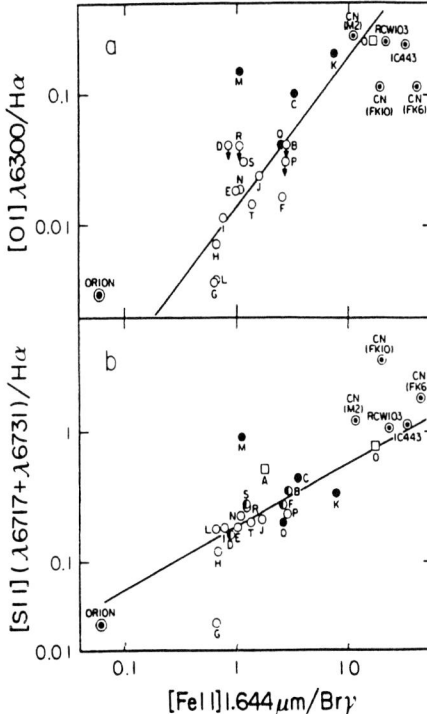

FIGURE 23. [O I]/Hα and [S II]/Hα versus [Fe II] λ1.644µm/Brγ for a sample of starbursts (open circles) and active galaxies. From Mouri et al. (1990).

diagnostics of the physical conditions in starbursts; when combined, observations spanning the whole range offer a unique way to explore sites of massive-star formation. For instance, [Fe II] λ1.26 µm and λ1.64 µm are excited in regions of fast shocks. Near-infrared [Fe II] lines are sometimes observed in starbursts. Mouri et al. (1990) compared the strengths of [Fe II] λ1.644µm/Brγ with those of shock-excited lines in the optical and found good correlation (Figure 23). The correlation can be understood in terms of shock heating by stellar winds and supernovae in the starburst. [Fe II] λ1.644 is increased as the result of an increased gas-phase abundance of Fe due to grain destruction in shocks. Strong [Fe II] emission is directly associated with young supernova remnants in the starburst galaxies M82 and NGC253 (Greenhouse et al. 1991; Forbes & Ward 1993). The correlation of [Fe II] with [O I] λ6300 suggests that a significant part of the [O I] emission can be attributed to shocks as well (see Section 4.5).

[Fe II] emission follows the evolution of individual supernova remnants, whose lifetime is on the order of 10^4 yr (Oliva, Moorwood, & Danziger 1989). Supernova remnants are immediately related to pre-supernovae, i.e. red supergiants. RSG's in turn evolve from hot stars so that a correlation is expected between the strength of recombination lines and [Fe II] if star formation occurs between the epoch of declining O-star numbers (after ∼5 Myr) and the formation of supernova remnants (after ∼10 Myr). Calzetti (1997) performed this test and found generally good agreement (Figure 24). The correlation in Figure 24 indicates that the current supernova rate (as measured by [Fe II]) is proportional to the future supernova rate (as measured by Pα). This is expected, given the physical size sampled by the observations (∼4.5 kpc), which encompasses numerous star-formation sites, and a quasi-continuous event is measured.

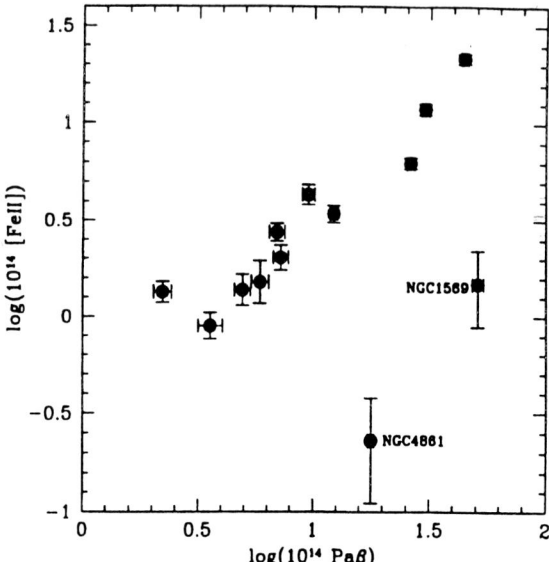

FIGURE 24. [Fe II] $\lambda 1.257\mu$m versus Pβ for a sample of starburst galaxies. The two lines correlate well in most galaxies. Exceptions are NGC1569 and NGC4861 whose light is dominated by exceptionally bright, compact clusters with no significant temporal spread of star formation. From Calzetti (1997).

4.5. *Evidence for non-thermal heating*

Emission from [Fe II] suggests significant non-thermal heating by massive stars, via shocks. Most of our prior discussion was restricted to the thermal energy output of massive stars, viz. their ionizing radiation. The observed spectral energy distribution, including most lines and continuum, and for stars and gas, can be reasonably well understood in terms of the radiant energy output. A few percent of the stellar luminosity is released as mechanical wind power (see Section 5.1). Does this non-thermal energy output produce other observable effects?

I am now returning to the bottom panel of Figure 20, where [O I] $\lambda 6300$/Hβ versus EW(Hβ) is plotted. There is a clear discrepancy between models and observations: even the most extreme models do not produce such a large proportion of observed cases with [O I] $\lambda 6300$/H$\beta > 0.02$. When faced with strong observed [O I] lines, one generally invokes the presence of shocks, for example from supernovae or stellar winds.

One can estimate how the shocks expected to occur in evolving starbursts may affect the intensities observed for diagnostic emission lines. Leitherer & Heckman (1995) have computed the total rate of mechanical energy deposited in starbursts by stellar winds and supernovae as a function of time (see Section 5.5). They find that the ratio of mechanical/ionizing luminosity increases with time as the burst ages, and is larger at higher metallicities. About 30% of this energy is converted into radiation emitted by shocked nebular gas (Abbott 1982). From the grid of shock models by Raymond (1979) — whose shock velocities range from 50 km s^{-1} to 200 km s^{-1} — it is possible to evaluate the proportion radiated in [O I] and other lines (see Stasińska & Leitherer 1996). The estimation suggests that shocks, *which are naturally expected to be present in starbursts as a result of stellar winds*, can possibly account for the observed intensities of the [O I] lines.

There is another indicator for significant non-thermal energy deposition by massive stars. Interstellar absorption lines found in starburst regions are anomalously strong. An example is the starburst region NGC4214#1 (see Table 3). Its interstellar lines are significantly stronger than in local Galactic H II regions (Leitherer et al. 1996c). The immediate conclusion is that the interstellar medium in NGC 4214#1 is different from that in the solar neighborhood. One can exclude different ionization states as an explanation for the different line strengths, as both low ionization states (e.g., C II $\lambda 1335$) and high ionization states (e.g., the interstellar contribution to C IV $\lambda 1550$) are stronger in NGC 4214#1. The enhanced interstellar lines in NGC 4214#1 are most likely a result of different velocities sampled by the sight lines toward NGC 4214#1. The observed profile will be the superposition of many individual interstellar absorption lines at different velocities. There is independent observational evidence for this suggestion. Numerous high-velocity shells are observed in 30 Dor, even though its stellar content is a factor of several less energetic than that of NGC4214# (Table 3). At the other extreme, the morphology of galaxies in the Hubble Deep Field is highly disturbed (see Figure 9) — possibly indicating a large-scale disruption of the ISM.

5. Release of mass and energy by massive stars

The previous sections primarily addressed the radiative properties of massive stars and stellar populations. The stellar "light" was utilized to detect massive stars and to derive their basic parameters. However, I already mentioned indications that this is an incomplete picture of the stellar population. Spectral lines, such as [Fe II] or [O I] can only be understood if there is an additional source of energy. In this section I will cover the other, "unseen" component of energy output: stellar winds. Strong winds are an ubiquitous feature of massive stars. They are evident in stellar ultraviolet spectra via pronounced P Cygni profiles (Figure 4), and they are known to be a major parameter in the evolution of massive stars. Mass-loss rates of 10^{-5} $M_\odot$ yr^{-1} and wind velocities on the order of 10^3 km s^{-1} result in mechanical energies around 10^{51} erg over the lifetime of massive stars (10^7 yr). This is comparable to the kinetic energy release during a supernova event. Therefore one expects stellar winds to be an important source for the mass and energy supply of the ISM. Reviews of the subject can be found in Bieging (1990) and Castor (1993). In this section I will focus on the "output" aspect: I will constrain the mass and energy output using predictions from models which have been calibrated via the observed *radiant* energy output, in other words via spectra. The model predictions will be compared to phenomena observed in the local universe without going into details on the resulting dynamics and geometry of the ISM. This "input" aspect is the topic of Section 6.

5.1. Wind properties of individual stars

One can identify four individual phases of stellar mass loss during the evolution of a massive star:
- the main-sequence and early post-main-sequence evolution;
- the LBV phase;
- the RSG phase;
- the Wolf-Rayet phase.

At the endpoint of the stellar evolution most (or even all) massive stars explode as supernovae which provide an additional source of mass loss and wind power. This mass-loss categorization is a natural consequence of the evolutionary channels introduced in Section 3.2. In the following I will discuss the observational and theoretical evidence for mass loss and its dependence on stellar parameters in each of these four phases. The discussion is valid for systems having ages less than $\sim 10^8$ yr since after that time evolutionary phases such as asymptotic giant branch (AGB) stars and the formation of planetary nebulae become important. Such late phases of low- and intermediate-mass stars are a major source for the mass and chemistry budget of old galaxies (Jura 1987).

5.1.1. *Mass-loss rates and wind velocities of OB stars*

The stellar-wind properties on the main-sequence and the early post-main-sequence (hereafter called OB phase) are determined by radiation pressure. Stellar winds of OB stars are best understood among all types of winds. As discussed in R. Kudritzki's contribution, observations and theoretical models result in consistent mass-loss rates and terminal velocities for a set of stellar parameters. A summary of the most pertinent stellar parameters is in Table 6, which makes use of the data in Vacca, Garmany, & Shull (1996). The entries in this table are valid for Galactic O stars within a few kpc from the Sun. Stars in this part of the Galaxy have a chemical composition that is close to, but somewhat below solar. Cunha & Lambert (1992) derived abundances of early-type stars in the Orion OB association which are about 0.2 dex below solar.

The values in Table 6 are the average for each spectral type. T_{eff} comes from spectral classification. Masses and luminosities are derived using atmospheres and evolution models. Note that the masses in column 3 are *current* masses. They are lower than $M_\circ$. N_L, the number of hydrogen ionizing photons (column 5) was derived from model atmospheres (see Vacca et al. for details). The values range between 10^{48} and 10^{50} s^{-1}. They should be compared to the ionization requirements of the H II regions in Table 4: Orion is ionized by a few O stars, whereas a few hundred are required for 30 Dor. Column 6 gives the radiative momentum flux, L/c.

Wind parameters are in columns 7 to 10 of Table 6. The mass-loss rates were derived from

$$\log \dot{M} = 1.738 \log L - 1.352 \log T_{\text{eff}} - 9.547, \quad (5.11)$$

which is the average relation for $\dot{M}$ in a sample of Galactic O stars studied by Lamers & Leitherer (1993). The standard deviation of individual mass-loss rates from eq. (5.11) is $\sigma = 0.23$. Wind velocities v_∞ were calculated with the theoretical relation

$$\log v_\infty = 1.23 - 0.30 \log L + 0.55 \log M + 0.64 \log T_{\text{eff}} \quad (5.12)$$

(Leitherer, Robert, & Drissen 1992b). The units in eqs. (5.11) and (5.12) are as in Table 6. The values for v_∞ agree with observations to within 20%. An extensive database for $\dot{M}$ and v_∞ exists in the literature. A few references are Howarth & Prinja (1989), Prinja, Barlow, & Howarth (1990), Lamers & Leitherer (1993), and Puls et al. (1996). The mechanical wind power

$$L_W = \frac{1}{2} \dot{M} v_\infty^2 \quad (5.13)$$

and the wind momentum flux

$$P_W = \dot{M} v_\infty \quad (5.14)$$

are in columns 9 and 10. The mechanical power and momentum can be compared with the corresponding radiative quantities in columns 4 (after conversion to cgs, with $\log L_\odot = 33.58$) and 6. In terms of the energy budget, O (and B) stars are dominated by radiation: winds carry less than 1% of the total energy. Winds do become significant when the momentum flux is considered: P_W is about 30% of L/c (cf. eq. (4.10)), independent of spectral type. This is a result of radiation pressure driving the winds, which operates with a 30% efficiency in converting radiative into kinetic momentum.

Spec. Type	T_{eff} (K)	M ($M_\odot$)	$\log L$ ($L_\odot$)	$\log N_L$ (s^{-1})	$\log \frac{L}{c}$ (cgs)	$\log \dot{M}$ ($M_\odot$ yr^{-1})	v_∞ (km s^{-1})	$\log L_W$ (cgs)	$\log P_W$ (cgs)
\multicolumn{10}{c}{Luminosity Class V}									
O3	51230	87	6.03	49.87	29.15	−5.43	3200	37.08	28.88
O4	48670	68	5.88	49.70	29.00	−5.66	3000	36.79	28.62
O4.5	47400	62	5.80	49.61	28.92	−5.78	2900	36.66	28.49
O5	46120	56	5.73	49.53	28.84	−5.90	2900	36.52	28.36
O5.5	44840	50	5.65	49.43	28.76	−6.02	2800	36.38	28.23
O6	43560	45	5.57	49.34	28.68	−6.14	2700	36.23	28.10
O6.5	42280	41	5.49	49.23	28.60	−6.27	2700	36.10	27.97
O7	41010	37	5.40	49.12	28.52	−6.39	2700	35.96	27.84
O7.5	39730	34	5.32	49.00	28.43	−6.52	2600	35.82	27.70
O8	38450	30	5.24	48.87	28.35	−6.65	2600	35.68	27.57
O8.5	37170	28	5.15	48.72	28.26	−6.78	2500	35.53	27.43
O9	35900	25	5.06	48.56	28.18	−6.91	2500	35.39	27.29
O9.5	34620	23	4.97	48.38	28.09	−7.04	2500	35.25	27.15
\multicolumn{10}{c}{Luminosity Class III}									
O3	50960	101	6.15	49.99	29.27	−5.22	3200	37.28	29.08
O4	48180	82	6.05	49.86	29.16	−5.37	2900	37.07	28.90
O4.5	46800	75	5.99	49.80	29.11	−5.45	2900	36.96	28.81
O5	45410	68	5.93	49.73	29.05	−5.53	2800	36.85	28.71
O5.5	44020	62	5.88	49.65	28.99	−5.61	2700	36.74	28.61
O6	42640	56	5.82	49.58	28.93	−5.70	2600	36.63	28.52
O6.5	41250	52	5.76	49.50	28.87	−5.78	2500	36.52	28.42
O7	39860	47	5.70	49.41	28.81	−5.87	2400	36.41	28.32
O7.5	38480	43	5.63	49.32	28.75	−5.96	2400	36.29	28.21
O8	37090	39	5.57	49.22	28.68	−6.05	2300	36.17	28.11
O8.5	35700	35	5.50	49.12	28.61	−6.14	2200	36.05	28.00
O9	34320	32	5.43	48.97	28.55	−6.24	2200	35.93	27.90
O9.5	32930	29	5.36	48.78	28.47	−6.34	2100	35.81	27.79
\multicolumn{10}{c}{Luminosity Class Ia}									
O3	50680	115	6.27	50.11	29.39	−5.00	3100	37.48	29.29
O4	47690	104	6.21	50.02	29.32	−5.08	3000	37.36	29.19
O4.5	46200	95	6.18	49.98	29.29	−5.12	2800	37.28	29.13
O5	44700	86	6.14	49.93	29.25	−5.16	2700	37.20	29.07
O5.5	43210	79	6.10	49.87	29.22	−5.21	2600	37.11	29.01
O6	41710	74	6.07	49.81	29.18	−5.25	2500	37.04	28.95
O6.5	40210	69	6.03	49.75	29.14	−5.30	2400	36.96	28.88
O7	38720	64	5.98	49.69	29.10	−5.35	2300	36.88	28.82
O7.5	37220	59	5.94	49.62	29.05	−5.40	2200	36.79	28.75
O8	35730	54	5.89	49.54	29.01	−5.46	2100	36.71	28.67
O8.5	34230	50	5.85	49.45	28.96	−5.52	2100	36.61	28.60
O9	32740	46	5.80	49.33	28.91	−5.58	2000	36.52	28.52
O9.5	31240	43	5.74	49.17	28.86	−5.64	1900	36.42	28.44

TABLE 6. Parameters for O stars, based on the tabulation of Vacca et al. (1996).

In terms of its consequences for stellar evolution, mass loss in the OB phase is not very important for stars with solar composition unless $M_\circ > 50$ $M_\odot$: as an example, the mass of a star of solar composition with $M_\circ = 40$ $M_\odot$ decreases only by about 10% during this phase (Table 6, together with a lifetime of ~ 5 Myr). Therefore uncertainties in the evolution models are relatively minor. The total amount of mass and energy returned to the interstellar medium by a stellar population during the OB phase, however, is significant. As I will show below, the OB phase, together with the WR phase is the dominant contributor to the total wind output.

5.1.2. WR stars

I will defer the discussion of LBV's and RSG's to the next section since their wind properties are different from those of hot stars, and will first present the end-point of stellar evolution: WR stars. WR stars are blue, luminous stars whose principal distinguishing property are strong emission lines in the ultraviolet, optical, and infrared spectral regions. The WR stars are classified in two main groups: those having strong nitrogen lines in their spectra (denoted WN stars), and those with strong carbon lines (WC stars). Smith, Shara, & Moffat (1990, 1996) have devised quantified classification schemes for WC and WN stars based on the line strengths and line ratios of optical emission lines. A few percent of all WR stars do not fit into either of the WC or WN categories, mostly due to their hybrid nature, or because they have strong oxygen lines (WO stars). The latest compilation of WR stars includes 189 individual WR stars in our Galaxy (van der Hucht 1995 and references therein). For a general review of this stellar species see Abbott & Conti (1987).

The strength of the emission lines in WR stars suggests the presence of very extended atmospheres. Abundance analyses indicate a likely overabundance of elements affected by stellar nucleo-synthesis with a significant helium-, nitrogen-, and carbon-enrichment in most objects (Crowther, Hillier, & Smith 1995; Hamann 1995). Optical, infrared and radio studies have shown the stars to have powerful stellar winds with mass-loss rates in excess of 10^{-5} $M_\odot$ yr^{-1} and wind velocities of more than 10^3 km s^{-1} (Crowther & Willis 1994). The strong stellar winds and probable chemical enrichment are consistent with WR stars being the evolved descendants of massive O stars (Maeder 1994). In this evolutionary scenario, massive stars lose a significant fraction of their original mass via stellar winds prior to and during the WR phase, eventually exposing their bare helium cores with typical masses of $10-20$ $M_\odot$. In some cases, interaction in close binary systems may also remove material from the outer atmospheres (Vanbeveren 1994), in particular at lower than solar metallicity, where mass loss by stellar winds is less effective in reducing the stellar mass. The fact that the majority of WR stars in the Small Magellanic Cloud are binaries (Moffat 1995) is consistent with this suggestion.

Table 7 gives typical stellar-wind parameters for each WR sub-type based on observations. The table is from Leitherer et al. (1992b). WR mass-loss rates are rather homogeneous, with little variation among subtypes. The velocities are similar to those of O stars. WR stars occupy roughly the same part of the HRD as O stars. They have luminosities around 10^5 to 10^6 $L_\odot$ and temperatures between 30,000 K and 100,000 K (Abbott & Conti 1987). O and WR stars have similar L and T_{eff} but $\dot{M}$ differs by 1 to 2 orders of magnitude. Therefore η of eq. (4.10) is a factor of 10 to 100 larger in WR stars than in O stars. The fact that $\eta > 1$ in WR stars is the main reason why no theory exists which could predict stellar-wind properties of WR stars as a function of stellar parameters. The origin of the strong stellar winds observed in WR stars is not known (Cassinelli 1991).

WR Type	log $\dot{M}$(M$_\odot$ yr^{-1})	v_∞ (km s^{-1})
WNL	−4.2	1650
WNE	−4.5	1900
WC6–9	−4.4	1800
WC4–5	−4.7	2800
WO	−5.0	3500

TABLE 7. Stellar-wind parameters of WR stars

5.1.3. Significance of cooler stars

Two additional evolutionary phases can be identified: LBV's and RSG's. The class of LBV's has originally been defined by Conti (1984). For a general discussion of the LBV phenomenon, the reader is referred to the conference proceedings edited by Nota & Lamers (1997). LBV's are early-type, luminous stars displaying dramatic photometric and spectroscopic variations over several distinct timescales, ranging between weeks and decades. LBV's as a class of stars are believed to represent a short-lived ($10^4 - 10^5$ yr) phase during the evolution of a massive star when a significant amount of mass is lost. It is largely the LBV phase which determines the evolutionary relation between O stars on the hydrogen-burning main-sequence and core-helium-burning WR stars in the theoretical HRD. Time averaged mass-loss rates are on the order of 10^{-4} M$_\odot$ yr^{-1} and typical wind velocities are around 200 km s^{-1} (Lamers 1989). The mechanism responsible for removing large amounts of mass from LBV's is not fully understood (e.g., de Koter, Lamers, & Schmutz 1996). LBV's are most noted for their episodic eruptions, when a few percent of the entire stellar mass can be shed. The most famous example is η Carinae (Figure 25). This LBV had a major outburst about 150 years ago, whose relics give rise to the spectacular bipolar nebula seen in this figure. A significant fraction of the time-averaged mass loss of LBV's is contributed by such eruptions.

RSG's have dense stellar winds with high mass-loss rates and low wind velocities. Observationally, outflow velocities between 10 and 100 km s^{-1} are derived (Drake 1986; Jura 1991). The mass-loss rates can be represented in terms of stellar parameters by

$$\dot{M} = 5.5 \times 10^{-13} \frac{LR}{M} \quad (5.15)$$

(Reimers 1975). $\dot{M}$ in M$_\odot$ yr^{-1}, and L, R, and M in solar units. A typical RSG has $M = 15$ M$_\odot$, $T_{\rm eff} = 3000$ K, and $L = 10^5$ L$_\odot$ (see the HRD in Figure 1), resulting in mass-loss rates of a few times 10^{-6} M$_\odot$ yr^{-1} from eq. (5.15). This is comparable to (or even higher than) O-star rates. Like the LBV phase, the RSG phase is crucial for stellar evolution since a large fraction of the stellar mass is lost in luminous red supergiants. In terms of the mass and energy return to the ISM, however, RSG's are insignificant (see Section 5.2). Therefore I will not discuss their wind properties any further.

5.1.4. Wind properties of hot stars at different metallicity

The winds are accelerated by momentum transfer resulting from absorption of photons in metallic lines. As the line strengths depend on the chemical composition, the wind properties should depend on Z. In fact, model calculations for O stars predict $\dot{M} \propto Z^\alpha$, with $0.5 < \alpha < 1$ (Kudritzki, Pauldrach, & Puls 1987; Leitherer et al. 1992b). For a metal-poor O star in the SMC, this means a decrease of $\dot{M}$ by about a factor of 3 with

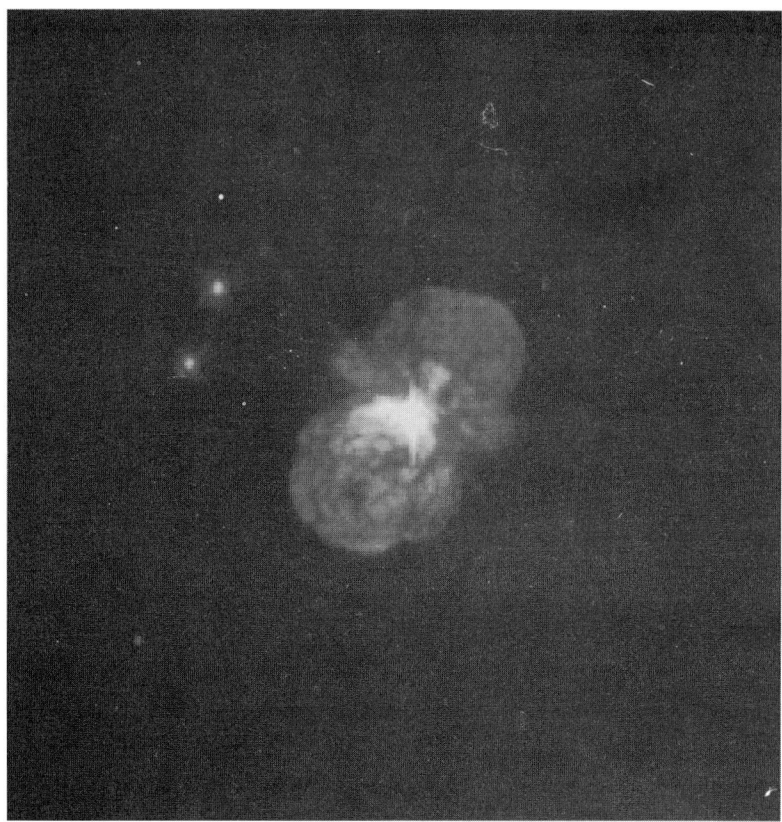

FIGURE 25. The shell around the LBV η Car, ejected in the mid-19th century. F410W WFPC2 image obtained with HST. The bipolar structure has a major axis of about $20''$ or 7×10^{17} cm.

respect to stars in the solar neighborhood. At the same time, v_∞ is predicted to be lower by a few hundred km s^{-1}. What is the observational evidence for a variation of $\dot{M}$ and v_∞ with Z? Stars in the Magellanic Clouds (and in the SMC in particular) have weaker wind lines in their ultraviolet spectra (Walborn et al. 1995), suggesting lower opacity in the corresponding ions. Lower wind velocities in the Clouds are well established, in agreement with the theory (Garmany & Conti 1985; Prinja 1987). Translating the observed opacity in, e.g., N V λ1240 into mass-loss rates requires detailed modeling of the radiative transfer for an extreme trace ion and in parallel solving the equation of motion of the outflow. The most recent efforts in this respect by Puls et al. (1996) are remarkably successful. Stellar winds observed in LMC and SMC stars have properties as predicted by the wind theory. Although a general survey of mass-loss rates in the Magellanic Clouds is not yet complete, the early results support the predictions of the theory for the metallicity dependence of the winds. The models for the mass and energy return by O stars in the following sections are based on the theoretical prediction $\dot{M} \propto Z^{0.8}$ and $v_\infty \propto Z^{0.13}$.

WR winds have been studied in detail in the Galaxy and in the Magellanic Clouds. No significant difference between the wind properties was found (e.g. Hamann 1995). The lack of correlation between Z and $\dot{M}$ and v_∞ does not necessarily indicate that radiation

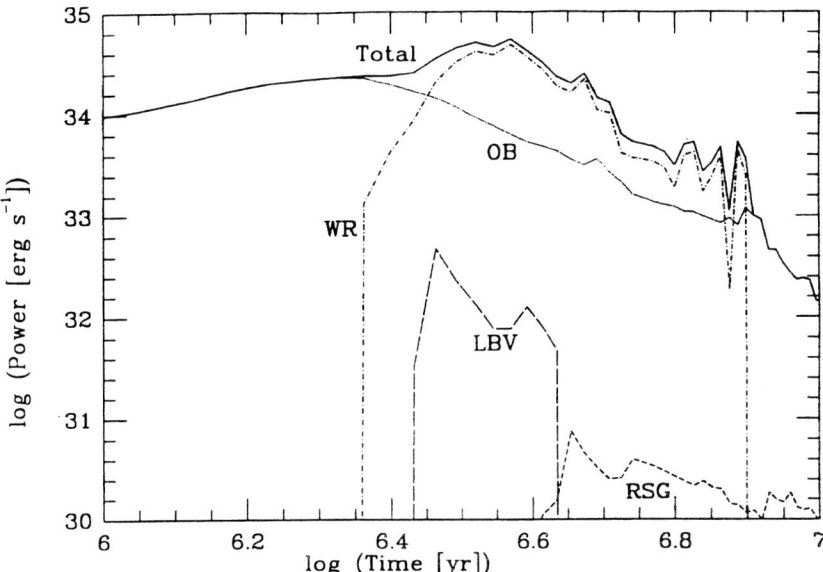

FIGURE 26. Wind power of a stellar population forming instantaneously, normalized to 1 $M_\odot$. $Z = Z_\odot$, $\alpha = 2.35$, $M_{up} = 120$ $M_\odot$, $M_{low} = 1$ $M_\odot$. Contributions from individual evolutionary phases to the total power are shown. From Leitherer et al. (1992b).

pressure is not responsible for driving the winds of WR stars. WR winds are 1 to 2 order of magnitude denser than O-star winds, causing a larger fraction of optically thick lines. If lines have higher optical depth, metallicity variations become less important. The lack of a theory for WR winds is a challenge for the field of stellar atmospheres but it is less of an issue for models of the mass and energy return by stellar populations. The observed properties of WR stars are simple and well enough established to suffice as a model ingredient for evolutionary synthesis.

5.2. Wind properties of an evolving population

I will begin with addressing the relative importance of individual stellar phases for non-radiant energy release by a population. The previous section covered wind properties of individual stars; now I will take into account the initial mass function and the stellar lifetimes to study the integrated wind properties of an entire population.

Let L_W be the total wind power of the population (calculated from $\frac{1}{2} \dot{M} v_\infty^2$ of each star) which is released into the ISM. The quantity is plotted versus time in Figures 26 and 27. The figures show the total wind power of a population of massive stars ($Z = Z_\odot$, $\alpha = 2.35$, $M_{up} = 120$ $M_\odot$, $M_{low} = 1$ $M_\odot$) formed during an instantaneous burst having a total mass of 1 $M_\odot$ (Figure 26) and with a continuous star-formation rate of 1 $M_\odot$ yr^{-1} (Figure 27), respectively. In the relevant epoch of the starburst, i.e. at $10^6 < t < 10^8$ yr, LBV's and RSG's are entirely negligible. This is a consequence of their short lifetimes and their low wind velocities. LBV's and RSG's gain in their importance relative to OB- and WR-stars in their output of the momentum flux P_W and $\dot{M}_W$ due to the weaker sensitivity of these quantities on v_∞. Even in those cases, however, LBV's and RSG's remain unimportant. Figures 26 and 27 suggest that the return of wind power (and of mass and momentum flux) is mostly dominated by OB stars and — if they are present — by WR stars. If stars form continuously, OB stars and WR stars are of roughly equal importance after the time an equilibrium between stellar birth and death has been reached ($t > 5 \times 10^6$ yr in

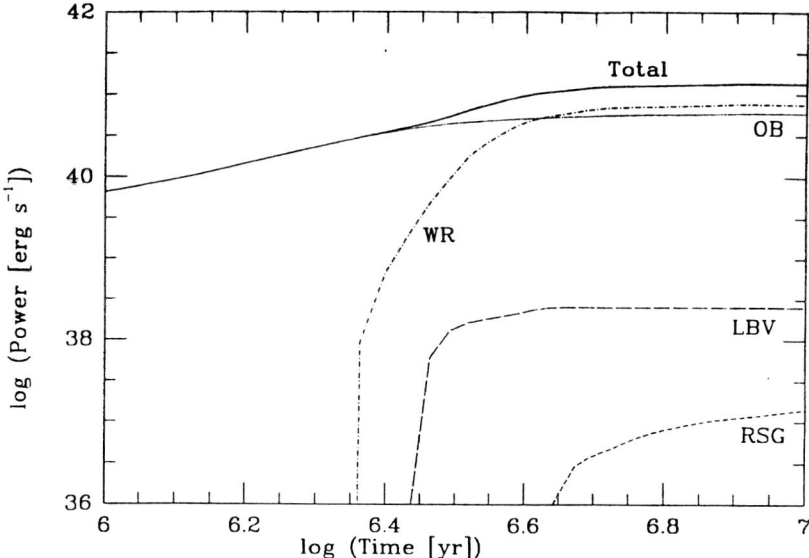

FIGURE 27. Total wind power of a population forming continuously, normalized to 1 M$_\odot$ yr^{-1}. $Z = Z_\odot$, $\alpha = 2.35$, $M_{\rm up} = 120$ M$_\odot$, $M_{\rm low} = 1$ M$_\odot$. Contributions from individual evolutionary phases to the total power are shown. From Leitherer et al. (1992b).

Figure 27). If, on the other hand, stars are formed instantaneously, WR stars dominate over OB stars for about 5×10^6 yr after the on-set of WR formation at $t \approx 3 \times 10^6$ yr. This time-span is controlled by the difference in the evolutionary timescale between the most and least massive O stars evolving into WR stars.

So far, the discussion of the mechanical luminosity output was restricted to stellar winds. How do stellar winds compare with supernovae? As a first example, I give an estimation of the mechanical energy release in our Galaxy. Observations suggest a total kinetic energy of 10^{51} erg associated with supernova explosions of massive stars (McKee 1990). There is no evidence for a strong dependence of this value on supernova type or on metallicity. I therefore assume that each massive star explosion releases a total energy of 10^{51} erg.

The Galaxy has experienced no strongly variable massive-star formation activity over the last 10^8 Myr so that the star-formation history in the Galactic disk can be approximated by a constant star-formation law with an equilibrium between birth and death of all stars with masses above 5 M$_\odot$. The injection of power into the ISM of the Galactic disk is completely dominated by supernovae (SN'e). The average total supernova rate in our Galaxy is approximately $SNR \approx 0.01$ yr^{-1} (van den Bergh 1991). This value includes type Ia, Ib, and II supernovae. Type Ia SN'e are probably related to explosions of a white dwarf in binary systems (Woosley 1986). The stellar progenitors of type Ib supernovae are being debated. They may or may not be related to massive stars. Supernovae of type II originate from massive single stars (Woosley 1986). Since SN'e of type II account for roughly 75% of all supernova events (van den Bergh 1991), I will adopt $SNR = 0.01$ yr^{-1} for the supernova rate due to massive stars in our Galaxy. The corresponding power injected into the ISM is 3.2×10^{41} erg s^{-1}. The total output due to winds in all stellar phases turns out to be only a few times 10^{40} erg s^{-1}. Stellar winds are only a minor contribution to the total power budget in the case of a constant star-formation scenario where all stars forming type II supernovae are in equilibrium. Abbott (1982) did an

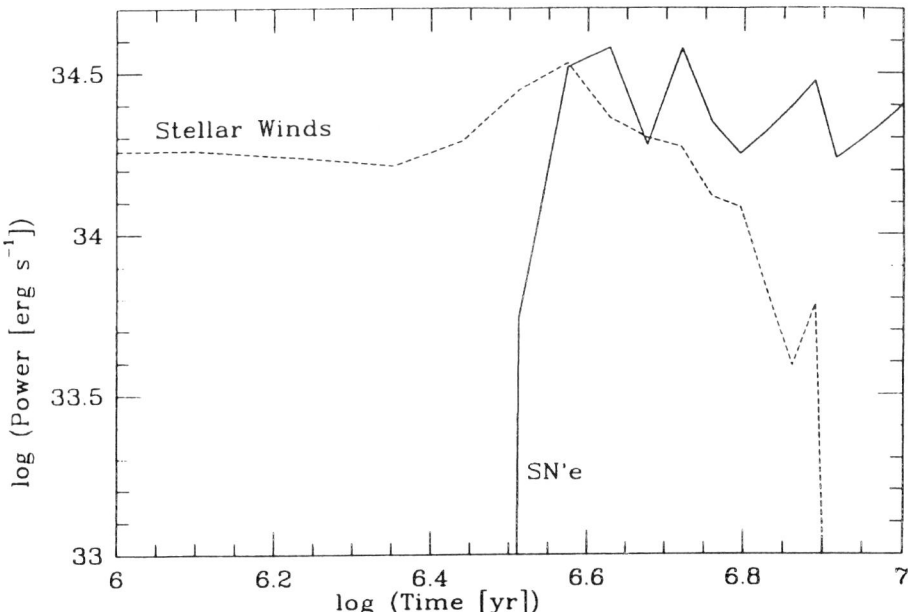

FIGURE 28. Contribution of stellar winds and supernovae to the total mechanical power in a single stellar population normalized to 1 $M_\odot$. $Z = Z_\odot$, $\alpha = 2.35$, $M_{\rm up} = 120\ M_\odot$, $M_{\rm low} = 1\ M_\odot$.

empirical estimation of the energy transfer rates in the solar neighborhood. His results agree with my simple estimate: winds play only a minor role for the heating of the local ISM. Stellar radiation dominates the energy budget: the total radiant energy output is 3×10^{40} erg s^{-1} kpc^{-2}, whereas the stellar-wind output is 2×10^{38} erg s^{-1} kpc^{-2} and the mechanical heating by supernovae is 1×10^{39} erg s^{-1} kpc^{-2}. The mechanical heating by stellar winds amounts to about 1% of the radiative heating (compare Table 6) and to about 20% of the supernova heating.

Stellar winds do become important as soon as young *bursts* of star formation are considered. The reason for the dominance of SN'e over winds in our Galaxy as a whole is the overwhelming contribution of stars below $M_\circ \approx 20\ M_\odot$ to the total supernova rate. The lifetime of stars in this mass range is larger than 10 Myr so that SN'e from these stars are not observed in a young starburst of smaller age. This is in agreement with Figure 28 where the power injected by SN'e and stellar winds is shown for a single stellar population. Three phases can be seen: during the first 3×10^6 yr of the burst SN'e are still absent and only winds are important. Between 3 and 6×10^6 yr winds and SN'e are of roughly equal importance. This time range is most relevant for the interpretation of Galactic and extragalactic H II regions. Therefore one expects both stellar winds and supernova explosions to be responsible for the high-velocity shells seen in 30 Dor (Section 2.2.2), for the exceptionally strong interstellar absorption lines in NGC4214#1, and for the large [O I]/Hβ ratios in H II galaxies (Section 4.5).

5.3. *Massive stars and the chemical enrichment of the ISM*

The surfaces of massive stars become enriched in elements affected by nuclear processes due to high mass loss and internal mixing. In Figure 29 the mass in individual elements returned to the ISM by winds is shown. The curves reflect the increasing helium and

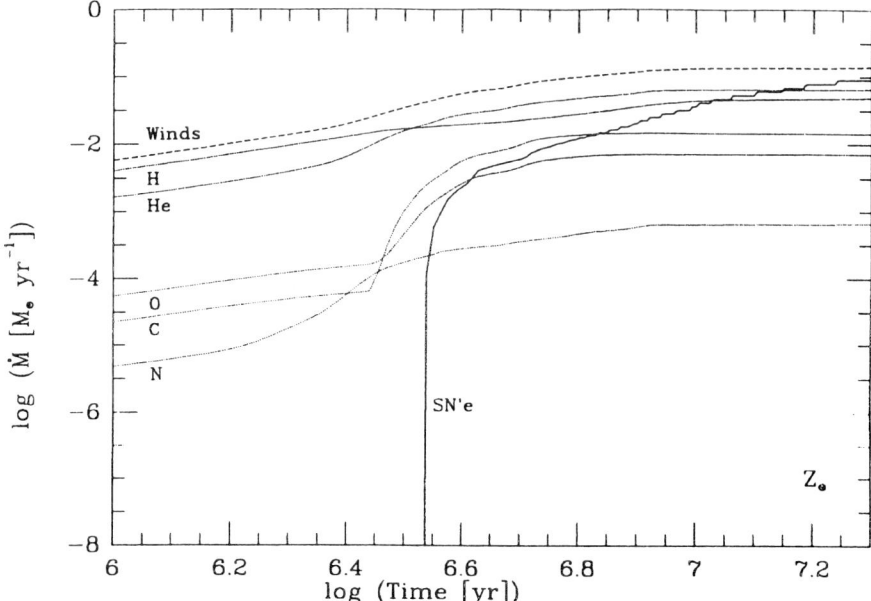

FIGURE 29. Mass deposition rate of stellar winds and supernovae in a system with a constant star formation rate of 1 $M_\odot$ yr^{-1}. Individual elements returned by stellar winds are plotted. $Z = Z_\odot$.

CNO abundances with later evolutionary state. The sharp increase in CNO around ~3 Myr is caused by the first occurrence of Wolf-Rayet stars. Initially, nitrogen-rich WN stars dominate and N/O increases. Observations of nebular ejecta around individual WR stars suggest significant enrichment of He and N, and depletion of O (Smith 1996). The occurrence of carbon-rich WC stars during the final stages of stellar evolution reverses the behavior of N/O: carbon and oxygen become more enhanced than nitrogen. This is a consequence of very strong mass loss in this particular evolutionary phase.

Do stellar winds and supernova explosions produce significant, i.e. observable enrichment of the *global* interstellar gas? The chemical evolution of H II galaxies can be understood if helium and secondary nitrogen pollute the ISM of the galaxies (Pagel 1987). The fraction of material not locked into stars but eventually returned to the ISM critically depends on the number of low-mass stars formed during the starburst. If $M_{\rm low} = 1$ $M_\odot$, about 10% of the gas originally locked into stars will be lost again by stellar winds. This implies the need for a high star-formation efficiency (= mass of all stars formed during the burst over the mass of the interstellar gas before the burst) if significant enrichment is expected. Otherwise the chemically-enriched wind material will be strongly diluted by left-over gas from the star-formation episode.

Model calculations by Cid-Fernandes et al. (1992) imply that efficiencies of about 50% lead to CNO overabundances by a factor of 2 in the interstellar medium of a starburst. Theoretical models for the formation of massive stars suggest much lower star-formation efficiencies of about 5% (Larson 1987). On the other hand, star-formation efficiencies around 70% have been derived in the centers of the starburst galaxies M82 and M83. Walker et al. (1993) used mm and sub-mm line observations in CO to derive gas masses and compared them to the masses of the newly formed stars. Their results suggest that the star-formation episode consumed most of the molecular gas and that stellar winds

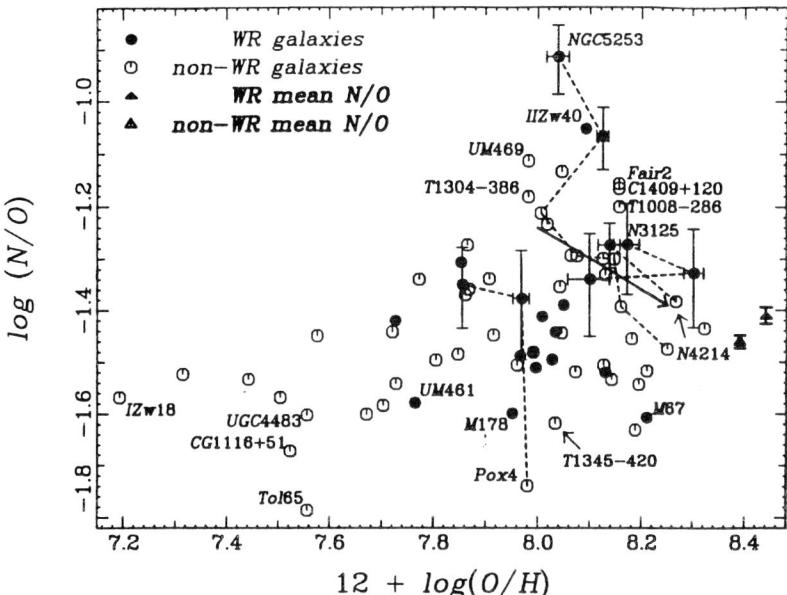

FIGURE 30. N/O versus O/H for a sample of 60 H II galaxies with and without WR stars. The two sub-samples show the same correlation, indicating that WR stars are not the main source of chemical enrichment. From Kobulnicky & Skillman (1996).

and supernovae (if they could be observed in the optical) may significantly enrich the ISM to measurable levels.

Observational evidence for chemical enrichment by stellar winds in H II galaxies exists but the theoretical interpretation is still controversial. Esteban & Peimbert (1995) modeled the expected chemical enrichment based on the observed stellar population. While abundance anomalies are observed in their sample of galaxies, the models could not satisfactorily reproduce He, N, and O in all galaxies. Kobulnicky & Skillman (1996) compared abundance anomalies in two samples of H II galaxies, one with and the other without WR stars. The sample containing WR stars is by definition skewed towards starbursts which have the strongest mass injection by winds and the highest chemical enrichment. Their result is in Figure 30. There is no significant difference between the two sub-samples. Abundance anomalies are observed both in galaxies with and without WR features. The best-studied example is NGC5253. Walsh & Roy (1989) and Pagel et al. (1992) found a region with clear nitrogen overabundance which also contains WR stars. The WR winds may be responsible for the N enrichment, and at the same time for the removal of the ambient ISM. Uncertain model predictions for the wind chemistry and mixing and dispersal process in the ISM are probably responsible for our lack of understanding of the observed abundance patterns.

5.4. Variations of metal abundance and IMF

The chemistry of the stellar birthplace affects the mass and energy return in a twofold way. First, the stellar-wind properties correlate with the metal abundance; second, higher metal-abundance favors the formation of WR stars with their powerful winds. Both

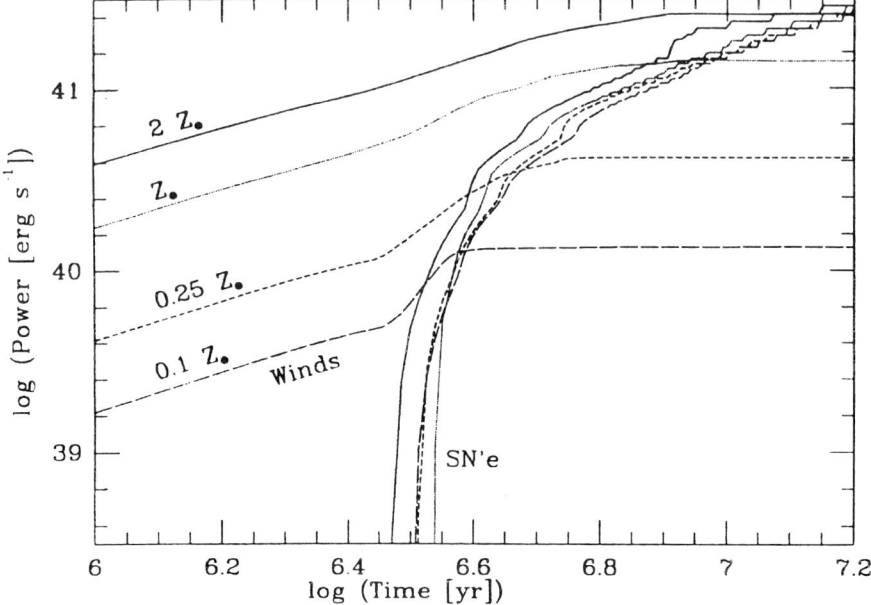

FIGURE 31. Energy deposition rates for a system forming stars continuously at a rate of 1 M$_\odot$ yr^{-1}. The same parameters as in the previous figures apply. Note the strong Z-dependence of the wind luminosity.

effects lead to the prediction of much higher deposition rates in a high-Z environment than at lower metallicities.

Figure 31 gives the energy deposition rate expected for a continuous star-formation rate of 1 M$_\odot$ yr^{-1} with $\alpha = 2.35$. There is no observational evidence for a Z-dependence of the total kinetic energy associated with a type II supernova event, which has been taken as 10^{51} erg in all cases. Except for small differences introduced by the pre-SN evolution, the mechanical energy release by supernovae alone is essentially independent of Z. In contrast, the wind luminosity varies by more than an order of magnitude over the metallicity range displayed in Figure 31. It is found that $L_W \propto Z$. This is simply a consequence of the functional dependence of $\dot{M}$ and v_∞ on Z (see Section 5.1.4). The age of the starburst hardly affects this result. If $Z \geq Z_\odot$, stellar winds are more important contributors to the total luminosity injection (and to the momentum and mass flux as well) than supernovae in starbursts younger than about 20 Myr. This is the situation in starburst galaxies with nuclear star formation. Conversely, low Z suppresses the output of matter and energy by winds, and stellar winds become negligible as soon as the first supernovae appear around ∼3 Myr.

The input of stellar material and related quantities into the surrounding interstellar environment depends on the relative frequency of massive versus less massive stars. This implies a strong dependence on the IMF slope and the cut-off masses. Figure 32 confirms this expectation. In this figure I show the total mechanical luminosity of a population of massive stars forming continuously with a star-formation rate of 1 M$_\odot$ yr^{-1} between 1 and 120 M$_\odot$. The metallicity is solar. A flatter IMF increases the power output due to the higher number of massive stars relative to low-mass stars. Varying the IMF slope α between 1.5 and 3.0 (the range measured in the local universe) modifies the wind power by nearly a factor of 50. This confirms my previous rough estimate that

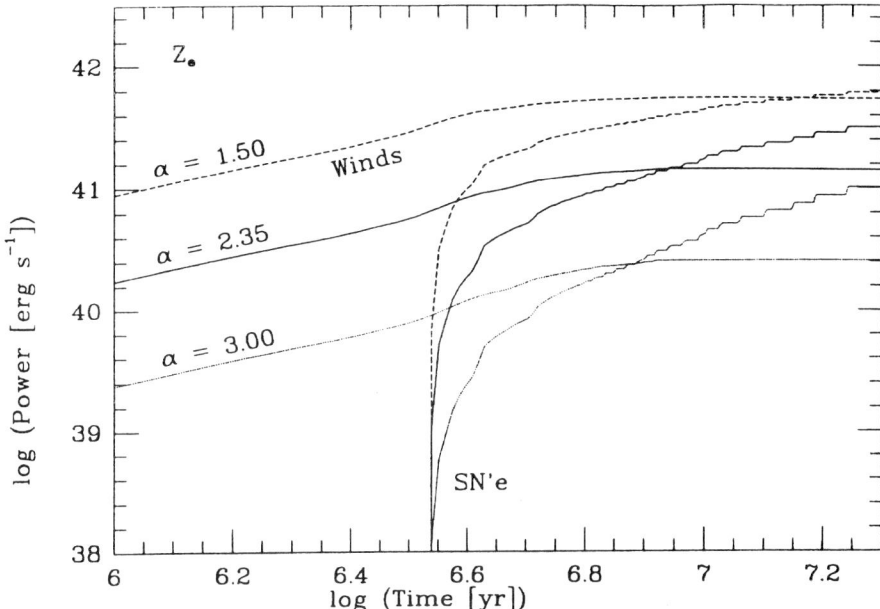

FIGURE 32. Mechanical luminosities for a system forming stars continuously at a rate of 1 $M_\odot$ yr^{-1}. The same parameters as in the previous figures apply. Note the strong IMF-dependence of the wind luminosity.

the output of mechanical luminosity is strongly weighted towards the most massive stars (Section 3.2.2). The relative contributions from winds and SN'e are affected in such a way that a flatter IMF favors winds over SN'e for a longer time until SN'e take over. If α is smaller, the typical SN progenitor has a lower mass, and the first SN explosions occur later due to the longer lifetimes of less massive stars. SN'e will always dominate at late phases of the evolution and in systems having low metallicity.

5.5. *The relative importance of mechanical and radiant luminosity*

The models I have discussed above permit an estimation of the rate at which radiant *and* mechanical energy are injected into the interstellar medium by massive stars and their evolutionary byproducts. These estimates are of considerable interest for the energy balance of galaxies. The ionization of the interstellar medium is regulated primarily by the Lyman continuum produced by massive stars, while the heating of the dust grains (whose infrared emission can sometimes comprise most of the bolometric luminosity, see Sanders & Mirabel 1996) is primarily provided by the non-ionizing radiation of the massive stars. The injected mechanical energy from massive stars and supernovae can power galactic-scale outflows ("superwinds" or "superbubbles"), such as those observed in NGC1569 (Figure 6). An estimate of the rate of mechanical energy deposition $L_{\rm mech} = L_{\rm W} + L_{\rm SN}$ is also important for understanding the global energy balance in the interstellar medium of normal galaxies like the Milky Way (cf. McKee & Ostriker 1977). In particular, the role of mechanical heating in maintaining the thermal and dynamical state of the hot and warm diffuse ionized gas in the disk and halo of the Milky Way and similar galaxies is a matter of on-going debate (Reynolds 1991; Dettmar 1993). I will first discuss the mechanical energy input, and will then consider the Lyman continuum, following the argumentation given in Leitherer & Heckman (1995).

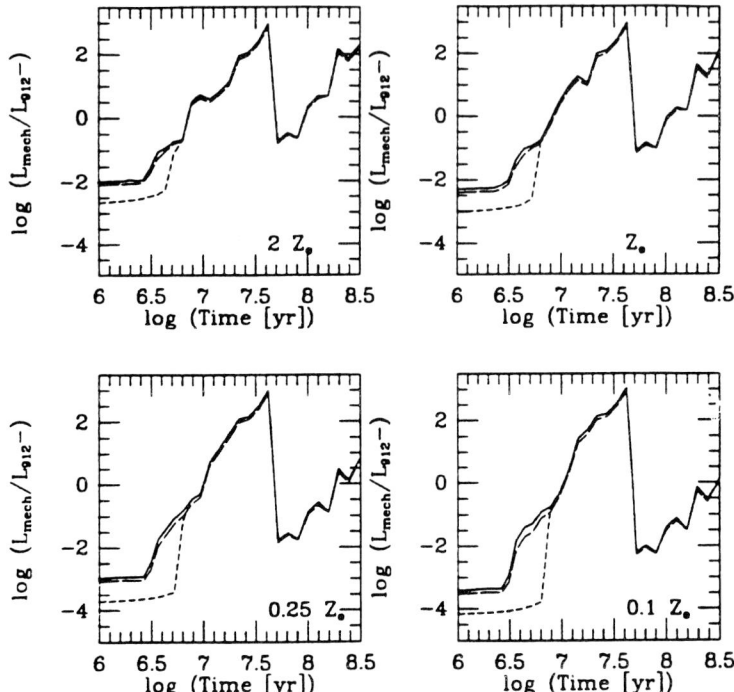

FIGURE 33. Ratio of the mechanical luminosity over the ionizing luminosity for an instantaneous burst having different IMF's. The four figures are for metallicities of 2 $Z_\odot$, $Z_\odot$, 0.25 $Z_\odot$, and 0.1 $Z_\odot$. Solid line: $\alpha = 2.35$; long-dashed: $\alpha = 3.3$; short-dashed: truncated Salpeter IMF with $M_{up} = 30$ $M_\odot$. The rise of the ratio after 50 Myr is an artifact and should be ignored. From Leitherer & Heckman (1995).

L_{mech} (and E_{mech}, its integral over some specified period of time) can be estimated if an assumption is made concerning the metallicity, the IMF, and the history of star formation in a galaxy. To see this, I will discuss the ratios of L_{mech}/L and L_{mech}/L_{912^-} for an instantaneous burst since the time evolution of these ratios are then most dramatic (see Figure 33). The bulk of the Lyman continuum is produced by the most massive stars (early O stars), and these have characteristic lifetimes of only a few million years. In contrast, the majority of the mechanical energy is supplied by supernovae (except during the first 6 Myr or so). Since the typical supernova progenitors are considerably less massive than early O stars, they have typical lifetimes of several tens of millions of years. Thus, between an instantaneous burst age of 1 Myr and 40 Myr, there is an overall strong increase in the ratio of L_{mech}/L_{912^-} by a factor between 10^5 and 10^7 (with this value depending on metallicity and IMF — see below). The evolution of the ratio L_{mech}/L shows a variation which is similar to, but less dramatic than the time evolution of the ratio L_{mech}/L_{912^-}: the dependence of L on stellar mass (and hence lifetime) is not nearly as steep as in the case of L_{912^-}.

The dependence of the ratios of L_{mech}/L_{912^-} and L_{mech}/L on metallicity and IMF follows from our earlier discussions. At early times (before supernovae occur) these ratios increase with increasing metallicity, primarily because the mechanical energy is then supplied by stellar winds at a rate that is strongly metallicity dependent (see Figure 31). Variations in the IMF have a similar effect during the early pre-supernova phase: during

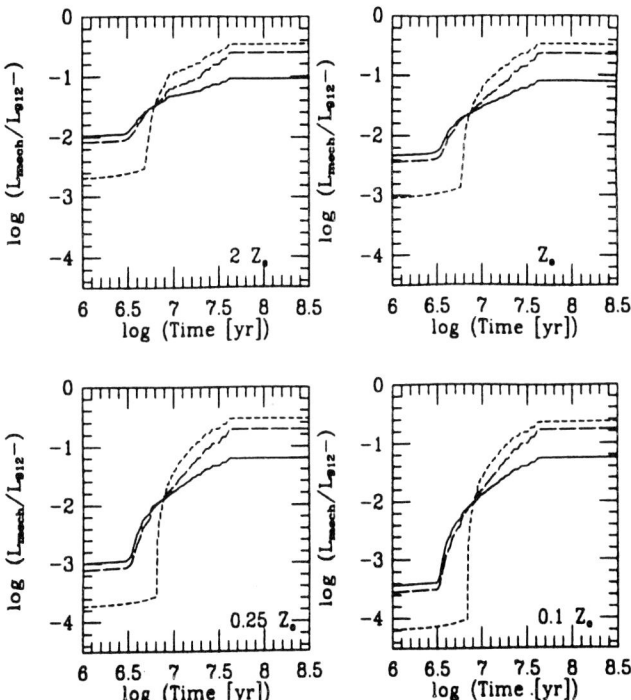

FIGURE 34. Ratio of the mechanical luminosity over the ionizing luminosity for stars forming continuously with three different IMF's. The four figures are for metallicities of 2 $Z_\odot$, $Z_\odot$, 0.25 $Z_\odot$, and 0.1 $Z_\odot$. Solid line: $\alpha = 2.35$; long-dashed: $\alpha = 3.3$; short-dashed: truncated Salpeter IMF with $M_{\rm up} = 30$ $M_\odot$. From Leitherer & Heckman (1995).

this stage, the ratio of $L_{\rm mech}$ is a rising function of stellar mass (see also Figure 3.2.2). Thus, a steeper IMF (larger α) results in a ratio of $L_{\rm mech}/L_{912-}$ and $L_{\rm mech}/L$ that is about a factor of two lower than that for our standard Salpeter IMF. The truncated Salpeter IMF (with an upper mass cut-off at 30 $M_\odot$) produces more dramatic results: compared to the standard Salpeter IMF, $L_{\rm mech}/L_{912-}$ and $L_{\rm mech}/L$ are decreased by about an order of magnitude during the first 3 Myr, but by about two orders of magnitude during the interval from 3 to 5 Myr. This is because the truncated IMF is very deficient in WR stars which dominate $L_{\rm mech}$ during this epoch. Finally, it is clear that once supernovae kick in and then dominate $L_{\rm mech}$ ($t > 6$ Myr), there is very little dependence of $L_{\rm mech}/L_{912-}$ or $L_{\rm mech}/L$ on IMF or metallicity.

Next the case of a constant star-formation rate is considered. $L_{\rm mech}/L_{912-}$ is in Figure 34. $L_{\rm mech}/L$ behaves similarly but is not shown for the sake of brevity. This case is far more relevant to normal star-forming galaxies, and even some starburst galaxies may be better approximated by a model of constant star formation for 10 to 100 Myr than by an instantaneous burst (e.g., Bernlöhr 1993; Rieke et al. 1993). These minimum timescales are plausible physically, since the minimum characteristic timescale for a starburst to "turn on" or "turn off" is (on simple grounds of causality) very unlikely to be much less than the characteristic dynamical timescale for the star-forming region (i.e. signals will propagate across the star-forming region no more rapidly than the free-fall or shockwave crossing time). These timescales are of-order a few times 10^8 yr for entire galaxies and 10^6 yr for extreme starbursts (Heckman 1994).

At early pre-supernova times, the ratio $L_{\rm mech}/L$ evolves in time and depends on IMF and metallicity in ways that follow immediately from the behavior of the instantaneous burst model discussed above. This behavior is largely due to the dependence of the mechanical luminosity of stellar winds on stellar mass and metallicity. At intermediate times (about 7 to 30 Myr) there is very little dependence of $L_{\rm mech}/L$ on IMF, but there is still a significant metallicity dependence. This can be understood because stellar winds are still making a very significant contribution to $L_{\rm mech}$ for the high-metallicity model at this evolutionary phase. However, at late stages (after about 30 Myr) when supernovae dominate, the ratio $L_{\rm mech}/L$ approaches a roughly constant value of about 1% for all metallicities.

The time evolution of $L_{\rm mech}/L_{912-}$ and its dependence on metallicity and IMF is particularly interesting (Figure 34). During the early pre-supernova phase ($t < 6$ Myr), the ratio $L_{\rm mech}/L_{912-}$ is larger for IMF's with relatively larger number of more massive stars. This is because during this stellar-wind dominated phase, the mass dependence of $L_{\rm mech}$ is steeper than that of L_{912-}. The effect is clearest in the model with an upper mass cut-off of 30 $M_\odot$, where $L_{\rm mech}/L_{912-}$ is about a factor of five lower than for the other two IMF's. In contrast, at later times when supernovae dominate $L_{\rm mech}$, the ratio of $L_{\rm mech}/L_{912-}$ is lower for IMF's favoring massive stars (the standard Salpeter IMF has $L_{\rm mech}/L_{912-}$ that is a factor of about three lower than for the other two cases). The reason is clear: the bulk of the mechanical energy supplied by supernovae comes from stars of lower mass than those that supply the bulk of the Lyman continuum. The general trend for $L_{\rm mech}/L_{912-}$ to be lower for the lower metallicity systems is due to the metallicity dependence of the stellar wind mechanical power. Thus, these differences are large only during the early pre-supernova phase.

In view of the above, how can one *measure* $L_{\rm mech}$ for a galaxy? For large systems, the models of constant star-formation over timescales of 10 to 100 Myr are probably the most appropriate. For these models and timescales, $L_{\rm mech}/L$ can be determined from L. That is, the ratio of $L_{\rm mech}/L$ is relatively constant at a value of about 1%. For more quiescent galaxies (e.g. systems forming stars at a constant rate over a Hubble time), one can use the models of Bruzual & Charlot (1993) to predict L. These models predict that L increases by only about 60% between $\log t = 8.5$ and $\log t = 10.0$. Thus, the models predict that $L_{\rm mech}/L \approx 0.6\%$ would be appropriate in such a quiescent system (excluding the significant contribution from type Ia supernovae).

In contrast to normal star-forming galaxies — where the duration of the star-formation is much longer than the lifetime of a typical supernova progenitor (so that the constant star-formation models are a reasonable approximation) — there are systems for which this is not the case. Giant H II regions are one clear example, and the small-scale starbursts that occur in dwarf galaxies may be another (see Section 3.2.1). In these cases, the instantaneous burst models are more appropriate, and the ratio of $L_{\rm mech}/L_{912-}$ exhibits strong time evolution. As can be seen in Figure 33, during the era from about 10 to 50 Myr, $L_{\rm mech}/L_{912-}$ increases from about unity to over 200, so that the heating and ionization of the ISM would be dominated by supernova-driven shocks. Galaxies in this phase would be called post-starbursts. Do we ever observe galaxies in this evolutionary phase? Bona fide examples of such systems may exist among the sample of IRAS-selected galaxies. For example, Koornneef (1993) has suggested that the compact infrared source in the center of the nearby edge-on galaxy NGC4945 may be such a post-starburst, and systems like NGC6240 and Arp220 (Beck 1994) may belong to this category.

Even in the "mundane" case of a constant star-formation rate, mechanical heating of the ISM is not entirely negligible compared to heating via photoionization by hot stars. As Figure 34 shows, the steady-state ratio of $L_{\rm mech}/L_{912-}$ (achieved after about 50 Myr) ranges from a few % (normal Salpeter IMF and 10% solar metallicity) to up to nearly 50% in extreme cases ("truncated" Salpeter IMF and twice solar metallicity). This is consistent with empirical estimations for the energy balance in the solar neighborhood (see Section 5.2). Abbott (1982) determined $L_{\rm mech}/L_{912^-} \approx 1\%$ from a direct census of the massive-star population within 3 kpc from the Sun.

6. Massive stars and the dynamics of the ISM

The goal of this section is to give an overview of the effects of stellar winds and supernovae on the ISM. This is the "input" aspect I mentioned before. How does the injected mass, momentum, and energy affect the structure and dynamics of the ISM? On a local scale, the interaction of winds with the ambient ISM leads to distinct, shell-like morphologies. If the processes become more energetic, for stellar populations on a galactic scale, the global ISM in a galaxy can be disrupted and driven out of the gravitational well. Numerous excellent reviews on the subject exist. A large concentration of relevant material is in the conference proceedings edited by Cassinelli & Churchwell (1993), Franco, Ferrini, & Tenorio-Tagle (1993), and Kunth et al. (1995).

6.1. Wind-blown bubbles around massive stars

The mechanical energy released by winds over the lifetime of a massive star is on the order of 10^{51} erg (Section 5), comparable to the energy in supernova shells. This led to the early suggestion that the cavity of the Rosette Nebula might be caused by strong stellar winds from the central stars (Matthews 1966). Subsequent theoretical studies, e.g. by Dyson (1973), demonstrated that strong winds can sweep up the surrounding ISM into thin, dense shells. In this section I will first describe the theory for wind-blown bubbles and then present several observed examples.

6.1.1. Theory of wind-blown bubbles

The first complete theoretical study of the interaction between a stellar wind and the surrounding interstellar gas was done by Castor, McCray, & Weaver (1975) and Weaver et al. (1977). See also the review by Shull (1993). The evolution of a wind-driven shell is analogous to that of a supernova shell (Woltjer 1972; see also below). Initially the stellar wind is in a brief phase of free expansion that lasts until a sufficiently large mass of interstellar gas has been swept up to stop the expansion. The duration of this phase is a few hundred years, at which time a bubble of radius 0.1 to 1 pc has formed. The resulting bubble has a characteristic 3-component structure:

• The interior is filled with a supersonic wind at a velocity of about 2000 km s^{-1} (Table 6). The wind expands freely out to a radius R_1.

• The region between R_1 and R_2 (farther out) is filled with shocked wind material. The temperature is between 10^6 and 10^7 K due to shock heating. This regions contains most of the mass released by the wind over the time t the bubble has formed: $M = \dot{M}t$. Since the region is hot, it expands with a velocity $v_2 = dR_2/dt$.

• Outside R_2 (called Region 3), a shell of swept-up interstellar gas has formed as a result of the expansion of Region 2. It has an outer radius R_3 and consists of pure interstellar material. This shell is highly supersonic, creating strong shocks due to interaction with the ISM, and is heated well above 10^6 K. Radiative cooling is negligible at such high temperatures (Raymond, Cox, & Smith 1976) so that this shell (and the region between R_1 and R_2 filled with shocked wind material) expand adiabatically. This phase is therefore called adiabatic phase.

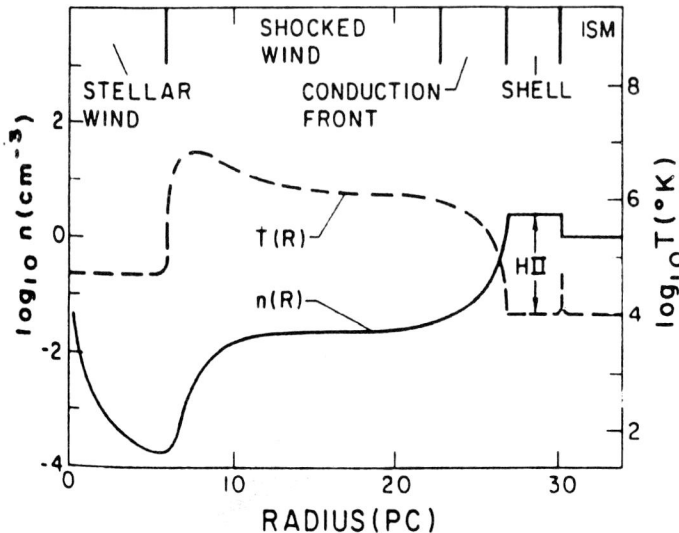

FIGURE 35. Temperature and density structure of an interstellar bubble for which $L_W = 1.27 \times 10^{36}$ erg s^{-1}, $N_\circ = 1$ cm^{-3}, and $t = 10^6$ yr. From Weaver et al. (1977).

The expansion of Regions 2 and 3 is slowly decelerating at a rate $\propto t^{-2/5}$. The model of Weaver et al. predicts that 45% of the available wind energy is used to heat Region 2 and the remaining 55% are used to heat and drive Region 3. The adiabatic phase will come to an end when its age becomes comparable to the timescale for radiative cooling, which is at

$$t \approx 1.7 \times 10^3 \sqrt{\frac{L_W}{N_\circ}}. \tag{6.16}$$

t is the age in years, L_W is the wind luminosity in 10^{36} erg s^{-1} (cf. Table 6), and $N_\circ$ is the ISM density in cm^{-3}. Inserting typical numbers, it is evident that the adiabatic phase lasts only a few thousand years before radiative cooling becomes important. At that time the bubble has a radius R_3 of somewhat less than 1 pc. The probability to observe this phase is negligibly small.

As soon as the shell of swept-up interstellar material has reached a temperature of somewhat below 10^6 K, radiative cooling becomes important and the shell quickly cools down to 10^4 K. The bubble has reached the radiative phase, also called the "snow-plow" phase, which is long enough and which produces a bubble large enough to be detectable by observations. The density and temperature structure of the bubble are in Figure 35. The boundary between the cold ($\sim 10^4$ K) shell and the ISM is an isothermal shock. The high compression rate leads to a very thin structure. The hot shocked wind region is still too hot for radiative cooling and continues to expand adiabatically. The temperature gradient at the boundary between the hot wind region and the cold shell is so large that thermal conduction by electrons transfers heat to the shell. As a result, the inner surface of the shell evaporates and material flows into the hot wind region, increasing the mass in the hot region. Notice that the density and the mass in the shell are orders of magnitude higher than in the shocked wind region (see Figure 35). The schematic in Figure 35 suggests that there exist five distinct regions:

• the freely expanding stellar wind;

- the hot, shocked wind;
- the conduction front;
- the swept-up shell;
- the ambient ISM.

The evolution of the structure can be described by solving the equations of momentum and energy. The most interesting quantities are the radius of the bubble (roughly R_2), its expansion velocity (v_2), and the thickness of the swept-up shell (d_3). Weaver et al. (1977) found:

$$R_2 = 28 \left(\frac{L_W}{N_o}\right)^{1/5} t^{3/5} \tag{6.17}$$

$$v_2 = 17 \left(\frac{L_W}{N_o}\right)^{1/5} t^{-2/5} \tag{6.18}$$

$$d_3 = 3 \times 10^{-4} T_{shell} \mu^{-1} \left(\frac{L_W}{N_o}\right)^{-1/5} t^{7/5} \tag{6.19}$$

R_2 in pc, L_W in 10^{36} erg s^{-1}, N_o in cm^{-3}, t in Myr, v_2 in km s^{-1}, d_3 in pc, T_{shell} in K. μ is the mean molecular weight in the shell. A typical bubble has a size of several tens of pc, expands at velocities of tens of km s^{-1}, and has a shell thickness of a few pc. The thickness of the shell depends strongly on its ionization state. It shrinks from a few pc to a fraction of a pc if it recombines from ionized to neutral, with a corresponding drop of T_{shell} from 10^4 K to 10^2 K.

The numerical example assumed typical O-star parameters. Similar structures will develop around WR stars, the only difference being somewhat larger bubble sizes and higher expansion velocities. L_W of WR stars is 1 to 2 orders of magnitude larger but since $R_2 \propto L_W^{1/5}$, the bubble has nearly the same size. After a few 10^6 yr the mass losing star will evolve into a RSG if its initial mass was less than $\sim$50 M$_\odot$. The wind velocity drops drastically in this stage. The bubble can continue its expansion until it stalls by the external pressure of the ISM. However, the RSG will explode as a supernova and repressurize the bubble. If the star has an initial mass above 50 M$_\odot$, it will likely form a supernova immediately following the hot-star phase. Hydrodynamical models for wind-blown bubbles around stars evolving through different evolutionary stages have been published by García-Segura, Mac Low, & Langer (1996a) and García-Segura, Langer, & Mac Low (1996b).

The physics of the interaction between supernovae and the ISM is analogous to the stellar wind case. Review articles are those of Woltjer (1972), Chevalier (1977, 1991), and Shull (1993). The main difference between the two processes is the timescale of the energy release. The stellar-wind phase last for several Myr and provides a constant supply of mechanical luminosity during that period, while a supernova event occurs essentially instantaneously. Since the total luminosities associated with a stellar-wind-blown bubble and a supernova bubble are similar, supernova bubbles have properties similar to those described above. An instructive comparison between both types of bubbles has been made by Lamers (1983a,b).

6.1.2. *Observed properties of wind-blown bubbles*

Several examples of wind-blown bubbles around Galactic hot stars are known. A catalog is given in Chu's (1991) review. The two best-studied cases among O stars are the bubbles around BD+60°2522 and HD148937.

BD+60°2522 is an O6.5 III star surrounded by a spherical structure, whose name

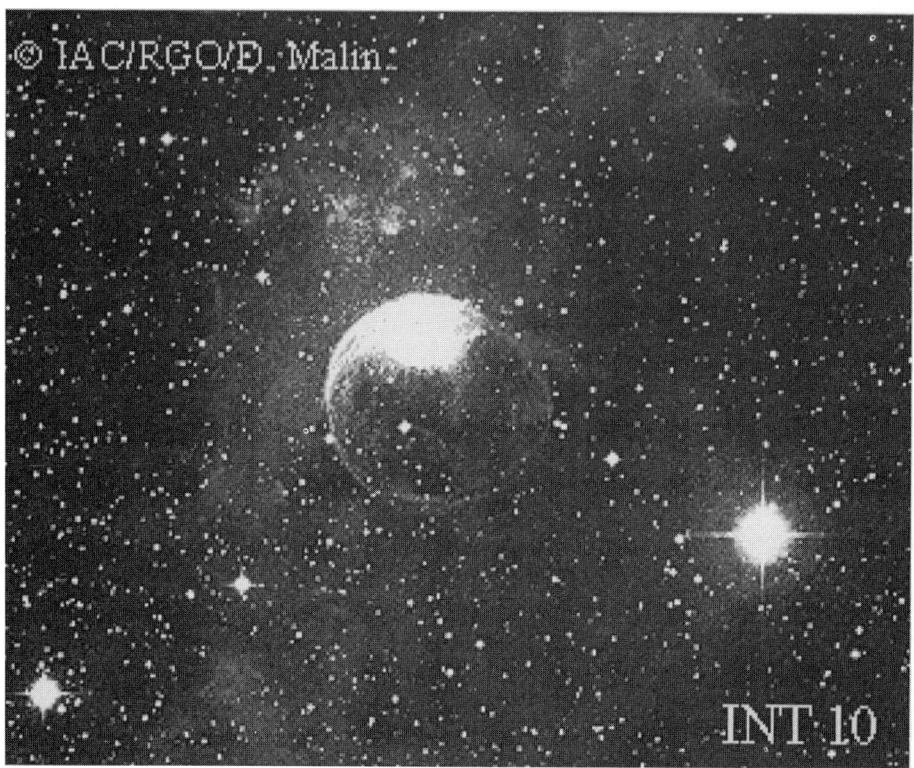

FIGURE 36. Optical image of the "Bubble Nebula" NGC7635 obtained with the Isaac-Newton Telescope. The central star BD+60°2522 is in the northern region of enhanced emission inside the bubble.

"Bubble Nebula" becomes immediately clear from an inspection of Figure 36. The nebula has a radius of a few pc (which is on the low side when compared to the prediction of the theory) and an expansion velocity of 25 km s^{-1} (Dufour 1989). The low expansion velocity and small size lead to an age estimate of only about 1 Myr. The mass of the bubble material is a few M$_\odot$. The high [O III]/Hβ ratio ($\sim$7) suggests non-radiative excitation due to shocks.

HD148937 (O6.5f?p) and its surrounding nebulosities are a spectacular example of an O star creating a Strömgren sphere via its ionizing radiation, forming a thin, wind-blown bubble, and ejecting processed material in the form of a bipolar flow. All three phenomena are visible in Figure 37. The most relevant structure for our discussion is the thin, filamentary nebulosity in the lower left panel of Figure 37. Deep images show this to be a complete ellipsoidal structure around HD148937. The star is surrounded by bipolar ejecta that are known to be chemically enriched (Leitherer & Chavarría 1987; Dufour, Parker, & Henize 1988). This suggests that HD148937 is relatively evolved and that the filamentary nebulosity resulted from the wind interaction in the past evolutionary history of HD148937.

The majority of wind-blown bubbles are detected around WR stars (e.g., Chu 1982; Treffers & Chu 1982). One of the most prominent examples is the ring nebula around the WN6 star HD192163. Detailed studies were performed, e.g., by Treffers & Chu (1982), Marston & Meaburn (1988), and Esteban et al. (1992). The nebula is found to be a

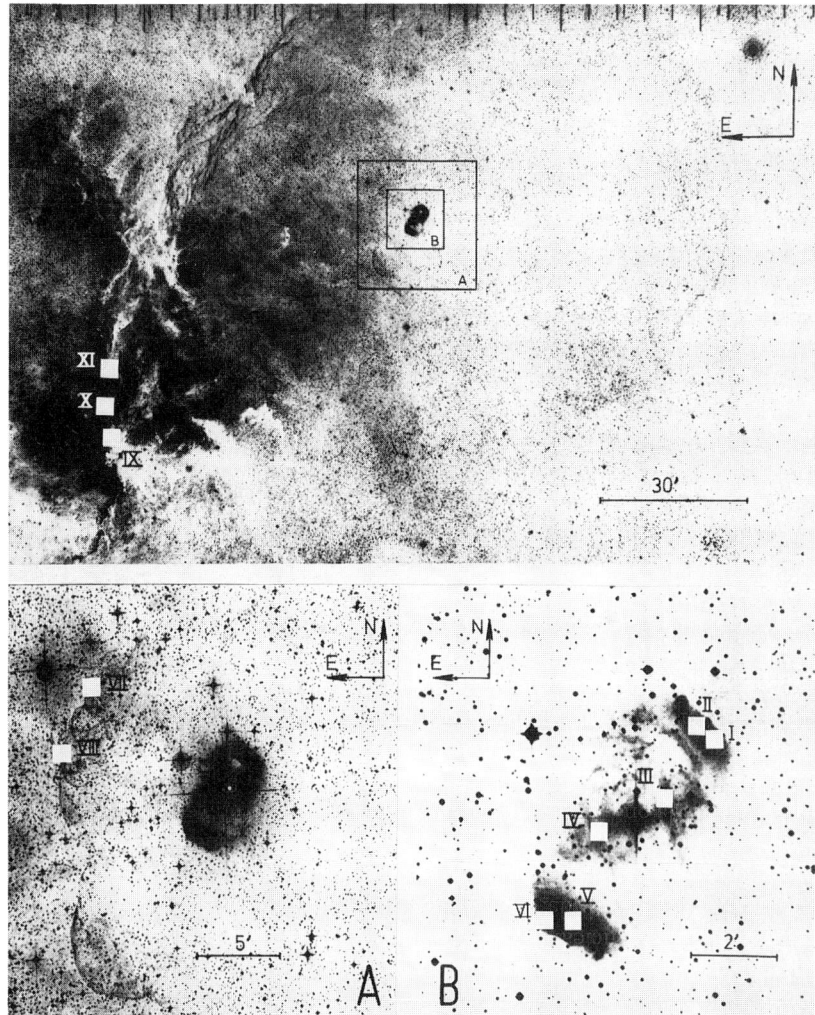

FIGURE 37. Reproductions from an ESO-SRC plate showing the different nebular complexes around HD148937. A and B are enlargements from the upper plate. White boxes indicate locations where spectra were obtained. From Leitherer & Chavarría (1987).

pressure-driven bubble expanding at a velocity of 80 km s^{-1}. An important result is the general presence of chemically processed material in bubbles surrounding WR stars. This suggests significant mixing between the chemically enriched WR winds and the ambient ISM, as well as a high fraction of WR material in the dense shell. The bubble model of Weaver et al. (1977) predicts the shell to be pure interstellar material. The clumpy structure of the winds and the ISM may contribute to the mixing. High resolution images of WR nebulae (e.g. RCW58; Chu 1982) show pronounced substructure, such as filaments and clumps.

6.2. Superbubbles around massive-star populations

The original theory by Weaver et al. (1977) provides the basic description for a wind-blown bubble. The stellar wind in their model is assumed to remain unchanged in the course of the bubble evolution. In reality, hot-star winds do change significantly with stellar lifetime, and massive stars form in clusters. Therefore very few stars are isolated enough that a bubble can form around a single star. A typical wind-blown bubble is powered by a stellar population. An obvious modification to the original theory is the inclusion of a time-dependent wind luminosity, provided by a stellar population following an initial mass function.

The generalization of the bubble theory from single stars to populations was done by McCray & Kafatos (1987), who followed the bubble evolution around a typical OB association. The dynamics of the "superbubble" can be approximated by scaling up the wind-bubble theory for a continuous energy input. Associations typically contain 10 to 100 OB stars with powerful winds and the potential to explode as supernovae. The radius and the velocity of a superbubble powered by a time-dependent wind can be derived by combining the equations for energy and momentum conservation into a single, third-order dynamical equation:

$$\frac{d}{dt}\left[R_2^3 \frac{d^2}{dt^2}(R_2^4)\right] = \frac{6R_2^2}{\pi\rho_o} L_{\mathrm{mech}}(t), \qquad (6.20)$$

where $L_{\mathrm{mech}}(t)$ denotes the time-dependent wind input (see Shull & Saken 1995). During the supernova phase (after about 10^7 yr) this energy input can be expressed as the product of the IMF, the mass-age relation, and the supernova energy. The mass-age relation for massive stars is $t(m) \propto m^{-\gamma}$ with $\gamma \approx 1.2$ (Schaerer et al. 1993). Therefore we have:

$$L_{\mathrm{mech}}(t) \propto m^{-x}\frac{dm}{dt} \propto t^{(x/\gamma)-1}. \qquad (6.21)$$

For a Salpeter IMF with $x = 1.35$ we find that $L_{\mathrm{mech}}(t) \approx$ constant. This essentially says that there are more and more supernova events for lower masses since there are more progenitors available (due to the IMF), but at the same time it takes longer for a star to turn into a supernova (due to the lifetime). The net effect is a nearly constant mechanical luminosity for the next 50 Myr.

For a constant wind luminosity $L_{\mathrm{mech}} = L_o$ this equation has a self-similar solution,

$$R_2 = \left(\frac{125 L_o t^3}{154\pi\rho_o}\right)^{1/5} \qquad (6.22)$$

$$v_2 = \frac{3R_2}{5t}. \qquad (6.23)$$

These expressions are identical to those for a bubble around a single star, except for constants. Superbubbles are simply scaled-up versions of single bubbles. After the last supernova went off, the structures dissipate, either because radiative cooling becomes important and the interior pressure is lost, or because the bubble bursts through the Galactic disk and a blow-out occurs (see below).

Superbubbles in the Milky Way are difficult to study in the optical because of severe extinction in the Galactic plane. The Cygnus Superbubble is one of the better studied cases (Abbott, Bieging, & Churchwell 1981; Saken et al. 1992).

The Magellanic Clouds offer a better view. Oey & Massey (1995) and Oey (1995, 1996) examined the stellar populations within a sample of LMC superbubbles. Their goal was to derive the stellar content, predict the stellar-wind parameters, and relate them to the dynamics of the superbubbles. In the following I am presenting their discussion on the superbubble DEM152, which is in the OB association LH47/48. The observed HRD is in Figure 38. Comparison of the luminosities with evolutionary models suggests masses somewhat above 40 $M_\odot$ for the most massive members. The stars inside the superbubble

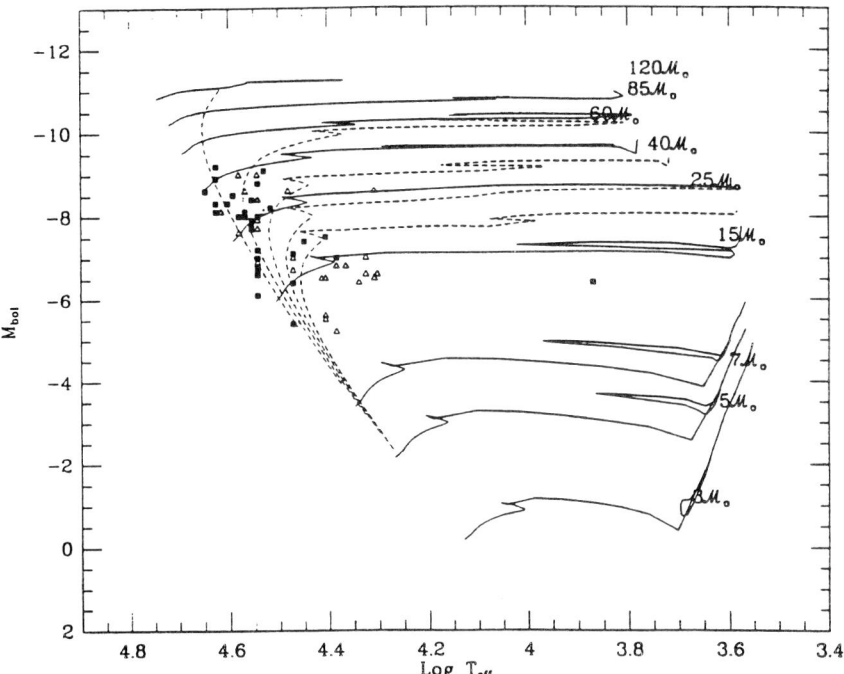

FIGURE 38. HRD of the LMC association LH47/48. Stars interior to the superbubble are denoted by open triangles and those exterior with filled squares. The dashed lines correspond to isochrones of 2, 4, 6, 8, and 10 Myr. From Oey & Massey (1995).

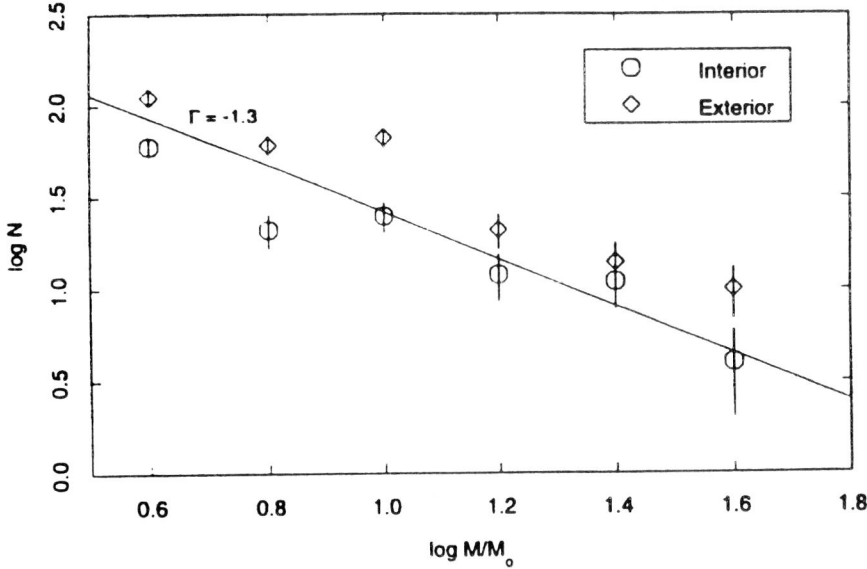

FIGURE 39. Comparison of the IMF for stars inside and outside the superbubble DEM152. From Oey & Massey (1995).

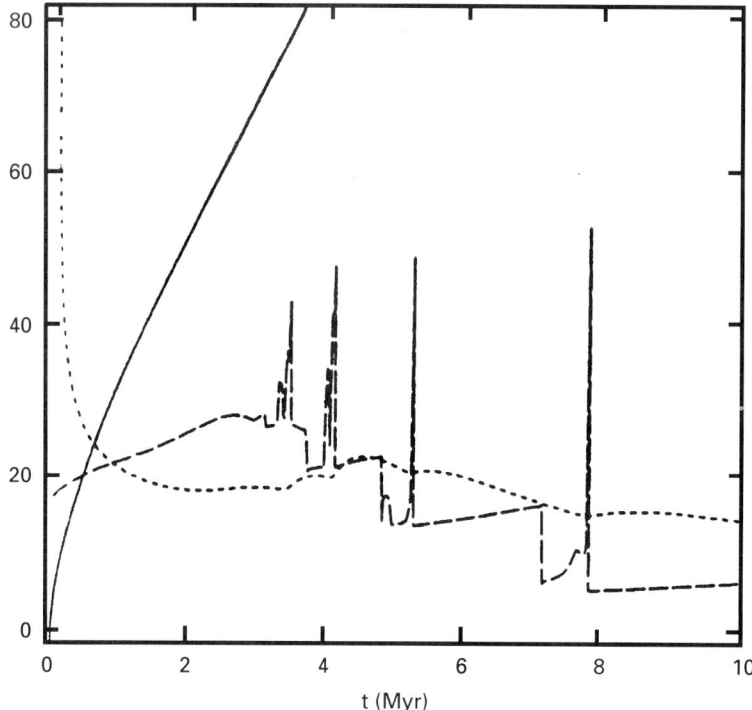

FIGURE 40. Superbubble model for DEM152 assuming a coeval burst of the currently observed population. Solid line: R_2 in pc; short-dashed: v_2 in km s^{-1}; long-dashed: $\log L_{\mathrm{mech}}(t)$ in 10^{35} erg s^{-1}. From Oey & Massey (1995).

are older than their exterior counterparts. The interior stars are already past the 10 Myr isochrone, while the exterior stars have ages around 5 Myr. The corresponding IMF is in Figure 39. The population in- and outside the superbubble is consistent with a Salpeter IMF.

Having derived the basic stellar properties, Oey & Massey (1995) modeled the superbubble dynamics using $L_{\mathrm{mech}}(t)$ as input. The results are in Figure 40. $L_{\mathrm{mech}}(t)$ reflects the wind properties varying with stellar evolution. O-star winds increase in their mechanical luminosity with time during the first 2.5 Myr, as L increases. The double-peak at $\sim$2.8 Myr is due to WR stars with their powerful winds ($M_\circ = 85$ M$_\odot$), followed immediately by a supernova event. Subsequent peaks are due to WR winds and supernovae from less massive stars. The measured major axis diameter and the expansion velocity of DEM152 are 60 pc and 40 km s^{-1}, respectively. There is an inconsistency between the model and the observations: the observed shell radius is too small and the velocity is too large. Other models having a wide range of stellar-population parameters were investigated, but none could resolve this discrepancy. Possible reasons for the break-down of the models include particular geometries and the requirement for a more complicated modeling of the supernova blast wave.

The study of Oey & Massey (and other studies as well) indicate *qualitative* agreement between the bubble theory and observations but discrepancies in the bubble properties by a factor of a few are present. Nevertheless, the formation and evolution of superbubbles and supershells are basically understood, and there is manifold observational evidence for their existence. Spectacular examples were already shown in Figure 7. The H I holes in Holmberg II are cavities carved into the ISM by multiple winds and supernova events.

6.3. Galactic superwinds

If superbubbles are in a medium with a density gradient, such as galactic a disk, they will expand preferentially along the direction of the vertical pressure gradient. Let H_{eff} be the effective vertical scale height of the galaxy, defined by

$$H_{\text{eff}} = \frac{1}{\rho_\circ} \int_0^\infty \rho(z)dz. \tag{6.24}$$

$\rho_\circ$ is the ambient gas density. After the superbubble reaches several vertical scale heights, Rayleigh-Taylor instabilities develop, and the wall of the superbubble will dissipate. This allows the interior hot gas to blow out of the disk and under certain conditions form a galactic superwind. The characteristic *minimum* wind luminosity L_{min} that determines whether a blow-out will occur is the luminosity at which the velocity of the superbubble at $H_{\text{eff}} = 1$ is equal to the isothermal sound speed (Koo & McKee 1992). This is the case for

$$L_{\text{min}} = 18\rho_\circ H_{\text{eff}}^2 c_\circ^3. \tag{6.25}$$

$c_\circ$ is the isothermal sound speed. All units in eq. (6.25) are cgs. Typical values for galaxies with scale heights of a few hundred pc are $L_{\text{min}} \approx 10^{38}$ erg s^{-1}. More detailed hydrodynamical simulations indicate that the actual luminosity requirements are about a factor of five higher than this minimum value (Mac Low, McCray, & Norman 1989). This simple argument suggests that a blow-out should frequently occur. In this section I will present the observational evidence for blow-outs leading the galactic superwinds and discuss relevant theoretical work.

6.3.1. *Observational evidence for superwinds*

Lynds & Sandage (1963) were the first to draw attention to a large-scale outflow from the nucleus of the nearby starburst galaxy M82. At that time the "starburst" phenomenon was yet unknown, and Lynds & Sandage speculated about an explosive event responsible for the outflow. NGC253 is at the same distance as M82 ($d \approx 3$ Mpc) and rivals M82 in the status of the proto-type starburst. Demoulin & Burbidge (1970) found evidence for an outflow from the nucleus of NGC253 on the basis of long-slit spectroscopy. Again, the true nature of the starburst nucleus of NGC253 was unknown, but Demoulin & Burbidge suggested that the expanding gas was produced by a "violent event" in the center of the galaxy. Ulrich (1978) did an extensive kinematic study of the nebulosity around NGC253, confirming the existence of a massive galactic wind. At that time the relevance of massive stars for the properties of the nucleus of NGC253 had been realized, and Ulrich proposed an "outburst of star formation" as a trigger of the flow. A total galactic-wind mass $\geq 10^4$ M$_\odot$ was derived.

The interpretation of the expanding nebulosities around M82 and NGC253 as a large-scale outflow did not remain unchallenged. Solinger, Morrison, & Markert (1977) proposed an alternative model for M82, which invoked scattering of photons by dust to *mimic* the observed kinematics. Strong support for the outflow model, however, came from Einstein HRI detections of M82 (Watson, Stanger, & Griffiths 1984) and NGC253 (Fabbiano & Trinchieri 1984). The hot X-ray emitting, extra-nuclear gas coincides spatially with the cold gas observed in the optical emission lines. The morphology of the optically detected gas suggest that the gas is located on the walls of cylinders, with the interior being filled by the X-ray emitting gas. Fabbiano (1988) substantially improved the spatial and spectral resolution of the available X-ray data of M82 and NGC253 on

the basis of Einstein IPC data. The work was extended to the starburst galaxy NGC3628 (Fabbiano, Heckman, & Keel 1990). Two major results emerged:
- The X-ray emitting gas is extended and aligned along the minor axis of the galaxies.
- The extended emission is present in soft (0.2 − 1.4 keV) X-rays, but absent at higher energies (1.7 − 4.5 keV).

The most detailed study of the optical morphology, physical conditions, and kinematics of M82 and NGC253 has been performed by McCarthy, Heckman, & van Breugel (1987). Heckman, Armus, & Miley (1987) extended this work to a sample of powerful far-IR galaxies. They found extended gas with physical and kinematic properties very similar to those of M82 and NGC253.

Other examples of galactic superwinds are the outflow from NGC1569, which was already mentioned before (Figure 6), and from IZw18. Extended diffuse emission and a partial shell extending well beyond the main body of the blue compact dwarf galaxy IZw18 has been detected by Martin (1996). A kinematic study suggests that a superbubble is observed. The superbubble may blow out of the galaxy, and a fraction of the ISM will leave IZw18 as a galactic superwind. Complete reviews of the observational aspects of galactic superwinds were given by Heckman, Lehnert, & Armus (1993) and Heckman (1995).

6.3.2. *A stellar population powering the superwind?*

It seems natural to invoke the massive-star population in the nuclei of galaxies as the powering sources of the observed superwinds: for instance, the bright starburst knot NGC1741B1 (see Table 3) contains about 10^4 O stars. According to Table 6, they can provide about 10^{40} erg s^{-1} of mechanical luminosity. Only a fraction of the available wind luminosity is thermalized. Efficiencies of 30% are typical (Weaver et al. 1977; Abbott 1982). Even so, comparison with eq. (6.25) suggests that the observed stellar populations should produce blow-outs of the ISM in starburst galaxies.

Heckman, Armus, & Miley (1990) and Leitherer et al. (1992b) did a quantitative assessment of the energy injected by winds and supernovae and compared it to the energy requirements of the superwinds observed in sample of galaxies ranging from low-luminosity dwarfs to infrared-luminous starbursts. The properties of the galactic superwinds can be determined from the emission-line diagnostics alone, whereas the stellar parameters are obtained *independently* from the infrared luminosity of the starburst. The main uncertainties is due to the adopted IMF. However, the general evidence for a Salpeter IMF, based on observations in the Local Group, gives some confidence that the assumption of a universal, Salpeter-like IMF is reasonable.

The energy injected by winds and supernovae is compared to the energy content derived for the outflows in Figure 41. The energy due to stars and supernovae is based on the model calculations discussed in Section 5. The total energy released by stars depends on the assumed burst duration. The length of the burst can be constrained by the dynamical timescale of the superwinds, which is on the order of a few times 10^7 yr. Therefore three sets of models with ages of 1×10^7 yr, 1.5×10^7 yr, and 3×10^7 yr were computed. The star-formation rate is assumed to be constant within the duration of the burst. The kinetic energies measured for the superwinds can be understood if the energy due to stellar winds and SN'e is thermalized and used to drive the outflows. The required burst durations for most galaxies are well within the limits imposed by the dynamical timescales of the outflows and the gas depletion timescales of the ISM.

The galaxies in Figure 41 display a trend implying higher burst ages for more powerful superwinds. In fact, Arp220, the most luminous galaxy has an unusually small ratio of the number of ionizing photons over bolometric luminosity (Heckman, Lehnert, &

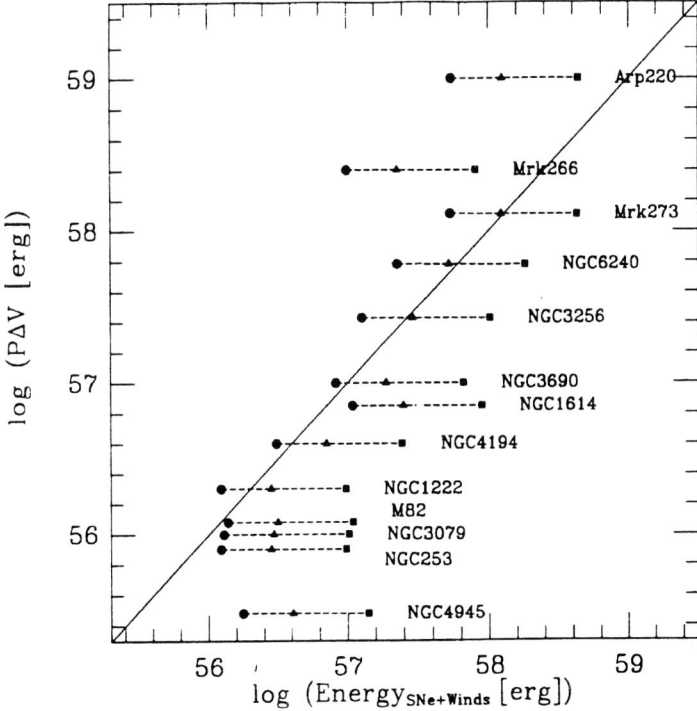

FIGURE 41. Energy released by winds and supernovae (abscissa) versus energy content measured for the superwinds (ordinate). The three tick marks denote burst durations of 1×10^7 yr (dots), 1.5×10^7 yr (triangles), and 3×10^7 yr (squares). From Leitherer et al. (1992b).

Armus 1993). This could be evidence for a relatively high burst age when the most massive stars have already disappeared (cf. the discussion on post-starburst galaxies in Section 5.5). Metallicity may also explain the trend visible in Figure 41. The energy due to winds and supernovae was derived for solar metallicities. Galaxies with the least powerful winds are mostly dwarf galaxies with $Z < Z_\odot$ whereas galaxies at the upper end are typically metal-rich infrared-luminous galaxies. Therefore these simple models tend to overestimate the energy injection of the galaxies in the bottom part of the figure and are an underestimate in the top part. Despite these uncertainties, the general agreement between stellar (horizontal axis) and galactic-wind (vertical axis) properties over three orders of magnitude is consistent with the assumption that superwinds are driven by the combined effect of winds and supernovae.

6.3.3. Theory of galactic superwinds

Correlation does not necessarily mean causation. Therefore the result of Figure 41 is not *proof* that winds and supernovae are indeed driving the outflow. Hydrodynamical models are required to investigate the physical properties and the morphology of the outflow with typical parameters of the galactic gaseous disk and the massive-star population. Chevalier & Clegg (1985) derived an analytical solution for a galactic wind flowing from a region of uniform mass and energy deposition. A wind is driven out of the nucleus if most of the energy input is thermalized. Their model provides the basic, qualitative description for the phenomenon of a galactic superwind. Mac Low (1995) gave a general review of the blow-out theory.

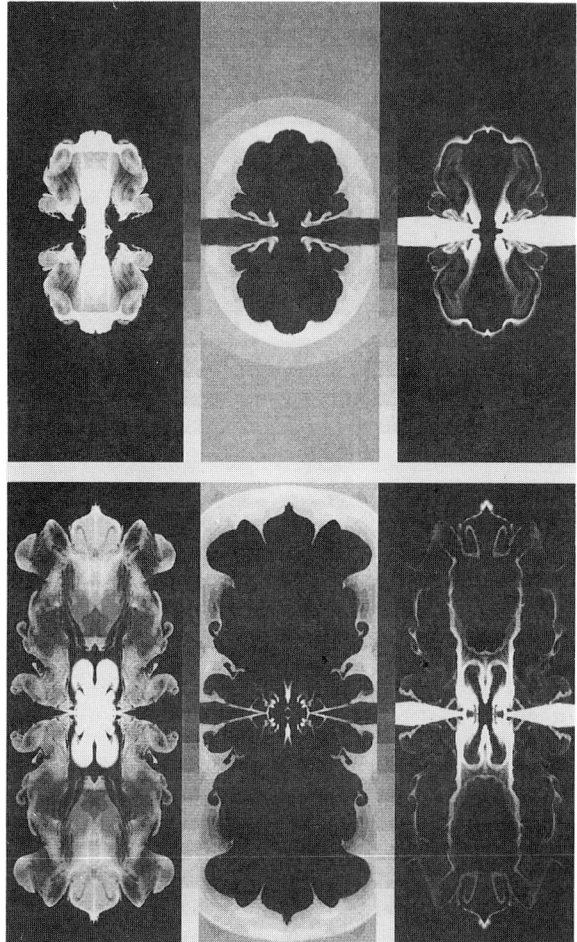

FIGURE 42. Density distribution of the disk gas (left), halo gas (center), and wind material (right). The upper panel is for $t = 8.3$ Myr when *stellar winds* are important. The lower panel is for $t = 16.6$ Myr when the energetics are dominated by supernovae. Each field measures 6×15 kpc. Salpeter IMF; $M_{\rm up} = 120$ $M_\odot$; $M_{\rm low} = 1$ $M_\odot$; continuous star formation at a rate of 2 $M_\odot$ yr^{-1}. From Suchkov et al. (1994).

Tomisaka & Ikeuchi (1988) did a pioneering study of this kind. They simulated the bipolar flow from the nucleus of M82 and found that the morphology and X-ray emission can be understood in terms of thermal emission from a hot wind if supernovae explode at a rate of 0.1 yr^{-1} and their energy is thermalized in the disk gas. Suchkov et al. (1994) revisited the hydrodynamics of galactic superwinds by addressing the "stellar" issue. They took into account the time-dependence of the mass and energy deposition in the starburst region as predicted from the properties of the stellar population.

Suchkov et al. did a two-dimensional hydrodynamical simulation of the wind interaction with a two-component disk-halo ambient ISM. The energy injection was treated as time-dependent, with self-consistent stellar-wind and supernova models. Parameters close to M82 were adopted. Results for the density distributions of the three gas components are shown in Figure 42. During the early phase of the starburst ($t < 5$ Myr) the energy deposition is dominated by stellar winds. Supernovae increase in their relative importance, and by $t = 8.3$ Myr winds and supernovae are roughly equally important. In the upper part of Figure 42 a bipolar outflow can be recognized that is collimated by the galactic disk whose material is partially dragged out. At $t = 16.6$ Myr (lower part of Figure 42) supernovae dominate the energy deposition. A substantial fraction of the galactic disk, which at $t = 8.3$ Myr provided the collimating walls, has now been entrained by the wind and removed from the disk into the halo. The geometry of the bipolar cones is remarkably similar to the observed outflows from starburst galaxies (Heckman et al. 1990).

The wind topology is crucially dependent on the history of the mass and energy deposition. The initial, moderate effect of stellar winds generates a cavity in the disk and halo gas without strongly affecting the disk itself. This determines the initial conditions for the escape of the hot gas during the subsequent supernova-dominated phase.

Tenorio-Tagle & Muñoz-Tuñón (1997) accounted for infall of material into the nuclear starburst in their hydrodynamical model of a galactic superwind. The infalling material is ejected along the maximum pressure gradient. The resulting flow closely resembles the observed extended material around starburst galaxies.

The models of Suchkov et al. (1994) suggest that the *soft* X-ray emission in the $0.1 - 2.2$ keV band of the wind material is negligible in comparison with that of the shocked halo and disk gas. The predicted soft X-ray luminosity of the shocked gas agrees with the observations of M82. An immediate prediction of these models is that the soft X-ray spectra will not necessarily exhibit signs of chemical enrichment. Only if the disk gas is sufficiently polluted due to a high star-formation efficiency, enhanced heavy-element abundances may be detectable.

The *hard* X-ray emission ($1.6 - 8.3$ keV) is produced by thermal emission from wind material at temperatures $> 2 \times 10^7$ K. However, the models predict that the hard X-ray luminosity is at least an order of magnitude below the soft X-ray luminosity. Observations of M82 and NGC253 (Fabbiano 1988) indicate comparable fluxes in the soft and hard X-ray bands. Therefore the observed hard X-ray emission is unlikely to be associated with the wind material but is probably due to massive X-ray binaries formed in the central starburst. This is consistent with the observed concentration of the hard X-ray emission towards the starburst nuclei such as observed, e.g., in NGC1808 with the ASCA satellite (Awaki et al. 1996).

6.3.4. *Superwinds and their cosmological implication*

I have summarized the manifold observational and theoretical evidence for the ubiquitous existence of large-scale outflows from starburst galaxies. Such outflows are not restricted to classical starbursts but are observed in active galaxies as well.

Baum et al. (1993) detected kpc-scale radio emission in 13 Seyfert galaxies. The flux density of the extra-nuclear radio emission agrees with the flux density predicted from the far-infrared flux density, under the assumption that the same radio-infrared correlation found in starburst galaxies also holds in Seyfert galaxies. It is therefore plausible to attribute the radio emission in Seyferts to the same mechanism responsible for the radio emission in starburst galaxies, i.e. synchrotron emission by cosmic-ray electrons. The existence of circumnuclear starbursts in Seyfert galaxies has been known before (see e.g. Bruhweiler, Truong, & Altner 1991 for the proto-type Seyfert2 galaxy NGC1068). Recent HST imagery and spectroscopy of the Seyfert2 galaxy Mrk477 by Heckman et al. (1997) has revealed a population of $\sim 10^5$ O stars near its nucleus. This makes this starburst an order of magnitude more powerful than the extraordinary starburst NGC1741B1. These results suggest that a significant fraction of Seyfert galaxies may have circumnuclear starbursts and associated outflows, similar to those observed in classical starbursts.

The universality of galactic superwinds has important cosmological consequences. A comprehensive review of the implications has been given by Heckman, Lehnert, & Armus (1993). Galactic superwinds play a crucial role in the chemical evolution of galaxies and the surrounding intergalactic medium, in the efficiency of galaxy formation in the early universe, and in the production of the soft X-ray background.

Continuing financial support of the starburst project at STScI, most recently through grants GO-3591.01-91A, GO-5444.01-93A, GO-5900.02-94A, GO-5954.01-94A, AR-5804.01-94A, AR-5809.01-94A, and NAG8-1075, is gratefully acknowledged. Nolan Walborn and Bill Vacca kindly provided electronic versions of parts of Tables 1 and 6. Carme Gallart, Barry Madore, Casiana Muñoz-Tuñón, Enrique Pérez, and Grażyna Stasińska made numerous comments on an earlier version of the manuscript.

REFERENCES

ABBOTT, D. C. 1982, ApJ, 263, 723
ABBOTT, D. C., BIEGING, J. H., & CHURCHWELL, E. 1982, ApJ, 250, 645
ABBOTT, D. C., & CONTI, P. S. 1987, ARAA, 25, 113
ABLES, H. D. 1967, Pub. US Naval Obs., 2nd Ser., 20, pt. 4, 61
APARICIO, A., ET AL. 1995, AJ, 110, 212
APARICIO, A., GALLART, C., CHIOSI, C., & BERTELLI, G. 1996, ApJ, 469, 97
ARMUS, L., HECKMAN, T. M., & MILEY, G. K. 1990, ApJ, 364, 471
ARP, H.C., & SANDAGE, A. 1985, AJ, 90, 1163
AWAKI, H., UENO, S., KOYAMA, K., TSURU, T., & IWASAWA, K. 1996, PASJ, 48, 409
BALICK, B., BOESHAAR, G. O., & GULL, T. R. 1980, ApJ, 242, 584
BALZANO, V. A. 1983, ApJ, 268, 602
BAUM, S. A., O'DEA, C. P., DALLACASSA, D., DE BRUYN, A. G., & PELAR, A. 1993, ApJ, 419, 553
BECK, S. C. 1994, in The Nearest Active Galaxies, ed. J. Beckman, L. Colina, & H. Netzer (Madrid: Consejo Superior de Investigaciones Cientificas), 169
BERNLÖHR, K. 1993, A&A, 268, 25
BESSELL, M. S., & STRINGFELLOW, G. S. 1993, ARAA, 31, 433
BIEGING, J. 1975, in H II Regions and Related Topics, ed. T. L. Wilson & D. Downes (Berlin: Springer), 443
———. 1990, in The Evolution of the Interstellar Medium, ed. L. Blitz (San Francisco: ASP), 137
BINNEY, J., & TREMAINE, S. 1987, Galactic Dynamics, ed. J. P. Ostriker (Princeton: PUP)
BLUM, R. D., DEPOY, D. L., & SELLGREN, K. 1995, ApJ, 441, 603
BRESSAN, A., FAGOTTO, F., BERTELLI, G., & CHIOSI, C. 1993, A&AS, 100, 647
BRINKS, E., & BAJAJA, E. 1986, A&A, 169, 14
BRUHWEILER, F. C., TRUONG, K. Q., & ALTNER, B. 1991, ApJ, 379, 596
BRUZUAL A., G., & CHARLOT, S. 1993, ApJ, 405, 538
CALZETTI, D. 1997, AJ, 113, 162
CALZETTI, D., KINNEY, A. L., & STORCHI-BERGMANN, T. 1994, ApJ, 429, 582
———. 1996, ApJ, 458, 132
CAMPBELL, A. 1988, ApJ, 335, 644

CASSINELLI, J. P. 1991, in IAU Symp. 143, Wolf-Rayet Stars and Interrelations with Other Massive Stars in Galaxies, ed. K. A. van der Hucht & B. Hidayat (Dordrecht: Kluwer), 289

CASSINELLI, J. P., & CHURCHWELL, J. P. 1993, Massive Stars: Their Lives in the Interstellar Medium (San Francisco: ASP)

CASTOR, J. I. 1993, in Massive Stars: Their Lives in the Interstellar Medium, ed. J. P. Cassinelli & E. B. Churchwell (San Francisco: ASP), 297

CASTOR, J. I., MCCRAY, R., & WEAVER, R. 1975, ApJ, 200, L107

CERVIÑO, M., & MAS-HESSE, J. M. 1994, A&A, 284, 749

CHEVALIER, R. A. 1977, ARAA, 15, 175

———. 1991, in Massive Stars in Starbursts, ed. C. Leitherer, N. Walborn, T. Heckman, & C. Norman (Cambridge: CUP), 169

CHEVALIER, R. A., & CLEGG, A. W. 1985, Nature, 317, 44

CHIOSI, C., BERTELLI, G., & BRESSAN, A. 1992, ARAA, 30, 235

CHIOSI, C., & MAEDER, A. 1986, ARAA, 24, 329

CHU, Y.-H. 1982, ApJ, 254, 578

———. 1991, in IAU Symp. 143, Wolf-Rayet Stars and Interrelations with Other Massive Stars in Galaxies, ed. K. A. van der Hucht & B. Hidayat (Dordrecht: Kluwer), 349

CHU, Y.-H., & KENNICUTT, R. C. 1994, ApJ, 425, 720

CID FERNANDES, R., DOTTORI, H. A., GRUENWALD, R. B., & VIEGAS, S. M. 1992, MNRAS, 255, 165

CLAYTON, C. A. 1986, MNRAS, 219, 895

CONTI, P. S. 1984, in IAU Symp. 105, Observational Tests of Stellar Evolution Theory, ed. A. Maeder & A. Renzini (Dordrecht: Reidel), 233

———. 1991, ApJ, 377, 115

CONTI, P. S., & VACCA, W. D. 1994, ApJ, 423, L97

COPETTI, M. V. F., PASTORIZA, M. G., & DOTTORI, H. A. 1986, A&A, 156, 111

CROWTHER, P. A., HILLIER, D. J., & SMITH, L. J. 1995, A&A, 293, 403

CROWTHER, P. A., & WILLIS, A. J. 1994, in Evolution of Massive Stars: A Confrontation between Theory and Observations, ed. D. Vanbeveren, W. van Rensbergen, & C. de Loore (Dordrecht: Kluwer), 85

CUNHA, K., & LAMBERT, D. L. 1992, ApJ, 399, 586

DA COSTA, L. N., PELLEGRINI, P. S., DAVIS, M., MEIKSIN, A., SARGENT, W. L. W., & TONRY, J. L. 1991, ApJS, 75, 935

DA COSTA, L. N., PELLEGRINI, P. S., SARGENT, W. L. W., TONRY, J., DAVIS, M., MEIKSIN, A., LATHAM, D. W., MENZIES, J. W., & COULSON, I. A. 1988, ApJ, 327, 544

DA COSTA, L. N., PELLEGRINI, P. S., WILLMER, C., DE CARVALHO, R., MAIA, M., LATHAM, D. W., & GEARY, J. C. 1989, AJ, 97, 315

DAVIDSON, K. 1989, in IAU Coll. 113, Physics of Luminous Blue Variables, ed. K. Davidson, A. F. J. Moffat, & H. J. G. L. M. Lamers (Dordrecht: Kluwer), 101

DE KOTER, A., LAMERS, H. J. G. L. M., & SCHMUTZ, W. 1996, A&A, 306, 501

DELLA CECA, R., GRIFFITHS, R. E., HECKMAN, T. M., & MACKENTY, J. W. 1996, ApJ, 469, 662

DE MARCHI, G., CLAMPIN, M., GREGGIO, L., LEITHERER, C., NOTA, A., & TOSI, M. 1997, ApJ, submitted

DEMOULIN, M.-H., & BURBIDGE, E. M. 1970, ApJ, 159, 799

DETTMAR, R. J. 1990, A&A, 232, L15

———. 1993, Rev. Mod. Astron., 6, 33

DEUL, E. R., & DEN HARTOG, R. H. 1990, A&A, 229, 362

DJORGOVSKI, S., & THOMPSON, D. J. 1992, in IAU Symp. 149, Stellar Populations, ed. B.

Barbuy & A. Renzini (Kluwer: Dordrecht), 337

DOHERTY, R. M., PUXLEY, P. J., LUMSDEN, S. L., & DOYON, R. 1995, MNRAS, 277, 577

DOYON, R., PUXLEY, P. J., & JOSEPH, R. D. 1992, ApJ, 397, 117

DRAKE, S. A. 1986, in Cool Stars, Stellar Systems, and the Sun, ed. M. Zeilik & D. M. Gibson (Heidelberg: Springer), 369

DREW, J. 1990, in Properties of Hot Luminous Stars, ed. C. D. Garmany (San Francisco: ASP), 230

DRISSEN, L., MOFFAT, A. F. J., WALBORN, N. R., & SHARA, M. M. 1995, AJ, 110, 2235

DRISSEN, L., ROY, J.-R., & MOFFAT, A. F. J. 1993, AJ, 106, 1460

DUFOUR, R. J. 1989, Rev. Mex. Astron. Astrof., 18, 87

DUFOUR, R. J., PARKER, R. A. R., & HENIZE, K. G. 1988, ApJ, 327, 859

DYSON, J. E. 1973, A&A, 23, 381

EBBELS, T. M. D., LE BORGNE, J.-F., PELLÓ, R., ELLIS, R. S., KNEIB, J.-P., SMAIL, I. & SANAHUJA, B. 1996, MNRAS, 281, L75

ECKART, A., & GENZEL, R. 1996, Nature, 383, 415

ECKART, A., GENZEL, R., HOFMANN, R., SAMS, B. J., & TACCONI-GARMAN, L. E. 1995, ApJ, 445, L27

ESTEBAN, C., & PEIMBERT, M. 1995, A&A, 300, 78

ESTEBAN, C., VÍLCHEZ, J. M., SMITH, L. J., & CLEGG, R. E. S. 1992, A&A, 259, 629

FABBIANO, G. 1988, ApJ, 330, 672

FABBIANO, G., HECKMAN, T. M., & KEEL, W. C. 1990, ApJ, 355, 442

FABBIANO, G., & TRINCHIERI, G. 1984, ApJ, 286, 491

FAGOTTO, F., BRESSAN, A., BERTELLI, G., & CHIOSI, C. 1994a, A&AS, 104, 365

———. 1994b, A&AS, 105, 29

FANELLI, M. N., O'CONNELL, R. W., BURSTEIN, D., & WU, C.-C. 1992, ApJS, 82, 197

FERLAND, G. J. 1991, HAZY, a brief introduction to CLOUDY V.80.08

FERGUSON, A. M. N., WYSE, R. F. G., GALLAGHER, J., & HUNTER, D. 1996, AJ, 111, 2265

FIGER, D. F., MCLEAN, I. S., & MORRIS, M. 1995, ApJ, 447, L29

FLIEGNER, J., LANGER, N., & VENN, K. A. 1996, A&A, 308, 13

FORBES, D. A., & WARD, M. J. 1993, ApJ, 416, 150

FRANCO, J., FERRINI, F., & TENORIO-TAGLE, G. 1993, Star Formation, Galaxies, and the Interstellar Medium (Cambridge: CUP)

FRITZE-V. ALVENSLEBEN, U. 1993, in Panchromatic Views of Galaxies, ed. G. Hensler, C. Theis, & J. S. Gallagher (Gif-sur-Yvette: Editions Frontieres), 245

GABLER, R., GABLER, A., KUDRITZKI, R. P., PULS, J., & PAULDRACH, A. 1989, A&A, 226, 162

GALLART, C., APARICIO, A., & VÍLCHEZ, J. M. 1996a, AJ, 112, 1928

GALLART, C., APARICIO, A., VÍLCHEZ, J. M., BERTELLI, G., & CHIOSI, C. 1996b, AJ, 112, 1950

GALLART, C., APARICIO, A., BERTELLI, G., & CHIOSI, C. 1997, AJ, in press

GARCÍA-SEGURA, G., LANGER, N., & MAC LOW, M.-M. 1996b, A&A, 316, 133

GARCÍA-SEGURA, G., MAC LOW, M.-M., & LANGER, N. 1996a, A&A, 305, 229

GARCÍA-VARGAS, M. L. 1996, in From Stars to Galaxies: The Impact of Stellar Physics on Galaxy Evolution, ed. C. Leitherer, U. Fritze-von Alvensleben, & J. Huchra (San Francisco: ASP), 244

GARCÍA-VARGAS, M. L., & DÍAZ, A. I. 1994, ApJS, 91, 553

GARCÍA-VARGAS, M. L., BRESSAN, A., & DÍAZ, A. I. 1995a, A&AS, 112, 13

———. 1995b, A&AS, 112, 35

GARMANY, C. D., & CONTI, P. S. 1985, ApJ, 293, 407

GARMANY, C. D., & STENCEL, R. E. 1992, A&AS, 94, 211

GENZEL, R., BECKLIN, E. E., WYNN-WILLIAMS, C. G., MORAN, J. M., REID, M. J., JAFFE, D. T., & DOWNES, D. 1982, ApJ, 255, 527

GENZEL, R., HOLLENBACH, D., & TOWNES, C. H. 1994, Rep. Prog. Phys., 57, 417

GENZEL, R., & TOWNES, C. H. 1987, ARAA, 25, 377

GALLAGHER, J. S. 1996, in From Stars to Galaxies: The Impact of Stellar Physics on Galaxy Evolution, ed. C. Leitherer, U. Fritze-von Alvensleben, & J. Huchra (San Francisco: ASP), 315

GOLDADER, J. D. 1996, in From Stars to Galaxies: The Impact of Stellar Physics on Galaxy Evolution, ed. C. Leitherer, U. Fritze-von Alvensleben, & J. Huchra (San Francisco: ASP), 461

GOLDADER, J. D., LEITHERER, C., & SCHAERER, D. 1997, in preparation

GOLDADER, J. D., & WYNN-WILLIAMS, C. G. 1994, ApJ, 433, 164

GONZÁLEZ-DELGADO, R. M., LEITHERER, C., HECKMAN, T. M., & CERVIÑO, M. 1997, ApJ, submitted

GREDEL, R. 1996, The Galactic Center, 4th ESO/CTIO Workshop (San Francisco: ASP)

GREENHOUSE, M. A., WOODWARD, C. E., THRONSON, H. A., RUDY, R. J., ROSSANO, G. S., ERWIN, P., & PUETTER, R. C. 1991, ApJ, 383, 164

GREGGIO, L., MARCONI, G., TOSI, M., & FOCARDI, P. 1993, AJ, 105, 894

GROENEWEGEN, M. A. T., LAMERS, H. J. G. L. M., & PAULDRACH, A. W. A. 1989, A&A, 221, 78

HAMANN, W.-R. 1995, in Wolf-Rayet Stars: Binaries, Colliding Winds, Evolution, ed. K. A. van der Hucht & P. M. Williams (Dordrecht: Kluwer), 105

HEAP, S. R., EBBETS, D., MALUMUTH, E. M., MARAN, S. P., DE KOTER, A., & HUBENY, I. 1994, ApJ, 435, L39

HECKMAN, T. M. 1991, in Massive Stars in Starbursts, ed. C. Leitherer, N. Walborn, T. Heckman, & C. Norman (Cambridge: CUP), 289

———. 1994, in Mass-Transfer Induced Activity in Galaxies, ed. I. Schlosman (Cambridge: CUP), 234

———. 1995, in The Interplay between Massive Star Formation, the ISM, and Galaxy Evolution, ed. D. Kunth, B. Guiderdoni, M. Heydari-Malayeri, & T. X. Thuan (Gif-sur-Yvette: Editions Frontieres), 159

HECKMAN, T. M., ARMUS, L., & MILEY, G. K. 1987, AJ, 93, 276

———. 1990, ApJS, 74, 833

HECKMAN, T. M., DAHLEM, M., LEHNERT, M. D., FABBIANO, G., GILMORE, D., & WALLER, W. H. 1995, ApJ, 448, 98

HECKMAN, T. M., GONZÁLEZ-DELGADO, R. M., LEITHERER, C., MEURER, G. R., KROLIK, J., WILSON, A. S., KORATKAR, A., & KINNEY, A. L. 1997, ApJ, submitted

HECKMAN, T. M., LEHNERT, M. D., & ARMUS, L. 1993, in The Environment and Evolution of Galaxies, ed. J. M. Shull & H. A. Thronson (Kluwer: Dordrecht), 455

HICKSON, P., KINDL, E., & AUMAN, J. R. 1989, ApJS, 70, 678

HILL, R. J., MADORE, B. F., & FREEDMAN, W. L. 1994, ApJ, 429, 204

HILLIER, D. J. 1988, ApJ, 327, 822

———. 1989, ApJ, 347, 392

HO, L. C., & FILIPPENKO, A. V. 1996, ApJ, 466, L83

HOWARTH, I. D., & PRINJA, R. K. 1989, ApJS, 69, 527

HUMPHREYS, R. M. 1978, ApJS, 38, 309

HUNTER, D. A., BAUM, W. A., O'NEIL, JR., E. J., & LYNDS, R. 1996, ApJ, 456, 174

HUNTER, D. A., SHAYA, E. J., HOLTZMAN, J. A., LIGHT, R. M., O'NEIL, JR., E. J., &

LYNDS, R. 1995, ApJ, 448, 179

JACOBY, G. H., HUNTER, D. A., & CHRISTIAN, C. A. 1984, ApJS, 56, 257

JURA, M. 1987, in Interstellar Processes, ed. D. J. Hollenbach & H. A. Thronson, Jr. (Dordrecht: Reidel), 3

———. 1991, in IAU Symp. 143, Wolf-Rayet Stars and Interrelations with Other Massive Stars in Galaxies, ed. K. A. van der Hucht & B. Hidayat (Dordrecht: Kluwer), 341

KAMPHUIS, J., & SANCISI, R. 1993, A&A, 273, L31

KENNICUTT, R. C. 1984, ApJ, 287, 116

———. 1991, in Massive Stars in Starbursts, ed. C. Leitherer, N. Walborn, T. Heckman, & C. Norman (Cambridge: CUP), 157

———. 1992, ApJ, 388, 310

———. 1995, in The Interplay between Massive Star Formation, the ISM, and Galaxy Evolution, ed. D. Kunth, B. Guiderdoni, M. Heydari-Malayeri, & T. X. Thuan (Gif-sur-Yvette: Editions Frontieres), 297

KENNICUTT, R. C., & CHU, Y.-H. 1994, in Violent Star Formation, From 30 Doradus to QSOs, ed. G. Tenorio-Tagle (Cambridge: CUP), 1

KINNEY, A. L., BOHLIN, R. C., CALZETTI, D., PANAGIA, N., & WYSE, R. F. G. 1993, ApJS, 86, 5

KINNEY, A. L., CALZETTI, D., BOHLIN, R. C., MCQUADE, K., STORCHI-BERGMANN, T., & SCHMITT, H. R. 1996, ApJ, 467, 38

KOBULNICKY, H. A., & SKILLMAN, E. D. 1996, ApJ, 471, 211

KOO, B.-C., & MCKEE, C. F. 1992, ApJ, 388, 93

KOORNNEEF, J. 1993, ApJ, 403, 581

KRABBE, A., ET AL. 1995, ApJ, 447, L95

KUDRITZKI, R.-P., HUMMER, D. G., PAULDRACH, A. W. A., PULS, J., NAJARRO, F., & IMHOFF, J. 1992, A&A, 257, 655

KUDRITZKI, R.-P., LENNON, D. J., HASER, S. M., PULS, J., PAULDRACH, A. W. A., VENN, K., & VOELS, S. A. 1996, in Science with the Hubble Space Telescope — II, ed. P. Benvenuti, F. D. Machetto, & E. J. Schreier (Baltimore: STScI), 285

KUDRITZKI, R. P., PAULDRACH, A., & PULS, J. 1987, A&A, 173, 293

KUNTH, D., GUIDERDONI, B., HEYDARI-MALAYERI, M., & THUAN, T. X. 1995, The Interplay between Massive Star Formation, the ISM, and Galaxy Evolution (Gif-sur-Yvette: Editions Frontieres)

KURUCZ, R. L. 1992, in IAU Symp. 149, The Stellar Population of Galaxies, ed. B. Barbuy & A. Renzini (Dordrecht: Kluwer), 225

LAMERS, H. J. G. L. M. 1983a, in Diffuse Matter in Galaxies, ed. J. Audouze, J. Lequeux, M. Lévy, & A. Vidal-Madjar (Dordrecht: Reidel), 35

———. 1983b, in Diffuse Matter in Galaxies, ed. J. Audouze, J. Lequeux, M. Lévy, & A. Vidal-Madjar (Dordrecht: Reidel), 45

———. 1989, in IAU Coll. 113, Physics of Luminous Blue Variables, ed. K. Davidson, A. F. J. Moffat, & H. J. G. L. M. Lamers (Dordrecht: Kluwer), 135

LAMERS, H. J. G. L. M., & CASSINELLI, J. P. 1996, in From Stars to Galaxies: The Impact of Stellar Physics on Galaxy Evolution, ed. C. Leitherer, U. Fritze-von Alvensleben, & J. Huchra (San Francisco: ASP), 162

LAMERS, H. J. G. L. M., & LEITHERER, C. 1993, ApJ, 412, 771

LANÇON, A., & ROCCA-VOLMERANGE, B. 1992, A&AS, 96, 593

LARSON, R. 1987, in Starbursts and Galaxy Evolution, ed. T. X. Thuan, T. Montmerle, & J. Tran Thanh Van (Gif-Sur-Yvette: Edition Frontieres), 467

———. 1991, in Fragmentation of Molecular Clouds and Star Formation, ed. E. Falgarone, F. Boulanger, & G. Duvert (Dordrecht: Kluwer), 261

LEHNERT, M. D. 1992, Ph. D. thesis, The Johns Hopkins University

LEITHERER, C. 1997, in Starburst Activity in Galaxies, ed. J. Franco, R. Terlevich, & G. Tenorio-Tagle, Rev. Mex. Astron. Astrofis. Conf. Ser., in press
LEITHERER, C., ET AL. 1996b, PASP, 108, 996
LEITHERER, C., & CHAVARRÍA, C. 1987, A&A, 175, 208
LEITHERER, C., FRITZE-V. ALVENSLEBEN, U., & HUCHRA, J. 1996a, From Stars to Galaxies — The Impact of Stellar Physics on Galaxy Evolution (San Francisco: ASP)
LEITHERER, C., GRUENWALD, R., & SCHMUTZ, W. 1992a, in Physics of Nearby Galaxies, ed. T. X. Thuan, C. Balkowski, & J. T. T. Van (Gif-sur-Yvette: Editions Frontieres), 257
LEITHERER, C., & HECKMAN, T. M. 1995, ApJS, 96, 9
LEITHERER, C., ROBERT, C., & DRISSEN, L. 1992b, ApJ, 401, 596
LEITHERER, C., ROBERT, C., & HECKMAN, T. M. 1995, ApJS, 99, 173
LEITHERER, C., VACCA, W. D., CONTI, P. S., FILIPPENKO, A. V., ROBERT, C., & SARGENT, W. L. W. 1996c, ApJ, 465, 717
LEJEUNE, T., BUSER, R., & CUISINIER, F. 1997, A&A, submitted
LIEBERT, J., & PROBST, R. G. 1987, ARAA, 25, 473
LILLY, S., LE FEVRE, O., CRAMPTON, D., HAMMER, F., & TRESSE, L. 1995, ApJ, 455, 50
LONGO, R., STALIO, R., POLIDAN, R. S., & ROSSI, L. 1989, ApJ, 339, 474
LUCY, L. B., & SOLOMON, P. M. 1970, ApJ, 159, 879
LYNDS, C. R., & SANDAGE, A. 1963, ApJ, 137, 1005
MADAU, P., FERGUSON, H. C., DICKINSON, M. E., GIAVALISCO, M., STEIDEL, C. C., & FRUCHTER, A. 1996, MNRAS, 283, 1388
MAC LOW, M.-M. 1995, in The Interplay between Massive Star Formation, the ISM, and Galaxy Evolution, ed. D. Kunth, B. Guiderdoni, M. Heydari-Malayeri, & T. X. Thuan (Gif-sur-Yvette: Editions Frontieres), 169
MAC LOW, M.-M., MCCRAY, R., & NORMAN, M. L. 1989, ApJ, 337, 141
MAEDER, A. 1990, A&AS, 84, 139
——. 1994, in Evolution of Massive Stars: A Confrontation between Theory and Observations, ed. D. Vanbeveren, W. van Rensbergen, & C. de Loore (Dordrecht: Kluwer), 349
MAEDER, A., & CONTI, P. S. 1994, ARAA, 32, 227
MALUMUTH, E. M., & HEAP, S. R. 1994, AJ, 107, 1054
MALUMUTH, E. M., WALLER, W. H., & PARKER, J. WM. 1996, AJ, 111, 1128
MARLOWE, A. T., HECKMAN, T. M., WYSE, R. F. G., & SCHOMMER, B. 1995, ApJ, 438, 563
MARSTON, A. P., & MEABURN, J. 1988, MNRAS, 235, 391
MARTIN, C. 1996, ApJ, 465, 680
MAS-HESSE, J. M., & KUNTH, D. 1991, A&AS, 88, 399
MASSEY, P., JOHNSON, K. E., & DEGIOIA-EASTWOOD, K. 1995a, ApJ, 454, 151
MASSEY, P., LANG, C. C., DEGIOIA-EASTWOOD, K., & GARMANY, C. D. 1995b, ApJ, 438, 188
MATTHEWS, W. G. 1966, ApJ, 144, 206
MCCARTHY, P. J., HECKMAN, T. M., & VAN BREUGEL, W. 1987, AJ, 93, 264
MCCRAY, R., & KAFATOS, M. 1987, ApJ, 317, 190
MCGAUGH, S. S. 1991, ApJ, 380, 140
MCKEE, C. F. 1990, in The Evolution of the Interstellar Medium, ed. L. Blitz (San Francisco: ASP), 3
MCKEE, C. F., & OSTRIKER, J. P. 1977, ApJ, 218, 148
MEHRINGER, D. M. 1994, ApJS, 91, 713
MELNICK, J. 1977, ApJ, 213, 15
——. 1985, A&A, 153, 235

———. 1992, in Star Formation in Stellar Systems, ed. G. Tenorio-Tagle, M. Prieto, & F. Sánchez (Cambridge: CUP), 253

MELNICK, J., MOLES, M., & TERLEVICH, R. 1985, A&A, 149, L24

MEURER, G. R. 1995, Nature, 375, 742

MEURER, G. R., FREEMAN, K., DOPITA, M., & CACCIARI, C. 1992, AJ, 103, 60

MEURER, G. R., HECKMAN, T. M., LEITHERER, C., ROBERT, C., KINNEY, A., & GARNETT, D. 1995, AJ, 110, 2665

MEYNET, G. 1995, A&A, 298, 767

MEYNET, G., MERMILLIOD, J. C., & MAEDER, A. 1993, A&AS, 98, 477

MEZGER, P. G. 1987, in Starbursts and Galaxy Evolution, ed. T. X. Thuan, T. Montmerle, & J. T. T. Van (Gif-sur-Yvette: Editions Frontieres), 3

MOFFAT, A. F. J. 1995, in Wolf-Rayet Stars: Binaries, Colliding Winds, Evolution, ed. K. A. van der Hucht & P. M. Williams (Dordrecht: Kluwer), 213

———. 1997, in Starburst Activity in Galaxies, ed. J. Franco, R. Terlevich, & G. Tenorio-Tagle, Rev. Mex. Astron. Astrofis. Conf. Ser., in press

MOFFAT, A. F. J., DRISSEN, L., & SHARA, M. M. 1994, ApJ, 436, 183

MOORWOOD, A. F. M. 1996, Space Sci. Rev., 77, 303

MORRIS, M., & SERABYN, E. 1996, ARAA, 34, 645

MORTON, D. C. 1967, ApJ, 147, 1017

MOURI, H., NISHIDA, M., TANIGUCHI, Y., & KAWARA, K. 1990, ApJ, 360, 55

MURRAY-HANSON, M., & CONTI, P. S. 1995, 448, L45

NAJARRO, F., HILLIER, D. J., KUDRITZKI, R. P., KRABBE, A., LUTZ, D., GENZEL, R., DRAPATZ, S., & GEBALLE, T. R. 1994, A&A, 285, 573

NOTA, A., & LAMERS, H. J. G. L. M. 1997, Luminous Blue Variables: Massive Stars in Transition (San Francisco: ASP), in press

O'CONNELL, R. W., GALLAGHER, J. S., & HUNTER, D. A. 1994, ApJ, 433, 65

OEY, M. S. 1995, ApJS, 104, 71

———. 1995, ApJ, 467, 666

OEY, M. S., & MASSEY, P. 1995, ApJ, 452, 210

OLIVA, E., MOORWOOD, A. F. M., & DANZIGER, I. J. 1989, A&A, 214, 307

OLIVA, E., ORIGLIA, L., KOTILAINEN, J. K., & MOORWOOD, A. F. M. 1995, A&A, 301, 550

OLOFSSON, K. 1989, A&AS, 80, 317

———. 1996, in From Stars to Galaxies: The Impact of Stellar Physics on Galaxy Evolution, ed. C. Leitherer, U. Fritze-von Alvensleben, & J. Huchra (San Francisco: ASP), 77

OSTERBROCK, D. E. 1989, Astrophysics of Gaseous Nebulae and Active Galactic Nuclei (Mill Valley: University Science Books)

PAGEL, B. E. J. 1987, in Starbursts and Galaxy Evolution, ed. T. X. Thuan, T. Montmerle, & J. Tran Thanh Van (Gif-Sur-Yvette: Edition Frontieres), 227

———. 1992, in The Feedback of Chemical Evolution on the Stellar Content of Galaxies, ed. D. Alloin & G. Stasińska (Paris: Obs. de Paris/Meudon), 87

PAGEL, B. E. J., SIMONSON, E. A., TERLEVICH, R. J., & EDMUNDS, M. G. 1992, MNRAS, 255, 325

PANAGIA, N. 1973, AJ, 78, 929

PARKER, W., & GARMANY, C. D. 1993, AJ, 106, 1471

PARKER, W., & GARMANY, C. D., MASSEY, P., & WALBORN, N. R. 1993, 103, 1205

PAULDRACH, A. W. A., KUDRITZKI, R. P., PULS, J., & BUTLER, K. 1990, A&A, 228, 125

PRINJA, R. K. 1987, MNRAS, 228, 173

PRINJA, R. K., BARLOW, M. J., & HOWARTH, I. D. 1990, ApJ, 361, 607

PRITCHET, C. J., & HARTWICK, F. D. A. 1990, ApJ, 355, L11

Puls, J., et al. 1996, A&A, 305, 171

Puche, D., Westpfahl, D., Brinks, E., & Roy, J.-R. 1992, AJ, 103, 1841

Rand, R. J., Kulkarni, S. R., & Hester, J. J. 1990, ApJ, 352, L1

Raymond, J. C. 1979, ApJS, 39, 1

Raymond, J. C., Cox, D. P., & Smith, B. W. 1976, ApJ, 204, 290

Reimers, D. 1975, Mem. Soc. R. Sci. Liège, 6e Série, VIII, 369

Reynolds, R. 1991, in IAU Symp. 144, The Interstellar Disk-Halo Connection in Galaxies, ed. H. Bloemen (Kluwer: Dordrecht), 157

Rieke, G. H. 1991, in Massive Stars in Starbursts, ed. C. Leitherer, N. Walborn, T. Heckman, & C. Norman (Cambridge: CUP), 285

Rieke, G. H., Lebofsky, M. J., Thompson, R. I., Low, F. J., & Tokunaga, A. T. 1980, ApJ, 238, 24

Rieke, G. H., Loken, K., Rieke, M. J., & Tamblyn, P. 1993, ApJ, 412, 99

Robert, C., Leitherer, C., & Heckman, T. M. 1993, ApJ, 418, 749 749

Rubin, V. C., Hunter, D. A., & Ford, Jr., W. K. 1990, ApJ, 365, 86

Rupen, M. P. 1991, AJ, 102, 48

Saken, J. M., Shull, J. M., Garmany, C. D., Nichols-Bohlin, J., & Fesen, R. A. 1992, ApJ, 397, 537

Salpeter, E. E. 1955, ApJ, 121, 161

Sanders, D. B., & Mirabel, I. F. 1996, ARAA, 34, 749

Sargent, W. L. W., & Searle, L. 1970, ApJ, 162, L155

Satyapal, S., et al. 1995, ApJ, 448, 611

Scalo, J. M. 1986, Fund. Cosm. Phys., 11, 1

Schaerer, D., & de Koter, A. 1997, A&A, in press

Schaerer, D., de Koter, A., Schmutz, W., & Maeder, A. 1996a, A&A, 310, 837

———. 1996b, A&A, 312, 475

Schaerer, D., Meynet, G., Maeder, A., & Schaller, G. 1993, A&AS, 98, 523

Schaerer, D., & Vacca, W. D. 1996, in Wolf-Rayet Stars in the Framework of Stellar Evolution, ed. J.-M. Vreux, A. Detal, D. Fraipont-Caro, E. Gosset, & G. Rauw (Liège: Inst. d'Astroph.), 641

Schaller, G., Schaerer, D., Meynet, G., & Maeder, A., 1992, A&AS, 96, 269

Schmutz, W., Hamann, W.-R., & Wessolowski, U. 1989, A&A, 210, 236

Schmutz, W., Leitherer, C., & Gruenwald, R. 1992, PASP, 104, 1164

Searle, L., Sargent, W. L. W., & Bagnuolo, W. G. 1973, ApJ, 179, 427

Sekiguchi, K., & Anderson, K. S. 1987, AJ, 94, 644

Shields, J. C. 1993, ApJ, 419, 181

Shields, J. C., & Kennicutt, Jr., R. C. 1995, ApJ, 454, 807

Shull, J. M. 1993, in Massive Stars: Their Lives in the Interstellar Medium, ed. J. P. Cassinelli & E. B. Churchwell (San Francisco: ASP), 327

Shull, J. M., & Saken, J. M. 1995, ApJ, 444, 663

Smith, L. F., Biermann, P., & Mezger, P. G. 1978, A&A, 66, 65

Smith, L. F., Shara, M. M., & Moffat, A. F. J. 1990, ApJ, 358, 229

———. 1996, MNRAS, 281, 163

Smith, L. J. 1996, in Wolf-Rayet Stars in the Framework of Stellar Evolution, ed. J.-M. Vreux, A. Detal, D. Fraipont-Caro, E. Gosset, & G. Rauw (Liège: Inst. d'Astroph.), 381

Soifer, B. T., Sanders, D. B., Madore, B. F., Neugebauer, G., Danielson, G. E., Elias, J. H., Lonsdale, C. J., & Rice, W. L. 1987, ApJ, 320, 238

Solinger, A., Morrison, P., & Markert, T. 1977, ApJ, 211, 707

STASIŃSKA, G. 1980, A&A, 84, 320

———. 1996, in From Stars to Galaxies: The Impact of Stellar Physics on Galaxy Evolution, ed. C. Leitherer, U. Fritze-von Alvensleben, & J. Huchra (San Francisco: ASP), 232

STASIŃSKA, G., & LEITHERER, C. 1996, ApJS, 107, 427

STEIDEL, C. C., GIAVALISCO, M., PETTINI, M., DICKINSON, M., & ADELBERGER, K. L. 1996, ApJ, 462, L17

STEIDEL, C. C., & HAMILTON, D. 1992, AJ, 104, 941

———. 1993, AJ, 105, 2017

SUCHKOV, A. A., BALSARA, D. S., HECKMAN, T. M., & LEITHERER, C. 1994, ApJ, 430, 511

TAMBLYN, P., RIEKE, G. H., MURRAY-HANSON, M., CLOSE, L. M., MCCARTHY, JR., D. W., & RIEKE, M. J. 1995, ApJ, 456, 206

TENORIO-TAGLE, G., & MUÑOZ-TUÑÓN, C. 1997, ApJ, 478, in press

TERLEVICH, E., DÍAZ, A., & TERLEVICH, R. GONZÁLEZ-DELGADO, R. M., PÉREZ, E., & GARCÍA-VARGAS, M. L. 1996, MNRAS, 279, 1219

TERLEVICH, E., SKILLMAN, E. D., & TERLEVICH, R. 1994, in Violent Star Formation, From 30 Doradus to QSOs, ed. G. Tenorio-Tagle (Cambridge: CUP), 182

TERLEVICH, R., & MELNICK, J. 1981, MNRAS, 195, 839

———. 1985, MNRAS, 213, 841

TERLEVICH, R., MELNICK, J., MASEGOSA, J., MOLES, M., & COPETTI, M. V. F. 1991, A&AS, 91, 285

THUAN, T. X. 1991, in Massive Stars in Starbursts, ed. C. Leitherer, N. Walborn, T. Heckman, & C. Norman (Cambridge: CUP), 183

TINSLEY, B. M. 1980, Fund. Cosm. Phys., 5, 287

TOLSTOY, E. 1995, in The Interplay between Massive Star Formation, the ISM, and Galaxy Evolution, ed. D. Kunth, B. Guiderdoni, M. Heydari-Malayeri, & T. X. Thuan (Gif-sur-Yvette: Editions Frontieres), 67

TOLSTOY, E., & SAHA, A. 1996, ApJ, 462, 672

TOMISAKA, K., & IKEUCHI, S. 1988, ApJ, 330, 695

TOSI, M., GREGGIO, L., MARCONI, G., & FOCARDI, P. 1991, AJ, 102, 951

TREFFERS, R. R., & CHU, Y.-H. 1982, ApJ, 254, 569

TULLY, R. B., & FISHER, J. R. 1987, Nearby Galaxies Atlas (Cambridge: CUP)

TYLENDA, R., ACKER, A., & STENHOLM, B. 1993, A&AS, 102, 595

ULRICH, M.-H. 1978, ApJ, 219, 424

VACCA, W. D., & CONTI, P. S. 1992, ApJ, 401, 543

VACCA, W. D., GARMANY, C. D., & SHULL, J. M. 1996, ApJ, 460, 914

VALLENARI, A., & BOMANS, D. J. 1996, A&A, 313, 713

VANBEVEREN, D. 1994, in Evolution of Massive Stars: A Confrontation between Theory and Observations, ed. D. Vanbeveren, W. van Rensbergen, & C. de Loore (Dordrecht: Kluwer), 327

VANBEVEREN, D., VAN BEVER, J., & DE DONDER, E. 1997, A&A, in press

VAN DEN BERGH, S. 1991, in Supernovae, ed. S. E. Woosley (New York: Springer), 711

VAN DER HUCHT, K. A. 1995, in Wolf-Rayet Stars: Binaries, Colliding Winds, Evolution, ed. K. A. van der Hucht & P. M. Williams (Dordrecht: Kluwer), 7

VAN DER HULST, T., & KAMPHUIS, J. 1991, in IAU Symp. 144, The Interstellar Disk-Halo Connection in Galaxies, ed. H. Bloemen (Dordrecht: Kluwer), 201

WALBORN, N. R. 1982, ApJ, 254, L15

———. 1991, in Massive Stars in Starbursts, ed. C. Leitherer, N. Walborn, T. Heckman, & C. Norman (Cambridge: CUP), 145

———. 1995, in The η Car Region: A Laboratory of Stellar Evolution, ed. V. Niemela, N. Morell,

& A. Feinstein, Rev. Mex. Astron. Astrofis. Conf. Ser., 2, 51

Walborn, N. R., & Blades, J. C. 1997, AJ, submitted

Walborn, N. R., Lennon, D. J., Haser, S. M., Kudritzki, R.-P, & Voels, S. A. 1995, PASP, 107, 104

Walborn, N. R., Nichols-Bohlin, J., & Panek, R. J. 1985, International Ultraviolet Explorer Atlas of O-Type Spectra from 1200 to 1900 Å (NASA RP-1155)

Walborn, N. R., & Parker, J. W. 1992, ApJ, 399, L87

Walker, C. E., Martin, R. N., Phillips, T. G., & Bash, F. N. 1993, in Star Formation, Galaxies, and the Interstellar Medium, ed. J. Franco, F. Ferrini, & G. Tenorio-Tagle (Cambridge: CUP), 197

Waller, W. 1991, ApJ, 370, 144

Walsh, J. R., & Roy, J.-R. 1989, MNRAS, 239, 297

Wang, Q., & Helfand, D. J. 1991a, ApJ, 370, 541

———. 1991b, ApJ, 373, 497

Watson, M. G., Stanger, V., & Griffiths, R. E. 1984, ApJ, 286, 144

Weaver, R., McCray, R., Castor, J., Shapiro, P., & Moore, R. 1977, ApJ, 218, 377

Weedman, D. W. 1987, in Star Formation in Galaxies, ed. C. J. Lonsdale (Washington: NASA), 351

Weedman, D. W., Feldman, F. R., Balzano, V. A., Ramsey, L. W., Sramek, R. A., & Wu, C.-C. 1981, ApJ, 248, 105

Williams, R. E., et al. 1996, AJ, 112, 1335

Woltjer, L. 1972, ARAA, 10, 129

Woosley, S. E. 1986, in Nucleosynthesis and Chemical Evolution, ed. J. Audouze, C. Chiosi, & S. E. Woosley (Sauverny: Geneva Observatory), 1

Yang, H., Chu, Y.-H., Skillman, E. D., & Terlevich, R. 1996, AJ, 112, 146

Yee, H. K. C., Ellingson, E., Bechtold, J., Carlberg, R. G., & Cuillandre, J.-C. 1996, AJ, 111, 1783

Zamorano, J., Rego, M., Gallego, J., Vitores, A., González-Riestra, R., & Rodríguez, G. 1994, ApJS, 95, 387

Zamorano, J., Gallego, J., Rego, M., Vitores, A., & Alonso, O. 1996, ApJS, 105, 343

Zinnecker, H. 1995, in The Interplay between Massive Star Formation, the ISM, and Galaxy Evolution, ed. D. Kunth, B. Guiderdoni, M. Heydari-Malayeri, & T. X. Thuan (Gif-sur-Yvette: Editions Frontieres), 249